Textiles and Fashion

The Textile Institute and Woodhead Publishing

The Textile Institute is a unique organisation in textiles, clothing and footwear. Incorporated in England by a Royal Charter granted in 1925, the Institute has individual and corporate members in over 90 countries. The aim of the Institute is to facilitate learning, recognise achievement, reward excellence and disseminate information within the global textiles, clothing and footwear industries.

Historically, The Textile Institute has published books of interest to its members and the textile industry. To maintain this policy, the Institute has entered into partnership with Woodhead Publishing Limited to ensure that Institute members and the textile industry continue to have access to high calibre titles on textile science and technology.

Most Woodhead titles on textiles are now published in collaboration with The Textile Institute. Through this arrangement, the Institute provides an Editorial Board which advises Woodhead on appropriate titles for future publication and suggests possible editors and authors for these books. Each book published under this arrangement carries the Institute's logo.

Woodhead books published in collaboration with The Textile Institute are offered to Textile Institute members at a substantial discount. These books, together with those published by The Textile Institute that are still in print, are offered on the Elsevier website at: http://store.elsevier.com/. Textile Institute books still in print are also available directly from the Institute's website at: www.textileinstitutebooks.com.

A list of Woodhead books on textile science and technology, most of which have been published in collaboration with The Textile Institute, can be found towards the end of the contents pages.

Woodhead Publishing Series in Textiles: Number 126

Textiles and Fashion

Materials, Design and Technology

Edited by

Rose Sinclair

The Textile Institute

AMSTERDAM • BOSTON • CAMBRIDGE • HEIDELBERG
LONDON • NEW YORK • OXFORD • PARIS • SAN DIEGO
SAN FRANCISCO • SINGAPORE • SYDNEY • TOKYO

Woodhead Publishing is an imprint of Elsevier

Published by Woodhead Publishing Limited in association with The Textile Institute
Woodhead Publishing is an imprint of Elsevier
80 High Street, Sawston, Cambridge, CB22 3HJ, UK
225 Wyman Street, Waltham, MA 02451, USA
Langford Lane, Kidlington, OX5 1GB, UK

British Library Cataloguing-in-Publication Data
A catalogue record for this book is available from the British Library

Library of Congress Control Number: 2014942758

ISBN 978-1-84569-931-4 (print)
ISBN 978-0-85709-561-9 (online)

For information on all Woodhead Publishing publications visit our website at http://store.elsevier.com/

Typeset by TNQ Books and Journals
www.tnq.co.in

Printed and bound in the United Kingdom

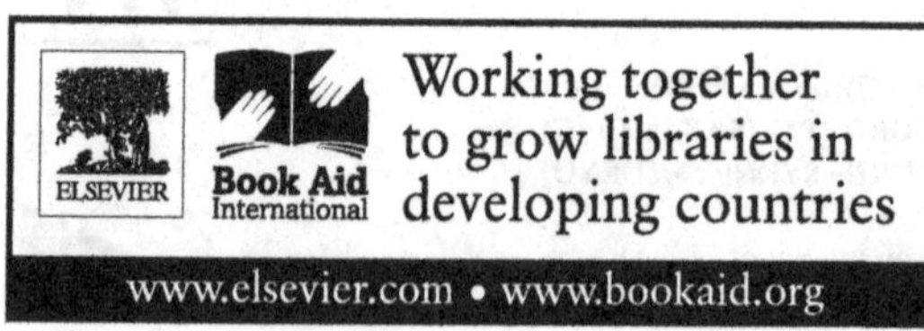

Contents

Contributors xxxi
Woodhead Publishing Series in Textiles xxxiii
Preface xli
Acknowledgements xlv
How to Use this Book xlvii

PART 1 FIBRE TYPES

CHAPTER 1 Understanding Textile Fibres and Their Properties: What is a Textile Fibre? 3

Learning Objectives 3
1.1 Introduction 3
1.2 Types of Textile Fibres 4
1.3 Fibres, Yarns and Fabrics 7
1.4 Fibre Properties 9
1.5 Fibre Length, Shape and Diameter 10
1.6 Fibre Colour and Lustre 12
1.7 Fibre Fineness 13
1.7.1 Fibres 13
1.7.2 Yarns 13
1.7.3 Fabrics 14
1.8 Fibre Strength, Flexibility and Abrasion Resistance 15
1.8.1 Tensile Strength and Extension (Elongation) 15
1.8.2 Flexibility or Stiffness 16
1.8.3 Elasticity 18
1.8.4 Resiliency 18
1.8.5 Abrasion Resistance 18
1.9 Moisture Absorbency 18
1.10 Electrical Properties of Fibres 19
1.11 Thermal Properties of Fibres 20
1.12 Chemical Reactivity and Resistance 22
1.13 Case Studies: From Fibre Properties to Textile Products 22
1.13.1 Case Study 1: Choosing Apparel Fibres 22
1.13.2 Case Study 2: Microfibres 24
1.14 Summary 24
1.15 Project Ideas 24
1.16 Revision Questions 25
References and Further Reading 25

CHAPTER 2 Natural Textile Fibres: Vegetable Fibres 29
Learning Objectives 29
2.1 Introduction 29
2.1.1 Molecular Composition of Cellulose 29
2.1.2 Common Properties of Vegetable Fibres 30
2.2 Cotton 31
2.2.1 Definitions and Types of Cotton 31
2.2.2 Cultivation and Ginning 31
2.2.3 Structure of Cotton 32
2.2.4 Composition of Cotton 32
2.2.5 Physical Properties of Cotton 34
2.2.6 Measurement of Fibre Properties 36
2.2.7 Cotton Application in Textile 39
2.3 Other Seed Fibres 40
2.3.1 Kapok 40
2.3.2 Coir 42
2.4 Bast Fibre 43
2.4.1 Flax 43
2.5 Other Bast Fibres 46
2.5.1 Ramie 46
2.5.2 Jute 48
2.5.3 Kenaf 50
2.5.4 Hemp 50
2.5.5 Sisal, Abaca, and Pineapple Fibre 51
2.6 Sustainability Issues/Eco Issues 52
2.6.1 Biotech Cotton 52
2.6.2 Organic Cotton 52
2.6.3 Naturally Coloured Cotton 52
2.6.4 Bast Fibres 54
2.7 Case Studies 54
2.8 Future Trends 55
2.9 Summary 55
2.10 Project Ideas 55
2.11 Revision Questions 56
References 56

CHAPTER 3 Natural Textile Fibres: Animal and Silk Fibres 57
Learning Objectives 57
3.1 Introduction 57
3.2 Wool Fibres 58
3.2.1 Structure of Wool 58
3.2.2 Amino Acid Composition 59

3.2.3 Properties of Wool Fibres 61
3.2.4 Applications 62
3.3 Silk Fibres 62
3.3.1 Sericulture and Cocoon Production 63
3.3.2 Silk Reeling 64
3.3.3 Silk Manufacture 65
3.3.4 Fine Structure of Silk 66
3.3.5 Amino Acid Composition 67
3.3.6 Properties of Silk Fibres 68
3.3.7 Applications 70
3.4 Other Specialty Hair Fibres 71
3.4.1 Cashmere Fibres 71
3.4.2 Camel Hair Fibres 72
3.4.3 Mohair Fibres 73
3.5 Applications of Natural Protein Fibres 74
3.6 Sustainability and Ecological Issues 75
3.7 Future Trends 75
3.8 Summary 76
3.9 Project Ideas 77
3.10 Revision Questions 77
References 77

CHAPTER 4 Synthetic Textile Fibers: Regenerated Cellulose Fibers 79
Learning Objectives 79
4.1 Introduction 79
4.2 Viscose Rayon 81
4.2.1 The History of Viscose Rayon 81
4.2.2 Viscose Rayon Production 82
4.2.3 Viscose Fiber Appearance 83
4.2.4 Viscose Fiber Mechanical Properties 85
4.2.5 Viscose Rayon Physical Properties 85
4.3 Lyocell Rayon 85
4.3.1 The History of Lyocell Rayon 85
4.3.2 Lyocell Rayon Production 86
4.3.3 Lyocell Fiber Appearance 87
4.3.4 Lyocell Fiber Mechanical Properties 87
4.3.5 Lyocell Fiber Physical Properties 89
4.4 Cellulose Acetate 89
4.5 Applications 90
4.6 Case Study 91
4.7 Future Trends 92
4.8 Summary 93

4.9 Project Ideas 93
4.10 Revision Questions 94
References 94

CHAPTER 5 Synthetic Textile Fibres: Polyamide, Polyester and Aramid Fibres 97

Learning Objectives 97
5.1 Introduction 97
5.2 Classification of Fibres 98
5.3 Polyamide Fibres 99
5.3.1 Production of Nylon 99
5.3.2 Structure and Properties of Nylon Fibres 100
5.3.3 Applications 101
5.4 Polyester Fibres 103
5.4.1 Production of PET Polyester 103
5.4.2 PET Fibre Formation 103
5.4.3 Structure and Properties of Polyester Fibre 104
5.4.4 Applications 105
5.5 Aramid Fibres 106
5.5.1 Production of Aramid Fibres 106
5.5.2 Structure and Properties of Aramid Fibres 108
5.5.3 Applications 109
5.6 Blended Fibres: Key Issues 109
5.7 Case Study: Polyester Fibres for Apparel and Clothing Applications 110
5.8 Future Trends 112
5.9 Summary 112
5.10 Project Ideas 113
5.11 Revision Questions 113
5.12 Sources of Further Information and Advice 113
References 114

CHAPTER 6 Synthetic Textile Fibres: Polyolefin, Elastomeric and Acrylic Fibres 115

Learning Objectives 115
6.1 Introduction 115
6.2 Polypropylene (PP) Fibres 116
6.2.1 Production of Polypropylene (PP) 116
6.2.2 Fibre Manufacture 116
6.2.3 Types of Yarns 117
6.2.4 Spin Finishes 118
6.2.5 Additives 119
6.2.6 Fibre Structure 119

6.2.7 Fibre Properties ... 122
6.2.8 Applications ... 122
6.3 Other Polyolefin Fibres ... 123
6.4 Acrylic Fibres ... 123
6.4.1 Production of Acrylic Fibres ... 125
6.4.2 Fibre Manufacture ... 126
6.4.3 Fibre Structure ... 127
6.4.4 Acrylic Fibre Variants ... 128
6.4.5 Fibre Properties ... 128
6.4.6 Applications ... 129
6.5 Modacrylic Fibres ... 129
6.6 Elastomeric Fibres ... 129
6.6.1 Elastane Fibres ... 130
6.6.2 Fibre Manufacture ... 131
6.6.3 Fibre Structure ... 131
6.6.4 Fibre Properties ... 131
6.6.5 Applications ... 132
6.7 Case Study: Why Are There So Many End-Uses for Polypropylene (PP) Fibres, But So Few in Apparel? ... 133
6.8 Future Trends ... 133
6.8.1 Polyolefin Fibres ... 134
6.8.2 Acrylic Fibres ... 135
6.8.3 Elastomeric Fibres ... 135
6.9 Summary ... 135
6.10 Project Ideas ... 136
6.11 Revision Questions ... 136
6.12 Sources of Further Information and Advice ... 137
References ... 137

CHAPTER 7 Synthetic Textile Fibres: Non-Polymer Fibres ... 139
Learning Objectives ... 139
7.1 Introduction ... 139
7.2 Carbon Fibres ... 139
7.2.1 Manufacture ... 140
7.2.2 Structure ... 142
7.2.3 Properties ... 143
7.2.4 Applications ... 147
7.3 Glass Fibres ... 147
7.3.1 Manufacture ... 147
7.3.2 Structure ... 148
7.3.3 Properties ... 148
7.3.4 Applications ... 148

7.4 Metallic Fibres 149
7.4.1 Manufacture 149
7.4.2 Basic Structure and Properties 150
7.4.3 Applications 150
7.5 Ceramic Fibres 152
7.5.1 Manufacture 152
7.5.2 Basic Structure and Properties 152
7.5.3 Applications 152
7.6 Case Study: The Use of CFRP in Sporting Goods 153
7.7 Future Trends 153
7.8 Summary Points 153
7.9 Project Ideas 154
7.10 Revision Questions 154
References and Further Reading 154

PART 2 MANUFACTURING TEXTILES: YARN TO FABRIC

CHAPTER 8 Conversion of Fibre to Yarn: an Overview 159
Learning Objectives 159
8.1 Introduction 159
8.2 Classification of Yarns 159
8.2.1 Staple Yarns 159
8.2.2 Continuous-Filament Yarns 159
8.2.3 Novelty Yarns 160
8.2.4 Industrial Yarns 161
8.2.5 High-Bulk Yarns 161
8.2.6 Stretch Yarns 161
8.3 Staple-Fibre Yarns 161
8.3.1 Spinning Methods 161
8.3.2 Operations in Staple-Fibre Spinning 162
8.3.3 Yarn Structure 162
8.3.4 Applications of Staple-Spun Yarns 164
8.4 Filament Yarns 164
8.4.1 Spinning Methods 164
8.4.2 Polymer Spinning Processes 165
8.4.3 Structures of Continuous Filament Yarns 166
8.4.4 Applications of Filament Yarn 168
8.5 Fancy Yarns 170
8.5.1 Marl Yarn 170
8.5.2 Spiral or Corkscrew Yarn 170

8.5.3 Gimp Yarn 170
8.5.4 Diamond Yarn 171
8.5.5 Boucle Yarn 171
8.5.6 Loop Yarn 171
8.5.7 Snarl Yarn 172
8.5.8 Knop Yarn 172
8.5.9 Slub Yarn 173
8.5.10 Fasciated Yarn 173
8.5.11 Tape Yarn 174
8.5.12 Chainette Yarn 174
8.5.13 Chenille Yarn 174
8.5.14 Ribbon Yarns 175
8.5.15 Composite Yarns 175
8.5.16 Covered Yarns 175
8.5.17 Metallic Yarns 176
8.6 Staple-Fibre Yarn Manufacturing 176
8.6.1 Ring (Conventional) Spinning 176
8.6.2 Hollow-Spindle Spinning 178
8.6.3 Combined Systems 178
8.6.4 The Doubling System 180
8.6.5 Open-End Spinning 182
8.6.6 Air-Jet Spinning 184
8.6.7 The Chenille Yarn System 185
8.6.8 Flocking 186
8.6.9 Mock Chenille 186
8.7 Future Trends 187
8.8 Summary 187
8.9 Project Ideas 188
8.10 Revision Questions 188
References 189

CHAPTER 9 Fibre to Yarn: Staple-Yarn Spinning 191
Learning Objectives 191
9.1 Introduction 191
9.2 Preparation of Cotton and Other Short Staple Fibres 192
9.2.1 Opening and Cleaning 192
9.2.2 Blending 193
9.2.3 Carding 195
9.2.4 Combing 195
9.2.5 Drawing 196
9.2.6 Roving 196

9.3 Preparation of Wool and Other Long Staple Fibres: The Woollen System 196
9.3.1 Opening 198
9.3.2 Scouring and Carbonising 198
9.3.3 Drying or Oiling 198
9.3.4 Blending 198
9.3.5 Carding 199
9.4 Preparation of Wool and Other Long Staple Fibres: The Worsted System 199
9.4.1 Carding 200
9.4.2 Gilling 200
9.4.3 Combing 201
9.5 Spinning Techniques for Staple Fibres 201
9.5.1 Ring Spinning 202
9.5.2 Twist-Spinning Methods: Open-End (Rotor and Friction) Spinning and Self-Twist Spinning 204
9.5.3 Open-End Spinning: Rotor Spinning 205
9.5.4 Open-End Spinning: Friction Spinning 206
9.5.5 Self-Twist Spinning 208
9.6 Wrap-Spinning Techniques 208
9.6.1 Air-Jet Spinning 208
9.6.2 Filament Wrapping Techniques 210
9.7 Future Trends 210
9.8 Summary Points 210
9.9 Project Ideas 211
9.10 Revision Questions 211
9.10.1 Cotton System 211
9.10.2 Wool System 212
References and Further Reading 212

CHAPTER 10 Fibre to Yarn: Filament Yarn Spinning 213
Learning Objectives 213
10.1 Introduction 213
10.1.1 Definitions 213
10.1.2 Classification of CF Yarns 214
10.1.3 Yarn Count System 215
10.2 Fibre-Extrusion Spinning 216
10.2.1 Melt-Spinning 216
10.2.2 Wet Spinning 219
10.2.3 Dry Spinning 221
10.3 Yarn Texturing 222
10.3.1 False-Twist Texturing 224
10.3.2 Air-Jet Texturing 229

10.4 Bulk Continuous Fibre (BCF) Technology 231
10.4.1 Twisting/Plying of Continuous-Filament Yarns 231
10.4.2 Metallised Yarns 233
10.5 Properties of CF Yarns 234
10.5.1 Morphology 234
10.5.2 Tensile Properties 237
10.6 Adding Functionality to Yarn 242
10.6.1 Moisture Absorption 242
10.6.2 Dyeability and Printability 243
10.6.3 Functional Additives 246
10.7 Applications 247
10.8 Future Trends 250
10.9 Project Ideas 251
10.10 Revision Questions 251
References 251

CHAPTER 11 Yarn to Fabric: Weaving 255
Learning Objectives 255
11.1 Introduction 255
11.2 Looms 255
11.2.1 Rigid Heddle Loom 256
11.2.2 Table Loom 256
11.2.3 Floor/Treddle Loom 257
11.2.4 Counterbalanced Loom 258
11.2.5 Dobby Loom 258
11.2.6 Computerised Loom 259
11.2.7 Jacquard Loom 260
11.3 Making a Warp and Dressing the Loom 260
11.3.1 Selecting a Warp Yarn 260
11.3.2 Calculating the Warp Yarns 260
11.3.3 Making the Warp 261
11.3.4 Making a Chain 261
11.3.5 Dressing the Loom 262
11.4 Documentation 266
11.4.1 Point Paper 267
11.4.2 Threading Plan 267
11.4.3 Lifting Plan 267
11.4.4 Reed Plan 268
11.5 Pattern Drafting 268
11.5.1 Straight Draft 268
11.5.2 Pointed Draft 268

11.5.3 Block Draft 270
11.5.4 Scattered Draft 271
11.6 Weave Structures 271
11.6.1 Balanced and Unbalanced Weave Structures 271
11.6.2 Balanced Plain Weave 271
11.6.3 Unbalanced Plain Weave 272
11.6.4 Basket Weave 272
11.6.5 Twill Weaves 272
11.6.6 Herringbone Twill 274
11.6.7 Satin Weave 275
11.7 Derivative-Weave Structures 276
11.7.1 Mock Leno 276
11.7.2 Double Weave 276
11.7.3 Honeycomb 278
11.7.4 Jacquard Weaves 278
11.8 Starting to Weave 279
11.9 Designing for Woven Textiles 279
11.10 Designing for the Jacquard Loom 280
11.11 Tapestry Weaving 281
11.12 Case Study: Honeycomb Woven Structures 283
11.13 Finishing 283
11.14 Tips for Weaving 284
11.15 Future Trends 284
11.16 Summary 285
11.17 Revision Questions 286
11.18 Sources of Further Information and Advice 286
11.18.1 Collections 286
Further Reading 286

CHAPTER 12 Yarn to Fabric: Knitting 289

Learning Objectives 289
12.1 Introduction 289
12.2 Loop Formation 290
12.3 Knitting Terminology 291
12.4 Weft-Knitted Structures 292
12.4.1 Weft-Knitting Machines 293
12.5 Warp Knitted Structures 294
12.5.1 Warp Knitting Machines 295
12.6 Knitting Developments 296
12.7 The Impact of Computers in Design and Technology 298
12.8 Quality Control 299

12.9 Case Study ... 300
12.10 Future Trends ... 301
12.11 Summary ... 302
12.12 Project Ideas ... 303
12.13 Revision Questions ... 303
12.14 Sources of Further Information and Advice ... 304
References ... 305

CHAPTER 13 Fibre to Fabric: Nonwoven Fabrics ... 307
Learning Objectives ... 307
13.1 Introduction ... 307
13.2 Technologies for the Formation of Nonwoven Fabrics ... 308
13.2.1 Fibrous Web Formation ... 309
13.2.2 Web Bonding Technologies ... 310
13.2.3 Nonwoven Fabric Finishing and Converting Techniques ... 312
13.2.4 Coating and Laminating ... 314
13.3 Characteristics of Nonwoven Fabric Structure and Properties ... 315
13.3.1 Characterisation of Fabric Bond Structure ... 315
13.3.2 Nonwoven Fabric Structural Parameters ... 322
13.4 Properties and Performance of Nonwoven Fabrics ... 327
13.5 Methods for the Evaluation of Nonwoven Fabric Structure, Properties and Performance ... 328
13.5.1 Standard Test Methods for the Evaluation of the Structure and Properties of Nonwoven Fabrics ... 328
13.5.2 Standards for the Evaluation of the Performance of Nonwoven Products ... 328
13.6 Nonwoven Fabrics and Their Applications ... 329
13.7 Nonwoven Fabrics in Fashion ... 330
13.8 Future Trends ... 332
13.9 Project Ideas ... 332
13.10 Revision Questions ... 332
13.11 Sources of Further Information ... 333
References ... 333

CHAPTER 14 Yarn to Fabric: Specialist Fabric Structures ... 337
Learning Objectives ... 337
14.1 Introduction ... 337
14.2 Triaxial Fabrics ... 337
14.3 Pile Fabrics ... 339
14.4 Knotted Fabrics ... 341
14.4.1 Nets ... 341
14.4.2 Macrame ... 341

14.4.3 Lace 342
14.4.4 Crochet 342
14.4.5 Knotting 342
14.5 Braided Fabrics 344
14.6 Three-Dimensional Fabrics and Future Developments 346
14.6.1 3D Solid Structures 346
14.6.2 Hollow Structures 347
14.6.3 Shell Structures 348
14.6.4 Knitted Structures 349
14.6.5 Nonwoven Structures 350
14.7 Summary 352
14.8 Project Ideas 352
14.9 Revision Questions 352
References and Sources of Further Information 353

CHAPTER 15 Yarn to Fabric: Intelligent Textiles 355
Learning Objectives 355
15.1 Introduction 355
15.2 What Are Intelligent Textiles Used For? 356
15.2.1 Smart Textile Applications 357
15.2.2 Research and Development of Smart Textiles 359
15.2.3 Phase Change Materials 360
15.2.4 Shape Memory Materials 363
15.2.5 Chromic and Conductive Materials 365
15.2.6 Stress-Responsive Materials 367
15.2.7 Wearable Electronics 367
15.3 Case Study: Biomimetics and Intelligent Textiles 368
15.3.1 Examples of Biomimetic Products 368
15.3.2 The Lotus Effect 368
15.4 Future Trends 369
15.4.1 Future Applications of Intelligent Textiles 369
15.4.2 Future Market Development 372
15.5 Summary 373
15.6 Project Ideas 374
15.7 Revision Questions 375
15.8 Sources of Further Information and Advice 375
References and Further Reading 375

PART 3 FABRIC FINISHING AND APPLICATIONS

CHAPTER 16 Fabric Finishing: Joining Fabrics Using Stitched Seams 379
Learning Objectives 379
16.1 Introduction 379

16.2 The Stitch....380
16.2.1 Class 100 Chain Stitches....380
16.2.2 Class 300 Lockstitches....381
16.2.3 Class 400 Multi-Thread Chain Stitches....381
16.2.4 Class 500 Overedge Stitches....382
16.2.5 Stitch Quality....382
16.3 The Seam....382
16.3.1 Class 1: Superimposed Seams....384
16.3.2 Class 2: Double-Lap Seams....385
16.3.3 Class 3: Bound Seams....385
16.3.4 Class 4: Flat Seams....385
16.4 Sewing Machines....385
16.4.1 The Needle....390
16.4.2 Machine Feeding Systems....394
16.4.3 Machines for Different Stitching Operations....400
16.5 Seam Quality Problems....402
16.5.1 Pucker....402
16.5.2 Thread Breakage....405
16.6 Future Trends....405
16.7 Summary....409
16.8 Case Study and Project Idea....409
16.9 Revision Questions....410
References and Further Reading....410

CHAPTER 17 Joining Fabrics: Fastenings....413
Learning Objectives....413
17.1 Introduction....413
17.2 Zips....413
17.2.1 Components of a Zip....415
17.2.2 Zip Functions and Applications....416
17.2.3 How to Measure the Correct Length of Opening for a Zip....416
17.2.4 Machinery and Attachments Used to Apply Zips....419
17.2.5 Continuous Zips....419
17.2.6 Safety Standards and Legislation for Selecting and Applying Zips....419
17.3 Buttons....422
17.3.1 Types of Buttons....423
17.3.2 Materials Used to Make Buttons....425
17.3.3 How to Measure Buttons....425
17.3.4 Machinery and Attachments Used to Apply Buttons....425
17.3.5 Safety Standards and Legislation for Selecting and Applying Buttons....431

17.4 Hook-and-Loop Fasteners....432
17.4.1 Types of Hook-and-Loop Tapes....433
17.4.2 Materials Used to Make Hook-and-Loop Tape....434
17.4.3 Machinery and Attachments Used to Apply Hook-and-Loop Tape....434
17.4.4 Safety Standards and Legislation for Selecting and Applying Hook-and-Loop Tape....434
17.5 Press Fasteners....435
17.5.1 Types of Press Fasteners....436
17.5.2 Materials Used to Make Press Fasteners....437
17.5.3 Machinery and Attachments Used to Apply Press Fasteners....439
17.5.4 Other Non-Snap Components....442
17.5.5 Safety Standards and Legislation for Selecting and Applying Press Fasteners....442
17.6 Cords, Ties and Belts....446
17.6.1 Materials Used to Make Cords, Ties and Belt Fastenings....446
17.6.2 Machinery and Attachments Used to Apply Cords, Ties and Belts....447
17.6.3 Safety Standards and Legislation for Selecting and Applying Cords, Ties and Belts....448
17.7 Hook-and-Eye Fasteners....448
17.7.1 Types of Hook-and-Eye Fasteners....449
17.7.2 Materials Used to Make Hooks and Eyes....449
17.7.3 Machinery and Attachments Used to Apply Hooks and Eyes....450
17.7.4 Safety Standards and Legislation for Selecting and Applying Hooks and Eyes....450
17.8 Hook-and-Bar Fasteners....451
17.8.1 Types of Hook-and-Bar Fasteners....451
17.8.2 Materials Used to Make Hooks and Bars....452
17.8.3 Machinery and Attachments Used to Apply Hooks and Bar/Fasteners....453
17.8.4 Safety Standards and Legislation for Selecting and Applying Hooks and Bar/Fasteners....453
17.9 Buckles and Adjustable Fasteners....454
17.9.1 Types of Buckles and Adjustable Fasteners....454
17.9.2 Materials Used to Make Buckles and Adjustable Fasteners....454
17.9.3 Machinery and Attachments Used to Apply Buckles and Adjustable Fasteners....455
17.9.4 Safety Standards and Legislation for Selecting and Applying Buckles and Adjustable Fasteners....455
17.10 Summary....456
17.11 Project Ideas....456
17.12 Revision Questions....457

17.13 Sources of Further Information 457

References and Further Reading 457

CHAPTER 18 Fabric Finishing: Pretreatment/Textile Wet Processing 459

Learning Objectives 459

18.1 Introduction 459

18.2 Processing Methods 460

18.3 Fabric Preparation Processes 461

18.3.1 Desizing 461

18.3.2 Scouring 462

18.3.3 Bleaching 463

18.3.4 Mercerization 464

18.3.5 Carbonization 465

18.3.6 Heat Setting 465

18.3.7 Drying 465

18.4 Quality Control in Fabric Preparation 466

18.5 Environmental Impact and Sustainability of Fabric Preparation 467

18.6 Research and Future Trends 467

18.7 Summary 468

18.8 Case Study 468

18.9 Project Ideas 469

18.10 Revision Questions 469

18.11 Sources of Further Information 469

References 469

CHAPTER 19 Fabric Finishing: Dyeing and Colouring 475

Learning Objectives 475

19.1 Introduction 475

19.2 Colour Theory 475

19.2.1 Light and the Human Eye 476

19.2.2 Colour Description and Measurement 476

19.2.3 Instrumental Colour Match and Shade Assessment 477

19.3 Selection of Dyes 477

19.3.1 Achieving the Required Shade 478

19.3.2 Compatibility of Dyes 479

19.3.3 Metamerism 479

19.3.4 Colour Fastness 479

19.3.5 Environmental Considerations 480

19.4 The Dyeing Process 480

19.4.1 Dyeing Conditions 481

19.4.2 Machinery for Dyeing 481

19.4.3 Further Textile Colouration 481

19.5 Classes of Dye for Different Fibre Types....485
19.5.1 Cellulosic Fibres....485
19.5.2 Protein Fibres....487
19.5.3 Polyamide Fibres....489
19.5.4 Polyester Fibres....491
19.5.5 Acrylic Fibres....492
19.5.6 Fibre Blends....493
19.5.7 Fluorescent Brightening Agents....493
19.6 Strengths and Weaknesses of Natural and Synthetic Dyes....494
19.6.1 Safety....494
19.6.2 Shade Range and Reproducibility....494
19.6.3 Colour Fastness....494
19.6.4 Availability of Natural Dyes....494
19.6.5 The Way Forward....495
19.7 Ensuring Quality and Effectiveness of Dyeing....495
19.7.1 Assessment of Shade....495
19.7.2 Assessment of Colour Fastness....496
19.7.3 Assessment of Overall Substrate Quality....496
19.8 Environmental Impact of Dyeing....496
19.8.1 Water Consumption....497
19.8.2 Energy Consumption....497
19.8.3 Air Emissions....497
19.8.4 Effluent Emissions....497
19.8.5 Occupational Safety....498
19.8.6 Safety of Dyed Products....498
19.9 Research and Future Trends....499
19.10 Summary....500
19.11 Case Study: Reactive Dyeing of Knitted Cotton Garments....500
19.11.1 Selection of Dyes....500
19.11.2 Selection of Process Method....501
19.11.3 Selection of Machinery....502
19.11.4 Post-Dye Operations....502
19.12 Project Ideas....503
19.12.1 Process Control to Reduce the Environmental Impact of Dyeing....503
19.12.2 Restricted Substances List....504
19.12.3 Controls Within a Dyeing and Finishing Operation....504
19.13 Revision Questions....504
19.14 Sources of Further Information....505
References....505

CHAPTER 20 Fabric Finishing: Printing Textiles....507

Learning Objectives....507

20.1 Introduction....507

20.2 Direct Printing....508

20.2.1 Pigment Printing....509

20.2.2 Reactive Dye Printing....510

20.2.3 Disperse Dye Printing....510

20.2.4 Vat Dye Printing....510

20.2.5 Acid Dye Printing....511

20.2.6 Digital Inkjet Printing....511

20.3 Other Printing Techniques....511

20.3.1 Resist Printing....512

20.3.2 Discharge Printing....512

20.3.3 Burn-Out (Devoré) Printing....512

20.4 Traditional Printing Methods....513

20.5 Screen Printing....515

20.5.1 Table Screen Printing....515

20.5.2 Automatic Flat-Bed Screen Printing....516

20.5.3 Rotary Screen Printing....517

20.5.4 Screen Design and Production....518

20.6 Transfer Printing....519

20.6.1 Gravure Printing....519

20.6.2 Digital Paper Printing....520

20.6.3 Heat Transfer Press....520

20.7 Digital Inkjet Printing....520

20.7.1 Technology and Characteristics....522

20.7.2 Design Application....523

20.8 Impact of CAD/CAM on the Design of Printed Textiles....523

20.9 Research and Future Trends....525

20.10 Summary....526

20.11 Case Study....526

20.12 Project Ideas....527

20.13 Revision Questions....528

20.14 Sources of Further Information and Advice....528

References....529

CHAPTER 21 Applications of Textile Products....531

Learning Objectives....531

21.1 Introduction....531

21.2 Apparel....532

21.3 Furnishing or Interior Textiles, Including Household Products....534

21.4 Technical Textiles....536
21.4.1 Industrial Textiles and Geotextiles....536
21.4.2 Smart Fabrics and Intelligent Textiles....536
21.4.3 Medical Textiles....537
21.4.4 Wearable Textiles and Protective Clothing....537
21.4.5 Eco Textiles....538
21.5 Textile Art....538
21.6 Textile Industry....539
21.7 Case Study: Traditional Bedouin al Sadu Hand-Woven Products and Contemporary Digital Applications....540
21.8 Future Trends....542
21.9 Summary....543
21.10 Revision Questions....543
21.11 Sources of Information....544
Further Reading....544

CHAPTER 22 Sustainable Textile Production....547
Learning Objectives....547
22.1 Introduction....547
22.2 Key Issues in Sustainability....548
22.3 The Textile Supply Chain....549
22.3.1 Supply Chain for Fabrics Made from Natural Fibres....550
22.3.2 Synthetic Fibres....552
22.4 Assessing the Environmental Impact of the Textile Supply Chain....553
22.5 Minimising the Environmental Impact of the Textile Supply Chain....554
22.6 Case Study: Creating Sustainable and Socially Responsible Fashion....556
22.6.1 Future Thinking in Sustainable Futures....556
22.7 Summary and Project Ideas....558
22.8 Sources of Further Information and Advice....559
References....560

PART 4 DEVELOPING TEXTILE PRODUCTS: THE CASE OF APPAREL

CHAPTER 23 Material Culture: Social Change, Culture, Fashion and Textiles in Europe....563
Learning Objectives....563
23.1 Introduction....563
23.2 Art and Society....564
23.2.1 Advances in Technology: The Industrial Revolution....564
23.2.2 Advances in Technology: Modern Developments....567

23.2.3 Travel and Discovery 568
23.2.4 Orientalism 569
23.2.5 The Space Race 572
23.2.6 Modern Day 575
23.3 Politics 575
23.3.1 Poster Art 575
23.3.2 Soviet Posters 576
23.3.3 Guerilla Art 578
23.3.4 T-Shirts 580
23.4 War 584
23.4.1 Bayeux Tapestry 584
23.4.2 Guernica 584
23.4.3 Fashion and World War II 585
23.4.4 Textile Design Following World War II 586
23.5 Impact of Culture on Design, Fashion and Textiles 589
23.5.1 Cubism and Delaunay 589
23.5.2 Surrealism and Schiaparelli 589
23.5.3 Pop Art 589
23.5.4 Op Art 593
23.5.5 Popular Culture/Pop Culture 594
23.5.6 Counter-Culture 595
23.5.7 Punk 595
23.6 Definitions of Textile Culture and Fashion Culture: Are They the Same? 597
23.7 Project Ideas 600
23.8 Revision Questions 600
23.9 Further Reading 600
References 602
Electronic Sources 603

CHAPTER 24 Fashion and Culture: Global Culture and Fashion 605
Learning Objectives 605
24.1 Introduction 605
24.2 Impact of Culture in European and Non-European Arenas 606
24.3 Case Study 607
24.3.1 Cultural Exchanges: Japan and the West 607
24.4 Future Trends 619
24.4.1 Globalisation and the Democratisation of Fashion 620
24.4.2 Sustainability in Fashion and Textiles: 'Green' Issues 624
24.4.3 Ethical Fashion 626
24.5 Summary Points 629
24.6 Project Ideas 630

24.7 Revision Questions 630
24.8 Further Reading....631
References....632
Electronic Sources....633

CHAPTER 25 Fashion and the Fashion Industry....635
Learning Objectives....635
25.1 Introduction....635
25.2 Emergence, Development and Change in Fashion 636
25.2.1 Fashion Is Evolutionary 636
25.2.2 What Is Fashion Style?....637
25.2.3 Fashion Moves in Cycles 637
25.3 The Standard Fashion-Trend Cycle 638
25.3.1 Style Regeneration 638
25.3.2 The Classic....639
25.3.3 The Fad....639
25.4 Why Fashion Changes?....639
25.4.1 Topman Case Study....639
25.4.2 Topman Case Study: Project Ideas....642
25.4.3 Asos Case Study....642
25.4.4 Asos Case Study: Project Ideas....644
25.5 Revision Questions 645
25.6 Summary Points....645
Magazines/Periodicals 645
References and Further Reading....646
Websites 647

CHAPTER 26 Visual Design Techniques for Fashion....649
Learning Objectives....649
26.1 Introduction....649
26.1.1 Research for Design....649
26.1.2 Planning the Collection....650
26.1.3 Developing the Samples....651
26.2 Why Consumers Buy New Designs....651
26.3 Market Research Methods for Identifying Emerging Consumer Needs....652
26.4 Finding Inspiration....653
26.4.1 Trade Shows....654
26.4.2 Fashion Forecasting 654
26.4.3 Copyright 654
26.4.4 Creative Thinking Techniques....654

26.5 Aesthetic Qualities in a Good Design 655
26.5.1 Shape and Silhouette 656
26.5.2 Proportion 656
26.5.3 Colour 656
26.5.4 Fabrics and Trimmings 656
26.5.5 Prints and Motifs 656
26.5.6 Details and Embellishments 657
26.5.7 Styling and Accessories 657
26.6 Design Tools 657
26.7 Moving from Sample to Production 658
26.8 Future Trends: Impact of New Technologies/Processes 658
26.9 Case Study: The Development of a Garment 659
26.10 Summary 666
26.11 Project Ideas 667
26.11.1 Analysing a Collection 667
26.11.2 Finding Trend Information 667
26.11.3 Analysing a Garment 667
26.12 Revision Questions 667
26.13 Sources of Further Information and Advice 668
References 668
Further Reading 668

CHAPTER 27 Computer-Aided Design (CAD) and Computer-Aided Manufacturing (CAM) of Apparel and Other Textile Products 671
Learning Objectives 671
27.1 Introduction 671
27.2 Fashion and Textile Software Programs 673
27.3 Using CAD to Design Fashion Products 674
27.3.1 Flats/Working Drawings 675
27.3.2 Specification Sheets (Specs) 675
27.3.3 Style Sheets 678
27.4 Other Uses of CAD in Fashion Design 678
27.4.1 Digital Design Library 678
27.4.2 Design Presentations 682
27.4.3 Digital Design Portfolio 682
27.4.4 The Place of the App 685
27.5 CAM in Fashion and Textiles 686
27.5.1 3D Digital and Virtual Fabrication in Textiles and Fashion 686
27.5.2 New 3D Printing and Fabrication in Textiles and Fashion 688

27.6 Case Studies: Fashion Designers Interviewed by Sandra Burke 691
27.6.1 Laura Krusemark....691
27.6.2 Alissa Stytsenko 693
27.7 Summary Points and Project Ideas....700
27.8 Revision Questions 700
27.9 Sources of Further Information and Advice....700
27.9.1 Books 700
27.9.2 Websites 701
27.9.3 Web Resources 702
27.9.4 Open Source Software 702
References and Further Reading....703

CHAPTER 28 Adding Functionality to Garments 705

Learning Objectives 705
28.1 Introduction....705
28.2 Factors Affecting Garment Function....706
28.3 Improving Fabric Handle and Tailorability....706
28.3.1 Fibre Properties 707
28.3.2 Yarn Properties 710
28.3.3 Fabric Properties 710
28.3.4 Dyeing and Finishing....710
28.3.5 Measurement of Fabric Handle and Making-Up 711
28.4 Reducing Wrinkling....712
28.4.1 Factors Affecting Wrinkling and Wrinkle Recovery During Wear 713
28.4.2 Fibre Properties....715
28.4.3 Yarn and Fabric Parameters....716
28.4.4 Mechanical and Chemical Finishing to Reduce Wrinkling 717
28.4.5 Measurement of Wrinkle and Crease Recovery....717
28.5 Reducing Pilling....719
28.5.1 How Pills Are Formed....720
28.5.2 Fibre Properties....720
28.5.3 Yarn Properties 722
28.5.4 Fabric Properties 722
28.5.5 Dyeing and Finishing....722
28.5.6 Measurement of Pilling....722
28.6 Reducing Bagging....724
28.6.1 Fibre Properties....724
28.6.2 Yarn Properties 725
28.6.3 Fabric Properties 725
28.6.4 Garment Construction....725
28.6.5 Finishing....725
28.6.6 Measurement of Bagging 725

28.7 Improving Fabric and Garment Drape 726
28.7.1 Fibre Properties 727
28.7.2 Yarn Properties 728
28.7.3 Fabric Properties 728
28.7.4 Dyeing and Finishing 729
28.7.5 Garment Construction 729
28.7.6 Measurement of Drape 729
28.8 Improving Fabric and Garment Durability 730
28.8.1 Fibre Properties 730
28.8.2 Yarn Properties 731
28.8.3 Fabric Properties 731
28.8.4 Garment Design and Fit 732
28.8.5 Dyeing and Finishing 732
28.8.6 Measurement of Fabric Durability 732
28.9 Research and Future Trends 735
28.10 Summary 735
28.11 Project Ideas 736
28.12 Revision Questions 736
References 736

CHAPTER 29 Improving the Comfort of Garments 739
Learning Objectives 739
29.1 Introduction 739
29.2 Tactile Comfort 740
29.3 Thermo-Physiological (Thermal) Comfort 742
29.3.1 Factors Affecting the Thermal Insulation of Fabrics and Clothing 742
29.3.2 Factors Affecting the Moisture (Vapour) Transmission Properties of Fabric and Clothing 743
29.3.3 Factors Affecting Liquid Water Transport Properties of Fabrics and Clothing 744
29.3.4 Factors Affecting Garment Fit and Ease of Body Movement 745
29.4 Measuring Physiological Comfort 746
29.4.1 Tactile Comfort 746
29.4.2 Thermal Contact 746
29.4.3 Thermal Insulation 746
29.4.4 Water Vapour Permeability 747
29.4.5 Liquid Water Transport Properties 749
29.4.6 Garment Fit and Ease of Body Movement 750
29.4.7 Pressure Comfort 750
29.4.8 Formaldehyde Content 750

29.5 Psychological Comfort.....751
29.5.1 Factors Affecting Psychological Comfort.....751
29.5.2 Assessing Psychological Comfort.....753
29.6 Improving Waterproofing and Breathability.....753
29.6.1 Factors Affecting Fabric and Garment Breathability.....755
29.6.2 Measuring Waterproofing and Breathability.....757
29.7 Research and Future Trends.....758
29.8 Summary.....759
29.9 Case Study.....759
29.10 Project Ideas.....760
29.11 Revision Questions.....760
References.....760

CHAPTER 30 The Marketing of Fashion.....763
Learning Objectives.....763
30.1 Introduction.....763
30.2 What is Marketing?.....763
30.2.1 The Four P's: Product, Price, Place and Promotion.....764
30.2.2 The Four C's: Consumer, Cost, Convenience and Communication.....764
30.3 The Marketing of Fashion.....764
30.3.1 Product Development.....766
30.3.2 Retailing Space.....766
30.3.3 Communication of the Product.....766
30.3.4 Public Relations.....766
30.4 Targeting a Market.....767
30.4.1 Customer Profiles.....768
30.4.2 Seasonal and Occasion Markets.....768
30.5 Branding.....768
30.5.1 Branding Case Study: Apple Inc......768
30.6 The Traditional Media Channels.....769
30.7 New Technologies as Media Channels.....770
30.7.1 Web 2.0 and Other Technological Developments.....770
30.7.2 Case Study: Wickedweb Digital Marketing Agency.....773
30.8 The Marketing Plan.....774
30.8.1 Case Study 1: 'Evolution' by Katie Lay.....774
30.8.2 Case Study 2: 'Stratagem' by Luke Anthony Richardson.....782
30.8.3 Case Study 3: Commercial Case Study – All Saints of Spitalfields.....790
30.9 Future Trends.....791
30.9.1 Fashion Forecasting.....791
30.9.2 New Technologies and Processes.....792
30.10 Summary Points.....794

30.11 Project Ideas and Revision Questions 795
30.12 Sources of Further Information 795
30.12.1 Books 795
30.12.2 Trend Forecasting Companies 795
30.12.3 Fashion Forecasting Companies 796
30.12.4 Magazines 796
30.12.5 Websites 796
References 797

CHAPTER 31 The Care of Apparel Products 799
Learning Objectives 799
Abbreviations 799
31.1 Introduction 800
31.2 Wear of Garments 801
31.2.1 Pilling 801
31.2.2 Abrasion 802
31.2.3 Colour Fading 803
31.2.4 Breaking of Yarns and Fabrics 803
31.2.5 Snagging 805
31.2.6 Seam Failure 805
31.2.7 Dimensional Change 806
31.2.8 Other Problems 807
31.3 Stains 808
31.4 Laundering 809
31.4.1 Laundering Chemicals 809
31.4.2 Laundering Aids 811
31.4.3 Laundering Equipment 812
31.5 Care Labelling 813
31.5.1 Care Label Requirements 814
31.5.2 Care Labelling Systems 815
31.6 Clothing Storage 818
31.7 Conclusions and Future Trends 818
31.8 Sources of Further Information 819
31.9 Summary Points 819
31.10 Project Ideas 819
31.11 Revision Questions 819
References 820

Glossary 823
Index 833

Contributors

R. Alagirusamy
Indian Institute of Technology Delhi, New Delhi, India

K.M. Babu
Bapuji Institute of Engineering Technology, Davanagere, India

S. Burke
Burke Publishing, London, UK

K. Canavan
University of Wales Institute Cardiff, Cardiff, UK

J. Chen
The University of Texas at Austin, Austin, TX, USA

A. Das
Indian Institute of Technology Delhi, New Delhi, India

B.L. Deopura
Indian Institute of Technology Delhi, New Delhi, India

L. Drew
The Glasgow School of Art, Glasgow, UK

I.A. Elhawary
Alexandria University, Alexandria, Egypt

J. Fan
Cornell University, New York, NY, USA

J. Gaimster
London College of Fashion, London, UK

R.H. Gong
University of Manchester, Manchester, UK

P. Hauser
North Carolina State University, Raleigh, NC, USA

L. Hunter
CSIR and Nelson Mandela Metropolitan University, Port Elizabeth, South Africa

T. Kikutani
Tokyo Institute of Technology, Tokyo, Japan

C. Lawrence
University of Leeds, Leeds, UK

N. Mao
University of Leeds, Leeds, UK

R.R. Mather
Heriot-Watt University, Edinburgh, UK

H. Mattila
Tampere University of Technology, Tampere, Finland

K. McKelvey
Northumbria University, Newcastle upon Tyne, UK

J. McLoughlin
Manchester Metropolitan University, Manchester, UK

A. Mitchell
Manchester Metropolitan University, Manchester, UK

R.K. Nayak
RMIT University, Melbourne, VIC, Australia

N.V. Padaki
Central Silk Technological Research Institute, Guwahati, India

R. Padhye
RMIT University, Melbourne, VIC, Australia

E.J. Power
University of Huddersfield, Huddersfield, UK

P.R. Richards
Richtex Textile Consultancy, Newark, UK

S.J. Russell
University of Leeds, Leeds, UK

C. Ryder
Liverpool John Moores University, Liverpool, UK

M. Shioya
Tokyo Institute of Technology, Tokyo, Japan

R. Sinclair
Goldsmiths, University of London, London, UK

S. Stankard
Universiti Teknologi MARA (UiTM), Selangor, Malaysia

M. Tomaney
University for the Creative Arts, Epsom, UK

H. Ujiie
Philadelphia University, Philadelphia, PA, USA

C. Yu
Dong Hua University, Shanghai, China

Woodhead Publishing Series in Textiles

1 **Watson's textile design and colour Seventh edition**
Edited by Z. Grosicki

2 **Watson's advanced textile design**
Edited by Z. Grosicki

3 **Weaving Second edition**
P. R. Lord and M. H. Mohamed

4 **Handbook of textile fibres Volume 1: Natural fibres**
J. Gordon Cook

5 **Handbook of textile fibres Volume 2: Man-made fibres**
J. Gordon Cook

6 **Recycling textile and plastic waste**
Edited by A. R. Horrocks

7 **New fibers Second edition**
T. Hongu and G. O. Phillips

8 **Atlas of fibre fracture and damage to textiles Second edition**
J. W. S. Hearle, B. Lomas and W. D. Cooke

9 **Ecotextile '98**
Edited by A. R. Horrocks

10 **Physical testing of textiles**
B. P. Saville

11 **Geometric symmetry in patterns and tilings**
C. E. Horne

12 **Handbook of technical textiles**
Edited by A. R. Horrocks and S. C. Anand

13 **Textiles in automotive engineering**
W. Fung and J. M. Hardcastle

14 **Handbook of textile design**
J. Wilson

15 **High-performance fibres**
Edited by J. W. S. Hearle

16 **Knitting technology Third edition**
D. J. Spencer

17 **Medical textiles**
Edited by S. C. Anand

18 **Regenerated cellulose fibres**
Edited by C. Woodings

19 **Silk, mohair, cashmere and other luxury fibres**
Edited by R. R. Franck

20 **Smart fibres, fabrics and clothing**
Edited by X. M. Tao
21 **Yarn texturing technology**
J. W. S. Hearle, L. Hollick and D. K. Wilson
22 **Encyclopedia of textile finishing**
H-K. Rouette
23 **Coated and laminated textiles**
W. Fung
24 **Fancy yarns**
R. H. Gong and R. M. Wright
25 **Wool: Science and technology**
Edited by W. S. Simpson and G. Crawshaw
26 **Dictionary of textile finishing**
H-K. Rouette
27 **Environmental impact of textiles**
K. Slater
28 **Handbook of yarn production**
P. R. Lord
29 **Textile processing with enzymes**
Edited by A. Cavaco-Paulo and G. Gübitz
30 **The China and Hong Kong denim industry**
Y. Li, L. Yao and K. W. Yeung
31 **The World Trade Organization and international denim trading**
Y. Li, Y. Shen, L. Yao and E. Newton
32 **Chemical finishing of textiles**
W. D. Schindler and P. J. Hauser
33 **Clothing appearance and fit**
J. Fan, W. Yu and L. Hunter
34 **Handbook of fibre rope technology**
H. A. McKenna, J. W. S. Hearle and N. O'Hear
35 **Structure and mechanics of woven fabrics**
J. Hu
36 **Synthetic fibres: Nylon, polyester, acrylic, polyolefin**
Edited by J. E. McIntyre
37 **Woollen and worsted woven fabric design**
E. G. Gilligan
38 **Analytical electrochemistry in textiles**
P. Westbroek, G. Priniotakis and P. Kiekens
39 **Bast and other plant fibres**
R. R. Franck
40 **Chemical testing of textiles**
Edited by Q. Fan
41 **Design and manufacture of textile composites**
Edited by A. C. Long

42 **Effect of mechanical and physical properties on fabric hand**
Edited by H. M. Behery
43 **New millennium fibers**
T. Hongu, M. Takigami and G. O. Phillips
44 **Textiles for protection**
Edited by R. A. Scott
45 **Textiles in sport**
Edited by R. Shishoo
46 **Wearable electronics and photonics**
Edited by X. M. Tao
47 **Biodegradable and sustainable fibres**
Edited by R. S. Blackburn
48 **Medical textiles and biomaterials for healthcare**
Edited by S. C. Anand, M. Miraftab, S. Rajendran and J. F. Kennedy
49 **Total colour management in textiles**
Edited by J. Xin
50 **Recycling in textiles**
Edited by Y. Wang
51 **Clothing biosensory engineering**
Y. Li and A. S. W. Wong
52 **Biomechanical engineering of textiles and clothing**
Edited by Y. Li and D. X-Q. Dai
53 **Digital printing of textiles**
Edited by H. Ujiie
54 **Intelligent textiles and clothing**
Edited by H. R. Mattila
55 **Innovation and technology of women's intimate apparel**
W. Yu, J. Fan, S. C. Harlock and S. P. Ng
56 **Thermal and moisture transport in fibrous materials**
Edited by N. Pan and P. Gibson
57 **Geosynthetics in civil engineering**
Edited by R. W. Sarsby
58 **Handbook of nonwovens**
Edited by S. Russell
59 **Cotton: Science and technology**
Edited by S. Gordon and Y-L. Hsieh
60 **Ecotextiles**
Edited by M. Miraftab and A. R. Horrocks
61 **Composite forming technologies**
Edited by A. C. Long
62 **Plasma technology for textiles**
Edited by R. Shishoo
63 **Smart textiles for medicine and healthcare**
Edited by L. Van Langenhove

64 **Sizing in clothing**
Edited by S. Ashdown
65 **Shape memory polymers and textiles**
J. Hu
66 **Environmental aspects of textile dyeing**
Edited by R. Christie
67 **Nanofibers and nanotechnology in textiles**
Edited by P. Brown and K. Stevens
68 **Physical properties of textile fibres Fourth edition**
W. E. Morton and J. W. S. Hearle
69 **Advances in apparel production**
Edited by C. Fairhurst
70 **Advances in fire retardant materials**
Edited by A. R. Horrocks and D. Price
71 **Polyesters and polyamides**
Edited by B. L. Deopura, R. Alagirusamy, M. Joshi and B. S. Gupta
72 **Advances in wool technology**
Edited by N. A. G. Johnson and I. Russell
73 **Military textiles**
Edited by E. Wilusz
74 **3D fibrous assemblies: Properties, applications and modelling of three-dimensional textile structures**
J. Hu
75 **Medical and healthcare textiles**
Edited by S. C. Anand, J. F. Kennedy, M. Miraftab and S. Rajendran
76 **Fabric testing**
Edited by J. Hu
77 **Biologically inspired textiles**
Edited by A. Abbott and M. Ellison
78 **Friction in textile materials**
Edited by B. S. Gupta
79 **Textile advances in the automotive industry**
Edited by R. Shishoo
80 **Structure and mechanics of textile fibre assemblies**
Edited by P. Schwartz
81 **Engineering textiles: Integrating the design and manufacture of textile products**
Edited by Y. E. El-Mogahzy
82 **Polyolefin fibres: Industrial and medical applications**
Edited by S. C. O. Ugbolue
83 **Smart clothes and wearable technology**
Edited by J. McCann and D. Bryson
84 **Identification of textile fibres**
Edited by M. Houck
85 **Advanced textiles for wound care**
Edited by S. Rajendran

86 **Fatigue failure of textile fibres**
Edited by M. Miraftab
87 **Advances in carpet technology**
Edited by K. Goswami
88 **Handbook of textile fibre structure Volume 1 and Volume 2**
Edited by S. J. Eichhorn, J. W. S. Hearle, M. Jaffe and T. Kikutani
89 **Advances in knitting technology**
Edited by K-F. Au
90 **Smart textile coatings and laminates**
Edited by W. C. Smith
91 **Handbook of tensile properties of textile and technical fibres**
Edited by A. R. Bunsell
92 **Interior textiles: Design and developments**
Edited by T. Rowe
93 **Textiles for cold weather apparel**
Edited by J. T. Williams
94 **Modelling and predicting textile behaviour**
Edited by X. Chen
95 **Textiles, polymers and composites for buildings**
Edited by G. Pohl
96 **Engineering apparel fabrics and garments**
J. Fan and L. Hunter
97 **Surface modification of textiles**
Edited by Q. Wei
98 **Sustainable textiles**
Edited by R. S. Blackburn
99 **Advances in yarn spinning technology**
Edited by C. A. Lawrence
100 **Handbook of medical textiles**
Edited by V. T. Bartels
101 **Technical textile yarns**
Edited by R. Alagirusamy and A. Das
102 **Applications of nonwovens in technical textiles**
Edited by R. A. Chapman
103 **Colour measurement: Principles, advances and industrial applications**
Fdited by M. L. Gulrajani
104 **Fibrous and composite materials for civil engineering applications**
Edited by R. Fangueiro
105 **New product development in textiles: Innovation and production**
Edited by L.Horne
106 **Improving comfort in clothing**
Edited by G. Song
107 **Advances in textile biotechnology**
Edited by V. A. Nierstrasz and A. Cavaco-Paulo

108 **Textiles for hygiene and infection control**
Edited by B. McCarthy
109 **Nanofunctional textiles**
Edited by Y. Li
110 **Joining textiles: Principles and applications**
Edited by I. Jones and G. Stylios
111 **Soft computing in textile engineering**
Edited by A. Majumdar
112 **Textile design**
Edited by A. Briggs-Goode and K. Townsend
113 **Biotextiles as medical implants**
Edited by M. W. King, B. S. Gupta and R. Guidoin
114 **Textile thermal bioengineering**
Edited by Y. Li
115 **Woven textile structure**
B. K. Behera and P. K. Hari
116 **Handbook of textile and industrial dyeing Volume 1: Principles, processes and types of dyes**
Edited by M. Clark
117 **Handbook of textile and industrial dyeing Volume 2: Applications of dyes**
Edited by M. Clark
118 **Handbook of natural fibres Volume 1: Types, properties and factors affecting breeding and cultivation**
Edited by R. Kozłowski
119 **Handbook of natural fibres Volume 2: Processing and applications**
Edited by R. Kozłowski
120 **Functional textiles for improved performance, protection and health**
Edited by N. Pan and G. Sun
121 **Computer technology for textiles and apparel**
Edited by J. Hu
122 **Advances in military textiles and personal equipment**
Edited by E. Sparks
123 **Specialist yarn and fabric structures**
Edited by R. H. Gong
124 **Handbook of sustainable textile production**
M. I. Tobler-Rohr
125 **Woven textiles: Principles, developments and applications**
Edited by K. Gandhi
126 **Textiles and fashion: Materials, design and technology**
Edited by R. Sinclair
127 **Industrial cutting of textile materials**
I. Viļumsone-Nemes
128 **Colour design: Theories and applications**
Edited by J. Best
129 **False twist textured yarns**
C. Atkinson

130 **Modelling, simulation and control of the dyeing process**
R. Shamey and X. Zhao
131 **Process control in textile manufacturing**
Edited by A. Majumdar, A. Das, R. Alagirusamy and V. K. Kothari
132 **Understanding and improving the durability of textiles**
Edited by P. A. Annis
133 **Smart textiles for protection**
Edited by R. A. Chapman
134 **Functional nanofibers and applications**
Edited by Q. Wei
135 **The global textile and clothing industry: Technological advances and future challenges**
Edited by R. Shishoo
136 **Simulation in textile technology: Theory and applications**
Edited by D. Veit
137 **Pattern cutting for clothing using CAD: How to use Lectra Modaris pattern cutting software**
M. Stott
138 **Advances in the dyeing and finishing of technical textiles**
M. L. Gulrajani
139 **Multidisciplinary know-how for smart textiles developers**
Edited by T. Kirstein
140 **Handbook of fire resistant textiles**
Edited by F. Selcen Kilinc
141 **Handbook of footwear design and manufacture**
Edited by A. Luximon
142 **Textile-led design for the active ageing population**
Edited by J. McCann and D. Bryson
143 **Optimizing decision making in the apparel supply chain using artificial intelligence (AI): From production to retail**
Edited by W. K. Wong, Z. X. Guo and S. Y. S. Leung
144 **Mechanisms of flat weaving technology**
V. V. Choogin, P. Bandara and E. V. Chepelyuk
145 **Innovative jacquard textile design using digital technologies**
F. Ng and J. Zhou
146 **Advances in shape memory polymers**
J. Hu
147 **Design of clothing manufacturing processes: A systematic approach to planning, scheduling and control**
J. Gersak
148 **Anthropometry, apparel sizing and design**
D. Gupta and N. Zakaria
149 **Silk: Processing, properties and applications**
Edited by K. Murugesh Babu
150 **Advances in filament yarn spinning of textiles and polymers**
Edited by D. Zhang

151 **Designing apparel for consumers: The impact of body shape and size**
Edited by M.-E. Faust and S. Carrier
152 **Fashion supply chain management using radio frequency identification (RFID) technologies**
Edited by W. K. Wong and Z. X. Guo
153 **High performance textiles and their applications**
Edited by C. A. Lawrence
154 **Protective clothing: Managing thermal stress**
Edited by F. Wang and C. Gao
155 **Composite nonwoven materials**
Edited by D. Das and B. Pourdeyhimi
156 **Functional finishes for textiles: Improving comfort, performance and protection**
Edited by R. Paul
157 **Assessing the environmental impact of textiles and the clothing supply chain**
S. S. Muthu
158 **Braiding technology for textiles**
Y. Kyosev
159 **Principles of colour appearance and measurement Volume 1: Object appearance, colour perception and instrumental measurement**
A. K. R. Choudhury
160 **Principles of colour appearance and measurement Volume 2: Visual measurement of colour, colour comparison and management**
A. K. R. Choudhury

Preface

Textiles and fashion are profoundly inter linked, and remain chained together by a series of interlocking processes, and an understanding of the elements that make up this interrelationship is explored through the individual chapters in *Textiles and Fashion*.

The Oxford dictionary defines 'Fashion as a currently popular style of clothing, or behaviour, etc., or the production and marketing of new styles of clothing and cosmetics', it defines textiles as 'a type of cloth or woven fabric or relating to fabric or weaving'.

In an ever increasing consumer-lead world, more weight is given to 'fashion' and its impact, on our everyday lives, yet the acknowledgement should be that textiles and fashion coexist and cannot exist without each other, and the need to understand how each of these areas coexist and the frameworks of operation are therefore an important part of the repertoire of the twenty-first-century textile or fashion designer, practitioner, technologist or those who engage with fashion and textiles on whatever level.

The mutual dependency of both these industries on each other, are even more relevant in a twenty-first-century design continuum. Teaching textiles as I have done in a 'Design' department, my textiles and fashion thinking has broadened, leading me to increasingly question how to deliver such specialist knowledge within a nonspecialist domain. It has also lead to surprising discoveries of material cognisance, and wider application of the textiles 'tacit' knowledge in other design disciplines. The application of textiles practises in smart materials, e-textiles and impact on areas of 3D printing, the utilisation of other technologies to explore 'textile craft practices', have all contributed to the development of chapters and thinking throughout this book. The development of a designer who has 'hybrid' skills and knowledge to imagine new design futures should be encouraged, not only in our design colleges, but also in our design studios, and making spaces.

This book has also been developed based on the ever growing need of the way in which fashion and textiles is taught. First, textiles and fashion is still for the most part still taught in specialist colleges, where students on fashion and textiles courses are often in the same building but not in the same lectures, or to those on other Design courses, often regarded as 'nonspecialists' who may not be specialising in fashion or textiles but want to utilise 'Textiles or Fashion' as the vehicle for output.

There is also the growing design 'materials' specialists who realise that it is the inherent properties of textile fibres or textile processes that could be utilised outside its normal domain, or those interested in the construction process who see the validity of the architectural nature of fashion construction as a process to do this through. The development of new smart and technical textiles, alongside the growth of areas such as digital and 3D printing, highlight the 'hybrid' nature of what is considered today to be textiles and fashion. The new 'designers' no longer come just from traditional textile and fashion schools, this text will allow even these students or professionals to engage with this subject.

This book provides a platform for an integrated study experience and researching additional information. Its aim is to provide a framework and platform from which to engage with a more holistic overview of the industry of textiles and fashion. This book seeks therefore to capture some of the diversity in textiles and fashion through the documentation of both the aesthetic and technical considerations that this topic entails, with the scope and the shape of each chapter shaped by the specialists that have contributed to the book.

The contributors of this book come from an international background, and as such the textiles and fashion terminology throughout have been made consistent with recognised conventions.

THE PURPOSE OF THIS BOOK

Textiles and Fashion therefore provides a comprehensive overview of the fundamental topics one might be expected to cover when teaching or researching textiles and fashion, ranging from types and properties of textile fibres to yarn and fabric, through to manufacture, fabric finishing, apparel production and fashion. It is ideal both for college and undergraduate students studying textiles or fashion courses. Texts on material culture allow readers to see that textiles and fashion studies do not operate in a vacuum, but are influenced by a range of internal and external factors.

For the lecturer, the book's 31 chapters can be divided into specialised parts to create your own bespoke core reading material for students.

For textile and fashion industry professionals, you have a single reference work that can be utilised to provide an instant guide to the key aspects of your professional practise.

To those who are not specialists in the industry, this text book will allow you to explore individual aspects of the industry, providing a platform for initial discovery, and allowing the opportunity to further build on that knowledge though further reading and supporting project activities.

OVERVIEW OF THE PARTS OF THE BOOK

The book is divided into 4 distinct parts.

PART 1

This part begins with an introduction to the types of fibres, with the first chapters reviewing the different kinds of natural and synthetic fibres and their properties, their uses and impact on textiles and fashion.

PART 2

This focuses primarily on the transition process from yarn to fabric, covering, spinning, weaving, knitting and nonwoven fabrics. This then moves on to discussing specialist fabric structures. The emergence of new smart intelligent textiles, e-textiles and their manufacture is discussed in this part.

PART 3

This reviews the importance of fabric finishes and applications, such as joining techniques, methods of applying colour to fabrics such as dyeing and colouring techniques, printing, which includes traditional techniques and the impact of digital printing is discussed in this part.

PART 4

This focuses on the development of apparel and clothing, with chapters on fashion, the impact of material culture on fashion, key issues in the fashion industry, including sustainability, visual design techniques, marketing, apparel and clothing manufacture and methods for improving apparel clothing, fabric functionality, testing and comfort.

GLOSSARY

All the key terms used in the book are included in the glossary.

Acknowledgements

I would like to thank all of my friends and colleagues, in the fashion and textiles industry who have given their time and expertise to make this book possible. To all the contributors, it is their knowledge and passion of the textiles and fashion industry that has provided invaluable.

To all the contributors who have provided the key elements for the book, that of the creative, technical and 'tacit' knowledge embedded in the practice of textiles and fashion, which contributes to such an ever-changing global industry.

The editorial team at Woodhead Publishing, especially Francis and Anneka, for their unending support and guidance, throughout all stages of this publication.

To my husband Audley, as my test reader, who now has an overview and knowledge of the textiles and fashion industry from all perspectives. To my mother, Bernice, who fuelled the love and passion of all things textiles and fashion, from making to understanding the value of good materials.

Personally I have learnt a lot from this process, including how the connectedness between textiles and fashion remains ever constant, whether it be through the materials, the process, or through the sharing with others who have the same passion.

How to Use this Book

The book is divided into 31 chapters. Each chapter is laid out in the following way, learning objectives, introduction, key information, discussion on emerging trends and sustainable issues, key summary points and references, also a glossary of terms is included at the end of the book. Illustrations are also provided throughout the chapters to assist understanding and the learning process. Each chapter can be used on its own as an introduction to a longer piece of research, as each chapter has activities that can be linked to independent research and further reading.

The case studies in each chapter demonstrate the different approaches that have been taken to explore aspects of creative, design and technological exploration.

PART 1

FIBRE TYPES

CHAPTER

1

UNDERSTANDING TEXTILE FIBRES AND THEIR PROPERTIES: WHAT IS A TEXTILE FIBRE?

R. Sinclair
Goldsmiths, University of London, London, UK

LEARNING OBJECTIVES

At the end of this chapter, you should be able to:

- Understand and know the key classifications of textiles fibres
- Understand key fibres and their properties
- Understand how fibres can be used to develop particular attributes in textile products
- Understand how the key characteristics and properties of fibres can affect end-user requirements

1.1 INTRODUCTION

Fibres are the foundation for all textile products and can either be natural (natural fibres) or man-made (manufactured or man-made rengenerated). Within these two types or groups, there are two main kinds of fibres:

- Fibres of indefinite (very great) length, called **filaments**
- Fibres of much shorter length, called **staple fibres**

Filaments are generally combined and twisted to form yarns, whilst staple fibres are spun to create yarns. Yarns are then typically woven or knitted into fabrics. A piece of a fabric contains a huge number of fibres. For example, a small piece of lightweight fabric may contain over 100 million fibres (Morton & Hearle, 2001). Individual types of fibres can be used on their own or combined with other types of fibres to enhance the quality of the end-product. The process for combining fibres is known as blending. There are many well-known blended fibres on the market, such as Viyella, which is made of a blend of cotton and wool (Corbman, 1983).

A fibre is defined as a small threadlike structure (Hearle, 2009). The American Society for Testing and Materials (ASTM) defines a ‘fibre’ (spelt ‘fiber’ in countries using American English) as ‘a generic term for any one of the various types of matter that form the basic element of a textile, and it is characterised by having a length at least 100 times its diameter’ (Anonymous, 2001). The Textile Institute defines a fibre as a ‘textile raw material, generally characterised by flexibility, fineness and high ratio of length to thickness’ (Anonymous, 2002). A similar industry definition is a ‘unit matter with a length

Textiles and Fashion. http://dx.doi.org/10.1016/B978-1-84569-931-4.00001-5

at least 100 times its diameter, a structure of long chain molecules having a definite preferred orientation, a diameter of 10–200 microns (micrometres), and flexibility' (Landi, 1998). All fibres have a molecular structure that contributes to their specific attributes and properties.

The common characteristics of fibres from these definitions are:

- The diameter of a fibre is small relative to its length
- Properties of 'flexibility' and 'fineness' (a way of describing the thickness of a fibre)

It is these key characteristics and qualities that make it possible to manipulate fibres to create the much larger structures that we are familiar with such as yarns and fabrics (Wilson, 2001). Key characteristics, such as fineness and flexibility, as well as length and diameter, have a profound effect on the properties of any textile product. Many of today's textile products are made up of blends or mixtures of different types of fibres which give a particular mix of properties that best fit how the product will ultimately be used. There are several reasons why different fibres may be blended or mixed (Bunsell, 2009; Erberle, 2004):

- To compensate for weaker attributes or properties of one type of fibre
- To improve the performance of the resulting yarn or fabric
- To improve or provide a different appearance
- To improve the efficiency of processing, especially of spinning, weaving and knitting
- To reduce costs

The advantages and disadvantages of blending and mixing fibres are discussed in Part 1 and Part 2 of this book, and include environmental issues such as recycling blended yarns (Fletcher, 2008).

1.2 TYPES OF TEXTILE FIBRES

Textile production up until the seventeenth century was predominantly a specialised domestic production system (cottage industry) mainly done by women (Gordon, 2011). The key fibres used were wool, cotton, silk, hemp and flax (for linen). The advent of the Industrial Revolution meant mechanisation of the production process, allowing new and much faster methods of manufacturing to emerge. Over the next 300 years, it would be both developments in processing and advances in engineered fibres that would change the textile landscape. Whilst the first man-made or manufactured fibres, namely regenerated cellulose fibres, were developed in the late nineteenth century, industrial production of these fibres only really started in the early twentieth century. Synthetic fibres were developed in the late 1930s and production took off after the Second World War. The quest in the 21st century is now to create fibres that are both functional and sustainable, along with smart fibres that can be adapted precisely to the changing needs of today's users (Aldrich, 2007; Clarke-Braddock & Harris, 2012). Designers, whether in fashion or textiles, seek also to push the boundaries of design, by exploiting the key characteristics of fibres and their properties, such as Japanese textiles company NUNO, or fashion designer Issey Miyake, as characterised in his formidable 'Pleats Please' fashion range.

There are three basic types of fibre groups:

- Natural fibres
- Regenerated fibres
- Synthetic fibres

Regenerated and synthetic fibres are collectively known as man-made or manufactured fibres. The various types of textile fibres are summarised in Figure 1.1. Images of some of these fibres are shown in Figure 1.2. Natural fibres are, as the name suggests, those which occur in nature, such as wool from

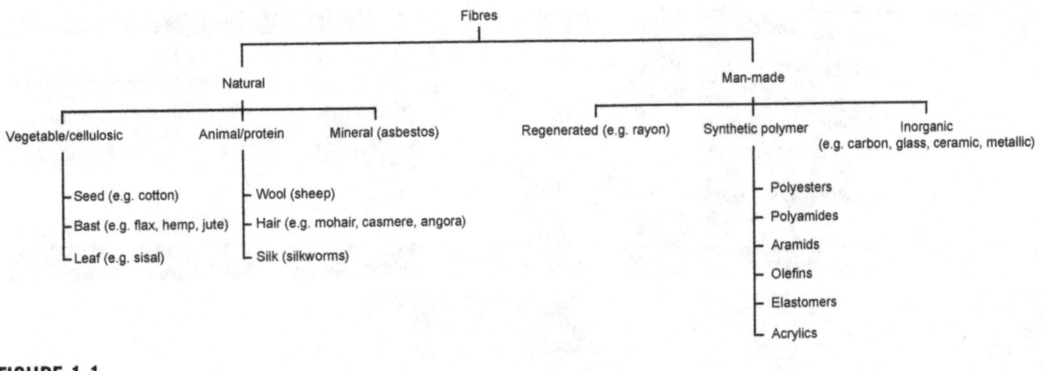

FIGURE 1.1

Types of textile fibres.

sheep or cotton from cotton plants (Kozlowski, 2012a, 2012b). Natural fibres are discussed in Chapters 2 and 3 of this book. Regenerated fibres are made from natural polymers that are not useable in their original form but can be regenerated (i.e. reformed) to create useful fibres (Woodings, 2001). One of first regenerated fibres was rayon, also referred to as viscose or viscose rayon, regenerated from wood pulp. Regenerated fibres are discussed in Chapter 4. The structure and properties of both natural and regenerated fibres are discussed in Eichhorn, Hearle, Jaffe, and Kikutani (2009b). In contrast, synthetic fibres are made by polymerising smaller molecules into larger ones in an industrial process (McIntyre, 2004). Synthetic fibres are discussed in Chapters 5–7.

Natural fibres can be divided into two main types:

- Vegetable or cellulosic fibres (discussed in Chapter 2)
- Animal or protein fibres (discussed in Chapter 3)

There is also one natural mineral fibre, asbestos, though this is no longer used because it has been found to be carcinogenic. Natural rubber is sometimes included in classifications of natural fibres. Other products, such as leather, are also sometimes classified as natural fibres, but fall outside the standard definitions of fibres given above.

Animal (protein)-based fibres can be divided into the following categories:

- Wool (from sheep)
- Hair (e.g. from goats, such as mohair and cashmere; or from rabbits, such as angora)
- Silk (from silkworms)

Wool fibres are discussed in Simpson and Crawshaw (2002) and Johnson and Russell (2008); silk is reviewed in Babu (2013) and Sonwalker (1993); whilst other animal fibres are covered in Franck (2001).

Based on which part of the plant they come from, vegetable fibres can be divided into:

- Seed (e.g. cotton)
- Bast (fibres derived from the outer, or bast, layers of plant stems, e.g. flax, hemp and jute)
- Leaf (e.g. sisal)

Cotton is reviewed in Gordon and Hsieh (2006) and Hallett and Johnston (2010, 2014), whilst bast and other plant fibres are discussed in Franck (2005).

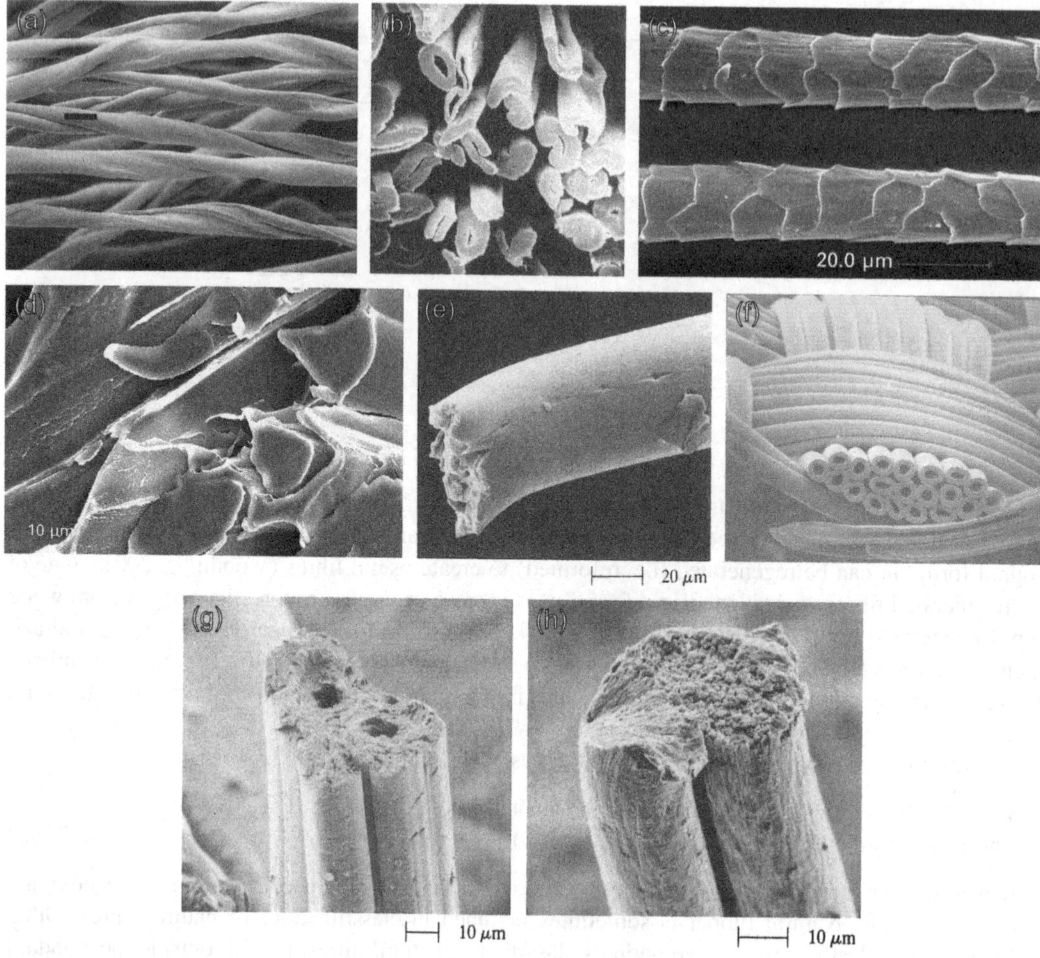

FIGURE 1.2

Images of a range of fibres: (a) longitudinal view of cotton fibres showing their characteristic twist (*Source: Gordon and Hsieh (2006).*); (b) view of cross-sections of cotton fibres showing their characteristic dog-bone shape (*Source: Kozlowski (2012a).*); (c) wool fibres showing scales on the fibre surface (*Source: Eichhorn et al. (2009b).*); (d) silk fibres showing a trilobal cross-section (*Source: Eichhorn et al. (2009b).*); (e) nylon fibre with round cross-section (*Source: Hearle, Lomas, and Cooke (1998).*); (f) nylon fibres with hollow cross-sections in a woven fabric (*Source: McIntyre (2004).*); (g) viscose rayon fibre with a multilobal cross-section (Source: *Hearle et al. (1998).*); (h) acrylic fibre showing a dog-bone shaped cross-section.

Source: Hearle et al. (1998).

Manufactured, or man-made, fibres can be classified as:

- Synthetic polymers, e.g. polyester, nylon (polyamide), acrylic, lycra
- Regenerated, e.g. viscose, modal, acetate
- Inorganic, e.g. carbon, glass, ceramic and metallic fibres

Non-polymer fibres are discussed in Chapter 7 and are also reviewed in Eichhorn et al. (2009b). Synthetic polymer fibres can be classified in a number of ways. One classification is as follows (included with their technical definitions):

- **Polyesters**: Defined as any long-chain synthetic polymer composed of at least 85% by weight of an ester of a substituted aromatic carboxylic acid, including, but not restricted to, substituted terephthalate units and parasubstituted hydroxybenzoate units (e.g. PET, PTT, PBT, PEN, PLA, high-modulus high-tenacity (HM-HT) fibres).
- **Polyamides**: Defined as polymers having in the chain recurring amide groups, at least 85% of which are attached to aliphatic or cyclo-aliphatic groups (e.g. nylon, PVA, PVC).
- **Aramids**: These are defined as polyamides, where each amide group is formed by the reaction of an amino group of one molecule with a carboxyl group of another (e.g. Kevlar, Nomex).
- **Olefins**: Defined as manufactured fibres in which the basic unit is any long-chain synthetic polymer composed of at least 85% by weight of ethylene, propylene or other olefin units (e.g. polypropylene, polyethylene).
- **Elastomers**: Defined as materials that, at room temperature, can be stretched repeatedly to at least twice their original length, and upon immediate release will return to approximately the original length (e.g. polyurethane, Lycra, Spandex).
- **Acrylics**: Defined as manufactured fibres in which the basic substance is a long-chain synthetic polymer composed of at least 85% by weight of acrylonitrile units.

These groups are discussed in Chapters 5 and 6 and are also reviewed in McIntyre (2004), Deopora, Alagirusamy, Joshi, and Gupta (2008), Ugbolue (2009) and Eichhorn, Hearle, Jaffe, and Kikutani (2009a).

1.3 FIBRES, YARNS AND FABRICS

Fibres are commonly classified in the following way:

- Staple fibres
- Filaments
- Tow

As noted earlier, a **staple** fibre is a fibre of relatively short length, as is the case with most natural fibres, which range from a few millimetres (e.g. the shortest cotton fibres, known as linters) to around a metre (e.g. fibres from bast plants). Staple fibres are typically between 3 and 20 cm in length. Given the differences in average fibre length, cotton fibres (2–3 cm) and wool fibres (5 cm or more) are, for example, sometimes referred to as 'short staple' and 'long staple' fibres, respectively.

A **filament** is a fibre of indefinite length. The various silks are the only natural filament fibres. Most regenerated and synthetic fibres are produced as filaments. These can be used in single or multifilament form. Some of these are also assembled to produce a 'tow' which is then cut or broken into required short lengths to produce staple fibres suitable for blending with other fibres, in particular with cotton or wool.

A **tow** can mean two different things:

- In the synthetic fibre industry, a tow is a large assembly of filaments that is destined to be cut into shorter (staple) fibres.
- In the processing of natural fibres (flax), tow is the shorter fibre produced when the stalks are processed to extract the fibres (the long fibres are called line flax).

A yarn has been defined as 'a product of substantial length with a relatively small cross-section, consisting of fibres and/or filaments with or without twist' (Anonymous, 2002; Elsasser, 2011). Another definition of yarn is 'groupings of fibres to form a continuous strand' (Cohen, 1997). Most staple fibres are made into yarn through a process of drawing, spinning and twisting that allows an assembly of fibres to hold together in a continuous strand (Briggs-Goode, 2011). There are different methods of spinning, depending on the fibre being spun, which are discussed in Chapters 9 and 10 and further identified and defined by El Mohaghzy (2009) and Cohen and Johnson (2010). Fibres can also be assembled into larger structures in other ways, e.g. felt and nonwoven fabrics. Nonwoven fabrics are discussed in Chapter 13.

Yarns can be categorised in various ways. Basic yarn types are:

- Monofilament
- Multifilament
- Staple or spun

These types are illustrated in Figure 1.3. As their name suggests, monofilament yarns contain a single filament. More commonly, many filaments are twisted together to form multifilament yarns.

As noted earlier, staple or spun yarns consist of staple fibres combined by spinning into a long, continuous strand of yarn. The key elements of a staple yarn are content, fineness and length, yarn ply and twist. There are many ways of creating a staple yarn from groups of fibres. Typical yarn formations include:

- Single (fibres combined into a single yarn)
- Ply/plied (two or more yarns twisted together)
- Cabled/corded (several plied yarns twisted together)
- Blended/compound (different fibre types combined in a yarn)
- Core spun (a yarn with one type of fibre, usually a filament, in the centre (core) of the yarn, which is usually covered (wrapped) by staple fibres)
- Fancy or effect yarns (yarns with special effects or deliberate irregularities, e.g. slubs (thicker portions) or loops occurring regularly or randomly along the length of the yarn)

Some of these structures and the process of spinning yarns are discussed in Chapters 9 and 10. The combination of different fibres and yarn structures can be used to engineer a particular set of properties (Gong, 2011; Lawrence, 2010; Lord, 2003). Sewing threads are an example of a yarn that is specifically engineered for a specific purpose. Additional finishes are often added to yarns to ensure they are fit for purpose.

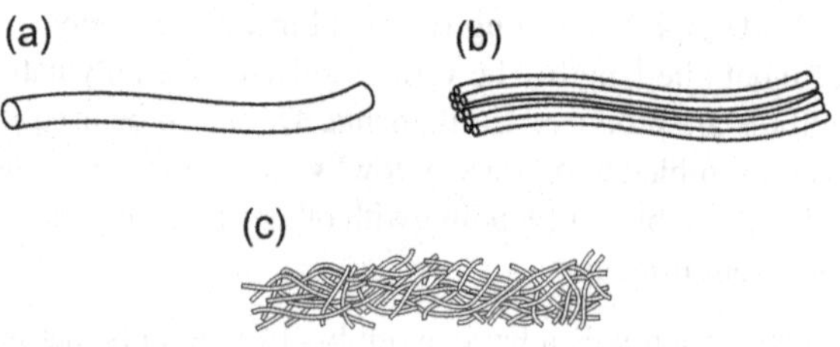

FIGURE 1.3

Types of yarn: (a) monofilament; (b) multifilament; (c) staple/spun.

Yarn fineness, thickness or size is characterised by two main types of numbering systems:

- Direct yarn numbering system: Based on weight or mass of a fixed length of yarn (mass per unit length).
- Indirect yarn numbering system: Based on the length of yarn of a fixed weight (length per unit mass).

Both systems provide a measure of the fineness (or thickness) of the yarn, which is important for its appropriate application in fabric construction techniques such as knitting and weaving.

Once yarns are processed, they then need to be assembled in some way to produce a fabric. A fabric is defined as 'a manufactured assembly of fibres and/or yarns that has substantial surface area in relation to its thickness, and sufficient cohesion to give the assembly useful mechanical strength' (Anonymous, 2002). There are many ways of combining yarns to create a fabric. Some of the most important are:

- Weaving (discussed in Chapter 11)
- Knitting (discussed in Chapter 12)
- Nonwoven fabric production (discussed in Chapter 13)

Some more specialised fabric types include lace, nets, braids and felts.

The different ways of constructing fabrics result in wide variations in texture, appearance, drape (the way a fabric hangs) and hand/handle (the feel of a fabric); as well as performance characteristics such as strength, durability (Schwartz, 2008), comfort and protection. The combination of different fibre types and yarn and fabric structures results in a huge range of products, with widely differing properties (Briggs-Goode, 2011). Weaving is discussed in more depth in various texts, such as Behera and Hari (2010) and Gandhi (2012), Cohen and Johnson (2010) and O'Mahoney (2011); knitting is reviewed in Au (2011) and Spencer (2001); whilst nonwoven production is covered in various texts such as Russell (2006) and Chapman (2010). The use of finishing techniques for fabric allows further refinement of these properties (Fan & Hunter, 2009; Gulrajani, 2013; Rouette, 2001; Wei, 2009). Finishing techniques are discussed in Chapters 17–19. Both fashion and textile designers will exploit the inherent properties and performance characteristics of different types of fabrics when designing and making products.

1.4 FIBRE PROPERTIES

Fibres have a wide range of physical, mechanical and chemical properties (Howes and Laughlin, 2012). Physical and mechanical properties include:

- Length, shape and diameter (fineness)
- Colour and lustre
- Strength and flexibility
- Abrasion resistance
- Handle (or feel), e.g. soft (cashmere), harsh (coir), crisp (linen), elastic (Lycra)
- Moisture absorbency
- Electrical properties

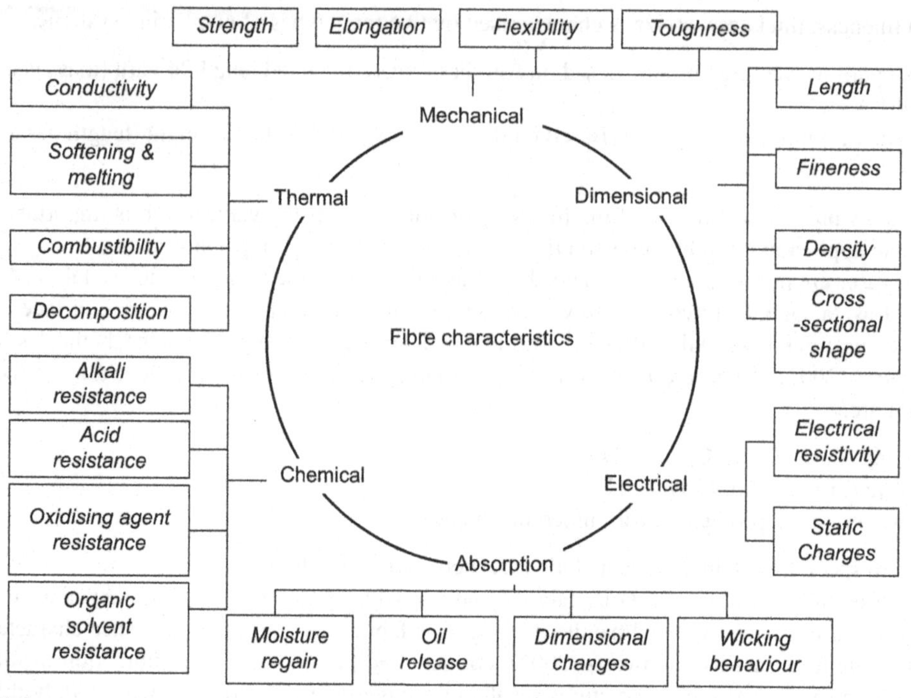

FIGURE 1.4

Key fibre properties.

Source: El Mogahzy (2009).

- Important chemical properties, including fire resistance
- Chemical reactivity and resistance (e.g. resistance to acids, alkalis, solvents, light, etc.)
- Antimicrobial properties

These are discussed in the following sections. Figure 1.4 suggests a way of categorising fibre properties related to performance (El Mogahzy, 2009).

1.5 FIBRE LENGTH, SHAPE AND DIAMETER

This chapter has already discussed the length of different fibre types. Fibre length affects many of the properties of a staple yarn, including strength, evenness and hairiness. Because they are continuous, filament yarns can be made into yarn with little or no twist, producing a smooth, bright appearance, particularly when no crimp (see below) is present. Staple fibres need to be twisted together to form a length of yarn with fibre ends protruding from the surface of the yarn. This produces a duller appearance and more uneven texture. This can be an advantage since it gives some fabrics a softer 'hand' or feel (Behery, 2005).

A characteristic feature of some fibres is 'crimp', which refers to the waviness of a fibre along its length. The only natural fibres that have significant crimp are animal fibres (wool and hairs). Crimp is often imparted to synthetic filament fibres, in order to make them more bulky and comfortable, and to man-made staple fibres for blending with wool or cotton and to facilitate processing. Crimp can be measured by counting the number of crimps or waves per unit length or the percentage increase in fibre length on removal of the crimp. Crimped fibres tend to be more bulky and cohere (cling) together more effectively when being spun into staple fibre yarn. Fibre 'cohesiveness' is an important factor in the successful spinning of staple fibres, and can produce stronger yarns. Crimped fibres, such as wool, have more bulk and better insulating properties, the latter being due to more entrapped air.

Natural fibres come in a range of shapes, whilst synthetic fibres can be manufactured in almost any shape required. Fibre shape can essentially be analysed in two ways:

- By looking at the cross-section of a fibre (i.e. cross-sectionally)
- By looking at the fibre lengthways (i.e. longitudinally)

Some common types of fibre cross-section are:

- Round
- Dog-bone shaped
- Trilobal
- Multilobal
- Serrated
- Hollow

Some common types of cross-section are illustrated in Figure 1.5. Amongst natural fibres, cotton fibres have a characteristic dog-bone (bean or kidney) shape, whilst silk fibres have a more rounded, trilobal shape. Whilst there is greater flexibility in selecting and producing a particular cross-section for synthetic fibres, some tend to have the same basic shape (mostly round or serrated).

Looking at fibres longitudinally, some have a smooth surface whilst others are rough and uneven. Wool fibres have scales similar to human hair, whilst cotton fibres have a characteristic twist. The shape and surface characteristics of a fibre can have an important effect on properties such as cohesion, wetting, wicking, how easily fibres can be cleaned, as well as fabric cover (discussed below) and lustre.

Synthetic filament yarns can be texturised (textured) to alter their surface and other properties. The purpose of texturising is to increase the bulkiness of the yarns and the comfort of the fabrics. Textured yarns can be knitted or woven into fabrics that have the appearance, drape and almost the handle of wool, cotton or silk fabrics, whilst retaining the synthetic fabrics advantages of better washability, ease of care and lower cost compared to natural fibre fabrics.

An important property of a fabric is 'cover', which describes the degree to which the yarn and fibres cover the space or area occupied by the fabric. The more open the fabric, with more open spaces or gaps, the lower the cover. Good cover in a fabric is important because it determines certain comfort-related and other properties such as degree of protection and transparency. The 'cover factor' of a woven fabric is a number indicating the area covered by the fibres and yarns in the fabric, relative to that covered by the fabric.

The diameter of a fibre is the distance across its cross-section. Because they are irregular, the diameter of natural fibres usually varies over their length, so an average is used. Fibre diameter is usually measured in millionths of a metre, known as 'microns' or 'micrometres' (using the symbol μm).

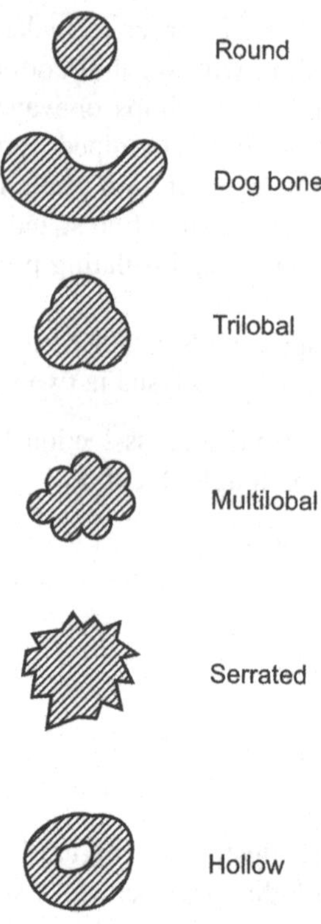

FIGURE 1.5

Fibre shapes.

Typical textile fibres have a diameter of between 10 and 20 μm, though some can reach 50 μm. Natural fibres range in diameter from silk (10–13 μm) to wool (up to 40 μm). Synthetic fibres can be manufactured in diameters from as small as 6 μm (known as microfibers) up to heavy-duty carpet fibres (over 40 μm). Nanofibres, with a diameter below 100 nm (nanometres), are also produced (Houck, 2009). A small diameter produces 'finer' fibres with a greater pliability, flexibility and softness. This results in a fabric with better or softer hand (i.e. feel) and drape (the way a fabric hangs). The fineness of fibres is discussed in more detail below. Apparel is typically manufactured from small-diameter fibres, whereas larger-diameter fibres are often used for heavy-duty applications such as carpets.

1.6 FIBRE COLOUR AND LUSTRE

The colours of natural fibres vary. Wool fibres can, for example, vary in colour from black to white, usually being creamish in colour. Filaments are usually white when manufactured but can be coloured to almost any colour, either prior to manufacturing or subsequently. When synthetic fibres are mass dyed (also called dope dyeing, mass-pigmentation or mass-colouring), the colouring matter is incorporated in the polymer before the filaments are formed (i.e. extruded). The colour and surface characteristics of different fibres have a major effect on fabric appearance, including lustre (the amount and nature of light reflected by a fibre, yarn or fabric). A smooth surface and more regular cross-sectional shape (e.g. the smooth, trilobal shape of silk fibres) will reflect light more strongly and evenly, creating a high lustre. A fibre like cotton, with a rough surface and irregular, twisted cross-sectional shape, has a lower lustre. As well as reflecting light more strongly, smooth, round fibres tend to show soiling more easily than, e.g. multilobal fibres, which are preferred for products such as carpets where dirt and wear may be a significant problem.

Another method of colouring fibres is to dye them in fibre form, before they are spun into yarns. This is called stock dyeing, and is to all intents and purposes confined to wool and hair fibres. Stock dyeing permits the mixing of differently coloured fibres before spinning, and this in turn permits the production of yarns containing several intimately mixed colours. This enables designers to achieve colour effects that cannot be obtained in any other way. Harris tweeds are an example of these colour effects. Dyeing can also be carried out in yarn, garment or fabric form, the latter being the most common.

1.7 FIBRE FINENESS

In textiles, fineness refers to the thinness of the fibre. Thinner fibres have greater surface-to-weight ratios and are more flexible, giving a softer drape (flexibility) and handle than thicker fibres. Generally speaking, the thinner the fibre, yarn or fabric, the better its quality and the higher its price. Fineness is measured in textiles in several ways, depending, for example, on whether it is in fibre, yarn or fabric form.

1.7.1 FIBRES

There are two usual ways of measuring the fineness of textile fibres. The one usually used for wool and animal hairs is to measure the diameters of a sample of the fibres and express the result as the average of these diameters. The most usual unit used, bearing in mind the smallness of these diameters, is the micron (μm), which is one millionth of a metre. As an example, fine Merino wool would have a diameter of between 18 and 20 μm.

The second common way of expressing the fineness of textile fibres, which applies particularly to filaments such as silk and certain synthetic fibres such as polyester and nylon, is in terms of weight per unit length, such as denier or tex. A denier is the weight in grams of 9000 m of filament or fibre. An example is ladies 15 denier nylon stockings, which vary from about 1.3 denier to 2.5 denier. The tex system, expressed in grams per 1000 m (see below), is also sometimes applied to fibres, especially in scientific and technical circles. Since fibres are usually very fine, millitex is more often used (1000 millitex = 1 tex).

1.7.2 YARNS

The systems of yarn fineness used in the textile industry vary according to the fibres used and the traditions of the areas where the yarns are produced. Yarn fineness measurement systems can be either direct or indirect.

In the direct system, the weight of a given length of yarn is used, the denier and tex systems being the most commonly used. The denier system is based on the weight in grams of 9000 m of filament or yarn. This is the traditional system used for silk and often for synthetic filaments. Therefore, 9000 m of a 20 denier silk yarn will weigh 20 g. The same length of 40 denier yarn will weigh 40 g. In the tex system (discussed in more detail below), 1000 m of a 3 tex yarn will weigh 3 g, and 1000 m of a 30 tex yarn will weigh 30 g. In direct systems, the greater the number of the yarn, the thicker the yarn. Direct systems are sometimes called linear density systems, since they represent the mass per unit length of yarn.

Over the last 50 years or so, as textiles became a truly global business, (Diamond, 2002) there was a growing need to have common systems of reference and measurement, including yarn fineness. This resulted in the development of the tex direct system of yarn fineness or linear density measurement, which has now attained worldwide recognition. The unit in this system is called the 'tex' and it corresponds to the number of grams per 1000 m (1 km) of yarn. Sometimes, for yarns finer than 1 tex, to avoid using fractions or decimals, the yarn number is expressed in decitex, where 10 decitex equals 1 tex (to convert from tex to decitex, multiply by 10).

Indirect systems are based on the length of yarn per unit mass. Different indirect measurement systems have developed over time. A traditional British name for these indirect measurement systems, also termed counts, is 'grist' (see Figure 1.6). Examples are:

- **The English cotton system.** Although called English cotton count, this system is also used in the United States, Europe, Asia and other places for describing cotton (100% cotton and cotton/polyester blended) spun yarns. This count is based on the number of 840 yard hanks in one pound (454 g). The number of 840 yard lengths of yarn together weighing one pound is its cotton count. For example, for a 30s cotton count (cc), one pound of yarn will consist of 30 × 840 yards = 25,200 yards (23,043 m) of yarn, whilst 15s cc yarn would be half that length. In indirect systems, the higher the count, the finer (thinner) the yarn.
- **Worsted count.** Number of 560 yard hanks per pound. There are also many woollen (as opposed to worsted) counts.
- **Flax count.** The number of 300 yard 'leas' (hanks) or yarn lengths per pound. Also called the 'lea' or linen system.

1.7.3 FABRICS

The usual way of measuring the 'fineness' or 'lightness' of fabric (normally one would use the phrase 'the weight of the fabric') is in grams per square metre, although sometimes it is expressed in terms of grams per running metre. A running metre is 1 m of the full width of the fabric, which is often, but not always, about 150 cm, depending, for woven fabrics, on the type and width of the loom used to weave the cloth.

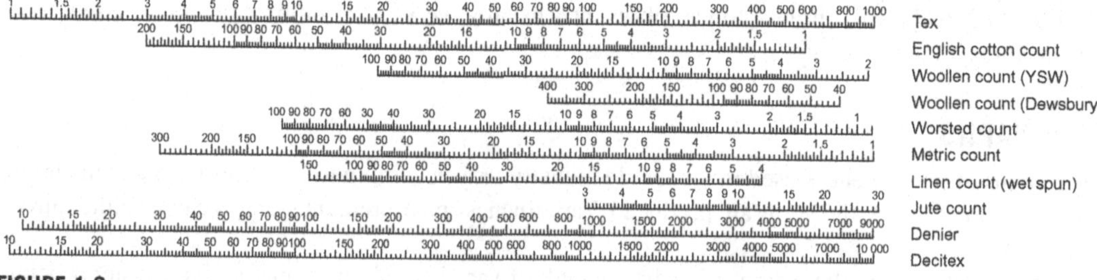

FIGURE 1.6

Comparing common systems of yarn counts.

Source: Slide rule reproduced courtesy of Bludell Harling.

1.8 FIBRE STRENGTH, FLEXIBILITY AND ABRASION RESISTANCE

This section reviews a number of mechanical properties: tensile strength, extension (elongation), flexibility (stiffness), elasticity, resiliency and abrasion resistance.

1.8.1 TENSILE STRENGTH AND EXTENSION (ELONGATION)

Tensile strength and extension are important characteristics of fibres since they influence the performance characteristics of the yarns and fabrics produced from them. As an example, if you are making climbing ropes, you will need a rope that stretches if you fall during a climb, as the stretch (extension) of the rope will absorb the force exerted (impact) when the rope reaches its full length at the end of your fall, thus preventing the sudden shock that might otherwise cause serious injury. You would use an extensible fibre, such as nylon filaments, for such a rope. You would not use flax, hemp or sisal (traditional rope fibres), which have practically no extensibility. These fibres, on the other hand, make excellent marine ropes, where extensibility would be a disadvantage. In the case of the example of a climbing rope, you would also need a rope that will not break when subjected to your full weight (plus that of your back-pack). Nylon, for example, has both high extensibility and high strength, making it a good fibre for climbing purposes.

The resistance to breaking, when a fibre is stretched, is called its 'tensile strength'. It is measured by exerting force, in the form of weights or springs, on the fibre or yarn, and noting both the extension of the fibre when it reaches breaking point (this is called 'elongation at break' or 'breaking elongation') and the force required to achieve this. The elongation at break of different fibres is shown in Figure 1.7.

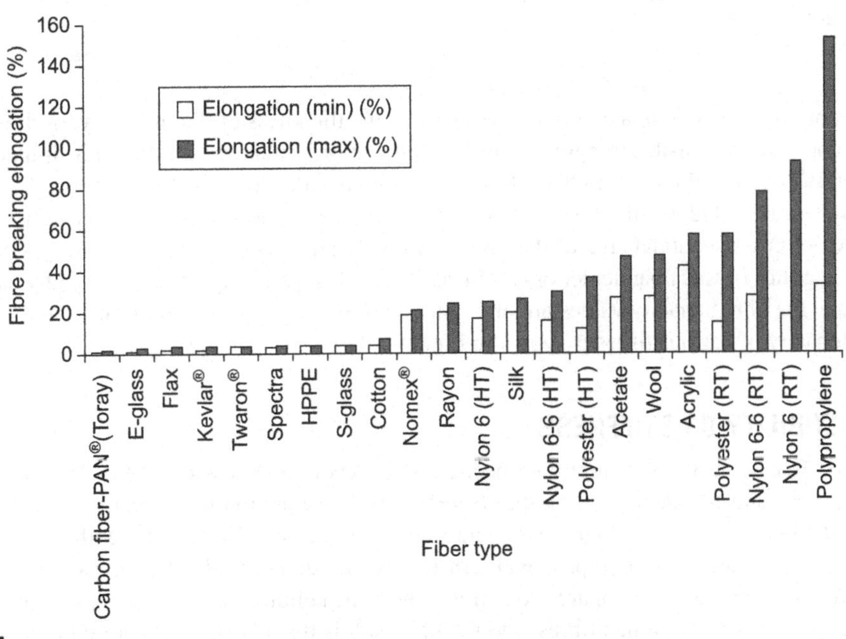

FIGURE 1.7

Elongation of various fibres.

Source: El Mogahzy (2009).

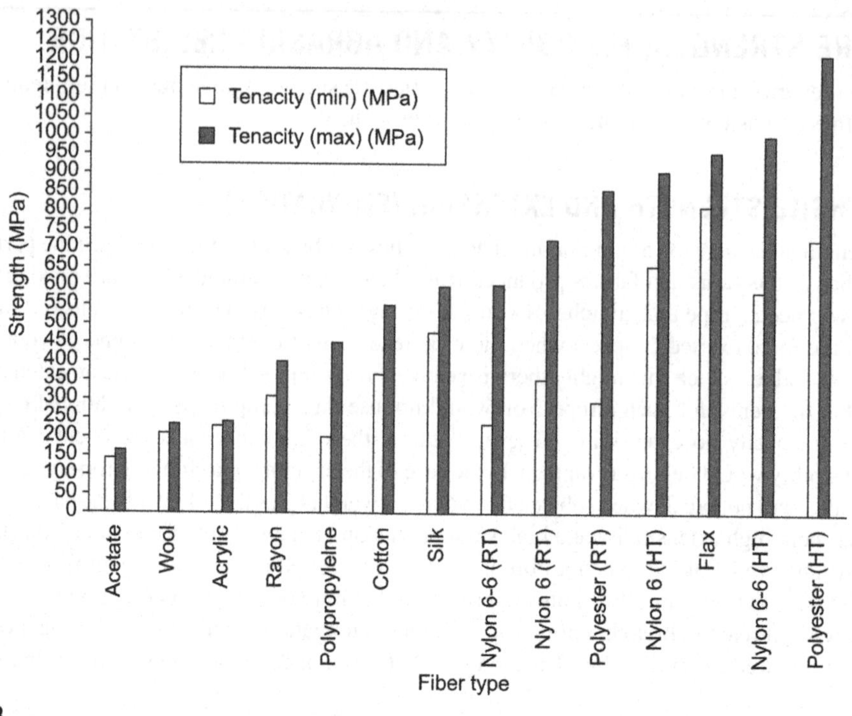

FIGURE 1.8

Tenacity of various fibres.

Source: El Mogahzy (2009).

In terms of tensile strength, a complicating factor is the thickness of the fibre or yarn. It is obvious that if you compare the tensile strength of a nylon filament of 15 denier with that of a similar filament of 30 denier, the latter will have a higher (close on double) tensile strength. To overcome this problem, the 'tenacity' is calculated, by dividing the tensile strength by the linear density (fineness, usually measured in tex), or cross-sectional area of the fibre or yarn. Tenacity is expressed in Newtons per tex[1] or sometimes in grams of force per denier or tex. In addition, because tenacity is influenced by the amount of moisture in the fibre, tenacity values are often measured in both dry (or conditioned) and wet conditions. The tenacity of various fibres is shown in Figure 1.8.

1.8.2 FLEXIBILITY OR STIFFNESS

Flexibility (stiffness) in fibres is important because it affects the processing of the fibre into yarn, as well as the drape and handle of fabrics. Fibre flexibility is a function of the chemical and physical structure of the fibre, and in particular the thickness and cross-sectional shape of the fibre. For example, bast fibres (e.g. flax) are composed principally of highly oriented crystalline cellulose molecules and are therefore fairly rigid when compared to cotton, where the cellulose is much more amorphous. Cotton fibres are also thinner than flax fibres, and the net result is that cotton fibres are more flexible than

[1] Newton: the unit of force that imparts an acceleration of 1 m/s^2 to a mass of 1 kg.

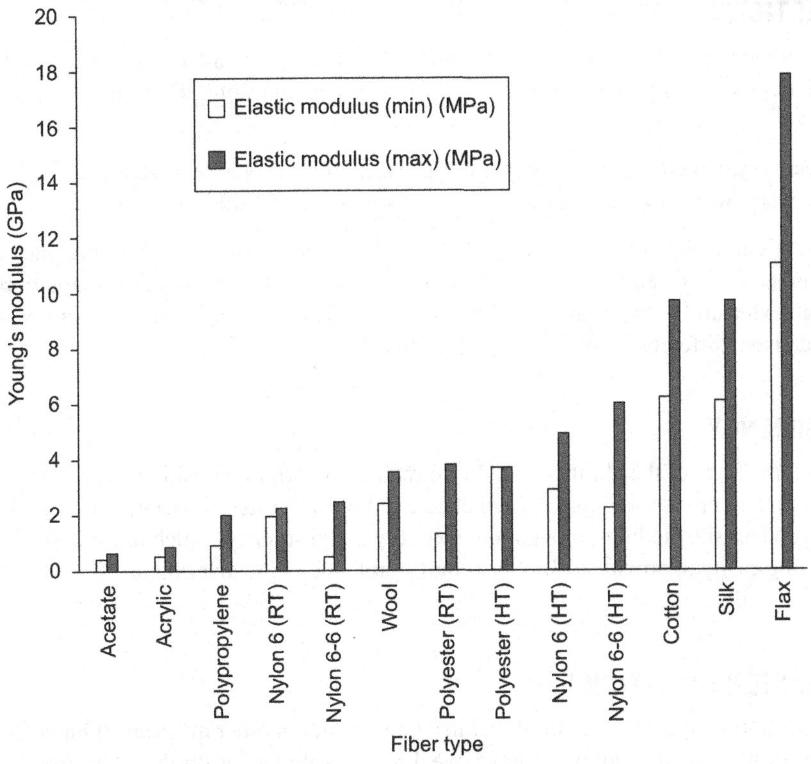

FIGURE 1.9

Modulus of various fibres.

Source: El Mogahzy (2009).

flax fibres. These differences in flexibility mean that fabrics of similar construction, made from the two different fibres, drape differently. Their handle is also different.

There are a number of measures of fibre flexibility, such as bending or tensile modulus. The fibre cross-sectional area, and therefore linear density, has a major effect on the force required to extend or bend the fibre. To compare fibres of different cross-sectional areas or linear density, there must be a correction for the differences in cross-sectional area or linear density. This is normally done by calculating the bending or tensile modulus of the fibre.

'Modulus' is a way of assessing the stiffness or flexibility of a fibre by measuring the fibre's initial resistance to a tensile force. It is measured in units of stress, i.e. force per unit area. Units of measurement include N/tex, GPa or MPa.[2] GPa and MPa stand for gigapascals and megapascals. A Pascal is a measure of force per unit area (pressure) and is the pressure exerted on 1 m^2 of surface by a force of 1 N. Fibres with low initial resistance, or modulus, will easily stretch or bend as the force is applied. Low-modulus materials therefore bend or stretch more easily than high-modulus materials. Most fibres have a modulus within a range of 0.08–10 GPa. Elastomeric fibres, like Spandex, have a low modulus, whilst para-aramid and flax have a high modulus. The modulus of various fibres is shown in Figure 1.9.

[2] GPa and MPa: gigapascal, megapascal. These are units of pressure. A pascal is the pressure exerted on 1 m^2 of surface by 1 N.

1.8.3 ELASTICITY

If a fibre is stretched and then released, it will tend to recover some, or all, of its original length, depending upon the degree of stretch and the inherent elasticity of the fibre. There are two basic types of 'recovery':

- Elastic recovery, which occurs virtually instantly after the stress is removed
- Creep recovery, which occurs gradually after the stress is removed

Elastic recovery can be quantified by measuring the reduction in extension of a fibre once a particular amount of applied force is removed, e.g. 50% recovery after 5% stretch. An elastomeric fibre is defined as having high extensibility, with rapid and substantially complete elastic recovery. Fibres with similar elongation can have different degrees of elastic recovery.

1.8.4 RESILIENCY

Resiliency is a measure of the ability of a fibre to recover its original position after it is distorted, e.g. by bending, twisting or compression. A high elastic recovery is often associated with good resiliency. Nylon has a good tensile modulus, elongation, recovery and resiliency, which make it ideal as a hosiery fabric, stretching easily to provide a close but comfortable fit, and recovering its original shape without wrinkling or bagging.

1.8.5 ABRASION RESISTANCE

Abrasion is the rubbing or friction of fibres against one another or other materials (Gupta, 2008). Fibres with poor abrasion resistance will wear down, weaken and splinter (or fibrillate), weakening the yarns and producing signs of wear in fabrics. Abrasion resistance is related to 'toughness', which is a function of the tenacity and elongation to break of a fibre. Fibres and fabrics can be subjected to three main types of abrasion:

- Flat abrasion, as a result of surface rubbing
- Flex abrasion, as a result of bending, flexing or folding
- Edge abrasion, such as wear of the fabric edges at collars, cuffs, and so on

Some fibres have better abrasion resistance than others. Nylon and wool, for example, have good abrasion resistance, while cotton does not.

1.9 MOISTURE ABSORBENCY

The way fibres respond to moisture is both an important property in itself and one that has a major influence on other properties (Pan & Gibson, 2006). 'Moisture regain' is the ability of a dry fibre to absorb moisture under set conditions of humidity, and can be determined by using the following equation:

$$\text{Regain}\ (\%) = \frac{\text{Mass of conditioned specimen} - \text{Mass of dry specimen}}{\text{Mass of dry specimen}} \times 100$$

Fibres can be divided into two groups:

- Those with a high regain, which absorb water or moisture easily and are known as 'hydrophilic' (i.e. water loving) fibres.
- Those with a low regain, which absorb water or moisture less easily, or not at all, and are known as 'hydrophobic' (i.e. water hating) fibres.

Differences in moisture regain have an important effect on fibre functionality. Hydrophilic/high-regain fibres, like cotton, may, for example, be more comfortable in warm weather because they absorb moisture, in the form of perspiration, from wearers, keeping them dry and comfortable. Hydrophobic/low-regain fibres, like polyester, may be less comfortable, though this may be counter-balanced by properties of adsorption and wicking (see below). Absorption of moisture generally causes fibres to swell, changing their size, shape and weight, as well as their mechanical and frictional properties and stiffness. Since water is an excellent conductor of electricity, its absorption also affects fibre electrical properties. The rate of absorption and desorption of a fibre also affects properties such as drying. Fibres with high moisture absorption can also tend to result in fabrics that change dimensions during use, because they swell and will tend to be more easily deformed during processing and use when moist.

One particularly important aspect of absorbency is that, when a fibre absorbs water or water vapour, heat is released. This is known as 'heat of sorption':

- Fibres with a high regain have a high heat of sorption
- Fibres with a low regain have a low heat of sorption

High-regain fibres, with a high heat of sorption, like wool, will both absorb moisture and generate heat. However, once they are saturated, this heating process stops. Under the same humid conditions, different fibres absorb different amounts of moisture from the atmosphere, depending upon their regain. This enables fibres to be compared in this respect and is called their 'natural regain', or standard regain under standard atmospheric conditions.

It is important to be aware that fibres take in water in two different ways:

- Absorption, in which water is drawn into the fibre
- Adsorption, in which water is held on the surface of the fibre

Adsorption is linked to 'wicking', where the water travels along the surface of the fibre. Some fibres have low absorbency but good adsorption/wicking properties. Examples include some synthetic fibres such as polypropylene, which wicks moisture or perspiration away from the skin and allows it to evaporate, under certain conditions, from the surface. This can be more effective than fibres with good absorption, which can become saturated.

1.10 ELECTRICAL PROPERTIES OF FIBRES

'Electrical conductivity' is the ability of a fibre to carry or transfer an electrical charge. Fibres with low conductivity (e.g. many synthetic fibres) tend to generate and store static electricity. This can cause fibres with the same charge to repel each other and balloon out, or fibres with opposite charges to attract each other, and cling together creating problems in spinning, for example. Similarly they can also cling to, or be repelled by, other surfaces encountered during processing. Because fibres with low conductivity store electricity, they can also cause a static electrical shock when touched. As has been noted, since water is an excellent conductor of electricity, fibres with high regain generally conduct electricity better, allowing it to dissipate. Antistatic finishes, often applied during processing, generally act by retaining moisture, thereby increasing conductivity. Electrical conductivity can also be affected by temperature.

1.11 THERMAL PROPERTIES OF FIBRES

The response of fibres to heat depends on their chemical composition. Fibres can be divided into two broad categories:

- Thermoplastics, which soften and then melt at higher than certain temperatures, the actual temperature depending on the fibre.
- Non-thermoplastics, which tend to char and become brittle, rather than soften or melt, at high temperature.

The application of heat to a fibre results in a loss of strength in the fibre. This can be measured by relating fibre strength to temperature and/or time:

- Thermoplastics: An example would be 'retains 50% of its strength at 150 °C'.
- Non-thermoplastics: An example would be 'retains 50% of its strength after 50 h at 150 °C'.

Aramid fibres, for example, can resist temperatures of 250 °C for 1000 h with only 35% decrease in breaking strength. Fibres that do not degrade readily under the influence of heat are said to have high 'thermal stability'. Glass fibres have a particularly high heat resistance, i.e. thermal stability up to 450 °C. In contrast, heat-sensitive thermoplastic fibres may both lose strength and shrink when heated. A special heat treatment (called 'heat setting') may be needed to make the fibre more resistant to changes in dimensions and other properties when encountering temperatures below that used during heat setting.

Thermal conductivity is the ability of a fibre to conduct heat. Most fibres exhibit low thermal conductivity, making them good insulators, but generally not as good as air. The thermal insulation or conductivity of textile fabrics, notably apparel fabrics, is influenced more by the amount of air trapped in the fabric than by the type of fibre or the thermal insulation or conductivity of the particular fibre. The thermal conductivity of various fibres is shown in Figure 1.10.

A critical property of fibres is their 'flammability', i.e. the tendency of a fibre to ignite and burn with a flame, creating a fire hazard. Fibres that char or smoulder, but do not ignite, are described as 'flame resistant' or 'flame retardant' (Horrocks & Price, 2001, 2008; Kilinc, 2008). A standard measure of flammability is the Limiting Oxygen Index (LOI) which represents the percentage of oxygen required to allow burning to occur. The lower the LOI, the more easily the fibre can burn. Natural fibres like cotton and certain synthetic fibres like polyester, nylon and olefin have relatively low LOIs of between 18 and 21; whereas other natural fibres like wool, and other synthetic fibres like the aramids, are more flame resistant, with LOIs between 25 and 30. The thermal properties and behaviour of a number of fibres are shown in Table 1.1.

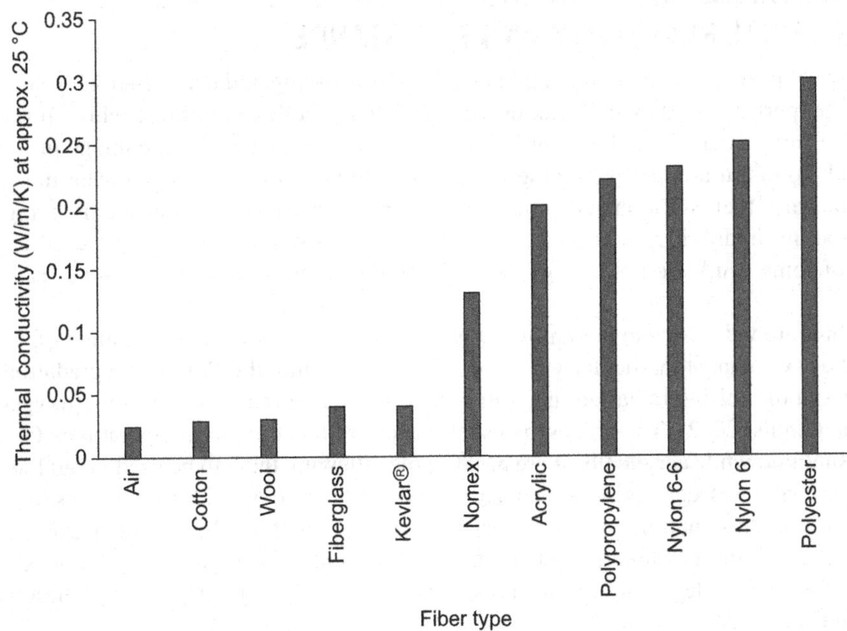

FIGURE 1.10

Thermal conductivity of various fibres.

Source: El Mogahzy (2009).

Table 1.1 Thermal behaviour of some fibres

Fibre	T_g (°C) (softens)	T_m (°C) (melts)	T_p (°C) (pyrolysis)	T_c (°C) (ignition)	LOI (%)
Wool			245	600	25
Cotton			350	350	18.4
Viscose			350	420	18.9
Nylon 6	50	215	431	450	20–21.5
Nylon 6-6	50	265	403	530	20–21.5
Polyester	80–90	255	420–447	480	20–21
Acrylic	100	>220	290 (with decomposition)	>250	18.2
Polypropylene	−20	165	470	550	18.6
Modacrylic	<80	>240	273	690	29–30
PVC	<80	>180	>180	450	37–39
Oxidised acrylic	–	–	≥640	–	–
Meta-aramid (e.g. Nomex)	275	375	410	>500	29–30
Para-aramid (e.g. Kevlar)	340	560	>590	>550	29

Source: Horrocks and price (2001).

1.12 CHEMICAL REACTIVITY AND RESISTANCE

Fibres will be exposed to a range of chemicals during processing and use, which may substantially affect their properties. Acids will break down and damage cellulosic fibres, whilst protein fibres are more resistant to acids. On the other hand, alkalis can break down and damage protein fibres, like silk and wool, but are less damaging to natural cellulosic fibres, and can even in some cases be used to improve their strength and appearance, as in the case of cotton (such a finishing process, involving sodium hydroxide, is known as 'mercerization'). Alkalis can also be used to alter the properties of some synthetic fibres, e.g. improving the fineness of polyester by the 'caustication' process.

Some fibres are vulnerable to oxidation, either from chemicals during processing or from oxygen present in the environment, particularly when combined with sunlight. Polymer degradation can take place as a result of ionising radiation (e.g. ultraviolet light), free radicals, oxidisers (oxygen, ozone) and chlorine (Annis, 2012). The process is usually accelerated at higher temperatures. Olefin fibres degrade in sunlight, while acrylic fibres are more stable, allowing them to be used in outdoor applications, such as awnings. Degradation or damage can also be caused by micro-organisms (e.g. bacteria and fungi), which can sometimes present a serious problem in natural fibres. With a growing concern about the cost and impact of textile waste on the environment, and an increasing emphasis on recycling, the controlled biodegradability of fibres has become an important property. These issues are discussed in Chapter 21.

1.13 CASE STUDIES: FROM FIBRE PROPERTIES TO TEXTILE PRODUCTS

Table 1.2 compares the following properties of a range of synthetic and natural fibres:

- Tensile strength
- Elongation
- Specific gravity
- Young's modulus
- Moisture regain

Comparing these properties helps manufacturers to decide which fibre or combination of fibres to use for a particular application, as the following case studies illustrate.

1.13.1 CASE STUDY 1: CHOOSING APPAREL FIBRES

The importance of understanding fibre properties, and their implications in use, can be seen, for example, in selecting the best fibres to use in apparel (Bubonia, 2012; El Mogahzy, 2009). An example is the choice between fibres, such as cotton and polyester. In the case of cotton, some key properties are:

- Tenacity: 2.6–4.3 cN/dtex in dry conditions; 2.9–5.6 cN/dtex in wet conditions
- Stiffness: 60–82 cN/dtex (Young's modulus)
- Moisture regain: 8.5% (under standard conditions: 65% humidity and 21 °C)

Table 1.2 Properties of a range of fibres

	Tensile strength (cN/dtex)		Elongation (%)	Specific gravity	Young's modulus (cN/dtex)	Moisture regain (%)
	Dry	Wet	Dry			
Polyester (PET)	3.8–5.3	3.8–5.3	20–32	1.38	79–141	0.4
Polyester (PBT)	2.6–4.4	2.6–4.4	20–40	1.31	18–35	0.4
Polyester (PTT)	3.7–4.4	3.7–4.4	20–40	1.34	23	0.4
Polyacrylonitrile	2.2–4.4	1.8–4.0	25–50	1.14–1.17	34–75	2.0
Nylon 6	4.2–5.7	3.7–5.2	28–45	1.14	18–40	4.5
Nylon 6-6	4.4–5.7	4.0–5.3	25–38	1.14	26–46	4.5
Vinylon (PVA)	3.5–5.7	2.8–4.6	12–26	1.26–1.30	53–79	5.0
Polypropylene	4.0–6.6	4.0–6.6	30–60	0.91	35–106	0
Polyethylene	4.4–7.9	4.4–7.9	8–35	0.94–0.96		0
Polyurethane	0.5–1.1	0.5–1.1	450–800	1.0–1.3		1.0
Viscose rayon	1.5–2.0	0.7–1.1	18–24	1.50–1.52	57–75	11.0
Triacetate	1.1–1.2	0.6–0.8	25–35	1.30	26–40	3.5
Silk	2.6–3.5	1.9–2.5	15–25	1.33	44–88	11.0
Wool	0.9–1.5	0.7–1.4	25–35	1.32	10–22	15
Cotton	2.6–4.3	2.9–5.6	3–7	1.54	60–82	8.5

Source: Houck (2009).

These properties give cotton such advantages as good handle and moisture absorbency, but also certain disadvantages, such as relatively low strength and durability. Typical parameters for polyester are:

- Tenacity: 3.8–5.3 cN/dtex (in both dry and wet conditions)
- Stiffness: 80–140 cN/dtex
- Moisture regain: 0.4%

These properties mean that polyester fibres are stronger and stiffer than cotton. In addition, they can be produced in a wide range of fibre lengths and levels of finess (Kitamura, 2012). This, for example, makes them easier to spin and produce a stronger yarn. They also have better wrinkle resistance and easy care properties. Polyester fibres are also cheaper to produce than cotton fibres. On the other hand, polyester fibres are stiffer and less absorbent than cotton, making them potentially less comfortable. A manufacturer therefore has a number of options, including:

- Using either 100% cotton or 100% polyester
- Using a blend of cotton and polyester
- A more advanced solution, such as using a core/sheath yarn, with a polyester fibre core and cotton fibre sheath

The final choice will depend on the precise mix of fabric properties, such as comfort, durability and easy care, required for the apparel range, and how these relate to such factors as cost and complexity of production operations. Wider Issues, such as sustanability, (Fletcher & Gross, 2011), are also becoming more important in the choice of fibre. Areas of concern include the 'afterlife' of the product (Fletcher, 2008). The end choice by the consumer is one of the parameters that needs to also be considered by the designer. A good choice overall for shirting fabric could, for example, be a 50/50 blend of cotton and polyester, combining in equal measure the properties of cotton and polyester.

1.13.2 CASE STUDY 2: MICROFIBRES

An example of the importance of understanding not only fibre properties but also how they affect processing can be seen in the case of microfibers (El Mogahzy, 2009). Microfibres were developed in the 1990s, allowing manufacturers to produce fibres that were finer than silk. Microfibres (sometimes known as microdenier or microfilament fibres) are synthetic fibres (made from a range of polymers such as nylon, polyester, acrylic, etc.) with very small diameters, and fineness typically less than 1 denier, though some have been created with a denier as low as 0.001. Their high aspect ratio results in greater fibre flexibility and a higher packing density, which are important in terms of softness, drape and air and water permeability in apparel. They are also used in non-apparel applications, such as filtration or sound insulation materials, where fineness is used to good effect.

The extreme fineness of microfibers, however, poses a range of processing challenges, since they are relatively weak and very flexible, and can break and also entangle easily during processing, particularly when in staple form. They are particularly prone to break and form neps during mechanical processing operations, such as carding and opening, making it critical to set and maintain very precise processing conditions. As a result, their higher performance characteristics need to be balanced against more complex and costly processing requirements.

1.14 SUMMARY

Fibres form the basic building blocks for the yarns and fabrics that are used in the fashion and textile industry today. Fibres can be divided into several categories depending on their origin and type, and can also be blended with each other, depending on the end product and user requirements. Understanding the properties, qualities and characteristics of fibres plays an important part in aligning fibres, yarns and fabrics for specific end-uses and users. Environmental and sustainability issues increasingly need to be considered. The development of new fibres, such as PLA, soy and bamboo, demonstrates the potential for a more holistic manufacturing process for textiles. Growth areas for fibre development include nanotechnology, high performance, multifunctional, smart and technical applications and further developments in microfilament production, as engineered fibres are refined and further developed.

1.15 PROJECT IDEAS

Starting right at the beginning with the choice of fibre or fibre blend, describe in detail, step by step, how you would produce fabrics that are most suitable for each of the following apparel products:

1. A formal suit
2. A formal shirt or blouse
3. A T-shirt

1.16 REVISION QUESTIONS

1. What are the key natural fibre groups, and what methods can be applied to identify them?
2. What are the key synthetic fibre groups?
3. What are the main differences between natural and man-made and synthetic fibres?
4. How are fibres turned into yarns? What particular yarn is specifically engineered for use in the fashion industry, and why does it need specific traits?
5. What are the different sources of information that a designer can use to identify fibres?
6. What are the main issues to be considered in the choice of fibres and their blends?
7. What are the key elements and features of microfibres, and how can these be exploited by designers?
8. What numbering systems are used to denote the fineness of a yarn? Explain how they are determined.
9. What is the main difference between filament and staple fibres, how are they made and how might these features be used to develop fabrics?

REFERENCES AND FURTHER READING

Aldrich, W. (2007). *Fabric form and flat pattern cutting* (2nd ed.). Oxford, UK: Blackwell Books.

Annis, P. (Ed.). (2012). *Understanding and improving the durability of textiles*. Cambridge, UK: Woodhead Publishing Limited.

Anonymous. (2001). *Compilation of ASTM standard definitions*. West Conshohocken, Pennsylvania, USA: American Society for Testing and Materials.

Anonymous. (2002). *Textile terms and definitions* (11th ed.). Manchester, UK: Textile Institute.

Au, K. (Ed.). (2011). *Advances in knitting technology*. Cambridge, UK: Woodhead Publishing Limited.

Babu, K. (2013). *Silkproperties, processing and applications*. Cambridge, UK: Woodhead Publishing Limited.

Behera, B., & Hari, P. (2010). *Woven textile structure*. Cambridge, UK: Woodhead Publishing Limited.

Behery, H. (Ed.). (2005). *Effect of mechanical and physical properties on fabric hand*. Cambridge, UK: Woodhead Publishing Limited.

Briggs-Goode, A. (Ed.). (2011). *Textile design: Principles, advances and applications*. Cambridge, UK: Woodhead Publishing Limited.

Bubonia, J. E. (2012). *Apparel production terms and processes*. USA: Fairchild Publications.

Bunsell, A. (Ed.). (2009). *Handbook of tensile properties of textile and technical fibres*. Cambridge, UK: Woodhead Publishing Limited.

Chapman, R. (Ed.). (2010). *Applications of nonwovens in technical textiles*. Cambridge, UK: Woodhead Publishing Limited.

Clarke-Braddock, S., & Harris, J. (2012). *Digital visions for fashion and textiles, made in code*. London, UK: Thames and Hudson.

Cohen, A. C. (1997). *Beyond basic textiles* (3rd ed.). USA: Fairchild Publications.

Cohen, A. C., & Johnson, J. (2010). *J.J. Pizzuto's fabric science* (9th ed.). USA: Fairchild Publications.

Corbman, B. P. (1983). *Textiles, fiber to fabric* (6th ed.). USA: McGraw Hill International Editions.

Deopora, B., Alagirusamy, R., Joshi, M., & Gupta, B. (Eds.). (2008). *Polyesters and polyamides*. Cambridge, UK: Woodhead Publishing Limited.

Diamond, E., & Diamond, J. (2002). *The world of fashion* (3rd ed.). USA: Fairchild Publications.

Eichhorn, S., Hearle, J., Jaffe, M., & Kikutani, T. (Eds.). (2009a). *Handbook of textile fibre structure. Fundamentals and manufactured polymer fibres: Vol. 1*. Cambridge, UK: Woodhead Publishing Limited.

Eichhorn, S., Hearle, J., Jaffe, M., & Kikutani, T. (Eds.). (2009b). *Handbook of textile fibre structure. Natura, regenerated, inorganic and specialist fibres: Vol. 1*. Cambridge, UK: Woodhead Publishing Limited.

El Mogahzy, Y. E. (2009). *Engineering textiles: Integrating the design and manufacture of textile products*. Cambridge, UK: Woodhead Publishing Limited.

Elsasser, V. H. (2011). *Textiles, principles and concepts* (3rd ed.). Fairchild Publications.

Erberle, H. (Ed.). (2004). *Clothing technology* (4th ed. (Eng.)) Germany: Verlag Europa.

Fan, J., & Hunter, L. (2009). *Engineering apparel fabrics and garments*. Cambridge, UK: Woodhead Publishing Limited.

Fletcher, K. (2008). *Sustainable fashion and textiles, design journeys*. London, UK: Earthscan.

Fletcher, K., & Grose, L. (2011). *Fashion and sustainability, design for change*. London, UK: Laurence King Publishing.

Franck, R. (Ed.). (2001). *Silk, mohair, cashmere and other luxury fibres*. Cambridge, UK: Woodhead Publishing Limited.

Franck, R. (Ed.). (2005). *Bast and other plant fibres*. Cambridge, UK: Woodhead Publishing Limited.

Gandhi, K. (Ed.). (2012). *Woven textiles: Principles, technologies and applications*. Cambridge, UK: Woodhead Publishing Limited.

Gong, H. (Ed.). (2011). *Specialist yarn and fabric structures*. Cambridge, UK: Woodhead Publishing Limited.

Gordon, B. (2011). *Textiles the whole story, uses, meanings and significance*. London: Thames and Hudson.

Gordon, S., & Hsieh, Y. (Eds.). (2006). *Cotton: Science and technology*. Cambridge, UK: Woodhead Publishing Limited.

Gulrajani, M. (Ed.). (2013). *Advances in the dyeing and finishing of technical textiles*. Cambridge, UK: Woodhead Publishing Limited.

Gupta, B. (Ed.). (2008). *Friction in textile materials*. Cambridge, UK: Woodhead Publishing Limited.

Hallett, C., & Johnston, A. (2014). *Fabric for fashion: The complete guide*. London, UK: Laurence King Publishing.

Hallett, C., & Johnston, A. (2010). *Fabric for fashion*. London, UK: The Swatch Book Laurence King Publishing.

Hearle, J. (2009). Fibre structure: its formation and relation to performance. In S. Eichhorn, J. Hearle, J. Jaffe, & T. Kikutani (Eds.), *Handbook of textile fibre structure. Fundamentals and manufactured polymer fibres: Vol. 1*. Cambridge, UK: Woodhead Publishing Limited.

Hearle, J., Lomas, B., & Cooke, W. (1998). *Atlas of fibre fracture and damage to textiles* (2nd ed.). Cambridge, UK: Woodhead Publishing Limited.

Horrocks, A., & Price, D. (Eds.). (2001). *Fire retardant materials*. Cambridge, UK: Woodhead Publishing Limited.

Horrocks, A., & Price, D. (Eds.). (2008). *Advances in fire retardant materials*. Cambridge, UK: Woodhead Publishing Limited.

Houck, M. (2009). Ways of identifying textile fibers and materials. In M. Houck (Ed.), *Identification of textile fibres*. Cambridge, UK: Woodhead Publishing Limited.

Howes, P., & Laughlin, Z. (2012). *Material matters, new materials in design*. London, UK: Blackdog Publishing.

Johnson, N., & Russell, I. (Eds.). (2008). *Advances in wool technology*. Cambridge, UK: Woodhead Publishing Limited.

Kilinc, S. (Ed.). (2008). *Handbook of fire resistant textiles*. Cambridge, UK: Woodhead Publishing Limited.

Kitamura, M. (Ed.). (2012). *Pleats Please Issey Miyake*. Germany: Tashen.

Kozlowski, R. (Ed.). (2012a). *Handbook of natural fibres. Types, properties and factors affecting breeding and cultivation: Vol. 1*. Cambridge, UK: Woodhead Publishing Limited.

Kozlowski, R. (Ed.). (2012b). *Handbook of natural fibres. Processing and applications: Vol. 2*. Cambridge, UK: Woodhead Publishing Limited.

Landi, S. (1998). *The textile conservator's manual* (2nd ed.). Oxford, UK: Butterworth-Heinemann.

Lawrence, C. (Ed.). (2010). *Advances in yarn spinning technology*. Cambridge, UK: Woodhead Publishing Limited.

Lord, P. (2003). *Handbook of yarn production*. Cambridge, UK: Woodhead Publishing Limited.

McIntyre, J. (Ed.). (2004). *Synthetic fibres: Nylon, polyester, acrylic and polyolefin*. Cambridge, UK: Woodhead Publishing Limited.

Morton, W. E., & Hearle, J. W. S. (2001). *Physical properties of textile fibres* (4th ed.). Cambridge, UK: Woodhead Publishing Limited.

O'Mahoney, M. (2011). *Advanced textiles for health and wellbeing*. London, UK: Thames and Hudson.

Pan, N., & Gibson, P. (Eds.). (2006). *Thermal and moisture transport in fibrous materials*. Cambridge, UK: Woodhead Publishing Limited.

Rouette, H.-K. (Ed.). (2001). *Encyclopaedia of textile finishing: 3 volume set*. Cambridge, UK: Woodhead Publishing Limited.

Russell, S. (Ed.). (2006). *Handbook of nonwovens*. Cambridge, UK: Woodhead Publishing Limited.

Schwartz, P. (Ed.). (2008). *Structure and mechanics of textile fibre assemblies*. Cambridge, UK: Woodhead Publishing Limited.

Simpson, W., & Crawshaw, G. (Eds.). (2002). *Wool: Science and technology*. Cambridge, UK: Woodhead Publishing Limited.

Sonwalker, T. (1993). *Handbook of silk technology*. USA: John Wiley Publishers.

Spencer, D. (2001). *Knitting technology* (3rd ed.). Cambridge, UK: Woodhead Publishing Limited.

Ugbolue, S. (Ed.). (2009). *Polyolefin fibres: Industrial and medical applications*. Cambridge, UK: Woodhead Publishing Limited.

Wei, Q. (2009). *Surface modification of textiles*. Cambridge, UK: Woodhead Publishing Limited.

Wilson, J. (2001). *Handbook of textile design*. Cambridge, UK: Woodhead Publishing Limited.

Woodings, C. (Ed.). (2001). *Regenerated cellulose fibres*. Cambridge, UK: Woodhead Publishing Limited.

CHAPTER

NATURAL TEXTILE FIBRES: VEGETABLE FIBRES

2

C. Yu
Dong Hua University, Shanghai, China

LEARNING OBJECTIVES

At the end of this chapter, you should be able to:

- Provide a definition of natural fibres, specifically natural vegetable fibres
- Understand the cultivation, dimension, and appearance of the main vegetable fibres
- Understand the composition and structure of the main vegetable fibres in relation to their properties
- Understand the application of the main vegetable fibres in textiles
- Understand the common properties of fibre and how they are tested

2.1 INTRODUCTION

Natural fibres have been used to make textiles since prehistoric times and are still used today. Nowadays, natural fibres including animal (protein) fibres and vegetable (cellulose) fibres make up almost 50% of the textile fibres produced annually in the world. Vegetable fibre is extracted from plants. Cellulose is the main content of vegetable fibre; therefore, vegetable fibre is usually referred to as plant fibre or natural cellulosic fibre. The categories of natural cellulose fibre include:

- seed fibres (such as cotton, kapok, milkweed, etc.)
- bast fibres (such as flax, ramie, jute, kenaf, hemp, etc.)
- leaf fibres (such as sisal, pineapple, abaca, etc.) and
- nut husk fibre (such as coir).

In addition to cotton, the most commonly used natural vegetable fibres include flax, ramie, jute, kenaf, and sisal. Nowadays, more and more new natural fibres, especially vegetable fibre sources, are being exploited, for example, kapok, pineapple, and apocynum.

2.1.1 MOLECULAR COMPOSITION OF CELLULOSE

Cellulose is a linear polymer, or long chain molecule, combining several 1000 anhydroglucose units. Although glucose, a simple sugar, is soluble in water, cellulose is not due to its huge polymolecule size. Cellulose is a carbohydrate, composed of carbon (44.4%), hydrogen (6.2%), and oxygen (49.4%); the cellulose molecule with the basic units of cellobiose is shown in Figure 2.1.

Textiles and Fashion. http://dx.doi.org/10.1016/B978-1-84569-931-4.00002-7

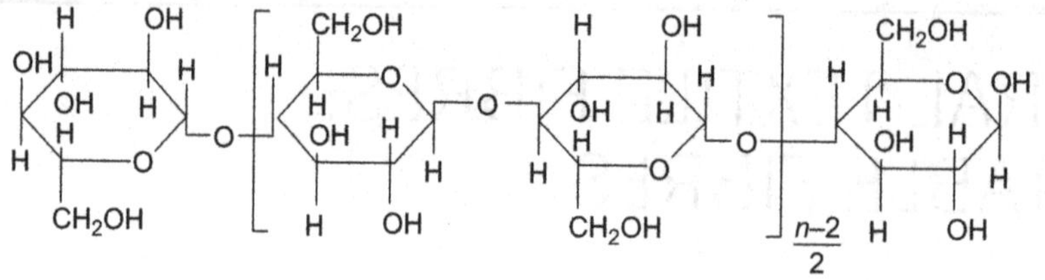

FIGURE 2.1

Diagram of cellulose molecule.

Cellobiose is formed by two combined glucose units; many cellobiose units combine to form cellulose. The number of anhydroglucose units in the cellulose molecule is referred to as the degree of polymerization. Each repeating unit equals 2*n*, or two anhydroglucose units. These units are flipped over as they combine together, as shown in Figure 2.1.

2.1.2 COMMON PROPERTIES OF VEGETABLE FIBRES

Cellulose molecules form into fibrils, or bundles of molecular chains that combine in groups to form the cellulose fibres. Each fibre is composed of many cellulose molecules. These are not arranged in a completely parallel manner; or rather although certain portions of the fibre may have the molecules lying parallel other areas are characterized by a somewhat random molecular arrangement. The parts of the fibre where molecules lie side by side and are held together by many associated forces are called crystalline. If the molecules, as well as lying side by side, are parallel to the longitudinal axis, there is a high degree of molecular orientation. Usually, high orientation and crystallinity imply strength, rigidity, low elongation, and low pliability.

The strength of cellulose fibres is also influenced by the degree of polymerization (dp) as well as by the molecular arrangement. The higher the dp, the stronger the fibre. A typical degree of polymerization for a native cellulose fibre is about 10,000; for regenerated cellulose, such as rayon, it is only about 500. The molecules within the fibre are usually held in place by hydrogen bonding. When cellulose fibres are bent, the hydrogen bonds are broken, and new ones form, causing creases or wrinkles that do not hang out. This is why all products made from cellulose fibres have the same tendency to crease. Removal of the hydroxyl unit may result in cross-linking of the cellulose molecules so that they are more stable, helping to form cellulose fibres that have a higher resistance to and recovery from creasing.

Because of the hydroxyl (–OH) groups in cellulose, these fibres usually have a high attraction for water. This means that in hot weather, perspiration from the body will be absorbed in fabrics made from cellulose fibres, transported along the yarns to the outer surface of the cloth, and evaporated into the air. Thus, the body maintains its temperature and feels comfortable.

Cellulose fibres also tend to burn easily and fast with a yellow flame, giving off a smell like burning paper or leaves, then depositing a light, fluffy, greyish residue or ash. Cellulose is usually decomposed by acid solutions, especially strong mineral acids, but it has excellent resistance to alkaline solutions. In general, cellulose is low in elasticity and resilience, so it is prone to creasing unless treated. The fibres are laundered readily and can withstand strong detergents, high temperatures, and bleaches (if used properly). This group of fibres is seldom damaged by insects, but fungi, such as mildew, will destroy cellulose or at least stain it severely.

2.2 COTTON

2.2.1 DEFINITIONS AND TYPES OF COTTON

Cotton is the most popular natural fibre, accounting for around 90% of all natural fibres. Cotton is one of the most important natural textile fibre crops, both from the agricultural and manufacturing sectors' points of view. It is the biggest source of clothing as well as being used to produce apparel, home furnishings, and industrial products. The main countries producing cotton in the world are China, United States, India, Pakistan, Uzbekistan, Turkey, and Brazil, which together account for over 80% of the world's cotton production.

There are many varieties of cotton, and each variety has different characteristics both in planting and processing performance. Cotton fibres may be classified roughly into three large groups based on staple length (average length of the fibres making up a sample or bale of cotton) and appearance.

1. Long-staple cotton is a fine, lustrous fibre with typical staple length ranging from about 30 to 40 mm, and includes types of the highest quality, such as Sea Island, Egyptian, and Pima cottons. Due to the difficulty in its cultivation and its limited production, long-staple cottons are costly and are used mainly for fine fabrics, yarns, and hosiery.
2. Medium-staple cotton is plentiful and standard, such as American Upland, with staple length from about 25 to 33 mm. Medium-staple cotton provides about 90% of the current world production of raw cotton fibre. This fibre is widely used for apparel, home furnishings, and industrial products.
3. Short-staple is coarse cotton, ranging from about 10 to 25 mm in length, which makes textile processing very difficult and consequently it is not commonly used in textiles, except for very low-grade products.

The colour of cotton fibre varies from almost pure white to a dirty grey. However, the standard cotton grades classify colour as white, light spotted, spotted, tinged, yellow stained, and light grey. High-quality cotton is usually very light or almost white.

2.2.2 CULTIVATION AND GINNING

Cotton for commercial purposes is grown as an annual crop. Cotton is a warm-weather plant cultivated in both hemispheres, mostly in North and South America, Asia, Africa, and India (in tropical latitudes). The period from planting to harvesting is usually 5–7 months. Planting time for cotton varies with locality, i.e. from February to June in the Northern Hemisphere. The time of planting in the Northern Hemisphere is harvest time in the Southern Hemisphere.

After harvesting, the seed cotton (consisting of cotton fibre attached to cottonseed and foreign plant matter) is transported to the ginning plant.

Ginning is the separation of the fibres from the seed and foreign plant matter such as dirt, twigs, leaves, and parts of bolls. Cotton essentially has no commercial value or use until the fibre is separated from the cottonseed and foreign matter at the gin. Ginning operations, which are regarded as part of the harvest rather than the textile process, normally include:

- conditioning (to adjust moisture content)
- seed–fibre separation

- cleaning (to remove foreign plant matter) and
- packaging.

There are two kinds of ginning: saw ginning and roller ginning.

In saw gin, the saw-teeth removes seed and short fibre efficiently with a high production rate, but greater damage to the fibres. In roller ginning, rollers gently separate the fibres from the seeds with less damage to the fibres, but the short fibres and foreign matter are not removed as efficiently reducing productivity. Usually, only the long–staple and high-quality cotton are ginned by roller gin where it is important to prevent fibre damage. Most cottons, like Upland cotton, are ginned on saw gin.

The long lint fibres are removed at the cotton gin and then packed into large bales and transferred to spinning mills for further processing (spinning). The seed with about 8% short linter remaining is used for seed oil after the short linter fibres are removed by the delintering process, producing fibres of different lengths.

2.2.3 STRUCTURE OF COTTON

Each cotton fibre is a single complete cell that develops in the surface layer of the cottonseed. Cotton fibres are composed of a cuticle, a primary wall, a secondary wall, and a central core or lumen (Figure 2.2(b)). The cuticle is the 'very outside' or 'skin' of the cotton fibre. It is composed of a waxy layer (cotton wax). The lumen is a hollow canal running the length of the fibre, which provides the nutrients while the plant is growing. Depending on the maturity of the fibre, the dimensions of the lumen vary enormously. Mature fibres will have a thick layer of cellulose in the secondary wall that results in a very small lumen, whereas an immature fibre has a very thin wall structure and a large lumen.

The longitudinal view of cotton fibre appears as a ribbon-like structure with twists at regular intervals along its length (Figure 2.3(b)). The twists, also called convolutions, give cotton an uneven fibre surface that increases interfibre friction and enables fine cotton yarns of adequate strength to be spun. The convolutions differentiate cotton fibres from all other forms of seed hair fibres and are partially responsible for many of cotton's unique characteristics.

Cotton is soft, absorbent, strong, and machine-washable. Its absorbency comes from the fibre surface (which has a strong affinity for water), the fibre core microstructure (which resembles millions of tiny sponge-like tubes), and the hydroxyl (–OH) groups in its molecules.

The width of cotton fibre is fairly uniform, varying between 12 and 20 μm wide. The cross-section of cotton fibre is generally referred to as being kidney-shaped (Figure 2.2(a), Figure 2.3(a)), and some are elliptical. Mercerization (a process using caustic soda and tension) can cause the fibre to swell, straighten, and become more cylindrical, promoting lustre, dyeability, and increased strength. Mercerized cotton is almost round in cross-section and more lustrous than untreated fibres.

The degree of crystallite orientation, the microstructural parameter obtained by X-ray diffraction, is directly related to the bundle strength. In general, fibres with more highly oriented crystallinity are stronger and more rigid. Usually, the crystallinity of cotton ranges from 60% to 70%.

2.2.4 COMPOSITION OF COTTON

Raw cotton fibre, after ginning and mechanical cleaning, contains approximately 90% or more cellulose (Table 2.1) and noncellulosics. Both the structure and composition of the cellulose and noncellulosics depend on the cotton variety and growing conditions.

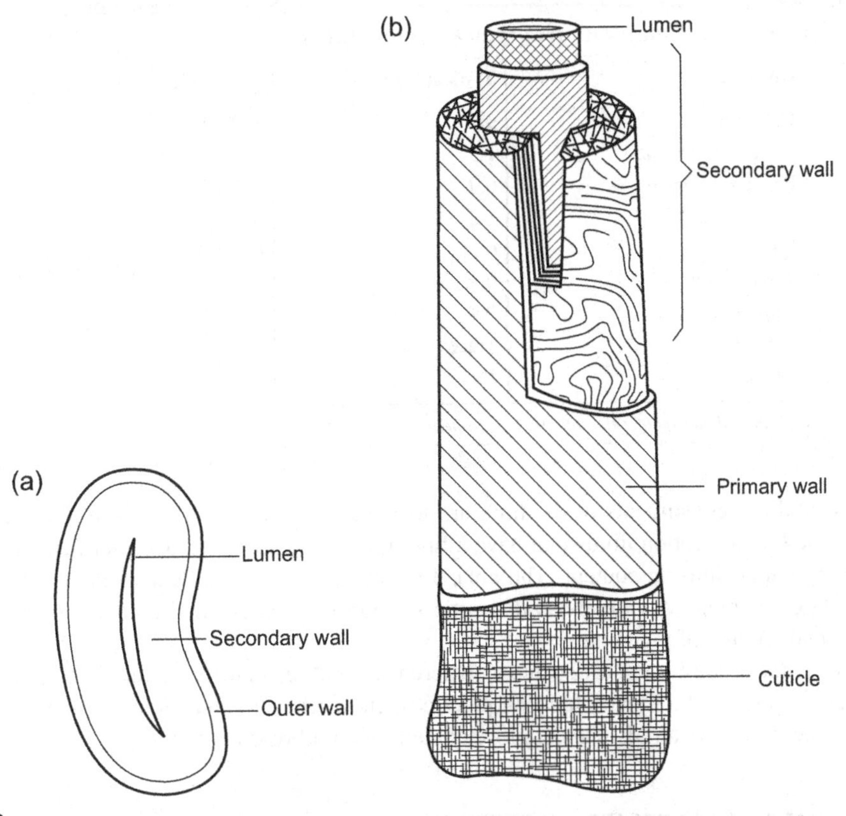

FIGURE 2.2

Microstructure of cotton. (a) Cross-section of cotton, (b) vertical section of cotton.

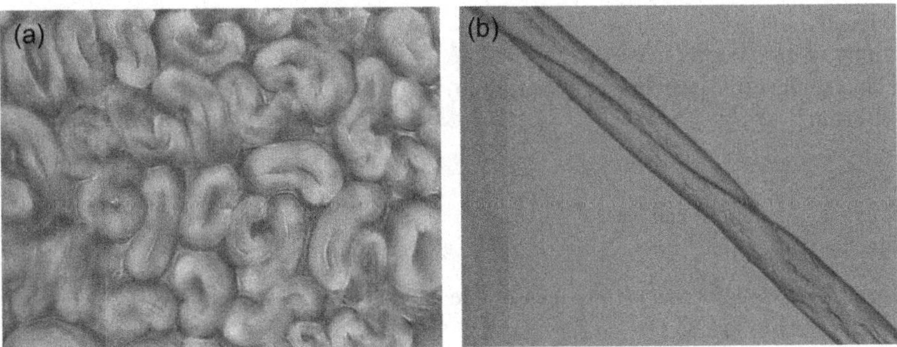

FIGURE 2.3

Microscopic view of cotton fibre. (a) Cross-sectional view of cotton, (b) longitudinal view of cotton.

Table 2.1 Composition of typical cotton fibres

Constituent	Typical (%)	Range (%)
Cellulose	95	88.0–96.0
Protein (%N×6.25)*	1.3	1.1–1.9
Pectic substances	0.9	0.7–1.2
Ash	1.2	0.7–1.6
Wax	0.6	0.4–1.0
Total sugars	0.3	0.1–1.0
Organic acids	0.8	0.5–1.0
Pigment	Trace	–
Others	1.4	–

**Standard method of estimating percent protein from nitrogen content (%N).*

The noncellulosic constituents of the fibre are located principally in the cuticle, the primary cell wall, and in the lumen. Cotton fibres that have a high ratio of surface area to linear density generally exhibit a higher noncellulosic content. The noncellulosic constituents include proteins, amino acids, other nitrogen-containing compounds, wax, pectic substances, organic acids, sugars, inorganic salts, and a very small amount of pigments.

Variations in these constituents arise due to differences in fibre maturity, variety of cotton, and environmental conditions (soil, climate, farming practices, etc.). After treatments to remove the naturally occurring noncellulosic materials, the cellulose content of the fibre is over 99%.

2.2.5 PHYSICAL PROPERTIES OF COTTON

High-quality cotton lint can produce high-quality yarn, fabric, and therefore high-quality end-products. High-quality cotton lint has a number of physical properties, some of which are measurable, whereas others are not. High-quality lint also implies the absence of certain harmful substances.

The properties of cotton lint that are important and measurable include the following:

- staple length
- uniformity of fibre length
- short fibre content (the percentage of short fibre)
- fibre strength
- fineness
- maturity
- elasticity (or elongation) and
- for some cottons, colour.

The properties of cotton lint that are not measurable include the following items:

- style
- silkiness and
- lustre.

These are the main properties considered when cotton is classified or graded, and also affect the subsequent processing and quality of end-products. High-quality cotton is also characterized by low trash content. Cotton with a high trash content needs more drastic action during processing to clean the fibres, which may damage it, decreasing the quality of end-products.

2.2.5.1 Length and fineness of cotton fibre

The average length of cotton fibre used for textiles ranges from 25 to 37 mm. The longer the fibre, the easier it is to process and the better the product's quality. Therefore, long-staple cotton has higher value in textile processing and is used for high-quality yarn or end-products. Being a natural product, the length of each cotton fibre may be different, and uniformity of fibre length is very important as the higher the uniformity, the better the fibre length. Short fibre content is the ratio of short fibre weight to the total tested fibre weight, which is also a key factor affecting processing. Short fibre usually refers to fibre shorter than a certain length. In HVI (high volume instrument) testing fibre less than 12.7 mm is considered to be short fibre and the higher the short fibre content the worse the fibre processing ability and lower the quality of end-products. The fineness of fibres or yarns is usually expressed as the linear density of them (fibres or yarns), with the units of tex, dtex, or metric count. The tex corresponds to the fibre or yarn weight in grams of 10,000 m length, whereas the metric count corresponds to the fibre or yarns in metres of the weight of a gram; the dtex (decitex) is 10th tex.

The fineness of cotton fibre is usually in the range of 5000–7000 metric count, which is the finest of all the natural vegetable fibres.

2.2.5.2 Tensile behaviour of cotton

The strength of fibre or yarn is commonly referred to the force to break or rupture the fibre or yarn, usually with the unit of grams-force (gf), kilograms-force (kgf), newtons (N), centinewtons (cN), or millinewtons (mN), whereas the tenacity is commonly referred to the resistance of break or rupture per unit of fineness of fibre or yarn; that is, the breaking force divided by the linear density of the unstrained material (fibre or yarn) usually with the unit of gf/tex, gf/dtex, cN/tex, N/tex.

The strength of single medium-staple cotton fibre is in the range of 3.5–4.5 g, whereas for long-staple the strength is in the range of 4–6 g. The higher the strength the more action the fibre can withstand during processing without damage which affects the tenacity, durability, and quality of the end-product. As yarn strength is determined not only by individual fibre strength but also by fibre to fibre interactions caused by length, friction and degree of twist it has been found that breaking bundles of parallel fibres gives a better prediction of yarn strength or tenacity by simulating the combination of fibre strength or tenacity and interaction. Cotton bundle tenacity ranges from about 17 cN/tex for short coarse cottons to approximately 43 cN/tex for long fine cottons. Generally, cotton strength or tenacity increases with moisture content and decreases with temperature.

The elongation of cotton is expressed as percent elongation taken at the point of breaking. For most cotton, elongation at break, or just elongation, is in the range of 6–9%. Moisture has the most pronounced effect on elongation. An elongation of about 5% at low relative humidity will increase to about 10% when the relative humidity is almost at saturation point.

2.2.5.3 Maturity and micronaire value

Maturity of cotton is an important property; the higher its maturity the stronger and thicker the fibre or higher its linear density which is usually related to better dyeability, ease of processing and ultimate quality of the product. Micronaire is generally used as a measure of maturity assessing both maturity and fineness. The importance of micronaire lies in its ability to predict obstacles may meet in processing. Lower micronaire fibres break more easily during mechanical action and being generally more flexible entangle more easily to form neps. Short fibre content affects spinning performance, yarn and fabric quality, dyed fabric appearance and neppiness in particular. It is generally considered that both too-low (immature) and too-high (over matured) micronaire cottons should be avoided, the ideal range being between about 3.8 and 4.2 for American Upland type cotton.

2.2.5.4 Elasticity, stiffness resilience, and rigidity

Compared to other fibres the elasticity of cotton is relatively low with elongation of only 3–7%. Like other vegetable fibres, cotton fibre has low resilience so products made from pure cotton crease easily and do not recover well from wrinkling Cotton fibre being finer than most other vegetable fibres is soft with low rigidity.

2.2.6 MEASUREMENT OF FIBRE PROPERTIES

2.2.6.1 Measurement of fibre length

Historically, fibre length was measured using the Baer diagram or Suter–Webb array method. Both methods are based on sorting fibres within a defined sample according to length and/or weight.

The fibres in each length group are accurately weighed. The resulting length–weight distribution is used in calculating various fibre length properties including the mean fibre length, upper quartile length by weight, (the fibre length exceeded by 25% of the fibre lengths by weight in the test specimen), and short fibre (less than a certain length, for example, 12.7 mm) content.

Nowadays, more instruments are used in measuring and computing the length and its relative results for cotton fibre, such as HVI and AFIS (advanced fibre information system).

2.2.6.2 Estimating fibre strength

There are two kinds of tensile testing of cotton fibre: one is for single fibre, and the other is for bundle fibre. A single fibre is clamped by two clamps (top and bottom). During testing the top clamp is fixed and the bottom clamp moves down imparting a force to the clamped fibre. The force is gradually increased while the clamp goes down and as the force increases the fibre elongates until it breaks. The force and elongation at the point when the fibre breaks is, therefore, the strength and elongation of the single fibre.

In the case of bundle fibre testing, a bundle of fibres is combed parallel and secured between two clamps. A force is applied to try and separate the clamps and gradually increased until the fibre bundle breaks. Fibre tensile strength is calculated from the ratio of the breaking load to the bundle mass. Due to the natural variation within a population of cotton fibres, bundle fibre selection and bundle construction such bundle mass measurements are subject to considerable experimental error.

In textile literature fibre strength is referred to as *breaking tenacity* or grams of breaking load per tex, where tex is the fibre linear density in grams per kilometre.

HVI and other instruments are used to measure the tensile property of cotton fibre.

2.2.6.3 Estimating fibre fineness and maturity (micronaire)

Direct measurement of *biological* fineness in terms of fibre or lumen diameter and cell-wall thickness are precluded by the high costs in both time and labour, the noncircular cross-sections of dry cotton fibres, and the high degree of variation in fibre fineness. Although advances in image analysis have improved measurements of biological fineness and maturity, fibre image analyses remain too slow and limited for fibres with irregular cross-sections so that this method is only really suitable for wool fibre with its almost regular circular cross-section.

Instead, indirect indicators such as linear density are widely used to express the fineness of fibre. There are two commonly used methods to estimate the indirect fineness of cotton fibre. One is cut-middles method, and the other is airflow method.

The cut-middles method

The cut-middles method involves weighing and counting a number of fibres of a certain length so that the weight per length can be obtained giving the linear density of the fibre.

Firstly, the fibre bundle, which has been weighed, is clamped at one end and then combed several times to remove the short fibres in the middle of the bundle. The fibre bundle is then cut between the neat and middle section into a certain fragment (for example, 10 or 20 mm) which is weighed and the number of fibres in this section counted. The fibre fineness, referring to the definition, is then calculated. This method is also used for measuring many other types of fibre, only the length cut is different depending on the tested fibre.

The air flow method

The airflow method is based on the resistance of a bundle of fibres to airflow. Air-permeability of the sample depends on its surface area. Basic fluid-flow theory states that air permeability is inversely dependent on the square of the fibre surface area; therefore, empirical formulae were developed for calculating the approximate weight per length (linear density) of the fibre. Due to its speed and ease of operation, this method is widely used to test the fineness of many kinds of fibres.

The maturity of cotton fibre is closely related to its fineness, the immature fibre being finer. Therefore, micronaire, a comprehensive indicator of both the maturity and fineness of cotton fibre is the most commonly used instrumental fibre-quality test. Micronaire is mostly tested using the airflow method. HVI, AFIS, and some other instruments can be used to estimate the fineness and micronaire of the cotton.

2.2.6.4 The new era of fully automatized high volume instruments (HVI)

A great deal of work has been and continues to be done developing devices and methods to ensure more comprehensive, objective cotton assessments. The first step in the transition to instrumental classing began with the introduction of the HVI line. Progress in the field of measuring systems is still important due to the growing competition of man-made fibres on the natural cottons. As a result of these developments, we have entered the new millennium with many fully automatized measurement systems like HVI Spectrum, AFIS, FiberLab, and Premier ART. Modern instruments allow measurement not only of the physical–mechanical parameters by the bundle or individual fibre methods, but also intrinsic properties like maturity, stickiness, color, and external factors influencing the cotton quality such as the cultivation, harvesting and ginning conditions like trash content, seed coat fragment content, and nep content. Now that all the measurement systems are fully automatized and computerized, we have a full database of fibre properties together with their images including other fibres, as well as cotton. These results are presented in Table 2.2.

Table 2.3 shows the properties of cotton in detail.

Table 2.2 The fully automatized and computerized measurement systems

Basic fibre properties measured	STI 2020 data products	Shaffner tech. Inc. ISO tester 2003	Uster technologies Inc. HVI spectrum 2003	Uster technologies AFIS	Uster technologies, IntelliGin	Lintronics fibrelab 2003	Premier polytronics premier ART (Automatic Rapid Tester) 2003	Premier HFT9000
Micronaire	Mic	Mic	Mic	–		Mic	Mic	Mic
Length	UHM LU SFC	UHM, inch, mm LU	UHML LU SFI	L(w), L(n) UQL(w), SFC(n) SFC(w), L(n)CV LCV(w)	–	L(w) UI SFI	CL(w) SFC Premier aQura	SL, L(w)
Strength	Strength, g/tex Elongation, %	Strength, g/tex	Strength, g/tex Elongation, %	–	–	Strength Elongation	Strength Elongation	Strength Elongation
Colour	CIE (fibre only) Colour grade	Rd, +b Colour grade	Colour +b, Rd Colour grade	–	Colour	Colour +b, Rd Colour grade	Rd, +b	Rd, +b
Trash	% Area Leaf grade Size dist. Bark and grass extraneous matter preparation	% Area Leaf grade	% Area Leaf grade Trash	Trash Cnt/g Dust Cnt/g VFM Mean size	Trash	Trash Cnt/g Trash size Trash area Dust Cnt/g	Trash	–
Moisture content	%	%	%	–	%	–	–	–
Neps, seed coat fragments	Neps/g Size dist. Seed coat frag-ments	Neps/g	Neps	Nep Cnt/g Nep size SCN Cnt/g SCN size	–	Nep Cnt/g Nep area Npe size SCN/g SCN area SCN size	Neps (Premier aQura)	–
Maturity and fineness	Mat ratio Immature fibre fraction		Maturity index	Maturity ratio Fineness IFC	–	Maturity index	Maturity estimate	–
Stickiness	Sticky points/g Size Sugar types	Sticky points/g	–	–	–	Sticky mass Cnt Stickiness grad Stickiness size	–	–

"Source: Iwona (2003)".

Table 2.3 Properties of cotton

Molecular structure	Cellulose
Macroscopic feature	
Length	0.3–5.5 cm
Cross-section	Kidney-shaped
Colour	Generally white, may be cream-coloured or brown
Light-reflection	Low lustre, dull appearance
Physical properties	
Tenacity (cN/dtex)	2.6 to 4.4 (dry), 3.2 to 5.3 (wet)
Stretch and elasticity	3–7% elongation at break. At 2% elongation, recovery is 70%
Resiliency	Low
Abrasion resistance	Fair to good
Dimensional stability	Fabrics may shrink during laundering
Moisture regain	8.5%
Specific gravity	1.54 g/cm^3
Chemical properties	
Effects of bleaches	Highly resistant to all bleaches
Acids and alkalies	Highly resistant to alkalies. Strong acids and hot dilute acids will cause disintegration
Organic solvents	Resistant to most organic solvents
Sunlight and heat	Withstands high temperatures well. Prolonged exposure to light will cause yellowing due to oxidation
Resistance to stains	Poor resistance to water-born stains
Dyeability	Good affinity for dyes. Dyed with direct, vat, and basic dyes Vat dyeing produces excellent wash and lightfastness
Biological properties	
Effects of fungi and moulds	Highly susceptible to attack by mildew. Mildew will promote odour and discolouration and results in rotting and degradation
Effects of insects	Starched cotton are attacked by silverfish
Flammability behaviour	Burns rapidly. Smouldering red afterglow
Electrical and thermal conductivity	Good heat conductor

2.2.7 COTTON APPLICATION IN TEXTILE

The versatility of cotton has made it one of the most valuable and widely used of all textile fibres. Cotton is inherently strong because the convolutions create friction within the fabric, preventing the fibres from slipping. Wet cotton fabric is stronger than dry cotton fabric, therefore cotton can withstand repeated washings and is ideal for household goods and garments that need to be washed regularly. Like other celluloses cotton cellulose is not unduly affected by moderate heat so that cotton fabrics can be ironed with a hot iron without damage.

Cotton fibres are able to absorb appreciable amounts of moisture and then evaporate it readily to the air which adds to the comfort of cotton garments. There are literally thousands of uses (about 100 major uses) for cotton in textile items, ranging from nappies to the most fashionable dresses, coats, and jackets. These uses can be classified into three main categories: apparel, home furnishings, and industrial.

1. **Apparel market**: Due to cotton's comfort and easy laundering, trousers and shirts, especially for leisure wear, account for the greatest use of cotton, followed by underwear. Jeans and denim fabrics utilize more cotton than any other single clothing item.

2. **Home furnishings**: The absorbency of cotton makes it an excellent material for household fabrics such as sheets and towels. Towels and flannels account for the largest amount of cotton used in home furnishings, followed by sheets and pillowcases.
3. **Industrial products**: The uses of cotton in industrial products are as diverse as tarpaulins, bookbindings, and zipper tapes. Cotton products are also used to clean up agrochemical spills and oil spills. Some of the major industrial markets for cotton are medical supplies and industrial threads.

2.3 OTHER SEED FIBRES

2.3.1 KAPOK

2.3.1.1 The physical properties of kapok fibre

Kapok refers to the seed fibre obtained from its seedpods. This fibre is botanically quite similar to cotton, belonging to closely related families. The kapok fibre is a single-cell natural cellulose fibre. Kapok fibres look transparent with characteristic air bubbles in the lumen (Figure 2.4(a)) under the light microscope. It has a hollow body and a sealed tail (as shown in Figure 2.4(b) and (c)) slightly widened at the roots (Figure 2.4(c)) with a lattice-like condensed cell wall (Figure 2.4(d)) narrowing towards the top, all of which are desirable features for functional textiles of this nature.

Kapok has a thin cell wall which enables the fibre to be compressed more easily and the subtle structures of the cell wall prevent other small particles entering the lumen.

In their appearance and characteristic properties, kapok fibres are very similar to milkweed fibres produced by the plants Asclepias, Ceropegia, and Calotropis, only that the latter are much longer.

Like all natural cellulosic fibres, kapok contains mostly alpha cellulose (35–50%), hemicelluloses (22–45%), lignin (15–22%), about 10–11% of moisture (conventional moisture regain 10.9%), and 2–3% of waxes. It also contains smaller quantities of starch, about 2.1% of proteins and inorganic substances, notably iron (1.3–2.5%). As to the content of alpha cellulose, kapok is more woody than flax and other plant fibres. A high content of lignin ensures kapok has good antibacterial resistance. Kapok contains xyloses (about 23%) and 4-0-methyl-glucuronic acid (about 5.9%) as the main hydrolysis product of its hemicellulose. Due to a high content of inorganic substances in the primary cell, kapok fibres have a lower ability to absorb water and higher resilience. Raw fibres are extremely hydrophobic and highly absorbent of non-polar liquids (oleophyllic) Kapok contains 70–80% air and provides excellent thermal and acoustic insulation. Kapok is also extremely buoyant in water. The fibre is resistant to acid and alkali at room temperature.

The physical properties of kapok fibre are listed in Table 2.4.

2.3.1.2 The quality evaluation

The quality of kapok is evaluated on the basis of the percentage of lignin, diameter of fibres (kapok with a more uniform diameter has a higher value), buoyancy in an alcohol solution with a density of 0.928 g/cm^3, and relative velocity of the fibres wetting and submersion. The percentage of lignin is qualitatively evaluated by microchemical reaction by using an alcohol solution of fluoroglucinol and HCl. Good-quality fibres are only produced from ripened kapok seedpods. Immature fibres have low strength, inferior lustre and colour, and do not withstand stress during compression into bales.

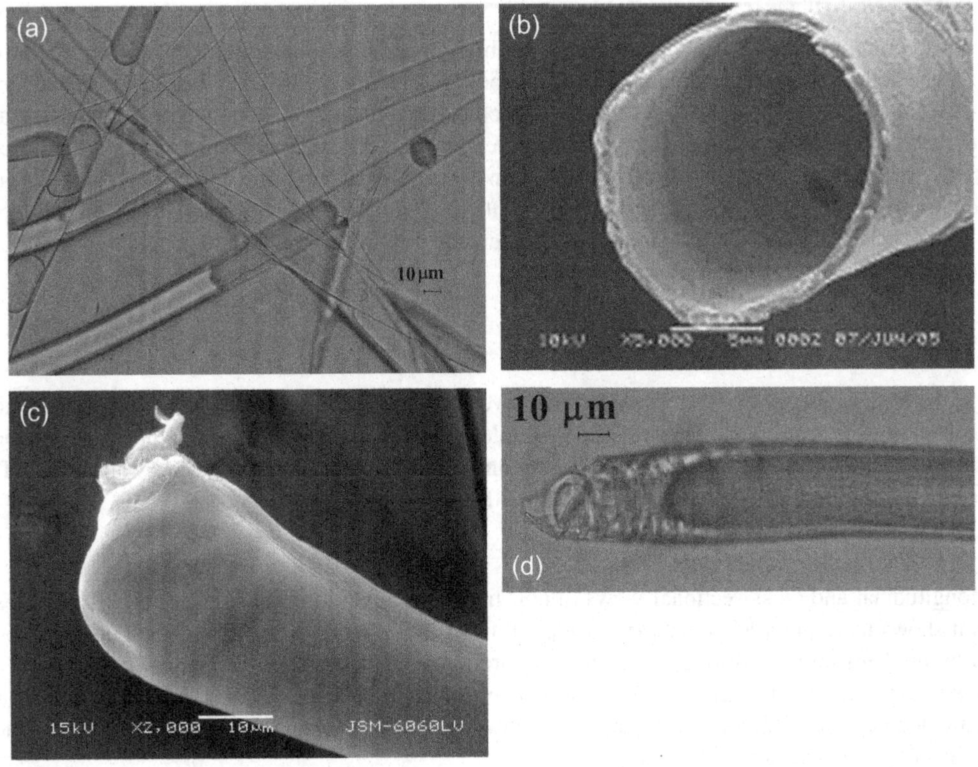

FIGURE 2.4

Morphology of kapok fibre: (a) kapok fibres observed under optical microscope *(Source: Foto, T. Rijavec (Rijavec, 2008))*, (b) the body of kapok fibre *(Source: Cui, Wang, Wei, & Zhao (2010))*, (c) characteristic widened root of a kapok fibre *(Source: Foto, T. Rijavec (Rijavec, 2008))*, (d) lattice-like end of a kapok fibre *(Source: Foto, T. Rijavec (Rijavec, 2008))*.

Table 2.4 The properties of kapok fibre

Fibre length (mm)	Linear density (dtex)	Crystallinity (%)	Moisture (%)
8–12	0.4–3.2	33	10.00–10.73
Breaking length (km)	Elongation at break (%)	Volume mass (g/cm^3)	Refractive index (%)
8–13	1.5–3.0	0.29	1.7176
Strength (cN/dtex) (93.3 MPa)	Young's module (Gpa)	The cell-wall thickness/µm	The length-to-diameter ratio
0.84	4	1–3	720

2.3.1.3 Applications of kapok fibre

The most commonly traded varieties of kapok are high-grade Java kapok and the lower-grade Indian kapok. The highest market share belongs to Java kapok. Kapok has traditionally been used as buoyancy aids, stuffing in mattresses and pillows due to its good thermal insulation, nonallergenic properties, and low density. Oil filters and composites are new potential uses. Machine spinning of kapok is difficult due to the short length, brittleness, and low cohesion of its fibres and is limited to coarse yarns or yarns blended with cotton. In the second half of the twentieth century the use of kapok drastically dropped with the advent of synthetic stuffing materials. However, during recent years, kapok, as a recyclable and biodegradable fibre, has become more attractive.

2.3.2 COIR

Coir is the fibre extracted from the fibrous outer covering of the fruit of the coconut palm and is native to the tropics. Coir is also regarded as a seed fibre, although its make up is similar to those of bast fibres with cellulose (about 44%), lignin (45%), pectin and related compounds (3%), and water (5%). The higher lignin content makes the fibre harder and stiffer.

2.3.2.1 The properties of coir

The longitudinal and cross-sectional views of coir fibre are shown in Figure 2.5(a) and (b); the cross-section shows there are a lot of cavities inside the fibre, and roughly one-third of the bulk of fibre is filled by air. This entrapped air gives rise to the pronounced springiness (resilience) of the fibre, its buoyancy in water, and increases the time water takes to penetrate the fibres. The properties of coir are less affected by wet conditions than other hard fibres. The thickness of coir fibre limits the products made by coir being coarser and heavier.

The tenacity of coir is not high, only around 1 cN/dtex, but its elongation is higher compared to other vegetable fibres, up to 15–40%; it is also less prone to wrinkle and crush because of the air-filled cavities. Coir can withstand exposure to all kinds of weather, making it a practical fibre for outdoor use.

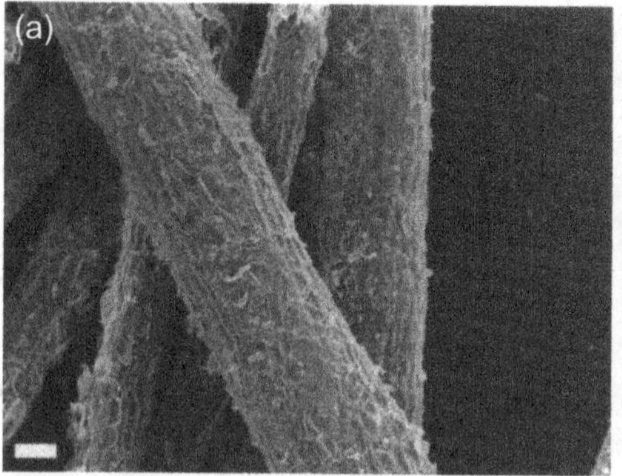

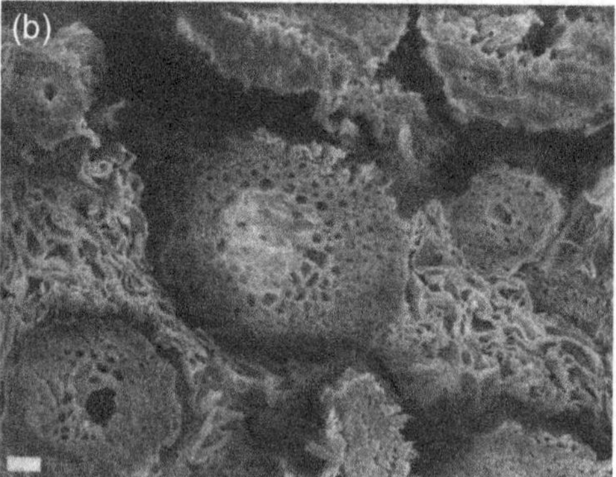

FIGURE 2.5

(a) Longitudinal and (b) cross section of coir fibre.

Source: P277, Bast and other plant fibres, Woodhead Publishing.

2.3.2.2 Applications of coir

The main applications of coir include:

- yarn
- rope
- carpet and
- mats.

2.4 BAST FIBRE

Bast fibre, or skin fibre, is regarded as one of the major natural vegetable fibre resources in the world and plays an important role in the textile industry. Such fibre is extracted from the phloem, or bast, surrounding the stem of certain plants (mainly dicotyledonic).

Most of the important bast fibres are obtained from plants cultivated in agriculture, for instance flax, ramie, jute, kenaf, and hemp. The valuable fibres are located in the phloem. Therefore, it is necessary to separate the fibres from the xylem and sometimes also from epidermis. The chemical components of main bast materials are shown in Table 2.5.

Normally, bast fibres are thick and boast high-tensile strength. Therefore, they are usually processed for use in coarse textiles such as ropes, carpet yarn, traditional carpets, geotextile, and hessian or burlap sacks. In addition, they can be used in composite technology industries manufacturing nonwoven mats and carpets, and composite boards such as furniture materials and car-door panels, for example.

2.4.1 FLAX

2.4.1.1 Definition and types of flax

Flax is native to the region extending from the eastern Mediterranean to India and was probably first domesticated in the Fertile Crescent. Cultivated flax is of two types: one is grown for the seed and the other for fibre production. Flax was extensively cultivated in ancient Ethiopia and ancient Egypt (Joseph, 1986) and has been grown since the beginning of civilization. In a prehistoric cave in the Republic of Georgia, dyed flax fibres have been found that date to 30,000 BC.

The word flax is used in connection with the plant and products that are made directly from it or that are closely associated with it. For example, flax fields, flax cultivation and production, line flax

Table 2.5 Chemical component of different bast materials (%)

Fibre varieties	Cellulose	Hemi-cellulose	Pectin	Lignin	Fat/Wax
Flax	62–71	16–18	1.8–2.0	3.0–4.5	1.5
Hemp	67–75	16–18	0.8	3.0–5.0	0.7
Ramie	68–76	13–14	1.9–2.1	0.6–2.0	0.3
Jute	59–71	12–13	0.2–4.4	11.8–12.9	0.5
Kenaf				12.0–15.0	

(long fibre flax), flax tow (short fibre flax) flax spinning and flax yarns, and also certain types of flax fabrics especially those of a heavier industrial kind. The word linen is used to refer to products that are further down the production line, lighter weight fabrics for household textiles, furnishings and garments and other consumer products made from these fabrics.

Flax is harvested for fibre production after approximately 100 days or a month after the plant flowers, and 2 weeks after the seed capsules form. Flax fibre is extracted from the bast or skin of the stem of the flax plant. Flax fibre is soft, lustrous, and flexible; bundles of flax fibre have the appearance of blonde hair, hence the description 'flaxen'. It is stronger than cotton fibre but less elastic and harder. The best grades are used for linen fabrics such as damasks, lace and sheeting. Coarser grades are used for the manufacture of twine and rope. Flax fibre is also a raw material for the high-quality paper industry and used for printed banknotes, rolling paper for cigarettes, and tea bags.

2.4.1.2 Fibre extraction

Flax fibres originate in the phloem (bast region) and provide an important food-conducting tissue for the plant. Flax plants potentially contain 25% textile grade fibre and 75% non-fibre material (cuticle and shive material) tightly associated with the plant stem.

After harvesting, the flax is allowed to dry, the seeds are removed, and then the fibre is separated from the rest of the stalk. The first step in this process is called retting. Retting is separating or loosening fibre bundles from the cuticularized epidermis and woody core cells and subdividing into smaller bundles and ultimately fibres. Microbial activity during retting causes a partial breaking down of the components binding the tissues together, so separating the cellulosic fibres from nonfibre tissues.

The commercial methods traditionally used to ret flax for industrial grade fibres are water-retting and dew-retting. Water and dew-retting depend upon colonization and partial plant biodegradation by microorganisms, and are influenced by environmental conditions that makes retting difficult to control.

Water-retting includes pond retting and stream retting.

- Pond retting is the fastest retting, but the retted flax is traditionally considered to be of lower quality. Here, the flax is placed in a pool of water for just a couple of days or weeks.
- Stream retting takes longer than pond retting, usually being left for 2 or 3 weeks in a stream or river; however, the quality of retted flax is better.

Dew-retting depends upon climatic conditions, characteristics of the sown flax, and fields. In dew-retting, the flax remains on the ground between 2 weeks and 2 months for retting. As a result of alternating rain and the sun, an enzymatic action degrades the pectins that bind fibres to the straw. The straw is turned over during retting to evenly ret the stalks. This process normally takes a month or more, but is generally considered to produce the highest quality flax fibres and the least pollution.

Currently, 'enzymatic' retting of flax is being researched as a retting technique to engineer fibres with specific properties. Enzyme-retting depends upon processing conditions using pectinase-rich enzymes and chelators to separate the fibres from the shive.

After retting, it is still necessary to remove the remaining straw from the fibres, which is dressing the flax. Dressing consists of three steps: breaking, scutching, and heckling. In breaking, the flax is 'broken', the straw is broken up into small, short bits, whereas the actual fibre is left unharmed, then fibres are 'scutched', where the straw is scraped away from the fibre, and finally pulled through 'hackles' which act like combs and comb the straw out of the fibre. The hackled fibre is ready for further processing.

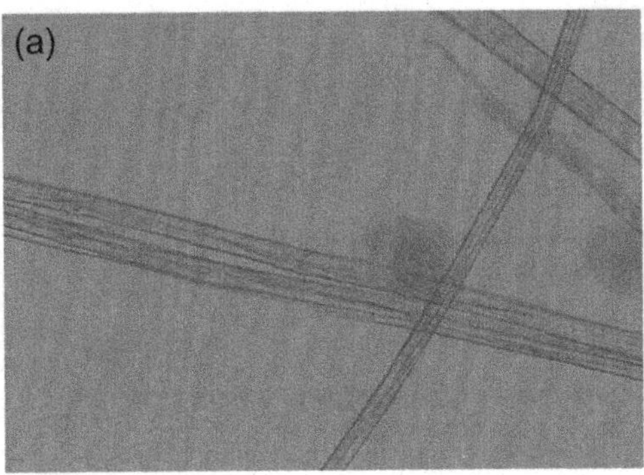
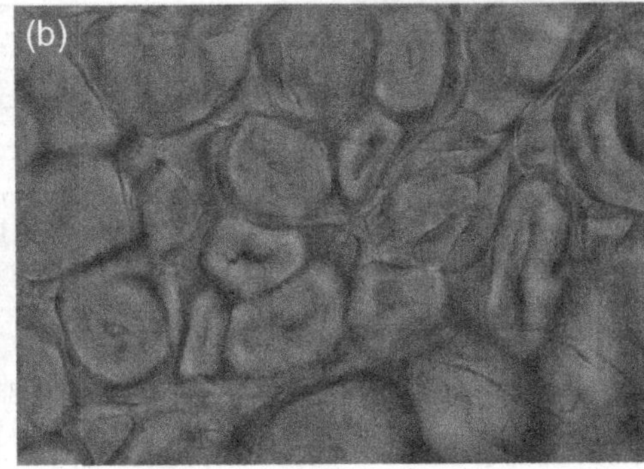

FIGURE 2.6

Microscopic view of flax fibre. (a) Longitudinal view, (b) cross-sectional view.

2.4.1.3 Fibre composition and appearance

The composition of flax includes cellulose, and gums, such as hemicellulose, pectins, lignin, and waxes; the precise makeup varying according to the variety and maturity of the flax, the cellulose is around 60–70%, and the lignin is roughly 3–5% as shown in Table 2.5.

Commercial flax consists of bundles of individual fibres held together by gum. The length of fibre bundles ranges from 200 to 1000 mm. Line fibre is usually longer than 300 mm, whereas the tow fibre is shorter. The bundle fibres are composed of single cells, from 15 mm to 80 mm, averagely in 25–30 mm long end to end.

As shown in Figure 2.6, the longitudinal view under the microscope shows that flax fibre, unlike cotton, has no convolutions, but longitudinal lines or striations, with protuberances, called nodes, spaced irregularly along the length.

A cross-sectional view reveals that the individual fibres are polygonal in shape with a thick, fleshy wall surrounding a central hollow core, or lumen. The mean diameter of an individual fibre is 15–20 μm.

However, the lumen in mature fibres is not as pronounced as that in cotton. The fibre ranges in colour from creamy white to yellowish brown to grey. Residual wax from the flax stem helps to give linen its lustrous appearance.

2.4.1.4 Properties of flax

Because of its short single cell, flax is usually used for textile processing in bundle fibres (several single cells bonded together by gums); therefore, flax bundle fibre is longer (usually 300–700 mm) and coarser (200–500 metric count) than cotton.

Flax fibre is stronger, more rigid, and has a lower elasticity than cotton, partially because of the high content of gums (such as lignin). Like cotton, wet flax is also stronger than dry; the linen cloth has low elasticity and resilience and therefore tends to wrinkle. Also, like cotton, if stored in highly humid conditions, flax is subject to mildew, although to a lesser extent. Flax has greater resistance to insect attack than cotton because of the greater level of gum in the fibre (Table 2.6).

2.4.1.5 Textile application of flax

Flax fibre is mostly used for clothing and household products (such as table coverings). Flax has a high moisture regain and releases heat easily so linen garments feel cool in warm weather.

Table 2.6 Properties of flax

Molecular structure	Cellulose
Macroscopic features	
Cross-section	Oval or polygonal
Colour	May vary from light ivory to dark tan or grey
Light reflection	Good lustre, almost silky in appearance
Physical properties	
Tenacity (cN/dtex)	4.8 to 5.8
Stretch and elasticity	2.7–3.3% elongation at break. At 2% elongation, recovery is 65%
Resiliency	Poor resiliency, creases and wrinkles badly
Abrasion resistance	Fair to good
Dimensional stability	Fibres will not stretch or shrink, but fabrics are subject to relaxation during laundering
Moisture regain	10–12% at 65% relative humidity
Specific gravity	1.50 g/cm^3
Chemical properties	
Effects of bleaches	Unaffected by common household bleaches
Acids and alkalies	Easily damaged by hot dilute or cold concentrated acids. Highly resistant to all alkalies
Organic solvents	Resistant to organic solvents
Sunlight and heat	Extended exposure to sunlight weakens fibres. Scorches at high temperatures
Resistance to stains	Is not as harmed by water-borne stains as cotton and will give up stain more readily
Dyeability	Linen does not have a good affinity for dyes. It is dyed with direct and vat dyes
Biological properties	
Effects of fungi and moulds	Very vulnerable to damage by mildew
Effects of insects	Not damaged by insects
Flammability behaviour	Burns rapidly. Smouldering red afterglow
Electrical and thermal conductivity	Good electrical and heat conductivity

In addition, the natural resistance of flax to chemicals, including detergents, bleaches, and dry-cleaning solvents provides a fabric that is easily maintained; therefore, linen is mostly used for summer clothing.

2.5 OTHER BAST FIBRES

2.5.1 RAMIE

2.5.1.1 Introduction

Ramie is one of the oldest fibre crops, having been used for at least 6000 years. It has been around for so long that it was even used in Mummy cloths in Egypt during the period 5000–3300 BC. Ramie has been grown in China for many centuries even before cotton entered in China in 1300 AD.

There are two types of ramie, one is ‘white ramie’, also known as ‘chinese grass’, and the other is ‘green ramie’, or Rhea.

2.5.1.2 Cultivation

Nowadays, ramie is cultivated in an area from south latitude 25° to north latitude 39°, particularly in the subtropics and tropics, in China, Formosa, Japan, India, and Malaya, and also in Queensland, Mauritius, the Cameroons, the West Indies, Brazil, Mexico, the southern states of North America, and southern Europe. The ramie production in China accounts for about 80% or more of the global ramie production. Green ramie only has a limited distribution in the tropics, whereas white ramie is commonly cultivated and is the ramie we normally consider. As a bast crop the bark (phloem) of the vegetative stalks is used for fibre extraction and the ramie fibre is principally used for fabric production.

Ramie grows best in a warm, moist climate. The planting period may extend from May to September depending on the local seasonal conditions. Normally, ramie is harvested two to three times per year. However, under good growing conditions, it can be harvested up to six times a year. Harvesting is carried out just before or soon after the beginning of flowering since at this stage the plant growth declines and the maximum fibre content can be achieved.

2.5.1.3 Extraction

The extraction of the fibre can be classified into three stages: First is decortication. In this stage, the cortex or bark is removed by hand or by machine. Secondly, the cortex is scraped to remove most of the outer bark, the parenchyma in the bast layer and some of the gums and pectins. Finally the residual cortex material is washed and dried leaving the raw ramie fibre which is then degummed to extract the spinnable fibre.

The raw ramie fibre is extracted from the freshly harvested green stalk (after defoliations) with the help of a decorticating machine which helps remove the outer bark and also crushes and removes the central woody portion, some gums and waxes. The product obtained through decortication is crude decorticated fibre still containing nearly 25–30% gum. However, it is generally free from the cortical tissues. This fibre must be properly degummed to obtain spinnable fibre. The gums of ramie primarily contain araban and xylans (hemicellulose) which are relatively insoluble in water but very soluble in hot alkaline solutions. Alkali helps break down the pectins into ribbons without attacking the cellulose in the fibres. Generally, in chemical degumming raw ramie fibre is initially boiled in aqueous alkaline solution and then washed in water. Factors such as concentration of the chemicals, temperature and the treatment time determine the quality of the degummed fibre.

The single cell of ramie fibre is very long, around 60–250 mm in length, so that the single cell is long enough for spinning processing and the fibre used for spinning is much thinner than the bundle fibres of flax, hemp, and jute, resulting in finer yarn and fabric being produced from ramie.

As shown in Figure 2.7, the longitudinal view of ramie fibre is similar to that of flax, straightened with striations and nodes, whereas the cross-sectional view of ramie fibre is similar to that of cotton, except for the width. The width of the cell is usually 20–35 μm, the fineness of fibre is 1500–2500 metric count; that means 3–4 times coarser than that of cotton fibre.

2.5.1.4 Properties

Ramie fibre is strong and durable, especially when wet, and has almost the highest tenacity of natural vegetable fibres. Like cotton and flax, ramie has poor resiliency but lower elongation than cotton. The absorbency and release of moisture from ramie is similar to that of flax and little better than cotton. The natural colour of ramie is white and lustrous, almost like silk in appearance. It is stiff and rigid, which contributes to the crispness of fabrics made from ramie.

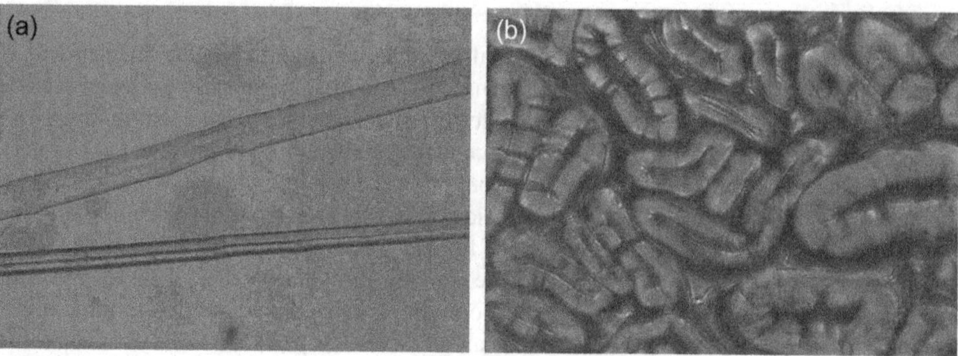

FIGURE 2.7

Microscopic view of ramie fibre. (a) Longitudinal view, (b) cross-sectional view.

2.5.1.5 Textile application

Ramie is widely used in clothing, especially summer apparel, also for household goods such as bed spreads, and for industrial uses such as high-quality packing cloth.

2.5.2 JUTE

2.5.2.1 Introduction

Jute is one of the most affordable natural fibres and is second only to cotton in the amount produced and its variety of uses. India and Bangladesh are the biggest producers of jute in the world.

2.5.2.2 Properties

The length of a single cell is short, only 3–6 mm on average, with a large width of 40–80 µm; therefore, jute, like flax, is usually used in the form of bundle fibre; the fibre bundle is usually 1–4 m long. As with other bast fibres such as flax and ramie, jute fibres are composed primarily of the plant materials cellulose (the major component of plant fibre) and lignin (the major component of wood fibre). However, the lignin content in jute is higher than that in flax and ramie so jute is much coarser and stiffer than flax and ramie making it harder to process and it can only produce coarser and heavier yarn and fabric. The longitudinal appearance of jute is also straightened with striations and gums, the cross-section is polygonal-like with a central lumen, as shown in Figure 2.8(a) and (b). The fibres are off-white to brown, with a golden, silky shine, hence its name 'The Golden Fiber' (Table 2.7).

2.5.2.3 Textile application

Advantages of jute include good insulating and antistatic properties, as well as having low thermal conductivity and moderate moisture regain. Jute fibre is the cheapest vegetable fibre with high tensile strength, low extensibility, and better breathability of fabrics. It is 100% biodegradable and recyclable, and is one of the most versatile natural fibres that has been used in raw materials for packaging, textiles, nontextile, construction, and agricultural sectors. Jute fibre is often called hessian; jute fabrics are also called hessian cloth, and jute sacks are referred to as gunny bags in some European countries.

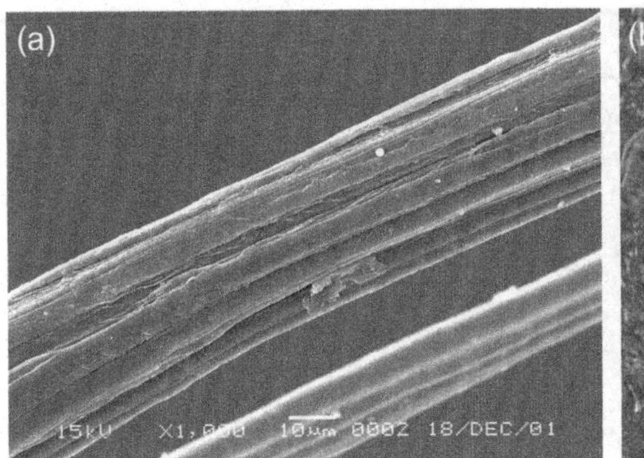

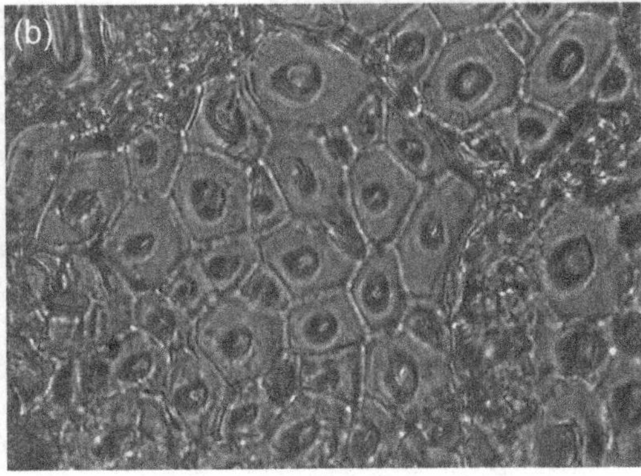

FIGURE 2.8

(a) Longitudinal and (b) cross-section of jute.

Table 2.7 Properties of jute

Molecular structure	Cellulose
Macroscopic features	
Cross-section	Polygonal with a central lumen
Colour	Yellow to brown to grey. May be bleached to white
Light reflection	Dull
Physical properties	
Tenacity	0.9–3.5 cN/dtex
Stretch and elasticity	2% Elongation at break. Very low elasticity
Resiliency	Poor
Abrasion resistance	Poor to fair
Dimensional stability	Good
Moisture regain	13.8%
Specific gravity	1.5 g/cm^3
Chemical properties	
Effects of bleaches	Not affected by oxidizing or reducing bleaches
Acids and alkalies	Easily damaged by hot dilute or cold concentrated acids. Resistant to alkalies
Organic solvents	Resistant to organic solvents
Sunlight and heat	Poor sunlight resistance. Scorches at high temperatures
Resistance to stains	Poor resistance to water-borne stains
Dyeability	Easily dyed, but light- and washfastness are poor
Biological properties	Scoured jute has good to excellent resistance to microorganisms and insects
Flammability behaviour	Burns rapidly. Smouldering red afterglow
Electrical and thermal conductivity	Moderate conductor of electricity and heat

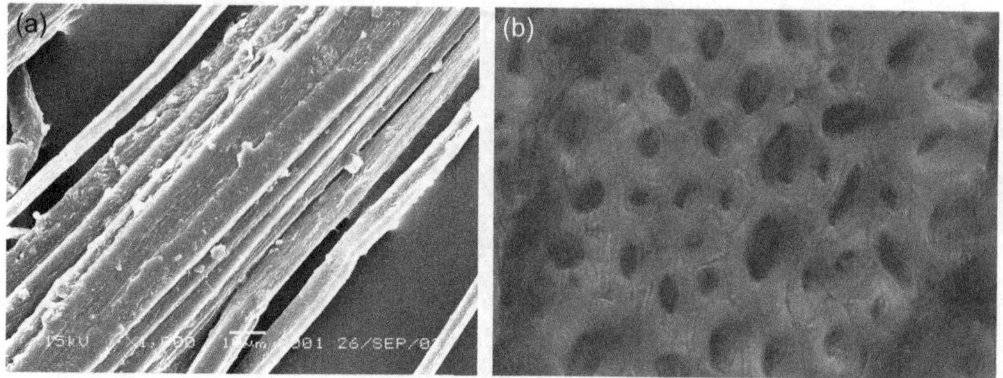

FIGURE 2.9

Microscopic view of kenaf fibre. (a) Longitudinal view, (b) cross-sectional view.

2.5.3 KENAF

2.5.3.1 Introduction

Kenaf is a plant in the Malvaceae family. Kenaf is one of the allied fibres of jute and shows similar characteristics. It is an annual plant growing to 1.5–3.5 m tall with a woody base. The stems are 10–20 mm in diameter, often, but not always, branched.

2.5.3.2 Properties

Kenaf fibre has a similar make up to other bast fibres, but a higher lignin content than jute, therefore, kenaf fibre is stronger, coarser and harder than jute and other bast fibres. The longitudinal and cross-section appearance of kenaf is similar to that of jute, and like jute and flax, is used in the form of bundle fibre because of its short single cell.

2.5.3.3 Textile application

Kenaf is used as a substitute for jute in low grade products such as packing cloth, ropes, cordage, and canvas because of its high production (Figure 2.9).

2.5.4 HEMP

2.5.4.1 Introduction

Hemp is the name of the soft, durable fibre that is cultivated commercially from plants of the *Cannabis* genus. Hemp is harvested and processed in much the same way as flax. It is coarser than flax and darker in colour. Hemp is also usually used in the form of bundle fibre.

2.5.4.2 Properties

The diameter of a single cell is from 15 to 50 μm with an average length of 30–40 mm, the length of the fibre bundle is about 1500–2500 mm. The breaking force of hemp fibre is a little higher than that of flax fibre, its elongation is low (1–2%). Among the bast fibres, hemp has better antibacterial ability.

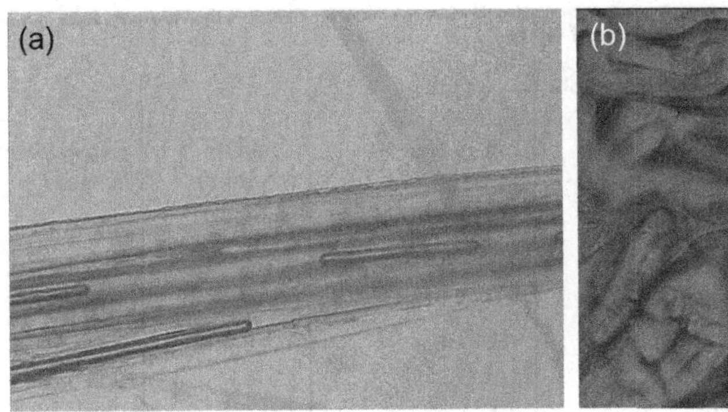

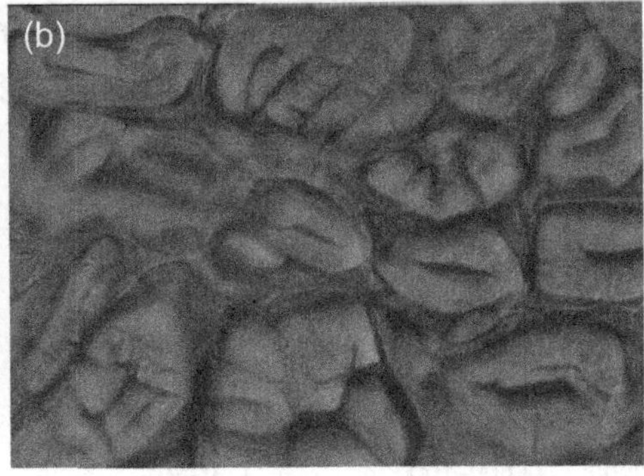

FIGURE 2.10

Microscopic view of hemp fibre. (a) Longitudinal view, (b) cross-sectional view.

The longitudinal and cross-sectional views of hemp are similar to other bast fibres, as shown in Figure 2.10.

2.5.4.3 Textile application

In modern times, hemp has been used for industrial purposes including paper, textiles, biodegradable plastics, construction, health food, fuel, and medical purposes with modest commercial success, as well as for clothing and household goods.

2.5.5 SISAL, ABACA, AND PINEAPPLE FIBRE

All of these fibres are obtained from the leaves of plants.

2.5.5.1 Sisal

Sisal is one group of fibres extracted from the leaves of plants belonging to the agave family. The sisal fibre scraped from the fresh leaves cut from the plant is stiffer and stronger, with low elasticity. It is important in the manufacture of such items as matting, rough handbags, ropes, cordage, especially marine rope (where good resistance to sea water is needed), and carpeting. Sisal cloth is also used to polish materials.

2.5.5.2 Abaca

Abaca, also known as Manila hemp, is stripped from the leaves of plants belonging to the banana family and grows mainly in the Philippines. Abaca fibre is processed in a similar manner to sisal, although the fibre shows a little more elasticity. The textile application of this fibre includes ropes, cordage and twine, and also marine ropes.

2.5.5.3 Pineapple fibre

Pineapple fibre is extracted from the leaves of the pineapple plant in a similar way to the extraction of sisal fibre, the difference being that pineapple leaves are narrower and shorter than sisal leaves. Pineapple fibre is also used in bundle form. The pineapple bundle fibre is finer and softer than sisal so large

Table 2.8 Mechanical properties of bast and leaf fibres

	Length of commercial fibre (mm)	Length of spinnable fibre (mm) (i.e. staple length)	Linear density (tex)	Tenacity (cN/dtex)	Extension at break (%)
Jute	750–1500	60–150	1.5–4.5	1.0–3.5	1.8–2.5
Kenaf	750–1500	80–150	1.9–6.0	3.0	1.6–2.3
Flax	700–900	60–150 (wet spun)	1.2–2.5	3.5–5.5	2.0–5.0
Ramie	800–1300	70–100	0.3–0.6	4.0–6.0	2.0–6.0
Hemp	2500 (long hemp)	50–150	0.3–2.2	3.5–5.8	2.0–5.0
Pineapple	600–1000	30–80	1.5–2.3	3.5	2–4.8
Sisal	800–1200	500–1000	12–20	2.0–5.8	1.8–3.5

amounts of pineapple fibres are used in the manufacture of clothing and accessories with elaborate embroidery. Some ropes and twines are also made from pineapple fibres.

The properties of the main bast and leaf fibres can be summarized as shown in Table 2.8.

2.6 SUSTAINABILITY ISSUES/ECO ISSUES

2.6.1 BIOTECH COTTON

Currently, in addition to the conventional breeding methods, research is underway to produce new hybrid varieties and modern biotechnology (recombinant DNA technology) is being used to produce biotech or transgenic cottons, which enhance production flexibility. Some countries, such as China, Australia, India, South Africa, and United States, already allow biotech cotton to be grown. More and more countries are considering approving the cultivation of biotech cotton.

2.6.2 ORGANIC COTTON

Organic cotton refers to naturally cultivated cotton without the use of any synthetic agricultural chemicals such as fertilizers or pesticides or transgenic technology. It was first planted in the 1980s as an attempt to secure sustainable, ecological, and biodynamic agriculture. Organic cotton promotes and enhances biodiversity and biological cycles and so is beneficial to human health and the environment. Therefore, even though the properties of organic cotton fibre are not as good as regular cotton fibre worldwide production of organic cotton is growing rapidly.

There has been some research to study the properties of organic and regular cotton and the fibre and yarn produced from them. Table 2.9 compares the properties of the fibre.

The comparison between the yarns made by organic and regular cotton are shown in Table 2.10.

2.6.3 NATURALLY COLOURED COTTON

Naturally coloured cotton is naturally pigmented fibre, and found in shades of red, green, and several shades of brown, as well as the yellowish, off-white colour typical of modern commercial cotton fibres.

Table 2.9 Comparison of the properties between organic cotton and regular cotton

Properties	Organic cotton	Regular cotton
Length (mm)	27.9	29.2
SFC (<16 mm, W%)	10.5	9.1
Uniformity ratio (%)	81.9	82.9
Fineness (dtex)	18.4	17.4
Micronaire (Mic)	4.9	4.5
Tenacity (cN/dtex)	2.7	2.8
Maturity ratio	1.65	1.8
Immature content (%)	6.8	6.3
Impurity (%)	1.7	1.1

Table 2.10 Comparison of the properties the yarn (18.2 tex) made by organic and regular cotton

	Tenacity (cN/tex)	Weight variation (%)	CV of weight (%)	CV (%)	Thin place (Cnt/1000 m)	Thick place (Cnt/1000 m)	Neps (Cnt/1000 m)
Organic yarn	13.3	1.39	2.28	14.44	3	21	19
Regular yarn	14.6	1.13	2.05	13.93	1	15	15

Table 2.11 Comparison among the natural coloured cottons

	2.5% Span Length (mm)	Tenacity (cN/tex)	Micronaire (Mic)	Uniformity ratio (%)	SFC (%)	Neps (Cnt/100 g)	Maturity ratio
White cotton	28–33	18–23	3.7–4.2	49–52	<12	80–200	1.50
Brown cotton	25–29	16–19	3.0–6.0	44–47	20–25	120–200	1.42
Green cotton	21–27	13–16	3.0–6.0	45–47	20–30	100–150	0.61

The natural colour is due to the plant's inherent genetic properties. Shades of coloured cotton vary over seasons and geographic location due to climate and soil variations.

Naturally coloured cottons are considered environmentally friendly because they have many insect and disease-resistant qualities, are drought and salt tolerant and do not have to be dyed artificially. Coloured cottons have been grown successfully with organic farming methods.

However, the majority of naturally coloured cottons are of a lesser quality (Table 2.11) than most conventional cottons and are currently only available in a limited range of colours. Even though breeders have improved the properties of naturally coloured cotton, the fibres are still shorter and weaker. Staple-core and filament-core spinning are two methods being used to produce composite yarns that blend coloured cotton fibre with stronger white cotton or synthetic fibres. These composite yarns are stronger, although still retaining the softness and appearance of coloured cotton yarns. The comparison of coloured cotton is shown in Table 2.11.

Naturally coloured cotton feels softer to the skin. This, combined with their unique nonfading and environmentally friendly properties, has helped to assure them a niche market. The future success of coloured cotton depends on continued improvement in fibre quality and the development of appropriate manufacturing procedures.

2.6.4 BAST FIBRES

As all bast fibres contain non-cellulose matter such as pectin, lignin, and hemicellulose, degumming or retting is needed to remove the non-cellulose matter. Degumming by chemicals and retting in water have negative environmental effects. However, recently enzyme-retting and enzyme or microbial degumming have been developed and put in to practice, resulting in less water pollution, less damage to fibre, and more efficient processing.

As the vegetable fibre is naturally decomposed which is better for the environment, more and more new applications are being found, for example, flax, ramie, jute, kenaf and hemp are now being used as suitable substitutes for man-made synthetic fibres such as heavier glass fibre.

2.7 CASE STUDIES

As discussed in this chapter, cotton has a range of properties that make it particularly suitable as an apparel fabric.

These properties include:

- Strength and durability
- High ability to absorb and desorb moisture, and
- Low resilience

As an apparel fibre, cotton has a lot of desired properties, combining durability with attractive wearing qualities and comfort, making it suitable for underwear and leisure wear.

The natural convolutions in the longitudinal aspect of cotton creates friction among the fibres, preventing them from slipping and contributing to the durability of the fabric. Furthermore, its characteristic of being stronger when wet means cotton can withstand repeated washing which is important for garments that must be washed regularly.

The comfort of a textile material depends upon its ability to absorb and desorb moisture. A garment that does not absorb moisture, such as those made from synthetic fibres, will tend to feel clammy as perspiration condenses on it from the skin. Cotton fibre can absorb appreciable amounts of moisture and transfer it readily to the air; therefore, cotton garments are comfortable and cool, passing on the perspiration from the body into the surrounding air no matter how tightly woven a cotton fabric may be.

Like all fabrics made from natural vegetable fibres, cotton fabric will wrinkle and crease. However, cotton cellulose is not affected unduly by moderate heat, so cotton fabrics can safely be ironed with a hot iron. Cotton fabric can also be treated to impart wrinkle resistance and dimensional stability.

Cotton is not very suitable for outerwear and coats because of its low resilience and tendency to crease. In the case of a dress shirt and other garments for which appearance is more important than comfort, cotton, at least pure cotton, is seldom used. Sometimes polyester fibre is blended with cotton to make the fabric more durable with increased resilience.

In the same way, flax, although having the same capacity as cotton to absorb and desorb moisture, being stronger, thicker, and less elastic is rarely used to make underwear because it is usually heavier and more rigid which is less comfortable on the skin. In most cases flax is blended with cotton to produce finer yarn, and consequently finer, softer fabric which is then suitable for underwear.

2.8 FUTURE TRENDS

Due to their short single cell, flax and hemp are usually used in the form of long, coarse bundle fibres, which limits the processing ability and quality of end-products. The wet processing widely used in flax and hemp spinning has the disadvantage of low productivity, and the wet spun yarn is too hard to be used for knitting. A new technology of deep degumming instead of traditional retting is being developed to produce a refined fibre. This refined fibre is short (but can be processed using the cotton spinning system) and fine (usually 2000 metric count or finer), so it can be spun easily, improving productivity. The resultant yarn and end-product are of high quality and can be used readily for more purposes such as knitting.

Another developing application of bast fibres is composite materials. Due to the stronger, coarser, higher rigidity and lighter weight of the fibre, more and more composite materials made from bast fibres are used in cars and decrocative boards in construction.

2.9 SUMMARY

Natural fibres include vegetable fibres and protein fibres. Among the natural fibres, cotton, which is finer and softer than the other vegetable fibres, has the largest production and widest application.

The main component of vegetable fibre is cellulose, so that all the natural vegetable fibres have similar characteristics of high water absorbency, low resilience, a tendency to wrinkle easily, resistance to alkalies and organic solvents, and are highly combustible.

All bast fibres have non-cellulose matters or gums besides the main components of cellulose; therefore, before being further processed, degumming or retting is needed to remove (at least part of) the gums. The high content of lignin in bast fibres means they are usually harder and less elastic than cotton.

2.10 PROJECT IDEAS

In this chapter, the most commonly used natural cellulose fibers are intruduced. The structure, compositons, property, and main application of these fibers are described, the basic and main test methods for the fibers are also mentioned. It is hope to make the learner better understand the characteristic and application of the natural fibers.

2.11 REVISION QUESTIONS

1. Which fibres are natural vegetable fibre?
2. What is the component of natural vegetable fibres?
3. What are the main properties of fibre? What are the characteristics of natural vegetable fibres?
4. Describe the kinds of cotton and the important properties of cotton.
5. How do the properties of fibre affect its processing and end-products?
6. Briefly explain the reason why natural vegetable fibres have high absorbance of water.
7. Why is retting or degumming needed for bast fibres?
8. What is the main difference between cotton and bast fibres?
9. Briefly describe the characteristics of organic cotton and naturally colored cotton.
10. List two applications of vegetable fibres and explain the advantages of using such fibre.

REFERENCES

Cui, P., Wang, F. -M., Wei, A., & Zhao, K. (2010). The performance of kapok/down blended wadding. *Textile Research Journal*, *80*(6), 516–523.

Iwona, F. (2003). Cotton quality evaluation: New possibilities. 62nd ICAC Plenary Meeting, Gdansk, Poland.

Joseph, M. L. (1986). *Introduction to textile science* (5th ed.). New York: Holt, Rinehart and Winston.

Rijavec, T. (2008). Kapok in technical textiles. *Tekstilec*, *51*(10–12), 319–331.

CHAPTER

NATURAL TEXTILE FIBRES: ANIMAL AND SILK FIBRES

3

K.M. Babu

Bapuji Institute of Engineering Technology, Davanagere, India

LEARNING OBJECTIVES

At the end of this chapter, you should be able to:

- Understand the cultivation, dimension, and appearance of the main natural protein fibres
- Understand the composition and structure of the main protein fibres in relation to their properties
- Understand the application of the main protein fibres in textiles

3.1 INTRODUCTION

Fibrous proteins have been exploited for many years. Natural protein fibres are formed by animal sources through condensation of α-amino acids to produce repeating polyamide units with various substituents on the α-carbon atom. The sequence and type of amino acids making up individual protein chains contribute to the overall properties of the resultant fibre. The two major classes of natural protein fibres are keratin (hair or fur) and secreted (insect) fibres, and of these two groups, the most important members are wool (derived from sheep) and silk (excreted by various moth larvae such as *Bombyx mori*), respectively. Wool is composed of an extremely complicated protein called *keratin*, which is highly cross-linked by disulphide bonds from cystine amino acid residues. By contrast, silk fibre is composed of much simpler secreted protein chains, arranged in a linear pleated structure with hydrogen bonds between amide groups on adjacent protein chains.

Common qualities of protein fibres are:

- moderate strength, resiliency, and elasticity
- excellent moisture absorbency
- anti-static
- fairly resistant to acids, but readily attacked by bases and oxidizing agents
- tendency of yellowing in sunlight
- comfortable under most environmental conditions
- excellent aesthetic qualities

Nature exhibits an abundance of structural materials in the form of protein fibres, which have attained remarkable levels of efficiency and performance through evolutionary selection. Alpha keratin

Textiles and Fashion. http://dx.doi.org/10.1016/B978-1-84569-931-4.00003-9

fibres, together with fibroin fibres such as silks and spiders webs, are all highly extensible fibrous proteins whose mechanical properties are of primary importance both to the animal from which they originate and their ultimate application by man. Similarly, collagens are highly inextensible fibrous proteins that form the major component of mammalian skin and connecting structures such as tendons. All these fibrous proteins are biological polymers of polypeptide chains whose mechanical and allied physical properties, such as water absorption are determined by their macrostructure and their molecular and near-molecular structure. The mechanical properties of α-keratin fibres are primarily related to the two components of the elongated cortical cells, the highly ordered intermediate filaments (microfibrils) which contain the α-helices, and the matrix in which the intermediate filaments are embedded. The matrix consists of globular proteins plus water, the content of the latter being dependent on the fibres' environment.

In this chapter, natural protein fibres such as wool, silk, and specialty hair fibres such as cashmere, mohair, and camel hair fibres are discussed with reference to their structure and physical properties. Various applications of these fibres are also discussed in brief.

3.2 WOOL FIBRES

Wool is a protein fibre chiefly composed of keratin. It is a natural, highly crimped protein hair fibre derived from different breeds of sheep such as Merino, Lincoln, and Sussex, amongst others. Worsted fabrics are made from highly twisted yarns of long fine wool fibres usually blended with polyester fibre whereas woollen fabrics are made from less twisted yarns of coarser wool fibres.

The fibre is made up of overlapping cuticle scales and an inner cortex and is slightly elliptical, unlike other animal fibres. Both the cortex and the cuticle influence the wool fibre's properties, as does the breed of sheep from which it originates. The particular fibre characteristic of specific breeds can be exploited by processing the fibre into appropriate end products. In general terms, wool varies from the super fine Merino producing a fibre similar to cashmere, very high lustre English breeds producing mohair-like fibre, to coarse hairy wools similar to the guard coat of some goats.

3.2.1 STRUCTURE OF WOOL

The structure of the proteins in wool differs between the various regions of the fibre. Some of the proteins in the microfibrils are helical, like a spring, which gives wool its flexibility, elasticity, resilience, and good wrinkle recovery properties. Other proteins, particularly in the matrix that surrounds the microfibrils have a more amorphous structure enabling wool to absorb relatively large amounts of water without feeling wet (up to around 30% of the mass of the dry fibre). The matrix proteins are also responsible for wool's ability to absorb and retain large amounts of dye.

In addition to its chemical complexity, wool also has a very complex physical structure, as shown schematically in Figure 3.1. The fibre is surrounded by cuticle cells that overlap in one direction which consist of at least four layers:

- the epicuticle
- the A-layer of the exocuticle
- the B-layer of the exocuticle
- the endocuticle

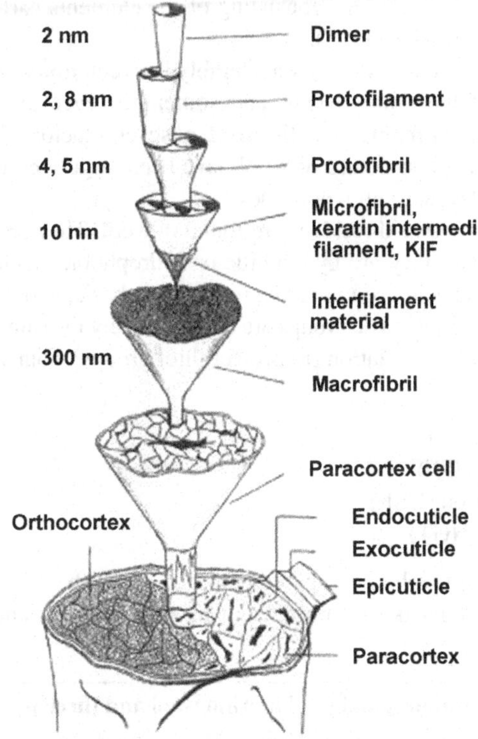

FIGURE 3.1

Fine structure of wool.

The cuticle surrounds a compacted mass of cortical cells in spindle form aligned with the fibre axis, their fringed ends interdigitating with each other (Rogers, 1959). Both cuticle and cortical cells are separated by the so-called cell membrane complex comprising internal lipids and proteins. This cell membrane complex is the component between the cells that guarantees strong intercellular bonding via proteins generally called desmosomes.

Australian merino wool fibres range in diameter, typically from 17 to 25 μm. They are composed of two types of cells:

- the internal cells of the cortex and
- the external cuticle cells that form a sheath around the fibre, as shown in Figure 3.1.

3.2.2 AMINO ACID COMPOSITION

Wool fibre is composed of extremely complex, highly cross-linked keratin proteins made up of over 19 different amino acids. The most important amino acid is cystine, which contains sulphur and forms cross-links between adjacent chains through disulphide bonds ($-CH_2-S-S-CH_2-$). Keratin differs from most other proteins because of its large sulphur content of about 3–4%; other than that, the elemental composition of wool is typical of proteins, consisting of the elements carbon, hydrogen, oxygen and nitrogen (Zahn, Wortmann, & Höcker, 1997).

The relative amounts of amino acids vary considerably between fibres from different sheep breeds, from different individuals of the same species and sometimes even along the length of one fibre (Rippon, 1992, Chap. 1). These differences are the result of several factors, including genetic origin and nutrition. Studies on the chemical structure of wool have been largely confined to fine merino fibres, although many aspects are relevant to all wool types.

Beside cystine, 20 other amino acid residues are found in wool. They are distinguished by their side chain, which determines whether they are hydrophilic or hydrophobic, acidic or basic. In their ionized state, a deprotonated carboxylic acid group may be regarded as basic, and a protonated amino group as acidic. The proportions of acidic and basic groups are approximately the same (800–850 µmol/g of each).

After reduction and carboxymethylation (to protect thiol groups), four fractions of proteins can be extracted from wool, namely:

- the low sulphur fraction (LSF)
- the high sulphur fraction (HSF)
- the ultrahigh sulphur fraction (USF)
- the high Gly/Tyr fraction (HGT)

The amino acid composition of wool, as compared with that of the above-mentioned fractions, is given in Table 3.1. Each fraction consists of a number of protein families and each one of them made

Table 3.1 Amino acid composition (µmol/g) of merino wool and three protein fractions extracted

Amino acid	Merino wool	LSF	HSF	USF
Ala	417	518	238	275
Arg	602	585	398	248
Asp	503	655	60	82
Cys*	943	546	1859	1734
Glu	1020	1138	772	905
Gly	688	709	497	702
Ile	234	295	215	330
Leu	583	826	144	151
Lys	193	326	38	1
Met	37	44	0	0
Phe	208	243	50	103
Pro	633	342	969	853
Ser	860	588	1163	1100
Thr	547	354	893	832
Tyr	353	345	164	151
Val	423	477	331	317

LSF = low sulphur fraction, HSF = high sulphur fraction, USF = ultra high sulphur fraction.
**Cysteine and half cystine.*
Source: Crewther (1975).

up of closely related members (Gillespie, 1965; Gillespie & Reis, 1966; Haylett, Swart, Parris, & Joubert, 1971; Jeffry, 1972; Lindley, Gillespie, & Haylett, 1967; Zahn & Biela, 1968).

3.2.3 PROPERTIES OF WOOL FIBRES

3.2.3.1 Fibre size and shape

Wool is usually harvested from sheep by annual shearing. Consequently, fibre length is determined largely by the rate of growth, which in turn depends on both genetic and environmental factors. Typical merino fibres are 50–125 mm long. They have irregular crimp (curvature), with the finer fibre generally showing lower growth rates and higher crimp. The fibre surface is rough due to the outer layer of overlapping cuticle cells. By far the most important dimension is the fibre diameter. Wool fibres exhibit a range of diameters, which like fibre length are dependent on both genetics and environment. Coarse wool fibres (25–70 μm diameter) are used in carpets, whereas fine merino fibres (10–25 μm) are used in apparels because of their soft texture. Fibres from an individual sheep also exhibit a range of diameters. The mean diameter is the prime dictator of price; however, the distribution of diameters is also important.

3.2.3.2 Tensile properties

The tensile properties of wool are quite variable but, typically, at 65% relative humidity (RH) and 20 °C, individual fibres have a tenacity of 110–140 N/ktex (140–180 MPa), a breaking elongation of 30–40%, and an initial modulus of 2100–3000 N/ktex (2.7–3.9 GPa) (Morton & Hearle, 1993). Although wool has a complicated hierarchical structure, the tensile properties of the fibre are largely understood in terms of a two-phase composite model (Feughelman, 1987; Hearle & Susutoglu, 1985; Wortmann & Zahn, 1994). In these models, water-impenetrable crystalline regions (generally associated with the intermediate filaments) lying parallel to the fibre axis are embedded in a water-sensitive matrix to form a semicrystalline biopolymer. The parallel arrangement of these filaments produces a fibre that is highly anisotropic. While the longitudinal modulus of the fibre decreases by a factor of three from dry to wet (Huson, 1998), the torsional modulus (a measure of the matrix stiffness) decreases by a factor greater than 10 (Mitchell & Feughelman, 1960). The longitudinal stress–strain curves for a wool fibre at different relative humidities are shown in Figure 3.2 (Morton & Hearle, 1993).

Three distinct regions can be discerned, especially for fibres at higher relative humidity. Once the fibre crimp is removed, a near-linear region up to about 2% strain is obtained (pre-yield region). For wet fibre, this is generally associated with stretching of the α-helices within the intermediate filaments. At lower (water) regain, the matrix phase plays an increasingly dominant role. Between 2 and 25% strain (yield-region) progressive unfolding of zones of α-helices occurs to form a β-pleat configuration. Very little increase in stress is observed during this stage and complete recovery is still possible, provided the fibre is allowed to relax in water. Beyond 25% strain (post-yield region), the fibre stiffens and breaks. At a molecular level, the reasons for this are still a matter of debate, but are believed to be partly due to resistance to the unfolding of a stabilized region of the intermediate filaments (Feughelman, 1987; Wortmann & Zahn, 1994) and the matrix's elastic response (Hearle & Susutoglu, 1985). A recent review looks critically at the different models (Hearle, 2000).

For a fibre immersed in water, the ratio of the slopes of the stress–strain curve in these three regions is about 100:1:10. While the apparent modulus of the fibre in the pre-yield region is both time and water dependent, the equilibrium modulus (1.4 GPa) is independent of water content and corresponds to the

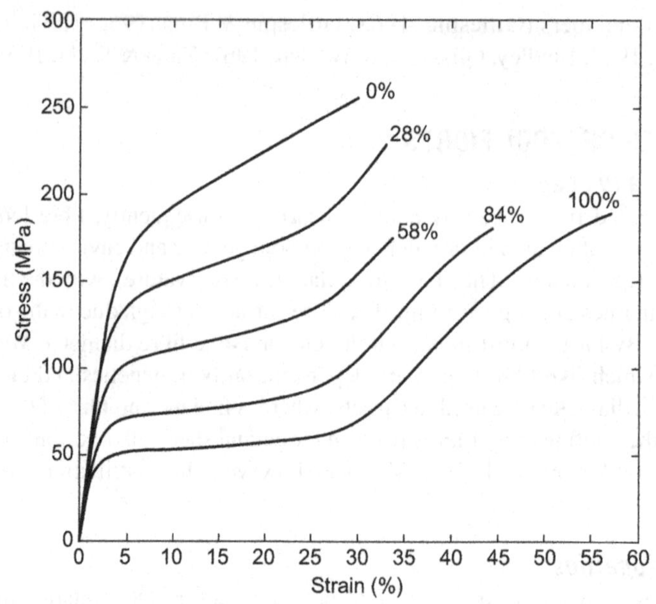

FIGURE 3.2

Stress–strain curves of typical wool fibres at different relative humidities.

modulus of the crystalline phase (Feughelman & Robinson, 1971). The time-temperature and water-dependence can be attributed to the visco-elastic properties of the matrix phase.

3.2.4 APPLICATIONS

Today, the emphasis of wool fibre and textile marketing has switched from quantity to quality. Product and process innovation has extended the appeal and applications of wool. The development of softer, lighter-weight fabrics based on finer yarns and fibres, innovative blends with other natural and man-made fibres and new finishing techniques has improved the technical performance range and *trans-seasonal* appeal of wool textiles. New developments have extended the application of wool textiles to include a wider range of casual and sportswear items. Another area of technical development has been the so-called single-stage processing enabling the production of fabric directly from scoured, dyed fibre in the form of felts or nonwoven products. This has seen applications in several areas including, blankets, building insulation, agro textiles, industrial felts, and performance sportswear.

3.3 SILK FIBRES

Silk is one of the oldest fibres known to man. Silk is an animal fibre produced by certain insects to build their cocoons and webs, and is the only natural fibre that occurs in filament form. Although many insects produce silk, only the filament produced by the larvae of the caterpillar from the cultivated *B. mori* moth and a few others in the same genus are used by the commercial silk industry

(Jolly, Sen, Sonwalker, & Prasad, 1979). Although there are several commercial species of silkworms, *B. mori*, commonly known as the mulberry silkworm because it feeds on the leaves of the white mulberry tree, is the most widely used. The silk produced by other insects, mainly spiders, is used in a small number of other commercial capacities, for example weapon and telescope cross hairs and other optical instruments (Chirs Spring & Julie Hudson, 2002).

Silk has been used and regarded as a highly valued textile fibre for over 4000 years. It is still considered a premier textile material in the world today due to its high (tensile) strength, lustre, and ability to bind chemical dyes (Zarkoob, Reneker, Ertley, Eby, & Hudson, 2000). Silk fibres are remarkable materials displaying unusual mechanical properties, such as being strong, extensible, and mechanically compressible (Matsumoto et al., 2006). Despite facing keen competition from man-made fibres, silk has maintained its supremacy in the production of luxury apparels and high-quality specialized goods (Robson, 1998) and has been referred to as the 'queen of textiles' for its lustre, sensuousness, and glamour (Manohar Reddy, 2009). Silk's natural beauty and properties of remaining cool in warm weather and providing warmth during colder months have ensured its continued use in high-fashion clothing. Silk fibres have outstanding natural properties that rival the most advanced synthetic polymers, yet their production does not require harsh processing conditions, encouraging widespread research into the possibility of artificially produced silk fibres (Chen et al., 2003).

Besides the growing of mulberry trees (mulberry culture), the production of silk can be viewed as a culmination of a number of separate stages:

- sericulture,
- silk reeling to obtain the raw silk filament thread, and
- throwing that converts the harvested thread into a useable yarn for fabric production.

3.3.1 SERICULTURE AND COCOON PRODUCTION

Sericulture is the rearing of silkworms and the production of cocoons. It initially involves the selection and separation of eggs into two categories; eggs to be used for reproduction of the silk moth and those used for actual silk production. The latter are incubated at a temperature of ca. 30 °C and hatch as caterpillars after about 10 days. They are then fed on mulberry leaves. Approximately 35 days after hatching, they will have increased in weight by 10,000, molted four times, and are ready to start spinning cocoons.

Below the mouth of each caterpillar, liquid silk (fibroin) and sericin from two internal glands is extruded through two small orifices (called *spinnerets*) to form liquid filaments coated in sericin that coalesce to a single thread, two to four denier in size, then solidify on immediate contact with air. After 2–3 days, a caterpillar will have spun about 1.5 km of solidified filaments to form a cocoon shell around its body. Sericin is a water-soluble gum (glycoprotein) that envelops the fibroin fibre with successive, sticky layers, helping to form a cocoon and constituting about 20–30% of the total cocoon weight (Robson, 1998). After harvesting the cocoons, they are immersed in heated water,[1] enabling the filaments to be unwound as yarns. The single thread from each cocoon is too fine to be used in subsequent fabric production, so typically the single threads from 4 to 20 cocoons are combined while being unwound onto a reel; during this process, the threads are slightly twisted together to form a single raw silk yarn.

[1] http://www.youtube.com/watch?v=uKaZaCpO138.

Sericulture is ideally suited for improving a country's rural economy, as it is a subsidiary industry to agriculture. Recent research has also shown that sericulture can be developed as a highly rewarding agro-industry.

3.3.2 SILK REELING

The unwinding of filaments from water softened cocoons, combining and winding them onto reels, is called silk reeling. Each cocoon has three parts:

- the floss, which are surface filaments that hold the cocoon to a support surface while it is being spun, and the actual outer surface of the cocoon;
- the shell of the cocoon, the coalesced filaments of which are to be unwound; and
- the dead body of the chrysalis.

When the cocoons are first softened in hot water, the objective is to loosen the floss so that it can be readily detached; the term used is *cooking* the cocoons. The detaching is achieved by a subsequent brushing (*beating* or *peeling*) action after which the cocoons are ready for unwinding. Originally, this would be performed manually by using a brush to drag off the floss; the modern practice is to use cocoon cooking and peeling machines, which minimizes damage to the cocoon shell and so reduces silk waste.

While the number of cocoons to be converted into a yarn are kept in a bowl of water continuously heated to 33–40 °C (*reeling bath* or *reeling basin*), the unwinding filaments pass through a series of guides and pulleys, so arranged that they twist together about 200 times during their passage to the reel, squeezing out any water adhering to the filament lengths and giving a circular cross-sectional shape to the yarn that gives its lustre. This is referred to as the *croissure*, and pulleys facilitating the coiling are called croissure pulleys. Before they reach the reel, the filaments move through an oscillating guide bar that spreads the yarn 75 mm wide across the reel to form a hank.[2] Modern machines would have up to 400 reeling positions per frame.

The quality of the water in the reeling bath influences the quality of the raw silk yarn. It should have a pH level between 6.5 and 7.0, as this removes the sericin molecules but leaves a film of sericin over the filament surfaces, assisting the croissure to bind (*agglutinate*) the collected filaments together, giving a neater appearance to the raw silk yarn. Although the sericin film retained must not be thick, neither must it be too thin, as this could lead to weak agglutination and poor mechanical properties adversely affecting subsequent manufacture.

As the reeling process progresses, the water becomes turbid owing to the increasing concentration of dissolved sericin, the pupal body, and other extraneous matter from the cocoon. The build up of these impurities will reduce cocoon reelability and/or the silk colour becomes dull as they adhere to the filament surface. A constant inflow and overflow of water to the bath is used to keep the concentration of these impurities as low as possible.

There are three main methods of raw silk reeling,

- direct reeling on a standard sized reel,
- indirect reeling on small reels, and
- the transfer of reeled silk from small reels onto standard sized reels on a re-reeling machine.

---

[2] Hank: an unsupported coil (135 cm in circumference) composed of wraps of yarn on a reeling machine.

The last technique is primarily applied in modern silk reeling processes. Reeling of mulberry cocoons is much more organized in the industry compared to nonmulberry silk reeling (Sonwalker, 1993, p. 181). Countries like China with advanced sericulture have adapted very sophisticated reeling processes with fully automatic reeling machines.

Following the reeling process, the hanks from the creel are then wound onto a spool by a spool-winding machine, subsequently degummed[3] and dyed followed by doubling, two-for-one twisting and steaming, for stress relation, ready for weaving or knitting. What is termed a thrown silk yarn is obtained by twisting together either two or more raw silk threads.[4] The dyeing of silk yarns and the principles of doubling and two-for-one twisting are widely reported in publications and the reader can refer to the two references cited here.

The innate pigmentation of raw silk is largely removed during degumming. However, the degummed filaments do retain yellowness. Silk is used in a wide range of textile applications, and white or light-dyed shades of silk are often required. The degummed silk filaments therefore need to be bleached to remove more pigmentation. These filaments are susceptible to tendering by oxidizing chemicals (bending), so the chemical agents used need to be chosen and applied carefully. Typical bleaching agents are nitric acid, sulphur dioxide, bromine, permanganate, sodium hydrosulphite, hypochlorites, salts of per-acids, and peroxides.

3.3.3 SILK MANUFACTURE

Silk has played an important role in the development of loom and weaving technology. Traces of primitive looms and woven fabrics have been found in excavations in Egypt, China, India, and Peru, but these tribal 2-bar bamboo devices, including later improved shaft looms – horizontal and vertical – were only suitable for plain or simple patterned coarse weaving or for carpets, tapestry, or floor coverings. The silk weavers of China invented the use of the heddle and draw loom, a revolutionary development over the traditional primitive loom, while India invented a foot treadle for silk weaving, another technical innovation (Datta & Nanavaty, 2005, p. 171).

In order to manufacture quality fabric and in widths that are acceptable in the international markets, silk weaving looms need to be standardized. Shuttle looms can weave silk fabrics efficiently, and a high speed machine ensures a high-quality warp and weft. The number of knots, cleanness, and cohesion are also important. In recent years, silk weaving has seen drastic changes both in the use of sophisticated looms and weaving technology. Today, silk fabrics are being successfully woven on modern shuttleless looms such as Rapier and Air jet, with electronic Jaquard and Dobby attachments for producing intricate designs. Despite this technology, sophisticated silk brocades and other intricate silks can still only be produced on traditional handlooms operated by master weavers in China, India, and other ancient silk-weaving countries (refer to Chapter 11 for more information on weaving).

---

[3] The degumming solution is made of two chemicals: an alkaline and a surfactant. The alkaline is washing soda, or sodium carbonate, also known as soda ash.

[4] Thrown silk: raw silk yarn that has been reeled and then two or more twisted into yarn. Thrown singles: a silk cord made by three processes of twisting (a single twisted, two or more singles twisted together, then several of these twisted together).

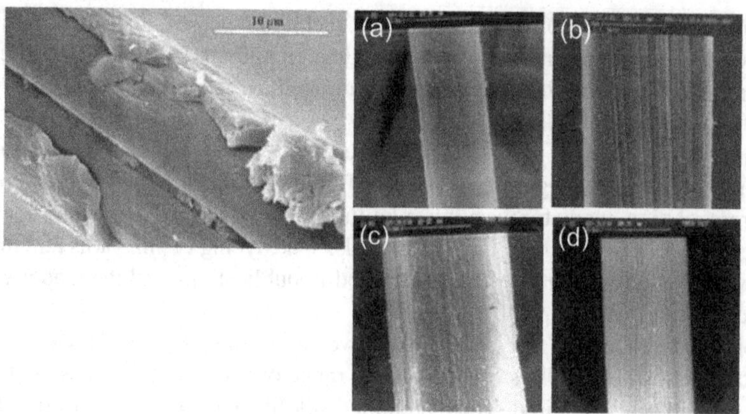

FIGURE 3.3

On the left, a longitudinal view of silk fibres (undegummed). On the right, longitudinal views of silk fibres (degummed): (a) mulberry; (b) tasar; (c) muga; (d) eri.

3.3.4 FINE STRUCTURE OF SILK

Silk fibres (*Bombyx mori*) spun from silkworm cocoons consist of fibroin in the inner layer and sericin in the outer layer. Each raw silk thread has a lengthwise striation consisting of two fibroin filaments of 10–14 lm each embedded in sericin. Generally the chemical composition is silk fibroin 75–83%, sericin 17–25%, waxes about 1.5%, and others about 1.0% by weight. Silk fibres are biodegradable and highly crystalline with well-aligned structure, higher tensile strength than glass fibre or synthetic organic fibres, good elasticity, and excellent resilience. Silk fibre is normally stable up to 140 °C and the thermal decomposition temperature is over 1500 °C. The densities of silk fibres range between 1320 and 1400 kg/m^3 with sericin and 1300–1380 kg/m^3 without sericin. Silk fibres are also commercially available in a continuous fibre type.

3.3.4.1 Longitudinal view

Scanning electron micrographs of longitudinal views of undegummed and degummed silk fibres are presented in Figure 3.3 (left) and 3.3(a-d), respectively. These show that mulberry silk has a more or less smooth surface (Figure 3.3(a)), whereas the nonmulberry silks such as tasar, muga, and eri (Figure 3.3(b)–(d)) all have striations on their surface.

3.3.4.2 Cross-sectional view

The scanning electron micrographs of silk fibre cross-sections are presented in Figure 3.4. This shows that two strands of fibroin filaments are enveloped by nonfibrous sericin. When a strand of fibroin filament is enlarged to show its inner structure, it appears like a large bundle of fibrils (Minagawa, 2000).

There are variations depending upon the variety of silkworms and also among individual cocoons. In this respect, the mulberry and nonmulberry silks show a very different cross-sectional morphology. The mulberry silks have a more or less triangular cross-section and smooth surface (Figure 3.4(a)), whereas the nonmulberry varieties, tasar and muga, exhibit an elongated rectangular or wedge-shaped cross-section, and a large cross-sectional area (Figure 3.4(b) and (c)). The eri silk has a more or less triangular shape (Figure 3.4(d)). Moreover, even in the same fibroin filament, there are variations in the cross-section depending upon the level of the cocoon layer.

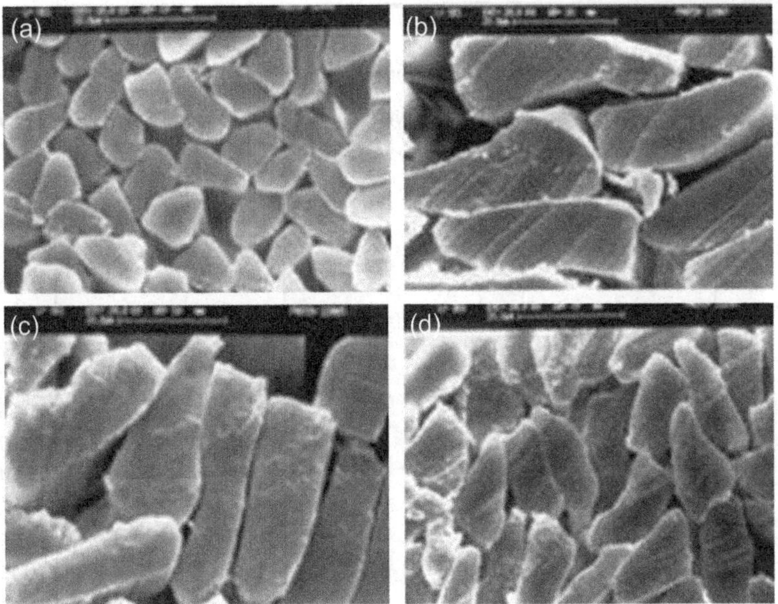

FIGURE 3.4

The cross-sectional view of silk fibres (degummed): (a) mulberry; (b) tasar; (c) muga; (d) eri.

3.3.5 AMINO ACID COMPOSITION

The amino acid composition varies in different varieties of silk. Three major amino acids, namely serine, glycine, and alanine, are found in mulberry and nonmulberry varieties. Among the other major amino acids present are tyrosine and valine. In mulberry silks, glycine, alanine, and serine together generally make up about 82%, of which about 10% is serine. Tyrosine and valine are the next largest constituents at about 5.5 and 2.5%, respectively. The overall composition of acidic amino groups (i.e. aspartic and glutamic acids) in mulberry silks is greater than that of the basic amino acids. The other important aspect is the presence of amino acids with bulkier side groups. These bulky side groups can hinder the crystallization process by restricting the close packing of molecules. However, in general, a large portion of the mulberry fibroin is made up of simple amino acids such as glycine and alanine, ensuring good crystallization (Sen & Murugesh Babu, 2004).

Compared to the mulberry silks, the total amount of glycine, alanine, and serine constitute about 73% in the nonmulberry variety, about 10% less. All the nonmulberry silks exhibit a high proportion of alanine compared to the mulberry variety. The proportion of alanine is about 34% in tasar, 36% in eri, and 35% in muga. On the other hand, the glycine content in these varieties is about 27–29%, which is lower than that found in the mulberry varieties at around 43%.

In addition, the nonmulberry varieties have a substantial proportion of amino acids with bulky side groups, especially aspartic acid (4–6%) and arginine (4–5%), which means that not only the acidic but also basic amino acid levels are greater. It is interesting to note the presence of sulphur-containing amino acids (i.e. cystine and methionine) in all the silk varieties. The methionine content in nonmulberry silks is slightly higher (0.28–0.34%) compared to that found in mulberry varieties (0.11–0.19%),

Table 3.2 Amino acid composition of silk fibres

	Amino acid composition (mol%)			
Amino acid	***Bombyx mori* (Mulberry)**	***Antheraea mylitta* (Tasar)**	***Antheraea assama* (Muga)**	***Phylisomia ricini* (Eri)**
Aspartic acid	1.64	6.12	4.97	3.89
Glutamic acid	1.77	1.27	1.36	1.31
Serine	10.38	9.87	6.11	8.89
Glycine	43.45	27.65	28.41	29.35
Hystidine	0.13	0.78	0.72	0.75
Arginine	1.13	4.99	4.72	4.12
Threonine	0.92	0.26	0.21	0.18
Alanine	27.56	34.12	34.72	36.33
Proline	0.79	2.21	2.18	2.07
Tyrosine	5.58	6.82	5.12	5.84
Valine	2.37	1.72	1.5	1.32
Metheonine	0.19	0.28	0.32	0.34
Cystine	0.13	0.15	0.12	0.11
Isoleusine	0.75	0.61	0.51	0.45
Leusine	0.73	0.78	0.71	0.69
Phenylalanine	0.14	0.34	0.28	0.23
Tryptophan	0.73	1.26	2.18	1.68
Lysine	0.23	0.17	0.24	0.23

whereas the cystine content is similar (Sen & Murugesh Babu, 2004). The amino acid composition of the different varieties of silk is presented in Table 3.2.

3.3.6 PROPERTIES OF SILK FIBRES

3.3.6.1 Tensile properties

Studies conducted on some mulberry and nonmulberry varieties by Iizuka et al., reveal that the tenacity, elongation, and modulus are all dependent on the linear density of the filament, and the linear density or mean size in turn depends on the silkworm race (Freddi, Gotoh, Mori, Tsutsui, & Tsukada, 1994; Iizuka, 1994, 1995; Iizuka & Itoh, 1997; Iizuka, Okachi, Shimizu, Fukuda, & Hashizume, 1993).

Tenacity is found to be related to the linear density of the filament (Figure 3.5). The correlation is negative, i.e. as the linear density increases, tenacity decreases. A similar trend has been observed for modulus. Elongation, on the other hand, increases with an increase in linear density.

Tenacity ranges between 2.5 and 4.82 g/d, for Japanese and Chinese mulberry varieties, 2.4–4.32 g/d for Indian mulberry varieties, and 3.74–4.6 g/d for Indian tasar varieties (Iizuka, 1995). A study on the chemical structure and physical properties of *Antheraea assama* (muga) silk found that the tenacity of muga varies between 3.2 and 4.95 g/d (Freddi et al., 1994; Iizuka et al., 1993). Another important non-mulberry variety, eri, showed the lowest tenacity value, ranging between 2.3 and 4.0 g/d (Iizuka & Itoh, 1997).

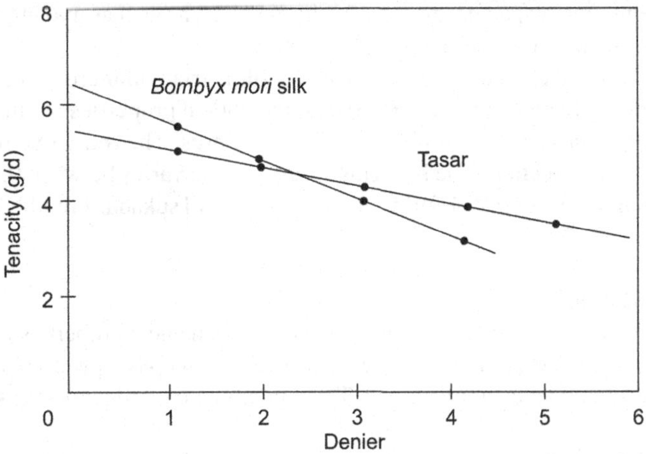

FIGURE 3.5

Tenacity versus denier thread relationship.

Table 3.3 Mechanical properties of different varieties of silk

Variety	Sex	Dynamic modulus (10^{10} dyn/cm^2)	Tan δ	Tenacity (g/d)	Elongation (%)
Shunreix shougetsu (mulberry)	M	1.847	–	5.265	20.36
	F	1.808	–	5.207	21.48
A. mylitta (tropical tasar)	M	1.132	0.030	3.412	31.36
	F	1.087	0.035	3.256	31.12
A. proylei (temp. tasar)	M	1.305	0.023	4.123	31.45
	F	1.087	0.025	4.128	31.48
A. assama (muga)	M	1.205	0.020	3.170	34.83
	F	1.230	0.023	3.823	34.10

Elongation-at-break, on the other hand, showed a higher value for all the nonmulberry silks compared to mulberry varieties. The values range between 31% and 35% for tasar, 34–35% for muga, and 29–34% for eri silks, respectively, while the elongation values for mulberry varieties ranged between 19% and 24%. Some of the mechanical properties of different varieties of silk are summarized in Table 3.3.

3.3.6.2 Optical properties

Silk fibroin extracted from silkworm cocoons is a unique biopolymer that combines biocompatibility with excellent optical properties. Silk may be used as an optical material for use in biomedical engineering, photonics, and nanophotonics. Silk can be nanopatterned with features smaller than 20 nm. This allows the manufacture of holographic gratings, phase masks, beam diffusers, and photonic crystals; for example, from a pure protein film. The properties of silk allow these devices to be 'biologically activated', offering new opportunities for sensing and biophotonic components. Many interesting bio-optical devices can be fabricated by doping silk films with fluorescent materials. Further possibilities are to enhance light emission by patterning the silk film surfaces, as well as making tunable wavelength devices and printing specific patterns on silk film surfaces.

Lustre associated with silk is due partly to the effect of its triangular cross-sectional shape on the pattern of light-reflection. In an attempt to understand the optical properties of silk, many researchers have determined the refractive index and birefringence of fibres. The refractive index of silk generally varies through its cross-section. The birefringence (n) value varies between 0.051 and 0.0539 for mulberry silk and from 0.030 to 0.034 for nonmulberry silks (Tsukada, Freddi, Minoura, & Allara, 1994).

3.3.6.3 *Visco-elastic behaviour*

Silk fibre exhibits visco-elastic behaviour. Time-dependent mechanical properties of silk fibres such as stress relaxation, creep, and creep recovery have been studied. Creep is a phenomenon associated with time-dependent extension under an applied load. The complimentary effect is stress relaxation under a constant extension.

It has been found that instantaneous extension and secondary creep are both higher for tasar silk compared to mulberry silk. The stress relaxation was also found to be higher in nonmulberry silks than in mulberry silk (Das, 1996).

Silk has also been shown to exhibit inverse stress relaxation phenomenon (Das, 1996). The inverse relaxation could be observed for both mulberry and tasar silks when the level of strain was maintained below a certain value. Inverse relaxation becomes higher with an increase in peak tension. Cyclic loading has been found to reduce the extent of inverse relaxation.

3.3.7 APPLICATIONS

Silk is one of the most beautiful fabrics available, with a long and colourful history and changing applications in the world today. Be it for dresses, mèdical use, home decor, and more, the use of silk is a wide and varied topic.

3.3.7.1 *Textile and apparels*

Silk's capacity to absorb water makes it comfortable to wear in warm weather and whilst engaged in activity. However, it is equally good in cold weather, as its low conductivity keeps warm air close to the skin. It is often used for clothing, but its elegant soft lustre and beautiful drape also make it perfect for many furnishing applications. It is used for upholstery, wall coverings, curtains (if blended with another fibre), rugs, bedding, and wall hangings.

Silk continues to be used as a material to produce fine dresses like traditional Chinese wedding Cheongsam dresses. Silk is chosen because it is one of the finest materials known in ancient Chinese culture. Delicately woven dragons, flowers, and butterflies are sewn into the silk dresses. The material is thick and shiny, which has a very flattering, slimming effect. Women's evening gowns are also often made from silk. It drapes well and, being slightly warmer, provides warmth, even for sleeveless gowns in winter.

Increasingly, bedding manufacturers have started to make silk sheets and pillowcases, as the health benefits of silk have become more widely known. Silk bedding is believed to prevent coughing and sneezing, especially for those allergic to dust mites, which do not like silk. Additionally, sleeping on silk sheets can be beneficial to women's hair, reducing tangles, and split ends.

3.3.7.2 Biomedical field

Silk, especially *B. mori* silk, has a very long history in biomedical applications. The unique mechanical properties of these fibres provide important clinical repair options. Silk from the silkworm, *B. mori*, has been used in native fibre form as sutures for wound ligation over the past 100 years and became the most common natural suture, surpassing collagen (Altman et al., 2003). Silk sutures are used in ocular, neural, and cardiovascular surgery due to its knot strength, handling characteristics, and ability to lie low to the tissue surface. In recent years, the reported exceptional nature of silk has lead to an increased interest in silk for biomedical applications (Hakimi, Knight, Vollrath, & Vadgama, 2007). Silk fibroin has been increasingly studied for new biomedical applications due to its biocompatibility, slow degradability, and remarkable mechanical properties. In addition, the ability to control molecular structure and morphology through versatile processability and surface modification options has expanded its use in a range of biomaterial and tissue-engineering applications. Silk fibroin in various formats (films, fibres, nets, meshes, membranes, yarns, and sponges) has been shown to support stem cell adhesion, proliferation, and differentiation *in vitro* and promote tissue repair *in vivo*. In particular, stem cell-based tissue engineering using 3D silk fibroin scaffolds has expanded the use of silk-based biomaterials to engineer a range of skeletal tissues such as bone, ligament, and cartilage, as well as connective tissues like skin (Wang, Kim, Vunjak-Novakovic, & Kaplan, 2006). More recent studies with well-defined silkworm silk fibres and films suggest that the core silk fibroin fibres exhibit comparable biocompatibility *in vitro* and *in vivo* with other commonly used biomaterials such as polylactic acid and collagen (Altman et al., 2003).

3.3.7.3 Fibre-reinforced composites

Although silk is extensively used as a valuable material for textile purposes in its own right, in recent years it has been increasingly used as a reinforcing material for composites made from epoxy and other biodegradable, biopolymeric resins. The organization of the silk fibres can contribute significantly to the impact resistance by ensuring both strength and good deformability in the composite. Also, silk yarn is readily available as a waste product in the textile industry, so the composite is cost-effective.

3.4 OTHER SPECIALTY HAIR FIBRES

There are many important specialty hair fibres having wool-like characteristics that are commercially valuable because combined with wool, they produce fabrics with interesting properties (Labarthe, 1975). Although these specialty hair fibres have similar properties to wool, they have some distinctive features too, like Mohair or Angora goat fibres where the epidermal scales are only faintly visible, making the fibre more lustrous and durable. It is widely used in men's clothing, often in combination with wool and silk.

3.4.1 CASHMERE FIBRES

Cashmere is a rare, natural fibre renowned for its softness. As a luxury fibre, cashmere commands some of the highest prices in the world of textiles. Only Vicuna and Musk Ox – neither of which is available in anything approaching commercial quantities – achieve a higher price than cashmere. The rarity, the geographical remoteness of cashmere production, the dependence on manual skills in the early stages of processing, and the fibre's association with exotic peoples following a traditional, rural way of life, add to the attraction of cashmere fibre in highly industrialized western markets.

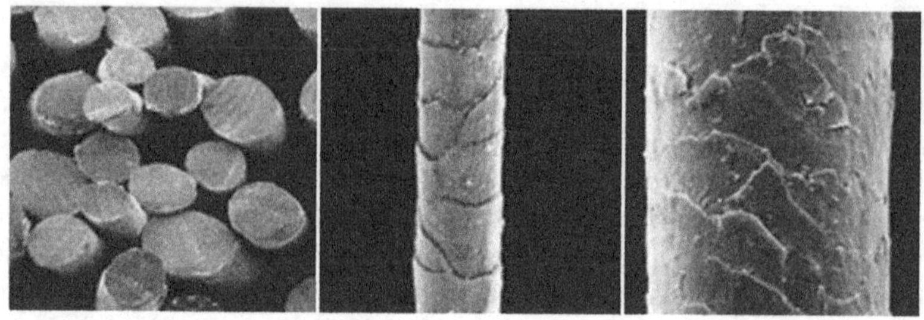

FIGURE 3.6

Cross-section and longitudinal view of cashmere fibres.

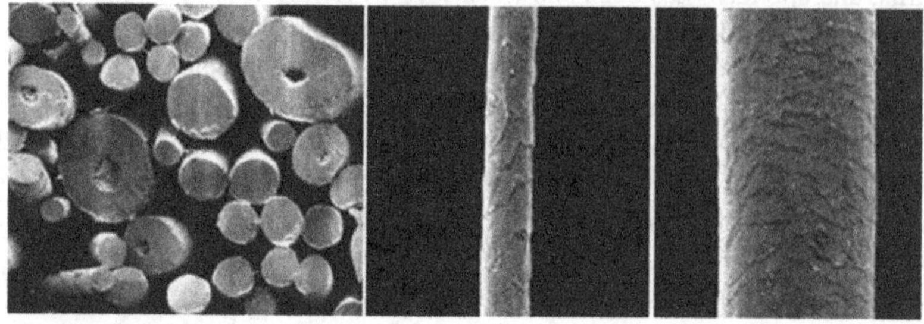

FIGURE 3.7

Cross-section and longitudinal view of camel hair fibres.

The cross-sectional and longitudinal view of cashmere fibres is given in Figure 3.6. Cashmere is one of the alternative (and lesser used) spellings of Kashmir, a region in the western Himalayas that bridges India and Pakistan. This wild and mountainous area gave its name to the fine, soft goat's wool, or down, which first came to the West in the form of intricately woven cashmere shawls. In fact, the fibre came from Tibet where it was gathered by herdsmen from their goats during the animals' spring moult; however, it was named after Kashmir, as it was there that it was spun, woven, and sold as a finished item. Today in Kashmir and Tibet, where the fibre is still produced and processed by hand, it is known locally as Pashmina.

3.4.2 CAMEL HAIR FIBRES

Camel hair, like cashmere, comprises relatively coarse outer hair and inner down fibre. The underdown produced by animals living in the hottest desert climates tends to be coarser and sparser than from those living in a more temperate climate. The best-quality fibre is found in Inner Mongolia and Mongolia. The fine down fibre varies from 19 to 24 μm – about 2 μm coarser than Iranian cashmere – and varies between 2.5 and 12.5 cm in length (Figure 3.7). This fibre is the result of many years of selective breeding. The outer hair of the camel is coarse and can be up to 37.5 cm in length with a diameter of 20–120 μm. The average weight of the adult female under hair is 3.5 kg, whereas that of the male is double this amount.

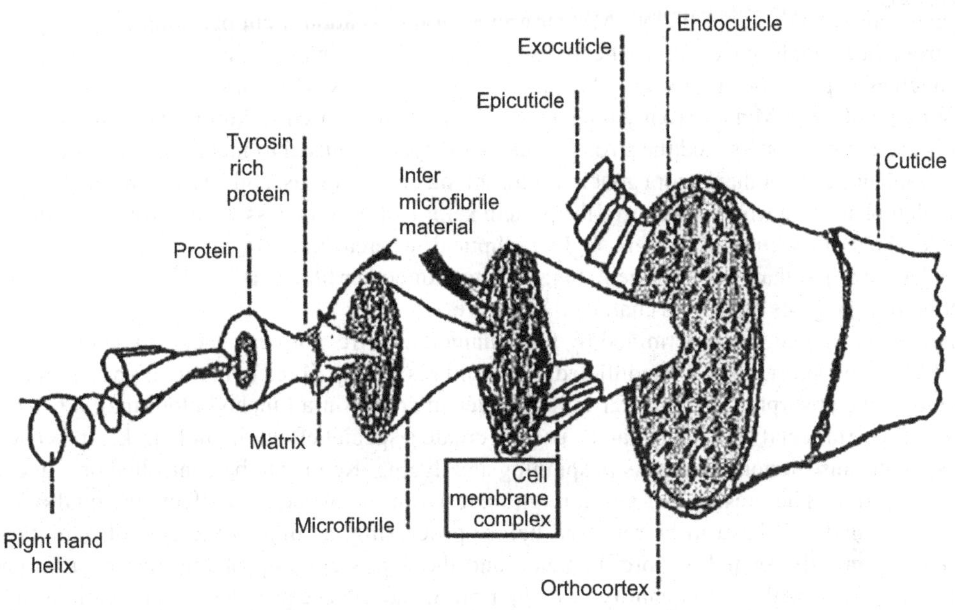

FIGURE 3.8

Structure of a mohair (adult) fibre.

Also, like cashmere, only the soft underwool or down hair is used to produce yarn. The moulting process takes place over a 6–8-week period starting in late spring, the camel losing first the neck hair, then the mane, and finally the body hair. The hair covering the humps is not shorn since the animals would be more susceptible to disease in the summer months without it. The hair is obtained by a number of methods: combing, shearing, and simply by collecting clumps of hair shed naturally during the moulting season. Varying from reddish to light brown, the hair is sorted according to shade and age of the animal. Baby camel hair, which has an average diameter of about 19 μm and a length of 2.5–12.5 cm, is the finest and softest.

3.4.3 MOHAIR FIBRES

Mohair, the lustrous fleece of the Angora goat, is one of the most important specialty animal fibres even though it represents less than 0.02% of total world fibre production. Mohair is a very beautiful, luxurious, and incredibly durable fibre. It is one of the warmest and most versatile natural fibres known. Angora goats took their name from Ankara an ancient Turkish city where they originated. Although the goats were farmed for their fibre from early times, it was not until the sixteenth century that export of the goats was permitted. The first exports were limited to France and Spain before spreading to many countries, reaching the Americas in 1849 and Australia in early 1900s. South Africa currently produces more than 60% of the total world production of mohair, whereas Australian production totals approximately 250,000 kg. The total global production is 5 million kg per year.

Mohair is a protein fibre. It has a smooth, overlapping, cuticular scale pattern on the surface that imparts lustre and has low felting capacity (Figure 3.8). This smooth scale is different to the wool fibre scale and consequently is not 'itchy'. Mohair grows rapidly at about 2 cm per month and is generally shorn from the animals twice a year. The fibres range from 23 μm in mean diameter at the first shearing to as much as 38 μm in older animals. This range of diameter makes the fibre incredibly versatile with a wide range of uses. Mohair from young goats (kid mohair) is used in knitwear, from intermediate age it is used to make suits, and the stronger 'fine hair' types are used to make coats and rugs.

The mohair fleece of the Angora goat is white, smooth, and lustrous, and has a high tensile strength. Although mohair (like wool) consists of the protein keratin, it nevertheless differs from wool in certain respects. The cross-section of wool is slightly elliptical, whereas the very fine mohair fibre is round. The scales are larger than wool and lie flatter, making a smoother fibre surface. The resultant increased reflection of light gives mohair its characteristic lustre.

The value of a fleece is determined by fibre diameter, lustre, softness, lack of kemps, and clean yield. Kemps are short, heavily medullated, coarse fibres. Kemp fibres contain air spaces (medulla) that reduce dye absorption and appear much lighter in colour in a finished cloth than other fibres. Although in certain end uses kemp can be used to create a special effect, in mohair, kemp is undesirable as it can cause serious problems in spinning and dyeing. Kemp can be controlled or reduced by genetic selection. The presence of any foreign material in the fleece also affects the quality of the final product and will have to be removed before processing, adding to the cost of manufacture. Grading is primarily related to fibre diameter, and the goat's age is probably the most important determinant in the quality and quantity of mohair produced. Fleece production increases from birth and peaks at approximately 3 or 4 years of age. The average peak production in South Africa is about 4–5 kg a year for females and 5–6 kg for males. Over the goat's lifespan, the fibre diameter increases from an average of 24 μm for kids up to 46 μm for strong adults. Kids normally produce fibre with a diameter of 24–28 μm at their first shearing, approximately 29–30 at 1 year, 31–34 at 18 months, and from 36 to 46 μm as adults. Staple length shows little change with age and grows at an average rate of 20–25 mm/month. The first three shearings produce the most sought-after fibre, and subsequently the fibre becomes coarser. Kid mohair fetches the highest price but only represents 16% of the clip.

3.5 APPLICATIONS OF NATURAL PROTEIN FIBRES

Wool, a fibre that has evolved over thousands of years to insulate and protect sheep, is the most complex and versatile of all textile fibres. It can be used to make products as diverse as cloth for billiard tables to the finest woven and knitted fabrics. The insulating and moisture absorbing properties of the fibre make fine wool products extremely comfortable to wear. The chemical composition of wool enables it to be easily dyed to shades ranging from pastels to full, rich colours. It is indeed justified to call wool 'Nature's Wonder Fibre'. In future, market trends may favour a shift back to wool products. This includes the prospect of a general fashion swing back towards smarter, more tailored apparel at work. In addition, apparel and furnishing markets are expanding quickly in developing nations due to their higher economic and population growth rates. The growing middle-class consumers in these nations, especially in China, are looking for higher-grade products.

There are two areas in which the use of silk could be developed more than it is today. The first area is in blends. Silk blends easily with other fibres, especially with other natural fibres. Silk–wool, silk–cashmere, silk–cotton, silk–linen blends are easier to dye than blends of silk and synthetic fibres. In addition, these blends are mutually enhancing. Silk benefits from the specific qualities of the other fibre it is blended with, whereas the addition of silk adds prestige and value to the blend. Although blends are already used frequently in clothing fabrics, their use could be developed and diversified, both by developing new fabrics based on fibre/fibre (intimate) blends of silk with wool, linen, cotton, or polyester and by the use of yarn/yarn blends.

Mohair is used to manufacture many products, including knitting yarn for hand or machine knitwear, lightweight suits, fabric for stoles, scarves and warm blankets, and durable upholstery velours. It is often blended with wool for top-quality blankets where the mohair content makes the fabric warmer, and at the same time lighter. Mohair is a lustrous fibre with a silky, luxurious appearance, but it is also very hard wearing. One of the advantages of this fibre is its tremendous versatility. It can be used both for clothing and furnishing and, within these, its end uses are wide ranging. Individual fibres are strong and make a finished fabric that is noncrushable, soil resistant, and does not pill. Mohair has an affinity for dyes and is able to absorb them completely and indelibly. It provides warmth during the winter months but also makes a cool suiting fabric for the humidity of summer, hence its popularity in East Asia, particularly Japan.

About 12% of total mohair production goes into furnishing fabrics, such as upholstery velours and moquettes. Such fabrics are often used for upholstery, particularly in prestige locations, including the first class areas of ships such as the QE2 because, although expensive, the fabric is extremely hard wearing and easy to clean while creating a feeling of quality.

One of the challenges facing the mohair industry – particularly the knitting sector – is that the fibre has been imitated by much cheaper acrylic yarns, developed to have similar properties, and described as 'mohair-like'. These products are seen to have done genuine mohair a disservice, adulterating its high-quality image.

3.6 SUSTAINABILITY AND ECOLOGICAL ISSUES

Sustainability, industrial ecology, eco-efficiency, and green chemistry are guiding the development of the next generation of materials, products, and processes. With increasing concerns regarding the effect the textile industry is having on the environment, more and more textile researchers, producers, and manufacturers are looking to biodegradable and sustainable fibres as an effective way of reducing that impact. New environmental regulations and social awareness have triggered the search for new products and processes that are compatible with the environment.

3.7 FUTURE TRENDS

With the current availability of technologies to produce 'designer' fibres based on genetic engineering strategies, new directions for protein fibre production can be considered. These new approaches might lead to the large-scale industrial production of protein fibres to be used as performance molecules in medical and technical applications. Studies of natural fibres promise a number of potentially useful lessons for materials chemistry and processing. Nature's range of functional materials represents the success of up to 4 billion years of development. Nature has optimized supramolecular self-assembly mechanisms, hierarchical microstructures, property combinations, and inservice durability resulting in fibrous materials that are not only damage-tolerant, but often self-repairing. They offer the attractions of biosynthesis (produced from renewable resources), benign processing conditions (assembled and shaped in an aqueous environment at mild temperatures), and biodegradability (breaking down into harmless components when exposed to specific natural environments).

Molecular biology techniques can be used to genetically engineer host cells or even multicellular organisms that are capable of synthesizing economic quantities of protein for possible processing into fibre producing proteins that already exist in nature, as well as entirely new materials.

Improved analytical techniques, together with biotechnology tools, enable a new generation of products to be envisioned with silk. The ability to tailor polymer structure to a precise degree leads to interesting possibilities in the control of macroscopic functional properties of fibres, membranes, and coatings, as well as improved control of processing windows. Biotechnology offers the tools to solve limitations in spider silk production that the traditional domestication and breeding approach used successfully with the silkworm has not been able to overcome. This is important because of the variety of silk structures available and the higher modulus and strength compared to silkworm silk.

Hybrid silk fibres have been synthesized with silk coextruded or grafted onto synthetic fibre cores. Cosmetics and consumer products, such as hair replacements and shampoos containing silk, have also been marketed. Sutures, biomaterials for tissue repairs, wound coatings, artificial tendons, and bone repair may be possible applications since immunological responses to the silks are controllable. It is also reasonable to speculate on the use of silk webbing for tissue and nerve cell growth, and brain repair applications such as temporary scaffolding during regrowth and reinfusion after surgery. Cell culture petri plates having genetically engineered silkworm silks containing cell binding or adhesive domains have already been produced and are sold commercially. Fibre spinning from resolubilized silkworm silk provides further opportunities in material fabrication by using native and genetically engineered silk proteins.

A range of technical applications in filter, membrane, paper, textile, and leather fabrication is being targeted to employ protein fibres. Recent technical applications for protein fibres include their use for patterning on the nanoscale. Applications of protein fibres are also being explored in the field of biosensors and in the medical and biomedical sectors, including the use of protein fibres as surgical threads and sutures and for the development of biological membranes and scaffolds to support cell growth and tissue function. Films, fibres, and matrices of proteins, such as collagen, are often used in clinical repairs, wound healing, ligament replacements, implants, cosmetic surgery, pharmaceutical delivery systems, tissue engineering, and in medical devices for soft tissue augmentation. Protein fibres can also be employed as carrier molecules in therapeutic applications to induce oral tolerance for certain drugs, and could find application in the food industry (e.g. as stabilizers). Protein fibres could also be utilized in the field of optometry for the production of contact lens material and in personal care products such as cosmetics.

Hydrogels formed from natural protein fibres, such as collagen, fibrin, and elastin have found numerous applications in tissue engineering and drug delivery. The disadvantage of such protein fibres is their limited range of mechanical properties. Therefore, silk, which provides impressive mechanical properties, compatibility, biodegradability, and cell interaction properties, has been tested as a new biomaterial.

3.8 SUMMARY

There is an incredible variety of natural protein fibres tailored-and tailorable-to purpose. Three examples, plus artificially manufactured fibres, illustrate the diversity. Wool, other animal hairs, and silk have a future as quality fibres when matched to commercial needs. This has implications for the industry in terms of linking growers to markets and producing fibres with the qualities needed for particular applications. Solution spinning of proteins may meet some special uses, e.g. medical applications, but is not a viable route for the production of fibres with good mechanical properties, although genetic engineering opens up new possibilities. Mechanics has lagged behind genetics, proteomics, and structural analysis in scientific advances of biological materials over the last 50 years. There is a need to understand structural mechanics in order to link formation to performance. This is of much wider biological and medical significance than just the study of fibres.

3.9 PROJECT IDEAS

1. Production of natural fibre composites using biodegradable resins for applications in construction and in automobile industry.
2. Production of natural fibre blended fabrics with other synthetic fibres for producing cost-effective blended fabrics for apparel and home furnishing fabrics.
3. Production of biomaterials using natural fibres like silk for biomedical applications.
4. Production of high-quality knitted fabrics for hosiery applications.
5. Development of natural antimicrobial textiles using natural bioactive antimicrobial agents for health care fabrics.

3.10 REVISION QUESTIONS

1. What are the two major classes of natural protein fibres?
2. What are the main chemical elements found in wool and typical of proteins?
3. What is the chemical compound that makes keratin different to other natural protein fibres?
4. What is the only natural fibre that occurs in filament form?
5. What are the main properties of silk fibre?
6. What are the three parts of the cocoon?
7. What are the two layers of the silk fibre?
8. Which three amino acids are the main components of mulberry silk?
9. Name three qualities of silk fibre that have led to it being considered a premier textile material.
10. Name three other examples of specialty hair fibre.

REFERENCES

Altman, G. H., Diaz, F., Jakuba, C., Calabro, T., Horan, R. L., Chen, J., et al. (2003). Silk-based biomaterials. *Biomaterials*, *24*, 401–416.

Chen, Z., Kimura, M., Suzuki, M., Kondo, Y., Hanabusa, K., & Shirai, H. (2003). Synthesis and characterization of new acrylic polymer containing silk protein. *Fiber*, *59*(5), 168–172.

Crewther, W. G. (1975). In *Proc. Int. Wool Text. Res. Conf., Aachen* (vol. 1).

Das, S. (1996). *Studies on tasar silk* (Ph.D. thesis). Delhi: IIT.

Datta, R. K., & Nanavaty, M. (2005). *Global silk industry: A complete source book*. USA: Universal Publishers.

Feughelman, M. (1987). In J. I. Kroschwitz (Ed.), *Encyclopedia of polymer science and engineering* (vol. 8) (p. 566).

Feughelman, M., & Robinson, M. S. (1971). *Textile Research Journal, 41*, 469.

Freddi, G., Gotoh, Y., Mori, T., Tsutsui, I., & Tsukada, M. (1994). *Journal of Applied Polymer Science, 52*, 775–781.

Gillespie, J. M. (1965). In A. G. Lyne, & B. F. Short (Eds.), *Biology of the skin and hair growth* (p. 377). Sydney: Angus and Robertson.

Gillespie, J. M., & Reis, P. J. (1966). *Biochemical Journal, 98*, 669.

Hakimi, O., Knight, D. P., Vollrath, F., & Vadgama, P. (2007). Spider and mulberry silkworm silks as compatible biomaterials. *Composites: Part B, 38*, 324–337.

Haylett, T., Swart, L. S., Parris, D., & Joubert, F. J. (1971). *Applied Polymer Symposium, 18*, 37.

Hearle, J. W. S. (2000). *International Journal of Biological Macromolecules, 27*, 123.

Hearle, J. W. S., & Susutoglu, M. (1985). In *Proc. 7th Int. Wool Text. Res. Conf.* (vol. 1) (p. 214). Tokyo, Japan: Society of Fiber Science and Technology.

Huson, M. G. (1998). *Textile Research Journal, 68*, 595.

Iizuka, E. (1994). *International Journal of Wild Silkmoth and Silk, 1*(2), 143–146.

Iizuka, E. (1995). *Journal of Sericultural Science of Japan, 65*(2), 102–108.

Iizuka, E., & Itoh, H. (1997). *International Journal of Wild Silkmoth and Silk, 3*, 37–42.

Iizuka, E., Okachi, Y., Shimizu, M., Fukuda, A., & Hashizume, M. (1993). *Indian Journal of Sericulture, 32*(2), 175–183.

Jeffry, P. D. (1971). *Journal of Textile Institute, 1972*(63), 91.

Jolly, M. S., Sen, S. K., Sonwalker, T. N., & Prasad, G. K. (1979). Non-mulberry silks. In G. Rangaswami, M. N. Narasimhanna, K. Kashivishwanathan, C. R. Sastri, & M. S. Jolly (Eds.), *Manual on sericulture* (pp. 1–178). Rome: Food and Agriculture Organization of the United Nations.

Labarthe, J. (1975). *Elements of textiles*. New York: Macmillan.

Lindley, H., Gillespie, J. M., & Haylett, T. (1967). In *Symposium on fibrous proteins* (p. 535). Sydney: Crewther, Butterworth.

Manohar Reddy, R. (2009). Innovative and multidirectional applications of natural fibre, silk a review. *Academic Journal of Entomology, 2*(2), 71–75.

Matsumoto, A., Kim, H. J., Tsai, I. Y., Wang, X., Cebe, P., & Kaplan, D. L. (2006). *Silk, hand book of fibre chemistry*. Taylor & Francis Group, LLC.

Minagawa, M. (2000). In N. Hojo (Ed.), *Structure of silk yarn* (vol. I) (pp. 185–208). New Delhi: Oxford & IBH Publishing Co. Pvt. Ltd.

Mitchell, T. W., & Feughelman, M. (1960). *Textile Research Journal, 30*, 662.

Morton, W. E., & Hearle, J. W. S. (1993). *Physical properties of textile fibres* (3rd ed.). Manchester, UK: The Textile Institute.

Rippon, J. A. (1992). In D. M. Lewis (Ed.), *Wool dyeing*. Bradford, UK: Society of Dyers and Colourists.

Robson, R. M. (1998). *Handbook of fibre chemistry* (p. 415). New York: Marcel Dekker.

Rogers, G. E. (1959). *Annals of the New York Academy of Sciences, 83*, 378–399 and (1959). *Journal of Ultrastructure Research, 2*, 309–330.

Sen, K., & Murugesh Babu, K. (2004). Studies on Indian silk. I. Macrocharacterization and analysis of amino acid composition. *Journal of Applied Polymer Science, 92*, 1080–1097.

Sonwalker, T. N. (1993). *Hand book of silk technology*. New Delhi: Wiley Eastern Limited.

Spring, C., & Hudson, J. (2002). *Silk in Africa*. Seattle: University of Washington Press.

Tsukada, M., Freddi, G., Minoura, N., & Allara, G. (1994). *Journal of Applied Polymer Science, 54*, 507–514.

Wang, Y., Kim, H.-J., Vunjak-Novakovic, G., & Kaplan, D. L. (2006). Stem cell-based tissue engineering with silk biomaterials. *Biomaterials, 27*, 6064–6082.

Wortmann, F.-J., & Zahn, H. (1994). *Textile Research Journal, 64*, 737.

Zahn, H., & Biela, M. (1968). *European Journal of Biochemistry, 5*, 567.

Zahn, H., Wortmann, F.-J., & Höcker, H. (1997). *Chemie in Unserer Zeit, 31*, 280–290.

Zarkoob, S., Reneker, D. H., Ertley, D., Eby, R. K., & Hudson, S. D. U.S. Pat. 6,110,590 (2000).

CHAPTER

SYNTHETIC TEXTILE FIBERS: REGENERATED CELLULOSE FIBERS

4

J. Chen

The University of Texas at Austin, Austin, TX, USA

LEARNING OBJECTIVES

At the end of this chapter, you should be able to:

- Review the fundamental characteristics of cellulose
- Understand how regenerated cellulose fibers are produced and what their properties features are
- Learn how to incorporate regenerated cellulose fibers into different apparel end uses

4.1 INTRODUCTION

Regenerated cellulose fiber is a type of manufactured or man-made fiber that uses cellulose (mainly from wood or plant fibers) as a raw material. Regenerated cellulose fiber was the first man-made fiber applied in the textile and apparel industry and in the early days of its development, during the 1850s, had the popular name "artificial silk" as manufacturers hoped to produce an artificial fiber to replace silk (Woodings, 2001).

In 1924, the generic name Rayon was adopted by the U.S. Department of Commerce and various industrial associations to label regenerated cellulose fiber. "Ray-" (ray of light) implied fiber brightness and "-on" represented the fibers cotton-like structure (Swicofil AG Textile Services, 2011).

Regenerated cellulose fiber has a smooth and lustrous appearance much like silk (although it is chemically completely different), and the excellent water absorption ability of cotton. Fabrics made of regenerated cellulose fiber are soft and display high drapability, leading them to be widely used for apparels such as blouses, jackets, skirts, slacks, lining, and suits.

Due to the development of new synthetic fibers, however, the market share of regenerated cellulose fiber has been shrinking. In the early 1980s, global production of regenerated cellulosic fiber reached its peak (about 2.96 million metric tons), and has been in decline ever since (American Fiber Manufacturers Association, 2011). Currently, world production of regenerated cellulose fiber is about 3 million metric tons per year, accounting for approximately 5% of global man-made fiber production.

World consumption of rayon fiber is illustrated in Figure 4.1. China is the largest rayon fiber consumer, demanding 51% of the world's supply. It is predicted that global rayon demand will grow at a rate of 3.8% per year from 2009 to 2014 (Blagoev, Bizzari, & Inoguchi, 2011).

Textiles and Fashion. http://dx.doi.org/10.1016/B978-1-84569-931-4.00004-0

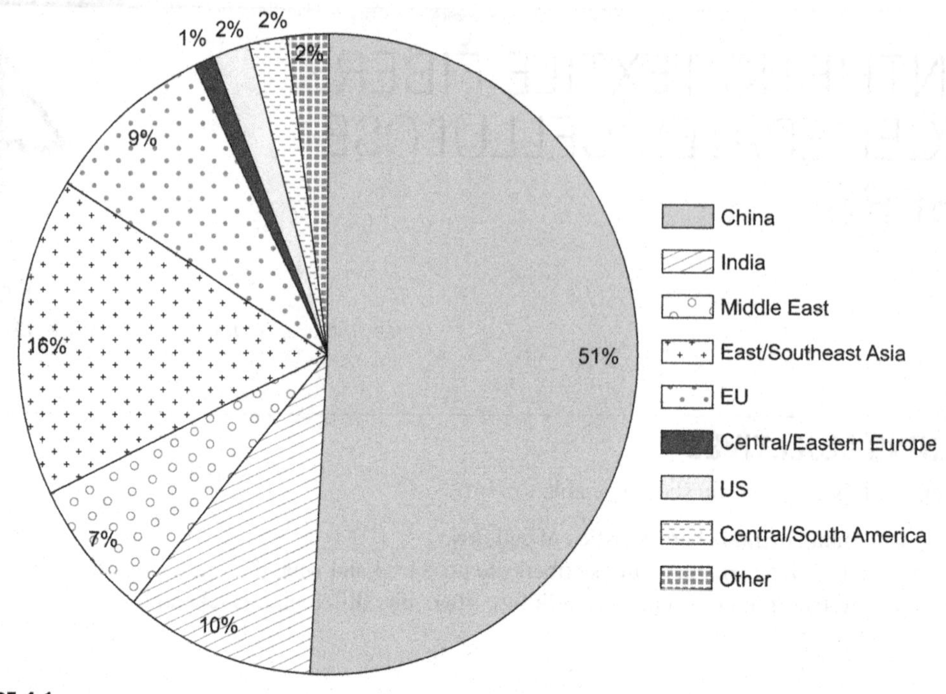

FIGURE 4.1

Global rayon fiber consumption in 2009.

There are four major types of regenerated cellulose or cellulose-derived fiber:

- viscose
- lyocell
- cupro
- acetate

The classification of these different regenerated cellulose fibers is based on fiber production method.

Viscose rayon fiber is produced by making alkali cellulose and reacting it with carbon disulfide to form cellulose xanthate. Viscose rayon is the dominant regenerated cellulose fiber accounting for more than 93% of the regenerated-cellulose and cellulose-derived fiber market.

Lyocell rayon fiber is produced by directly dissolving cellulose into the solvent N-methylmorpholine-N-oxide (NMMO). Lyocell rayon is a new generation of regenerated cellulose fiber with environmentally friendly processing and improved fiber properties. Current production of lyocell fiber however is still limited, at less than 5% of the rayon fiber market.

Cupro rayon fiber is produced by dissolving cellulose into cuprammonium solution and then wet-spinning to regenerate cellulose. Because this process needs to use high-price cotton cellulose and copper salts, cuprammonium rayon is not competitive with viscose rayon and ceased commercial development after World War I. Today, only a few manufacturers (such as Asahi in Japan and Bemberg in Germany) still produce cuprammonium rayon to supply a niche market for artificial silk and medical

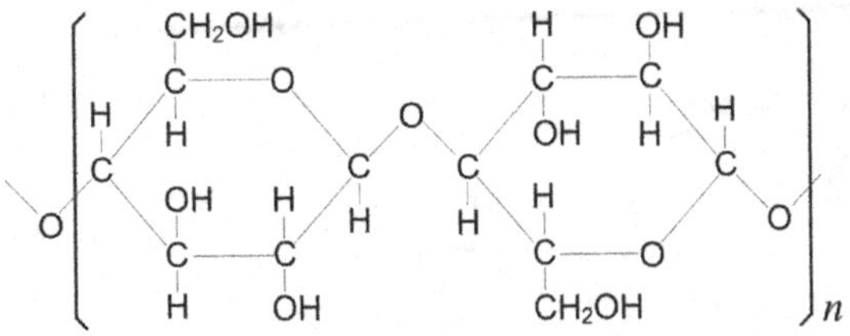

FIGURE 4.2

Cellulose polymer molecule unit.

filter materials (Kamide & Nishiyama, 2011). Cupro rayon fiber will not be discussed further in this chapter.

Acetate is a cellulose-derived fiber rather than a regenerated cellulose fiber. It is produced by acetylating cellulose using acetic anhydride liquid with a sulfuric acid catalyst. The resulting cellulose acetate is dissolved in acetone and spun into fiber through a dry spinning process. The application of cellulose acetate fiber is limited to cigarette filters only and it accounts for about 2% of the total market for cellulose fiber (Banks, 2011).

The production of regenerated cellulose fiber includes two essential steps:

- dissolution of the raw cellulose using either chemical or physical methods, and
- regeneration of cellulose through a spinning process (wet spinning, dry spinning, or dry-jet/wet-spinning).

This production approach causes no change in molecular content (Figure 4.2), although the dissolving-precipitating process does result in some changes in polymeric molecular structure after re-crystallization compared to raw cellulose. These changes manifest in cellulose crystallinity, crystal orientation, and crystallite size and benefit the fiber properties in terms of strength, absorbency, dimensional uniformity, drapability, dyeability, and apparel tailorability.

4.2 VISCOSE RAYON

4.2.1 THE HISTORY OF VISCOSE RAYON

According to the U.S. Federal Trade Commission (FTC) regarding generic names and definitions for manufactured fibers, rayon is defined as (16 CFR, 2011):

> *A manufactured fiber composed of regenerated cellulose, as well as manufactured fibers composed of regenerated cellulose in which substituents have replaced not more than 15 percent of the hydrogens of the hydroxyl groups.*

Viscose rayon fiber is regenerated cellulose fiber produced using the viscose technology and was originally patented by the British chemists Charles Cross, Edward Bevan, and Clayton Beadle in 1892 (Andrew, 2001). They discovered that cellulose xanthate could be formed by using raw cellulose from wood or cotton via reaction with an alkali and carbon disulphide. The resulting cellulose xanthate solution could also be precipitated in an ammonium sulfate solution and changed back to cellulose after neutralization using dilute sulfuric acid.

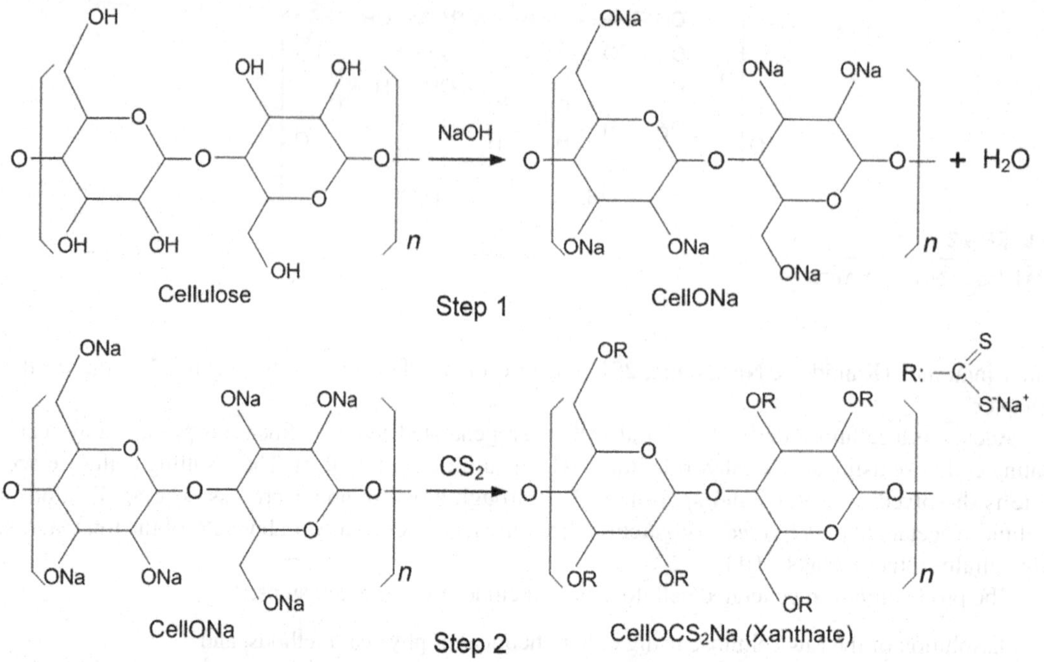

FIGURE 4.3

Chemistry of viscose formation.

The spinning method for producing viscose fiber was developed by Charles Henry Stearn in cooperation with Charles Cross in 1898, and was successfully commercialized in 1904 when Samuel Courtauld & Co. Ltd in England acquired the viscose process patents. The first U.S. manufacturer, the American Viscose Company, was registered in 1910.

4.2.2 VISCOSE RAYON PRODUCTION

Rayon fiber is made from pure cellulose, often derived from wood pulp. For spinning purposes, the wood pulp needs to be dissolved into a solution via a conversion that takes place in two steps (Figure 4.3):

1. caustic soda is mixed with wood pulp to make alkali cellulose, and
2. carbon disulfide is added to react with the alkali cellulose to form sodium cellulose xanthate.

The resulting sodium cellulose xanthate is then dissolved in a weak caustic soda solution to form a spinning solution called viscose. Using a wet spinning process (Figure 4.4), the viscose solution is extruded through a spinnerette into a sulfuric bath (H_2SO_4). After neutralization in the sulfuric bath, the cellulose from the viscose is regenerated into a continuous fiber (tow) drawn through the first and second drawing units for stretching.

Between these two drawing areas is a tow washing step that removes carbon disulfide. The second drawing unit feeds the washed tow into a cutter to cut the tow fiber into staple (shorter) fiber with a fiber length between 1 and 6 in (25–152 mm). The rayon staple fiber is then laid on a wash belt to pass

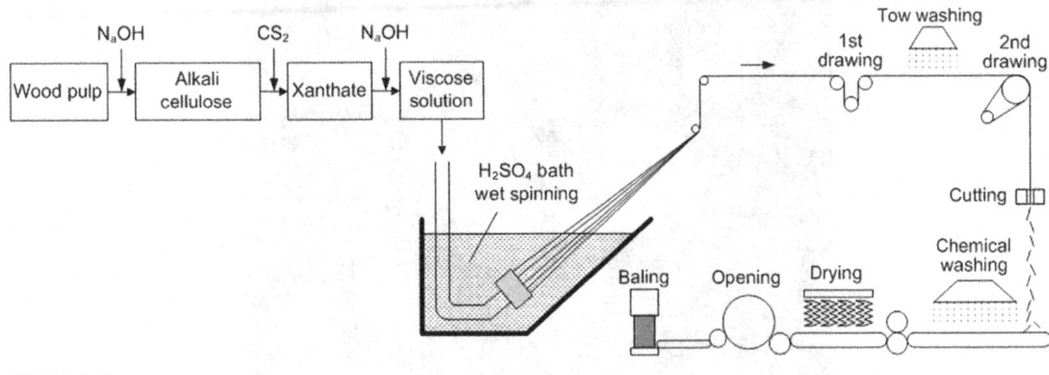

FIGURE 4.4

Schematic of rayon fiber production.

through a series of washing steps that include a hot water wash to remove residual acid; a sulfide wash to wash out residues from the desulfurizing bath such as sulfides; a bleaching bath; and finally, a hot water wash.

After washing, the rayon staple fiber is dried, opened, and sent to a bale press for baling. Rayon filament fiber can also be wound directly onto a bobbin if the cutting step is skipped. In today's viscose rayon market, however, 85% of rayon fiber production is staple fiber.

Rayon staple fiber is commonly used for blending with natural fibers or synthetic fibers to make blended yarns. Fabrics made of rayon blended yarns are highly desirable for a wide range of apparel applications. To enhance the spinnability of viscose rayon blended yarns, viscose rayon staple fiber is often crimped before cutting.

With changes to the viscose reaction conditions and control parameters, viscose fiber properties vary. This results in the production of modified viscose rayon fibers, among which high-wet-modulus (HWM) viscose and high-tenacity viscose are the two most important. To produce HWM viscose fiber, aging in cellulose alkalization and ripening in xanthation is no longer required and chemical concentrations are reduced, so that cellulose coagulation speed is reduced, allowing more time for fiber stretching. For the production of high-tenacity viscose fiber, cellulose regeneration speed needs to be reduced by increasing zinc sulfate concentrations in the spin bath.

HWM viscose fiber has two commonly used brand names: "Modal" and "Polynosic."

4.2.3 VISCOSE FIBER APPEARANCE

Viscose rayon fiber used in the textile and apparel industries can be staple fiber or multifilament fiber. Figure 4.5 shows a typical fiber shape for a commercial viscose rayon fiber product with a 1.5-denier fineness and 1.5-in (38 mm) length. The cross-section resembles a distorted circle with a serrated contour and the fiber surface is smooth but striated longitudinally, as shown in Figure 4.6. The luster of viscose rayon fiber can be bright, semi-dull, or dull. Commonly used fineness of viscose rayon fiber is in the range of 1.5–15 denier. Viscose rayon microfiber (fineness less than 1 denier) is also available for production of microfiber fabrics.

FIGURE 4.5

Viscose rayon fiber cross-sectional view (FEI QUANTA FEG 650).

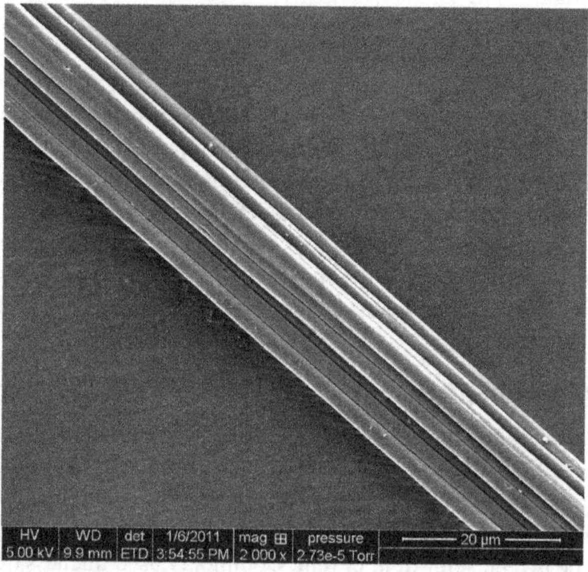

FIGURE 4.6

Viscose rayon fiber longitudinal view (FEI QUANTA FEG 650).

Table 4.1 Viscose rayon fiber tensile strength and elongation

Fiber	Tenacity (g/denier)		Elongation (%)	
	Dry	Wet	Dry	Wet
Viscose	2.6–3.1	1.2–1.8	20–25	25–30
HWM	4.1–4.3	2.3–2.5	13–15	13–15
Tencel	4.8–5.0	4.2–4.6	14–16	16–18
Cotton	2.4–2.9	3.1–3.6	7–9	12–14
Polyester	4.8–6.0	4.8–6.0	44–45	44–45

Courtaulds Fibers Inc., 1999.

4.2.4 VISCOSE FIBER MECHANICAL PROPERTIES

As with other textile fiber materials, tensile strength and breaking elongation are two of the most important mechanical properties for viscose rayon fiber. Fiber tensile strength is often expressed by tenacity with a unit of force per denier or tex. Table 4.1 lists tensile strengths and elongations of viscose rayon compared to HWM viscose, tencel, cotton, and polyester.

Viscose rayon fiber has a lower tensile strength under wet conditions (e.g., during washing) than under dry conditions. This difference can be greater than a factor of 2. Another mechanical feature of viscose rayon fiber is its substantially high elongation, up to 25% (dry) and 30% (wet). Therefore, viscose rayon or rayon blended fabrics tend to have good stretchability. In comparison, HWM viscose fiber shows a higher tensile strength in both dry and wet conditions, but a lower elongation at break.

4.2.5 VISCOSE RAYON PHYSICAL PROPERTIES

Viscose rayon fiber has a density of 1.52 g/cm^3, higher than all other natural fibers and also higher than most synthetic fibers. Thus, fabrics made of pure rayon fiber are usually heavier than those made of other fibers at fixed fabric thickness.

Ability to absorb water is often described by the moisture content (%) and moisture regain (%). Moisture content is calculated as the weight of absorbed water divided by fiber weight measured under a standard temperature and humidity. This fiber weight is equivalent to the fiber dry weight plus the weight of absorbed water. Moisture regain is defined as the weight of absorbed water divided by the fiber dry weight. The moisture regain for viscose rayon fiber is between 12% and 14% (Morton & Hearle, 1993). This value is higher than the moisture regains for all natural and synthetic fibers excepting wool, indicating that viscose rayon fibers possess superior water absorption. This is due to their lower crystallinity.

4.3 LYOCELL RAYON

4.3.1 THE HISTORY OF LYOCELL RAYON

Lyocell fiber (U.S. brand name Tencel) is another type of regenerated cellulose fiber made from wood pulp. The method to produce cellulose solution is totally different from that of viscose rayon fiber, however. The Federal Trade Commission defines lyocell as a cellulose fiber that is precipitated from an organic solution in which no substitution of the hydroxyl groups takes place and no chemical intermediates are formed. Lyocell fiber is classified as a subcategory of rayon.

Driven by environmental concerns, researchers have sought new methods for the preparation of cellulose solutions. NMMO was discovered to be a solvent which can directly dissolve cellulose pulp. This invention appeared first in a patent describing a basic process of dissolving cellulose by using the NMMO solvent (Mcorsley, 1981). Work on dissolution of different compounds including cellulose in NMMO was reported by D. L. Johnson of Eastman Kodak Inc., in the United States, during 1966–1968 (Mbe, 2001). In the following 10 years from 1969 to 1979, another U.S. company, American Enka, explored the spinning of regenerated cellulose fiber using the NMMO cellulose solution but failed to commercialize the process.

Eventually, an R&D team led by Pat White of Courtaulds in the United Kingdom developed a successful engineering approach for cellulose solution spinning. In 1982, Courtaulds built the first small pilot plant capable of making up to 100 kg per week of lyocell fiber in Coventry, England. In 1984, the production capacity of this pilot line was increased to 1 ton/week. A 25 ton/week semi-commercial production line went into operation in 1988 at Grimsby, England. In 1992, in Mobile Alabama, in the United States, Courtauds reached its full commercial production capacity of producing this new regenerated fiber with the trade name "TENCEL®."

Another major European company engaged in the production of lyocell fiber is Lenzing AG in Austria. Traditionally specializing in the manufacture of viscose rayon fiber, Lenzing established a pilot plant to start making lyocell fiber in 1990 (Lenzing AG, 2011b). Lenzing's full-scale production plant at Heiligenkreuz came into operation in 1997, with an annual capacity of 12,000 metric tons of lyocell staple fiber called Lenzing Lyocell®. In 2004, Lenzing completed an acquisition of the TENCEL® Group (one plant in Mobile, Alabama, and another in Grimsby). Today, Lenzing is the world's largest lyocell fiber manufacturer, capable of supplying about 130,000 metric tons of lyocell fiber for the global rayon market each year.

4.3.2 LYOCELL RAYON PRODUCTION

The production of lyocell rayon fiber includes all steps indicated in Figure 4.7. Raw cellulose (wood pulp) is mixed with the NMMO solvent and dissolved in NMMO by heating. The formed cellulose solution is called "dope." A solvent spinning technique (also called dry-jet and wet-spun) is used to press the dope through a spinnerette into a spin bath where regenerated cellulose fiber precipitates as the NMMO solvent is dissolved in the spin bath. The formed cellulose fiber is further processed by water washing, lubricant finishing, drying, and static removing. At this stage, lyocell filament fiber is produced.

To produce lyocell staple fiber, the regenerated cellulose fiber is crimped and cut to a certain staple length for press and packing. During the production of the Lenzing Lyocell® staple fiber, the fiber cutting is done before the washing and finishing steps. The NMMO solvent from the washing unit is recycled through the solvent recovery system where the dilute NMMO solvent is concentrated and then pumped into a mixing tank for dissolving new pulp (Woodings, 1995). In comparison with the method of viscose rayon production, the lyocell fiber spinning process is an environmentally friendly green technology that eliminates toxic chemical use and chemical reactions, and substantially reduces air and water emissions.

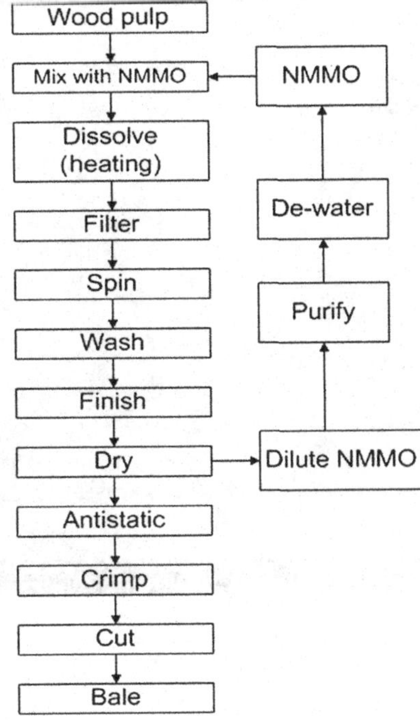

FIGURE 4.7

Tencel fiber production flowchart.

4.3.3 LYOCELL FIBER APPEARANCE

Lyocell fiber has a close to circular cross-section (Figure 4.8). Its longitudinal surface is very smooth and cylindrical without any striation (Figure 4.9). Lyocell rayon fiber is different from viscose rayon in fiber shape and appearance, and this differentiation allows lyocell rayon fabrics to exhibit better fabric feel and drape. Table 4.2 lists major lyocell rayon fiber products produced by Lenzing.

4.3.4 LYOCELL FIBER MECHANICAL PROPERTIES

Using the lyocell spinning technology, the regenerated cellulose fiber can be produced with a higher dry and wet tensile strength. Referring to Table 4.1, the dry tenacity value of lyocell fiber is larger than that of viscose and HWM rayon fibers, and almost equivalent to that of polyester fiber. Lyocell fiber is also capable of keeping 85% of its dry tenacity in the wet condition. It is the only regenerated cellulose fiber with a wet tensile strength exceeding the cotton wet strength. Compared to viscose rayon fiber, lyocell fiber has a significantly reduced elongation the value of which is slightly above that of HWM rayon fiber. Because of a high degree of cellulose crystallinity and crystal orientation, lyocell fiber features a fibrillar structure with microfibrils aligned parallel to the fiber axis. This allows lyocell fiber to easily develop a fibrillated surface under mechanical abrasion (Collier et al., 1999).

FIGURE 4.8

Lyocell rayon fiber cross-sectional view (FEI QUANTA FEG 650).

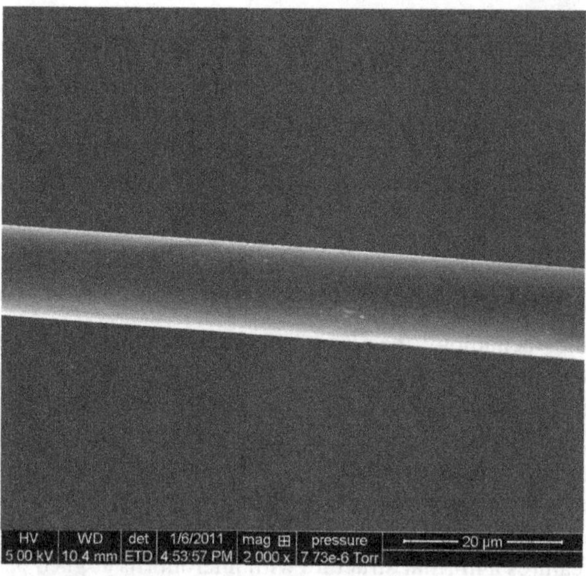

FIGURE 4.9

Lyocell rayon fiber longitudinal view (FEI QUANTA FEG 650).

Table 4.2 Major lyocell fiber products from Lenzing

TENCEL®	Fineness (dtex)	Length (mm)
Standard	1.3	38
	1.4	38
	1.7	38/51
	2.2	50
Micro	0.9	34
LF	0.9	34
	1.3	38
	2.2	50
A 100	1.4	38
	3.0	75/98
FILL	2.3	15
	6.7	22/32
	6.7	60

Lenzing AG, 2011a.

4.3.5 LYOCELL FIBER PHYSICAL PROPERTIES

Lyocell fiber is similar to viscose rayon fiber in many physical aspects, but exhibits enhanced properties in terms of softness, drapability, dimensional stability, dye uptake, and colorfastness. Moisture regain of lyocell fiber is around 11%, slightly lower than that of viscose rayon (Hongu & Phillips, 1997). This is mainly because the lyocell spinning process produces a higher cellulose crystallinity.

4.4 CELLULOSE ACETATE

Cellulose acetate fiber is another type of manufactured fiber related to cellulose. It is no longer considered a regenerated cellulose fiber because the polymer formula to form acetate fiber is acetate (cellulose ester) instead of cellulose.

The method for producing cellulose acetate was discovered by Paul Schützenberger in 1865, while trying to dissolve cotton in acetic anhydride liquid. When this solution was poured in water, white flakes called triacetate were precipitated. However, this invention did not directly lead to progress in acetate fiber spinning until the method of converting triacetate into secondary acetate (diacetate) was discovered by George Miles in 1904 (Hearle & Woodings, 2001). This invention solved the problem of directly dissolving triacetate for fiber spinning.

The first commercial production of acetate filament was undertaken by the British Cellulose Co. Ltd at Spondon, England, in 1921. One year later, Camille Dreyfus (one of the British Cellulose Co. Ltd owners) launched acetate fiber production in the United States. Acetate fiber is currently rarely employed in the textile and apparel markets.

There are two types of acetate fiber:

- regular acetate also called secondary acetate or diacetate
- triacetate

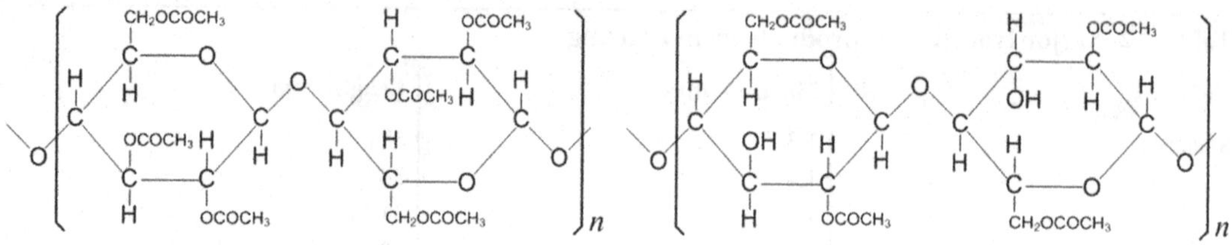

Triacetate (fully acetylated cellulose)

Diacetate

FIGURE 4.10

Molecular expression of diacetate and triacetate fibers.

The chemical formulae for these two fibers are indicated in Figure 4.10. These two fiber types are produced with two different fiber spinning approaches.

Making triacetate is a common starting point for the manufacture of acetate fiber. Raw cellulose (wood pulp or cotton lint) is purified and mixed with acetic acid and acetic anhydride liquid for acetylation. Sulfuric acid is added as a catalyst to accelerate the acetylating reaction. Upon completion of the acetylation, cellulose triacetate is formed as a thick gelatinous solution. To make triacetate fiber, the triacetate solution is poured into water to precipitate triacetate flakes. After washing and drying, the triacetate flakes are dissolved in methylene chloride to form the spinning dope. Triacetate fiber is produced by extruding the dope through the spinnerette into a chamber of heated air.

To make regular acetate, the acetylated cellulose (triacetate) needs to be hydrolyzed by adding water to be precipitated as triacetate flakes. Diluted access acetic acid (95% concentrations) triggers an acid hydrolysis with a ripening time of 20h. During the hydrolysis, triacetate is converted to diacetate as one-sixth of the acetate groups (CH_3COO–) are replaced by hydroxyl groups (–OH). Following the hydrolysis, of diacetate flakes is precipitated by adding excess water. To produce spinning dope, washed and dried diacetate flakes are dissolved in acetone. The same dry spinning method as used for the triacetate fiber production is applied to make regular acetate fiber.

The cross-sectional shape of regular acetate and triacetate fibers resembles that of viscose fiber with irregular lobed contours. There are also some longitudinal striations on the surface of regular acetate and triacetate fibers. The luster of regular acetate and triacetate fibers can be controlled to yield bright, dull, and semi-dull versions. Both acetate fibers have a volume density of 1.32 g/cm^3, lighter than the viscose fiber density. The strength of regular acetate fiber is between 1.2 and 1.4 g/denier (dry) and 0.9–1.0 g/denier (wet). For triacetate fiber, dry tenacity is 1.1–1.4 g/denier and wet tenacity is 0.8–1.0 g/denier. Dry break elongation is the same for both regular acetate and triacetate, ranging from 25% to 35%. Wet break elongation is 35–45% for regular acetate and 30–40% for triacetate. Regular acetate fiber has a moderate ability to absorb water with a standard moisture regain of 6.5% (close to that of cotton). In contrast, standard moisture regain for triacetate fiber is within 3.2–3.5% (close to that of nylon) (Joseph, 1986). Both regular acetate and triacetate fibers provide a good dimensional stability with some stretch and resistance to shrink. In terms of wrinkle resistance, triacetate performs slightly better.

4.5 APPLICATIONS

Viscose rayon fiber is used for a wide range of apparel applications. Blouses, skirts, dresses, and shirts are often made of pure viscose staple fiber for excellent drapability and comfortability. Viscose rayon fiber is also blended with wool, polyester, or acrylic for suits, jackets, and blazers for the purpose of enhancing appearance and moisture permeability. Viscose rayon filament fiber is often used for making linings with a silky surface.

Although viscose rayon fiber brings unique features of drapability, softness, and moisture absorption to both formal and casual apparel products, it can also have some drawbacks. These include easy stretching with poor elastic recovery, low abrasion resistance, wet shrinkage, and low wrinkle resistance. To eliminate the potential of developing these problems, evaluation of the following fabric properties is critical during garment design and make up.

1. *Tensile and shear properties.* Extensibility of viscose/lyocell rayon fiber fabrics is an influential parameter for the development of excessive wet shrinkage or a slack appearance for apparel products. The maximum extension of viscose rayon fabrics under a certain load should be controlled. Lining fabrics made of viscose rayon filament usually have a very low shear rigidity. This may easily result in difficult fabric spread or inaccurate pattern cut in apparel manufacture. So the shear rigidity of viscose lining fabrics also needs monitoring for the purpose of fabric purchase control. Fabric tensile properties can be tested using a strip method described in the standard ASTM D 5035. The Kawabata instruments and FAST instruments are capable of measuring fabric shear rigidity as well as tensile properties.
2. *Abrasion resistance.* Apparel fabrics made of viscose or lyocell rayon fiber tend to develop micro-fibril surfaces (known as fibrillation) after frequent wearing and laundering. Where this fibrillated appearance is not aesthetically required, evaluation of the abrasion resistance for these fabrics is needed to control a threshold level for fibrillation. Fabric abrasion tests can be performed using different instruments in accordance with the standard methods ASTM D 3884, ASTM D 3885, ASTM D 4966, and AATCC 93.
3. *Wrinkle resistance.* Rayon apparel fabrics are also prone to develop a wrinkled appearance after laundering. The wrinkle resistance of rayon fabrics should be evaluated and a minimum value specified for the fabric. The test methods of fabric recovery angle (AATCC 66) and fabric wrinkle recovery appearance (AATCC 128) are commonly used for the evaluation of fabric wrinkle resistance.
4. *Dimensional stability.* Because of wet shrinkage, rayon apparel fabrics can cause garment dimensional changes after laundering. Control of fabric dimensional stability is critical to ensure the production of high-quality garments. A maximum shrinkage should be defined in the fabric material specifications. Evaluation of rayon fabric dimensional changes can be implemented according to the standard method AATCC 135. For wool-rayon blended fabrics, the dimensional stability test can be referred to FAST-4, a method defined in the FAST system (Boos & Tester, 1994).

4.6 CASE STUDY

A wedding gown using viscose rayon fiber is shown in Figure 4.11. The garment design is based on a course project conducted by a fashion design student. Wedding gowns are very special and high-end garment products. Material drape and luster properties are essential considerations for fashion designers to achieve the required structural and luxurious appearance.

A typical bridal wedding gown is usually composed of two garment components: an upper part perfectly fitting the body contour and a lower part exhibiting elegant skirt drape. The drape style is highly customized ranging from a simple as an A-line silhouette to a Victorian toilette pleating. The fabric material selected for this wedding gown is a satin woven fabric, with 102 g/m^2 fabric weight and

FIGURE 4.11

Wedding gown made of viscose rayon.

Source: Photo courtesy of Crystal Colmenero, the wedding gown designer.

0.229-mm fabric thickness. This satin woven structure has a drape coefficient of 36.8% showing a fairly good drapery characteristic as, according to the Cusick drape test, the higher the drape coefficient the stiffer the fabric.

4.7 FUTURE TRENDS

The U.S. National Energy Policy sets goals to secure energy supplies by using a more diverse mix of domestic renewable resources (Biomass Technical Advisory Committee, 2002). Implementation of this energy policy is having a long-term impact on the manufacturing and consuming of downstream polymer and chemical products from petroleum refineries.

One challenge for achieving this energy goal is to develop biomass technologies that enable production of bioenergy and biobased products in a more efficient and economical manner. Regenerated cellulose fibers are as a result becoming more popular as the only fiber group among man-made fibers produced from cellulose biomass. However, the dominant fiber remains vicose rayon, the production of which consumes a large volume of toxic chemicals such that environmental pollution has been a serious concern. A likely trend for future production is to see a reduction in the production of viscose rayon and a steady increase in that of lyocell rayon.

Further developments could see the use of more diverse cellulose feedstock and enhancements to fiber performance for healthcare, personal care, and high-end functionality type applications.

Wood pulp is currently the major cellulose feedstock for the production of regenerated cellulose fiber. Other cellulose products or residues, such as bamboo, bast fibers, cotton linters, and sugarcane bagasse, may be increasingly used as additional cellulose feedstock.

Viscose rayon fiber with improved flame retardancy is already being used in military uniforms for burn protection, breathability, and comfort. Trilobal-shape viscose rayon is also being used for hygiene products with superior absorbency (Lenzing AG, 2011c). Antimicrobial lyocell rayon fiber is produced by new spinning technologies to load silver or copper salt components. This new lyocell rayon fiber can be used for underwear and sportswear, hygiene products, and healthcare textiles. Lyocell rayon with a thermal self-regulating function is has also been developed by integrating phase change materials such as paraffin. This specialty lyocell rayon is entering the end-use market of sportswear and bedding products (Smartfiber AG, 2011).

4.8 SUMMARY

- Viscose rayon fiber was the first man-made fiber invented for use in the textile and fashion apparel industries with the original name "artificial silk."
- New synthetic fibers have developed that have created many new industrial and consumer application fields.
- Viscose rayon, lyocell rayon, and acetate/triacetate are the three major types of regenerated cellulose or cellulose-derived fiber being used in the textile and apparel industries.
- Lyocell fiber is seen as "green" and sustainable.
- It can be envisaged that regenerated cellulose fiber will become a primary fiber material in the development of eco-friendly fashion products.

4.9 PROJECT IDEAS

In children's apparel product lines, there is an interest to produce children's school uniforms with antimicrobial function and children's sleepwear with flame retardancy. In the product line of ladies' blouse, excellent drape appearance and high wrinkle resistance may be the most important aesthetic quality to consider. These application potentials may be considered by apparel design students to develop projects in accordance with course requirements. Critical technical routes for these projects would be how to select and incorporate an antimicrobial rayon fabric in their design for children' school address (challenge: bioactive durability), how to select and incorporate a flame retardant rayon fabric in their design for children's sleepwear (challenge: comfort), and how to select and incorporate a micron rayon fabric in their design for high-end blouses (challenge: tailorability). Attention should be drawn to the following steps:

- fabric material sourcing,
- material quality specifications,
- material test methods,

- tailorability evaluation,
- make-up appearance assessment, and
- target cost estimation.

4.10 REVISION QUESTIONS

- What are the drawbacks relating to the use of viscose rayon in the apparel industry?
- Describe the difference in the physical properties of viscose rayon and lyocell rayon.
- By what characteristic are the different cellulose fibers classified?
- What is the most commonly produced cellulose fiber today?
- What is the main application of cellulose acetate fiber?
- Which type of cellulose fiber would you use for a highly colored garment?

REFERENCES

16 CFR. (2011). *U.S. Code of Federal Regulations Title 16—Commercial Practices (16 CFR), Part 303—Rules and regulations under the Textile Fiber Products Identification Act.* http://ecfr.gpoaccess.gov/cgi/t/text/text-idx?c=ecfr&tpl=/ecfrbrowse/Title16/16cfr303_main_02.tpl.

American Fiber Manufacturers Association. (2011). *Fiber facts.* http://www.fibersource.com/F-Info/fiber%20production.htm.

Andrew, G. W. (2001). The viscose process. In C. Woodings (Ed.), *Regenerated cellulose fibres* (pp. 37–61). Cambridge, England: Woodhead Publishing Limited.

Banks, T. H. (2011). *Cellulose acetate and triacetate fibers.* http://www.sriconsulting.com/CEH/Public/Reports/541.1000/.

Biomass Technical Advisory Committee. (2002). *The vision of bioenergy and biobased products in the United State.* http://www.brdisolutions.com/pdfs/BioVision_03_Web.pdf.

Blagoev, M., Bizzari, S., & Inoguchi, Y. (2011). *Rayon and lyocell fibers.* http://www.sriconsulting.com/CEH/Public/Reports/541.3000/.

Boos, A. D., & Tester, D. (1994). *SiroFAST fabric assurance by simple testing.* Geelong, Australia: CSIRO. WT92.02.

Collier, B. J., Negulescu, I. I., Romanoschi, M. V., Goynes, W. R., Hoven, T. V., Graves, E., et al. (1999). Effects of finishing and dyeing on lyocell and lyocell-blend fabrics. *Textile Chemist and Colorist & American Dyestuff Reporter, 1*(2), 40–45.

Courtaulds Fibers Inc. (1999). *Tencel technical overview.* New York, NY: Courtaulds Fibers Inc.

Hearle, J. W. S., & Woodings, C. (2001). Fibres related to cellulose. In C. Woodings (Ed.), *Regenerated cellulose fibres* (pp. 156–173). Cambridge, England: Woodhead Publishing Limited.

Hongu, T., & Phillips, G. O. (1997). *New fibres* (2nd ed.). Cambridge, England: Woodhead Publishing Limited.

Joseph, M. L. (1986). *Introductory textile science* (5th ed.). New York, NY: CBS College Publishing.

Kamide, K., & Nishiyama, K. (2011). Cuprammonium processes. In C. Woodings (Ed.), *Regenerated cellulose fibres* (pp. 88–155). Cambridge, England: Woodhead Publishing Limited.

Lenzing AG. (2011a). *TENCEL-the universal fiber.* http://www.tencel.at/index.php?id=75&L=1.

Lenzing AG. (2011b). *The Lenzing group history.* http://www.lenzing.com/en/concern/lenzing-group/history.html.

Lenzing AG. (2011c). *Viscose data sheet.* http://www.viscose.at/index.php?id=89&L=1.

Mbe, P. W. (2001). Lyocell: the production process and market development. In C. Woodings (Ed.), *Regenerated cellulose fibres* (pp. 62–87). Cambridge, England: Woodhead Publishing Limited.

Mcorsley, C. C. (1981). *Process for shaped cellulose article prepared from solution containing cellulose dissolved in a tertiary amine N-oxide solvent* US 4246221, Akzona Incorporated (patent).

Morton, W. E., & Hearle, J. W. S. (1993). *Physical properties of textile fibres* (3rd ed.). Manchester, England: The Textile Institute.

Smartfiber AG. (2011). *Smartcel™ bioactive*. http://www.smartfiber.de/index.php?option=com_content&view=article&id=72&Itemid=120.

Swicofil AG Textile Services. (2011). *Viscose rayon*. http://www.swicofil.com/viscose.html.

Woodings, C. (2001). A brief history of regenerated cellulose fibres. In C. Woodings (Ed.), *Regenerated cellulose fibres* (pp. 1–21). Cambridge, England: Woodhead Publishing Limited.

Woodings, R. (1995). The development of advanced cellulosic fibres. *Journal of Biological Macromolecules, 17*(6), 305–309.

CHAPTER

5 SYNTHETIC TEXTILE FIBRES: POLYAMIDE, POLYESTER AND ARAMID FIBRES

B.L. Deopura[1], N.V. Padaki[2]

[1]Indian Institute of Technology Delhi, New Delhi, India;

[2]Central Silk Technological Research Institute, Guwahati, India

LEARNING OBJECTIVES

At the end of this chapter, you should be able to:

- Describe polyamide, polyester and aramid fibres
- Explain the methods of preparation of these synthetic fibres
- Relate the structures of polyamide, polyester and aramid fibres to their properties
- Describe the main properties of these synthetic fibres
- Understand the applications of polyamide, polyester and aramid fibres

5.1 INTRODUCTION

The mid-nineteenth century witnessed the evolution of chemically synthesized fibres with enhanced performance compared to natural fibres. The first entirely synthetic fibres became widely available in the early twentieth century. In 1940, DuPont introduced the synthetic fibre nylon, invented by Wallace Carothers. John Whinfield and James Dickson patented polyethylene terephthalate (PET) polyester, based on the early research work of Wallace Carothers. Whinfield and Dickson, along with inventors Birtwhistle and Ritchie, created the first polyester fibre called Terylene in 1941.

Today it is estimated that the world's total textile fibre production is about 70.5 million metric tons of which 40.3 million metric tons are synthetic fibres, as per 'The Fibre Year 2010' published by Oerlikon Textiles, Germany (Engelhard, 2010). Polyamides (nylons), polyolefin (polyethylene and polypropylene), acrylic and polyesters constitute 98% of synthetic fibre production, and are used in almost every field of fibre and textile applications. Polyesters are now the main type of synthetic fibres produced (60%) and consumed worldwide (24 million metric tons).

Synthetic fibres have the following advantages:

- These fibres are strong and durable.
- They are able to retain crease, for a longer duration. They do not wrinkle easily.

Textiles and Fashion. http://dx.doi.org/10.1016/B978-1-84569-931-4.00005-2

- They are resistant to most chemicals.
- They are resistant to insects, fungi and rot.
- They have low moisture absorbency, and hence are easy to dry.
- They do not shrink when washed.

This chapter covers preparation, processes, fibre structure, important properties and applications of polyamides, polyesters and aramid fibres. The importance of fibre blending, the purpose of blending and other key issues will be discussed.

5.2 CLASSIFICATION OF FIBRES

The classification of textile fibres into natural and man-made types based on fibre source and manufacture is shown in Figure 5.1. Man-made fibres can be classified into organic and inorganic types based on the composition of the fibres. Organic man-made fibres are further categorized into regenerated and synthetic fibres.

Regenerated fibres are manufactured using raw materials available in nature. Typical examples of regenerated man-made fibres include viscose rayon, cellulose acetate, cellulose triacetate, cuprammonium rayon and lyocell fibres.

Synthetic fibres are made from chemically synthesized polymers. Amongst the most well known are polyester, acrylic and polyamide fibres, which have a wide variety of applications in fashion and apparel sectors.

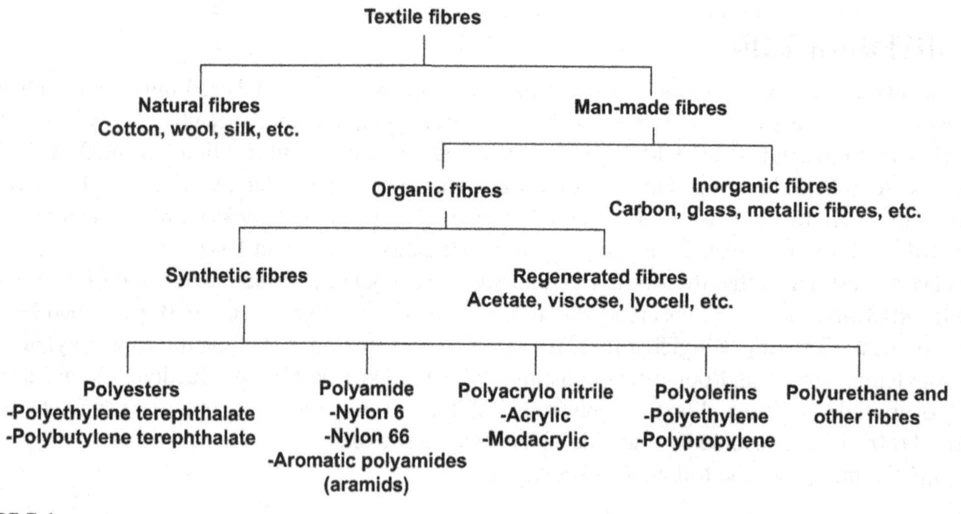

FIGURE 5.1

Classification of man-made fibres.

5.3 POLYAMIDE FIBRES

Polyamides were the first synthetic polymer developed through research by Wallace Carothers at DuPont. Nylon, the first synthetic polymer introduced to the market, was immediately successful; it is documented that four million pairs of nylon stockings were sold within the first few hours of sale in May 1940. During World War II, nylon fabrics were used by the military for waterproof tents and light-weight parachutes amongst other things.

The word 'nylon' was introduced to signify the fineness of the filament. A pound of nylon could be converted to a length equivalent to the distance between London ('lon') and New York ('ny'). 'Nylon' then became the generic word representing polyamide polymers.

There are two types of polyamides, which can be represented as nylon XY and nylon Z. In nylon XY, X refers to the number of carbon atoms in the diamine monomer and Y represents the number of carbon atoms in the diacid monomer. In nylon Z, the Z refers to the number of carbon atoms in the monomer. Nylon 6 (nylon Z type) and nylon 66 (nylon XY type) are the two most manufactured polyamides, which are commonly used in a wide range of applications from apparels, ropes, carpets, tyre cords to innumerable technical textile applications.

There are several other nylons such as nylon 46, nylon 610, nylon 612, nylon 10 and nylon 12, which are used for specialized applications. One such application is the use of nylon 610 monofilament for bristles and brushes. Nylon 46 has a superior ability to retain mechanical properties even at elevated temperatures of up to 220 °C, and thus finds applications in the automobile, electronics and electrical sectors.

5.3.1 PRODUCTION OF NYLON

5.3.1.1 Nylon 66

Nylon XY is synthesized from a diacid and a diamine. For nylon 66, hexamethylene diamine (H) and adipic acid (A) are reacted to form hexamethylene diadipate (H-A) salt. These two components are reacted in methanol at high temperature to obtain the salt that precipitates from the methanol solution. The H-A salt is dissolved in water at a concentration of 60%. The solution is then heated to around 250 °C for poly-condensation and production of nylon 66. The chemical reaction involved in the process is illustrated in Figure 5.2.

5.3.1.2 Nylon 6

Nylon 6 is produced from caprolactam. A catalyst is required to convert a small amount of caprolactam to ε-aminocaproic acid, which in turn aids the polymerization process. Water is usually used as a catalyst to control the polymerization reaction at temperatures between 225 and 285 °C. The chemical reaction involved in the production process of nylon 6 is given in Figure 5.3.

$$\underset{H}{NH_2\ (CH_2)_6\ NH_2} + \underset{A}{COOH\ (CH_2)_4\ COOH} \longrightarrow \underset{\text{H-A salt}}{H\ [-NH\ (CH_2)_6\ NHCO\ (CH_2)_4CO-]\ OH} + H_2O$$

$$n\,\underset{\text{H-A salt}}{H\ [-NH\ (CH_2)_6\ NHCO\ (CH_2)_4\ CO-]\ OH} \longrightarrow \underset{\text{Nylon 66}}{H\ [-NH\ (CH_2)_6\ NHCO\ (CH_2)_4\ CO]_n OH} + (n-1)\ H_2O$$

FIGURE 5.2

Reactions for production of nylon 66.

$$n\,[HN{-}(CH_2)_5{-}CO] \longrightarrow H\,[{-}HN{-}(CH_2)_5\,{-}CO{-}]_n\,OH$$

Caprolactum   Nylon 6

FIGURE 5.3

Poly-condensation reaction for production of nylon 6.

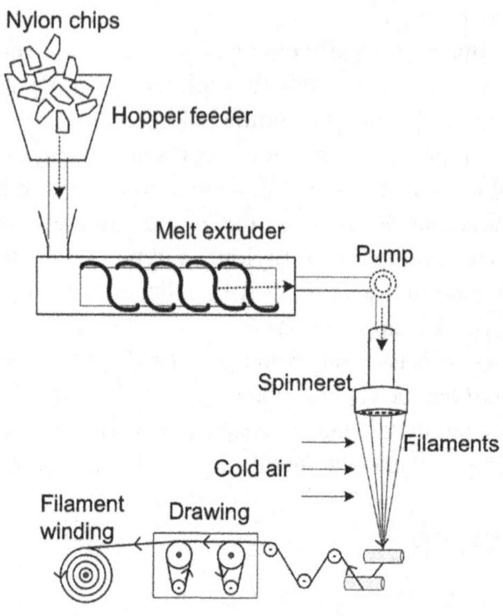

FIGURE 5.4

Production of nylon filament yarn by melt spinning.

The polyamide polymers nylon 6 and nylon 66 are produced with a number average molecular weight in the range of 18,000–20,000. The polymer is then subjected to melt spinning and drawing operations to manufacture nylon filaments, as shown in Figure 5.4.

The hopper feeder melts the nylon polymer chips and feeds them to the melt extruder. The extrusion temperature for nylon 6 is around 260 °C and for nylon 66 is around 280 °C. The molten polymer then passes through the extruder and is continuously pumped at a uniform pressure through the spinneret to form fine filaments. These filaments are subjected to a drawing process to improve the fibre properties such as tensile strength and stiffness.

5.3.2 STRUCTURE AND PROPERTIES OF NYLON FIBRES

Although both nylon 6 and 66 contain the same chemical groups in the same proportion, there exists a subtle difference in the arrangement of the molecules (Morton & Hearle, 2008, p. 58). These differences result in slightly lower crystallinity values for nylon 6 compared to nylon 66. Both nylons are semi-crystalline.

The morphological structure of high-strength nylon consists of microfibrilar and inter-microfibrilar regions (Prevorsek, Kwon, & Sharma, 1977), as represented by Figure 5.5. The microfibrilar regions

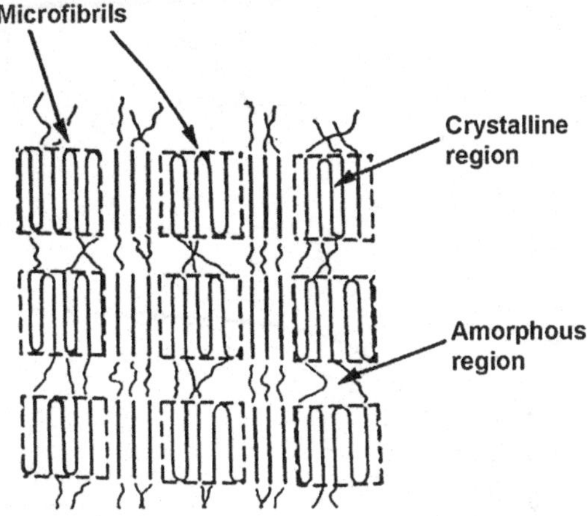

FIGURE 5.5

Microstructure of highly drawn polyamide fibres.

consist of crystalline and amorphous regions, whereas the inter-microfibrilar regions are formed of oriented molecular chains.

Nylons have a density of 1.14 g/cc. They are hygroscopic, but the high crystallinity of the fibre limits the moisture regain of nylons to 4%. Nylon fibres typically have a glass transition temperature (T_g) of 40–55 °C dependent on the moisture levels. The melting range of nylon 6 is 215–230 °C, while for nylon 66 it is between 250 and 265 °C and these differences result in a higher temperature stability of nylon 66 as compared to nylon 6.

Nylon fibres exhibit the following properties:

- Excellent tensile strength
- Good elastic recovery
- Low initial modulus
- Excellent abrasion resistance
- High work of rupture energy needed to break the fibre
- Excellent resistance to most chemicals

A typical stress–strain curve of nylon is shown in Figure 5.6. The tensile behaviour of polyester and aramid (Kevlar) fibres is also included for comparison. Initially nylon is quite extensible, followed by a yield point and a sigmoid type of deformation, as shown in the stress–strain curve. Compared with polyester fibres, nylons show similar strength values but are more extensible. Further, nylons have large extensions to break (about 25%) combined with relatively high strength (0.5 N/tex) and thus result in high work of rupture (energy needed to break the fibre).

5.3.3 APPLICATIONS

Nylon fibres are widely used in both the apparel and industrial sectors (Deopura, 2008). Lightweight and sheer garments are produced from nylon 6 and nylon 66, where extensibility, high strength and good abrasion resistance are of particular importance.

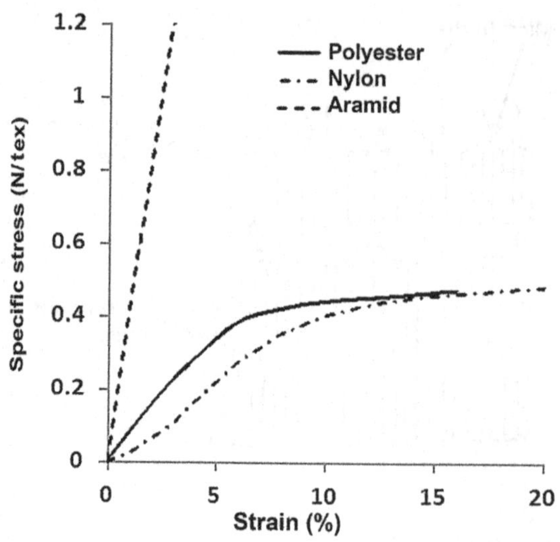

FIGURE 5.6

Stress–strain curves of synthetic fibres.

Nylon fabrics made from 15 denier monofilaments are used for hosiery products. Nylon fabrics also have excellent shape retention due to good elastic recovery behaviour. Setting in steam allows dimensional stability, as required for permanent pleating.

Fabrics made from fine nylon filament are extensively used for dress materials. Faux fur fabrics made from nylon are also popular due to their recovery behaviour and long life. Wool is blended with nylon to improve its durability, and this is particularly important when the application is for manufacture of outerwear or floor coverings.

In technical textiles, nylon fibres are widely used for a number of the following applications:

- In carpet fibres, because of resilience and excellent aberration resistance.
- Parachute fabrics, safety belts in cars, hoses and lightweight canvas for luggage are exclusively made from nylon filaments.
- Polyurethane-coated nylon fabrics are used for making hot air balloons.
- Multi-filament nylon yarns find extensive tyre cord applications for reinforcing rubber tyres. The twisted multi-strand tyre cords offer flex fatigue and a high surface area for easier bonding with latex. Critical application tyres for trucks and airplanes are made from nylon tyre cords.
- Nylon ropes and cordages have strength, durability and resistance to water.
- Fishing nets are mostly made from nylon twine due to excellent elastic recovery and high wet strength.
- Sailcloth made from nylon allows optimal deformation due to wind and recovery on reduction of wind speed.
- Other applications include ribbons for printers, bolting cloths, sutures and toothbrush bristles.

5.4 POLYESTER FIBRES

The term 'polyester' is applied to polymers containing ester groups in their main polymeric chain. Polyesters are derived from a poly-condensation reaction between dicarboxylic acids and diols. Polyesters can be broadly classified into two types:

1. Thermoplastic polyesters, which consist of fibre-forming, film-forming and engineering application polymers such as PET.
2. Thermoset polyesters, which are basically unsaturated polyester resins (liquid form) that upon curing form highly cross-linked structures (solid). They are widely used as a matrix for preparation of fibre-reinforced composite materials to bind the fibrous structure together.

The thermoplastic polyester PET is the most widely used synthetic polymer. Although many variants of polyesters are available, PET is very important for textiles and other commercial sectors. PET polyester is also used for the manufacture of plastic bottles, films, canoes, liquid crystal displays (LCD) and high-quality laminations. The major advantage of PET polyester is its recyclability.

Polyester fibres are the second largest type of fibres produced and consumed worldwide, second only to the cotton fibre.

5.4.1 PRODUCTION OF PET POLYESTER

PET is commercially manufactured using two methods. One is called dimethyl terephthalate (DMT) method and the other one is the Terephthalic acid (PTA) method (http://web.utk.edu/~mse/Textiles/Polyester fiber.htm).

5.4.1.1 DMT method

This method involves use of DMT, reacted with mono ethylene glycol (MEG) in the presence of a catalyst, typically a metal oxide, at 150–200 °C to obtain the monomer diethylene glycol terephthalate (DGT).

DGT then undergoes poly-condensation polymerization in the presence of a catalyst, usually a metal acetate, at a temperature between 265 and 285 °C and pressure of 1 mm Hg to produce PET polymer. The chemical reaction involved in the process is shown in Figure 5.7.

5.4.1.2 PTA method

In this method, terephthalic acid is reacted with ethylene glycol in the presence of metal oxides as catalysts, and at a temperature of 250–290 °C to obtain the monomer BHET. The DGT undergoes poly-condensation polymerization to produce PET polymer. The chemical reactions involved in the process are shown in Figure 5.8.

5.4.2 PET FIBRE FORMATION

Fibre-grade polyester polymers are formed in chips and fed into the fibre melt spinning process, as illustrated in Figure 5.4. The hopper feeder melts the polymer, which then feeds into the melt extruder. The extrusion temperature is usually kept at around 280–290 °C. The filaments are subjected to a drawing process to improve fibre properties such as tensile strength and stiffness. Depending on the end use, the polyester fibres are either retained in the filament form or converted into short staple fibre form.

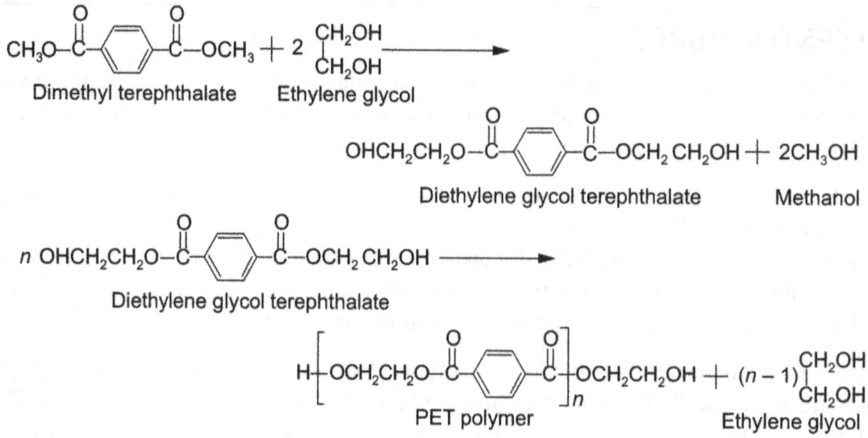

$$CH_3O{-}\overset{O}{\overset{\|}{C}}{-}C_6H_4{-}\overset{O}{\overset{\|}{C}}{-}OCH_3 + 2\begin{matrix}CH_2OH\\ |\\ CH_2OH\end{matrix} \longrightarrow$$

Dimethyl terephthalate   Ethylene glycol

$$OHCH_2CH_2O{-}\overset{O}{\overset{\|}{C}}{-}C_6H_4{-}\overset{O}{\overset{\|}{C}}{-}OCH_2CH_2OH + 2CH_3OH$$

Diethylene glycol terephthalate   Methanol

$$n\ OHCH_2CH_2O{-}\overset{O}{\overset{\|}{C}}{-}C_6H_4{-}\overset{O}{\overset{\|}{C}}{-}OCH_2CH_2OH \longrightarrow$$

Diethylene glycol terephthalate

$$H{-}\left[OCH_2CH_2O{-}\overset{O}{\overset{\|}{C}}{-}C_6H_4{-}\overset{O}{\overset{\|}{C}}\right]_n{-}OCH_2CH_2OH + (n-1)\begin{matrix}CH_2OH\\ |\\ CH_2OH\end{matrix}$$

PET polymer   Ethylene glycol

FIGURE 5.7

Production of polyethylene terephthalate (PET) polyester by dimethyl terephthalate method.

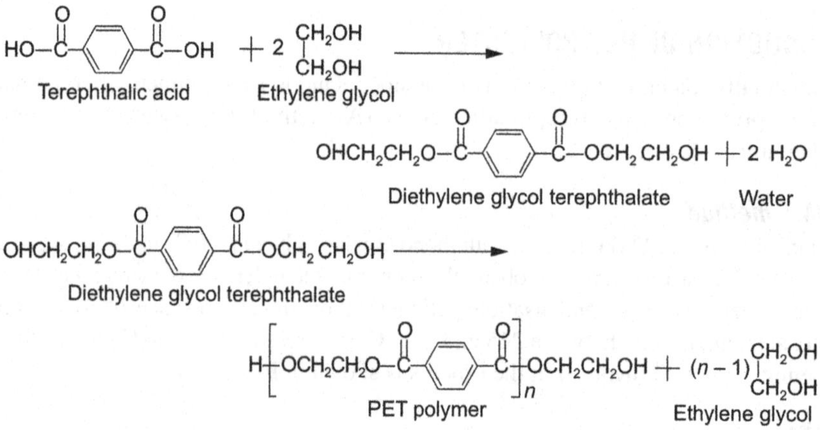

$$HO{-}\overset{O}{\overset{\|}{C}}{-}C_6H_4{-}\overset{O}{\overset{\|}{C}}{-}OH + 2\begin{matrix}CH_2OH\\ |\\ CH_2OH\end{matrix} \longrightarrow$$

Terephthalic acid   Ethylene glycol

$$OHCH_2CH_2O{-}\overset{O}{\overset{\|}{C}}{-}C_6H_4{-}\overset{O}{\overset{\|}{C}}{-}OCH_2CH_2OH + 2\ H_2O$$

Diethylene glycol terephthalate   Water

$$n\ OHCH_2CH_2O{-}\overset{O}{\overset{\|}{C}}{-}C_6H_4{-}\overset{O}{\overset{\|}{C}}{-}OCH_2CH_2OH \longrightarrow$$

Diethylene glycol terephthalate

$$H{-}\left[OCH_2CH_2O{-}\overset{O}{\overset{\|}{C}}{-}C_6H_4{-}\overset{O}{\overset{\|}{C}}\right]_n{-}OCH_2CH_2OH + (n-1)\begin{matrix}CH_2OH\\ |\\ CH_2OH\end{matrix}$$

PET polymer   Ethylene glycol

FIGURE 5.8

Production of polyethylene terephthalate (PET) polyester by PTA method.

5.4.3 STRUCTURE AND PROPERTIES OF POLYESTER FIBRE

The chemical structure of PET polyester is given in Figure 5.9. The molecular weight of the PET repeat unit is 192. The morphological structure of polyester fibres is similar to polyamide fibres, as both are polymeric fibres formed by a melt spinning and drawing process (Morton & Hearle, 2008, p. 63).

The polyester fibres are composed of a partially oriented and partially crystalline structure (see Figure 5.10) similar to that of polyamide fibres. Typically the degree of crystallinity of drawn polyester fibre is 55%. The glass transition temperature of polyester fibres is about 70 °C and the melting point is in the range of 255–270 °C. The density of the polyester fibre is 1.39 g/cc.

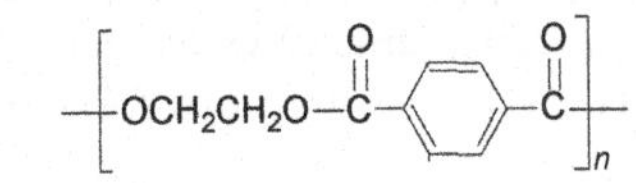

$$\left[-OCH_2CH_2O-\overset{O}{\overset{\|}{C}}-C_6H_4-\overset{O}{\overset{\|}{C}}- \right]_n$$

FIGURE 5.9

Chemical structure of polyethylene terephthalate polyester.

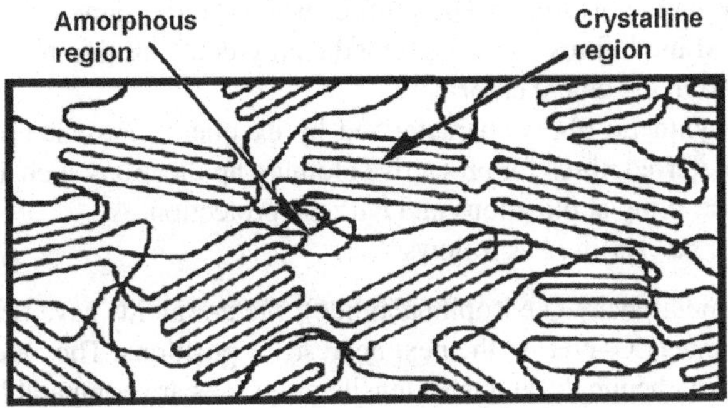

FIGURE 5.10

Structure of polyester fibre.

Properties of polyester fibres include the following:

- Polyester fibres are hydrophobic and have a low moisture regain value of 0.4%. Due to its hydrophobicity, the polyester fibres are water repellent and quick drying.
- Polyester fibres display excellent tensile strength.
- Resistance to stretching.
- Have negligible shrinkage.
- Are wrinkle resistant.
- Have excellent abrasion resistance.
- Easy care.
- Resistance to chemicals.
- Resistant to mildews.

Figure 5.2 displays the stress–strain behaviour of polyester fibres. The tenacity of polyester fibres is 0.4–0.5 N/tex with 15–25% elongation at break. The initial extensibility for polyester fibres is lower compared to that of nylon; otherwise the overall tensile behaviour of polyester is similar to that of the nylons.

5.4.4 APPLICATIONS

With its excellent tensile behaviour, resistance to stretch, anti-shrinkage and easy-care properties, polyester fibres tend to be used in a very large number of applications both in apparel and industrial sectors.

Polyester is the preferred fibre in the blending mix for cotton and wool. Apart from apparel applications, polyesters are used in home furnishings such as curtains, carpets, draperies, sheets, covers and upholsteries.

Polyesters have wide applications in technical textiles in such products as tyre cords, belts, ropes, nets, hoses, sails and automobile upholstery, and as fibre-fills for various products such as cushions and furniture, to name a few.

5.5 ARAMID FIBRES

'Aramid' is short for aromatic polyamide. These fibres belong to the same polyamide class as nylons, as shown in Figure 5.1. Aramid fibres are formed by the poly-condensation polymerization reaction of aromatic diamines and aromatic diacid chlorides.

Aramids are strong synthetic fibres characterized by excellent resistance to heat, chemicals and abrasion. They are the preferred fibres for protective clothing applications such as chemical protection, thermal protection, fire-proofing applications and ballistic protection.

The aramid fibres can be classified as follows:

1. Meta-aramids (poly meta-phenylene isophthalamide): Meta-aramids have benzene rings connected by amide linkage (CO–NH) in the next nearest 1,3 positions. They are resistant to temperature, abrasion and most chemicals, and their tensile properties are comparable to those of high-performance polyester and nylon fibres.
2. Para-aramid (poly para-phenylene terephthalamide): Para-aramids have benzene rings connected by amide linkage (CO–NH) in the opposite 1,4 positions. They are a highly oriented type of synthetic fibre, having high strength and excellent abrasion and chemical resistance (Kevlar aramid fibre technical guide, 2010).

5.5.1 PRODUCTION OF ARAMID FIBRES

5.5.1.1 Meta-aramid

Meta-aramid is synthesized by a poly-condensation reaction between *m*-phenylene diamine and isophthaloyl dichloride in an *n*-methyl pyrrolidone solvent (Jassal & Ghosh, 2002). The reaction is shown in Figure 5.11.

5.5.1.2 Para-aramid

Para-aramids are also synthesized by a poly-condensation reaction between *p*-phenylene diamine with terephthaloyl dichloride in an *n*-methyl pyrrolidone solvent. The reaction is shown in Figure 5.12.

Meta-aramid fibres are produced from a solution spinning process, which is also known as wet spinning (see Figure 5.13). The meta-aramid polymer is dissolved in 100% sulphuric acid to form aramid dope. This polymeric dope is forced through the spinneret immersed in a spin bath containing water to obtain fibres. Sulphuric acid solvent in the dope is removed in the water bath and the fibres formed are drawn, dried and heat-set.

Para-aramid fibres are made by the dry-jet, wet-spinning method (see Figure 5.14). The para-aramid polymer is immersed in 100% sulphuric acid (solvent) to form a liquid crystalline state and kept only partially liquid, which keeps the polymer chains together. The polymeric dope is forced through the spinneret at 100 °C. The fibre becomes highly oriented in the air gap before entering the spin bath containing water. Sulphuric acid solvent in the dope is removed in the water bath and the fibres formed are heat-set to obtain highly oriented fibres.

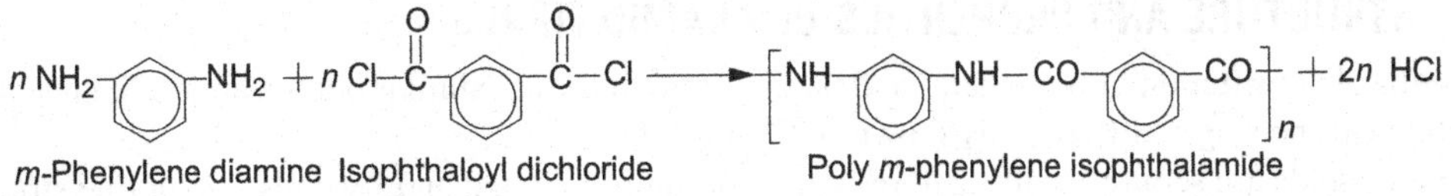

$$n\,H_2N{-}C_6H_4{-}NH_2 + n\,Cl{-}\overset{O}{\overset{\|}{C}}{-}C_6H_4{-}\overset{O}{\overset{\|}{C}}{-}Cl \longrightarrow \left[NH{-}C_6H_4{-}NH{-}CO{-}C_6H_4{-}CO\right]_n + 2n\,HCl$$

m-Phenylene diamine Isophthaloyl dichloride Poly *m*-phenylene isophthalamide

FIGURE 5.11

Production of meta-aramid polymer.

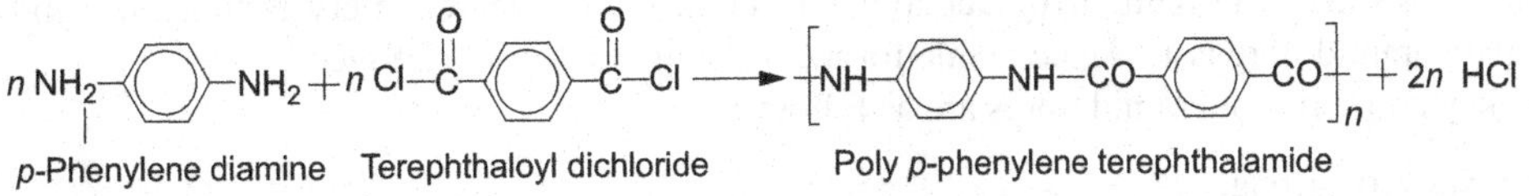

$$n\,H_2N{-}C_6H_4{-}NH_2 + n\,Cl{-}\overset{O}{\overset{\|}{C}}{-}C_6H_4{-}\overset{O}{\overset{\|}{C}}{-}Cl \longrightarrow \left[NH{-}C_6H_4{-}NH{-}CO{-}C_6H_4{-}CO\right]_n + 2n\,HCl$$

p-Phenylene diamine Terephthaloyl dichloride Poly *p*-phenylene terephthalamide

FIGURE 5.12

Production of para-aramid polymer.

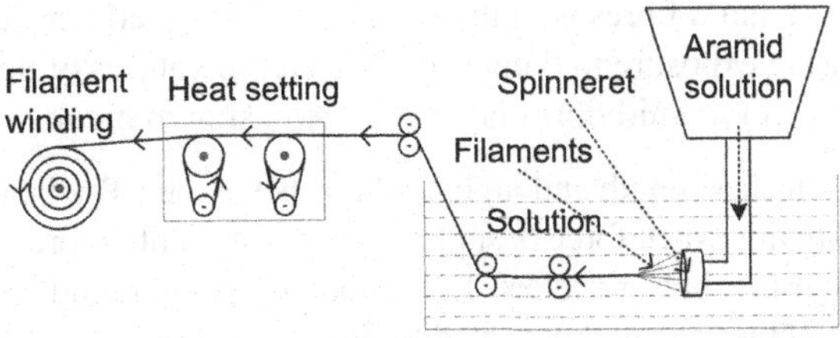

FIGURE 5.13

Production of meta-aramid filament yarn by solution spinning.

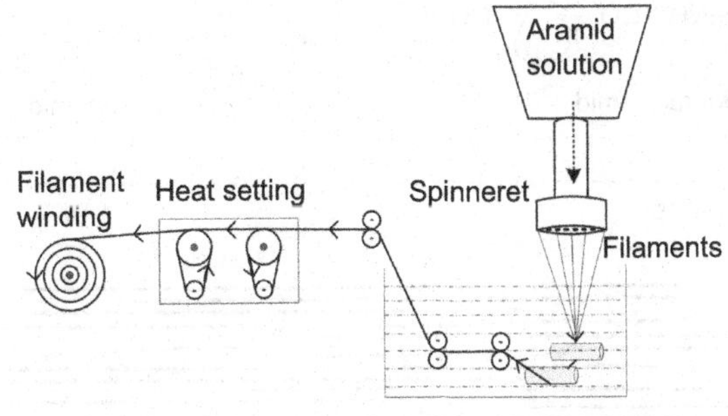

FIGURE 5.14

Production of aramid filament yarn by dry jet wet spinning.

5.5.2 STRUCTURE AND PROPERTIES OF ARAMID FIBRES

The chemical constitution of both meta- and para-aramid fibres are similar but the amide linkage in the aromatic benzene ring is different, as shown in Figure 5.15.

Meta-aramids are semi-crystalline fibres with a partially oriented and partially crystalline structure similar to polyester and nylon. The presence of amide linkage in the meta-aramid polymer restricts complete extension of the molecular chain, hence high orientation and high crystallinity is not obtained in the meta-aramid fibres (Morton & Hearle, 2008, p. 72).

For para-aramids, the presence of the amide linkage in the polymeric chain favours extended chain configuration assisting liquid crystal formation. Thus the resulting structure of the para-aramid fibre consists of fully extended chains, packed together with a very high degree of crystallinity and very high orientation, as shown in Figure 5.16. Due to the difference in the degree of crystallinity, the fibre density of para-aramid fibre is 1.44 g/cc, while for meta-aramid fibre it is 1.38 g/cc.

The key properties of aramid fibres are as follows:

- They are very strong.
- They have excellent resistance to heat, abrasion and chemicals.
- They are electrically nonconductive.
- They do not melt but commence to degrade beyond 500 °C.
- Meta-aramids (brands Nomex, Tenjinconex, New star, X-fiper, Kermel) in particular have a tensile strength of about 0.5 N/tex, comparable to high-strength synthetic fibres of nylon and polyester.
- The uniqueness of meta-aramid fibres is in their ability to withstand tensile stress, abrasion and chemical resistance during exposure to flames and high temperatures of up to 400 °C.
- The moisture regain of meta-aramid fibres is 5% and extension to break is 15%.

Para-aramids display high tensile strength and high modulus behaviour. The latter describes the tendency of a material to deform elastically when force is applied. Para-aramid fibres are available in low-modulus, high-modulus and very-high-modulus varieties. Low-modulus para-aramid fibres (Kevlar 29) usually have a tensile strength of 2 N/tex, a modulus of 490 N/tex and 3.6% elongation to break. The high-modulus para-aramid fibres (Kevlar 49, Twaron, Technora) have a tensile strength of 2.1 N/tex with a modulus of 780 N/tex and 2.8% elongation to break. Very-high-modulus para-aramid fibres (Kevlar 149) display similar tensile strength but the modulus is 1000 N/tex with a 2% elongation at break.

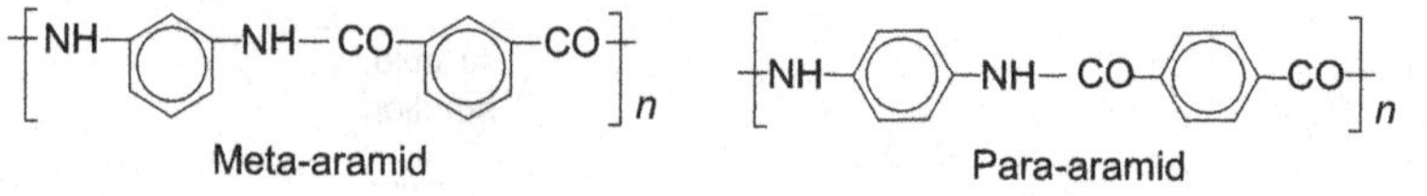

FIGURE 5.15

Chemical structure of aramid fibre.

FIGURE 5.16

Microstructure of para aramid fibre.

Moisture regain of para-aramid fibre ranges from 1.2% for very-high-modulus fibre to 7% for low-modulus fibres. Para-aramids have better abrasion resistance than meta-aramids and also the meta-aramid fibres tend to fibrillate and wear more easily during abrasion.

5.5.3 APPLICATIONS

Meta-aramid fibre, due to its excellent heat resistance, flame-proof properties, chemical resistance, electrical non-conductance and abrasion resistance, is established for applications involving protective apparel, specifically for fire-proof, cut-proof and abrasion-resistant clothing for automobile drivers. Meta-aramid fibres are also used for electrical insulation purposes.

Para-aramids, due to their high tenacity, excellent heat resistance and abrasion resistance properties, are the preferred fibres for ballistic protection wear such as bullet-proof armour vests and helmets. Para-aramids have extensive applications as fibre reinforcement for composite materials and can be used in technical textile applications such as automobile clutch plates, brake linings, aircraft parts, boat hulls and sporting goods. They are also extensively used for ropes, cables, optical cable systems, sail cloths, high-temperature filtration fabrics, drumheads and speaker diaphragms.

5.6 BLENDED FIBRES: KEY ISSUES

Fibre blending is a process of combining two or more fibres to form a yarn that comprises the best qualities of each. The selection of fibres for blending is based on the following objectives:

- To achieve functionality
- To improve process performance
- Economic factors
- For improved aesthetic qualities

When blending fibres, the functional properties usually targeted (Alagirusamy & Das, 2008) are:

- Improved tensile strength
- Uniformity
- Better appearance
- Increased wear life
- Crease resistance
- Crease recovery
- Shrinkage resistance
- Elasticity
- Better comfort properties

Fibre blending is carried out mainly for apparel end uses and for special yarns is carried out in staple fibre form before spinning during the fibre-processing stage. The advent of synthetic fibres such as polyester and nylons enabled the opportunity to create tailor-made yarns and fabrics. The key issues related to fibre blending are:

- Types of fibres
- Compatibility of the constituent fibres
- Fibre blend ratio

The types of fibres used for blending determine the final properties of the yarn or fabric to a significant degree. A single-component fibre such as 100% polyester can provide excellent strength, abrasion resistance, durability and easy-care properties. On the other hand, 100% polyester would be clammy, slippery and uncomfortable to the wearer. Similarly, 100% cotton or viscose fibres due to their hygroscopicity and thermal conduction properties provide excellent comfort properties to the wearer but have disadvantage of creasing, the need for repeated laundering, low colour stability and low abrasion properties.

Blending cotton or viscose rayon with polyester fibres would combine the advantages of both fibre types, thereby minimizing the weakness of each fibre. The synthetic fibres polyester and nylons are used along with natural fibres such as cotton and wool to produce fibre blends that now have established applications in the apparel sector.

The fibres selected for blending should be compatible with respect to several attributes such as fibre length, fineness (denier), strength, extensibility and dyeing behaviour. Although the purpose of blending is to achieve better properties in the final product, a mismatch of attributes would lead to quality and uniformity issues that would counteract the purpose of blending.

The ratio of each fibre in the blend is controlled to optimize the desired end-product property. The fibre blend proportions 80/20, 60/40, 55/45 and 50/50 of cotton/polyester, wool/polyester, wool/nylon, polyester/viscose rayon are often employed in the apparel industry, respectively. Some tertiary blends with three types of fibres such as cotton/polyester/polyurethane in 60/38/2 blend proportion are used for durability and comfort, while blends such as cotton/polyester/polynosics in 30/40/30 blend proportion are used in low-cost products.

5.7 CASE STUDY: POLYESTER FIBRES FOR APPAREL AND CLOTHING APPLICATIONS

A vast range of apparel types exist:

- Casual wear
- Formal wear
- Occasion wear
- Outdoor activities
- Indoor clothing
- Hosiery and lingerie
- Sports and extreme weather clothing

The range of products available has led to higher consumer demands, especially with respect to their functional properties. Polyester fibres are the main synthetic fibres used in the textile and apparel sectors worldwide. Polyester fibres:

- Are strong, durable and stretchable
- Resist abrasion and creasing
- Possess exceptional dimensional stability
- Are easy to wash, dry and use
- Are stable with respect to properties in wet and dry conditions
- Resist chemicals and mildew
- Are all-weather resistant and not damaged by sunlight

- Are economical and easy to blend with other fibres

Coupled with the above advantages, there are also disadvantages of polyester fibres for apparel applications. They are as follows:

- The fibres are uncomfortable to wear next to the skin in hot conditions due to low moisture absorption, poor vapour transmission and low thermal conductivity.
- They have static electricity issues due to hydrophobicity (a dislike of contact with water).
- Polyester has an unpleasant feel and poor aesthetic properties due to the smooth, round fibre surface.

The low moisture absorption, poor moisture transmission and low thermal conductivity can be overcome by blending polyester fibres with hygroscopic fibres such as cotton and viscose rayon. Cotton and viscose rayon if used alone would be prone to problems such as crease formation, low colour stability, low durability, dimensional stability and low abrasion resistance. Blending polyester with cotton/viscose brings out the best of the fibres' properties. These polyester/cotton and polyester/viscose blends have abundant apparel applications such as formal/casual shirts, trousers and outdoor garments.

Blending with hygroscopic fibres also solves the problems associated with the generation of static electricity. Many chemical finishes have also been developed for polyester fibres to impart a cotton-like feel, and hygroscopicity, improving the comfort characteristics.

Polyester fibres with different cross-sections such as triangular, trilobal, serrated, oval, band and hollow (see Figure 5.17) have greatly improved the aesthetic, tactile, comfort, physiological and technological properties in the end-products.

Hollow porous polyester fibres improve thermal insulation and moisture absorption. Modification of the polyester fibre surface during fibre production or by an after-treatment process to impart fibrillated, roughened, smoothened and peeled effects also improves the thermal insulation and moisture absorption properties.

Manufacturers have also used three-carbon glycol to make polyester fibres with a higher extensibility, making them softer, with improved stretch and elastic recovery, and also with improved comfort characteristics.

The characteristics introduced during fibre production have a greater impact on final properties than the modifications made after fibre production. Texturing is a process by which a luxurious feel, bulk, greater absorbency and improved stretch is introduced through the permanent

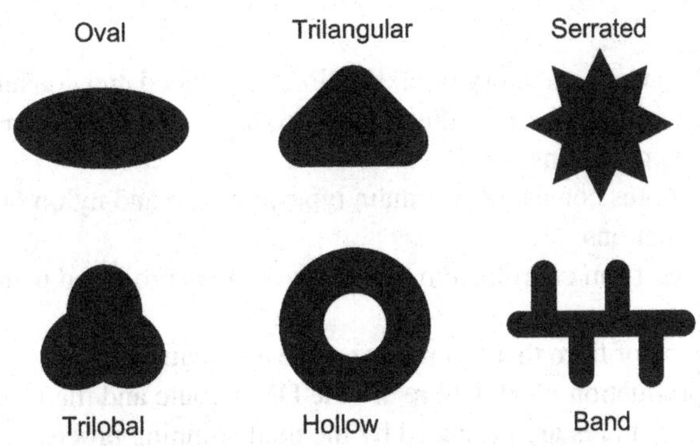

FIGURE 5.17

Different cross-sectional shapes of fibres.

introduction of crimp (waviness in the fibre), coils, loops or crinkles into an otherwise continuous filament yarn. Texturing also greatly influences the thermal insulation behaviour, tactile properties and moisture transport properties in apparel applications.

The production of ultra-fine micro denier polyester fibre with fineness of less than 0.1 denier has revolutionized the garment sector in general and the sportswear sector in particular. The ultra-fine fibres impart excellent softness, smoothness and flexibility to the textile, achieving a tactile feel without losing the tensile, durability, anti-shrinkage and other properties of polyester fibres.

Microfibre polyester clothes have been advertised as ideal travel apparel since they are comfortable, durable, look smart and retain their wrinkle-free appearance. The low absorbency of these fibres and their ability to be close packed inhibits the penetration of water, enabling their use in rainwear. In contrast, some jackets made with microfibers are fitted with a mesh lining to allow the fabric to wick away perspiration in the tiny channels between the fibres.

5.8 FUTURE TRENDS

Increased competition and shrinking profits for nylon and polyester production has driven the market towards specialist, niche, and thus more profitable products. It is possible to combine a multitude of fibre types and corresponding attributes to make unique textile products. For example, selection of square cross-section instead of circular cross-section fibres results in a significant increase in stiffness of the fabrics.

Material science research has enabled incorporation of composite and nanotechnologies into synthetic fibres. Attempts are being made to engineer material properties and to control design attributes in order to produce tailor-made material properties at an affordable cost. This has resulted in fibres having a range of properties such as anti-microbial, self-cleaning, thermo-sensitive, chemo-chromic, highly wickable, liquid repellency, etc.

Nylons, polyesters and aramid fibres are petrochemical products. Their conversion from crude petroleum into final product involves large energy costs and the use of many chemicals resulting in limited levels of environmental pollution. In this context, polyester, which is recyclable and biodegradable, is superior to nylon.

5.9 SUMMARY

- Synthetic fibres are the largest category of textile fibres produced and consumed worldwide.
- Polyester fibres are the main type of synthetic fibre produced and have been used extensively in apparel and industrial applications.
- Polyamides, or nylon, fibres consist of two main types, nylon 6 and nylon 66, which have wide apparel/industrial applications.
- Nylon 6 is manufactured from caprolactum, while nylon 66 is produced using hexamethylene diamine and adipic acid.
- PET is an important type of fibre that has many textile applications.
- The two methods of production of PET fibre are the DMA route and the PTA route.
- Both nylon and polyester fibres are produced by the melt spinning process and both have similar semi-crystalline morphological structures.

- Polyester fibres have a wide range of applications in the apparel sector due to the suitability of attributes such as strength, durability, crease resistance and easy-care properties.
- Polyamide fibres have a wide range of apparel/industrial applications due to excellent elastic recovery and abrasion resistance.
- Aramid fibres are aromatic polyamides, of which there are two types: meta- and para-aramids.
- Aramid fibres are produced by solution or the dry jet wet spinning method.
- Meta-aramids have exceptional resistance to heat, abrasion and chemicals, while para-aramids have excellent tensile properties.
- Para-aramid fibres have high tenacity, high modulus and high crystallinity in the fibre structure.

5.10 PROJECT IDEAS

- Study the properties and end uses of nylon fibres for naval (technical textile) applications.
- Using garment labels as a starting point, study various types of clothing and apparel (casual, formal, summer, winter, sports) noting their fibre blend and composition, and appropriateness for end-use application.
- Using the theme technical textiles as a starting point, study different types of ropes for their structure, fibre blend composition and properties (aerial ropes, construction ropes, mountaineering ropes, marine ropes, etc.).

5.11 REVISION QUESTIONS

1. Using a figure, outline the key man-made fibres.
2. Describe the production process for nylon 66 fibres.
3. List the most important applications of polyamide fibres.
4. Discuss the key elements of the structure and properties of polyester fibres.
5. What are the important properties of meta- and para-aramid fibres?
6. Describe the dry jet wet spinning process, and explain how it influences the structure and properties of para-aramid fibres.
7. What are the main applications of aramid fibres?
8. Why is fibre blending useful? What is the importance of polyester blend proportions with cotton and wool?
9. List the attributes of the polyester fibre that make it most suitable fibre for apparel applications.

5.12 SOURCES OF FURTHER INFORMATION AND ADVICE

1. Moncrieff, R. W. (1975). *Man-made fibres* (6th ed.). Newnes-Butterworths.
2. Gupta, V. B., & Kothari, V. K. (Eds.), (1997). *Manufactured fibre technology*. Chapman & Hall.
3. Deopura, B. L., Alagirusamy, R., Joshi, M., & Gupta, B. (Eds.), (2008). *Polyesters and polyamides*. The Textile Institute, Woodhead Publishing Limited.

4. Hongu, T., Takigami, M., & Phillips, G. O. (Eds.), 2005. *New millennium fibres*. The Textile Institute, Woodhead Publishing Limited.
5. Morton, W. R., & Hearle, J. W. S. (2008). *Physical properties of textiles fibres* (4th ed.). The Textile Institute, Woodhead Publishing Limited.
6. Lewin, M. (2008). *Hand book of fibre chemistry* (3rd ed.). Taylor & Francis.

REFERENCES

Alagirusamy, R., & Das, A. (2008). Property enhancement through blending. In B. L. Deopura, R. Alagirusamy, M. Joshi, & B. Gupta (Eds.), *Polyesters and polyamides* (p. 220). The Textile Institute, Woodhead Publishing Limited.

Deopura, B. L. (2008). Polyamide fibers. In B. L. Deopura, R. Alagirusamy, M. Joshi, & B. Gupta (Eds.), *Polyamides, in polyesters and polyamides* (pp. 41–61). The Textile Institute, Wood head Publishing Limited.

Engelhard, A. (May 2010). *The fibre year 2010*. Germany: Oerlikon Textile GmbH & Co. www.oerlikontextile.com.

Jassal, M., & Ghosh, S. (2002). Aramid fibres – an overview. *Indian Journal of Fibre and Textile Research, 27*, 290–306.

Kevlar aramid fibre technical guide. (2010). DuPont. http://www2.dupont.com/Kevlar/en_US/.

Morton, W. R., & Hearle, J. W. S. (2008). *Physical properties of textiles fibers* (4th ed.). The Textile Institute, Woodhead Publishing Limited.

Prevorsek, D. C., Kwon, Y. D., & Sharma, R. K. (1977). *Journal of Material Science, 12*, 2310–2328.

CHAPTER

SYNTHETIC TEXTILE FIBRES: POLYOLEFIN, ELASTOMERIC AND ACRYLIC FIBRES

6

R.R. Mather

Heriot-Watt University, Edinburgh, UK

LEARNING OBJECTIVES

At the end of this chapter, you should be able to:

- Understand the processing, structure and property relationships of polypropylene fibres
- Understand the processing, structure and property relationships of acrylic fibres
- Understand the processing, structure and property relationships of elastomeric fibres

6.1 INTRODUCTION

This chapter concerns three very important, but very different types of synthetic fibres: polyolefin, acrylic and elastomeric fibres. One aspect they have in common, however, is that each type of fibre embraces a wide range of constituent polymers.

A polymer consists of very large molecules constructed from many smaller structural units called monomers, chemically bonded together. According to *Textile Terms and Definitions* (11th edition), a polyolefin fibre is a manufactured fibre in which the fibre-forming substance is any long-chain synthetic polymer composed of at least 85% by mass of ethene (ethylene), propene (propylene) or other olefin units (Denton & Daniels, 2002).

By far the most prominent polyolefin fibre used commercially is polypropylene (PP), although polyethylene fibre also has important commercial applications. Some other types of polyolefin fibres are also produced, with limited commercial application.

Fibres attracting some commercial interest have been produced from copolymeric polyolefins, such as ethene–propene and ethene–propene–butene copolymers. Significant among these are elastomeric polyolefin fibres named 'Lastol' by the Federal Trade Commission in the United States and manufactured by the Dow Chemical Company as XLA fibres. These fibres have been used as minor components of a wide variety of stretch wool garments, denims, leisure shirts, swimwear and intimate apparel. It is notable too that the elastomeric fibres can be spun with an outer sheath of cotton fibre to give the appearance and feel of cotton.

With acrylic fibres, the constituent polymer chains must contain at least 85% of cyanoethane (acrylonitrile) groups (Denton & Daniels, 2002). The remaining 15% consists of other groupings that

Textiles and Fashion. http://dx.doi.org/10.1016/B978-1-84569-931-4.00006-4

assist fibre processing and confer several useful properties to the fibres. Elastomeric fibres embrace a wide range of constituent polymers: elastane, elastodiene, elastomultiester and polyolefins.

For each of the three types of fibre a considerable range of structures, and hence properties, can be produced at the molecular, and other, levels. For example, on a larger scale, the arrangements of the polymer chains within each fibre are significant, in that they strongly influence the fibre's mechanical properties, such as its strength and elasticity. The appearance of the fibre should also be considered, such as the size and shape of the fibre cross-section, hollow fibres and crimp (the waviness of the fibre).

The aspects of molecular architecture and fibre construction are critical in determining the physical and mechanical properties of the fibres. In turn, these aspects determine the commercial applications available for each type of fibre.

6.2 POLYPROPYLENE (PP) FIBRES

PP fibre in many ways can be considered the 'workhorse' amongst synthetic fibres. Despite its low cost, its key properties include its high strength, high toughness and good resistance to chemical attack. Consequently it finds a wide range of applications – from sacking and large industrial bags to high-tech medical applications. A major drawback, however, is that PP fibre (in common with all polyolefin fibres) cannot be dyed unless the fibre is modified with suitable additives (Shamey, 2009).

The commercial impact of PP fibre began in the 1960s, when it started to replace jute for carpet backings and bast fibre for rope. By 2009, the total annual world production of PP in textiles was well over 5 million tonnes, compared with 32 million and 3.3 million tonnes for polyester and polyamide fibre, respectively (Anonymous, 2010a). PP accounts for well over 90% of polyolefin fibre production.

6.2.1 PRODUCTION OF POLYPROPYLENE (PP)

PP is produced by the polymerisation of propene (propylene), itself produced as a by-product in the cracking of oil and from the fractionation of natural gas. Natural gas is a mixture of individual component gases, and fractionation involves their separation and isolation. Propene is generally polymerised by a process devised by Giulio Natta in 1954, following on from a similar process for producing polyethylene, invented by Karl Ziegler the previous year. The process involves special catalysts, the so-called Ziegler–Natta catalysts, which contain a mixture of a titanium chloride (or sometimes a vanadium chloride) and an organometallic compound of aluminium and a hydrocarbon. More recently, alternative catalysts, called metallocenes, have been developed. These catalysts, often based on the element zirconium, are more active and more specific than Ziegler–Natta catalysts, and provide better control of the structure of the PP chains produced. As a result, the filaments processed from metallocene PPs contain superior mechanical properties (Schmenk, Miez-Meyer, Steffens, Wulfhorst, & Gleixner, 2000). However, metallocene catalysts are much more expensive than Ziegler–Natta catalysts.

6.2.2 FIBRE MANUFACTURE

PP fibres are produced by melt extrusion of PP granules (melt extrusion is discussed in Chapter 10), and a number of different melt extrusion processes are used. The choice depends on the type of fibre required and the application for which the fibre is destined. There are different processes for producing multifilament yarns, monofilament yarns, staple, tape and non-wovens. Each of these is discussed briefly as follows.

Multifilament yarns of PP are produced in several forms: partially oriented (POY) yarns, fully oriented (FOY) yarns or bulked continuous filament (BCF).

- The POY process is low cost. It also allows for greater flexibility in subsequent processing such as drawing (stretching of the filaments), twisting and texturing. POY yarns are generally produced with linear densities of 40–200 dtex, with 0.5–4.0 dtex for the linear density of each constituent filament. An important feature of the POY process is the long length of the cooling unit through which the filaments pass after melt extrusion from the spinneret. The length of the cooling unit can be up to 10 m in order to ensure that, at the high production speeds used, the individual filaments are sufficiently cooled before being wound together as multifilament yarn. Winding speeds are in the range 2000–3000 m/min. The POY yarns are then processed further by drawing and possibly also by twisting and texturing.
- With the production of FOY yarn, spinning and drawing form consecutive parts of a continuous process. The FOY process is highly versatile, and depending on the control settings used and the grade of PP, yarns with a wide range of different mechanical properties can be produced. Final winding speeds are up to 5000 m/min. Both high-strength yarns, of linear density 5–10 dtex per filament for technical fabrics, and standard yarns, of linear density 1–2 dtex per filament, for more traditional applications can be produced by the FOY process. FOY yarns are generally stronger than yarns produced by the POY process, even after POY yarns have been drawn.
- The BCF (bulked continuous filament) process combines spinning, drawing and texturing in one process. The speed of filament production is 1500–4000 m/min. It resembles the FOY process, excepting that the filaments are textured after drawing.

6.2.3 TYPES OF YARNS

Monofilament yarns: PP monofilaments possess linear densities of >100 dtex and are much coarser than multifilaments. Monofilaments produced by the methods used to produce multifilaments tend to curl (Ahmed, 1982, p. 430) and are thus rendered unsuitable for many applications. Instead, PP monofilaments are usually formed at lower speeds and extruded into water for more efficient cooling. After leaving the spinneret, the monofilament travels through <5 cm air before it enters a water bath. Subsequent drawing may involve as many as three stages at elevated temperatures, typically 80–90 °C. PP monofilaments are produced for high strength in applications such as belts and ropes, and in hawsers (cables) used for mooring and towing ships.

Staple fibres: Staple fibres can be produced by two processes: either a two-stage discontinuous process or a single-stage compact continuous process. The two-stage process is adopted for very fine, high-quality staple fibre (~0.5 dtex per fibre), and the spinning speed is typically ~2000 m/min. The spun filaments produced from the first stage move onto the second stage, which comprises drawing, crimping and cutting into staple fibres of the required length. An alternative to cutting is converting, in which a sliver of parallel fibres is formed. These fibres are then used for worsted spinning, discussed in Chapter 9. A disadvantage of the two-stage process is the large amount of space needed to accommodate all the equipment required. In particular, the cooling zone is very long. A real advantage, however, is that each stage can be operated independently under its own optimum conditions.

The single-stage compact process combines all the production stages and is now considered more economical for small quantities for staple fibre. Although the spinning speeds are low (100–300 m/min), productivity is gained from the use of spinnerets with up to 100,000 holes, arranged in a grid structure.

The filaments produced from the spinning stage are combined into a tow, which is fed continuously into a drawing unit. The fibres produced possess a minimum linear density of 1–3 dtex, and are used for carpet yarns and non-woven products.

Non-wovens: PP non-wovens can be produced by a multistage process that incorporates the production of staple fibres. More often, a single-stage operation is utilised which integrates filament production and the formation of a non-woven fabric. The web comprising the fabric is usually held together by thermal bonding, because of the low melting point of PP fibres (160–165 °C). A thermally bonded non-woven fabric is composed of fibres containing heat-sensitive material, such as PP, bonded by the application of heat (Denton & Daniels, 2002).

There are two important versions of the single-stage process: the spun-bond process and the melt-blown process (Goswami, 1997). In the spun-bond process, the non-woven fabric produced consists of a web of randomly distributed, thermally bonded filaments. Spun-bonded PP webs are noted for their strength. In the melt-blown process, PP is extruded through numerous small spinneret holes situated close to one another. Just below the spinneret holes, a stream of molten polymer is caught in a current of hot air moving at high speed, and as a result is broken up into a network of very fine entangled fibres. These fibres are immediately deposited onto a moving conveyor belt. Although they are weaker than spun-bonded webs, melt-blown non-wovens possess textures that render them suitable for applications where filtration or absorption is important.

Production lines combining both processes are used for the formation of SMS (spun-bonded, melt-blown, spun-bonded) multilayer fabrics, which combine the desirable properties of spun-bonded and melt-blown materials. The external spun-bonded layers provide good mechanical properties, whilst the internal melt-blown web provides good filtration and absorption. SMS fabrics are used extensively in disposable hygiene and medical products.

Tapes: PP is quite often processed into tapes. Tapes are used predominantly for household products such as carpets and technical textile products, such as sacks, industrial bags and tarpaulins. PP tapes have also been used in sunhats, handbags and beach shoes.

There are two main methods for producing tapes (Schmenk et al., 2000). In the more common method, a film of PP is extruded, and then cooled in a water bath or on rollers. The film is then stretched in one direction to about 10 times its original length, when it is cut with knives into tapes. Alternatively, if stretched sufficiently, the film starts to split of its own accord.

In the alternative method, each tape is melt extruded separately through an individual slit-shaped orifice. This process is considerably more expensive, and is generally restricted to tapes for specialist uses, such as medical applications.

6.2.4 SPIN FINISHES

As with the processing of other synthetic fibres, PP filaments are treated with a spin finish to protect the filament surfaces and to dissipate any static electricity generated during processing. Static electricity can cause ballooning in multifilament yarns and may also distort or rupture the yarn during processing. In addition, it may cause electric shocks when the processing equipment is touched. Static electricity may also build up on the finished garment and cause it to cling to the body or to another garment. It can be discharged when the wearer touches an electrically conducting material, usually metal. The wearer may, as a consequence, experience discomfort.

An additional function of a spin finish is the reduction of friction during fibre processing, a function that assumes increasing importance as fibre production speeds rise. The heat generated by friction can cause softening or even melting of the fibres. The spin finish may contain an antimicrobial agent to kill or inhibit micro-organisms, although PP fibres are themselves quite resistant to micro-organisms.

PP fibres are highly hydrophobic: they very effectively repel water. To take account of this property, special wetting agents also have to be included in the formulation of the finish. In addition, some components of the spin finishes, commonly used in the processing of other synthetic fibres, cannot be applied to PP fibres because they migrate into the fibres and cause swelling. A spin finish contains a complex mixture of chemical components and is often applied as an emulsion.

6.2.5 ADDITIVES

Additives are used to assist the processing of PP fibres and to achieve the fibre properties required. They are either present in the PP granules supplied to the fibre producer or are incorporated into the PP melt before it is extruded.

Additives have a number of roles. Some additives act as heat stabilisers, to prevent thermal degradation of the polymer chains during fibre processing. Others act as stabilisers against light, and especially ultra-violet radiation, during the end-use of the fibres, or as flame retardants, to prevent ignition of the fibres. Antistatic additives are also likely to be present.

Since commercially produced PP fibres contain a variety of additives with a number of functions, care must be taken that the effectiveness of one additive is not compromised by another one. In some cases, however, there may be synergistic effects, in which the presence of one additive actually assists the function of another. Furthermore, all the additives must be able to withstand the fibre-processing conditions, the processes involved in the conversion of the fibre to the finished article, as well as the end-use of the article during its lifetime. Specific additive packages are under development (Crangle, 2009), and must be rigorously assessed against undesirable effects, both during fabric manufacture and end-use.

6.2.6 FIBRE STRUCTURE

The structure of a fibre can be considered on a number of levels. At the molecular level, the structure of the individual polymer chains is important. On a larger scale, the arrangements of these chains within the fibre are also significant. The appearance of the fibre is another factor, including fibre size and cross-section.

The chemical structure of PP is given in Table 6.1. PP chains can adopt several molecular configurations, which are shown schematically in Figure 6.1. In stretched-out projections of isotactic PP chains, all the methyl, $-CH_3$, side-groups, projecting from the main chain, are situated uniformly on the same side of the chain. In practice, however, the chains are not stretched out but generally adopt a three-dimensional helical formation, as illustrated in Figure 6.2. The isotactic form is the one used in commercial PP fibres, as the highly regular structure confers superior mechanical properties. In syndiotactic

Table 6.1 Structures of polypropylene and polyethylene

Polymer	Repeat unit
Polypropylene	$-(CH_2-CH)-$ I CH_3
Polyethylene	$-(CH_2-CH_2)-$

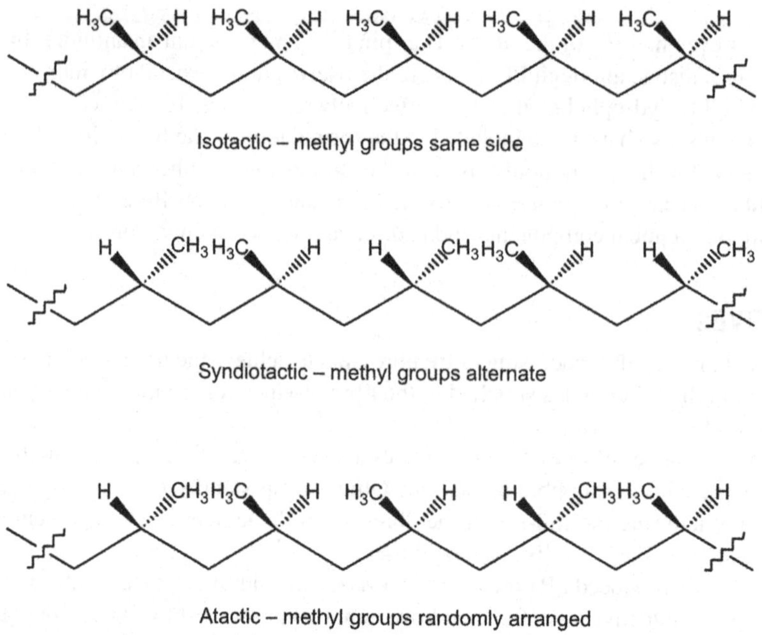

H3C H H3C H H3C H H3C H H3C H
Isotactic – methyl groups same side
H CH3 H3C H H CH3 H3C H H CH3
Syndiotactic – methyl groups alternate
H CH3 H3C H H3C H H CH3 H3C H
Atactic – methyl groups randomly arranged

FIGURE 6.1

Structures of polypropylene chains.

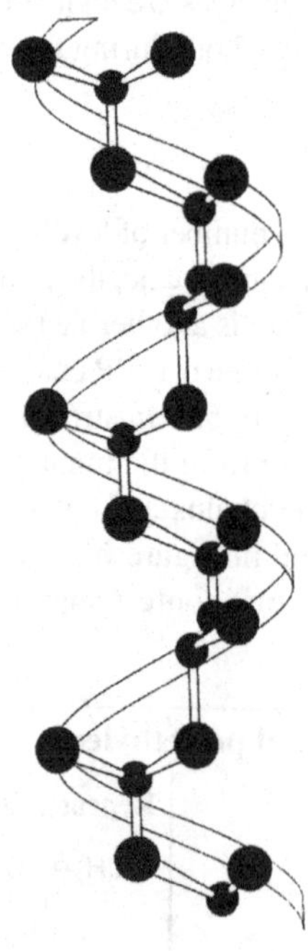

FIGURE 6.2

Three-dimensional helical configuration of isotactic polypropylene chains.

PP, the methyl groups alternate between the two sides of each chain (Figure 6.1). The chains in this case also adopt a helical formation. In atactic PP, the methyl groups are arranged randomly on the two sides of the chain.

The arrangements of the chains within a fibre may be complex and are strongly influenced by the fibre-processing conditions. At the simplest level, a fibre consists of crystalline regions, in which polymer chain segments are arranged into well-defined lattice structures; and amorphous regions, which consist of random interpenetrating polymer chains, as shown in Figure 6.3. An individual polymer chain may form part of more than one crystalline region, thus providing continuity along the length of the fibre. However, this concept is in practice too simplified, as several intermediate degrees of chain organisation are also likely to be present (Mather, 2005; Tomka, Johnson, & Karacan, 1992).

Three distinct crystalline forms of PP are known: α, β and γ. The α form is the most stable, and also the most important in the context of PP fibres. The α structure contains both left-handed and right-handed helical forms, and each individual helix lies for the most part next to helices of opposite handedness (Cheng, Janimak, & Rodriguez, 1995). A para-crystalline form is also known, in which there is some degree of chain order along the length of the fibre, but not at right angles to it.

Although many PP fibres are extruded with a circular cross-section, the filaments produced possess a waxy feel, which may not be desirable. However, it can be reduced, or even virtually eliminated, in filaments with some types of non-circular cross-sections, including triangular, multilobal and

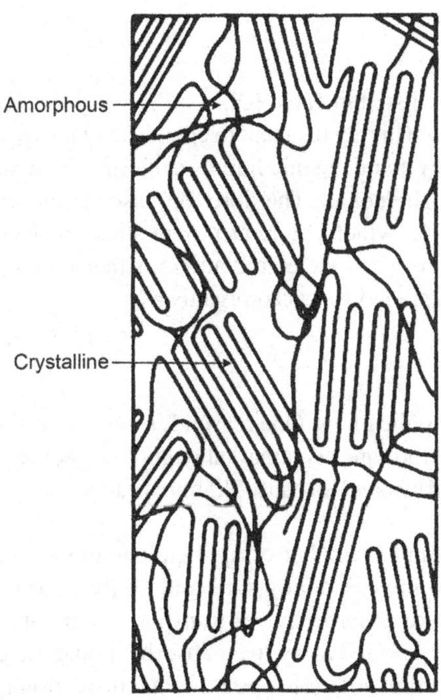

FIGURE 6.3

Amorphous and crystalline regions in fibres.

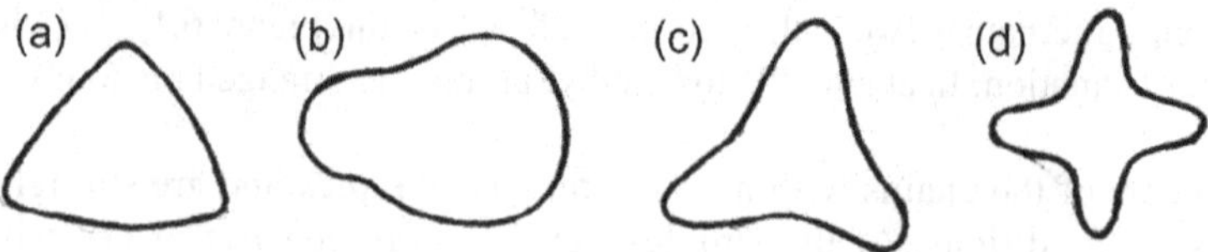

FIGURE 6.4

Examples of non-circular filament cross-sections in polypropylene fibres: (a) triangular; (b) bilobal; (c) trilobal; (d) cross-shaped.

cross-shaped, as illustrated in Figure 6.4. Hollow filaments are also commercially produced. Melt extrusion of such filaments requires more complex profiles for the spinneret holes.

Other properties influenced by the cross-sectional profile include bulkiness, effectiveness in providing heat insulation, resilience and extent of soiling. For example, fibres with more exotic cross-sectional shapes often pack more loosely than those with circular cross-sections. Fibre bulk is then increased, and insulation is improved.

6.2.7 FIBRE PROPERTIES

In common with other synthetic fibres, the properties of PP fibres are influenced by a number of factors:

- The grade of PP used in the formation of the fibre
- Fibre-processing conditions
- The additives present

Some properties of PP fibres are listed in Table 6.2.

PP fibres, for the most part, are highly resistant to chemical attack by acids, alkalis and most organic liquids. The fibres are swollen by some organic liquids at elevated temperatures and may even dissolve in them if the temperature is high enough. The fibres are also susceptible to the action of strong oxidants, such as hydrogen peroxide, which can reduce fibre strength and cause discolouration. In addition, PP fibres are weakened by ultra-violet radiation so that commercial PP fibres contain light stabilisers. The fibres are not degraded by micro-organisms.

6.2.8 APPLICATIONS

PP fibres have a wide range of applications. One reason for this is that, for a fibre that is so inexpensive to produce, its technological performance is remarkably good. Moreover, the production of PP fibres is comparatively straightforward. However, because PP fibres cannot be readily dyed, their application in clothing is much more limited.

PP fibres are, however, finding a number of applications in sportswear and activewear, such as walking socks, cycle shorts, swimwear, diving suits and lightweight wear for climbers. Because PP fibres possess almost no moisture absorbency yet transport water very easily, sweat generated from the body is not absorbed but transported away to an absorbent outer layer. PP garments therefore feel very comfortable next to the skin. In addition, good thermal insulation is retained, as the fibres do not become wet.

Table 6.2 Properties of polyolefin fibres

	Polypropylene	Gel-spun polyethylene
Melting point (°C)	160–165	135–150
Density (g/cm)	0.90	0.97
Moisture regain (%)	0.04	0
Tenacity (cN/tex)	30–80	260–370
Elongation at break (%)	15–35	3–4
Abrasion resistance	High	High

6.3 OTHER POLYOLEFIN FIBRES

Although several other polyolefins have been considered for textile applications, only polyethylene enjoys any real commercial interest in fibre form. The polymer is produced by the polymerisation of ethene (ethylene), principally by the Ziegler–Natta process described in Section 6.2.1. Most commercial polyethylene fibres are gel spun, although there are a few applications for melt spun fibres, such as artificial grass for sports fields. In gel spinning, polyethylene filaments are formed in a gel state, containing a solvent such as tetralin or paraffin oil (kerosene). The filaments are then drawn before the solvent is removed. This process gives rise to exceptionally strong fibres (Van Dingenen, 2001).

The chemical structure of polyethylene is shown in Table 6.1. The polymer chains tend to adopt a zig-zag formation, as shown schematically in Figure 6.5. As with PP chains, the polymer chains adopt various degrees of organisation – from crystalline through to amorphous. In contrast to PP chains however, there are no side-groups projecting from the main chain structure so that polyethylene chains can pack together more closely.

Some properties of gel-spun polyethylene fibres are shown in Table 6.2. It can be seen that their strength (as measured by fibre tenacity) far exceeds that of melt spun PP fibres. By contrast, they are also far less extensible than PP fibres.

Gel-spun polyethylene fibres are used extensively for protective clothing. Not only does the clothing provide protection, it is also very light to wear because of the low density of polyethylene. The fibres possess high cut and puncture resistance and therefore clothing made from them provides good protection against knife attacks. They are also used in cut-resistant gloves, chain-saw protective wear and fencing suits. Gel-spun polyethylene fibres are also widely used for ballistic protection, such as the vests and helmets used by police and military personnel. Figure 6.6 shows an example.

6.4 ACRYLIC FIBRES

Acrylic fibres have been produced on a commercial scale for over 50 years. At the height of its popularity, the fibre was used extensively for knitted clothing, such as jumpers and socks. An acrylic jumper is shown in Figure 6.7. Acrylic fibre was also used for blankets, and as a component in carpet facings. It resembles wool in terms of bulkiness and handle, but also possesses superior resistance to many chemical compounds and to micro-organisms.

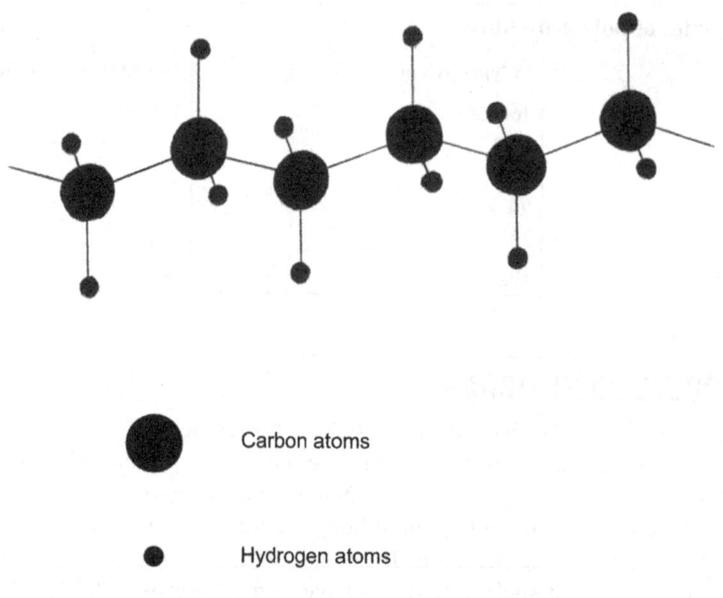

FIGURE 6.5

Schematic illustration of a polyethylene chain.

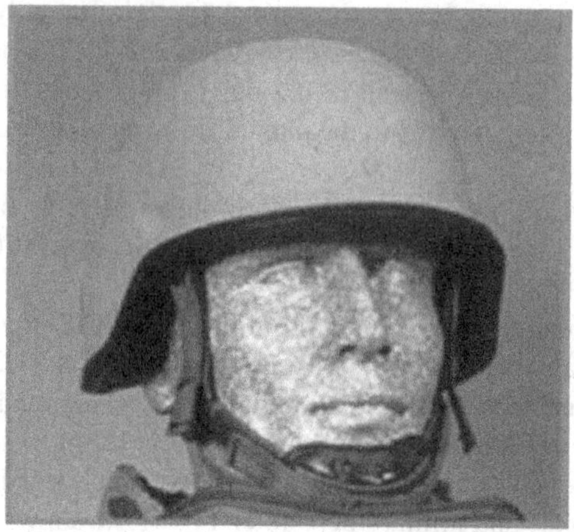

FIGURE 6.6

Military helmet made using gel-spun polyethylene fibres.

Global production of acrylic fibres peaked in the 1980s. Since then they have tended to become commercially less attractive in the clothing and household sectors. This decline is partly attributable to a trend by consumers towards textiles fabricated from natural fibres, and partly to commercial developments, such as microfibres, in polyamide and polyester fibres. Nevertheless, acrylic fibres are still important in a variety of industrial applications. In 2009, the total production of acrylic fibres in textiles was just under 2 million tonnes (Anonymous, 2010b).

FIGURE 6.7

Acrylic jumper.

Source: Courtesy of KSJ Knitwear Limited, published under the Creative Commons License.

6.4.1 PRODUCTION OF ACRYLIC FIBRES

Pure polyacrylonitrile is produced from the polymerisation of acrylonitrile (cyanoethene), the structure of which is shown in Table 6.3. In its pure form, polyacrylonitrile is chiefly used as a precursor for the manufacture of carbon fibres. For the production of suitable fibre for other applications, notably clothing and household products, other monomers are incorporated into the acrylic polymer chains. Common amongst these so-called comonomers are methyl acrylate, methyl methacrylate and vinyl acetate, whose structures are shown in Table 6.3. One of their functions is to yield a more open fibre structure to enable easier fibre processing. Furthermore, ionic comonomers may be added in order to provide further sites for the uptake of dyes to facilitate dyeing to deep shades. A halogen-containing comonomer, such as vinyl bromide (Table 6.3), may also be incorporated to confer good flame resistance.

Table 6.3 Some compounds used in the formation of acrylic and modacrylic polymers

Acrylonitrile	$CH_2{=}CH{-}CN$
Methyl acrylate	$CH_2{=}CH{-}CO{-}O{-}CH_3$
Methyl methacrylate	$CH_2{=}C(CH_3){-}CO{-}O{-}CH_3$
Vinyl acetate	$CH_2{=}CH{-}O{-}CO{-}CH_3$
Vinyl bromide	$CH_2{=}CH{-}Br$
Vinyl chloride	$CH_2{=}CH{-}Cl$
Vinylidene chloride	$CH_2{=}CCl_2$

A number of techniques are available for polymerising acrylonitrile and its comonomers. Solution polymerisation and aqueous dispersion polymerisation are common commercial approaches (Cox, 2005). Solution polymerisation is attractive, in that there is no need for the polymer to be isolated prior to fibre formation. Solvents commonly used are dimethyl formamide, dimethyl sulfoxide and aqueous sodium thiocyanate solution (45–55% by weight). However, there are some disadvantages with solution polymerisation. Polymers of high molar mass are difficult to achieve. Unreacted monomers that are not volatile are difficult to remove, and some monomers, such as vinyl acetate, are not readily incorporated because of their low reaction rates. The most widely used commercial process is aqueous dispersion polymerisation in which the polymer chains become insoluble in water as they grow, and the polymer is precipitated forming a slurry.

6.4.2 FIBRE MANUFACTURE

Unlike polyolefin fibres, acrylic fibres cannot be melt spun because the polymer tends to decompose below its melting point. Instead, the polymer is dissolved in a suitable solvent and fibres are formed by means of solution spinning (discussed in Chapter 10). The solvent used is, to some extent, determined by the nature of the comonomers in the polymer chains. Common solvents are aqueous sodium thiocyanate solution, dimethyl acetamide and dimethyl formamide. Most acrylic fibres are produced by wet spinning into a coagulation bath. Some are produced by dry spinning, nearly always as solutions in dimethyl formamide, whereby the solvent is removed from the fibres by evaporation.

Acrylic filaments produced by wet spinning are highly porous and are therefore passed over heated rollers in order to induce drawing. As the filaments dry, so the voids creating the porous structure collapse. The fibres are then washed to remove solvent, drawn, dried and relaxed in steam or hot water. Relaxation is a particularly important stage for stabilising the fibres, as the fibres can shrink by as much as 40% in length. Moreover, without relaxation, whilst the fibres may possess adequate strength, their extensibility and abrasion resistance are low, and they easily fibrillate. Fibrillation is the splitting of a fibre or tape into a network of smaller interconnected fibres (Denton & Daniels, 2002). Finally, crimp is imparted to the fibres, to promote bulkiness, and they are then most usually converted into staple fibres.

6.4.3 FIBRE STRUCTURE

Because the predominant monomer unit in acrylic polymers is formed from acrylonitrile, there is scope for hydrogen bonding between adjacent polymer chains, as shown in Figure 6.8. Another factor governing acrylic fibre structure is the repulsion between the nitrile, –CN, groups of adjacent chains – or even between groups in the same chain.

In a manner similar to PP chains (Section 6.2.6), isotactic, syndiotactic and atactic configurations of acrylic polymer chains are possible. However, in contrast to PP fibres, it is difficult to influence the type of structure that develops during the synthesis of the chains. The repulsion between nitrile groups also influences chain configuration, such that each chain tends to be twisted into an irregular helical structure.

Unlike many synthetic polymer fibres, such as polyamide, polyester and PP, separate crystalline and amorphous regions within acrylic fibres are not clearly identified. However, there is good evidence for the existence of two phases in acrylic fibres. Overall, the structure of acrylic fibres is particularly complex.

A wide variety of fibre cross-sectional shapes can be formed, as illustrated in Figure 6.9. In the case of acrylic fibres produced by wet spinning, the shape is strongly influenced by the shape of the spinneret holes and the conditions in the coagulation bath. Kidney-bean cross-sections are common (Figure 6.9(b)). With dry spinning, dog-bone shaped cross-sections (Figure 6.9(c)) can occur, even after spinning through spinneret holes of circular shape.

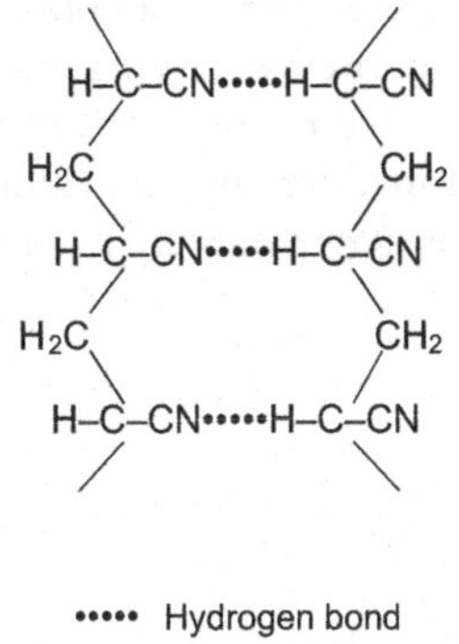

H–C–CN•••••H–C–CN

H_2C CH_2

H–C–CN•••••H–C–CN

H_2C CH_2

H–C–CN•••••H–C–CN

••••• Hydrogen bond

FIGURE 6.8

Hydrogen bonding between acrylic polymer chains.

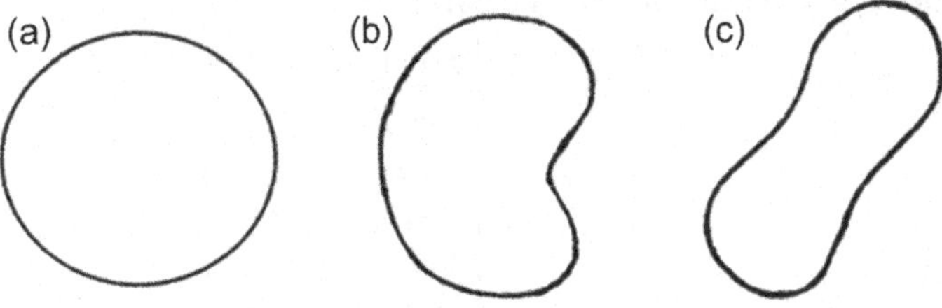

FIGURE 6.9

Acrylic fibre cross-sectional shapes: (a) circular; (b) kidney-bean; (c) dog-bone.

6.4.4 ACRYLIC FIBRE VARIANTS

Acrylic fibres are notable for the many variants that have been produced over the past decades. A range of variants is possible depending on the nature and proportion of comonomer in the polymer chains. Thus, some types of acrylic fibre contain basic comonomers that attract dyes containing acidic functional groups. These fibres are important for blends of acrylic fibres and wool. Other types of acrylic fibre contain a high proportion of halogen-containing comonomers. Many of these are modacrylic fibres and are discussed in Section 6.5.

Fibres with specifically designed properties have been produced by blending different types of acrylic polymer prior to fibre formation (Cox, 2005). Bicomponent fibres are also produced in which there are different acrylic polymers on each side of a fibre. On heating of the fibres, the component with the higher content of comonomer shrinks more than the other component and the fibres thus develop a helical crimp. Moreover, in wet spinning a wide range of fibre structures are possible depending on coagulation conditions.

A number of functional additives are often incorporated into the polymer before fibre spinning. Apart from additives such as titanium dioxide, which confer protection against ultra-violet radiation, and those that protect against micro-organisms, microencapsulated additives of c. 2 μm in size may also be present. These latter additives are phase change materials possessing melting points at around body temperature. They respond to body temperature either by changing phase from a solid to a liquid through absorption of heat, or vice versa (Cox, 2005). In this way, clothes incorporating these fibres provide comfort to the wearer. A good example is surgical clothing that is designed to prevent penetration of liquids and solid particles carrying micro-organisms, and for which comfort is improved markedly by coating the inside of the garment with microcapsules of phase change materials. Other examples are gloves for water sports, cycling and golf. Variants of acrylic fibres for improved aesthetic appeal, such as those conferring natural fur aesthetics, have also been marketed.

6.4.5 FIBRE PROPERTIES

In view of the many variants of acrylic fibres, it is difficult to provide representative values for their physical and mechanical properties. These properties also reflect the fact that acrylic fibres are most commonly produced as staple (length, 2.5–150 mm), which is converted to yarn by the methods used for natural fibres (see Chapter 9). Some properties of acrylic fibres are listed in Table 6.4.

Table 6.4 Properties of acrylic and modacrylic fibres

	Acrylic	Modacrylic
Density (g/cm^3)	1.14–1.19	1.28–1.37
Moisture regain (%)	1.5–2.5	1.5–3.5
Tenacity (cN/tex)*	10–35	10–25
Elongation at break (%)	25–60	25–45
Abrasion resistance	Moderate	Moderate

**For wet fibres, tenacity is reduced by 15–20%.*

Acrylic fibres possess good resistance to most types of chemical attack. They withstand acids and weak alkalis, although strong alkalis will rapidly degrade them at elevated temperatures. They are also resistant to oxidants. Most organic liquids are incapable of swelling acrylic fibres, the chief exceptions being dimethyl formamide, dimethyl acetamide and dimethyl sulfoxide. Some of these liquids are used as solvents in the formation of acrylic polymer and for spinning acrylic fibres.

Acrylic fibres readily withstand temperatures of 150 °C, although they may become slightly discoloured. The fibres also possess good resistance to ultra-violet radiation and to micro-organisms.

6.4.6 APPLICATIONS

As noted in Section 6.4, acrylic fibres have lost something of their prominence in clothing and household furnishings. Nevertheless, they are still important in a number of more technical applications. They are used extensively in outdoor fabrics, where good stability against light and weather and resistance to mould are essential. Such fabrics include awnings, retractable car tops and boat covers. A number of specialised acrylic fibres have been developed for the sportswear market over recent years.

6.5 MODACRYLIC FIBRES

Modacrylic fibres contain at least 35%, but not more than 85% by mass, of acrylonitrile. Although a wide variety of modacrylic fibres are possible under this classification, they are in practice notable for their substantial flame-retardant properties. These properties are attributable to the nature of the comonomers in the modacrylic polymer chains. Examples of these comonomers are vinyl bromide, vinyl chloride and vinylidene chloride (see Table 6.3). Some of the properties of modacrylic fibres are given in Table 6.4.

Modacrylic fibres can be processed by wet spinning in a manner broadly similar to that used for producing acrylic fibres. Like acrylic fibres, they require heat stabilisation after spinning and are usually cut into staple. The number of modacrylic fibres in the market today is quite small. They find use as specialist fibres for applications where flame-retardant properties are essential. In blends with other fibres, they are used in carpets. They also find extensive application in aircraft seating, protective clothing, children's nightwear, dolls' hair and other products where non-flammability is critical.

6.6 ELASTOMERIC FIBRES

Clothing that stretches with the human body, supports it and conforms to it, almost certainly contains elastomeric fibres. Elastomers are polymers possessing high extensibility, with rapid and substantially complete recovery (Denton & Daniels, 2002).

At one time, fibres made from rubber were the predominant type of elastomeric fibre, but these have largely been superseded and elastomers are produced from several different polymeric types. By far the most important in commercial terms are elastanes, which contain 85% or more by mass of segmented polyurethane (Denton & Daniels, 2002). Polyurethanes are polymers containing urethane groups, –NH–CO–O–, along the length of each constituent chain. These fibres are known for possessing up to 99% elastic recovery. They also possess high extensibility (up to 500%), the actual value of which is determined by the composition of the elastane polymeric chains and the method used to produce them. In Europe, these fibres are often known as Lycra fibres (though other brands such as Roica and Dorlastan also exist), and in the United States they are known as Spandex. Other types of elastomeric fibres include:

- Elastodiene fibres
- Elastomultiester fibres
- A few types of polyolefin fibres

These are commercially less important than elastane fibres, however, and are not discussed further. Information about these fibres can be obtained in Mather and Wardman (2011).

6.6.1 ELASTANE FIBRES

The polymeric chains comprising elastane fibres each consist of alternate blocks of 'hard' and 'soft' segments, as shown schematically in Figure 6.10. The soft segments comprise either polyethers or polyesters. Polyether elastane fibres are more resistant to detergents and mildew, whereas polyester elastane fibres are more resistant to oxidation (e.g. by ultra-violet radiation) and the absorption of oils, such as suntan oils.

In the relaxed state, the soft segment structures are loosely arranged, but when the fibre is extended, the segments unwind and become more aligned with the fibre axis (Figure 6.10). On release of tension,

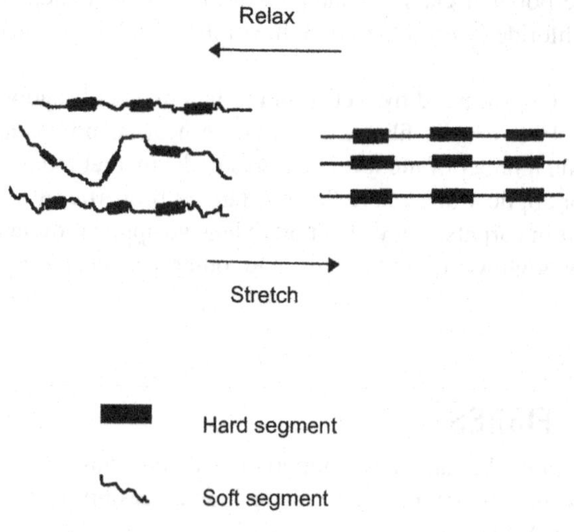

FIGURE 6.10

Hard and soft segments in elastane polymeric chains.

the segments revert to their original state. The hard segments, by contrast, contain rigid aromatic structures throughout. Each hard segment can interact with hard segments in adjacent polymeric chains, chiefly through hydrogen bonding. A strong network is, therefore, provided throughout the elastomeric fibre and slippage between adjacent chains is inhibited. Hard and soft segments in each chain are joined to one another through urethane linkages.

The balance between the proportion of soft and hard segments is important for determining the properties of the fibre. If the interaction between polymeric chains is too great, then the fibre will not achieve a sufficient degree of stretch. If the interaction is too small, the chains will slip relative to one another and the fibre will not recover properly after being stretched. Judicious control of the structure and length of each segment and of the ratio of hard to soft segments can however provide good control of the mechanical properties obtained.

6.6.2 FIBRE MANUFACTURE

Elastane fibres can be spun in several ways:

- Solution dry spinning
- Solution wet spinning
- Melt spinning

All of these methods are described in Chapter 10. The commonest method is dry spinning. The solvent used is either dimethyl formamide or dimethyl acetamide. The solid filaments formed when the solvent evaporates are sticky and come together on touching. This property has the advantage that yarn of required linear density can often be readily produced. Once the required linear density is attained, yarns are treated with a finishing agent such as magnesium stearate or polydimethyl siloxane, to prevent further yarn adhesion.

6.6.3 FIBRE STRUCTURE

Elastane fibres possess either round or square cross-sections (Hatch, 1993, p. 247). Linear density may range from 20 to 6000 dtex, depending on the application for which the fibres are required. When adjacent filaments come together during the spinning process, they are not joined along all of their length.

Elastane filaments can be used in combination with yarns of other types of fibre. In some cases, elastane filaments form the inner core of a covered yarn. In these covered yarns, the elastane cores are each individually covered with yarn of another fibre type, as illustrated in Figure 6.11. The elastane is then protected from agents that may degrade it such as oils and ultra-violet radiation. The covering yarn also provides greater control of elastane filament extension.

6.6.4 FIBRE PROPERTIES

The properties of elastane fibres are given in Table 6.5. They are quite weak in comparison to other textile fibres, but they possess good resistance to chemical attack. They can, however, be damaged by extended exposure to chlorine. Elastane fibres based on polyether are more resistant to chlorine and are

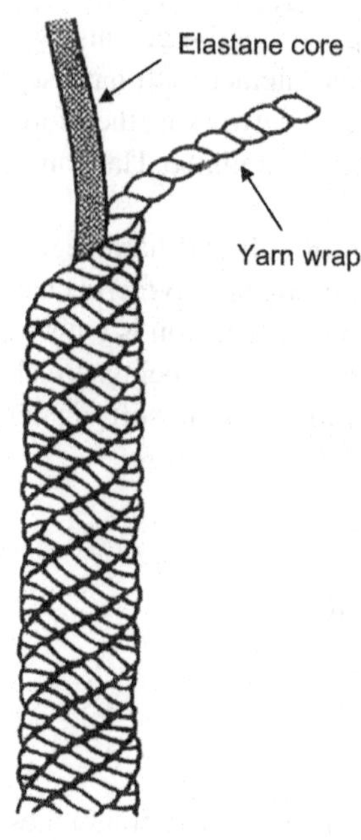

FIGURE 6.11

Schematic illustration of a covered elastane yarn.

Table 6.5 Properties of elastane fibres	
Density (g/cm^3)	1.21
Moisture regain (%)	1.3
Tenacity (cN/tex)	5–14
Elongation at break (%)	400–650
Abrasion resistance	Medium

therefore preferred for swimwear. Elastane fibres are resistant to ultra-violet radiation and to micro-organisms.

6.6.5 APPLICATIONS

Elastane fibre is used in lightweight support hosiery (20–250 dtex), men's hosiery (150–600 dtex) and swimwear and foundation garments (80–2500 dtex). Usually, only a small proportion of elastane is present in a fabric (2–5%), to provide the elasticity required. In foundation garments, however, as much as 45% elastane may be present. Covered elastane yarns are used extensively in cotton and woollen garments, often at a level of 2–5%.

6.7 CASE STUDY: WHY ARE THERE SO MANY END-USES FOR POLYPROPYLENE (PP) FIBRES, BUT SO FEW IN APPAREL?

As discussed in Section 6.2, PP fibre is low cost, yet possesses many good mechanical properties. These properties include:

- High strength
- Good toughness
- Good resistance to chemical and microbial attack

In addition, propene is still perceived as a secure raw material for the production of PP. PP fibre is formed by melt spinning, the least complex spinning process for synthetic fibres. The fibre is also a good substitute for many other materials and can be readily recycled. It is considered to have little environmental impact because it is chemically inert.

The applications of PP fibre are diverse. The fibre is used for carpet backings and in carpet pile yarns owing to the texturing processes that are available for BCF yarns. The pile can be safely subjected to many kinds of spillage. PP fibre is also applied in household furnishings, such as upholstery cover fabrics. It is used in bed linen, bedspreads and net curtains. Mattress ticking made from PP fibres can help to alleviate the problems suffered by those with allergies such as asthma and eczema.

PP fibres are widely used in geotextile fabrics for reinforcing roads and stabilising embankments. The fibres are also used in ropes and netting. They have medical and healthcare applications: surgical gowns, plasters and bandages, sutures (for wound closures) and vascular grafts (for repairing blood vessels) and artificial lungs.

In view of such a wide range of household, technical and medical applications, why does the fibre have such limited appeal to the fashion industry? A major disadvantage of PP fibres is the difficulty of dyeing them and printing on them. The inherent inertness that confers resistance to so many chemical compounds also confers resistance to dyes. Instead PP fibres are coloured by tiny pigment particles mixed with the polymer prior to melt spinning. In comparison with other types of fibres, it is therefore more difficult to respond quickly to the needs of the apparel industry and to current trends in fashionable colours. A lot of ingenious approaches have been devised to modify the fibre itself and the process for producing it in order that it becomes receptive to dyes. However, the advantages of low cost and ease of production are generally lost by such efforts.

PP textiles are, however, used for some types of garments. PP garments are worn next to the skin because they absorb little water. Consequently, perspiration passes from the skin through the garment to an absorbent outside layer. The wearer is kept warm and dry. PP fibres are thus used in some types of underwear and socks. They are used in sports and activewear, such as lightweight outerwear for climbers, cycle shorts, swimwear and garments for divers. An example of a competition swimsuit made from PP fibres is shown in Figure 6.12.

6.8 FUTURE TRENDS

It will be evident from the preceding sections in this chapter that polyolefin, acrylic and elastomeric fibres have quite different structures and properties. It follows that most new processes for developing these fibres for use in clothing are likely to be specific to the individual types of fibres. These processes will be

FIGURE 6.12

Polypropylene competition swimsuit.

discussed in turn as follows, but it is worth noting that some emerging fibre technologies could shape the commercial futures of all of them. A particularly exciting development is the growing application of gas plasma treatments to textiles on a commercial scale (Mather, 2009; Neville, Mather, & Wilson, 2007).

Gas plasmas are energetic sources of radiation and chemical species that interact in a complex fashion both with one another and with the substrate being treated. The plasma atmosphere consists of a mixture of ions, electrons, free radicals and ultra-violet radiation. Fibre surfaces can be modified both physically and chemically, yet the fibre bulk remains unaffected.

This technology has implications for assisting dyeing, oil and water absorption, abrasion resistance, fire-proofing and a number of other processes. Its application provides a 'clean' technology in an environmental sense, in contrast to the more traditional techniques for fibre surface modification (Mather, 2009). Indeed, it is likely to provide more modifications than are currently achievable. Plasma technology could be exploited by designers in the future to influence aesthetic properties and colour tone (Mather et al., 2003).

6.8.1 POLYOLEFIN FIBRES

The future of polyolefin fibres in clothing will be largely restricted to specialist applications such as protective clothing (gel-spun polyethylene) and sports clothing (PP), unless a commercially successful dyeing process can be found. PP fibres, modified so that they can absorb and retain dyes, have found only limited use. These modifications often lead to reduced fibre strength and lower thermal stability, and can also cause difficulties in fibre production (Shamey, 2009). The key will, therefore, be to dye unmodified

PP fibres. Although no commercial process yet exists for achieving this goal, the chemical company, BASF, has announced the production of a PP fibre, branded MOOO> (Anonymous, 2008), which contains anchor points for the retention of disperse dyes, normally manufactured for dyeing polyester fibres.

One report has described the successful application of some disperse dyes to PP fibres that have been melt spun but not yet drawn (stretched) to strengthen them (Ahmed, 2006). This approach may perhaps offer a commercial opportunity for PP fibre. Also, gas plasma technology can improve the printability of PP fabrics (Tsai, Wadsworth, & Roth, 1997), as the fabric surfaces are rendered more receptive to dye molecules.

6.8.2 ACRYLIC FIBRES

The past success of acrylic fibres can be attributed in great measure to their wool-like properties, most notably bulkiness and fabric handle. However, in commercial terms, the fibre has been in decline in both clothing and household furnishings sectors. Some observers see a more assured future for the fibre as a technical textile. It is a common precursor for the production of carbon fibres (Frushour & Knorr, 2007; Mather & Wardman, 2011, pp. 196–199), is used to replace asbestos fibre in wall cladding, to reinforce cement, and in a number of other products (Frushour & Knorr, 2007). Continuous acrylic fibre is used for outdoor applications such as tents and awnings, because of the fibre's superior resistance to sunlight.

A problem with acrylic fibres is their inability to withstand hot, wet conditions. Thus, hot water is not desirable for laundering acrylic garments. Acrylic fibre technology would greatly benefit from improved resistance to these conditions. However, as noted in Section 6.4.4, acrylic fibres come in a great number of variants. These variations may result from the constitution of the acrylic polymer, the bicomponent nature of the fibre, the nature of the blended fibre, or the coagulation conditions during fibre processing. Judicious handling of such a wide range of variants may provide a key to overcoming the susceptibility of acrylic fibres to hot, wet environments without prejudicing other desirable fibre properties and hence provide new opportunities in the clothing sector.

6.8.3 ELASTOMERIC FIBRES

The development of elastomeric fibres, and in particular elastane fibres, has been highly beneficial for garments requiring good stretch and elasticity. Adjusting the ratio of the proportions of hard and soft segments in the elastane polymer chain can in turn closely govern fibre stretch and elasticity. This capability will undoubtedly be valuable in the design and development of stretch garments in the future. Moreover, other fibre properties might be improved by incorporating suitable additives prior to the spinning process. Different types of covered elastane yarns may also play a part in the future development of these fibres.

6.9 SUMMARY

- The most important commercial polyolefin fibre is PP, with an isotactic polymer chain structure.
- Commercial PP fibres exist in a variety of forms:
  - Multifilament yarns
  - Monofilament yarns
  - Staple fibres

  - Non-wovens
  - Tapes
- PP fibres are produced with a variety of cross-sectional profiles, including circular, triangular, multilobal, cross-shaped and hollow.
- PP fibre is used in a wide variety of technical applications but its use in apparel is limited because it cannot be readily dyed.
- Gel-spun polyethylene fibres possess high strength and are used extensively for protective clothing.
- Acrylic fibres are produced in many variants by solution spinning.
- Acrylic fibres are today less prominent in apparel and furnishings, although they are used extensively in outdoor fabrics and sportswear.
- Modacrylic fibres are notable for their substantial flame-retardant properties.
- The most important commercial elastomeric fibre is elastane (Spandex).
- The elastic properties of elastane are derived from soft and hard segments in their constituent polymer chains.
- Elastane fibres are used extensively in hosiery, swimwear and foundation garments.

6.10 PROJECT IDEAS

Analyse the influence of fibre cross-sectional profile in PP and acrylic fibres on their properties and end-uses.

Consider how different methods of blending elastane fibres with other types of fibres will influence the aesthetics of stretch apparel.

Survey the modifications made to PP fibres to render them dyeable.

Consider how gas plasma treatments of fabrics could be exploited by designers to influence aesthetic properties and colour tone.

6.11 REVISION QUESTIONS

1. Describe the configurations that the polymer chains in PP may adopt, and explain why only one of these configurations is adopted in commercial PP fibres.
2. Describe the different processes used for producing PP fibres.
3. Describe the various applications of PP fibre, and explain why its use in apparel is restricted.
4. Summarise the properties of gel-spun polyethylene, and outline its applications.
5. Discuss the variations in the structure of the polymer chains in acrylic fibres, and explain why so many variants are produced commercially. (Include modacrylic fibres in the answer.)
6. Describe the types of garment that include elastomeric fibres, and explain why these fibres are present in them.
7. Explain why elastane fibres possess such high stretch and elasticity.
8. Describe the variety of cross-sectional shapes that are found in PP, acrylic and elastane fibres, and explain how these different cross-sections are formed in the production of these fibres.
9. Discuss how PP, acrylic and elastomeric fibres may be developed in the future for use in clothing.

6.12 SOURCES OF FURTHER INFORMATION AND ADVICE

Detailed accounts of polyolefin and acrylic fibres are given in *Synthetic fibres: Polyamide, polyester, acrylic, polyolefin*, edited by J. E. McIntyre, Woodhead Publishing, Cambridge, 2005.

A book devoted to polyolefin fibres has recently been published: *Polyolefin fibres industrial and medical applications*, edited by S. C. O. Ugbolue, Woodhead Publishing, Cambridge, 2009.

Gel-spun polyethylene fibres are discussed in *High-performance fibres*, edited by J. W. S. Hearle, Woodhead Publishing, Cambridge, 2001.

There are chapters on the structures of polyolefin, acrylic and elastomeric fibres in *Handbook of textile fibre structure.* volume 1: *Fundamentals and manufactured polymer fibres*, edited by S. J. Eichhorn, J. W. S. Hearle, M. Jaffe and T. Kikutani, Woodhead Publishing, Cambridge, 2009.

Information on the tensile properties of polyolefin and acrylic fibres can be found in *Handbook of tensile properties of textile and technical fibres*, edited by A. R. Bunsell, Woodhead Publishing, Cambridge, 2009.

Finally, there are sections on the chemistry of polyolefin, acrylic and elastomeric fibres in Chapter 5 of *The chemistry of textile fibres*, by R. R. Mather and R. H. Wardman, The Royal Society of Chemistry, 2011.

REFERENCES

Ahmed, M. (1982). *Polypropylene fibers – science and technology*. Amsterdam, Oxford and New York: Elsevier.

Ahmed, S. I. (2006). *Coloration of polypropylene: prospects and challenges*. PhD thesis. Edinburgh: Heriot-Watt University.

Anonymous. (2008). Award for dyeable polypropylene. *International Dyer* June edition, 6.

Anonymous. (2010a). Worldwide olefin fibers production and producing capacity by product type: 2006–2011. *Fiber Organon* June 2010 edition, 110–111.

Anonymous. (2010b). Worldwide synthetic fiber production and producing capacity by fiber except olefin: 2006–2011. *Fiber Organon* June 2010 edition, 104–105.

Cheng, S. Z. D., Janimak, J. J., & Rodriguez, J. (1995). Crystalline structures of polypropylene homo- and copolymers. In J. Karger-Kocsis (Ed.), *Polypropylene structure blends and composites.* Volume 1: *Structure and morphology* (pp. 31–35). London: Chapman and Hall.

Cox, R. (2005). Acrylic fibres. In J. E. McIntyre (Ed.), *Synthetic fibres: Nylon, polyester, acrylic, polyolefin* (pp. 167–234). Cambridge: Woodhead Publishing Ltd.

Crangle, A. (2009). Types of polyolefin fibres. In S. C. O. Ugbolue (Ed.), *Polyolefin fibres: Industrial and medical applications* (pp. 3–34). Cambridge: Woodhead Publishing Ltd.

Denton, M. J., & Daniels, P. N. (2002). *Textile terms and definitions* (11th ed.). Manchester: The Textile Institute.

Frushour, B. G., & Knorr, R. S. (2007). Acrylic fibers. In M. Lewin (Ed.), *Handbook of fiber chemistry* (3rd ed.) (pp. 811–973). Boca Raton, London and New York: Taylor & Francis.

Goswami, B. G. (1997). Spunbonding and melt-blowing processes. In V. B. Gupta, & V. K. Kothari (Eds.), *Manufactured fibre technology* (pp. 560–594). London: Chapman and Hall.

Hatch, K. L. (1993). *Textile science*. Minneapolis/St. Paul: West Publishing Company.

Mather, R. R. (2005). Polyolefin fibres. In J. E. McIntyre (Ed.), *Synthetic fibres: Nylon, polyester, acrylic, polyolefin* (pp. 235–292). Cambridge: Woodhead Publishing Ltd.

Mather, R. R. (2009). Surface modification of textiles by plasma treatments. In Q. Wei (Ed.), *Surface modification of textiles* (pp. 296–317). Cambridge: Woodhead Publishing Ltd.

Mather, R. R., Robson, D., Fotheringham, A. F., Neville, A., Wei, Q., & Warren, J. M. (2003). Effects of gas plasma treatments of textiles on their technological and aesthetic properties. In *Proceedings of the international textile design and engineering conference (INTEDEC)*. Edinburgh: Heriot-Watt University, Section 7B.

Mather, R. R., & Wardman, R. H. (2011). *The chemistry of textile fibres*. Cambridge: The Royal Society of Chemistry. pp. 196–199.

Neville, A., Mather, R. R., & Wilson, J. I. B. (2007). Characterisation of plasma-treated textiles. In R. Shishoo (Ed.), *Plasma technologies for textiles* (pp. 300–315). Cambridge: Woodhead Publishing Ltd.

Schmenk, B., Miez-Meyer, R., Steffens, M., Wulfhorst, B., & Gleixner, G. (2000). Polypropylene fiber table. *Chemical Fibers International*, *50*, 233–253.

Shamey, R. (2009). Improving the colouration/dyeability of polyolefin fibres. In S. C. O. Ugbolue (Ed.), *Polyolefin fibres: Industrial and medical applications* (pp. 363–397). Cambridge: Woodhead Publishing Ltd.

Tomka, J. G., Johnson, D. J., & Karacan, I. (1992). Molecular and microstructural modelling of fibres. In S. K. Mukhopadhyay (Ed.), *Advances in fibre science* (pp. 181–206). Manchester: The Textile Institute.

Tsai, P. P., Wadsworth, L. C., & Roth, J. R. (1997). Surface modification of fabrics using a one-atmosphere glow discharge plasma to improve fabric wettability. *Textile Research Journal*, *67*, 359–369.

Van Dingenen, J. L. J. (2001). Gel-spun high-performance fibres. In J. W. S. Hearle (Ed.), *High-performance fibres* (pp. 62–92). Cambridge: Woodhead Publishing Ltd.

CHAPTER

SYNTHETIC TEXTILE FIBRES: NON-POLYMER FIBRES

7

M. Shioya, T. Kikutani
Tokyo Institute of Technology, Tokyo, Japan

LEARNING OBJECTIVES

At the end of this chapter, you should be able to:

- Understand the basic properties of inorganic man-made fibres
- Describe the methods of manufacture of the fibres
- Outline the main uses of the fibres
- Understand the net environmental impact of the fibres

7.1 INTRODUCTION

Textile fibres can be classified into natural fibres and man-made fibres, or alternatively into organic and inorganic fibres. Most natural fibres such as cotton, silk and wool consist of polymers of organic compounds. One exception is asbestos, which can be defined as a natural inorganic fibre. Most man-made fibres, including natural-polymer fibres such as rayon or acetate, and synthetic-polymer fibres such as polyesters and polyamides, are also composed of polymers of organic compounds.

Carbon fibres, glass fibres, ceramic fibres and metallic fibres can be classified as non-polymer, man-made, inorganic fibres, some of which are actually produced from organic polymer fibres, i.e. carbon fibres from polyacrylonitrile fibres and silicon carbide fibres from polycarbosilane. Some fibres included in this chapter, such as carbon fibres and glass fibres, can be regarded as inorganic polymer fibres.

7.2 CARBON FIBRES

Carbon fibres are composed almost entirely of carbon atoms. They are black in colour and have diameters in the range of several micrometres. Carbon fibres possess extremely high tensile modulus and strength and low density. They were originally developed as a reinforcement phase in composite materials and typically used in carbon fibre-reinforced plastics (CFRP), produced by combining carbon fibres and a polymer matrix (bulk) phase.

Textiles and Fashion. http://dx.doi.org/10.1016/B978-1-84569-931-4.00007-6

7.2.1 MANUFACTURE

Carbon-based materials possess high thermal and chemical resistances. Graphite, for example, does not change its state until heated to about 3650 °C at atmospheric pressure. This makes it difficult to produce carbon fibres directly from the bulk carbon materials by melt or wet spinning.

Chemical vapour deposition (CVD) is one method to produce carbon fibres from hydrocarbons such as ethylene, acetylene, benzene and toluene with the aid of catalyst particles such as Fe, Co and Ni. In a reactor heated at 500–2200 °C, hydrocarbons are decomposed at the surface of the catalyst particle and the resulting carbon atoms dissolve into the particles and then deposit in the form of a fibre. The carbon fibres produced with this method, which are called vapour-grown carbon fibres (VGCF), have diameters in the range from 15 to 150 nm.

Continuous carbon fibres with diameters in the range of several micrometers are usually produced from organic compounds by first forming fibres and then heat-treating them. A high-temperature heat treatment in an inert (unreactive) atmosphere up to about 1500 °C is called carbonization, with which carbon atoms form carbon layer stacks with relatively large disorder (turbostratic structure). An additional high-temperature heat treatment in an inert atmosphere above 1500 °C is called graphitization, in which graphite is formed.

It is difficult for any material to possess both spinnability for melt or wet spinning and the ability to retain shape during carbonization. A heat treatment, called stabilization, may be applied to the fibre before carbonization to develop chemical structures that do not suffer combustion, extensive volatilization or melting during carbonization. Stabilization is usually carried out in air at relatively low temperature and is usually a lengthy process.

Not all organic compounds are suitable precursors (starting materials) for carbon fibre. An appropriate precursor needs to meet several requirements such as high yield of conversion to carbon fibre (i.e. high process efficiency), the ability to keep fibre shape without causing volatilization or fusion during carbonization and the ability to develop highly orientated carbon layers if high-performance carbon fibre is to be produced. Organic compounds used as carbon fibre precursors include polyacrylonitrile (PAN), pitch, cellulose and phenolic resin.

Carbon fibre was originally used as the filament of incandescent lamps through a process of thermal decomposition of bamboo and cotton by Thomas Edison and Joseph Swan. Afterword, rayon-based carbon fibres have been developed for the aerospace industry. Today, the PAN-based carbon fibres are seeing the largest demand.

PAN-based carbon fibres are produced through the following steps, with the changes in chemical structure during the process illustrated in Figure 7.1.

1. Wet spinning
2. Stretching at an elevated temperature
3. Stabilization at 200–300 °C
4. Carbonization at 1000–1500 °C
5. Optional graphitization at 2000–3000 °C

Wet spinning is used, as the decomposition of PAN on heating renders melt spinning difficult. Stretching of PAN fibres before and during stabilization is important to develop the preferred orientation of carbon layers in the final carbon fibre structure.

Pitch-based high-performance carbon fibres are produced in the following ways.

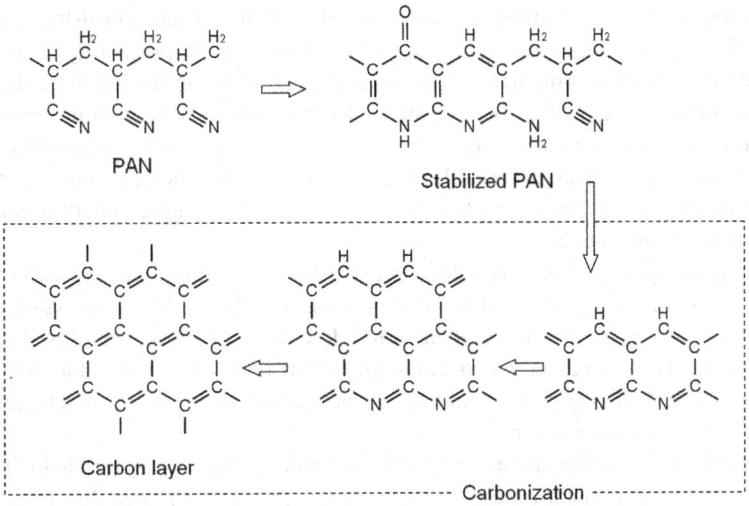

FIGURE 7.1

Changes in chemical structure during conversion of PAN fibre into carbon fibre.

1. Polymerization of pitch to mesophase pitch
2. Melt spinning
3. Stabilization at 200–350 °C
4. Carbonization at 1000–1500 °C
5. Graphitization at 2000–3000 °C

Pitches are a mixture of organic compounds composed of condensed benzene rings and alkyl chains. The composition of a pitch varies depending on its source, for example, petroleum asphalt or coal tar. Upon heating to a temperature above 300 °C, pitch molecules polymerize, forming a liquid crystal called mesophase pitch.

In composite materials, interfacial bonding between the reinforcement (fibre) phase and the matrix is critical to the transference of an external load from the matrix to the reinforcement phase. Without sufficient interfacial bonding, the excellent fibre properties such as high modulus and high strength cannot be fully utilized. As-received carbon fibres show weak bonding to polymers, and this is improved by surface treatment.

The processes to fabricate carbon fibres into CFRP are diverse depending on a number of factors, including:

- Whether the matrix is a thermoplastic or thermosetting resin.
- Whether the carbon fibre is continuous or staple.
- How the fibre is arranged.

There are two types of production processes, one of which uses a moulding material. The moulding material is prepared by incorporating fibres within the resin.

A prepreg is a moulding material composed of a thin sheet of unidirectional or woven fibres impregnated with a thermoplastic or thermosetting resin. A unidirectional fibre/thermosetting resin prepreg is produced by passing an array of continuous fibres through a resin bath and a heat-treatment chamber to cure the resin to B-stage. Curing refers to the hardening of the resin by promoting chain extension and cross-linkage. B-stage is an intermediate stage of curing before a continuous three-dimensional network structure is formed. The resin at B-stage shows light stickiness at room temperature and fluidity on reheating. Prepregs using thermosetting resins are usually stored under refrigeration conditions in order to suppress curing until ready to use. To produce a finished product, prepregs are cut and laid up onto a mould surface by hand or machine, and the resin is cured by heating under pressure for consolidation in an oven called an autoclave.

Sheet moulding compound (SMC) is a thin sheet of short fibres impregnated with a thermosetting resin and used as the moulding material for compression moulding. In compression moulding, the required amount of SMC is placed in a preheated mould cavity, which is then closed and the resin is cured under pressure. Thermoplastic pellets containing short fibres are used as the moulding material with injection moulding. With injection moulding, the pellets are melted in a heated barrel, injected into the mould and cooled for solidification.

Production methods for composite materials without using moulding materials include pultrusion, filament winding and resin-transfer moulding. In pultrusion, fibres are pulled through a thermosetting resin bath and a heated die with the desired cross-sectional shape and the resin is cured. Composite materials in the form of round and square tubing, channel, I-beam and rods are produced with this method.

With filament winding, fibres are pulled through a thermosetting resin bath and wound onto a mandrel. On the mandrel, fibres are laid down in a geometric pattern designed to most effectively utilize the tensile properties of the fibres. The mandrel is removed after the resin is cured or it becomes a part of the structure. Cylindrical structures and pressure vessels are produced with this method. Prepreg tapes can also be processed with filament winding.

In resin-transfer moulding, dry fibres are formed into a three-dimensional structure called perform, which is placed into a mould cavity, filled with a low-viscosity liquid resin and then cured.

In summary, CFRP are produced through fibre spinning, stabilization, carbonization, graphitization, impregnation with resin and arranging of the fibres into the desired pattern followed by curing. This process is a very effective way to make use of the potential performance of the material. Spinning and drawing not only effectively change the shape of the material but simultaneously facilitate the development of a highly orientated structure.

Fibrous form is of particular importance in the stabilization of PAN fibres, where oxygen is necessary for producing carbon fibres with a high yield. Fibrous form helps oxygen to diffuse homogeneously to the inner portion of the material. The flexibility of fibres and adjustability of fibre direction in the composite structure also provides a way to process a stiff and strong material into the desired shape, and helps to control the anisotropy of the properties. Specially designed fibre orientation angles allow a layered composite material to deform against stress in a very different way from isotropic materials. Further, the fibre/matrix interface can be controlled to promote toughening mechanisms such as crack arrest and deflection, interfacial debonding and fibre pull-out.

7.2.2 STRUCTURE

Carbon atoms can form molecules with various configurations such as carbyne, fullerene, nanotube, graphene, graphite, diamond and amorphous carbon.

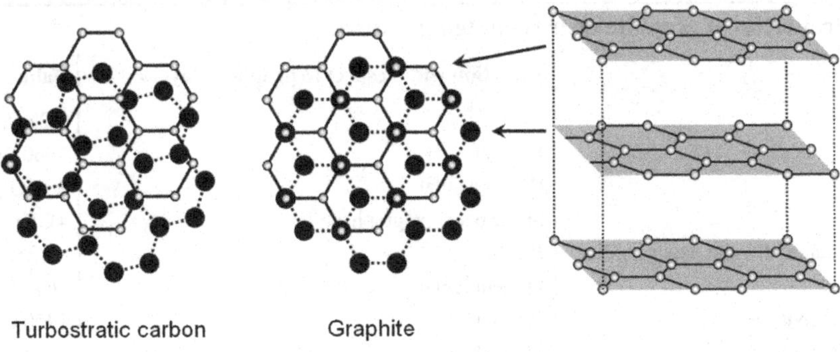

FIGURE 7.2

Graphite and turbostratic carbon.

When graphite is viewed in cross-section, graphene layers are shifted alternately in an A-B-A-B type stacking sequence as shown in Figure 7.2. With stacking of this type, a three-dimensional lattice of a hexagonal type is formed. Carbon layers developed in some types of carbon fibres resemble a graphene layer, but unlike graphene contain many defects. These imperfect carbon layers tend to stick to each other, but the stacking of the layers is largely disordered, as shown in Figure 7.2. This type of carbon layer stack is called turbostratic carbon. The interlayer spacing of turbostratic carbon is larger than that of graphite and can be in the range of 0.35–0.38 nm. Conversely, amorphous carbon in carbon fibres completely lacks any long-range order of structure.

The properties of graphite and turbostratic carbon are extremely anisotropic (different in different directions). The high modulus and high strength properties of carbon fibres rely on the orientation of the carbon layer stacks, which are nearly parallel to the fibre axis. In addition to the orientation relative to the fibre axis, carbon layers show various alignments in the cross-section perpendicular to the fibre axis such as circumferential, radial and random alignment.

7.2.3 PROPERTIES

The physical properties of graphite are shown in Table 7.1 (Blakslee, Proctor, Seldin, Spence, & Weng, 1970; Gray, 1972; Holland, Klein, & Straub, 1966; Matsubara, Sugihara, & Tsuzuku, 1990; Morgan, 1972). The properties of carbon fibres depend on such factors as fractions, sizes and orientation of carbon layer stacks, amorphous regions and microvoids. In turn, these factors depend on such elements as the type of precursor, heat-treatment temperature and the stretching of fibres during heat treatment. As shown in Figure 7.3 (Johnson, 1969), the tensile modulus of PAN-based carbon fibres tends to increase with increasing heat-treatment temperature. The tensile strength of PAN-based carbon fibres also increases with increasing heat-treatment temperature up to about 1300–1500 °C, and then decreases at higher temperatures.

In Figures 7.4–7.6, the tensile strength, tensile strain at break, density, coefficient of linear thermal expansion, thermal conductivity and the electric conductivity are plotted against the tensile modulus for commercially available PAN- and pitch-based carbon fibres of various types (Otani, Okuda, & Matsuda, 1983; Japanese Carbon Fiber Manufacturers Association). The relation between the tensile modulus and the tensile strength for different commercial PAN-based carbon fibres is shown in Figure 7.4 and resembles

Table 7.1 Properties of graphite near room temperature

Properties	Direction with respect to graphene layers	Values
Density		2.25 g/cm^3
Stiffness	Parallel	1060 GPa
	Perpendicular	36.5 GPa
	Shear between graphene layers	4 GPa
Coefficient of linear thermal expansion	Parallel	-1.2×10^{-6}/K
	Perpendicular	26×10^{-6}/K
Thermal conductivity	Parallel	1950 W/m K
Electric conductivity	Parallel	2×10^{6} S/m
	Perpendicular	1×10^{5} S/m

Source: Blakslee et al. (1970); Gray (1972); Morgan (1972); Holland et al. (1966); Matsubara et al. (1990).

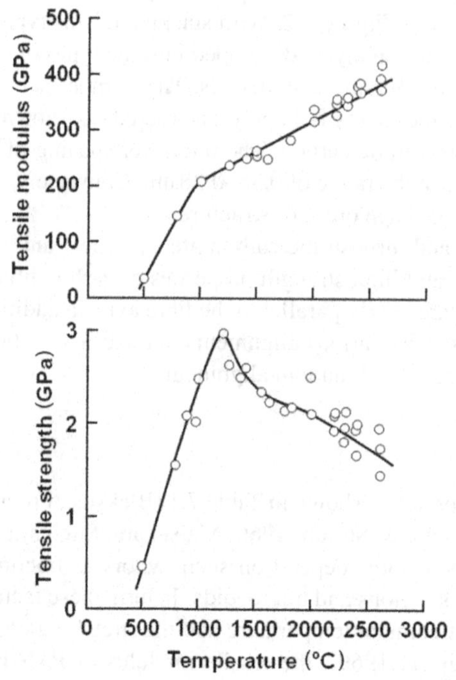

FIGURE 7.3

Changes of tensile modulus and strength of a PAN-based carbon fibre with heat-treatment temperature (Johnson, 1969).

those shown in Figure 7.3. Extremely high-modulus carbon fibres can be derived from pitch, whereas extremely high-strength carbon fibres can be derived from PAN. A high modulus and high strength in combination with a low density are highly desirable in the aerospace and automotive industries.

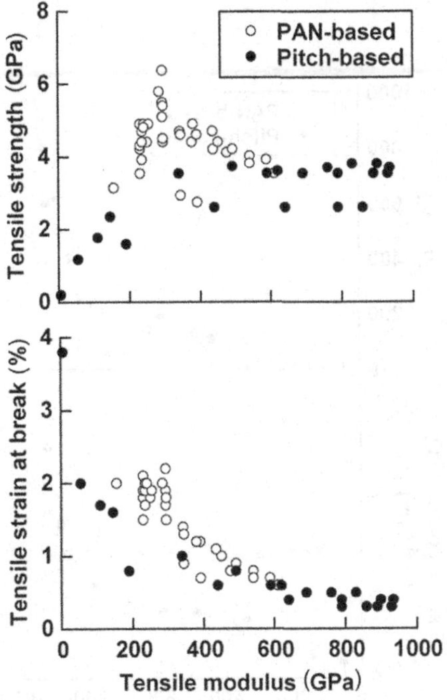

FIGURE 7.4

Tensile strength and tensile strain at break plotted against tensile modulus for commercially available PAN- and pitch-based carbon fibres (Otani et al., 1983; Japanese Carbon Fiber Manufacturers Association).

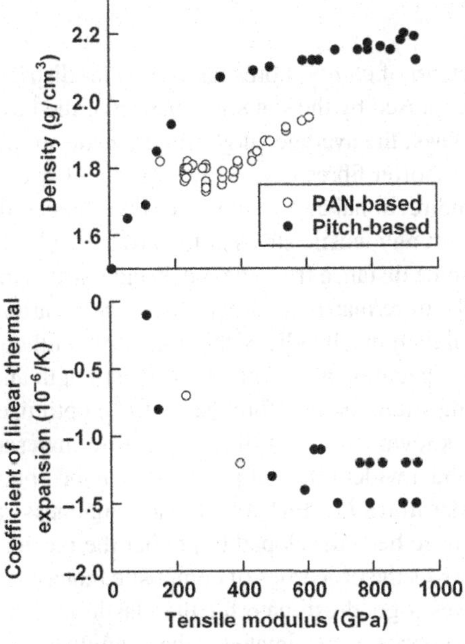

FIGURE 7.5

Density and coefficient of linear thermal expansion plotted against tensile modulus for commercially available PAN- and pitch-based carbon fibres (Otani et al., 1983; Japanese Carbon Fiber Manufacturers Association).

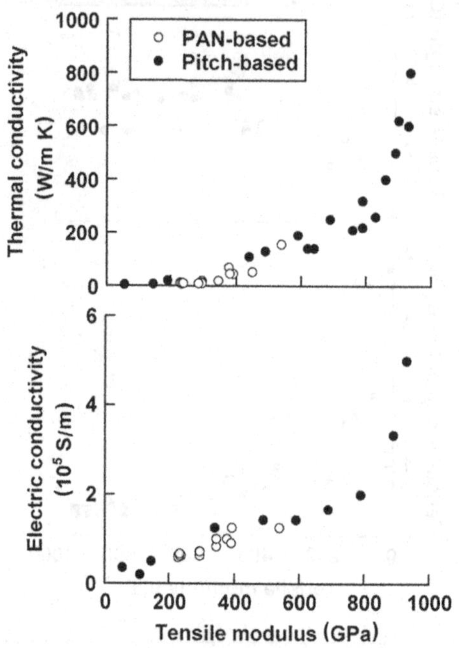

FIGURE 7.6

Thermal conductivity and electric conductivity plotted against tensile modulus for commercially available PAN- and pitch-based carbon fibres (Otani et al., 1983; Japanese Carbon Fiber Manufacturers Association).

The single-fibre tensile strength of carbon fibres has a broader distribution than those of polymeric fibres. The scatter in strength is caused by the statistical nature of the flaws contained in the fibre. Due to the statistical nature of the flaws, the average single-fibre tensile strength of carbon fibres increases with decreasing fibre length, as shorter fibres have a lower probability of having a flaw.

When a carbon fibre in a unidirectional composite material is loaded in tension it breaks at the flaw. The broken fibre does not support any tensile stress at the broken end. The tensile stress in the broken fibre, however, increases over some distance from the break and reaches the stress born by the unbroken fibres due to stress transfer at the fibre/matrix interface. As a result, carbon fibres show a higher tensile strength in a composite material than in a bundle without a matrix, although interfacial bonding that is too strong leads to easy crack propagation and reduces tensile strength and toughness of the composite material. The interfacial bonding strength, therefore, needs to be optimized.

In the composite materials known as hybrid fibre composite materials, carbon fibres are used in combination with other fibres and a wider range of properties can be obtained. For example, by adding a small amount of glass or Kevlar fibres in CFRP, the impact toughness can be enhanced.

Various theoretical models have been developed to predict the mechanical properties of composite materials. A simple estimate derives the properties of composite materials via a weighted average of the constituent properties. This gives a good estimate for the elastic properties of continuous fibre-reinforced composite materials. In contrast, the estimate for the strength of the composite materials derived in this manner can strongly deviate from the measured values. This can be understood by considering an extreme case where the composite material breaks due to the fracture of the matrix and debonding at the interface. In such a case, the exact strength value of the fibre is not explicitly reflected in the strength of the composite material. The ratio of the measured strength to the strength estimated is the utilization efficiency of the fibre strength.

7.2.4 APPLICATIONS

Carbon fibre-reinforced composites are used largely in those industries where their high strength-to-weight ratio can be exploited. These include:

- Aerospace industry
- The automotive industry
- Turbine blades for wind-derived power generation

The ability to control the mechanical properties of the composites in any particular direction also renders them attractive in areas where such variation (called anisotropy) is useful, for example, with pressure vessels designed to contain gases at high pressure (and which thus exert the force in the hoop direction twice as much as the force in the longitudinal direction of the vessels).

7.3 GLASS FIBRES

Glass fibres are composed mainly of silica (SiO_2), like window glasses, and have diameters in the range of about 3–25 μm. Glass fibres are commonly used for reinforcement of polymers. Glass wool is another type of glass fibre about 10–50 mm long and entangled in the bulky cotton-like form, which is used for thermal and acoustical insulation. Glass fibres have a lower production cost than carbon fibres, which is of advantage. Carbon fibres are black in colour, highly anisotropic and electrically conductive, while glass fibres are transparent, isotropic and electrical insulators.

7.3.1 MANUFACTURE

Glass fibres are produced by melt spinning. Various raw materials are mixed and melted at a temperature above 1300 °C (Nittobo Glass Fiber Note, 107). The molten glass is extruded through many orifices and wound up on a drum.

Glass fibres are supplied in various forms, as shown below, and processed into composite materials.

- Filament: a fibre whose length is long enough so that it cannot be determined precisely.
- Strand: a bundle of the filaments.
- Roving: a group of several strands wound on a cylinder.
- Yarn: one or several strands twisted together.
- Cloth: a woven fabric of rovings and yarns.
- Chopped strand: short fibres produced by cutting continuous strand into lengths of about 3–50 mm.
- Milled fibres: short fibres produced by grinding continuous strand into lengths of about 0.8–3 mm.

7.3.2 STRUCTURE

Glass, as an amorphous solid, lacks long-range order of structure as opposed to crystalline structures. The glass transition temperature is the temperature at which discontinuous changes in thermal properties, like the thermal expansion coefficient, take place. When the glass transition temperature is reached during the cooling process, viscosity significantly increases so that the material becomes rigid. The glass implies the state of a material below the glass transition temperature that possesses both the structure of a liquid and the rigidity of a solid. The network structures in a crystal and a glass of silica are illustrated in Figure 7.7. Commercial glasses contain various oxides in addition to silica, which make network structure more irregular.

7.3.3 PROPERTIES

Properties of glass vary widely with composition. The principal component of all glass fibres is SiO_2. A glass fibre of pure SiO_2 shows superior properties such as high thermal and chemical resistance, low thermal expansion and high transparency. It is, however, only used where its special properties are required, such as for optical fibres, since a high-temperature manufacturing process is costly. At present, glass fibres called E-glass are widely used for reinforcement in polymers.

7.3.4 APPLICATIONS

Glass fibres are used without matrix as filters and fibrous blankets for thermal and acoustical insulation. Glass fibres are used as reinforcement of polymers in various fields such as aerospace, automobile, marine, sporting and leisure goods, and construction and civil engineering. One of the principal advantages

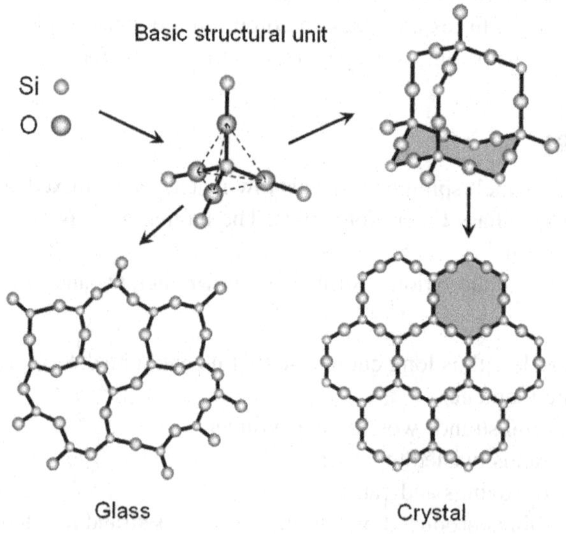

FIGURE 7.7

Crystal and glass of silica.

of using glass fibres for reinforcement of polymers is their high performance per cost ratio. An example of the application of glass fibres in the form of membrane is the architectural membrane made of poly(tetrafluoroethylene) (PTFE)-coated glass fibre for ceilings of stadiums and airports.

7.4 METALLIC FIBRES

Metallic materials consist of a single metallic element or combination of several such. These materials have good electrical and thermal conductivities as well as a lustrous appearance. In addition, metals are quite strong yet deformable. In general, metallic fibre (or metal fibre) is defined as manufactured fibres made from any metal. A manufactured fibre composed of metal and another material, such as polymer fibres coated with metal or metal fibres coated with polymer, are also defined as metallic fibres. In this chapter, metallic fibres composed only of metals will be dealt with.

7.4.1 MANUFACTURE

From the view of the fibre length, metallic fibres can be categorized as short fibres and long fibres. Long fibres may be divided into relatively long discontinuous fibres and continuous-filament fibres. Depending on the required length of fibres, there are various ways to manufacture metallic fibres. This is due to high processability of metals; metals can be cut, deformed and melt-processed.

Continuous metallic fibres can be produced by the drawing of single metal wire or bundle of metal wires through a die. Fine fibres may break because the stress evolved by squeezing the wire through a die exceeds the fibre strength. A bundle of long or continuous fibres also can be produced by machining a roll of metal foil from the end of the roll and winding up the produced fibres continuously as shown in Figure 7.8 (Goto, 1994a; Kaneko, Jun, & Yanagisawa, 2005).

The melt process can also be applied for producing either discontinuous or continuous metallic fibres. Continuous metallic fibres can be produced at a low cost adopting the melt-spinning process. Spinning conditions need to be adjusted carefully because a thin thread of molten metal tends to break up into droplets because of the combination of its low viscosity and high surface tension.

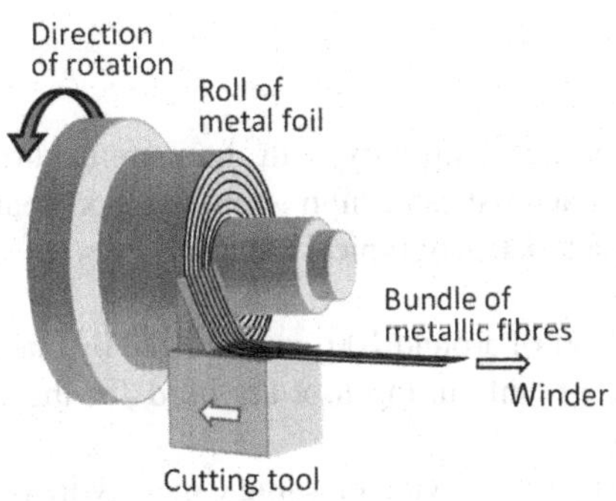

FIGURE 7.8

Schematic of continuous metallic fibre production through machining of a roll of metal foil (Goto, 1994a; Kaneko et al., 2005).

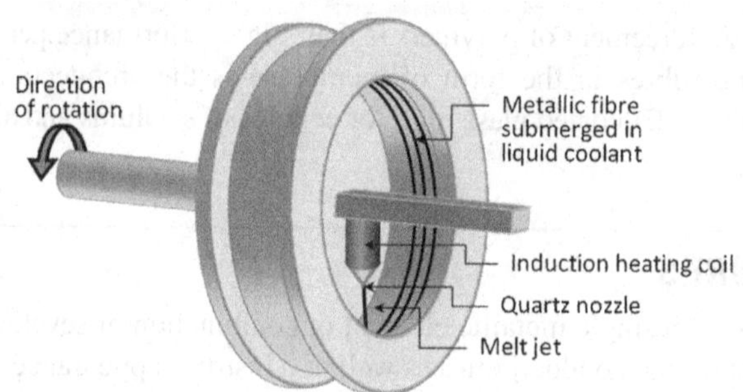

FIGURE 7.9

Schematic of "in-rotating-liquid-spinning process" (Goto, 1994a; http://www.unitika.co.jp/shinki/E/home.htm).

A unique "in-rotating-liquid-spinning process" (http://www.unitika.co.jp/shinki/E/home.htm) was developed to produce continuous metallic fibres stably utilizing the melt-spinning process. In this process, metal is heated in a melting pot utilizing an induction heating coil. The melt jet of metal extruded from a quartz nozzle is immediately quenched in the liquid coolant layer held in the groove of a rotating drum by centrifugal force. The quenched fibre can be pulled out from the bottom of the groove utilizing magnet rolls. A schematic of this process is shown in Figure 7.9 (Goto, 1994a; http://www.unitika.co.jp/shinki/E/home.htm).

7.4.2 BASIC STRUCTURE AND PROPERTIES

Various metals such as tungsten, stainless steel, copper, brass, aluminium, titanium, etc., are used as the material of metallic fibres and are produced in the forms of short fibre, continuous fibre, single fibre, tow, felt, sheet, woven fabric, etc.

Quenching of metals in melt processing possibly leads to the development of fibres with a unique crystalline form or amorphous structure, which may not appear in ordinary cooling conditions. Amorphous metals generally show anti-corrosion properties and soft magnetic properties because of the absence of grain boundaries.

7.4.3 APPLICATIONS

Application of metallic fibres correlates strongly with the manufacturing method, while most of the applications are based on the characteristics of high strength, good heat resistance and electrical conductivity of metallic fibres. Characteristics of typical metallic fibres are summarized in Table 7.2 (Goto, 1994b; Goto, 2004).

Short steel fibres with a diameter of around 100 μm produced by machining of solid metals utilizing the inherent vibration of a single-point cutting tool are used for the reinforcement of concrete and shielding of electromagnetic waves.

The tungsten fibres produced by the drawing of a single metal wire or bundle of metal wires through a die are used as the filament in a light bulb. Wires of copper, gold, silver and aluminium are also produced by applying the drawing method. In the case of piano wires, i.e. carbon steel fibres, production of fibres with a diameter of less than 100 μm by the drawing method becomes costly; however, piano wires of 40 μm diameter reach high strengths.

Table 7.2 Characteristics of metallic fibres (Goto, 1994b; Goto, 2004)

Materials	Manufacturing method	Crystal form	Density (g/cm^3)	Melting temperature (°C)	Diameter (µm)	Tensile modulus (GPa)	Tensile strength (GPa)
Tungsten	Drawing	bcc	19.4	3400	13	407	4.02
Molybdenum	Drawing	bcc	10.2	2630	25	329	2.16
Piano wire	Drawing	–	7.8	1400	80	199	3.43
Stainless steel	Multi-wire drawing	fcc	7.98	1450	15	186	2.5
Aluminium	Drawing	fcc	2.8	660	10	63	0.2
Gold	Drawing	fcc	19.3	1094	25	80	–
Fe-Co-Cr-Si-B	Melt spinning	Amorphous	7.6	535*	100	163	3.67
Fe-Co-Cr-Mn-Si-B	Melt spinning	Amorphous	7.7	545*	100	166	3.72

**Crystallization temperature.*
Source: Goto (1994b); Goto (2004).

7.5 CERAMIC FIBRES

Ceramics are non-metallic inorganic materials consisting of metallic and non-metallic elements. A wide range of materials composed of clay minerals, cement and glass are included in this classification. Ceramic materials are often prepared through a burning or sintering process. Therefore they are more resistant to high temperatures and harsh environments in comparison to metals and organic materials. In addition, ceramics are good insulators for both electricity and heat.

Ceramic fibres are categorized into non-oxide fibres and oxide fibres. Silicon carbide and alumina are the most commonly used non-oxide and oxide ceramic fibres, respectively. Boron fibres and silicon fibres are normally classified as ceramic fibres even though boron and silicon are the metalloid elements. Short fibres, long fibres and continuous fibres of ceramics are available in the market. Continuous ceramic fibres are produced mainly through either the melt-spinning or dry-spinning processes.

7.5.1 MANUFACTURE

The manufacturing processes of ceramic fibres can be divided into two types (Cooke, 1991; Isoda, 1994).

Direct formation of ceramic fibres can be achieved through various methods such as (1) the melt-spinning process, (2) deposition through cooling of the melt of ceramics, (3) crystallization after sublimation and (4) deposition or growth from a seed crystal after a gas-phase chemical reaction. Melt-spinning technologies developed for the production of glass fibres can be applied for the direct formation of ceramic fibres.

In the production of ceramic fibres through the two steps, i.e. fibre formation and metamorphosis processes, green fibres can be prepared either by the melt-spinning or dry-spinning processes.

The metamorphosis of green fibres to develop ceramic fibres can be accomplished by applying heat to induce thermal decomposition, oxidization decomposition, chemical reaction, extraction and sintering. Thermal decomposition in inert gas to form cross-linked structure is utilized for the production of non-oxide fibres, whereas oxidization decomposition is utilized for the production of oxide fibres.

7.5.2 BASIC STRUCTURE AND PROPERTIES

The mechanical properties of ceramic fibres can be characterized by their high tensile modulus, high tensile strength and brittleness. Brittleness originates from its low elongation at break and is sometimes considered a negative aspect of ceramic fibres. This fact also implies that the tensile strength of ceramic fibres is governed by the defects existing mainly on the fibre surface.

In general, tensile strength and creep behaviour of ceramic fibres deteriorate only at extremely high temperatures. In comparison with metals, ceramics have lower density, lower thermal conductivity and better heat resistance. In addition, ceramics are chemically stable at elevated temperatures. Accordingly, ceramic fibres are widely applied as reinforcing materials in composites for use at high temperatures. Their low thermal conductivity leads to the use of ceramic fibres as heat-resistant thermal insulators.

7.5.3 APPLICATIONS

Ceramic fibres have a high heat resistance and are therefore attractive for use in areas where this characteristic is important. For example, ceramic fibres are used in heat-resistant curtains and cushions, and as the covering material for electrical wires.

The fibres are also applied for parts of gas turbines for power generation and in parts for spacecraft and aircraft. In the latter case, heat resistance is of particular importance and is in fact a regulatory requirement for certain components.

7.6 CASE STUDY: THE USE OF CFRP IN SPORTING GOODS

Although carbon fibres possess superior mechanical properties, their drawback is the high production cost. One of the key areas of application for carbon fibres is the sporting and leisure goods market, which includes products made using CFRP such as tennis rackets, baseball bats, golf club shafts, fishing rods and canoes.

Until the early 1980s, the frames of tennis rackets had been made from wood, but this was replaced by metal frames using steel and aluminium. The increased stiffness and strength offered by metals allowed production of tennis rackets with oversize heads. The success of metal frames led to testing of other non-traditional materials for racket frames. Among these materials, CFRP was recognized as a prospective material for the further development of racket frames due to the extremely high modulus and high strength in spite of the high production cost. CFRP tennis rackets with large head size, higher stiffness and reduced weight have advantages including higher ball velocity. Current CFRP rackets are 40% larger in head size, 3 times stiffer and 30% lighter than the most highly developed wooden rackets (International Tennis Federation).

7.7 FUTURE TRENDS

The environmental friendliness of a material is evaluated using life-cycle assessment (LCA), see Chapter 22 for more information on LCA, which is an assessment of the environmental impacts of a product associated with all stages of production from raw material, through manufacture and use, to waste management. The carbon fibre production process involves a high-temperature heat treatment and is energy-consuming and causes the emission of large amounts of carbon dioxide. However, LCA has shown that the use of CFRP greatly contributes to the reduction of carbon dioxide emissions. According to an LCA report by the Japanese Carbon Fiber Manufacturers Association, carbon fibre production causes emission of 20 tons of carbon dioxide per 1 ton of carbon fibres. However, weight reduction of automobiles by 30% through the use of carbon fibres contributes to the reduction of 50 tons of carbon dioxide emissions per 1 ton of carbon fibres. Further, weight reduction of aircraft by 20% through the use of carbon fibres contributes to the reduction of 1400 tons of carbon dioxide emissions per 1 ton of carbon fibres.

7.8 SUMMARY POINTS

- Utilization of a material in the form of fibre and fibre-reinforced composite materials has advantages that include excellent mechanical properties of the material.
- High-strength and high-modulus fibres are used for reinforcement in composite materials for various applications such as aerospace, transportation, energy industry, civil engineering, medical fields, and sporting and leisure goods.
- Carbon fibres consist of highly oriented carbon layer stacks, as a result of which carbon fibres are extremely stiff and strong and low in density.

- Metallic fibres are produced in various ways including solid-state manufacturing and melt-state manufacturing, since metals can be cut, deformed and melt-processed.
- Ceramic fibres have characteristic of low density, low thermal and electrical conductivities, good heat resistance and high chemical stability at elevated temperatures.

7.9 PROJECT IDEAS

- Discuss some of the potential uses of fibre-reinforced composite materials that have not been discussed in this chapter (clue: think about their high strength-to-weight ratio).
- Research the health and safety aspects of carbon, glass, metal and ceramic fibres.
- Consider whether glass fibres might have application in the apparel industry and explain your conclusion.

7.10 REVISION QUESTIONS

1. What are the advantages of fibre-reinforced composite materials?
2. List and describe the typical production processes of fibre-reinforced composite materials.
3. What are the starting materials for carbon fibres?
4. What are the main applications of metallic fibres?
5. What are the main applications of ceramic fibres?
6. What are the advantages of non-polymer fibres from the viewpoint of environmental friendliness and energy saving?

REFERENCES AND FURTHER READING

Blakslee, O. L., Proctor, D. G., Seldin, E. J., Spence, G. B., & Weng, T. (1970). *Journal of Applied Physics*, *41*, 3373–3382.

Cooke, T. F. (1991). Inorganic fibers – a literature review. *Journal of American Ceramic Society*, *74*, 2959–2978.

Data of the Japanese Carbon Fiber Manufacturers Association. http://www.carbonfiber.gr.jp/english/index.html.

Goto, T. (1994a). Fundamentals on manufacturing of metal and ceramic fibres, metal fibres. In *Handbook of fibres* (2nd ed., pp. 63–66). Japan: The Society of Fiber Science and Technology (in Japanese), Maruzen.

Goto, T. (1994b). Structure and properties of various fibres, metal fibres. In *Handbook of fibres* (2nd ed., pp. 130–131). Japan: The Society of Fiber Science and Technology (in Japanese), Maruzen.

Goto, T. (2004). Manufacture, structure and properties of various fibres, metal fibres. In *Handbook of fibres* (3rd ed., pp. 194–197). Japan: The Society of Fiber Science and Technology (in Japanese), Maruzen.

Gray, D. E. (Ed.). (1972). *American institute of physics handbook* (3rd ed.). American Institute of Physics, McGraw-Hill.

Holland, M. G., Klein, C. A., & Straub, W. D. (1966). *Journal of Physics and Chemistry of Solids*, *27*, 903–906.

International Tennis Federation. http://www.itftennis.com/technical/.

Isoda, T. (1994). Fundamentals on manufacturing of metal and ceramic fibres, ceramic fibres. In *Handbook of fibres* (2nd ed., pp. 66–71). Japan: The Society of Fiber Science and Technology (in Japanese).
Johnson, J. W. (1969). *Applied Polymer Symposia*, *9*, 229.
Kaneko, M., Jun, M., & Yanagisawa, A. (2005). *Journal of the Textile Machinery Society of Japan*, *58*, T136.
Matsubara, K., Sugihara, K., & Tsuzuku, T. (1990). *Physical Review*, *B41*, 969–974.
Morgan, W. C. (1972). *Carbon*, *10*, 73–79.
Nittobo Glass Fiber Note 107.
Otani, S., Okuda, K., & Matsuda, S. (1983). *Carbon fibre*. Kindai Henshu Ltd.

PART 2

MANUFACTURING TEXTILES: YARN TO FABRIC

CHAPTER

CONVERSION OF FIBRE TO YARN: AN OVERVIEW

R. Alagirusamy, A. Das
Indian Institute of Technology Delhi, New Delhi, India

LEARNING OBJECTIVES

At the end of this chapter, you should be able to:

- The classification of yarns
- Types of spinning processes and the spinning process variables
- Different types of fancy yarns, their properties and end-uses
- The relationships between the structures and properties of yarns

8.1 INTRODUCTION

The term yarn may be defined as a linear collection of filaments or fibres in a twisted state or bound by other means, and possessing good tensile strength and elasticity properties. From the various types of commercially manufactured yarns, it may be seen that there are a great many functional and design possibilities. Fibres are processed in both pure and blended states. Considerable variation in yarn is made from a particular fibre or filament is possible. The classification of yarns (Table 8.1) is based on their physical characteristics as well as performance properties.

8.2 CLASSIFICATION OF YARNS

8.2.1 STAPLE YARNS

Staple yarns are defined as assembled strands of fibres twisted together to form a continuous strand according to the properties required. There are four staple-yarn manufacturing systems commercially available. These are carded, combed, woollen and worsted systems. Synthetic and regenerated fibres are stapled and treated like natural fibre (cotton or wool) with different fibre lengths, diameters and crimps, which enables processing to be carried out with few difficulties.

8.2.2 CONTINUOUS-FILAMENT YARNS

Silk was the first natural continuous-filament yarn in use before the introduction of man-made fibres. Man-made filaments are manufactured by extruding polymer solution through a spinneret, at which

Textiles and Fashion. http://dx.doi.org/10.1016/B978-1-84569-931-4.00008-8

Table 8.1 Yarn types and their properties

Yarn type	General yarn properties
Staple yarns Combed cotton Carded cotton Synthetic and blends Worsted Woollen	Good hand, cover, comfort and textured appearance Average strength and uniformity
Continuous-filament yarns Natural Man-made or synthetic	High strength, uniformity and possibility for very fine yarns Fair hand and poor covering power
Novelty yarns Fancy Metallic	Decorative features and characteristics
Industrial yarns Tyre cord Rubber or elastic core Multiply coated	Functional; designed and produced to satisfy a specific set of requirements
High-bulk yarns Staple Continuous filament (Taslan)	Great covering power with less weight, high loftiness or fullness
Stretch yarns Twist-heat set-untwist Crimp heat-set Stress under tension Knit-deknit Gear crimp	High stretchability and cling without high pressure, good handle and covering power

point the solution solidifies by coagulation, cooling or evaporation. The number of orifices in the spinneret dictates the number of filament in the bundle. The diameter and amount of drawing provided will subsequently decide the diameter of the filament. The filaments are cut and crimped according to the required length for conversion to staple fibre, which will undergo further staple-yarn production.

8.2.3 NOVELTY YARNS

These are designed and produced for decorative purposes and are seldom used to make an entire fabric, except in drapery applications. Most novelty yarns are of the fancy or metallic type.

Fancy yarns are generally produced by the irregular plying of staple fibre or continuous filaments and are characterised by the presence of abrupt and periodic effects. The periodicity of these effects may be irregular or constant. The novelty effect is brought about by a programmed difference in twist level or input rate in one or more components during the plying of the yarns. This results in differential bending or wrapping between the components or in segments of buckled yarn that are permanently entangled in the composite yarn structure.

8.2.4 INDUSTRIAL YARNS

Industrial design requires special end-use yarns with specific functional characteristics. These yarns are engineered for performance under specified conditions. Many industrial yarns do not have the visual and tactile properties of yarns used for apparel and home-furnishing applications. Examples are tyre cord, asbestos and glass yarns, twine, rubber or elastic threads, core-spun yarns, wire yarn, sewing thread, heavy monofilaments and split-film yarns.

8.2.5 HIGH-BULK YARNS

A high-bulk yarn may be a staple or continuous-filament yarn with normal extensibility but an unusually high level of loftiness or fullness. These yarns retain bulkiness in both relaxed and stressed conditions. High covering power with lesser weight is possible in fabrics made of high-bulk yarns.

8.2.6 STRETCH YARNS

Textured yarns preset for high extensibility are termed stretch yarns. While most can be stretched up to twice their normal or relaxed length, some can be stretched up to even three or four times their relaxed length. These yarns are highly extensible and highly elastic. Most stretch yarns are produced by texturising thermoplastic continuous-filament yarns. This results in reasonably good nonlinearity or crimp in the individual filaments. The nonlinear structure of the filaments is heat-set but not entangled as in the case of the high-bulk yarns.

8.3 STAPLE-FIBRE YARNS

A staple-spun yarn is a linear assembly of fibres, usually held together by the insertion of twist, to form a continuous strand that is smaller in cross-section but of a particular specified length. It is used for making fabrics in processes such as knitting, weaving and sewing. The amount of strength or the quality of handle and appearance will depend on the way the fibres are assembled in the yarn system (Eric, 1987; Klein, 1995). The fibres can be natural or man-made. Man-made staple-fibre yarns are manufactured by using staple fibre cut from continuous filament before spinning. Staple-fibre yarns can be subdivided and classified on the basis of fibre length, spinning method and yarn construction. Each of these categories can be further subdivided.

Staple-spun yarn can be classified as either **short staple or long staple**. A staple fibre has a length of between 10 and 500 mm. **Short staple fibre** has a maximum length of 60 mm (cotton fibre is a short staple at about 25–45 mm). **Long staple fibre** has a length of more than 60 mm (wool fibre is a long staple at about 60–150 mm).

8.3.1 SPINNING METHODS

Different spinning methods are available in making yarns, including ring-spun, rotor-spun, twistless, wrap-spun and core-spun yarns.

- **Ring-spun yarns:** This is the most widely used method of staple-fibre yarn production. The fibres are twisted around each other to give strength to the yarn.

- **Rotor-spun yarns:** These are similar to ring-spun yarns and usually made from short staple fibres. They produce a more regular and smoother, though weaker, yarn than ring spinning.
- **Twistless yarns:** The fibres are held together by adhesives, not by the twist, and are often laid over a continuous filament core.
- **Wrap-spun yarns:** These yarns are made from staple fibres bound by another yarn, which is usually a continuous man-made filament yarn. The yarns can be made from either short or long staple fibres.
- **Core-spun yarns:** Core-spun yarns have a central core that is wrapped with staple fibres, and are produced in a single operation at the time of spinning. For example, a cotton sheath for handle and comfort, with a filament (often polyester) core for added strength; or cotton over an elastomeric core.

8.3.2 OPERATIONS IN STAPLE-FIBRE SPINNING

There are a number of operations.

- **Mixing:** The term 'mixing' refers to the bringing together of more than one variety of the same basic fibre. For example, Egyptian cotton fibre can be combined with American cotton fibre and the final yarn remains 100% cotton.
- **Blending:** Blending refers to the bringing together of fibres of different types. For example, wool and silk or cotton and polyester fibres.
- **Cleaning and fibre separation:** Bales of raw fibres contain a variety of impurities, which must be removed. The first process divides and splits the bales into loose bunches of fibres of diminishing size to remove dust, seeds and other debris. Some fibre types are then washed or scoured and others can be combed or carded to further separate and clean the fibres.
- **Fibre alignment:** This process follows carding and combing. Several slivers or groups of carded or combed fibres are combined and attenuated to form a single sliver of straightened fibres. This process is known as drawing.
- **Drafting and twisting:** Drafting is the process of gently stretching out the slivers to reduce their linear density or thickness. The exact method and machinery used will depend upon the required yarn quality and count. In the final process, the required amount of twist is inserted into the single yarn.

8.3.3 YARN STRUCTURE

The major aspect of yarn structure is its visual appearance, which is determined by the peripheral layer of fibres. The second aspect is its internal structure. Yarn structures are very variable. The differences are caused partly deliberately, according to the intended use of the yarn, but for the most part they are predetermined by the means available. For example, it is difficult to produce a yarn having properties similar to a ring-spun yarn by new spinning processes, and ring-spun yarn still represents the standard reference for comparison (Table 8.2) (Lord, 2003).

Yarn structure depends primarily upon the properties of the raw materials, the spinning process and parameters, the spinning unit conditions, machine parameters and settings and twist levels, and so on. The yarn structure can be open or closed; bulky or compact; smooth, rough or hairy; soft or hard; circular or flat; thin or thick, and so on.

Table 8.2 Yarn structures from different spinning methods

	Ring-spun yarn		Open-end yarn		Air-jet yarn		Wrap yarn
	Classic	**Compact**	**Rotor spun**	**Friction spun**	**Jet spun, two nozzles, false twist process**	**Vortex spun, one nozzle**	**Filament wrapped**
Fibre deposition							
In the core	Parallel, helical	Parallel, helical	Less parallel, helical	Less parallel, helical	Parallel w/o twist	Parallel w/o twist	Parallel w/o twist
In the sheath	Parallel, helical	Parallel, helical	More random, less twisted	Less parallel, helical	6% of fibres twisted around the core in spirals	20% of fibres twisted around core in spirals	Filament windings
Fibre orientation							
Parallelism	Good	Very good	Medium	Low	Medium	Good	Very good
Compactness	Compact	Very compact; round	Open	Compact to open	Compact	Compact	Compact
Handle	Soft	Soft	Hard	Hard	Hard	Medium to hard	Soft
Hairiness	Noticeable	Low	Very low	Low	Some	Low to medium	Very low
Stiffness	Low	Low	High	High	High	Fairly high	Low

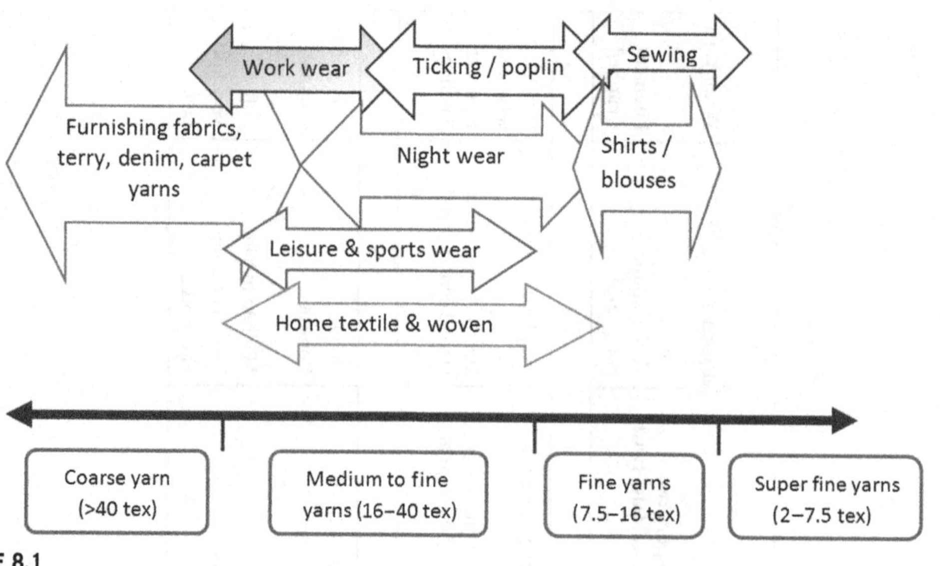

FIGURE 8.1

Application chart for staple spun yarns.

In addition to appearance, yarn structure also influences the following: handle; strength; elongation; insulating capacity; covering power; ability to resist wear, damage, strains, and so on; abrasion resistance; dyeability; tendency towards longitudinal bunching of fibres; and wearing comfort, and so on (Lawrence, 2003).

8.3.4 APPLICATIONS OF STAPLE-SPUN YARNS

Figure 8.1 shows the count range for the different end-uses of staple-fibre yarns. In addition to the very fine yarn count range of 2–7.5 tex meant for hosiery, staple-fibre yarns have almost similar market areas, where fine to medium yarn counts of 7.5–40 tex are largely used to make textiles for clothing and apparel. Spun staple yarns hold a major position in the market for items such as shirts, blouses, home textiles, bed linen, trousers and suits (Lord, 2003).

8.4 FILAMENT YARNS

A filament yarn is a collection of parallel bundles of filaments lying close together along the whole length of the yarn. Man-made yarns are made by extruding the required number of filaments in a single operation at the desired linear density. Yarns with only one filament are known as **monofilaments** and those with more than one filament are known as **multifilaments**.

8.4.1 SPINNING METHODS

Most synthetic fibres are extruded using polymers derived from by-products of petroleum and natural gas, which include polyethylene terephthalate (PET) and nylon as well as compounds such as acrylics, polyurethanes and polypropylene. The polymers are first converted from solid to fluid state by means of melting, dissolving using solvent, and so on. The fluid polymer is then extruded through a spinneret to convert the solution into filaments. There are four major types of synthetic fibre production techniques: Dry Spinning, Wet Spinning, melt spinning and gel spinning (Deopura, Alagirusamy, Gupta, & Joshi, 2008; Gupta & Kothari, 1997; McIntyre, 2004).

The spinneret is a metal component having one to several hundred small holes. The fluid polymer is injected through these tiny openings to produce filaments from polymer solution. This process of extrusion and solidification of innumerable filaments is known as the spinning of polymers. There are two types of extrusion: single-screw and twin-screw extrusion.

- **Single screw-extrusion:** This is one of the elementary tasks of polymer processing. The single-screw extrusion process builds pressure on a polymer melt and forces it through a die or injects it into a mould. Most single-screw extrusion machines are plasticating, taking solids in pellet or powder form and melting them while simultaneously building pressure.
- **Twin-screw extrusion:** This is commonly used for mixing, compounding or reacting polymeric materials. The flexibility of the twin-screw extrusion tool allows the operation to be designed specifically for the formulation being processed. For example, the two screws may be co-rotating or counter-rotating, inter-meshing or non-inter-meshing. The design and configuration of the screws themselves may also be changed using forward conveying elements, reverse conveying elements, kneading blocks, and other designs that can assist in obtaining particular mixing characteristics.

8.4.2 POLYMER SPINNING PROCESSES

There are a number of spinning processes.

- **Wet spinning:** Of the four processes, the oldest is wet spinning, as shown in Figure 8.2. This is used for polymers that need to be dissolved in a solvent before they can be spun. The spinneret remains submerged in a chemical bath, which causes the polymer to precipitate and then solidify as it emerges from the spinneret holes. (The name of the process is derived from the use of this 'wet' bath.) Acrylic fibre, Rayon fibre, Aramid fibre, Modacrylic fibre and Spandex fibres are all manufactured by the wet spinning process.

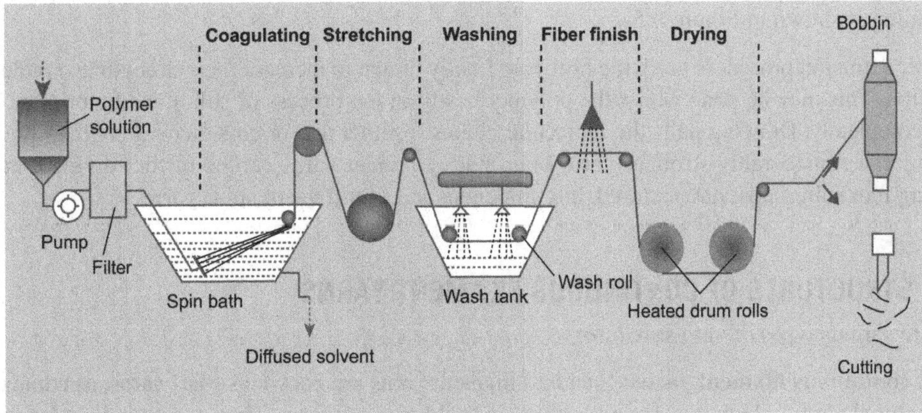

FIGURE 8.2

Schematic of wet-spinning process.

- **Dry spinning:** Dry spinning is used for polymers that need to be dissolved in a solvent. However, solidification is obtained from evaporation of the solvent. After the polymer is dissolved in a volatile solvent, the solution is pumped through a spinneret. As the fibres emerge from the spinneret, air or inert gas is used to evaporate the solvent from the fibre. This results in solidification of the fibres, which can then be collected on a take-up wheel. The fibres are drawn to provide orientation to the polymer chains along the fibre axis. This technique is used only for polymers that cannot be melt-spun due to the safety and environmental concerns associated with solvent handling. Dry spinning may be used for manufacturing acetate fibre, triacetate fibre, acrylic fibre, modacrylic fibre, PBI, Spandex fibre and vinyon. The process flow of dry spinning is shown in Figure 8.3.
- **Melt spinning:** In this process, the polymer is melted and then extruded through a spinneret. The cooled and solidified molten fibres are then collected on a take-up wheel. The fibres are stretched in the molten and solid states, which assists the orientation of the polymer chains along the fibre axis. Melt-spun fibres can be extruded through a spinneret in different cross-sectional shapes, including circular, trilobal, pentagonal and octagonal. Trilobal-shaped fibres are capable of reflecting more light, which gives a sparkle to the fabrics. Pentagonal-shaped and hollow fibres are soil and dirt resistant and are used in making carpets and rugs. Octagonal-shaped fibres offer glitter-free effects, while hollow fibres trap air, creating better insulation. Polymers such as polyethylene terephthalate and nylon 6-6 are melt-spun in high volumes. Nylon fibre, olefin fibre, polyester fibre, saran fibre, and so on are also manufactured through melt spinning. The process flow of melt spinning is shown in Figure 8.4.
- **Gel spinning:** Gel spinning is also known as dry-jet-wet spinning, because the filaments first pass through air and are then cooled further in a liquid bath. Gel spinning is used to make very strong fibres with special characteristics. The polymer is in a partially liquid or 'gel' state, which keeps the polymer chains bound together to some extent at different points in a liquid crystal form. This bond results in strong inter-chain forces in the fibre, which increases its tensile strength. The polymer chains within the fibres also have a large degree of orientation, which further increases its strength. The strength is still further enhanced by the filaments emerging with an unusually high degree of orientation relative to each other. High-strength polyethylene and aramid fibres are manufactured by this process. The process flow of gel spinning is shown in Figure 8.5.

Whatever extrusion process is used, the fibres are finally drawn to increase their strength and molecular orientation. This may be done whilst the polymer is still in the process of solidification or after it has cooled completely. Drawing pulls the molecular chains together and orients them along the fibre axis, resulting in a considerably stronger yarn. Man-made filament yarns can be further divided into the following four subgroups: flat, textured, bi-component and film (tape or split) yarns.

8.4.3 STRUCTURES OF CONTINUOUS FILAMENT YARNS

There are various types of yarn structure:

- **Flat continuous filament yarns:** Standard filament yarns are known as 'flat' yarns, as compared to textured yarns, which may have filaments in curly or wavy form. They can be dull and matt or bright and lustrous as required.

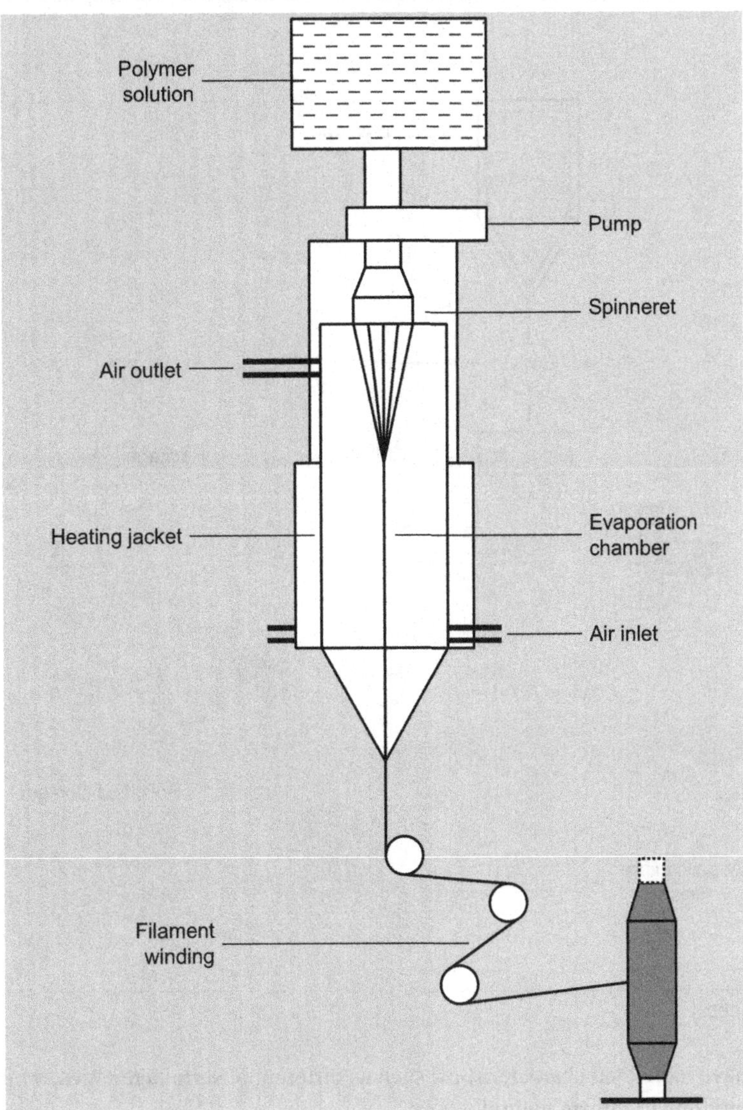

FIGURE 8.3

Dry-spinning process.

- **Textured continuous filament yarns:** These are not to be confused with fancy or decorative yarns. Textured filament yarns are man-made yarns in which the filaments have been altered for a higher degree of bulkiness. Distortions such as crimps, loops and knots are introduced during the texturing process. Various methods are used to introduce these deformations.
- **Bi-component continuous filament yarns:** These yarns are produced from two different polymer components that are brought together at the fibre extrusion stage. Each component can be

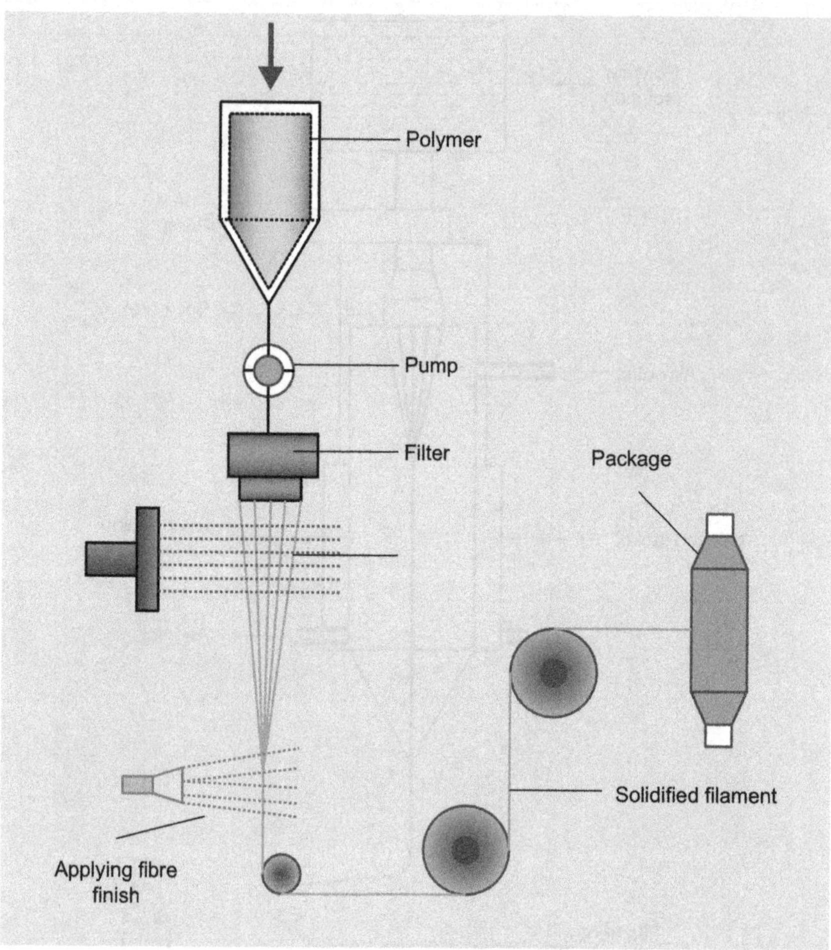

FIGURE 8.4

Melt-spinning process.

designed to have individual characteristics, such as differential shrinkage ratios, which can cause kinking or spiralling to imitate natural wool.

- **Tape or split-film yarns:** To make film, the polymer is extruded in a thin, wide sheet. The film is cut into narrow strips or ribbons to produce tape. Slits can then be made along the length of the ribbon to produce a split-film yarn.

8.4.4 APPLICATIONS OF FILAMENT YARN

Figure 8.6 illustrates the wide count range for the different end-uses of filament yarns. These yarns are highly competitive in the carpet-yarn and sportswear applications and in industrial yarn applications for technical textiles.

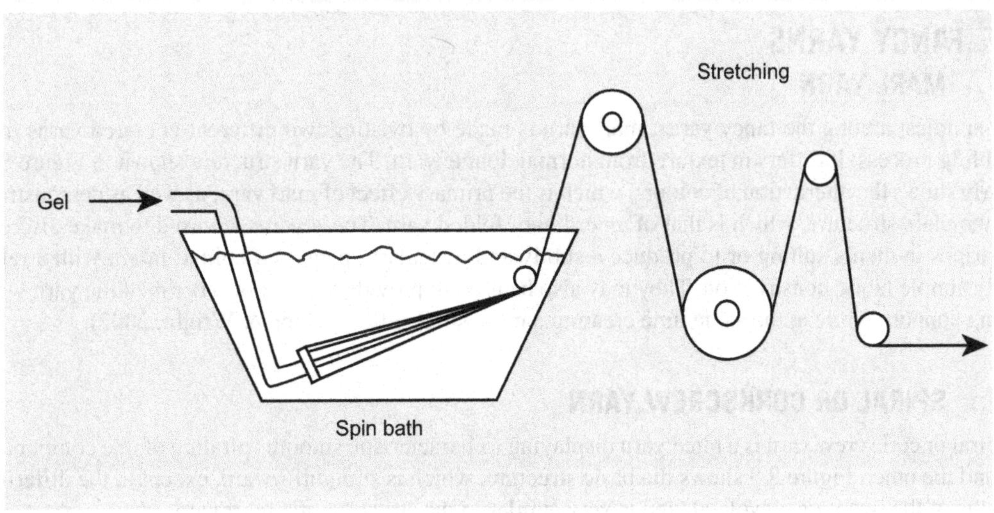

FIGURE 8.5

Schematic of gel-spinning method.

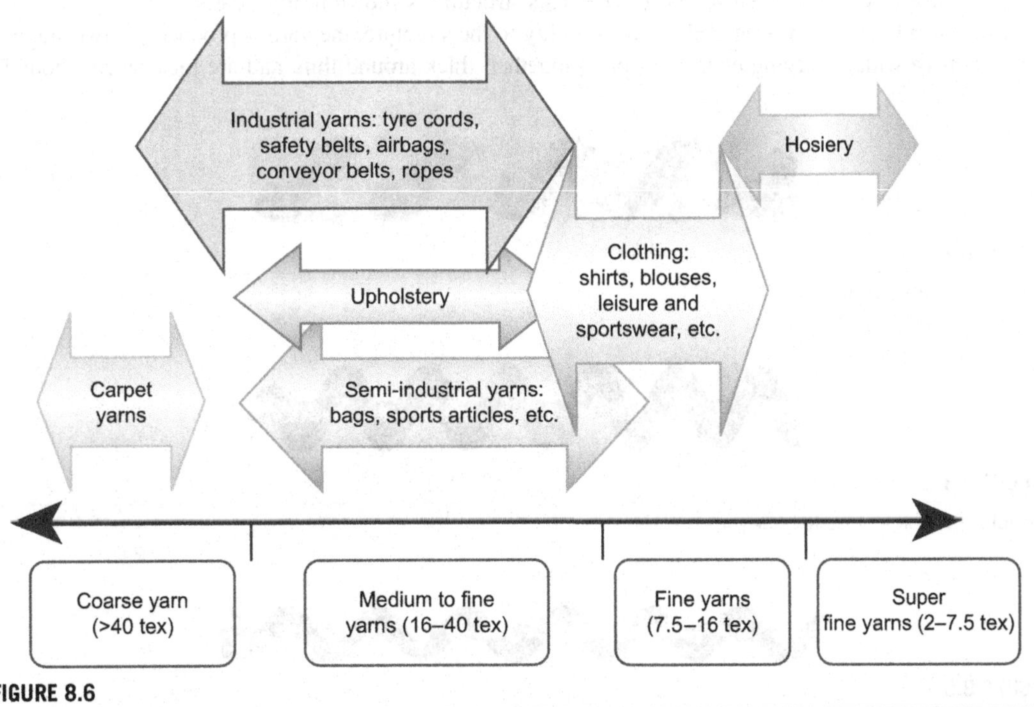

FIGURE 8.6

End-uses of filament yarns.

8.5 FANCY YARNS

8.5.1 MARL YARN

The simplest among the fancy yarns, marl yarn is made by twisting two different-coloured yarns in a doubling process. It differs in texture from normal double yarn. The yarn structure shown in Figure 8.7 clearly shows the alternation of colours, which is the primary effect of marl yarn, as well as demonstrating the plain structure, which is that of an ordinary folded yarn. These yarns are used to make discreet pinstripes in men's suiting or to produce a subtly and irregularly patterned knitted fabric with a relatively simple fabric construction. They may also be used to provide a Lurex or other metallic yarn with strong support, while at the same time creating a more subtle effect (Gong & Wright, 2002).

8.5.2 SPIRAL OR CORKSCREW YARN

A spiral or corkscrew yarn is a plied yarn displaying a characteristic smooth spiraling of one component around the other. Figure 8.8 shows the basic structure, which is straightforward, except in the differing lengths of the two yarns involved, and is very similar to the structure of a marl yarn.

8.5.3 GIMP YARN

A gimp yarn is a compound yarn consisting of a twisted core with an effect yarn wrapped around it so as to produce wavy projections on its surface. This structure is shown in Figure 8.9.

Because a binder yarn is needed to give stability to the structure, the yarn is produced in two stages. Two yarns of widely varying counts are plied together, thick around thin, and are then reverse bound.

FIGURE 8.7

Structure of marl yarn.

FIGURE 8.8

Structure of spiral or corkscrew yarn.

FIGURE 8.9

Structure of gimp yarn.

Reverse binding removes the twist that creates the wavy profiles as it makes the effect yarns longer than the actual length of the completed yarn. The texture properties of a gimp are clearly better than those of a spiral yarn. The finer of the two gimps shows that the effect is less regular and perhaps less well-defined.

8.5.4 DIAMOND YARN

A diamond yarn is produced by folding a coarse single yarn or roving with a fine yarn or filament of contrasting colour using S-twist. This is cabled with a similar fine yarn using Z-twist. Multifold 'cabled' yarns may be made by extending and varying this technique to produce a wide range of effects. A true diamond yarn would show some compression effect upon the thick yarn from the thin ones, but in the interests of clarity this is not shown in Figure 8.10.

Diamond yarn is very useful to designers in the creation of subtle effects of colour and texture, particularly in relatively simple fabric structures.

8.5.5 BOUCLE YARN

This type of yarn is characterised by tight loops projecting from the body of the yarn at fairly regular intervals, as shown in Figure 8.11. Some of these yarns are made by air-jet texturing but most are of three-ply construction. The three components of the yarn are the core, the effect and the tie, or binder. The effect yarn is wrapped in loops around a core or base yarn, and then the third ply, or binder, is wrapped over the effect ply in order to hold the loops in place. The individual plies may be filament or spun yarns. The characteristics of these yarns determine the ultimate design effect.

8.5.6 LOOP YARN

A loop yarn consists of a core with an effect yarn wrapped around it and overfed to produce a nearly circular projection on its surface. Figure 8.12 shows the structure of a loop yarn, in this case somewhat simplified by showing the core as two straight bars. In reality, the core always consists of two yarns twisted together, which entraps the effect yarn. As a general rule, four yarns are involved in the construction. Two of these form the core or ground yarns. The effect yarn(s) are formed with an overfeed of about 200% or more. It is important for these to be of the correct type and of good quality. Even,

FIGURE 8.10

Structure of diamond yarn.

FIGURE 8.11

Structure of boucle yarn.

low-twist, elastic and pliable yarn is required. The effect yarn is not completely entrapped by the ground threads and therefore a binder is necessary. The size of the loops may be influenced by the level of overfeed, the groove space on the drafting rollers, the spinning tension or the twist level of the effect yarn. Loop yarns may also be made with slivers in place of yarns for effect.

8.5.7 SNARL YARN

Snarl yarn has a similar twisted core-to-loop structure. Again for the sake of simplicity, the core is shown in Figure 8.13 as two parallel bars. A snarl yarn displays 'snarls' or 'twists' projecting from the core. It is produced by similar method to the loop yarn, but uses a lively, high-twist yarn and a somewhat greater degree of overfeed as the effect yarn. The required size and frequency of the snarls may be obtained by careful control of the details of overfeed and spinning tension, and by the level of twist in the effect yarn.

8.5.8 KNOP YARN

A knop yarn contains prominent bunches of one or more of its component threads, which are arranged at regular or irregular intervals along its length (Figure 8.14). It is normally produced by using an apparatus with two pairs of rollers, each capable of being operated independently. This makes it possible to deliver the base threads intermittently, while the knopping threads that create the effect are delivered continuously. The knopping threads join the foundation threads below the knopping bars. The insertion of twist collects the knopping threads into a bunch or knop. The vertical movement of the knopping

FIGURE 8.12

Structure of loop yarn.

FIGURE 8.13

Structure of snarl yarn.

threads forms a bunch or knop. The vertical movement of the knopping bars determines whether the knop is small and compact or spread out along the length of the yarn.

8.5.9 SLUB YARN

This is a yarn in which slubs are deliberately created to make the desired effect of discontinuity. Slubs are thick places in the yarn that may take the form of a very gradual change, with only a slight thickening of the yarn at its thickest point. Alternatively, a slub may be three or four times the thickness of the base yarn and the increase in thickness may be achieved within a short length of yarn. The yarn pictures in Figure 8.15 should give a clear impression of the structure of the yarn itself.

8.5.10 FASCIATED YARN

This is a staple-fibre yarn that consists of a core of parallel fibres bound together by wrapper fibres. Yarns made by the air-jet spinning method are structured in this way. Yarns produced by the hollow-spindle method are also frequently described as fasciated, as the binder is applied to an essentially twistless core of parallel fibres. The fasciated yarn shown in Figure 8.16 is produced using the hollow-spindle process. It is possible to see fibres that have escaped the dark binding thread and contrast with one of the two slivers used as feedstock in making the yarn.

FIGURE 8.14

Structure of knop yarn.

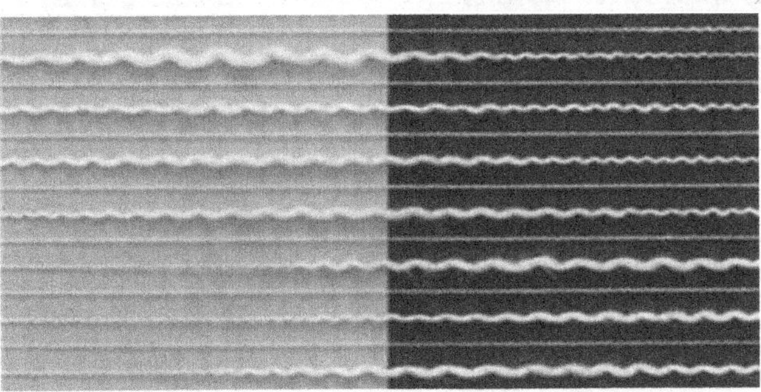

FIGURE 8.15

Structure of slub yarn.

FIGURE 8.16

Structure of fasciated yarn.

8.5.11 TAPE YARN

Tape yarns may be produced using various processes including braiding, warp knitting and weft knitting (Figure 8.17). In recent years, these materials have become better known, especially in fashion knitwear. It is also possible to use narrow woven ribbons, narrow tapes of nonwoven material, or slit film in the same way.

8.5.12 CHAINETTE YARN

Chainette yarn, shown in Figure 8.18, is produced in a miniature circular weft knitting process, often using a filament yarn and a ring of between 6 and 20 needles. The process has been used on a small scale for many years and is now used extensively in fashion knitwear.

8.5.13 CHENILLE YARN

True chenille yarns are produced from a woven leno fabric structure that is slit into narrow, warp-wise strips to serve as yarn. These are pile yarns, and the pile length may be uniform throughout the length of the yarn or may vary in length to produce a yarn of irregular dimensions. Chenille yarns are used in furnishings and apparel. Chenille yarns, as shown in Figure 8.19, have a soft, fuzzy cut pile that is bound to a core. These yarns can be spun, but the machinery required is very specialised. For this reason, these yarns are usually woven on a loom. The effect yarn forms the warp, which is bound by a weft thread. The weft thread is spaced out at a distance of twice the required length of pile. The warp is then cut halfway between each weft thread.

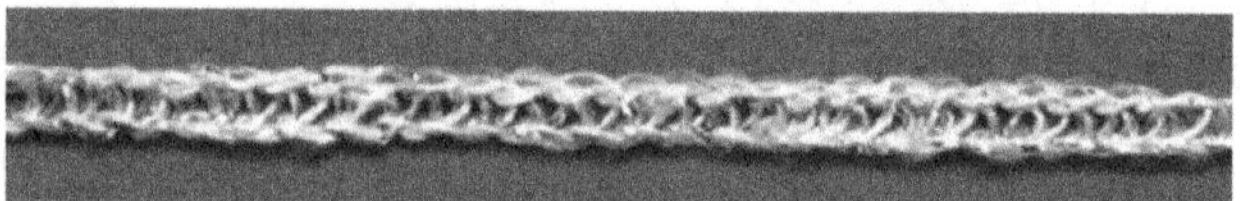

FIGURE 8.17

Structure of metallic tricot tape yarn.

FIGURE 8.18

Structure of chainette yarn.

FIGURE 8.19

Structure of chenille yarn.

8.5.14 RIBBON YARNS

These yarns are not produced by spinning and consist of finely knitted tubes, pressed flat to resemble ribbon or tape. The ribbons are usually soft, shiny and silky.

8.5.15 COMPOSITE YARNS

Also known as compound yarns, these consist of at least two threads. One forms the core of the composite yarn, and the other strand forms the sheath component. One thread is a staple-fibre yarn and other a filament yarn. Compound yarns are even in diameter, smooth and available in the same count range as spun and filament yarns.

8.5.16 COVERED YARNS

Covered yarns have a core that is completely covered by fibre or another yarn. Figure 8.20 shows different types of covered yarns. The core might be an elastomeric yarn, such as rubber or Spandex, or other yarns, such as polyester or nylon. Covered yarns may have either a single or double covering. The second covering is usually twisted in the opposite direction to the first.

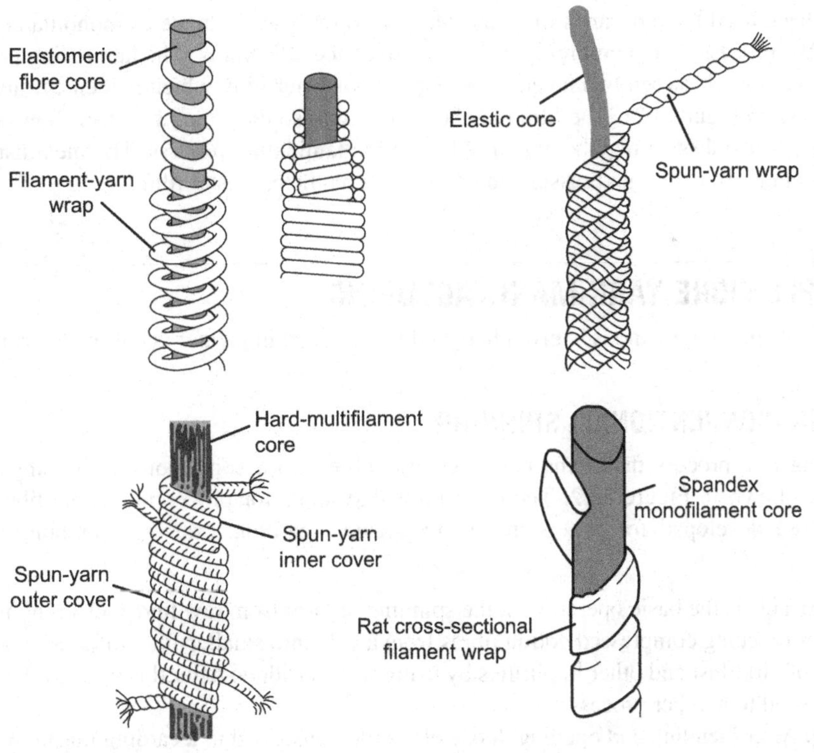

FIGURE 8.20

Different types of covered yarns.

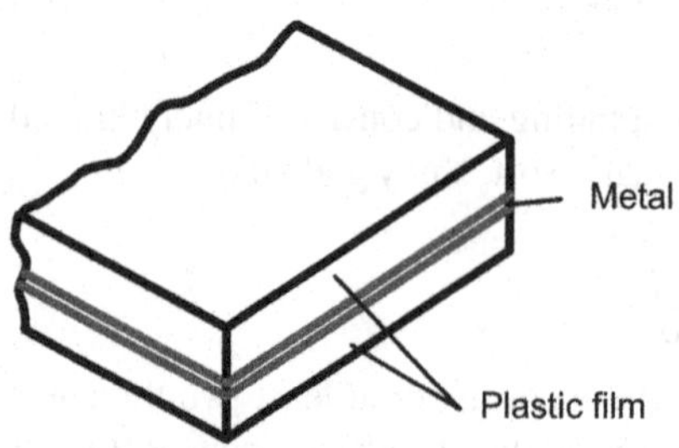

FIGURE 8.21

Cross-section of metallic yarn.

Single-covered yarns have a single yarn wrapped around them. They are lighter, more resilient and more economical than double-covered yarns and can be used in satin, batiste, broadcloth and suiting as well as for lightweight foundation garments. Most ordinary elastic yarns are double-covered to give them balance and better coverage. Fabrics made with these yarns are heavier.

8.5.17 METALLIC YARNS

These have been used for thousands of years. Metallic yarns may be made of monofilament fibres or ply yarns. Two processes are commonly used to produce metallic yarns. The laminating process seals a layer of aluminium between two layers of acetate or polyester film, which is then cut into strips for yarns, as shown in Figure 8.21. The film may be transparent, so the aluminium foil shows through, or the film and/or the adhesive may be coloured before the laminating process. The metallising process vaporises the aluminium at high pressure and deposits it on the polyester film.

8.6 STAPLE-FIBRE YARN MANUFACTURING

An overview of spinning methods is given here and is presented in greater detail in Chapter 9.

8.6.1 RING (CONVENTIONAL) SPINNING

Ring spinning is a process that spins the short, raw fibre into a continuous yarn using a series of machines, as shown in Figure 8.22. The conventional systems for processing staple fibre into spun yarns are those developed for cotton and wool: opening, carding, drawing, combing, roving and spinning.

- **Opening:** This is the basic operation in the spinning of yarn from raw fibres. Opening is the process of reducing compressed cotton fibres from a bale into smaller-fibre tufts. It removes the particles of dirt, dust and other impurities by using spiked rollers. After this process the fibre will be transferred to another process.
- **Carding:** After blending and opening, loose fibres are transferred to a carding machine. Carding is performed by opposing sets of teeth or small wire hooks known as card clothing, which cover the

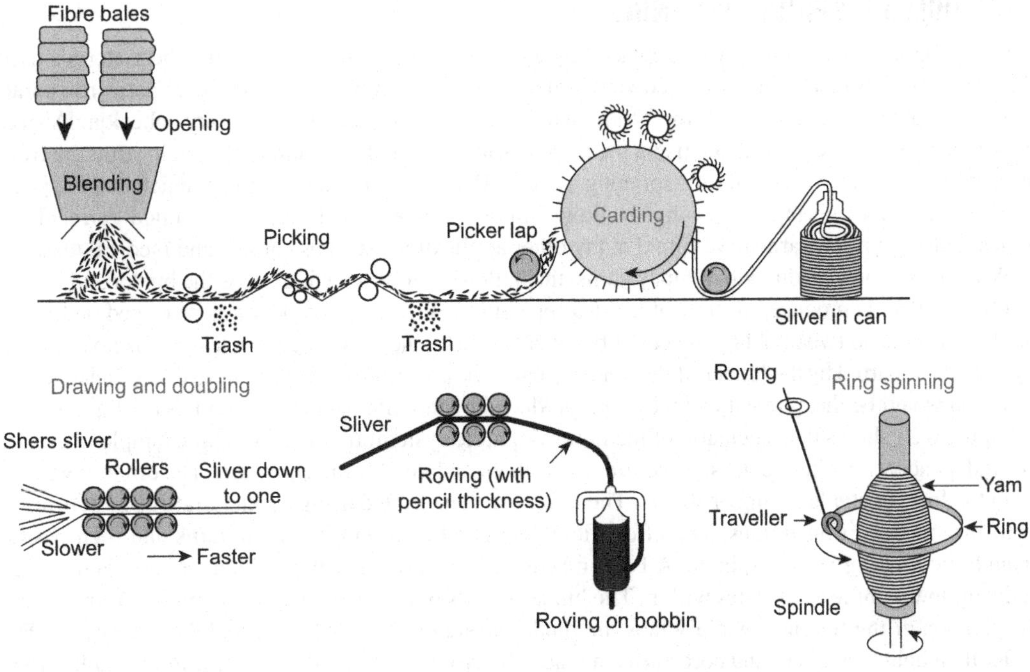

FIGURE 8.22

Process flow chart of conventional ring-spinning process.

machine parts and include a licker-in, a cylinder, revolving flats and a doffer. The cylinder and the flats may rotate in the same or opposing directions but at different speeds to tease the fibre tufts into a thin, filmy web, which is then collected into a loose rope-like structure called a sliver, which is often coiled, and deposited in cans. Carding further opens the fibre tufts and extracts any fine particles, neps and short fibres enclosed by the fibre aggregates. The drawing frame uses a series of rollers arranged in pairs and rotating at different speeds. Slivers are passed between the rollers and combined. The fibres will be well parallelised and mixed after going through this process.

- **Combing:** Combing is the process used to remove short fibres and neps from sheets of cotton fibres (lap). A roller with fine-toothed elements fixed on a half-lap is used. The amount and length of the short fibres extracted will depend on the combing parameters selected. The fibres will be straightened and paralleled during this process.
- **Roving:** In this process, slivers are reduced to around one-eighth of their original diameter by three pairs of rollers, rotating at different speeds. The required level of twist is also imparted to keep the rovings stable under the stretching caused by winding and unwinding.
- **Ring spinning:** The conversion of roving in to yarn is called the spinning process. This is usually done in a roller drafting system that will have some means of fibre control, such as a double apron. Twist is imparted to the fibre strands to prevent slippage through the ring and traveller. The yarn is then wound onto suitable bobbins known as ring cops for further processing.

8.6.2 HOLLOW-SPINDLE SPINNING

In this process, the twist in a yarn is replaced by wrapping a filament binder around the materials used (Figure 8.23). This results in a fasciated yarn structure in which most of the core fibres/yarns run parallel to one another along the axis of the strand, while the binder imparts the necessary cohesion. Despite superficial similarities, yarns made using the hollow-spindle system are quite different in structure from those made by the conventional ring-spinning system. They are also likely to differ in details of appearance and behaviour during processing. Hollow-spindle yarns are used mainly in knitted garments or fabrics, although plain yarns have found many other applications such as carpets and medical textiles.

When used in the production of fancy yarns, the hollow-spindle technique adds the binder and immediately the effect is produced instead of using a separate second operation. In yarns produced on hollow spindles, there is no twist holding the core fibres/yarns of the fancy yarn together, so there is no cohesion beyond that imparted by the binder. If the binder breaks, the core fibres fall apart more freely and dramatically than would be the case with a fancy yarn produced by the ring-spinning system (Salhotra, 1992).

Figure 8.23 shows the schematic of the hollow-spindle system. In this particular example, there are four independent feeding devices, three for effect fibres and one for the core yarn. The effect fibres are fed in the form of staple roving or slivers. The fibres are then drafted using a roller drafting system similar to that used on ring frames. The effect fibres are combined with the core yarns and then passed through the rotating hollow spindle. A bobbin bearing the binder, usually a filament yarn, is mounted on the hollow spindle and rotates with it. The binder yarn is pulled into the hollow spindle from the top. The rotation of the hollow spindle wraps the binder around the staple strand and the core yarns. The binder then holds the effect and core yarns in place. To avoid the possibility of the drafted staple strand disintegrating before it is wrapped by the binder, the spindle usually generates a false twist in the staple strand. The staple strand does not therefore pass directly through the hollow spindle but is first wrapped around a twist regulator, which is usually placed at the bottom of the spindle.

A very wide range of fancy effects can be produced with the hollow-spindle system. Many of these effects can be controlled by controlling the speeds of the corresponding feeding devices. It is also possible to use the hollow-spindle system to create fancy yarns that include yarns in their effect. Many more effects can be produced by controlling the final yarn delivery speed. Since the effect fibres do not have a real twist, hollow-spindle yarns differ from ring yarns in both their appearance and their performance characteristics. The former tend to be bulkier and have lower wear resistance.

8.6.3 COMBINED SYSTEMS

Combined systems were first established in order to unite the benefits of the ring and hollow-spindle systems in a single machine, as it was thought that a yarn with twist had a more stable and reliable structure than one with a fasciated structure. Later, it was recognised that two hollow spindles could also be assembled in series and that this would offer a variety of yarns and a different range of benefits. This is illustrated in Figure 8.24, which depicts two hollow spindles, arranged one above the other, which wrap the staple strand with two binders applied in opposing directions.

This technique is used to produce special-effect yarns that have a more stable structure, as the effect fibres are trapped by two binders instead of one.

Figure 8.25 shows the original combined system in which the hollow spindle and ring spindle are combined in a single machine. In this case, the wrapped yarn is provided with some true twist by the

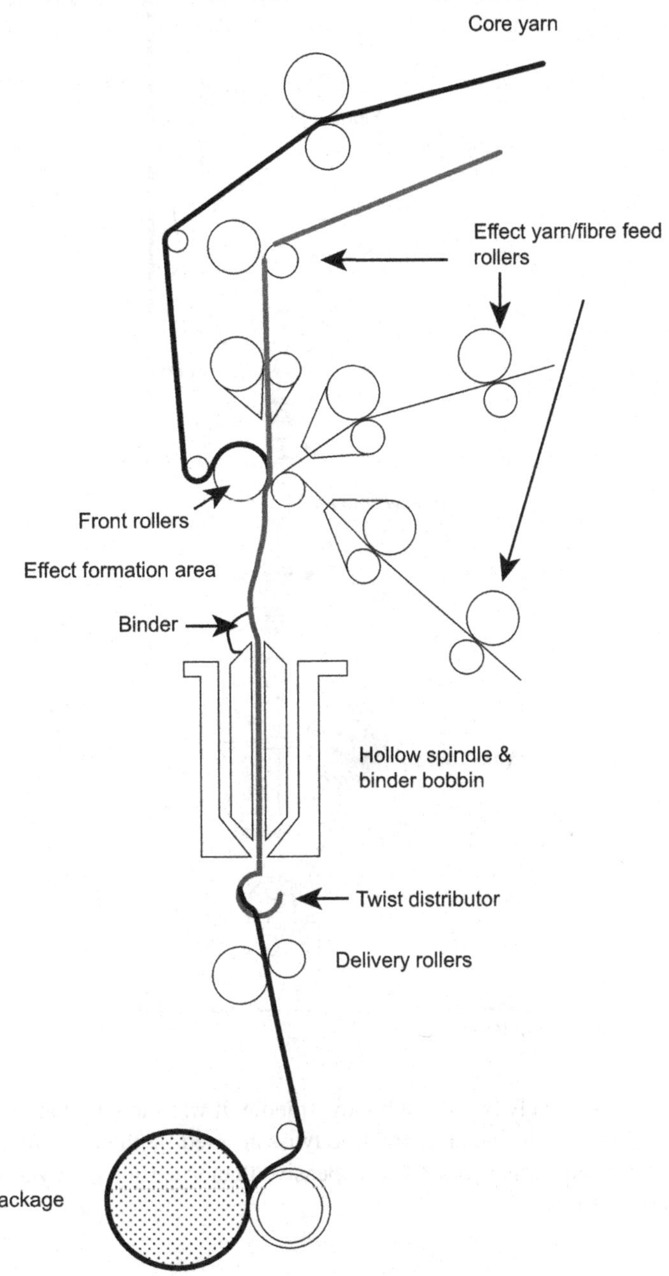

FIGURE 8.23

Schematic of hollow-spindle spinning method.

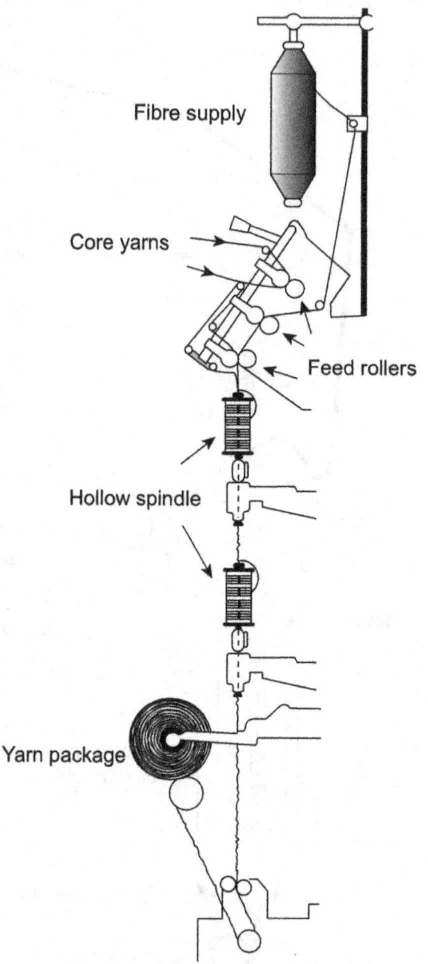

FIGURE 8.24

Schematic of two-spindle wrap system.

ring spindle placed immediately below the hollow spindle. It was thought that the speed of the hollow-spindle assembly would be enhanced by the true twist inserted by the ring spindle, and that it would therefore be able to create yarns that are less expensive than true ring-spun yarns while still retaining some of their characteristics.

8.6.4 THE DOUBLING SYSTEM

The conventional doubling system is based on ring spinning. The arrangement provides two or more yarns that can be fed independently at controlled speeds. These may include uniform, fluctuating or intermittent feeds as required, so permitting a simple means of producing spiral or marl-type yarns,

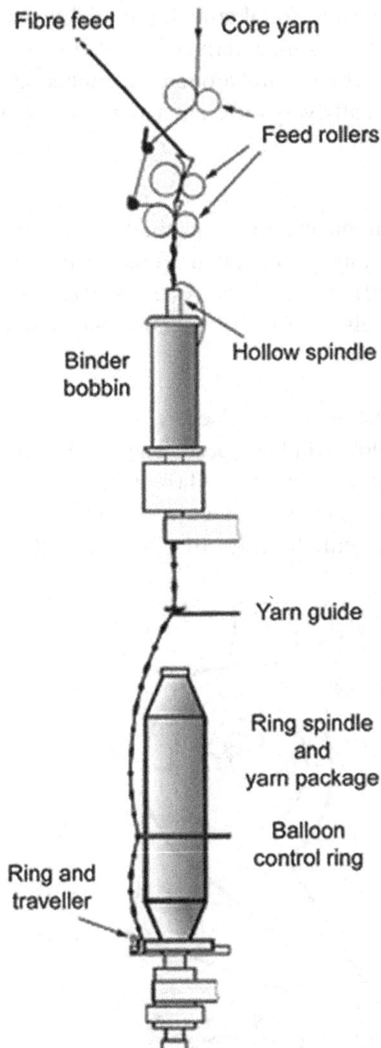

FIGURE 8.25

Combination of ring and hollow spindles.

although obviously requiring the feed material to be in yarn form. This method allows spinners who do not specialise in fancy yarn production to manufacture some of the simpler fancy yarn structures. The doubling frame can produce some interesting effects, particularly when it is used to combine two existing fancy yarns to produce another.

It is also possible to manufacture spiral effects using an ordinary doubling system. This can be done by combining two yarns of very different counts and opposite twist. If the doubling twist is in the same direction as that of the thicker single-yarn twist, the thicker yarn contracts while the thinner yarn expands, thus causing the thinner yarn to spiral around the thicker one. If the doubling twist is in the opposite direction to that of the thicker single-yarn twist, the thicker yarn expands while the thinner yarn contracts, causing the thicker yarn to spiral around the thinner one. Although their basic structure is identical, these yarns are aesthetically very different and will be used in different ways.

8.6.5 OPEN-END SPINNING

There are two commonly used yarn manufacturing methods operating on the open-end spinning principle: rotor and friction. The rotor system is mainly used for the production of coarse to medium count short staple yarns (Lawrence, 2010). The friction system is used mainly to make coarser industrial yarns. However, both systems may also be used for making some fancy yarns.

8.6.5.1 Rotor system

In open-end spinning, the yarn twisting action is separated from the winding action and the package needs to rotate only at a relatively low winding speed. The process may be divided into the following steps: opening, transport, alignment, overlapping and twist insertion.

In the rotor-spinning process, as shown in Figure 8.26, individual fibres are carried into the rotor on an air stream and laid in contact with the collecting surface so that a strand of fibres is assembled

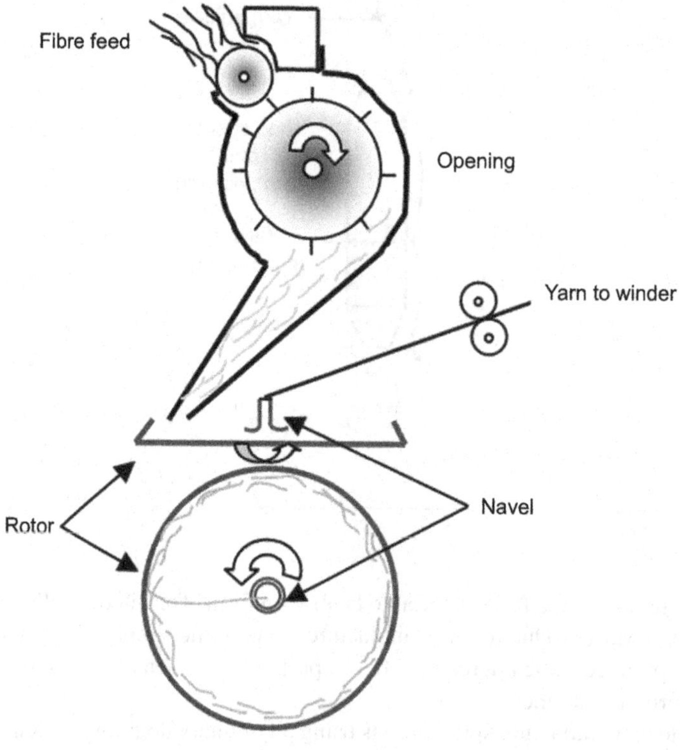

FIGURE 8.26

Principle of rotor spinning.

around the circumference. As the fibres are drawn off, twist is imparted by the rotor, to produce a yarn. Rotor spinning is most suitable for spinning short staple-fibre yarns. Recent developments in electronic control have allowed the development of rotor spinning machinery that is also capable of producing slub yarns. These yarns are used in furnishings and drapes, rather than in apparel fabrics, although they are sometimes used in denim fabrics. They are manufactured using attachments to ordinary open-end spinning devices, which usually incorporate an electronically controlled device for the brief acceleration of the drawing-in roller. As a result of the back doubling action inside the rotor, it is not possible to produce slubs shorter than the circumference length of the rotor because any variation in the fibre feed material is spread over a minimum length of the rotor circumference.

There have also been attempts to alter the fibre flow and thus introduce variations in the yarn appearance by the use of injecting pressurised air into the fibre-transportation tube. However, the effects created using this approach are very limited, as the fibre flow in the transportation tube is slight and the variation in the yarn caused by changes in the airflow is therefore very small.

8.6.5.2 Friction system

Friction spinning is an entirely different open-end spinning technique. Instead of using a rotor, two friction rollers are used to collect the opened-up fibres and to twist them into yarn. The principle of DREF-2 is shown in Figure 8.27. The company also produces the DREF 3 machine, which has an extra drafting unit in the machine in order to feed drafted staple fibres to form a core component.

The fibres are fed in sliver form and are opened by the opening roller. The opened fibres are then blown off the opening roller by an air current and transported to the nip area of two perforated friction drums. The fibres are drawn onto the surfaces of the friction drums by air suction. The two friction drums rotate in the same direction, and twist is imparted to the fibre strands because of the friction with the two drum surfaces. The yarn is withdrawn in the direction parallel to the axis of the friction drums and is delivered to a package-forming unit. A high twisting speed can be obtained even while using a relatively low speed for the friction drums, because the friction drum diameter is much larger than that of the yarn.

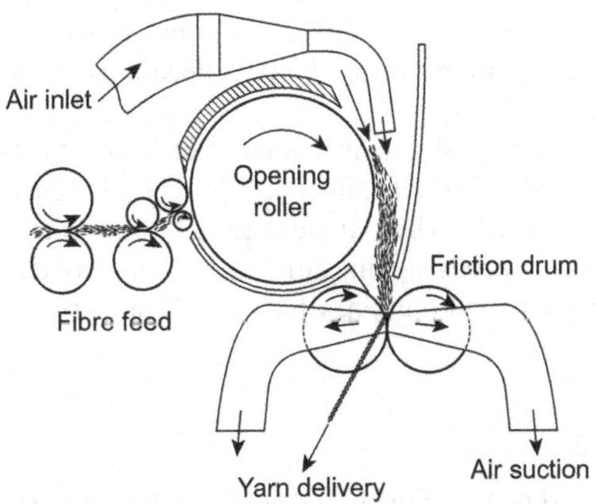

FIGURE 8.27

Principle of friction spinning.

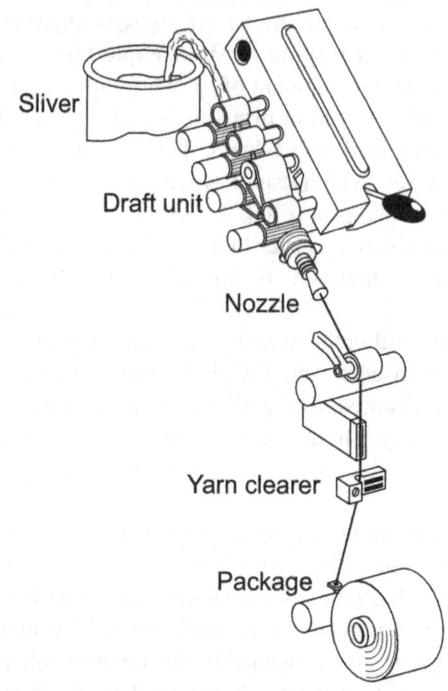

FIGURE 8.28

Feed material passage in Vortex spinning.

8.6.5.3 Vortex spinning

Vortex open-end spinning technology has recently emerged as a revolutionary development in the field of fasciated yarn technologies. But the earlier impression of low quality and processing ability has hindered acceptance of these technologies. The most advanced fasciated yarn-spinning technology that has gained momentum during the past decade is the Vortex-spinning technology (Figure 8.28).

Vortex spinning was introduced by Murata machinery Ltd Japan at OTEMAS'97. This technology is best explained as a development of air-jet spinning, especially designed to overcome the limitation of fibre types in Murata air-jet spinning. The main feature of Murata Vortex spinning (MVS) is its ability to spin carded cotton yarns at speeds significantly higher than any other system currently in existence. The machine produces yarn at 400 m/min, which is almost 20 times that of ring-spinning frame production (Figure 8.29).

8.6.6 AIR-JET SPINNING

This is a pneumatic method and is not an open-end spinning process. The drafted fibre strands pass through one or two air nozzles located between the front drafting roller and the take-up system (Figure 8.30). The roller drafting system drafts the input sliver into a ribbon-like parallel fibre

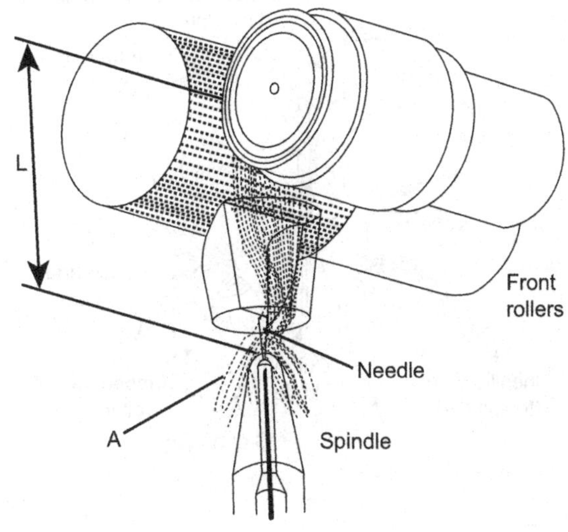

FIGURE 8.29

Expansion of fibre edges due to whirling force of the jet air stream.

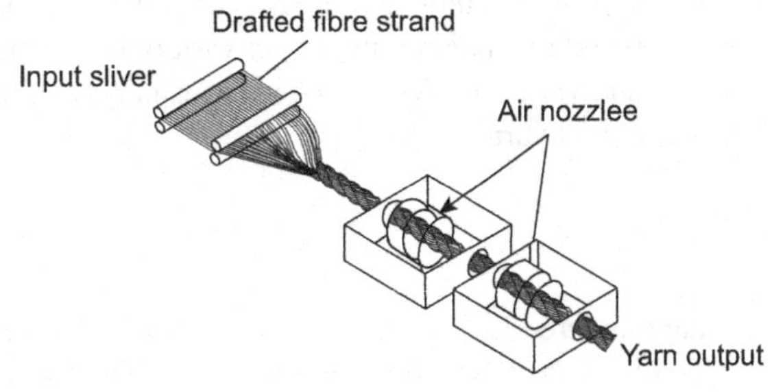

FIGURE 8.30

Principle of air-jet spinning.

strand. High-pressure air is injected into the nozzles, causing swirling airstreams inside the nozzle. This results in the insertion of false twist into the drafted fibre strands. The edge fibres wrap onto the surface of core strand and form the yarn.

8.6.7 THE CHENILLE YARN SYSTEM

A method of producing a chenille yarn that forms two ends at each unit is illustrated in Figure 8.31. The effect yarns are wrapped around a gauge or former which is triangularly shaped at the top, narrowing towards the base to allow the effect yarn coils to slide downwards onto the cutting knife. The width at the bottom of the gauge determines the effect length, by maintaining the depth of the pile, or 'beard', in the final yarn. Although, for the sake of simplicity, the cutting knife is shown in Figure 8.31 as a straight knife edge, modern machines all use a circular cutting knife.

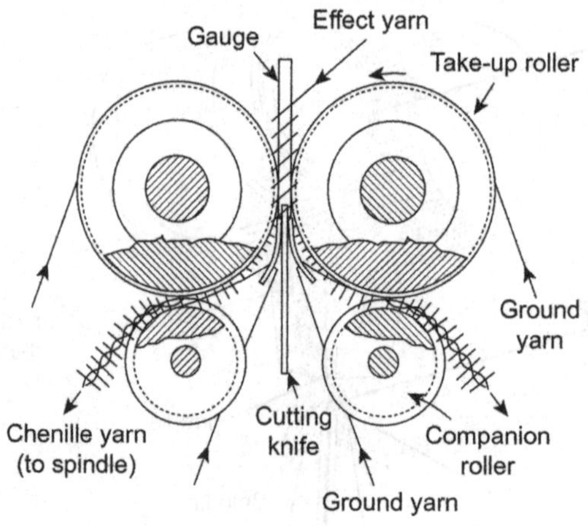

FIGURE 8.31

Chenille yarn production process.

On each side of the cutting knife there are two ground yarns, which may be either single or twofold One ground yarn is guided by the take-up roller while the other is guided by the companion roller. The take-up roller is pressed against the profiled guide and inter-meshes with the companion roller, allowing the two ground yarns to trap the pile created by the effect yarn in between them at right angles to the ground yarn axis. The two ground yarns are twisted together, usually by a ring spindle at the lower part of the machine, to produce the final yarn.

8.6.8 FLOCKING

Chenille effects can also be manufactured by a flocking process in which a ground yarn coated with adhesive is flocked electrostatically with loose fibres (Figure 8.32). The loose fibres and the ground yarn are charged with opposite electrostatic charges. As a result of this, the loose fibres are attracted to the ground yarn and are bonded to it by the adhesive. The loose fibres have the same electrostatic charge and repel each other, leading to good fibre separation and also forcing them to 'stand' on the ground yarn rather than lie flat on its surface. This is a very economical manufacturing method, but the yarn has poor abrasion resistance because the anchor of the loose fibres onto the ground yarn is weak and they can therefore by easily be worn off, leaving the ground yarn bare.

8.6.9 MOCK CHENILLE

A mock chenille effect can be manufactured by plying two gimp, boucle, or loop yarns with dense effects (two loop yarns with large numbers of small loops). The yarn may not look like chenille, but when it is made into fabric, the large number of small loops in the fabric results in a fabric surface resembling a chenille effect.

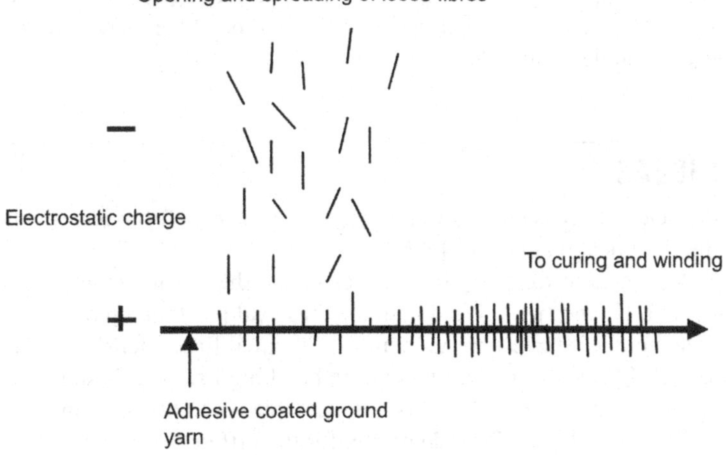

FIGURE 8.32

Chenille yarn production by flocking method.

8.7 FUTURE TRENDS

New uses for yarns beyond the traditional apparel and fancy applications are emerging. These include technical applications in which the properties of the yarns have to be engineered. In these cases, composite or hybrid yarns, which combine different types of fibres and structures to produce the required functional properties, are produced with techniques such as core spinning.

Although many yarn manufacturing techniques are very well established, the technical application of textiles in fields such as industry, civil use, medicine and sport require ongoing development in processing methods in order to obtain specific types of yarn structures able to meet stringent property requirements.

8.8 SUMMARY

Different types of yarns from staple fibres and continuous filaments from both natural and synthesised materials are described in terms of their structures, properties and applications. Appropriate yarn structures and manufacturing technologies are selected for required applications and properties. Detailed definitions and manufacturing principles for different types of yarns (e.g. simple, ply, cord, fancy and core-spun) are provided. Particular emphasis is given to various types of fancy yarns, which are described with suitable diagrams and detailed explanations of their manufacturing methods.

In the case of synthetic continuous-filament yarns, manufacturing methods including melt spinning, wet spinning, dry spinning and gel spinning techniques are explained and different forms of continuous-filament yarns are dealt with, such as flat yarns, textured yarns, bi-component yarns and split tape or film yarns. In case of staple-fibre yarns, manufacturing technologies including ring, rotor, air-jet spinning and Vortex spinning are explained together with the type of yarn structures obtained from the respective technologies and their end-uses.

8.9 PROJECT IDEAS

A. Go to the market and collect as many different yarns as possible. Try to identity the type of yarn according to the classification given in Table 8.1.

B. With the help of a sensitive weighing balance, calculate the linear density of these yarns in (i) English count, N_e and (ii) Tex and tabulate these values. Pick up one continuous filament yarn and another staple fibre yarn almost of equal linear density. If an Universal Testing Machine (UTM) is available, find out the breaking loads of these two yarns and see which one is taking higher load. If UTM is not available try to break them with hand and see which is more difficult to break. Provide reason for the difference based on their structure.

C. Take staple fibre yarns of almost same linear density produced with ring spinning, rotor spinning and air-jet spinning. Repeat the same experiment mentioned in B and see which type of yarn is taking maximum load to break. Give reason for the difference based on the structure of these yarns. Observe the surface of these yarns under a microscope. Comment about the surface structure of these yarns.

D. Take two yarns of same linear density one made with ring spinning and the other made with rotor spinning. Cut 30 pieces of 1 m length from each of these two types of yarns. Measure the weights of these cut pieces accurate up to mg and calculate the CV% for each of these yarn types. Find out which yarn type has higher CV% and try to reason out based on the manufacturing procedures of these yarns that you studied in this chapter.

E. Collect different types of fancy yarns available in the market. Looking at the structure, try to find out the name of these fancy yarns. From your collection of yarns (both A and E put together) locate fancy and normal yarns of almost same linear density and check which is taking more load to break using an UTM or with hands. Give reason for the differences based on the differences in the structures of these yarns.

8.10 REVISION QUESTIONS

1. How are the different types of yarns classified based on their properties in general?
2. What are the different methods available to manufacture yarns from staple fibres?
3. How do the structures of staple fibre yarns vary based on their method of manufacturing?
4. What are the typical yarn counts used for (a) furnishing fabrics (b) shirts (c) night wear and (d) carpets?
5. What are the principles involved in melt spinning, dry or solution spinning and gel spinning?
6. Give examples of fibres spun through spinning, dry or solution spinning and gel spinning.
7. What are the different applications for synthetic filament yarns?
8. Mention at least four types of fancy yarns and explain how they are made?
9. What are the process steps involved in manufacture of staple fibre yarns?

10. What is the principle of rotor spinning? Why this method is preferred to spin only coarse counts of yarns?
11. How does a friction spinning machine work? Explain how core spun yarns are made with these machines.

REFERENCES

Deopura, B. L., Alagirusamy, R., Gupta, B., & Joshi, M. (Eds.). (2008). *Polyesters and polyamides*. Cambridge: Woodhead Publishing.

Eric, O. (1987). *Spun yarn technology* (pp. 138–212). London, Boston, MA: Butterworth & Co (Publishers) Ltd.

Gong, R. H., & Wright, R. M. (2002). *Fancy yarns: their manufacture and application*. Cambridge: Woodhead Publishing.

Gupta, V. B., & Kothari, V. K. (Eds.). (1997). *Manufactured fibre technology*. Springer.

Klein, W. (1995). *New spinning systems, short staple spinning series. Manual of textile technology* (Vol. 5). Manchester: The Textile Institute International.

Lawrence, C. A. (2003). *Fundamentals of spun yarn technology*. Boca Raton: CRC Press.

Lawrence, C. A. (Ed.). (2010). *Advances in yarn spinning technology*. Cornwall: Woodhead Publishing.

Lord, P. R. (2003). *Hand book of yarn production: technology, science and economics* (pp. 184–211). Cambridge: Woodhead Publishing.

McIntyre, J. E. (Ed.). (2004). *Synthetic fibres nylon, polyester, acrylic, polyolefin*. Cambridge: Woodhead Publishing.

Salhotra, K. R. (1992). An overview of spinning technologies: possibilities, application and limitations. *Indian Journal of Fibre and Textile Research, 17*(4), 255–262.

CHAPTER

FIBRE TO YARN: STAPLE-YARN SPINNING

9

I.A. Elhawary
Alexandria University, Alexandria, Egypt

LEARNING OBJECTIVES

At the end of this chapter, you should be able to:

- Understand the preparation of short and long staple fibres for spinning
- Understand the cotton, woollen and worsted systems, their differences and similarities
- Understand the different techniques for spinning staple yarns, their advantages and disadvantages
- Understand the properties and end-uses of yarns produced by each method

9.1 INTRODUCTION

Fibres can be classified as either filament fibres or staple fibres. Filament fibres are long, continuous fibres; they are usually synthetic, although silk is an exception in that it is a natural filament fibre. The production of yarns from filament fibres is discussed in detail in Chapter 10.

This chapter will review the production of yarn from staple fibres. A staple fibre is a non-continuous fibre of relatively short length. Because of their short length, staple fibres must be twisted together to form a long, continuous yarn (hence the term 'spun yarns'). Staple fibres are usually natural fibres, although synthetic fibres can be cut into similarly short lengths to be blended with natural staple fibres or used on their own to produce yarns with a natural feel. Natural staple fibres include cotton, wool and flax.

The natural staple fibres used in the textile industry are classified by the typical length of the fibre and are described as 'short staple' or 'long staple':

- Short staple fibres have a maximum length of 60 mm. Cotton fibres are short staple fibres, having a length of 25–45 mm. Cotton linters (the fibres that remain adhered to the cotton seed after the first ginning) are even shorter, at just a few millimetres long; these are used for the manufacture of lower-quality products such as lint and cotton wool.
- Long staple fibres have a length of more than 60 mm. Wool fibres are long staples, with a length of about 60–150 mm.

As a result of their short length and non-uniform nature, staple fibres require greater processing before a satisfactory yarn can be produced; this obviously adds to production costs. However, the desirable characteristics of staple fibre yarns – such as comfort, warmth, softness and appearance – often compensate for these increased costs.

Textiles and Fashion. http://dx.doi.org/10.1016/B978-1-84569-931-4.00009-3

The system used to prepare staple fibres for spinning depends on the type. Short staple fibres are prepared using the cotton system, while longer wool fibres may be prepared using either the woollen system or the worsted system (also called the long staple system). It should be noted here that shorter wool fibres (<40 mm) can also be processed using the cotton system. These systems will be discussed in turn in the following sections. The chapter will then review the various different spinning techniques that are used to produce staple fibre yarns and the characteristics of the yarns produced by the different methods.

9.2 PREPARATION OF COTTON AND OTHER SHORT STAPLE FIBRES

Cotton fibres arrive at the cotton spinning mill in the form of cotton bales that are packed densely, typically with wrapping secured with polyproblene ties. At this stage, the cotton contains 1–15% impurities (e.g. dust, dirt, vegetable matter) which must be removed as the cotton is processed. In order to convert the raw cotton into fibres that are separated and aligned in a suitable manner for yarn production, the cotton passes through the following processing stages:

1. Opening and cleaning
2. Blending
3. Carding
4. Combing (optional)
5. Drawing
6. Roving

Stages 1 to 3 take place in the blowroom, so called because the cotton is transferred from stage to stage using pneumatic transport, i.e. it is blown from one machine to the next. The fibres that leave the blowroom are separated into disentangled individual fibres which can then be further processed ready for yarn production. These six stages will now be discussed in turn.

9.2.1 OPENING AND CLEANING

Figure 9.1 shows the various operations that occur in the blowroom. The term 'opening' refers to the process whereby the high-density bale is broken down into large clumps or tufts of fibres; these tufts are subsequently further broken down into smaller tufts ('tuftlets'). At this stage the individual fibres are not separated. In modern production mills, numerous cotton bales are unwrapped and arranged in lines in the blowroom for the start of processing. The automatic bale opener, which consists of sets of opposing points or spikes, then works its way along a line of bales (termed 'bale lay downs'), plucking tufts from each bale. The tufts are then transported pneumatically to the next stage, which is the cleaning operation. At this stage a 'pre-mixing' operation is also carried out in which bales of different cotton grades or even fibre types (cotton + viscose, etc.) can be positioned in the bale lay-down to achieve a certain blended specification.

Natural fibres such as cotton will inevitably contain impurities such as leaf, seed, trash and dust, which must be removed if high-quality yarns are to be produced. The opening and carding operations of the blowroom machines disentangle these impurities from the fibres, which can then be removed using air currents that simultaneously transport the cotton tufts. Figure 9.2 shows a typical

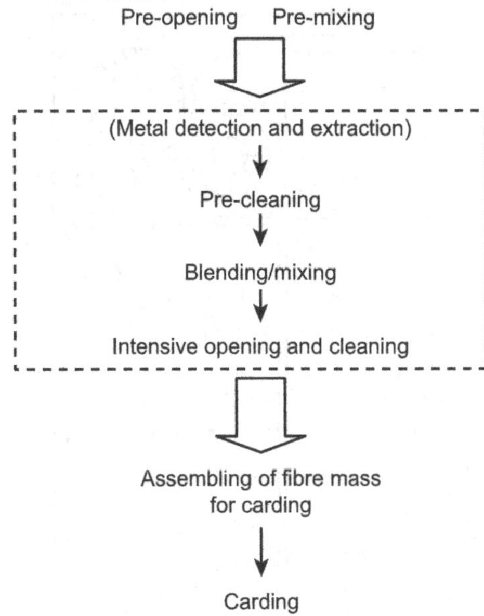

FIGURE 9.1

Blowroom operations.

opening/pre-cleaning machine that uses a complex system of beaters, suction and inter-fibre friction to remove further dust and impurities.

9.2.2 BLENDING

Natural fibres such as cotton can have noticeable variations in properties, such as maturity, length, strength and elongation. In order to avoid processing problems further downstream in the cotton mill, it is essential that fibres are well blended to produce a homogeneous mass that should result in a consistent yarn quality (in terms of, for example, strength and evenness). Blending may also be used to reduce production costs, i.e. higher (more expensive) grades of cotton may be blended with lower (cheaper) grades of cotton to reduce raw material costs per kilogram. The blending of cotton with man-made fibres often takes place further downstream from the blowroom. Manufacturers may, for example, use cotton/polyester blends to create easy-care fabrics. Some mills, however, will carry out blending of cotton and man-made fibres in the blowroom stages as it may lead to better blending.

As the tufts arrive at each machine in the blowroom, they are collected in a hopper before being processed by the machine. As the tufts are collected, there is the opportunity for more mixing or blending to occur, so that tufts from different bales and different parts of the same bale can be mixed together to reduce any variations in quality. Blowroom mixing takes place principally in a stack blender. Vertical compartments (usually 4–10) of a storage bin are filled in sequence and then layers are removed from consecutive stacks to form a sandwich of layers, which mixes the tufts from different bales, or parts of bales, together (see Figure 9.3).

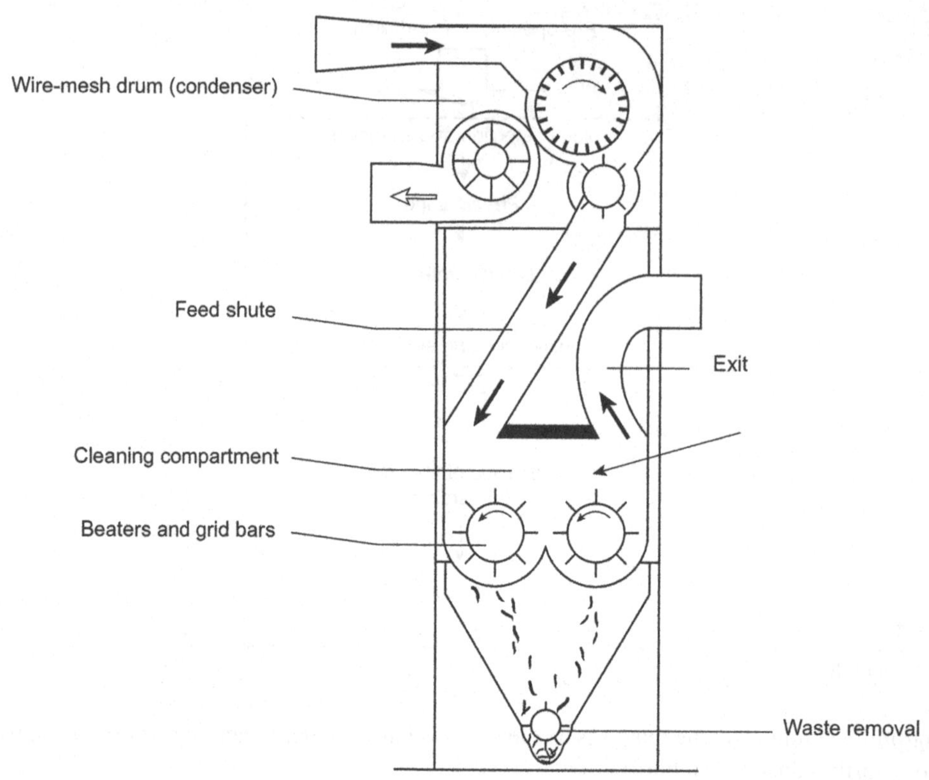

FIGURE 9.2

Opening/pre-cleaning machine.

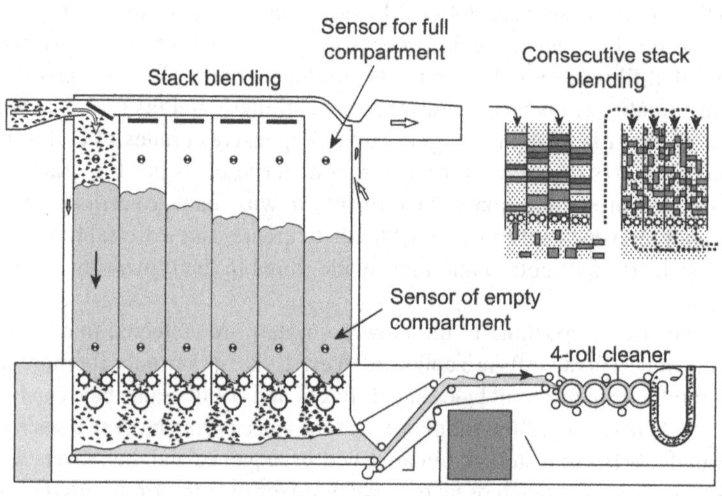

FIGURE 9.3

Stack blending operation.

9.2.3 CARDING

The final stage in the blowroom process is carding. In this stage, the cleaned and blended short tufts are converted into individual fibres. For the first time in the blowroom, fibre orientation is important. Fibres begin to be straightened and to become oriented in a common direction. These oriented fibres are then reassembled into a twistless rope of disentangled fibres held together by inter-fibre friction. This twistless rope is called a carded sliver, and is coiled into large cans ready for use in the subsequent processing steps.

The carding machine consists of an inner cylinder and an outer belt (the 'flats'), both of which are covered in hundreds of fine wires (Figure 9.4). These wires intermesh and act to separate the fibres and arrange them in a more or less parallel form. The wires also remove any remaining trash and entangled lumps of fibres (termed 'neps'). The carding cylinder forms a thin web of fibre. The fibre is removed from the cylinder in the 'doffing' process, whereby a second cylinder, the doffer, also covered with saw-tooth wire, converts the fibres from a sheet-like web to the rope-like sliver. The sliver is held together by the friction between the parallel fibres. The carding process also provides further opportunity for blending different fibres or different qualities of the same fibre.

9.2.4 COMBING

Combing is an optional step in fibre production and is used when a smoother, finer yarn is required. Fibres undergo further orientation by means of a comb-like device that arranges the fibres into an even stricter parallel form. The fibres are fed into the machine so that a fringe of fibres (held in place by nip jaws) can be combed by rotating pins. The combing process also removes short fibres (<0.5 in), fibre hooks and any neps or impurities that might remain. Fibre hooks are fibres with hooked ends, created during the carding process as the fibres are moved along by the carding machinery; if not removed, they will result in the production of a weaker yarn.

Combed yarns have a superior appearance compared to carded yarns, having smoother surfaces and finer diameters. The removal of short fibres means that fewer short ends show on the surface of the fabric, and the lustre of the fabric is also increased. The extra processing stage means that combed yarns

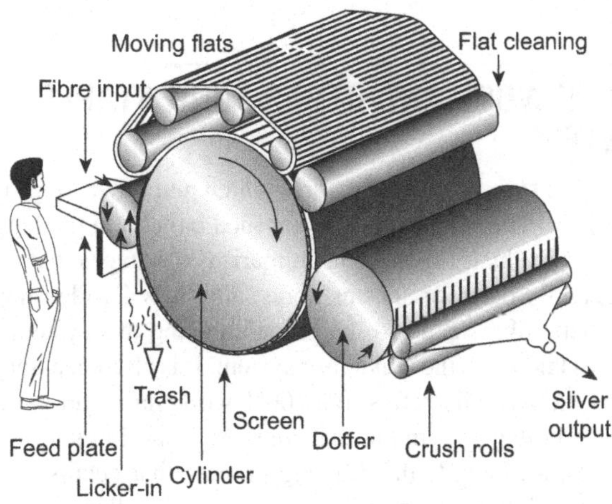

FIGURE 9.4

Carding machine.

are more expensive to produce, and therefore fabrics made form combed yarns are higher in price. As a result of this extra cost, the majority of cotton yarn is only carded, rather than carded and combed.

9.2.5 DRAWING

Drawing involves the processes of doubling and drafting. 'Doubling' refers to the fact that several of the slivers formed during carding (or carding and combing) may be combined during drawing. 'Drafting' refers to the attenuation (lengthening) and straightening of the slivers. However, the terms 'drafting' and 'drawing' are often used interchangeably. Again, blending can take place at this stage, and in fact this is where cotton is usually blended with a man-made fibre.

The carded sliver may contain up to 30,000 fibres in its cross-section. The function of the drawing process is to elongate (or 'draw out') the sliver to reduce the linear density to a level that is suitable for spinning (about 100 fibres in a cross-section). This elongation is achieved on a drawframe by a series of rollers rotating at different speeds to produce a single, uniform strand, which is then fed into large cans. This uniformity is vital to downstream production processes, for example, a thin area in a wide sliver could become a very thin and weak area in the final yarn.

The drawing process also removes fibres with hooked ends from the carded sliver. The drawing process is repeated twice for carded slivers, while combed slivers are drawn once before combing and twice after combing – hence the improved quality of fibres that have undergone both carding and combing operations.

9.2.6 ROVING

The drawn sliver is fed into a machine called a roving frame, the final step before spinning. In the roving frame, the strands of the fibre are lengthened still further by a series of rollers. As the fibres are wound onto the bobbins they are twisted slightly, and this twisted strand is called the roving. The twist results in improved cohesion of the strand and also condenses it so that the strand can be handled effectively in the subsequent spinning process.

9.3 PREPARATION OF WOOL AND OTHER LONG STAPLE FIBRES: THE WOOLLEN SYSTEM

Two different fibre-preparation systems are used for woollen and other long staple fibres: the woollen system and the worsted system. The system chosen depends largely on the length of the staple and the properties required in the end product. Figure 9.5 summarises the stages that occur in the two systems. The woollen system is generally used for shorter fibres, although a wide range of wool types can be prepared in this way. Synthetic fibres can also be prepared using this system, and for this reason the woollen system is sometimes known as the condenser system. It is also necessary to distinguish between the terms 'woollen yarn', which usually refers to a 100% wool yarn, and 'woollen-spun yarn', which refers to a yarn spun on the woollen system but that contains other fibres.

The woollen system does not include the drawing process that occurs in the cotton system. This means that the fibres are not straightened, and as a consequence woollen yarns are soft and bulky. They also have many fibre ends on the surface of the yarn, which gives them a fuzzy appearance and hand.

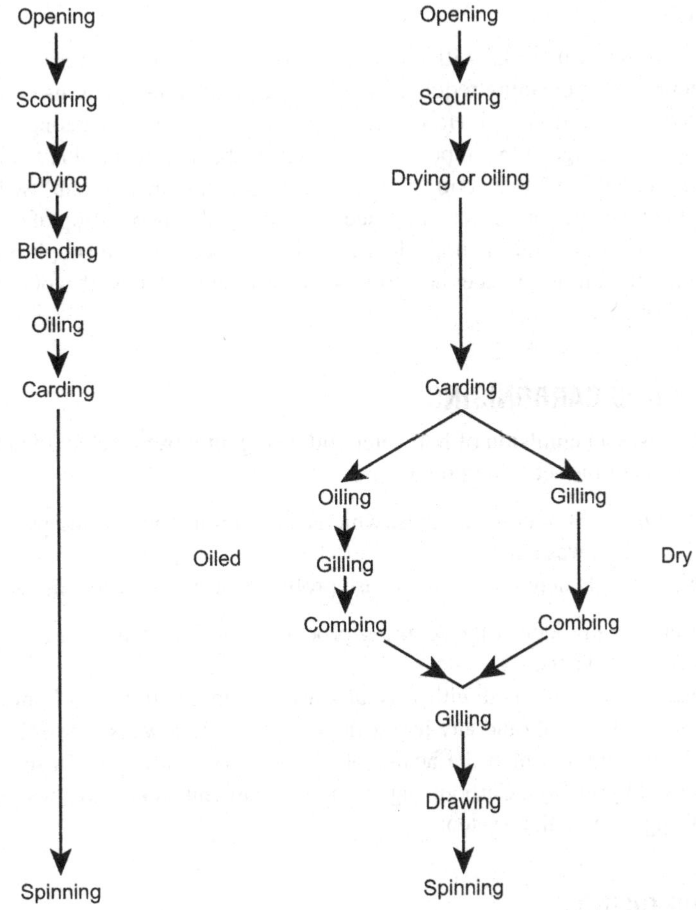

FIGURE 9.5

woollen (left) and worsted (right) systems.

Because of the air within the yarn they have excellent insulation properties. However, the fibres are relatively weak and have poor abrasion resistance.

The main steps in the woollen system are as follows:

- Opening
- Scouring
- Drying
- Blending
- Oiling
- Carding

However, because of the wide range of raw materials processed using this system, the exact sequence of machines used by individual manufacturers varies widely.

9.3.1 OPENING

The condition of the wool raw material will vary in terms of cleanliness and grease content, and hence the amount of opening and/or cleaning required will also vary. The equipment used for wool opening is similar to that used in the cotton mill. However, because the wool fibres are longer, the teeth on the sets of opposing spikes are longer. One important difference is that the material is fed into the machine and left there for a specified time; this contrasts with the continuous-feed system used in opening cotton. The wool-opening machine (or 'teaser') also needs to deal with any build-up of grease that occurs; again, this is not the case in cotton opening. If the opening process is too rough it may damage or entangle fibres; it may therefore be necessary to process different batches of wool in different ways before any blending can occur.

9.3.2 SCOURING AND CARBONISING

The scouring process uses an emulsion of hot water and detergent to remove contaminants and grease from the wool. The aims of the scouring process are:

- To ensure effective removal of contaminants, which can result in poor dyeing performance, weakened yarns and high wastage.
- To minimise fibre entanglement, which can cause problems during subsequent carding.

Lanolin is an important by-product of the scouring process. It is extracted from the scouring effluent and used in the cosmetics industry.

If the wool is heavily contaminated with vegetable matter, an additional carbonising step may be required. In this process, the wool (and any impurities) is soaked in a weak acid solution; the wool is then heated to a temperature at which the acid-soaked vegetable matter carbonises. The carbonised material is then removed by suction. Carbonising is a harsh treatment and results in short fibres that can only be processed using the woollen system.

9.3.3 DRYING OR OILING

Because the scoured fibres now contain no grease, a small amount of oil (5–10% of oil or 50/50 oil/water emulsion) may be added to lubricate the fibres; the lubricant may also contain a bactericide. The lubrication is applied to minimise fibre breakage, reduce static electricity, increase fibre cohesion and lubricate the fibres for drawing and twisting. In some systems, however, the fibres are simply dried without the addition of oil.

9.3.4 BLENDING

To achieve consistency in terms of cleanliness and fibre properties, several blending processes may be required before carding. Because the carded material goes directly to the spinning process, there is no further opportunity to correct any unevenness. Initial blending and quality control are therefore of the utmost importance. Modern blending methods involve pneumatic handling of the fibres, which are mechanically spread out in layers. Vertical slices are then removed from these layers. Other fibres such as cotton or linen can be blended at this stage to impart desirable properties to the finished yarns or to reduce the cost of the finished yarn by blending with cheaper materials.

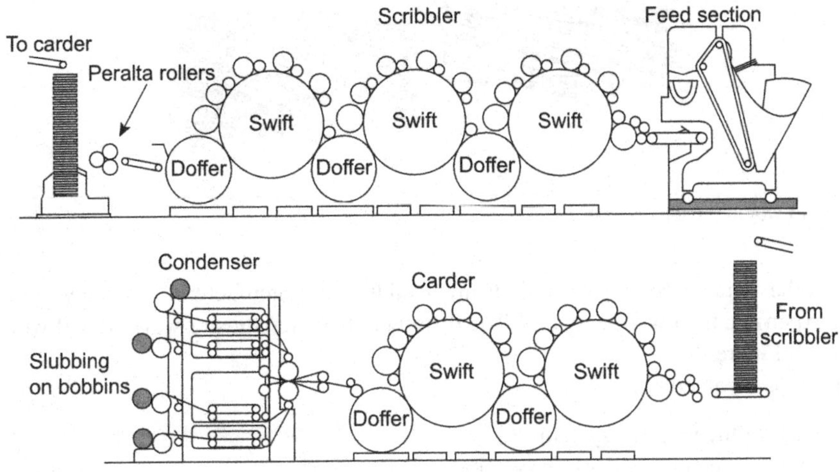

FIGURE 9.6

Carding (woollen system).

9.3.5 CARDING

Wool fibres are longer and have better cohesion than cotton fibres. The wool-carding process is also more effective than the cotton-carding process. For these reasons, the carded material can be used directly in spinning processes without further drawing. The exact specification of the woollen card will vary according to the types of fibres that will be processed and the range of yarn linear densities required. All systems consist of a series of rollers (swifts) surrounded by smaller rollers; the fibres are transferred from one swift to the next by a doffer, which has angled pins to pick up the fibres. All the rollers are covered in metal pins that open the fibre tufts. The density of these pins increases as the fibres progress through the carding machine, thereby increasing the opening and disentangling that occurs. Figure 9.6 shows the two main sections of the woollen card: the scribbler (breaker card) and the carder (finisher card).

A major difference between the woollen system and the worsted system is that the woollen system delivers 'slubbings' rather than slivers (as in the worsted system). Slubbings are similar to rovings but have no twist. The web of fibres produced by the carding machine is split into a number of ribbons by a condenser; these ribbons are then rubbed together to produce the fibre cohesion needed to form the slubbing.

9.4 PREPARATION OF WOOL AND OTHER LONG STAPLE FIBRES: THE WORSTED SYSTEM

The worsted system is used for longer and finer wool fibres and also synthetic staple fibres of appropriate lengths. The name comes from the village of Worstead in Norfolk, England, where several processes for producing finer cloths were first developed. In contrast to the woollen system described earlier, the worsted system includes a drawing process (termed 'gilling') that straightens out the fibres. Worsted yarns are therefore smoother, sleeker and more compact in appearance than woollen yarns. They are also stronger than woollen yarns and have a crisper hand (feel).

FIGURE 9.7

Woollen (left) and worsted (right) yarns.

Because of the extra processing operations involved in their manufacture, worsted yarns are significantly more expensive than woollen yarns. The different characteristics of worsted and woollen yarns are illustrated in Figure 9.7.

The main steps involved are:

- Scouring – as for the woollen system
- Drying or oiling – as for the woollen system
- Carding
- Gilling
- Combing

Only the processes that differ from the woollen system will be described in the following section. It should be noted that there is no distinct blending stage in the worsted system, since this can be carried out during gilling and combing.

9.4.1 CARDING

The principle of carding in the worsted system is similar to that of the woollen system, but there are fewer elements in the worsted carding machinery. The aim is to open up and disentangle the scoured wool fibres and to remove dirt and vegetable matter. The fibres are formed into a continuous web, which is then condensed into a card sliver. The continuous web may contain small, tight clusters of fibres (termed 'neps'); these can result from inadequate opening or can be created during carding if the machine settings are incorrect or the steel pins on the rollers have become blunt. These neps have to be removed before spinning.

9.4.2 GILLING

Gilling is comparable to the drawing process in the cotton system. Fibres pass through gill boxes in which pins control the movement of short fibres, remove fibre hooks, straighten and blend the fibres, and improve the evenness of slivers. These actions reduce fibre breakage during subsequent combing. A typical intersecting gill box is shown in Figure 9.8. The machine consists of two sets of rollers with the slivers being dragged from one to another through a bed of moving pinned combs. The feed rollers pick up the slivers and the delivery rollers, that rotate far faster, draw out the combined slivers into a lighter weight single sliver. The comb pins are pushed into the sliver, which is still gripped by the feed rollers, and are drawn through the fibres in a combing action.

At the other end, the delivery rollers pull the fibres faster than the pins are moving, drawing the sliver through the pins, giving another combing effect. A single sliver comes out of the gill box and is

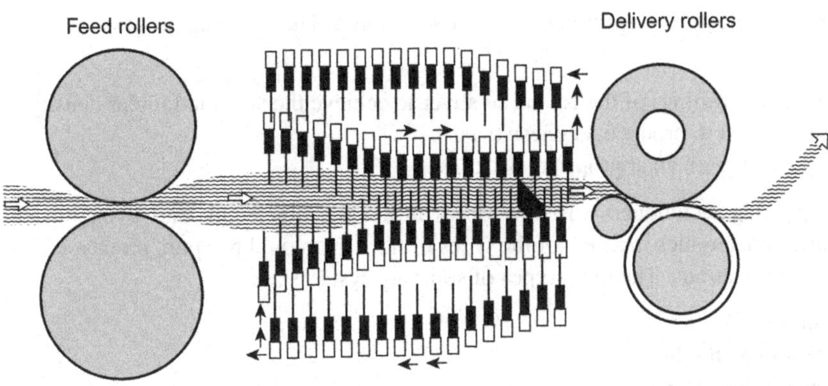

FIGURE 9.8

Fill box.

coiled into a can. Often there are three gill passages between the cards and the combs, partly so that any hooks formed during carding are presented to the comb with the hooks in the trailing position (trailing hooks are more likely to be straightened by the combs; leading hooks are often broken by the combs, which results in a decrease in top length). The gilling stage also allows extra processing oil and moisture to be added prior to combing. Because of the speed of modern processing machines, regain loss can be as high as 2% on each passage, so moisture needs to be added when possible.

9.4.3 COMBING

The combing process separates out the short fibres (or 'noil'), which can then be used in the manufacture of woollen yarns. Once the noil is removed, a rope-like strand of fibres, called the 'top', remains. The market value of the noil is about 40% of that of the top. Combing is today largely carried out on a rectilinear comb. The slivers are fed into the combing machine, where the fibres are held in place by a nipper. The free ends of the fibre are combed by a cylinder covered in progressively finer pins. Any short fibres/vegetable matter not being held by the nipper are combed out as noil. The fringe of now-parallel fibres is grabbed by detaching rollers, which pull the fibres through a top comb for a final combing stage. The combed fringe is laid onto previously combed fringes to form a continuous combed sliver.

Combed tops are then processed further by gilling, combing and finally drawing into roving before they are spun into yarns. An alternative 'semi-worsted' system is also used by some manufacturers. In this system, the combing step is omitted. This results in cheaper processing but the semi-worsted yarns are less smooth and lustrous compared to worsted yarns.

9.5 SPINNING TECHNIQUES FOR STAPLE FIBRES

Once the staple fibres have undergone preparation as described earlier for cotton, woollen and worsted systems, the end product is the roving (or slubbing in the woollen system), a continuous yarn that has just enough twist imparted to hold the fibres together. In order to produce yarns with enough strength to allow fabric manufacture, a spinning step is now required. The spinning process involves three main operations:

- Attentuation (or drafting) of the roving or sliver to achieve the specified linear density.
- Insertion of a twist to produce a cohesive yarn.
- Collecting the yarn on an appropriate package.

A range of spinning techniques was developed during the nineteenth and twentieth centuries, all with different characteristics in terms of the types of fibre that could be spun, process economics, yarn properties and applications. The main types of spinning systems are:

- Ring spinning
- Twist-spinning methods
- Rotor spinning (open-end)
- Friction (open-end)
- Self-twist
- Wrap-spinning methods
- Surface fibre wrapping (air-jet spinning)
- Filament wrapping (selfil spinning and hollow-spindle spinning)

The majority of staple yarn manufactured worldwide is produced either by ring spinning (70%) or rotor spinning (23%), with ring spinning dominating because of the superior quality (particularly of strength and evenness) in yarns produced using this method. However, rotor spinning is much faster, with one rotor spindle having the same output as five ring spindles. The main factors determining the process economics of the spinning systems are the number of processes needed to prepare the raw material for spinning, the production speed, the wound package size and the degree of automation possible. Table 9.1 shows the production characteristics of the major spinning systems. Table 9.2 shows the fibre properties that have a significant effect on spinning outcomes, and also shows that the importance of certain fibre properties varies according to the spinning system under consideration. The various spinning systems used for staple-yarn production are discussed in more detail in the following sections.

9.5.1 RING SPINNING

The ring-spinning frame was invented in 1832 by John Thorp and remains the most commonly used spinning technique in the textile industry. The popularity of this technique can be attributed to its versatility in terms of the range of fibres that can be processed, the range of yarn linear densities that can be produced, its

Table 9.1 Characteristics of different spinning methods

		System limited by		
Spinning methods	**Actual twist-insertion rate per minute**	**Twist-insertion rate**	**Drafting and fibre-transport speed**	**Delivery speed (m/min)**
Ring	15,000–25,000	Yes	No	20–30
Rotor	80,000–150,000	Yes	Partly	100–300
Air-jet	150,000–250,000	No	Yes	150–450
Friction	200,000–300,000	No	Yes	150–400

Table 9.2 Fibre properties affecting spinning methods

Order of importance	Ring	Rotor	Air-jet	Friction
1	Length and length uniformity	Strength	Fineness	Strength
2	Strength	Fineness	Cleanliness*	Strength
3	Fineness	Length and length uniformity	Strength	Fineness
4		Cleanliness*	Length and length uniformity	Length and length uniformity
5			Fineness	Cleanliness*

**Trash, dust, etc.*

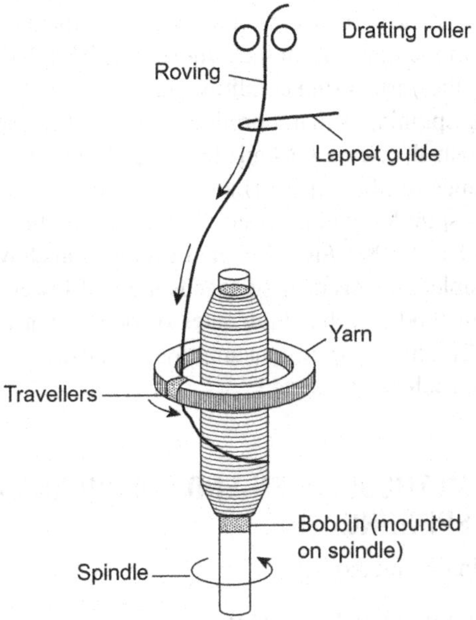

FIGURE 9.9

Ring spinning.

ability to produce fine yarns, and the yarn structure and properties. Because ringspinning produces a twist structure that is homogeneous across the length and cross-section, ring-spun yarns are strong, even and fine with low stiffness and high tenacity (a measure of the tensile strength of a fibre or yarn before it breaks).

During ring spinning, fibres are twisted around each other to produce a strong yarn. The ring spinner is made up of the following components (see Figure 9.9):

- Spools on which the roving (the strand of fibres for spinning) is wound
- A series of drafting rollers through which the roving passes

- A guiding ring or eyelet (also known as the lappet guide)
- A stationary ring around the spindle
- A traveller – a small, C-shaped clip that rotates around the ring
- A spindle
- A bobbin

The roving is fed from the spool through the drafting rollers, which attenuate (or elongate) the roving and control fibre orientation. The roving passes through the guiding ring and then through the traveller. The traveller moves freely around the stationary ring. The spindle turns the bobbin at a constant speed, and it is this turning of the bobbin and the movement of the traveller that result in the yarn being twisted. The yarn is twisted and wound onto the bobbin in one operation. Spinning costs are proportional to the amount of twist inserted, i.e. the minimum twist to produce acceptable spinning performance (i.e. low breakage rate) and acceptable yarn properties (in terms of hairiness, stiffness and tensile properties).

When the bobbin is full, it is removed from the machine. The removal process is called 'doffing'. Once removed, the bobbins are transferred to a winding machine, where the yarn is wound onto packages. During the winding process, the yarn is also 'conditioned' so that the moisture content of the yarn is brought into equilibrium with the moisture in the atmosphere. Wax and other coatings that facilitate weaving may also be added to the yarn during conditioning.

The main advantage of ring spinning is that it produces finer and stronger yarns than other spinning techniques due to good fibre control, orientation and alignment during spinning. It is also a very flexible system that can spin a wide range of fibre types. There are, however, a number of disadvantages. Yarn production rates are limited by spindle rotation speed. When ring spinning was first developed, spindle rotation speeds were of the order of 4000 r/min. Recent developments have meant that rotation speeds of up to 21,000 r/min are feasible. However, these speeds are still lower than those of other spinning methods. The ring spinning method also has high rates of power consumption, traveller wear, heat generation and yarn tension. The high power consumption is caused by the necessity of rotating the bobbin at a rate of one turn for each twist inserted.

9.5.2 TWIST-SPINNING METHODS: OPEN-END (ROTOR AND FRICTION) SPINNING AND SELF-TWIST SPINNING

There are three main twist-spinning methods:

- Rotor spinning (an open-end spinning technique)
- Friction spinning (an open-end spinning technique)
- Self-twist spinning

Open-end spinning was developed in an attempt to overcome the speed limitations of ring spinning. In this method, a sliver of fibres is fed into the spinner. The preparatory step that involves the formation of a roving is therefore omitted, with considerable savings in production costs. The input material is drafted until individual fibres are drawn out. These fibres are then collected onto the exposed end of the existing yarn (i.e. the 'open-end') and then rotated to twist the fibres to form a yarn. In contrast to ring spinning, open-end processes are continuous, i.e. material is continuously fed into the spinner and the formed yarn is continuously removed. Open-end spinning is now second only to ring spinning in terms of short staple-yarn production.

Yarns formed by open-end spinning techniques have a structure that is different from those formed by ring spinning. The yarn consists of:

- A core.
- A sheath of fibres wrapped around the core.
- Wrapper fibres.

These three components have a significant effect on the yarn's characteristics. The core of the rotor-spun yarn is similar to that of ring-spun yarns and is responsible for the strength of the yarn. The sheath affects the yarn bulk and handle. The wrapper fibres are wrapped tightly around the yarn, almost at right angles to its axis. They have an adverse effect on the handle (feel) of the yarn: as the number of wrapper fibres increases, the harshness of the yarn increases. This means the yarns produced are not so fine as those produced by ring spinning. The tensile strength of rotor-spun yarns is lower than that of ring-spun yarns. Because of their structure, open-end spun yarns are also more absorbent. Open-end yarns are often used in fabrics such as towelling, denim and heavy-weight bed linens.

As has been noted, a key advantage of open-end spinning is increased speed of production compared with ring spinning. Other advantages include the fact that energy costs are lower, because there is no need for the whole package to be rotated, and that finished yarns can be wound on any sized bobbin or spool. Disadvantages include the fact that open-end spinning is less versatile for producing complex/fancy yarns. Recent advances have extended the range of yarn sizes possible.

The two main open-end spinning methods currently in use in the textile industry are:

- Rotor spinning
- Friction spinning

These are discussed in the following sections.

9.5.3 OPEN-END SPINNING: ROTOR SPINNING

The principle of rotor spinning is similar to the production of cotton candy/candy floss. Cotton candy is produced by melting sugar in a drum rotating at high speed. Centrifugal force pulls the molten sugar out of the drum, allowing it to solidify and form fibres. The fibres spin round a stick to create the cotton candy. Figure 9.10 shows the basic elements of open-end rotor spinning. A sliver of fibres is fed into the spinner by a stream of air. The fibres then come into contact with a rotating pin-covered beater, which creates a thin stream of fibres. These fibres are then deposited in the v-shaped groove on the outside edge of the rotor. The rotor spins rapidly, imparting twist to the fibres. Fibres fed to the rotor are incorporated into the rapidly rotating 'open-end' of a previously formed yarn that extends out of the delivery tube. The yarn is withdrawn through a doffing tube (also termed a 'nave').

The fineness of the yarn is determined by the rate at which it is drawn out of the rotor relative to the rate at which fibres are being fed into the rotor. The twist is determined by the speed of the rotor in relation to the withdrawal speed of the yarn (i.e. the higher the speed of the rotor, the greater the twist). One turn of twist is inserted for each revolution of the rotor. The yarn is wound directly onto a package (cone). At this stage, the quality, e.g. evenness, of the yarn can be assessed; treatments such as waxing (for knitting yarns) can also be applied.

Great advances have been made in the automation of stages of the rotor-spinning process (e.g. rotor cleaning and doffing). Other advances include the development of online quality monitoring

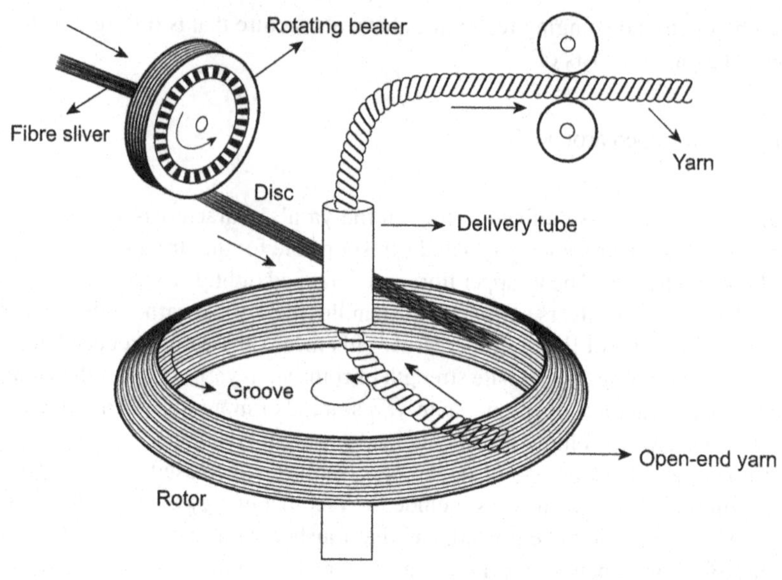

FIGURE 9.10

Rotor spinning.

using optical or capacitance systems, so that characteristics such as linear density can be assessed continuously and faults (e.g. the presence of foreign fibres) can be identified and removed automatically.

Recent developments in electronic control have allowed the development of rotor-spinning machinery that is also capable of producing slub yarns. These yarns are used in furnishings and drapes, rather than in apparel fabrics, although they are sometimes used in denim fabrics. They are manufactured using attachments to ordinary open-end spinning devices, which usually incorporate an electronically controlled device for the brief acceleration of the drawing-in roller. As a result of the back-doubling action inside the rotor, it is not possible to produce slubs shorter than the circumference length of the rotor, because any variation in the fibre feed material is spread over a minimum length of the rotor circumference.

There have also been attempts to alter the fibre flow and thus introduce variations in the yarn appearance by the use of injecting pressurised air into the fibre transportation tube. However, the effects created using this approach are limited, as the fibre flow in the transportation tube is slight and the variation in the yarn caused by changes in the airflow is therefore very small.

9.5.4 OPEN-END SPINNING: FRICTION SPINNING

Friction spinning is an alternative open-end spinning method in which friction, rather than a rotor, is used to insert twist. An opening roller attenuates a feed sliver, which is then fed into the groove formed by two perforated drums rotating in opposite directions (Figure 9.11). The fibres are held against the surface of the drums by suction. The tail of the forming yarn is also held in the groove by

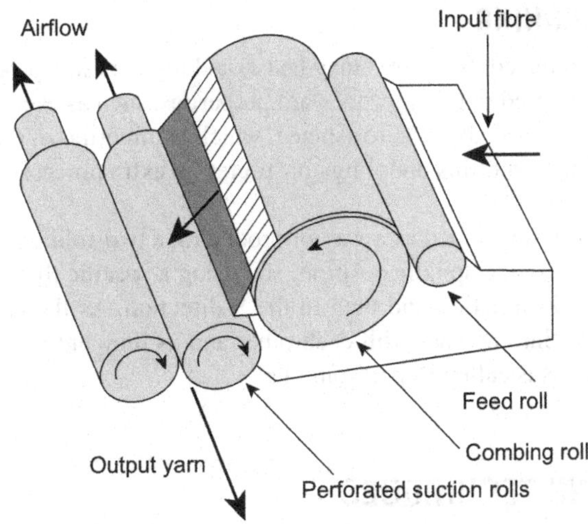

FIGURE 9.11

Friction spinning.

suction, and as the individual fibres pass through the drums, they are bound to the tail as a result of frictional forces generated. Twist is inserted because of the friction with the two drum surfaces, and the completed yarn is continuously drawn away in the direction parallel to the axis of the friction drums and delivered to a package-forming unit. A high twisting speed can be obtained even while using a relatively low speed for the friction drums, because the friction drum diameter is much larger than that of the yarn.

Because the yarn is withdrawn from the side of the machine, fibres fed from the machine end away from the yarn delivery point tend to make the yarn core, while fibres fed from the machine end closer to the yarn delivery point tend to make the sheath. This characteristic can be conveniently used to produce core-sheath yarn structures for a specific purpose, such as a yarn with the strength of a polyester core and the natural feel of a cotton sheath. Additional core components, filaments or drafted staple fibres, can also be fed from the side of the machine, while fibres fed from the top of the machine, the normal input, form the sheath. The fibre configuration in friction spun yarns is quite different from that of other yarns. When the fibres come to the friction drum surfaces, they decelerate sharply from a high velocity to become almost stationary. This causes fibre blending and disorientation. Due to the very low tension in the yarn formation zone, fibre binding within the yarn is also poor.

Slub yarns, which are potentially important for decorative effects, can be manufactured on the friction system by changing the feed rate of one or more of the slivers, or by injecting the fibres directly into the friction zone. However, the yarn tends to offer poor performance in processing and use, as a result of the poor binding of the fibres in the yarn indicated earlier. Friction spinning is not often used for cotton spinning because the structure of the yarn (in terms of fibre orientation, fibre packing, etc.) results in lower strength. It is used for producing coarse yarns for industrial applications, and particularly technical textiles.

9.5.5 SELF-TWIST SPINNING

Once the finished yarn is produced, two yarns may be twisted together in a process known as doubling or plying. This doubling is carried out to improve yarn properties, such as evenness. Doubling will also reduce yarn hairiness and increase abrasion resistance, which is important if the yarns are destined for weaving worsted fabrics. However, this doubling constitutes an extra processing step, which therefore adds to the cost of production.

The self-twist method was developed as a way of producing a two-fold yarn effect using ring spinning. Two rovings are fed into each spindle position, and using a specific threading geometry, the rovings are twisted first in the Z-direction and then in the S-direction. As the twist is released, the two strands are allowed to come into contact with each other and as they untwist they form a two-strand yarn. The commercial process is called Repco spinning.

9.6 WRAP-SPINNING TECHNIQUES

The previous sections have described spinning techniques that use twist to convert parallel fibres into a yarn. In contrast, the wrap-spinning techniques described in this section make use of the surface fibres protruding from the sliver, wrapping them around the parallel fibres to form a yarn; alternatively, a continuous filament can be wrapped around the parallel fibres. These two techniques are known as:

- Surface fibre wrapping (air-jet spinning).
- Filament wrapping (selfil and hollow-spindle spinning).

These techniques are described in the following sections.

9.6.1 AIR-JET SPINNING

Air-jet spinning is also known as Vortex or fasciated yarn spinning. It was introduced in the 1980s. A basic air-jet spinning system is shown in Figure 9.12. Before the sliver from the drawframe is supplied to the air-jet spinner, combing is often used, as it is imperative to get rid of any dust or trash that could obstruct the spinning jets. Twist is inserted to the fibres, mostly on the yarn surface, by the vortex created in one or two air-jet nozzles. The resulting yarn consists of a core of parallel fibres and a sheath of wrapped (twisted) fibres. The yarn produced by air-jet spinning resembles a ring-spun yarn but is not as strong. The yarns are also inclined to shrink. High delivery rates of 150–450 m/min are possible with this technique.

The feed material, i.e. the slivers, is fed to the four-roller drafting arrangement. The fibres come out of the front rollers, are sucked into the spiral orifice at the entrance of the air-jet nozzle, and are then held together more firmly as they move towards the tip of the needle protruding from the orifice. At this stage, the fibres are twisted by the force of the air-jet stream. This twisting motion tends to flow upwards. The needle, which acts as a guide on the needle holder, projects towards the inlets of a hollow spindle. A nozzle block provides a swirling air current, which acts on the drafted fibre bundle. The needle protruding from the orifice prevents this upward propagation (twist penetration). Therefore, the upper portions of some fibres are separated from the nip point between the front rollers, but are kept 'open'. After the fibres have passed through the orifice, the upper portions of the fibres begin to expand due to the whirling force of the air-jet stream and twine over the hollow spindle.

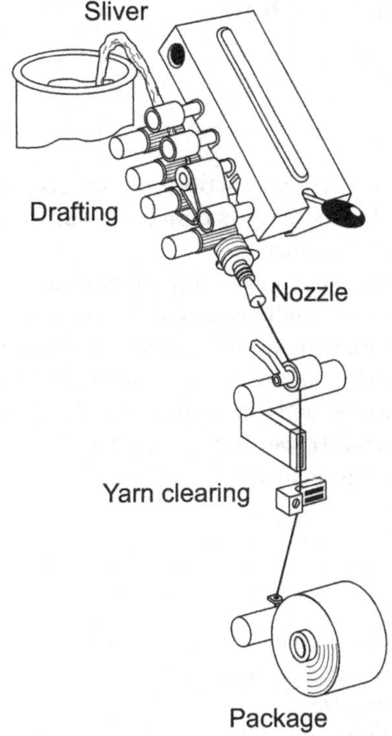

FIGURE 9.12

Air-jet spinning.

The fibre bundle is then sucked into the hollow spindle, and twist insertion starts when the fibre bundle is subjected to a compressed air vortex as it enters the spindle. Twist propagation towards the nip of the front rollers is prevented by the guide members acting as the centre of the fibre bundle. On leaving the guide member, the whirling action of the air current separates the fibres randomly towards the bundle.

The leading ends of all the fibres are held into the body of the yarn being formed, while the trailing ends leaving the front roller nip set are whirled up in the air and are inverted and separated from each other at the inlet of the hollow spindle. The leading ends are less subject to the air current as they are moved around the guide member towards the inlet of the spindle and converted into spun yarn. The fibres twined over the spindle are whirled around the fibre core and made into vortex yarn as they are drawn into the spindle.

The finished yarn is wound onto a package after any defects have been removed. The final package is then removed automatically. The leading ends of the fibre bundle are drawn into the hollow spindle by the fibres of the preceding portion of the bundle being twisted into a spun yarn. The trailing ends of the fibres are inverted at the inlet, separated from each other, and exposed to the swirling air blown through the nozzles. The trailing ends of the fibres are thereby caused to twist around the porting of the fibre bundle being converted into a spun yarn.

9.6.2 FILAMENT WRAPPING TECHNIQUES

There are two types of filament wrapping techniques:

- Selfil spinning
- Hollow-spindle spinning

Selfil spinning is based on the Repco system described earlier. Instead of feeding two twisted strands of rovings into the spindle, one of the strands is replaced by a twisted continuous filament. As the filament and strand untwist, they ply together.

Hollow-spindle spinning is the more commonly employed method of filament wrapping. A continuous filament yarn on a hollow spindle is wrapped around an untwisted staple core. The core constitutes about 80–95% of the yarn composition; the filament wrapping yarn is usually a fine yarn. Delivery speeds of 200 m/min can be achieved. The yarn has a soft handle and is used for coarse-count knitting rather than weaving. Compared with ring-spun yarns, these yarns are less hairy and bulky and have improved strength and evenness. However, the fine filament wrapper required is expensive and hence limits the widespread use of this method.

9.7 FUTURE TRENDS

- The increase of the running speeds of the different machines inside the spinning mills for the sake of productivity with keeping an acceptable quality to price ratio. Therefore the masses of the machines movable parts are minimized.
- Shortening the spinning lines with emphasizing on FCE i.e. fibre-control elements in the drafting systems either roller drafting, pin drafting and toothed drafting.
- The incorporation of the autolevellers in the production units of the yarn production train to substitute the production lines shortening.
- The application of automation inside the spinning mills (will lead to the reduction of labour power and to the increase of products quality).
- More and more applications of the ROBOTS in the cotton or wool spinning mills.

9.8 SUMMARY POINTS

- Fibers can be classified as either filament fibres or staple fibres.
- Chapter 9 will review the production of yarn from staple fibres (short or long), because of their short length, they must be twisted together to form spun yarn.
- Short staple fibre means cotton fibres while long staple fibre means wool fibres.
- The staple artificial fibre can be processed either by the cotton system or by the wool system.
- To convert the raw cotton into fibres that are separated and aligned in a suitable manner for yarn production, the cotton passes via the following stages; opening and cleaning, mixing (blending), carding, combing (optional), drawing and roving.

- The first three stages take place in the blow room where the cotton is transferred from stage to stage via pneumatic transport. The fibres that leave the blow room are separated into individual fibres that can be further processed ready for yarn production.
- For wool industry, two different fibre preparation systems are used foe woollen & other long-staple fibres as worsted and the chosen system depends largely on the length of the staple and the required properties. in the end product.
- The main procedures in the woollen system preparation are opening, scouring, drying, blending, oiling and carding. The end product is slubber.
- The main procedures in the worsted preparation are: scouring as woollen, drying or oiling as woollen, carding, gilling and combing. The end product is roving.
- To produce yarns with enough strength to allow fabric production, the spinning step is required. Any spinning process involves three steps: drafting, twisting and packaging. The main systems are ring spinning, twisting spinning (rotor, friction& self-twist) and wrap spinning (air jet, self and hollow spindle).
- The majority of staple yarn production world wide is either by using ring spinning (70%) or rotor spinning (23%). The ring spinning produces finer and stronger yarns more than the other systems.
- Yarns formed by open-end spinning techniques (rotor and friction) have a core, a sheath and wrapper fibres. Air jet spinning is known as vortex or fasciated yarn spinning.

9.9 PROJECT IDEAS

1. Study the effect of the ring or compact spinning machines (cotton or wool) parameters (traveller mass, spindle speed, break draft, etc.) on the quality of the ring-spun or compact-spun yarns, reviewing the effects on factors such as tenacity, extension at break, mass variation, Uster value, count strength product, IPI-total yarn imperfections, etc.
2. Explore the interaction between the worsted card-set variables and the quality of the produced sliver (top) and neps potential.
3. Study how different cotton comber lap preparation systems could affect the quality of the combed spun yarns.
4. Investigate the transfer function, using the Laplace transform function, for the autoleveller of a cotton draw frame, i.e. the relationship between the fed variability function of the fed sliver and the output drawn sliver variability function.

9.10 REVISION QUESTIONS

9.10.1 COTTON SYSTEM

- What are the technological functions for each of the following processes?
  - Cotton blowroom
  - Cotton card
  - Cotton comber

- Explain the phenomenon of the hooked fibres in both the cotton and worsted spinning industries. What is the role of the combers in the treatments with them?
- What is meant by 'floating-swimming fibres' in the cotton industry? How do they affect the quality of the spun products? How could the technologist control them?
- Using a sketch, explain the concept of rotor spinning. How can the turns per metre be calculated in the rotor-spinning machine?
- How does the cotton type and its quality affect the technological design of the blowroom?

9.10.2 WOOL SYSTEM

- Explain the mechanism of hooked fibre formation in the worsted card.
- What is the difference between screw-driven gills and chain-driven gills?
- What is meant by 'fibre-control elements'?
- Outline the factors affecting wool fibre breakage during the combing process.
- State the relative merits of the rubbing index, attenuation degree and irregularity index.
- By using Simpson and Crawshaw's line diagram, explain the differences between ring spinning and compact spinning.

REFERENCES AND FURTHER READING

Carr, H., & Latham, B. (1998). *The technology of clothing manufacture*. Oxford, UK: BSP Professional Books.

Elhawary, I. A. (2006). *Quality upgrading of Egyptian cottons*. San Antonio, TX: BWCC.

Elhawary, I. A. (2010). *Technology of cotton yarns plying*. Lectures, Egypt: PGS: Alex. University.

Elhawary, I. A., et al. (1974). *A comparison of open - end spun & ring spun yarns produced from two Egyptian cottons*. Lankester, UK: JTI.

Harrison, P. W. (1978). *Textile progress, the production & properties of staple-fiber yarns made by recently devolped techniques* (Vol. 10, #1/2). Manchester, UK: Textile Inst.

Hunter, L. (2002). Mechanical processing for yarn production. In W. Simpson, & G. Crawshaw (Eds.), *Wool: Science and technology*. Cambridge, UK: Woodhead Publishing Limited.

Hunter, L. (2007). Cotton spinning technology. In S. Gordon, & Y.-L. Hsieh (Eds.), *Cotton: Science and technology*. Cambridge, UK: Woodhead Publishing Limited.

Klein, W. (1994). *Man-made fibers & their processing* (Vol. 6). Manchester, UK: Textile Inst.

Lawrence, C. (2007). The opening, blending, cleaning and carding of cotton. In S. Gordon, & Y.-L. Hsieh (Eds.), *Cotton: Science and technology*. Cambridge, UK: Woodhead Publishing Limited.

Lawrence, C. (2010). Overview of developments in yarn spinning technology. In C. Lawrence (Ed.), *Advances in yarn spinning technology*. Cambridge, UK: Woodhead Publishing Limited.

Lord, P. (2003). *Handbook of yarn production*. Cambridge, UK: Woodhead Publishing Limited.

Mccreight, D. J., Feil, R. W., Booterbavgh, J. H., & Backer, E. E. (1977). *Short staple manufacturing*. Durham, NC: Carolina Academic Press.

Shanuan Wang, X. (2006). *Textile yarns (English ed.)*. China: Donghna University Press.

CHAPTER

FIBRE TO YARN: FILAMENT YARN SPINNING

10

C. Lawrence
University of Leeds, Leeds, UK

LEARNING OBJECTIVES

At the end of this chapter, you should be able to:

- Understand the classification of filament yarns
- Understand the natural and synthetic polymers used for the production of filament yarns
- Understand the principles of the basic manufacturing processes used to produce continuous filaments
- Understand the principles of false-twist and air-jet texturing of continuous filaments
- Understand the structure–property relationship of continuous filaments
- Understand the end uses for filament yarns in textiles and apparel
- Understand the future trends

10.1 INTRODUCTION

This chapter describes the basic principles of the production processes used in producing continuous-filament yarns from widely used natural and synthetic raw materials for applications in conventional textiles and apparel. In doing so, it is useful to begin by considering what is meant technically by continuous-filament yarns, and to outline the commonly used raw materials that are converted into such yarns.

10.1.1 DEFINITIONS

There are various published definitions (Composite, 2012; McIntyre & Daniels; Miniknittingstuff, 2012; Robinson & Marks, 1993, chap. 1, p. 1; Textileglossary, 2012) of a continuous-filament yarn, but before giving the more descriptive of these, it is useful to consider the terms **filament and continuous filament**. The Textile Institute's publication *Textiles Terms & Definitions* (McIntyre & Daniels) refers to a filament as a fibre of indefinite length[1]. A continuous-filament yarn can then be defined as 'A yarn composed of one or more filaments of length/lengths equal to the specified yarn length'. Thus, if we have a bobbin with, say, 10 m of continuous-filament yarn, all the filaments comprising the yarn would

[1] A fibre being a textile raw material, generally characterised by flexibility, fineness and a high ratio of length to thickness (diameter) aspect ratio.

Textiles and Fashion. http://dx.doi.org/10.1016/B978-1-84569-931-4.00010-6

be 10 m long. If there were 100 m of yarn on the bobbin, all the filaments would be 100 m in length. For the sake of brevity, we will from now use 'CF yarn(s)' to mean continuous-filament yarn(s).

10.1.2 CLASSIFICATION OF CF YARNS

In a similar manner to spun yarns, commercial CF yarns may be classified as **natural** or **manufactured** yarns (i.e. man-made filaments, or mmf) based on their constituent polymeric materials. However, as illustrated in Figure 10.1, the only natural filament yarn is silk (see Chapter 3), whereas the mmf category can be divided into CF yarns made from natural raw materials (Chapter 4) and synthetic polymers (Chapters 5 and 6). To convert these materials into suitable CF yarns for textiles and clothing they must first be extruded into continuous filaments.

Cellulose is the most abundant natural organic material and is found as a naturally synthesized polymer. However, it is a non-meltable polymer and has to be dissolved into a solution in order to be extruded into filaments. Chapter 4 describes the various processes used to solubilize cellulose raw materials for the production of viscose and Lyocell filaments. Natural rubber – caoutchouc – is an elastomer that is obtained as a milky colloidal suspension (latex) from the sap or 'milk' of various plants. It is compounded with colloidal emulsions and dispersions, and subsequently extruded into filaments (Daily Finance, 2012; Natural Rubber, 2012).

The synthetic polymers listed in Figure 10.1 are thermoplastics and therefore can be extruded into filament when in their molten state. The aromatic polyesters are by volume the most used of the CF yarns, followed by nylon. Among many available polyesters, the three popular for textiles and

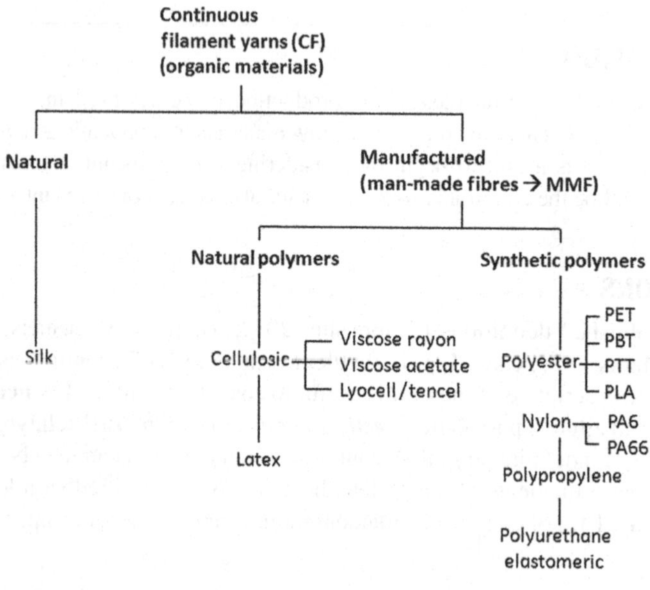

FIGURE 10.1

Classification of continuous-filament yarns.

apparel applications are polyethylene terephthalate (PET), polybutylene terephthalate (PBT) and polytrimethylene terephthalate (PTT). These polyesters are synthesised from man-made chemicals, as is described in Chapters 5 and 6. Polyesters can, however, also be produced with naturally occurring chemicals from renewable resources such as corn starch. Polylatic acid (PLA) is an aliphatic polyester obtained from such natural chemicals, and the resultant fibre, like the cellulosics, has the advantage of being biodegradable (Farrington, Hunt, Davis, & Blackburn, 2005) and can be placed into the sustainable materials category.

Nylon or polyamide (PA) was the first synthetic fibre to be commercialized (introduced to the textile industry in 1939). Today there are various commercial nylon polymers, for example nylon 6, 11, 12, 6/6, 6/10 and 6/12. Of these, the most widely used for textile and clothing applications are nylon 6 and nylon 6/6. The others are mainly employed as engineering materials (Fambre, Pegoretti, Mazzurana, & Migliares, 1994). Polypropylene (PP), also known as Polypropene, is also a synthetic thermoplastic polymer and is used in a wide range of applications (Galanty & Bujtas, 1992, pp. 23–30), including, thermal underwear, outerwear, blankets and carpets. Spandex is a synthetic polymer elastomer, made up of a long-chain polymer called polyurethane (PUR or PU), and over 90% of the world's spandex fibres are used to make stretchable clothing such as sportswear (Lewin & Preston, 1985; Ulrich, 1996).

Although Figure 10.1 depicts the CF groupings that are mainly used for conventional textiles and apparel, there are, however, many other materials that are used to make CF yarns but largely for more technical or engineering applications, collectively termed **technical textiles** (Horrocks & Anand, 2000; McIntyre & Daniels; Technicaltextile, 2012; Teonline, 2012). The focus of this chapter is on CF yarns for conventional textiles and apparel, and readers wishing to obtain information on CF yarns for technical textiles may do so by accessing the cited references.

In our classification of CF yarns, two further points must be considered. Firstly, when a CF yarn is in the form of one filament, it is referred to as a **mono-filament** CF yarn (or in short a **monofilament**), and when it consists of several filaments, a multifilament CF yarn (or just a **multifilament**). The second point is that when CF yarns are formed from their raw materials, they are straight along their lengths and have little significant bulkiness. In this form they are called **flat** or **untextured yarns**. Manufactured CF yarns, in particular synthetic CF yarns, are available not only in the flat form, but may be further processed to impart various degrees of bulkiness and as such are termed as **textured** CF yarns. The texture may be visualised as being either of two forms; a wavy, continuous crimp, or a profusion of loops along the lengths of the filaments. Both flat and textured CF yarns give visual and tactile aesthetic to fabrics for apparel end-uses.

10.1.3 YARN COUNT SYSTEM

The yarn count system is the means by which the principal dimension of CF yarns is expressed, the principal dimension being the yarn thickness or **yarn fineness** – the latter term is the accepted convention. Intuitively, filaments may be visualised as being of round cross-section and the measured diameter thought of as the means of expressing fineness. Although filament diameters are measured and used in the design and engineering of some technical textile products, there are two main reasons why this approach is not widely practised. Firstly, filaments can have non-circular cross-sections, and secondly, the practicality of diameter measurements would be restrictively cumbersome if a sufficient number of measurements for quality assurance purposes were to be performed when producing many hundreds of kilos of yarns per day. Secondly, it would also be highly impracticable to employ yarn diameter as a specified parameter for textured CF yarns. A more useful approach has been to use a measure of the linear density (mass per unit length) of a CF yarn; this measure is referred to as the **yarn count**.

The yarn count is defined as the mass that a standardised length of a CF yarn weighs. The unit of mass is the gram, and the standardised length can be either 1 or 9 km. If 1 km is used, then the count of the CF yarn is referred to as the yarn tex (i.e. the number of grams that a 1000 m length of the CF yarn weighs). If 9 km is used, then the reference is to the yarn denier (i.e. the number of grams that a 9000 m length of the CF yarn weighs). Thus a 10 tex yarn can be also be written as a 90 denier yarn. Evidently, for a given polymer, the larger the tex or denier value, the coarser the yarn, the coarseness being directly proportional to the increase in tex or denier value. For example, 20 tex yarn (180 denier) is twice the size of a 10 tex yarn (90 denier). With multifilament yarns, the number of constituent filaments is also given in addition to the yarn count, so 33 tex/150 (297 denier /150) indicates the yarn is comprised of 150 filaments. Each filament would therefore be 0.22 tex or 2.2 dtex (2 denier). With the tex unit, multiples and decimal fractions of the tex can be referred to in terms of the base-10 scale. Thus, 1000 tex = 1 kilotex (ktex), 0.1 tex = 1 decitex (dtex) and 0.001 tex = 1 millitex (mtex).

10.2 FIBRE-EXTRUSION SPINNING

There are various fibre-extrusion spinning methods, but the most commercially popular is melt-spinning for thermoplastic polymers, followed by wet spinning (**solution wet spinning**) and dry spinning (**solution dry spinning**), which are used mainly for non-meltable polymers.

10.2.1 MELT-SPINNING

Melt spinning may be defined as a continuous process in which a thermoplastic polymer is heated to form a molten solution that is extruded by pumping it through an array of very small diameter holes to form molten filaments. These are subsequently cooled to their solid state while being simultaneously attenuated by stretching, thereby producing what is termed as 'as-spun' or 'partially drawn' CF yarn.

With the exception of PP, the thermoplastic polymers listed in Figure 10.1 have some degree of moisture absorbency and therefore the polymer chips have to be dried prior to melt spinning. Figure 10.2 depicts the basic vertical arrangement of the melt-spinning process, called the spinning line. The dried polymer chips or pellets are supplied to a screw extruder via a hopper. The screw extruder has several hot zones to progressively heat the polymer chips to a few degrees above their melting point, T_m (**melt temperature**). The molten solution is fed to the metring pump, also kept at a temperature above the T_m. The pump is connected to a spin pack (Figure 10.3). This comprises a stack of circular plates with metal filters sandwiched between them. Each plate, subsequent to the inlet plate, has a series of holes through which the polymer flows until it reaches the last plate, the spinneret. The spinnerets may have between 500 and 4000 holes per disk. The holes in the spinneret largely govern the diameter of the molten filament streams; these are stretched while being cooled to their solid form. Stretching occurs as the partially solidified, semi-molten filaments are pulled down from the spinneret holes at a faster speed than the molten flow through the holes. The ratio of the former to the latter speed is called the draw-down ratio. The pump is

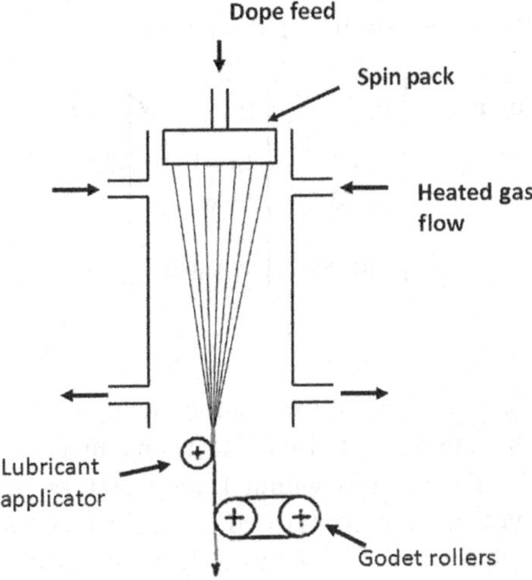

FIGURE 10.2

Melt-spinning process: spinning line.

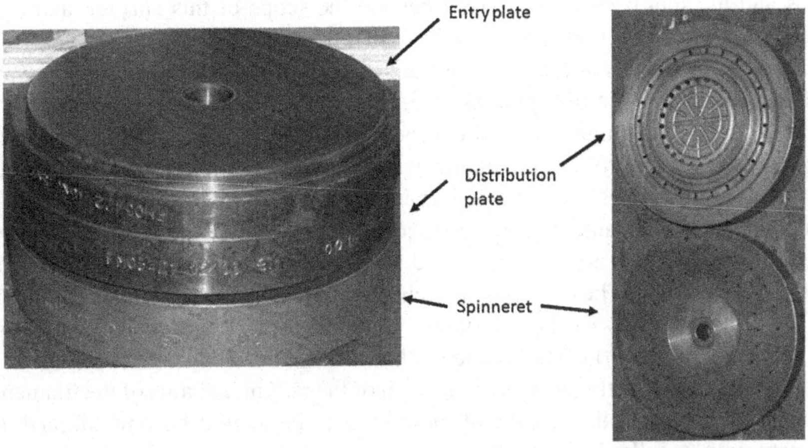

FIGURE 10.3

Spin pack plates.

designed to maintain a constant flow through the holes of the spinneret, hence the term 'metring pump'. This enables a relatively consistent molten filament diameter. The resulting filament diameter is therefore largely dependent on the stretching by the godet rollers, i.e. the draw-down ratio.

The T_m of the polymer and the flow of the molten polymer are important factors in the process. The ease of flow is directly related to the viscosity of the polymer, which in turn is a function of its molecular weight. Polymers suitable for textile fibre production have high molecular weights, and therefore these molten polymer solutions have high viscosities, which means they do not easily flow. However,

Table 10.1 Process temperatures for synthetic polymers

Temperatures (°C)	PET	PBT	PTT	PLA	PA 6	PA 6.6	PP	PU
Melting point (T_m)	265	225	228	150–160	220	265	168	280
Glass transition point (T_g)	80	66	45–65	55–60	40–87	50–90	−17/−4	−38/−35
Spinning (T_s)	280–300	250–280	240–270	220–240	240	280	220	300

viscosity decreases with increasing temperature and consequently extrusion commonly occurs at temperatures higher than the T_m. For example, with PET, extrusion temperatures (**spinning temps, T_s**) may be within 280–290 °C for molecular weights within 15,000–20,000. Table 10.1 summarises typical spinning temperatures for the synthetic polymers listed in Figure 10.1, along with their melting points.

The study of the melt-flow behaviour is referred to as the rheology of the molten liquid, and enables determination of the pump pressure required for a steady flow of material through the spinneret. Pump pressures can be up to 70 MPa[2].

The ratio of length to diameter of the spinneret holes is also important to achieve a steady flow. This can range from 2 to 5, depending on the polymer. A more detailed consideration of the rheology of molten liquids and the spin pack parameters is beyond the scope of this chapter, and the interested reader is referred to (Fourne', 1999; Salem, 2000).

The structure and properties of the spun yarn are strongly influenced by the cooling and attenuation of the extruded filaments in the spinning line. The action of attenuation by stretching is called drawing. Drawing tends to straighten and align the constituent polymer chains with the fibre axis; this improves the tensile properties of the filaments. Increasing the speed of draw-down godet rollers therefore increases the polymer orientation. Depending upon the way in which the melt-spinning process is implemented, the filaments may be subjected to low, moderate or high draw-down ratios. When low and moderate ratios are used, a second stage of drawing is applied by reheating the filaments whilst stretching them. The filaments are reheated to just above the temperature at which they are still in a solid state but very easily deformable (**plastic state**). This is called the glass transition temperature (T_g) (see Table 10.1). The filaments that result from this stage are referred to as fully drawn. Although drawing directly induces orientation of the polymer chains of the filaments, because of chain entanglement, not all the lengths of the polymer chains will become aligned and parallel with the filament axis. A fully drawn filament therefore, will be orientated to the limit that may be obtained with the particular polymer. For a detailed technical explanation of the developed molecular structure of filaments during melt spinning, the reader is referred to Salem (2000).

Fully dawn filaments can be achieved in one of two ways. Heated godets may be fitted to the melt-spinning machine and positioned after the pull-down godet. Each godet is set to rotate at a faster surface speed than the previous one. While the filaments are being heated, the speed differences impart the stretching

[2] Pascal (pronounced pass-KAL and abbreviated Pa) is the unit of pressure equivalent to one newton (1 N) of force applied over an area of one metre squared (1 m^2). MPa (pronounced megapass-KAL) is one million Pascals. 70 MPa is equivalent to 690 × atmospheric pressure.

or drawing and the ratio of the faster to the slower speed gives the draw-ratio. It is usual for hot drawing to be carried out progressively so a sequence of godets is employed, the number depending on the total draw ratio required.[3] The second approach for fully drawn filaments is to conduct hot drawing on a separate machine. The low or medium orientated as-spun filaments would first be wound onto bobbins, which are then transported to a separate drawing machine.

The structure–property relation of the polymers listed in Figure 10.1 will be discussed later when dealing with the properties of CF yarns. At this point, it is useful to consider in more detail the classification of the melt-spun CF yarns. From the above description of the melt-spinning process, there are four basic melt-spun CF yarns, which will be described next.

Where extrusion is done separately from orientational drawing, i.e. the yarn is first extruded and then transferred to a separate machine for hot-drawing, the extruded filaments possess little or no orientation, and such yarns are classified as LOY (**low oriented yarns**). Relatively low melt-spinning speeds are used for producing these yarns. For example, PET LOY yarns would be spun at a speed of around 1500 m/min. After hot drawing, the yarns are classified as FDY (**fully drawn yarns**). FDYs are mainly used in weaving for furnishing fabrics.

The drawing system can be incorporated into the spinning line. The yarn produced would then be drawn to and have the appropriate degree of orientation. These yarns are referred to as SDY (**spin-drawn yarns**). The output or production speed would be approximate to the product of the extrusion speed and the total draw ratio.

Where the godet rollers below the spinneret are run at a much higher speed than is the case for LOY yarns, i.e. within the range of 2500 to 4000 m/min, the draw-down will be sufficient to partially orientate or pre-orientate the polymers parallel to the filament axis. This type of yarn is referred to as POY (partially oriented yarn), and is mainly used in one of a range of subsequent processes known as texturing or texturising, POY can also be used in draw warping for the weaving and warp knitting of fabrics (Draw-warping Apparatus, 1989)

Higher draw-down speeds of 6000–7000 mm/in may be used to obtain highly oriented filaments that are not subjected to further drawing. These yarns are termed HOY (highly oriented yarns) when produced at a speed of 4000 to 6000 m/min and FOY (fully oriented yarns) at speeds above 6000 m/min. FOY are usually made for technical textile applications (McIntyre, 1998).

Although FDY/SDY yarns are used in textile and fashion fabrics, the majority of yarns for such applications are POY.

10.2.2 WET SPINNING

Not all polymers can be melt-spun, as some will thermally degrade rather than melt. In these cases, the polymer may be dissolved into a solution of sufficient viscosity to permit extrusion through a spinneret. The viscous solution (**termed the dope**) is extruded into a bath containing a second solution (**called the spin bath or coagulation bath**), which precipitates the polymer by diffusion and coagulates the polymer chains into continuous solid filaments.

[3] The total draw ratio is equal to the speed at which yarn leaves the output godet of the drawing section, divided by the speed of the godet at beginning of the drawing stage. The total draw ratio is therefore equal to the product of the draw ratios of each stage in the drawing sequence.

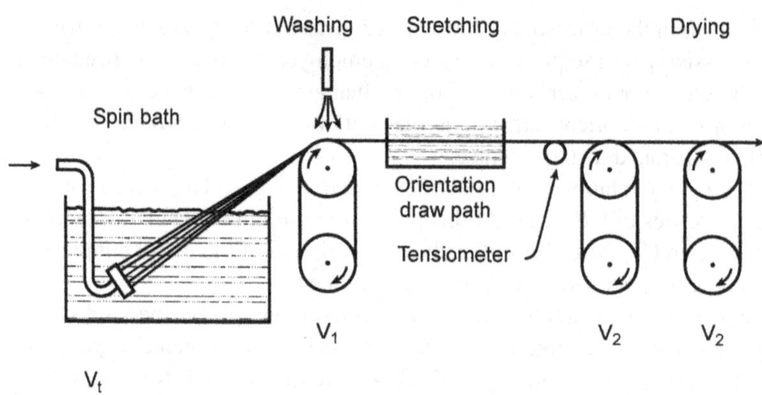

FIGURE 10.4

Wet-spinning process.

The dope is usually prepared by dissolving and stirring the polymer in a heated solvent. The amount of polymer addition should be sufficient to give the dope adequate viscosity for the continuous extrusion of liquid filaments. The dope is de-aired, removing air bubbles that may occur during the mixing stage, and filtered to remove any impurities or partially dissolved polymer. The wet-spinning process is illustrated in Figure 10.4. The dope is forced by a metring pump through the holes of a spinneret immersed in the coagulating solution. The filaments are drawn by the first set of godets as they coagulate, initially into a gel state and then into a more solid state when the solvent diffuses into the solution.

The partially drawn or as-spun filaments are washed as they are driven around by the godets. On leaving the first set of godets, the filaments pass through a heated bath, where they are further attenuated in their solid state to straighten and align (i.e. orientate) the polymer chains with the fibre axis. They are then dried prior to winding into a package.

Of the list of polymers given in Figure 10.1, viscose rayon is one example of a material converted into a CF yarn by the wet-spinning method. Figure 10.5 illustrates the full production process.

The viscose dope – cellulose xanthate – is pumped by a 'metering' gear pump at 2.6 to 5 times atmospheric pressure. This is a much smaller pressure than in melt spinning, because the viscosity of the dope is significantly lower. Similar to melt spinning, the pump must deliver the dope to the spinneret at a constant rate to obtain a consistent filament diameter. The spinneret holes are usually 50–100 μm in diameter. The spinneret is submerged in an acidic coagulation/spin bath of sulphuric acid, sodium sulphate, zinc sulphate and water. The acid neutralises the cellulose xanthate by diffusion and precipitates the cellulose polymers, coagulating them into filaments.

To prevent corrosion by the acids used in the process, the spinnerets may be made of non-reactive metals such as platinum, tantalum or gold. The spin bath is made from lead sheets. The temperature and chemical concentration of the bath in relation to the dope are important factors in the actions of diffusion, polymer precipitation and coagulation. (For further technical details, the reader is referred to Salem, 2000; Peters, 1963.) On leaving the spin bath, the filaments are passed through a drawing (*stretching*) zone in which stretching may range from 20% to 200%, depending on the spinning conditions, and are then washed, bleached and rewashed in a finishing stage to remove acid, salts and occluded sulphur. The spun yarns can then be dried and prepared for winding into a package.

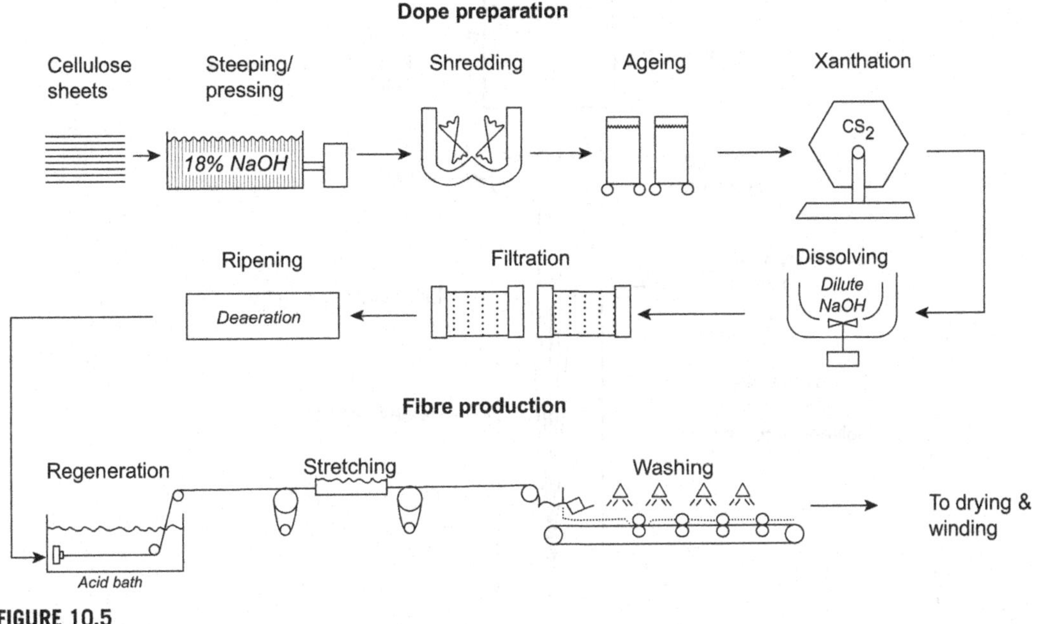

FIGURE 10.5

Wet spinning of viscose rayon.

Source: Dyer and Daul (1998a).

10.2.3 DRY SPINNING

A volatile solvent may be used to dissolve non-meltable polymers. Dry spinning is a process for producing CF yarns by dissolving a polymer in a volatile solvent to make a viscous solution that is extruded into a heated atmosphere to coagulate the polymer through solvent evaporation while stretching the filaments by means of drawing.

The polymers cellulose acetate and polyurethane (**spandex**)[4] are converted into CF yarns by the dry-spinning method. The basics of the process are similar to that illustrated in Figure 10.6. After the polymer is dissolved and the dope (which, again, is of much lower viscosity than with melt spinning) is filtered, de-aired and pre-heated, it is pumped through further filters and through the spinneret into a heated cabinet and gas flow. This may be air or an inert gas. Air is used for acetate. For polyurethane, nitrogen (N_2) plus a solvent gas cause the polymer to react chemically to form solid strands. As the viscous filament streams enter the gas flow, there is an instantaneous evaporation of solvent from the surface of each stream, which initially forms a solid skin. The filaments then solidify with further evaporation as they pass through the gas flow. Simultaneous stretching by down-drawing can be applied as required, similar to melt spinning, so enabling orientation of the polymer chains along the fibre axis. At the bottom of the heated cabinet, the filaments converge into CF yarns and are wound onto packages by a ring and traveller system (Schildknecht, 1956), which inserts a small degree of twist to hold the filaments together. Usually the evaporated solvent is recovered to meet the requirements of safety, environmental regulation and economical processing.

[4] Spandex fibres can be produced by melt spinning and wet spinning; however, dry spinning is used to produce over 90% of the world's spandex fibres (http://www.madehow.com/Volume-4/Spandex.html).

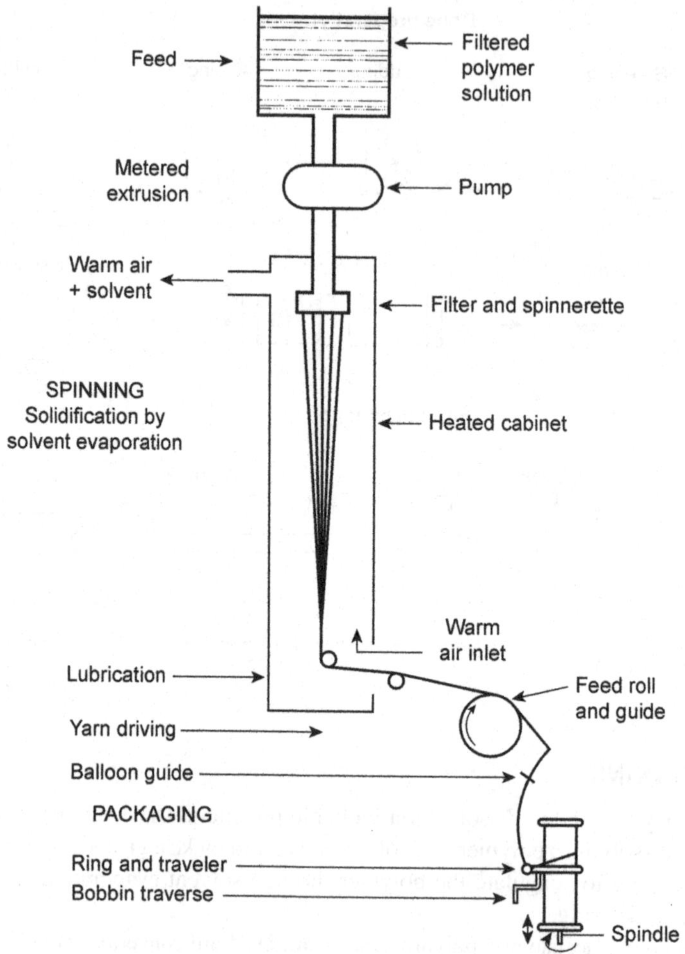

FIGURE 10.6

Dry spinning.

Source: Schildknecht (1956).

10.3 YARN TEXTURING

As obtained from the extrusion spinning processes, CF yarns are often described as flat yarns, meaning that they are inherently smooth and have no significant bulk or loft. With the exception of elastomers, they have little stretch at low applied forces and are lustrous unless altered by additives, as will be explained later. In this form they have a limited range of applications. Therefore CF yarns are generally modified to make them more opaque and to add bulk and stretch at low tension, thereby imparting such fabric aesthetics as increased thickness, cover, softness, warmth, stretch and water vapour permeability (i.e. breathability). Usually modification is performed only on multifilament CF yarns. The filaments are subjected to a major change in their physical form by being crimped, coiled or looped along their lengths (see Figure 10.7).

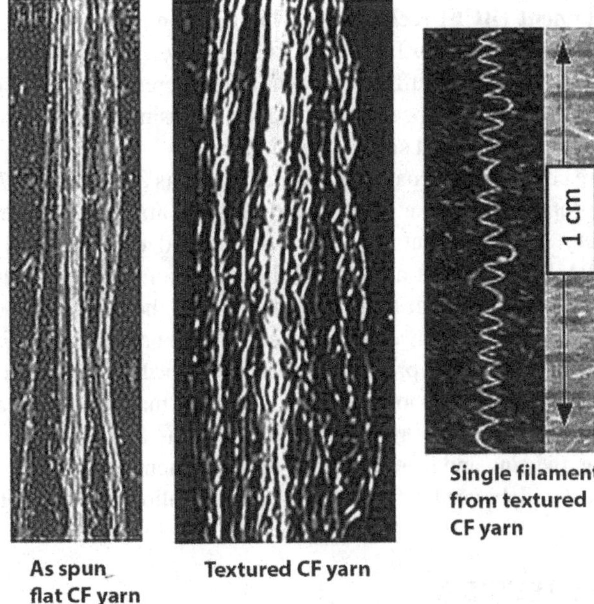

FIGURE 10.7

Flat and textured CF yarns.

The definition for a textured yarn can therefore be written as:

> a multifilament yarn that has been processed after extrusion spinning to introduce durable crimps, coils, random loops or other fine distortions along the lengths of the filaments, primarily giving the yarn increased bulk with or without increased stretch.
>
> **McIntyre and Daniels**

The processes used to make the change in physical form are collectively called texturing. Various texturing methods are practised, most of which are only suitable for thermoplastic CF yarns as they involve heating the filaments to achieve the texture profile and cooling to retain it. The methods commonly referred to are the following (Smith, Pieters, & Morrison, 1972).

- **Knife edge**: The filament yarn is heated and pulled across a blade at an acute angle. When the yarn is cooled and released, the retained internal stresses in the filaments cause them to collectively adopt a spring shape or curled ribbon appearance, i.e. the profile is heat-set.
- **Stuffer box**: The filaments pass through a heated box at a faster rate than they are removed (known as overfeed). This forces them to adopt a random wavy, crimped pattern while heated. The textured form is set by subsequent cooling.
- **Air jet**: High-speed overfeeding is also employed, but instead of using heat to effect the texture profile, compressed air is blown into the chamber, which causes the loose lengths of filaments in the yarn to separate and form entangled random loops. The entanglement retains the texture of random loops.
- **False twist**: The CF yarns are twisted and heated simultaneously and then untwisted when cooled, thus loosely retaining the heat-set helical shape of the twist.
- **Knit-deknit**: Filament yarns are knitted into the shape of a small-diameter tube and heat-set (i.e. heated and then cooled). The yarns are then de-knitted, giving them a wavy configuration.

Bulk continuous-filament (BCF) technology: This is essentially an integrated process of melt spinning and texturing, using either a stuffer box or a hot-fluid jet. The filaments leaving the extruder is tightly bunched up into the heated stuffer box with hot compressed air or driven by the hot compressed air through an air jet at high speed and a high plasticising temperature. The yarn becomes irregularly kinked and is then cooled and set.

BCF yarns usually are comprised of coarse individual filaments of 8–30 dtex (7–27 denier) for nylon (PA and PA 6.6) or polypropylene (PP), i.e. PABCF and PPBCF yarns. Their main areas of applications are carpets and upholstery. Finer filaments of 3.3–5.6 dtex (3–5 denier) in nylon (PA and PA 6.6) are possible for producing PA-BCF yarns of around 600 dtex, which may be suitable for furnishings and outerwear. PET has a more complex heat transfer to the core of the filaments, but development work has enabled PET-BCF yarns to become available with significant growth in the US market. Since BCF technology is basically an integration of processes earlier described, it will not be considered further.

From these brief descriptions of the above methods, texturing may be defined as a process by which continuous multifilament yarns are converted from their flat state to a greater bulked form, either by crimping and heat setting, twisting and heat setting, or the entanglement of loops. Although these methods are all in use, the most widely used is false-twist texturing followed by air-jet texturing.

10.3.1 FALSE-TWIST TEXTURING

False twist is the action by which twist is inserted into a yarn at the position of contact with a twisting device, and is then removed by the same device as the yarn leaves contact with it. A more detailed understanding can be acquired by comparing the actions involved in real and false twisting.

10.3.1.1 Real twist

When twist is inserted into a yarn, starting at its free end, the filaments adopt a helical shape or spiral, which may be clockwise or counter-clockwise in accordance with the direction of rotation of the twisting device. In a clockwise spiral, the inclination of the helix mirrors the inclination in the letter Z, and is therefore called Z-twist. In a counter-clockwise spiral, the twist is referred to as S-twist (see Figure 10.8). After the action of twisting, the inserted twist remains in the yarn[5].

10.3.1.2 False twist

If the twisting device through which the yarn passes is located so that the Z-twisting action (the clockwise torque) occurs as the yarn runs into the device, and the S-twisting action (the counter-clockwise torque) occurs as it runs out, then no twist will remain in the yarn. This is because the Z-twist will be removed by the counter-clockwise torque. This means that within a short time of starting the twisting process, Z-twist will be seen in the yarn as it runs into the device, but no twist will be present as it leaves. This is the false-twisting action, as illustrated in Figure 10.9.

The figure depicts the situation in which a CF yarn, nipped by two pairs of rollers at positions A and B, is driven at a linear speed of V_d m/min. The yarn is constantly twisted at the location X as it moves through the distance AB. If the twisting device is rotating at N_s rpm in the direction shown, then when viewed from A, along the length AX, it will appear to be turning clockwise.

[5] There may be a slight loss of twist as the fibres will try to recover from the strain of being twisted. This will depend on the fibre stiffness (torsional modulus) and the amount of twist inserted. The twist loss increases with both factors.

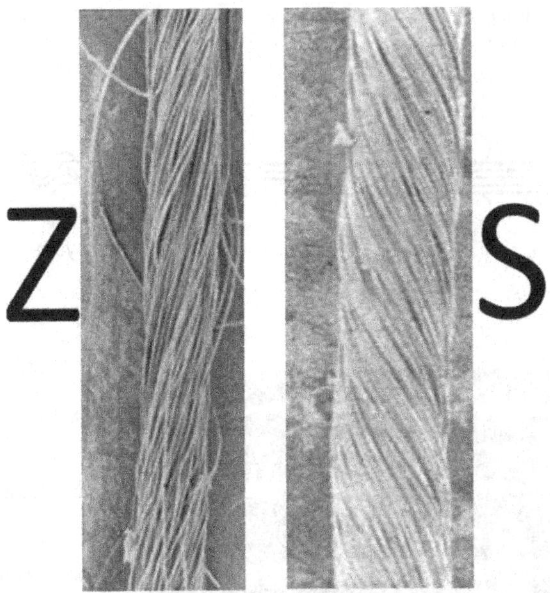

FIGURE 10.8

Z- and S-twisted yarns.

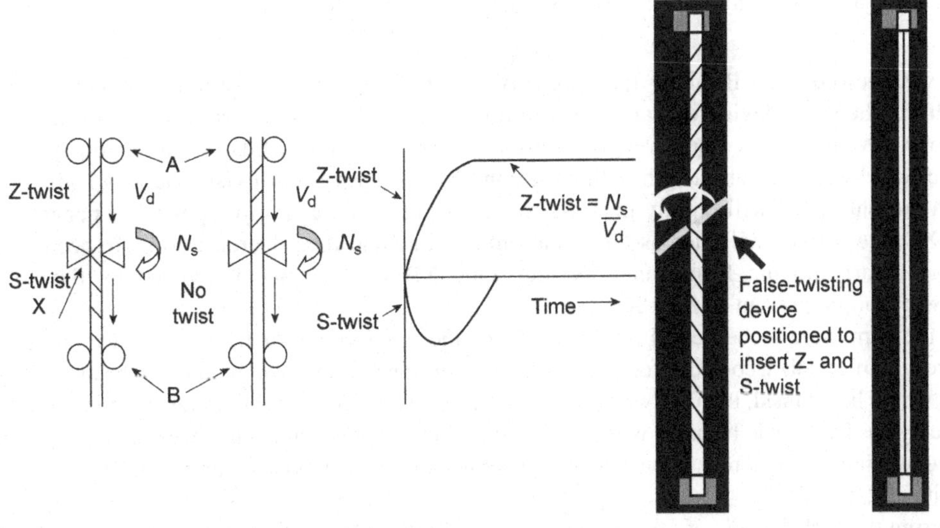

FIGURE 10.9

Illustration of the false-twisting action.

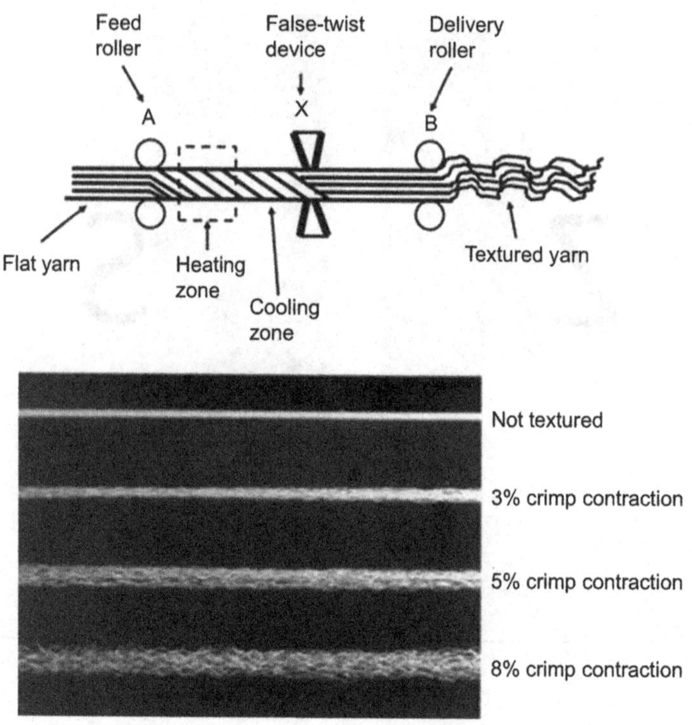

FIGURE 10.10

Illustration of false-twist texturing principle.

However, viewed from B, along the length BX, it will appear to be turning counter-clockwise. Therefore, the same device will at any given time effectively twist the yarn length present within the zone AX in a clockwise direction, thus inserting a Z-twist whilst simultaneously twisting counter-clockwise the yarn length within the zone XB, inserting an S-twist. The Z-twist in the yarn length present in AX will rapidly increase to a constant value, equal to V_d/N_s turns per metre[6]. In zone XB, the S-twist will increase to a maximum value and then decrease to zero, because each length of yarn moving from zone AX into zone XB will become untwisted by the counter-clockwise torque present in zone XB.

If the yarn were to be heated above the T_g of the polymer whilst being Z-twisted in zone AX (Figure 10.10), and then cooled before being untwisted in zone XB, the spiral shape of the individual filaments would be retained, but they would become free of the twist compaction, resulting in a false-twist textured yarn. This would be bulky with considerable stretch at low applied force, depending on the level of false twisting. For a more in-depth study of the principles of false-twist texturing, the reader is referred to Smith et al. (1972).

During the earlier years of false-twist texturing, FDY/HOY yarns were texturised using a rotating pin as the false-twist device. Figure 10.11 illustrates the pin twister as a hollow spindle with a

[6] Twist is usually referred to, and measured, in turns per metre (tpm).

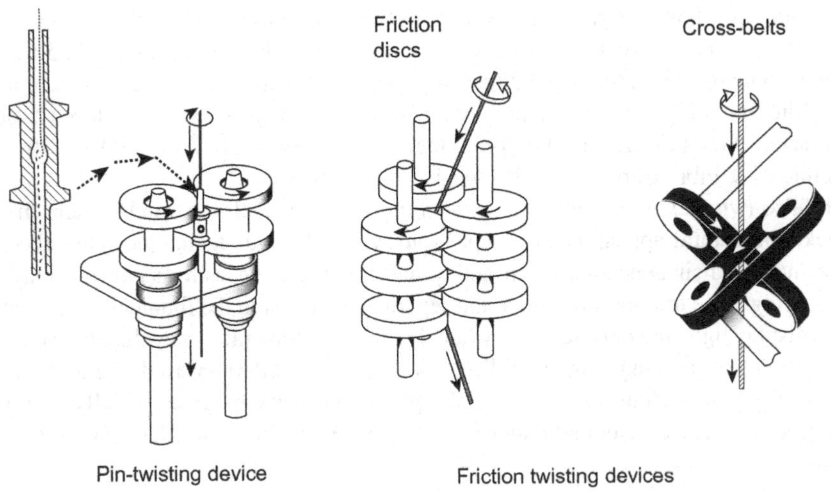

FIGURE 10.11

Illustrations of friction-twisting devices.

central pin positioned across the tubular interior. The flat CF yarn passes down the tube and around the pin. The pin is rotated by a driven disc assembly.

Partly oriented yarn may be utilised by the addition of a hot-draw zone before the heating/twisting/cooling zone, thus giving better process economics and product variety in terms of bulk and stretch (depending on the pre-drawing). The resulting yarn is then referred to as draw-textured yarn, or DTY. The pin-twister was later replaced by a friction-twisting device (Hearle, Hollick, & Wilson, 2001, Chap. 1, pp. 1–13). As shown in Figure 10.11, an assembly of overlapping discs enables the CF yarn to be directly twisted by utilising the frictional contact between the yarn and the disc surface, thus replacing the need for a hollow spindle. Another type of friction-twisting device is based on driven cross-belts. These devices enable higher twisting speeds: 20×10^6 rpm as opposed to 8×10^5 for pin twisting. Depending on the polymer and yarn count, friction twisting enables production speed of up to 12,000 m/min to be attained.

DTYs are considered to be highly stretchable. In addition to the intrinsic extension of the constituent filaments, there is added extension attributable to the crimp imparted by the texturing process. Highly stretchable yarns can be made by applying high false-twist levels that give crimp extensions of 150–300%. However, where DTYs are required to have relatively low crimp extension while retaining a high bulk, an additional heating stage is incorporated after the roller-pair B (Figure 10.11). A third pair of rollers would then be used as output rollers. This heating stage gives an additional heat-setting treatment that reduces the intrinsic elasticity and crimp extension of the DTY. The yarn is then described as stabilised or *set-DTY*.

An example of the set-DTY process is illustrated in Figure 10.12. This shows the melt spinning of nylon POY, which is then transferred to a draw texturing process. Here, the yarn is first drawn at a draw ratio of the order of 2.3–3.5 and a temperature of at least 50 °C (T_g=40 °C). It is then subjected simultaneously to false twisting (friction disc device), heating and further drawing at a draw ratio of 1.1–1.5. After false twisting, the DTY passes through the additional heating zone to become a set-DTY. The process is particularly useful for the manufacture of nylon textured yarn with a linear density in the range of 10 dtex/9 denier to 50 dtex/45 denier (Reimschuessel, 1998)

Thermoplastic multifilament yarns made from polyesters, nylons and polypropylene can usually be converted into DTYs. Polyester yarns are normally within the count range of 55.6 to 333 dtex (50–300 denier), with the emphasis on 83.3 to 166.7 dtex (75–150 denier). Nylon yarns are in the range of 16.7–122.2 dtex (15–110 denier); the majority of fine hosiery yarns are 24.4 dtex (22 denier) and coarser yarns are 77.8 dtex (70 denier). Polypropylene yarns are typically 7.7 to 100 tex (70–900 denier) (Acelon Chemicals & Fibre Corporation; Brown & Chuah, 1997).

Although false-twist texturing gives CF yarns a more natural feel, bulk and stretch, their uniform geometry lacks the natural appearance of staple-spun yarns. Various techniques have therefore been developed to modify their appearance. An additional drawing zone using a hot pin may be placed before the texturing stage to impart irregular and repetitive changes to the filaments so that after texturing, the yarns display light and dark sections when dyed. If the draw ratio is cyclically varied from high to low during the main drawing stage and the twist insertion similarly varied, the textured yarn will have thick and thin places along its length. Although this property may be partially removed during downstream processing, a sufficient amount of the property is retained, say, during weaving to give the fabric a linen-like appearance.

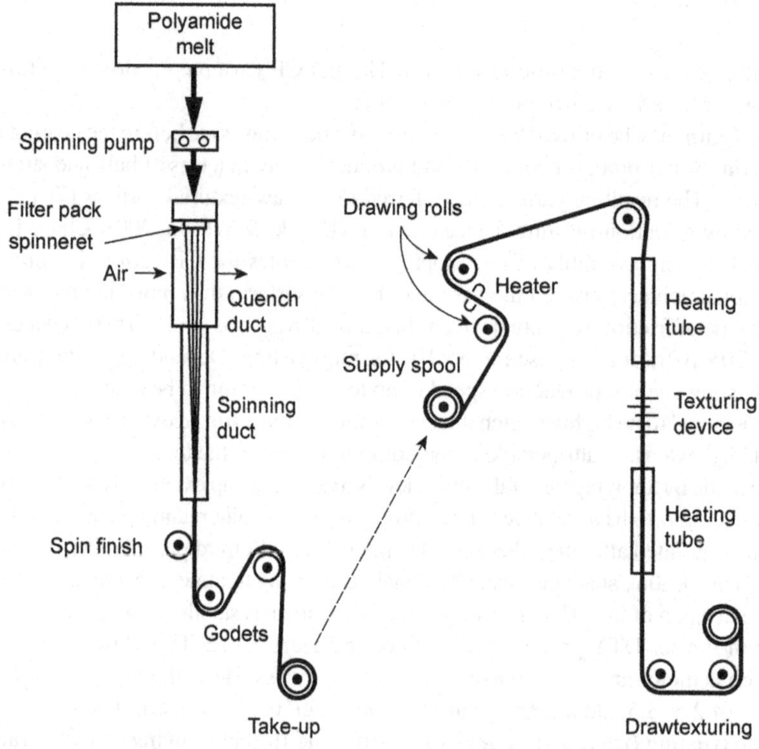

FIGURE 10.12

Illustration of the set-draw-textured yarn process.

Source: Reimschuessel (1998).

10.3.2 AIR-JET TEXTURING

Air-jet texturing has become increasingly important because of its capacity to process CF yarns of any polymer type. For example, cellulosic rayon, which is not a thermoplastic, can be air-jet texturised. A typical air-jet textured polyester/viscose yarn blend would comprise a FOY polyester of 80/75 tex (72/75 denier) and a viscose CF of 80/24 dtex (72/24 denier).

Air-jet texturing is also capable of providing products with aesthetic characteristics (tactile handle) superior to most other texturing processes. It can be used to process FDY/FOY (to give air-textured yarn, or ATY) or with a pre-drawing stage in an integrated draw-texturing system to convert POY to a drawn-air textured yarn (DATY).

Various air-jet designs may be used to make ATY and DATY. Figure 10.13 depicts one of the widely used jets, referred to as an axial-jet (or venturi jet). The filament lengths to be texturised are fed into the jet without tension. This is achieved by ensuring that the speed at which they are fed into the jet is faster than the speed at which they leave (overfeed >1). The compressed air entering the jet then acts on the loose filament lengths. The compressed air is accelerated towards the outlet and pulls the filaments along. A spherical barrier is positioned close to the outlet, which causes the compressed air to become turbulent as it exits. The filament lengths in this flow become separated and form a profusion of loops of various sizes, which entangle as the filaments are pulled together by the removal rollers to deliver the air- jet textured CF yarn.

A vortex-jet design may also be used. In this device, the compressed air has a spiralling turbulent flow to the outlet in which loose filament lengths are separated and form different-sized entangled loops as they exit the jet to become the ATY. These jets usually operate at pressures of about 10 bar (145 psi), and the actual texturing occurs directly at the exit of the jet within the turbulent air stream. To assist in the formation of loops, the filaments may be wetted prior to entering the jet, either by spraying or immersion in a water bath. This aids separation of the filaments and acts as a lubricant inside the jet.

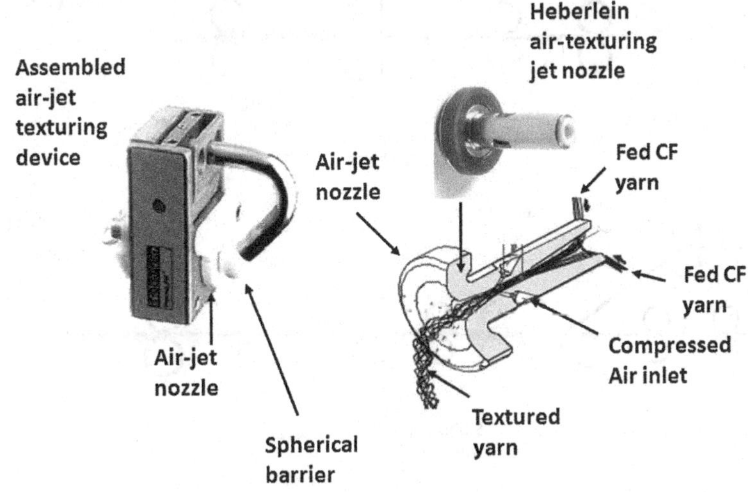

FIGURE 10.13

Air-jet texturing device.

POY is preferred for air-jet textured apparel yarns, largely because it is less expensive than FOY. The operating stages of a DATY machine would therefore be of a similar configuration to that of a DTY machine, but with the false-twisting device replaced by an air-jet system, as illustrated in Figure 10.14.

Two types of yarn can be produced with air-jet texturing: parallel yarns and core-effect yarns. The production of parallel yarns involves one or more POY being fed into the air-jet device with precisely the same overfeed for all the yarns used (Figure 10.14). Usually, an overfeed of 18–30% may be applied, depending on the end-use requirements. Parallel yarns are commonly used for apparel and cut-pile plush fabrics.

Core-effect yarns (or fancy yarns) are produced with two main components: a core yarn and an effect yarn. Both may consist of one or more POYs. The overfeed used for the core yarn(s) is always lower than that for the effect yarn(s): the former is normally between 5% and 15% whereas the latter can be up to 400%, depending on the end-use. For example, overfeeds of 8% core and 40% effect would be used for an apparel end-use such as nylon sportswear; while 120% effect yarn overfeed would be appropriate for upholstery fabrics for domestic use and car seats.

Figure 10.14 shows that the drawing zones for both yarn types have a heating element, which may be a heated pin, godet or plate. The heating element is essential for polyester POY, but nylon and polypropylene POY may be cold drawn. Note that in the case of core-effect type yarns, the two yarn paths must have their own drawing zone and heating element.

Although CF yarns of virtually any polymer can be air-jet textured, the number of filaments and the dtex/denier per filament are very important; a high number of filaments of low dtex/denier per filament

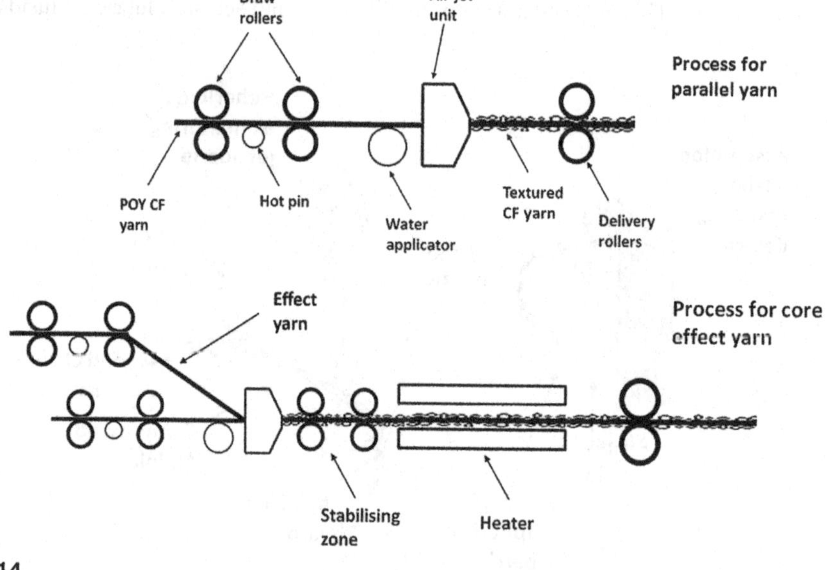

FIGURE 10.14

Processes for parallel and core-effect air-textured yarn.

is preferred. Yarns with a higher than 3.3 dtex (3 denier) per filament are rarely air-jet textured. Currently, the practise is to use between 1.1 and 2.2 dtex (1 and 2 denier) per filament (dpf), the most popular yarns tending towards 1 dpf or finer.

Air consumption is a significant cost factor in the process and is kept as low as possible. Therefore a variety of air-texturing jets are available to cover the different ranges of dpf. Vortex-type jets are generally used for products that require less than 100% overfeed, whereas the venturi type of jet may be used with overfeeds of up to several hundred percent.

Loop size and stability are important factors, especially in finer yarns. Smaller loops are more stable and give a bulkier feel than larger loops. As shown in Figure 10.14, the textured yarn can be heat-treated to increase the stability and shrink the loops as they leave the air jet. The temperatures must be sufficient to ensure that the constituent filaments are adequately treated. For finer yarns, sufficient setting temperatures are usually 230–240 °C. If required, an additional drawing zone placed between the air jet and the heater may also be used, ensuring small-sized loops are obtained.

A further addition to the process line can be made that converts small loops into hairs, thus simulating the structure of staple-spun yarns, which have short protrusions of fibre ends from the yarn surface. The process used for achieving this staple fibre-like yarn is called Texspun and has been developed by the machine manufacturer Barmag (Air Jet Texturing and Fabrics). The Texspun device is located in the heat-stabilising zone. As the yarn passes through the device, the loops are torn and free fibre ends, similar to staple fibres, are produced.

10.4 BULK CONTINUOUS FIBRE (BCF) TECHNOLOGY

10.4.1 TWISTING/PLYING OF CONTINUOUS-FILAMENT YARNS

A single multifilament FDY yarn (flat yarn) can be twisted so that the twist holds the filaments together and imparts some degree of stretch as a result of the twist helix in combination with the intrinsic stretch of the spun polymeric material. When a small amount of twist (40 tpm or less) is used to hold together the filaments of a multifilament FDY yarn, it is referred to as producer twist, since it is usually inserted at the yarn-production stage. The more common practice is to intermingle the filaments with a special air jet (Fourne', 1998).

Twist levels can range from that of producer twist to over 1000 tpm, and the higher the value, the more monofilament-like the multifilament yarn will become. Two or more yarns (mono- or multifilaments or combinations) may also be twisted together to produce greater bulk, or in the case of textured yarns, to alter the stretch and mechanical performance.

When two or more yarns are twisted together, the action is called **doubling, folding, plying**, or **ply twisting**. The resulting yarn is termed a **doubled, folded** or **plied** yarn (e.g. 2-ply, 8-ply). For example, PET yarns as fine as 49 dtex and up to 110 tex are often labelled as PET 49 dtex 1 × 2 (i.e. 2 yarns plied) and PET 1100 dtex 1 × 3 (i.e. 3 yarns plied) (Trougott Baumann). In plied yarns, each yarn may be individually Z-twisted, and then twisted together in the S-direction, thus alleviating a tendency for the plied yarn to snarl. Plied CF yarns are used in embroidery and for sewing threads.

There are three basic methods used in filament yarn twisting: up-twisting, down twisting and two-for-one twisting. Figure 10.15 shows the principal features of each method.

In up-twisting and down-twisting, a ring (R) and traveller (C) arrangement may be used. The traveller is the 'C' metal or plastic clip that is loosely clipped onto the ring profile. The traveller circles the

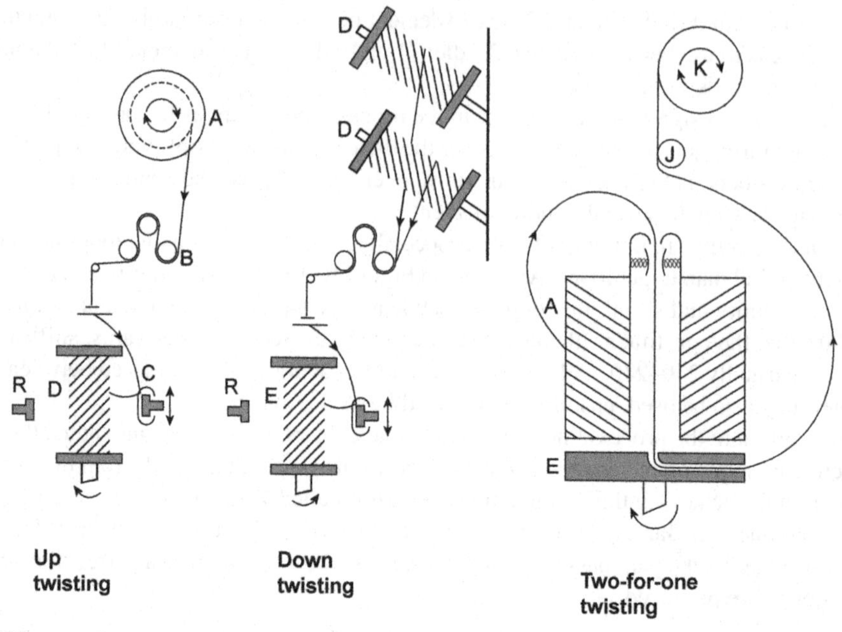

FIGURE 10.15

Yarn-twisting processes.

ring, and effectively each circulation will insert one turn of twist into a single yarn or a combination of two or more yarns.

In up-twisting, the circulation of the traveller around the ring is obtained by placing a bobbin (D) of CF yarn on a spindle (S) concentric with the ring, threading the yarn from the bobbin through the gap between traveller and ring and via a set of delivery rollers (B) onto a take-up bobbin (A). The rollers remove yarn from (D) and deliver it to (A). As the spindle rotates in the direction indicated, the yarn is easily removed and the take-off rollers simultaneously pull the yarn through the traveller/ring gap, causing the traveller to be dragged around the ring. The circulation of the traveller causes the length of yarn between traveller and take-off rollers to move radially outwards. This is termed ballooning, because the speed of the circulating length gives the optical illusion of an inflated balloon. For each rotation of the traveller, one turn of twist is inserted in the length. The delivery rollers then feed forward the twisted yarn for it to be wound onto the take-up bobbin by a drive system, the process being continuous until all the yarn is removed from the bobbin (D).

Up-twisting is suited to single CF yarn, but if more than one yarn is to be twisted, they may be initially wound together onto the up-twisting supply bobbin (D). This is referred to as assembly winding.

In down-twisting, two or more yarns on separate supply bobbins are pulled by the delivery rollers, which operate in the opposite way to the up-twisting action. The yarns pass via the ring and traveller onto a bobbin mounted on the spindle. The spindle rotates in the opposite direction to that of up-twisting. The yarn is now wound onto the spindle bobbin, instead of being unwound from it. As before, the yarn drags the traveller around the ring as it is being wound onto the bobbin, and the twist is inserted within the yarn length between the traveller and the delivery rollers. The winding action occurs because the friction between the ring and traveller causes the traveller to always move slightly more slowly than the surface speed of the rotating bobbin. The theory of ring and traveller combination is given in Lawrence (2003).

Figure 10.15 also illustrates the principle of two-for-one twisting. The yarn or yarns are withdrawn from a stationary supply package or packages (A), passed through the centre of the package(s) and through the centre of a rotating mechanism (E), driven by the spindle (S). The yarn(s) then balloons around the supply package(s) to be delivered by a pair of rollers (D) and wound onto the take-up package (K). The yarn (or yarns) is twisted once as it passes down the package centre (A), and a further turn is inserted between E and the delivery rollers (D), resulting in two turns of twist for each turn of the spindle (E).

Wrapping one yarn around another to produce a covered yarn may be considered as a special case. This process is normally used to produce elasticated filament yarns, which are PU-CF yarns covered with a textured CF yarn or staple-fibre yarn[7], making them easier to weave or knit and providing them with better tactile characteristics and fabric drape. Such yarns are used in woven fabrics, narrow fabrics, flat knitting and circular knitting, and can be found in gloves, sweaters, intimate apparel, swimwear, hosiery and socks (BeYarns). With wrapping, snarling should not occur. Therefore wrapping with a single-yarn covering can be carried out with either a Z- or an S-twist. It is common to use a Z-twist for elasticated fabrics. Double covering may be employed where the cover yarns are wrapped in the opposite directions, one being the outer covering (S-twist), and the other the inner covering (Z-twist). The different twist directions improve the tactile properties. Double covering may be used for socks and underwear to enable a more accurate body profiling.

10.4.2 METALLISED YARNS

A less conventional method of making filament yarns is to first extrude (using long narrow-gap dies instead of spinnerets) a thermoplastic polymeric material into a thin, wide sheet or film similar to that employed for packaging. Metering pumps are generally not used. The film is normally quenched in a water bath, or on chilled rolls. On leaving the bath, the film is blow-dried. It is then cut into narrow ribbons, drawn and annealed. Annealing is where the film is re-heated for a short period of time below the glass transition temperature of the polymer, in order to relieve internal stresses and allow the polymer chains to partially relax. The film is then gradually cooled to room temperature. The oriented polymer chains readily allow slits to be made along the length of the ribbons, i.e. the ribbons become fibrillated. These can then be twisted to make split film (or fibrillated) yarns. Although the majority of these yarns are produced for ropes and other technical textile applications, the process can be adapted for metallised yarns that are used for their visual aesthetics. Typically, a PET film is metallised by depositing a thin layer of aluminium or gold, and then lacquered with a protective epoxy resin that maintains the shiny look of the metallised surface. If a coloured shiny look is required, the requisite dye is added to the resin prior to coating. The film can then be cut into ribbons that are further reduced to strips of around 0.2–0.3 mm width (M-Type Metallised Yarns) and can be plied with other CF yarns or staple-fibre yarns for decorative end-uses.

[7] Staple-fibre yarns (or staple-spun yarns, or spun-staple yarns) are made from fibres of discrete lengths, such as cotton, wool or mmf cut to specified lengths. These fibres are assembled as a ribbon of fibres that is then twisted to make the staple-fibre yarn.

10.5 PROPERTIES OF CF YARNS

10.5.1 MORPHOLOGY

Filaments may have various cross-sectional shapes. Figure 10.16 shows cellulose fibres produced with serrated, multilobal, round and flat (not shown) cross-sections. Most differences in cross-sectional shapes are obtained with the melt-spinning process, including round, trilobal, pentagonal and octagonal. Figure 10.17 illustrates examples for PA 6.6 and PET.

The orifices of spinnerets are normally circular and produce filaments of circular cross-section. However, non-circular orifices are extensively used to produce filaments of non-circular cross-section for special fabric characteristics such as lustre, opacity, air permeability, thermal insulation and resistance to soiling. Modified filament cross-sections can be formed either by coalescing/fusing the molten streams issuing from the spinneret of single or multiple orifice designs, or by direct extrusion through profiled capillaries.

In addition to their effect on dyeing, caused by the change in surface area, different cross-sectional geometries can enhance certain physical properties. Triangular/trilobal-shaped fibres are capable of reflecting more light, which gives a sparkling effect to fabrics, whereas octagonal-shaped fibres offer glitter-free effects. Both give attractive tactile characteristics. Pentagonal-shaped and hollow fibres are soil and dirt resistant and impart easy-care properties. Hollow fibres

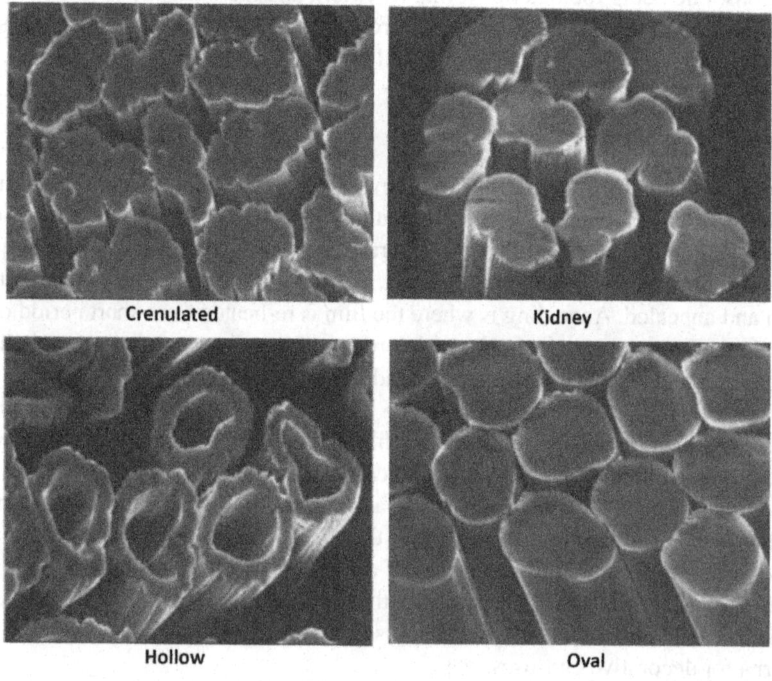

FIGURE 10.16

Cross-sectional shapes for rayon (2000× magnification).

Source: Dyer and Daul (1998b).

also trap air, which provides good insulation and reduced weight per unit area. The effect of different geometries on fibre packing and hence on air permeability/breathability is illustrated in Figure 10.17.

It is widely acknowledged that fibre fineness strongly influences the tactile aesthetics of fabrics, and fibres of less than 1 dtex (0.9 denier.) in fineness are therefore of importance. These are referred to as microfibres. Fabrics made from microfibres are very drapable and silk-like in hand and appearance. For example, rayon microfibres have been successfully produced at 1 dtex, and polyester CF microfibre yarns can be made with individual filaments of 0.4 dtex (0.36 denier) fineness. Both give good fabric-tactile qualities.

These fine fibres can be obtained directly by conventional spinning using a combination of spinnerets with smaller orifices and lower throughputs of polymer per hole. However, the preferred method for synthetics is to produce fibres near 1 dtex (0.9 denier) POY that can then be draw-textured. Microfilaments can also be produced from PET by treating coarser filaments with hot alkali (NaOH) to reduce their diameters. This process can be used to reduce microfibres to finer diameters. Even finer fibres can be made using what is known as the bi-component spinning technique. This is essentially a modification of the melt-extrusion stage of the melt-spinning process, and involves spinnerets from which two differing polymers are extruded together. Each filament of the CF yarn is comprised of the two polymers, coalesced in such a way as to enable their separation in a subsequent process to form very fine microfibres. The types of coalescences are referred to by their cross-sectional appearance, as illustrated in Figure 10.18.

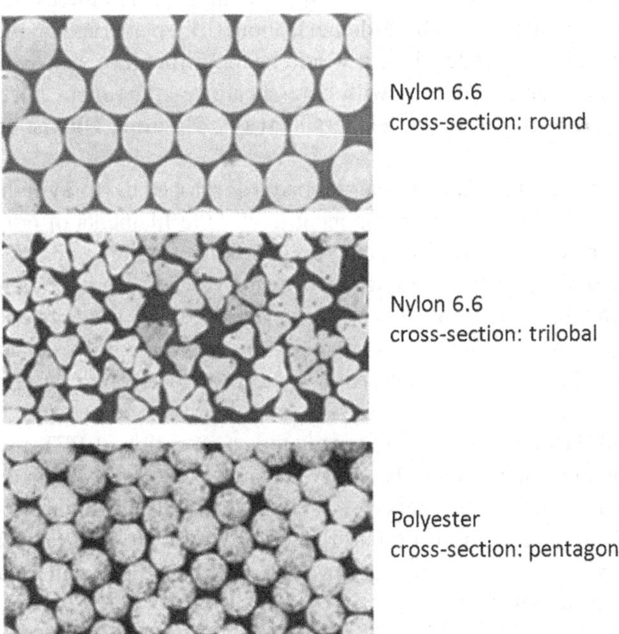

FIGURE 10.17

Examples of cross-sectional shapes of melt-spun filaments.

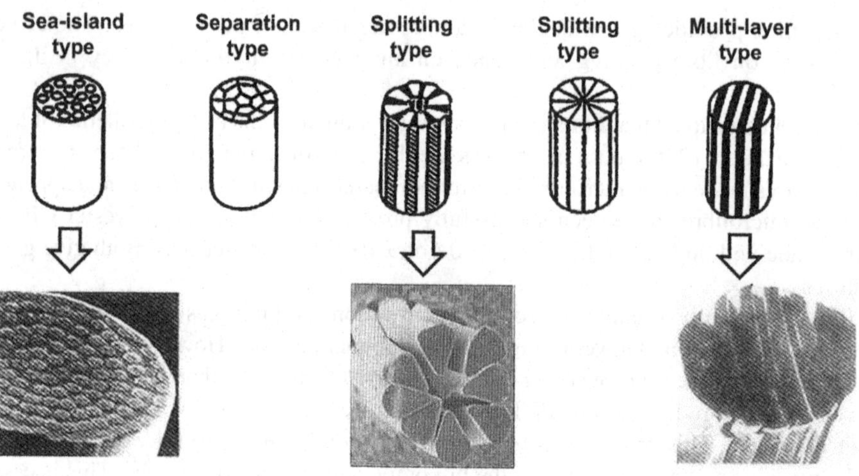

FIGURE 10.18

Bicomponent fibres.

An example of 'island-in-the-sea fibres' is 145 PET filaments (island) in a matrix (sea) of polystyrene.[8] The latter is then dissolved away using dimethyl formamide as the solvent. This process produces filaments of 0.001 dtex (0.0009 denier), about 0.1 μm average diameter. The dissolving is usually carried out in the fabric state, since it would be too difficult to weave or knit such fine filaments. A similar approach can be adopted with core-sheath bi-component fibres. Applications for such materials include suede-like artificial leathers (**Toray's Ecsaine, Ultrasuede, Alcantara**) and fine woven fabrics.

Alcantara® fabrics are made from a proprietary process developed by Toray Industries. This process involves producing 0.04 or 0.14 denier bi-component microfibre filaments of polyester core/polystyrene sheath, using 16- or 36-hole extruder spinnerets. The filaments are subsequently cut to 50-mm staple fibres and carded to form a web or veil, which is then layered to produce a felt. The felt is needle-bonded into a nonwoven fabric and a solvent (100% recycled) is used to remove the polystyrene. The fabric is then passed through a buffing process that slightly raises the polyester microfibres to give a velvety texture and silk-like shine. The material is used for a wide range of applications including apparel.

An example of a multilayer, splittable fibre type is the co-extrusion of PET and PA 6, which gives poor interfacial bonding and can therefore be easily separated mechanically or chemically (in fabric form) to make very fine filaments of unusual cross-sections, particularly flattened cross-sections. Alkaline hydrolysis (NaOH) may be used on PET to reduce up to 40–50% of the polymer. This type of modification produces what are called Shingosen fabrics, which have enhanced tactile aesthetics (Fukuhara, 1993; Matsudaira, 1994).

[8] Polystyrene is a petroleum-based thermoplastic polymer made from styrene monomers.

ZEPHYR 200 RTM, developed by Kanebo Ltd, is a highly soft nylon fabric made from a radially splittable bi-component CF yarn in which each filament consists of 75% PA and 25% PET by weight. After being woven, the PET is removed by alkali treatment to leave a fabric composed of microfibre PA filaments. The yarn is initially spun to a count of 111 dtex (100 denier), comprising of 50 bi-component filaments, 2.2 dtex (2 denier) per filament, with eight triangular nylon segments. When the PET is removed, the yarn becomes 100% PA with 400 filaments, c. 0.21 dtex (0.19 denier) per filament.

A technically interesting feature that can be achieved through fibre morphology is iridescence that simulates the Brazilian Morpho Hecuba butterfly's changes in colour. The iridescence occurs as a result of light interference, depending on the angle of view. Two approaches have been developed, both using bi-component materials. One approach (Hongu & Phillips, 1997a) attempts to copy the scale structure of the butterfly, by extruding bi-component filaments of a flat cross-section from materials having different thermal properties. When thermally treated, each filament becomes twisted, forming vertical and horizontal sections running consecutively along its length. The alternating orientations cause repeated reflection and absorption of incident light and diminish direct reflection. The resulting fabrics have deep brilliant colours with subtle changes depending on the angle of the incident light.

The second approach uses a three-layer film technology (Hongu & Phillips, 1997b) in which a drawn polymer film is sandwiched between two undrawn polarising films. The incident light wave is first vertically polarised then turned into elliptically polarised light by the drawn film, orientated 45°, and finally horizontally polarised via the third film to give the iridescent effect. The effect can be also obtained by using a reflecting metallised film as the final component that produces the iridescence. The film combination may be converted into slit-film yarns, using a similar process to that described earlier for metallised yarns. These iridescent yarns are used in woven or knitted garments.

10.5.2 TENSILE PROPERTIES

The tensile properties of CF yarns are principally a combination of the tensile properties of the individual filaments. The combination will depend on the arrangement of the filaments within the CF yarn. In a flat CF yarn, where all the filaments are straight and parallel to each other, the combination would be additive. If a flat CF yarn is twisted or textured, the combination would clearly be more complex and more difficult to determine mathematically. Whichever combination is used, the practical approach to obtaining full knowledge of CF-yarn tensile properties is to measure them for the filaments and for the yarns themselves. This is done by using a tensile testing machine that stretches each filament and the CF yarn to the point of break, while recording the continually increasing force required to extend the material at a specified rate (e.g. 10 mm/min) to the point of rupture. The recorded values are then plotted and the graph characterises the tensile behaviour of the particular filament or CF yarn. The actual recorded values would be in units of Newtons (N) for the force applied and millimetres (mm) for the extension of the yarn. However, to make possible the comparison of different filaments and CF yarns, the unit of force is normalised by dividing it by the tex of the filament or yarn tested. This gives what is termed the specific stress in units of N/tex. The extension (i.e. the value of the increase in length) divided by the value of the original length, gives what is referred to as the strain. This has no units but the calculated value may be multiplied by 100 to give the percentage strain (% strain). The reader wishing to learn further details of the tensile test procedure is advised to read Hongu and Phillips (1997b). However, the following simple equations summarise the main points:

$$\varepsilon = e/L_o \tag{10.1}$$

or

$$\%\ \varepsilon = [e/L_o]\ 100 \tag{10.2}$$

Also called % elongation at break:

$$\delta = \text{N/tex} \tag{10.3}$$

where δ is the specific stress (unit: N/tex, N is the symbol for Newtons); $\mathcal{E}$ is the strain; F is the force (units: N); L_o is the length of the test specimen prior to testing, i.e. the original length (unit: mm) and e is the amount by which the specimen extends at a certain F value.

Figure 10.19 shows the typical graph obtained when δ is plotted against $\mathcal{E}$, and it depicts the tensile behaviour of the material tested. It can be seen that with continuously stretching the material, the applied force initially increases rapidly and linearly up to a certain extension, after which further extension requires smaller increases in the applied force. This point of deflection of the graph is called the yield point, i.e. the point at which the initial resistance to stretching the specimen is overcome. However, when a much larger value of extension is reached, the applied force rapidly increases again (i.e. the material's resistance to stretching increases) up to the point at which the specimen ruptures (i.e. the breaking point).

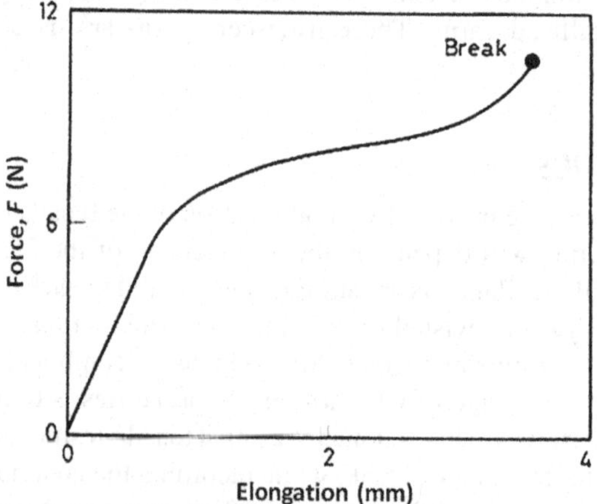

If the yarn or filament count is *C* tex, e.g. 10 tex, then for a Force of 6 N
δ - specific stress (N/tex) = *F/C* = 0.6 N/tex
The specific stress at break is the yarn strength, called the yarn tenacity.
If the test specimen length is L_o = 50 mm and it is extend *e* = 2 mm, the %Strain, % $\mathcal{E}$, or % elongation = $\mathcal{E}$ / L_o = 4%. The elongation at break can be used to calculate the breaking strain.

FIGURE 10.19

Force–elongation tensile characteristic of CF yarns.

A number of important property values can be determined from the force–elongation (or stress v strain) graph. These are:

- The yield stress: specific stress at the yield point
- The yield strain: strain at the yield point
- The elastic modulus: the resistance to stretching up to the yield point
- Tenacity (or tensile strength): the maximum specific stress
- The work of rupture

With regard to the elastic modulus, the stress–strain relationship is linear up to the yield point. If the applied stress was removed just before the yield point, the test specimen would quickly return to its original length L_0; thus $\varepsilon=0$, i.e. no residual or permanent strain would be left in the specimen. This is similar to elastic behaviour, and this region is often referred to as the elastic part of the graph. The slope or gradient of this linear region is termed the elastic modulus or the initial modulus, as it is the initial part of the graph. It is given the symbol E and calculated:

$$E=\delta/\varepsilon \text{ (units:N/tex)} \tag{10.4}$$

Table 10.2 shows that filaments made from differing polymers will have different E values, and even with the same polymer, the process parameters used in manufacturing can be altered to produce different E values.

The maximum specific stress for a single filament is likely to be its point of rupture, but for a CF yarn it may not be the breaking point of the yarn. This is because the variation in properties among the constituent filaments of a CF yarn may result in some filaments breaking before others, thereby weakening the yarn. Rupture will therefore occur at a lower stress value with further strain after the point of maximum specific stress. A flat or textured form will also influence the value of the breaking stress. Since the strength of the yarn (i.e. its tenacity) is the point of maximum stress, the strain at break may not necessarily be the strain at maximum stress.

As indicated in Table 10.2, the different polymers give different CF yarn tenacities. Even with the same polymer, the conditions used in extrusion spinning and texturing can be chosen to obtain different CF yarn tenacities, as is indicated in the table by the range given for a particular polymer. CF yarns can be therefore produced with tenacities to suit differing end-use requirements.

When extending a length of yarn, mechanical work is said to be done. By definition, the work done equals the product of the applied force required to stretch the length, and the actual amount of stretch.

Table 10.2 Mechanical properties of CF yarns from synthetic polymers

Fibre type property	Silk	Rayon	PET	PBT	PTT	PLA	PA 6	PA 66	PP	PU (spandex)
Tenacity (cN/dtex)	4.0	1.2–5.0	3.7–4.4	2.8–3.5	3.4–3.7	2.0–3.2	4.0–7.2	4.1–4.5	3.5–13.1	0.5–1.0
% Elongation at break	20	15–30	30–38	25–40	36–42	40–80	45–85	32–44	12–100	700
E, initial modulus (cN/dtex)	8.9 GPa/ density = 1.25	54	97	23	23	35	40	31	88–110	0.04

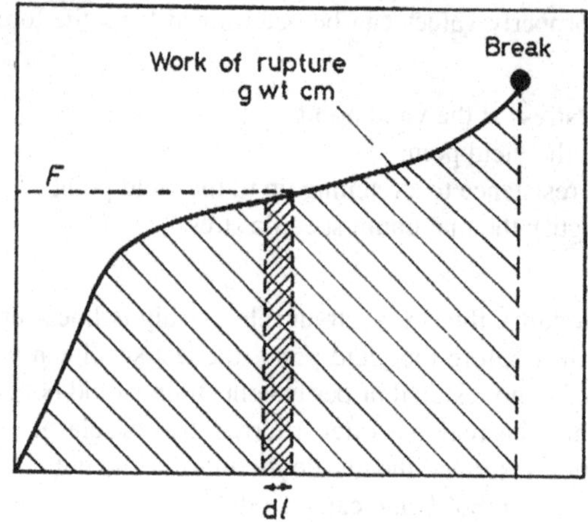

FIGURE 10.20

Work to rupture.

Consider the small elongation and the force required to achieve it, shown in Figure 10.20. The work done equals F·dl, where:

$$F = \delta/\text{tex} \tag{10.5}$$

and

$$dl = d(\%\,\varepsilon) \cdot L_o/100 \tag{10.6}$$

Since dl is small, F·dl would equal the area of the rectangle. Now consider the stretch from the start to the point of break. The series of the work done for small amounts of strain to the point of break would be equal to the area under the curve. Mathematically this is obtained by the following integration:

$$\text{Work of Rupture} = \int_0^b \sigma d \tag{10.7}$$

This is called the work of rupture (units: joules) because it is the amount of work done in stretching the specimen to the point of break or rupture.

Whereas the differences in the mechanical properties of filaments spun from different polymers are principally attributable to differences in the polymer structures, the ranges shown in Table 10.2 for filaments spun from essentially the same polymer type are deliberately produced by adjusting the process parameters during manufacturing, particularly in the filament drawing. The greater the degree of filament stretch in the drawing operation, the higher will be the tenacity and the lower the % strain to break. This can be explained as follows.

When linear semi-crystalline polymers are extruded through the die holes of a spinneret, there is a tendency for some parts of each chain length to become folded on themselves, forming localised layered structures (lamellar microstructures) in which the atoms are arranged into identifiable geometrical forms similar to crystals, i.e. crystalline regions. Other parts of the chain are randomly buckled and can be interlaced with parts of other chain lengths to form amorphous regions (i.e. regions without clearly defined shape or form). Where some parts come close together and lie parallel, crystalline regions of another type are formed.

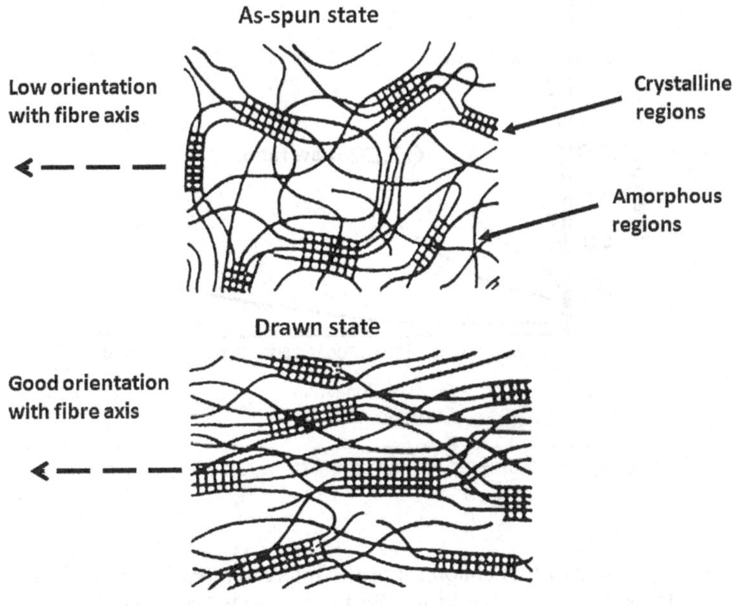

FIGURE 10.21

Fringe micelle structure of filaments of semi-crystalline polymers.

Generally these different regions will have a small degree of orientation along the direction of the filament length, which is induced by the draw-down of the first godet rollers. Therefore, as shown in Figure 10.21, the microstructure of the as-spun filaments will be comprised of crystalline regions and amorphous regions. This is referred to as the fringed micelle structure. Micelles are the crystalline regions, each having a fringe of lengths forming amorphous regions. A polymer chain may therefore pass through both crystalline and amorphous regions. The volume ratio of crystalline to amorphous regions (i.e. the degree of crystallinity), and the degree of preferred orientation, will be low. Consequently, in the as-spun state, filament tenacity and modulus will be small, while the strain to break will be large.

When a filament is hot-drawn, the polymer chain lengths of the amorphous regions become straightened with a much higher degree of preferred orientation. The lamellar crystalline regions are pulled more into the preferred orientation, as shown in Figure 10.21. The figure also illustrates that some parts of individual chains in the amorphous region will become parallel and closely spaced with each other, thus forming new crystalline regions (micelles) that may be of larger volume than the lamellar crystalline regions. Thus, the degree of crystallinity and chain orientation increases with the degree of drawing (i.e. the draw-ratio).

Annealing after drawing will further increase the volume of newly formed micelles. The chain lengths that are not part of a crystalline region (lamellar or micelle) will remain part of the amorphous region, even though more orientated. It should be noted that high draw-down obtained with high speeds for first godet rollers will form micelle regions as well as lamellar regions.

There are various descriptive theoretical models cited in the literature, but the description given here is an attempt to include the essential elements of the chain folding/lamellar and fringed-micelle models. A more detailed account of the science of fibre microstructure formation would be extensive

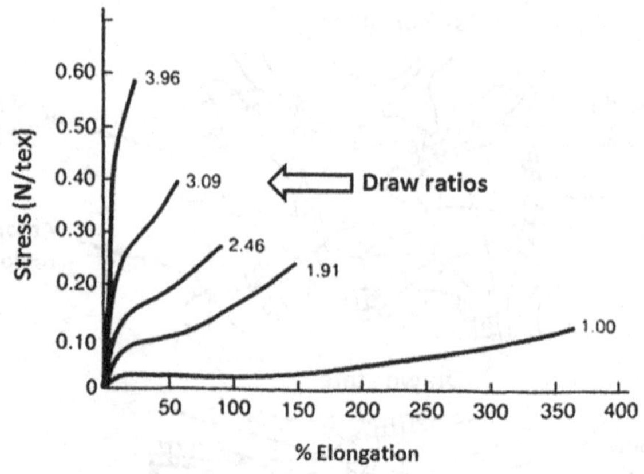

FIGURE 10.22

Stress-strain characteristics of nylon 6 filaments for increasing draw ratios.

Source: Rwimschuessel (1998).

and therefore beyond the scope of this chapter. For the interested reader, much has been written and published (McIntyre, 1998; Meares, 1965; Mukhopadhyay, 1992; Peters, 1963; Salem, 2000), including information on the analytical techniques, such as X-ray diffraction, infrared spectroscopy and birefringence, used to determine the type of crystal geometry, the degree of crystallinity and chain orientation.

Figure 10.22 shows how the draw ratio, by causing these changes in filament microstructures, alters the stress–stain curve with a tendency to eliminate the yield point, and to increase the elastic region, modulus and tenacity, but to reduce the % strain, causing the filaments and therefore the CF yarn to become stiffer. While extreme drawing may produce CF yarns suitable for technical textile applications, it is unlikely to result in desirable apparel fabrics. CF yarns having tenacities within the low to middle range of the values given in Table 10.2 are therefore used for conventional textiles clothing and apparel.

10.6 ADDING FUNCTIONALITY TO YARN

10.6.1 MOISTURE ABSORPTION

The moisture absorption of fibres is governed by their micromolecular and chemical structures. Moisture absorption depends on the diffusion of water molecules into configurations of the polymer chain by the attracting forces of certain molecules forming part of the chemical structure. These are referred to as functional groups that react with the water molecule. For example, Table 10.3 shows that among the CF yarns considered earlier, viscose rayon has the highest absorbance. The constituent cellulose molecules contain three hydroxyl group (OH) for each glucose part (see Chapter 4), and hydrogen bonds can therefore be formed between these groups and the water molecule (H_2O or HOH).[9] The number of functional groups that

[9] Hydrogen bonding is where, typically, hydrogen (an electropositive atom) links with strongly electronegative atom such as oxygen, nitrogen or fluorine. Hydrogen bonds are responsible for the bonding of water molecules to solids. They are one of several chemical bonds.

Table 10.3 Moisture regain at 65 RH 21 °C

Filament type	Function groups	% Moisture regain
Silk	Amide groups (CONH)	11
Viscose rayon	Hydroxyl groups (OH)	11.5–15
Cellulose acetate	Hydroxyl groups (OH)	5.4–6.5*
Polyesters	Ester group (COOR)	0.2–0.8
Nylons	Amide groups (CONH)	4–4.5
PP	–	–
PU elastomers	Amide groups (CONH)	0.3–1.2

**The moisture absorption is substantially lower than viscose rayon because many of the OH groups are prelaced by comparatively inert acetyl groups (CH_3COO) which do not strongly attract water (Morton & Hearle, 1962).*

can bond with water molecules will determine the amount of water absorbed, and this is influenced by the micromolecular structure; the crystalline and amorphous regions. In the crystalline regions, the molecules of the chemical structure are so closely packed together that water molecules cannot as easily penetrate and react with the functional groups as they can in the amorphous regions. This means the degree of crystallinity influences the moisture absorption. The high moisture absorbency of regular rayon results in a low wet modulus, and as a result, fabrics made from regular rayon are weak and are easy to stretch and deform in their wet state. The low wet strength is likely, however, to cause shrinkage if rayon fabrics are hand or machine washed. Such fabrics are usually labelled **dry clean only**. Higher-modulus rayons overcome this disadvantage and may be machine washed and tumble dried satisfactorily.

As can be seen from Table 10.3, polypropylene is at the other end of the water absorbance scale. It is highly crystalline and has no functional groups. Table 10.3 summarises the functional groups and the % moisture regain typical of the various CF yarns.

The amount of moisture in a material is expressed in terms of % moisture regain, which is used in commercial transactions to ensure that when sold by weight, the correct amount of the actual material and not added moisture is being obtained. Moisture regain is calculated simply by:

$$= \frac{\text{Weight of water absorbed}}{\text{Bonedry weight}} \cdot 100 \qquad (10.8)$$

Morton and Hearle (1962, Chaps 7–12) give a detailed description of the physics and chemistry of moisture absorbance in textile materials.

10.6.2 DYEABILITY AND PRINTABILITY

The subject of colouration of textile materials is extensive in its own right and many textbooks are dedicated to various theories and practises of dyeing and printing (Sjostrom, 1993; Trotman, 1975). Only a very brief consideration is therefore given here to illustrate the importance of the fibre micromolecular structure and the chemical structure of the constituent polymer.

Since moisture absorption is often involved, the colouration of fibres is strongly dependent on their physical and chemical properties in relation to this action. During the process of colouration, a dye molecule moves from the liquid phase (for dyeing) or paste phase (for printing; in heat transfer printing this would be a sublimation phase) towards the fibre, is then absorbed through diffusion into the fibre and held within it, preferably by chemically bonding with polar functional groups or by being physically locked within the microstructure; this represents the fixation of the dye molecule. Fibres made from polymers that have polar functional groups in their chemical structure are more easily coloured because the polar groups attract moisture and serve as active sites for combination with dye molecules by chemical bonds.

Polar groups are molecular groups such as –OH (hydroxyl) and –CONH (amide) within the chemical structure of the polymer that have positive and negative charges and thereby attract other polar molecules, such as water and polar dyes, bonding the polar dyes into the fibre structure. The bonds formed may be primary covalent bonds and/or secondary bonds (by means of van der Waals forces, dipole association and hydrogen bonding). The hydrogen bond is stronger than the van der Waals bond, but weaker than a covalent or ionic bond.

The diffusion of dye molecules into the fibre/filament is an important action in the colouration process. The ease or rate of diffusion will depend on the size of the dye molecule able to penetrate the micromolecular of the fibre/filament, and since the amorphous region can be more easily penetrated, the degree of crystallinity will affect the ease of colouration. The rate of aggregation of the dye molecules and the pore size of the amorphous regions are therefore important factors in controlling the rate and depth of penetration. If these two factors are incompatible, then the dye molecules will at best only remain attached to the surface of the fibre/filament and effective colouration will not occur. Steps have to be taken to inhibit dye aggregation and to temporally expand the pore size of the amorphous regions. Heat is used in transferring dye molecules from the solution to the fibre as well as assisting in swelling the fibre/filament to permit the dye molecules to diffuse more easily into the fibre.

Different types of dye can be used for the colouration of natural and synthetic CF yarns; however, the prevalent types are acidic dyes for natural protein filaments (e.g. silk) and reactive dyes for cellulosic filaments. Although nylon 6 and nylon 6.6 are synthetic filaments, their chemical structures (polyamide) enable them to be easily dyed with the same acid dyes commonly used for natural protein filaments. Acidic dyes are water-soluble anionic chemical substances that are applied to fibres via a neutral to acid bath ($pH<7$). Their attachment to filaments is partly attributed to the bonding between anionic (negatively charged) groups in the dyes and the cationic (positively charged – carboxylic and amino) groups in filaments. Nylon and natural protein filaments have many cationic sites to which the anionic dye molecules are attracted. Acids may be also added to the dye bath to increase the number of protonated amino-groups present. The hydrogen of the acid dyes attracted to the cationic sites accounts for the fixation of the coloured anions in the resulting dyed material.

Reactive dyes are best for cellulosic CF yarns. Rayons dye well with cold water reactive dyes, giving bright, long-lasting colours. These dyes are a class of highly coloured organic substances containing reactive groups that, when applied in a weakly alkaline dye bath, attach themselves to filaments by a chemical reaction that forms a covalent bond between the dye molecules and those of the filaments. The dyestuff then becomes a part of the filament and is much less likely to be removed by washing than are dyestuffs that only adhere by adsorption. Reactive dyes are the most permanent of all dye types and significantly improve the colour stability and washability of the resulting fabric. They were originally developed for cellulose materials, and this is still their main use. However, there are also commercially available reactive dyes for natural protein and polyamide filaments, which are applied under weakly acidic conditions.

Polyester CF yarns, which are fairly crystalline with polymer chains tightly packed together, tend to be hydrophobic. This means that polyesters have very little affinity for large ionic dyes, which simply cannot distribute between the chains or form satisfactory intermolecular interactions. The dyeing of polyesters therefore requires the use of disperse dyes, as other dye types leave the natural colour of the fibre unchanged. Although disperse dyes are mainly used for dyeing polyesters, they were originally developed for the dyeing of cellulose acetate/triacetate and can also be used to dye nylon.

Disperse dyes are water insoluble. They are finely ground in the presence of a dispersing agent and sold as a paste, or spray-dried and sold as a powder. The very fine particle size gives a large surface area that aids dissolution to allow uptake by the fibre/filament. The dyeing is done also with a suitable dispersing agent in a pressurised dye bath at high temperatures ranging from 130 to140 °C. At these high temperatures, thermal agitation causes the polymer structure to become more open, providing gaps for the dye molecules to enter.

The general structure of disperse dyes is small, planar and non-ionic, with attached polar functional groups such as $-NO_2$ and –CN. The shape makes it easier for the dye to slide between the tightly packed polymer chains, and the polar groups give some water solubility and dipolar bonding between dye and polymer. The main interactions between the dye and polymer are thought to be van der Waals and dipole forces. Their small size enables them to diffuse and disperse within the filament structure but also means that post dyeing, they are quite volatile at elevated temperatures and tend to sublime out of the filament. The volatility of the dye may cause, at elevated temperatures, a loss of colour density and staining of other materials in contact. This can be counteracted by using larger molecules or making the dye more polar (or both), but the drawback is that larger, more polar molecules will need more extreme conditions of temperature and pressure to dye the polymer.

Polyurethane can be dyed with 1:2 metal-complex monoazo dyes, such as the acid dyes that make up part of the range of dyes for wool. However, fabrics that contain spandex can be easily damaged by heat; therefore when dyeing fabrics of protein fibre (or filament)/spandex blends, consideration has to be given to the need for gentle treatment. Cellulosic/spandex blends dye suitably with cold-water reactive dyes. The reactive dye does not actually dye the spandex. However, this is not a problem if the blend consists predominantly of staple-fibre cellulose (e.g. cotton or cut rayon), with only 3–12% spandex, as this would usually be a core-spun yarn with spandex as the core, covered by the cellulose. Both nylons and polyesters require significant heat for dye fixation, so blends with spandex cannot be dyed, as the spandex will be damaged by the heat applied. The nylon or polyester must be dyed prior to combination with the spandex.

CF yarns made from polypropylene are lightweight, hydrophobic and highly crystalline. The latter two characteristics mean they have a high resistance to wetting, which gives them good anti-staining properties. However, they cannot therefore be chemically dyed, and dye pigments must be incorporated into the polymer chips prior to melt spinning. The dye pigment is mixed at high concentration into a quantity of molten polypropylene and the material extruded, cooled and chopped into chips or granules collectively called a master batch (BASF). The production process is that of compounding (Applied Market Information Ltd). During the melt spinning, a predetermined amount of the master batch is fed into the screw extruder along with virgin (un-pigmented) polymer chips; the amount depends on the depth of shade required and may be from 1% to 5% *w/w*. During their passage through the screw extruder, the two molten materials are blended to distribute the pigment throughout the molten liquid, giving a coloured melt that is then extruded through the spinneret to form pigmented filaments. This technique is called 'solution dyeing' or 'dope dyeing', and the CF yarn is referred to as 'spun-dyed' or 'dope-dyed'. An alternative approach is to add dye-accepting chemicals to make the PP filaments dyeable downsteam of spinning (Applied Market Information Ltd). Although mainly used for polypropylene, the master batch process can be used with other polymers. POY and FDY are often dope dyed as this is more efficient than first making fabrics with raw-white yarns and then dyeing them. Doped-dyed yarn is also more evenly coloured throughout its volume.

10.6.3 FUNCTIONAL ADDITIVES

Functional additives can be incorporated into CF yarns using the compounding process described above. Metal oxide particles termed **delusterants,** e.g. TiO_2 (titanium dioxide) may be added to the polymer to impart a semi-dull or dull appearance. POY and FDY yarns can be made to have bright or semi-dull appearance. The bright appearance is due to the cross-sections in the filaments, which can be circular or trilobal. CF yarns having trilobal filaments of a bright lustre are widely used in making curtains, bed-sheets and carpets.

Although predominantly applied in the dyeing and finishing of fabrics, florescent whitening additives, referred to as optical brighteners, e.g. 2,2′-(1,2-ethenediyldi-4,1-phenylene) bis benzoxazole, may be used in melt spinning to give the spun filaments a resistance to any subsequent yellowing caused by daylight and to induce a bright whiteness on which printed patterns become more vivid. The absorption of short-wavelength ultraviolet (UV) light by the untreated filaments produces a yellow appearance on the resulting fabrics. A fluorescent whitening additive absorbs the invisible UV energy of the daylight spectrum and re-emits most of the absorbed energy as blue fluorescent light in the 400–500 nm wavelength portion of the spectrum, thereby giving the fabric a whiter appearance (McElhone, 2009).

Some polyurethane materials can be vulnerable to damage from heat, light, atmospheric contaminants and chlorine. For this reason, stabilisers are added to protect the polymer. One type of stabiliser that protects against light degradation is a UV screener called hydroxybenzotriazole. Antioxidants are used to protect against oxidation reactions. Various antioxidants are available, such as monomeric and polymeric hindered phenols. For certain applications such as swimwear, anti-mildew compounds may be added to the polyurethane. Compounds that inhibit discolouration caused by atmospheric pollutants are another type of added stabiliser. These are typically compounds of tertiary amine functionality that can react with oxides of nitrogen in air pollution.

The CF yarns used in the production of fabrics for applications such as nightwear, curtains and upholstery have to meet national/international standards on flammability. Polymers used to produce CF yarns for textiles and clothing have a low inherent resistance to combustion, so steps are often taken to bring these materials up to an acceptable standard. Flame retardancy may be imparted to fabrics through special finishing treatments, but flame-retardant additives compounded into the polymers are also widely used. The advantage of the latter approach is that the retardant can be uniformly distributed throughout the mass of the filament, whereas finishing treatments concentrate it at the surface of the filaments.

Viscose rayon can be made flame retardant by using chemical additives, examples being silicic acid, aryl phosphates, phosphazenes, phosphonates and polyphosphonates. There are also various compounds, in particular phosphorous-based chemicals, which can be added to synthetic CF yarns during melt spinning to impart flame retardancy to the spun filaments (Stackman, 1982; Weil & Levchik, 2008).

Bioactive/antibacterial functionality is today widely demanded by consumers. Again, there is the choice between finishing treatments or additives; the later gives a more inherent antibacterial property to the resulting fabrics. Typical additives are certain elemental metals and metal oxides, the most popular being nanosilver and silver oxide (Duran, Marcto, De Souza, Alves, & Esposito, 2007). Most strains of bacteria are unable to develop a resistance to silver despite repeated exposure. Silver's lethality to micro-organisms is not based upon the metal itself but on the positively charged silver ion, Ag^+. The mechanism by which Ag^+ kills microorganisms is twofold. Silver ions effectively kill bacteria due to their adsorption onto the negatively charged bacterial cell wall, which initially deactivates cellular enzymes and disrupts membrane permeability, i.e. the silver ions block the respiratory enzyme system (energy production) of the cells. Secondly, they alter the microbial DNA and the cell wall, ultimately leading to cell lysis and cell death. This can occur at silver concentrations as low as 1 ppm (part per million). Importantly, these lethal actions are seen to have no toxic effect on mammalian cells at such low concentrations. With regard to inherent bactericidal or bacteriocidal properties in natural polymers, bamboo-based rayon filaments are reported[10] to have good bacterial inhibition and germicidal properties. Bamboo has properties of bacteriostasis which, it is claimed, remains in the filament yarn (Bamboo Filament Yarn).

Synthetic materials generally have poor moisture absorbency and are electrically non-conducting. Electrostatic buildup can therefore become problematical. For example, synthetic carpets can initiate electrostatic shocks; static charges generated by clothing made of synthetic fibres can be a cause for discomfort as the clothing clings to the skin. Dust and soil particles will, through static charges, readily adhere to synthetic surfaces. One approach to alleviating these problems is to include synthetic fibres/filaments made with antistatic properties within the fabric. These properties are usually obtained by introducing carbon particles in the melt-spinning process (Brown & Pailthorpe, 1986; Li et al., 2004).

Much attention is given to producing fabrics that enhance comfort, health and well-being. One approach towards achieving this has been to develop filament yarns that generate infrared radiation for maintaining muscular energy. The filaments are spun with ceramic particles incorporated in them, e.g. zirconium, magnesium or iron oxides, and when these particles reach body temperature (~36 °C) they emit far-infrared energy of 8–14 μm wavelengths and ~60 mW. Typical applications for such heat-regenerating CF yarns are sportswear, leisure wear, bed sheets and covers.

The use of scent essences is also a means of developing CF yarns for improved well-being. Hollow filaments are spun with the essence embedded in the inner wall of the hollow structure. The filaments have four essence-containing zones radially positioned around the central cavity. This enables the scent to permeate the ends and the annular wall of each filament.

10.7 APPLICATIONS

As explained earlier in this chapter, CF yarns can be produced in untextured and textured forms: the texturing being draw- (false) twist for thermoplastic polymers and air jet for natural and synthetic polymers. The untextured yarn market is substantially smaller than that for textured yarns. Untextured CF yarns for apparel are used mainly in tricot warp-knitted fabrics for garments such as shirts and lightweight women's wear and in woven fabrics for end-uses such as interlinings.

[10] Tests conducted at the Testing Center of China Textile Academy according to the standard JISL 1902–2002 of the Japan Textile Association.

The two most popular textured yarn structures are draw-twist texturing and air-jet texturing. The major apparel applications for draw-twist are in knitted garments, especially double jersey for women's outerwear and woven fabrics. Air-jet yarns are used in menswear, as they have more closely the appearance and handle of spun yarns.

Yarns used in apparel and home furnishings usually have a count range of 80–160 deniers. However, for apparel the trend has been towards fine-count plied yarns, which often consist of a combination of textured and untextured yarns. Fine-count CF yarns provide better aesthetics, a silk-like handle and good drape, although this can result in much reduced crease resistance.

Table 10.4 gives an overview of the applications in textiles and clothing for the natural and synthetic polymers used in CF yarns. Viscose rayon has a silk-like appearance and feel and with its good moisture absorbency, superior to that of cotton; it does not build up static electricity and therefore is comfortable to wear. It absorbs perspiration, allowing it to evaporate away from the skin. Rayon conducts heat more readily than, for example, silk and so gives a cooler feel next to the skin. A wide variety of fabrics can be made from viscose rayon staple fibres, but in filament form it is considered suitable for lightweight summer wear and coat linings, which are major end-uses.

Largely because of its easy-care properties, PET has a wide range of applications from suit fabrics to hosiery as well as textiles for home furnishings. CF semi-dull PET yarns are used for various types of apparel, including lingerie, while the dull version is used for shirts and blouses. Fabrics made from PET have good resistance to, and recovery from, creasing or wrinkling in both their dry and wet states. They also have a high resistance to stretch, abrasion, sunlight, heat ageing and chemicals. PET fabrics are therefore often used for curtains and outerwear. The high strength and durability of this polymer enables garments to withstand repetitive abrasion, and its low moisture absorbency, when aided by a water-resistant finish, makes PET garments suitable for use in wet or damp climates.

Although PU-CF yarns are widely used to impart stretch in many apparel applications (see Table 10.4), PBT and PTT have similar end-uses but often combined with other filament yarns or staple-fibre yarns, such as PET, acetate, acrylic, cotton, wool or nylon. PBT can be used for the development of knitted and woven elastic fabrics for underwear, socks, sportswear, elastic jeans and wool sweaters. PTT is more suited to power stretch fabrics, which are based on knitted interlock structures. Stretch fabrics can also be obtained from weaving, where the stretch may be designed to be present in the warp, weft or both.

Nylon combines good mechanical properties, abrasion and chemical resistance with light weight, making it suitable for a variety of applications, including technical as well as conventional textiles and apparel. Hosiery and knitted garments are popular end-uses. Characteristically, nylon fabrics are soft and bulky with a cotton-like touch, which makes them acceptable for apparel applications such as tight-fitting lingerie, panty hose and some sportswear requiring the stretch effect that is obtained with nylon textured yarns. Major end-uses for nylon CF yarns are carpets and upholstery.

In carpets, nylon gives good appearance and texture retention, owing to its resistance to abrasion, as well as having good durability, recovery from compression and dye fastness. Similar characteristics apply to upholstery, but it is the combination of strength, dyeability, abrasion resistance and light fastness that is of specific importance.

Polypropylene is also widely used for carpets, initially for carpet backing, but due to developments in spun-dyed technology, the CF yarns are now used for more aspects of a carpet. PP offers a combination of properties that enables it to compete with other polymers. It is light and strong, has good resistance to abrasion, and because of its hydrophobicity, possesses anti-soiling and anti-staining characteristics. PP-CF yarns can be processed by all the various texturing methods to be given loft and dimensional stability. They have therefore made steady progress in apparel knitwear applications.

Table 10.4 Overview of applications in textiles and clothing

CF yarns	Applications (examples)	General characteristics
Silk	Outerwear (dresses, blouses, skirts, jackets, scarves, ties); underwear (including thermal); nightwear	Good drape; mainly hand wash; good resistance to pilling and abrasion; good colouration (dye/print); poor light fastness; creates static charges
Viscose rayon	Lining fabrics; upholstery; embroidery; crocheted and knitted fabrics; outerwear and underwear (imitation silk fabric); crepe fabric; blends with natural and synthetics	Soft, high moisture absorbance, low wet strength, stretches and shrinks, poor resistance to abrasion and damage to weak acids, biodegradable, good drape, good lustre, poor crease recovery, good anti-pilling
Viscose acetate	Women's outerwear (coats, blouses, sweaters, scarves, etc.); neckties; home furnishings and bedding	Acetate/Tricacetate – lighter than cotton and other cellulosics, lustre and tactile feel as silk, good colouration, acceptable moisture absorption, good drape, good dimensional stability, biodegradable, hypoallergenic
PET	General clothing and apparel; home furnishings and beddings; sport and leisure wear; casual wear	Good strength (wet and dry), good dimensional stability, low moisture absorbance/easily washed/quick drying, wrinkle resistant, mildew resistant, abrasion resistant
PBT	Used for the development of knitted and woven elastic fabrics; underwear (foundation garments; lingerie); socks; sportswear; comfort wear (elasticated garments)	Good strength, elongation between nylon and PU, quick drying, resistant to humidity/perspiration, good chemical resistance, good colouration, soft handle
PTT	Power stretch fabrics (knitted interlock); woven fabrics made in combination with PET, acetate, acrylic, cotton, wool or nylon in the warp or weft directions; sport and leisurewear; casual wear; underwear; upholstery	High stretch and recovery, soft touch
PLA	Sport and leisurewear; underwear; upholstery and furnishings (blankets, sheeting, duvets, decorative fabrics); carpet floor tiles	Biodegradable; unstable, good loft/resilience, resistant to bacterial growth; dyeing problematical; poor abrasion resistance; thermally sensitive low moisture absorption/high wicking; low density
Nylon	Outerwear (blouses, dress, raincoats, blouson, thermal windbreakers, parkas); lining fabric; underwear (foundation garments, lingerie); hosiery; sportswear (swimsuits, cycle wear, etc.); medical and healthcare (nappy/diapers, feminine care products)	Good abrasion resistance; good colouration; soft, flexible
Polypropylene (PP)	Sportswear; thermal underwear and durable nonwoven (interlinings); carpets (incl. backings); upholstery and wall coverings	Hydrophobic, stain/soil resistant, abrasion resistant, high wet and dry strength, thermally bondable, very low density, excellent resistance to deterioration from chemicals, mould, insects, perspiration, rot and weather, thermally sensitive
Spandex	Sport and leisure wear (ski pants, jeans, trousers, belts); medical and healthcare (bandages, orthopaedic brace, diaper/nappy); underwear (bodysuits, bra straps and side panels); hosiery, leggings, socks	Generally mixed with other filaments or acting as the elastic core of core-spun yarns with staple fibre sheath, e.g. cotton, nylon, polyester, etc., and accounts for a small percentage of the final fabric

10.8 FUTURE TRENDS

Global production of the raw materials for textile and clothing may be broadly divided into two main sectors: natural fibres (mainly cotton, wool and silk) and manufactured or man-made fibres. During the last 5–10 years, the world trade in textiles has almost doubled, growing from $480 billion to $700 billion (Global Textiles and Clothing Industry). Global fibre consumption has reflected this growth with annual increases of up to 4.2%. Reported figures show 70.5 million tonnes, of which manufactured or man-made fibres account for about 63% and natural fibres 37% (The Fibre Years 2009/10, 2011). These percentages are for recent years, but would seem indicative of a general trend for future years (Fibres, Yarns & Threads Industry Overview; Overview of Textile Industry; Textiles & Apparel: Market Opportunities). However, the issues surrounding increasing oil prices, sustainability, environmental concerns relating to manufacturing technologies and the increasing demand for food (both meat and grain) will have an impact in the coming years and could alter the future weighting of man-made to natural fibres.

Among manufactured fibres, CF yarns for textiles and apparel, excluding carpets and industrials, account for 20.7 million tonnes as compared with 37 million tonnes for short- and long staple fibres in the production of spun yarns. It can therefore be seen that, although at a lower volume than spun yarns, CF yarns hold a significant base line as a raw material in the production of textiles and apparel.

Of the various types of CF yarns described in this chapter, polyester (PET), nylon (PA 6 and 6.6), polypropylene (PP) and polyurethane (PU) are currently the most important polymers used for filament yarns: 99.5% of the global CF yarn market is based on these polymeric types.

With the ever-growing consumer demand for improved comfort and functionality (e.g. antimicrobial, microclimate control, static and electrical conductivity, etc.) in textiles and clothing, and for what are termed smart textiles, these filament yarns will gain an increased market share in areas of application hitherto catered for by spun yarns. In developed and newly industrialised countries, spun yarns are experiencing increasing competition from CF yarns. This may be attributed to their greater potential for producing customised end-products (particularly with the use of new nano-additives) combined with a more favourable cost–benefit ratio resulting largely from shorter production processes. As the development of production processes continues, extrusion spinning and false-twist texturing may become the next process integration, producing fine-count CF yarns for apparel. One indication of this is the development of a false-twist texturing system that uses rotating heated cylinders instead of a conventional heater (Recent Developments in the Field of Texturing Technologies). This has enabled the integration of POY spinning with spin-draw texturing to be demonstrated as technically feasible. Continued development in the extrusion spinning of bi-component CF yarns could also play a role in reducing process stages by establishing self-crimping CF yarns for apparel.

Future production and development of CF yarns will need to take greater account of issues such as sustainability, recycling and biodegrability, which may lead to polymers including PTT, PBT and PLA becoming more dominant. However, the current consumption of these newer raw materials is only 20–30 ktonnes/year, compared with about 25 million tonnes for mainstream polymers. It is therefore also possible that the greater volume usage of the current mainstream synthetics will be a major driving force in their further development aimed at addressing concerns about sustainability and clean technology. One example of this is where issues surrounding viscose rayon gave rise to the production of Lyocell. A defining factor will be the economic cost of the available polymeric materials, and at present this has led to the dominance of PET, which can be produced over the full range of fibre finenesses and is used in nearly all areas of application.

10.9 PROJECT IDEAS

- Undertake market research for establishing a specification of a CF yarn suitable for lightweight breathable outerwear garments.
- Determine the most appropriate CF yarn to use for ultra-comfort lingerie.

10.10 REVISION QUESTIONS

- Draw a chart that shows a classification of CF yarns according to the types of polymeric materials used.
- Explain the yarn count system used to specify CF yarns.
- Describe the three most used processes for making CF yarns, giving examples of the type of polymers used in each process and the reason(s) for this.
- Describe the following processes:
  - False-twist texturing
  - Air-jet texturing
- Illustrate the typical tensile characteristic of a melt-spun multifilament CF yarn that has been drawn, but not fully drawn, and explain the parameters used to quantify its tensile properties.
- Describe ways by which added functionalities can be imparted to CF yarns.

REFERENCES

Acelon Chemicals & Fibre Corporation, http://www.allproducts.com/textile/acelon/12-split_micro_filament.html.

Air jet texturing and fabrics, http://www.polyspintex.com/education/ayt-air-jet-textured-yarn/.

Applied Market Information Ltd, Compounding/Masterbatch, http://www.amiplastics.com/cons/markets/compounding.aspx.

Bamboo filament yarn. http://www.swicofil.com/bambrotexenduses.html.

BASF, Masterbatch colour solutions. http://www.luvitec.de/portal/basf/ien/dt.jsp?setCursor=1_417479.

BeYarns. http://www.beyarns.com/.

Brown, N. S., & Chuah, H. H. (1997). Texturing of textile filament yarns based on poly(trimethylene terephthalate). *Chemical Fibres International*, *47*, 72–74.

Brown, D. M., & Pailthorpe, M. T. (1986). Antistatic fibres and finishes. *Coloration Technology*, *16*, 8–15.

Composite. (January 2012). http://composite.about.com/library/glossary/c/bldef-c1269.htm.

Daily Finance. (2012). http://www.dailyfinance.com/2010/02/03/bamboo-zled-ftc-says-retailers-fibbed-about-bamboo-product-clai/.

Draw-warping apparatus. (1989). United States Patent 4852225.

Duran, N., Marcto, P. D., De Souza, G. I. H., Alves, O. L., & Esposito, E. (2007). Antibacterial effect of sliver nanoparticles produced by fungal process on textile fabrics and their effluent treatment. *Journal of Biomedical Nanotechnology*, *3*, 203–208.

Dyer, J., & Daul, G. C. (1998a). Rayon fibres. In M. Lewin (Ed.), *Handbook of fibre chemistry* (pp. 150/750). Marcel Dekker (Chapter 10).

Dyer, J., & Daul, G. C. (1998b). Rayon fibres. In M. Lewin (Ed.), *Handbook of fibre chemistry* (pp. 786/787). Marcel Dekker (Chapter 10).

Fambre, L., Pegoretti, A., Mazzurana, M., & Migliares, C. (1994). Biodegradable fibres. Part I: Poly-L-lactic acid fibres produced by solution spinning. *Journal of Materials Science: Materials in Medicine, 5*, 679–683.

Farrington, D. W., Hunt, J., Davis, S., & Blackburn, R. S. (2005). Poly(lactic acid) fibres. In R. S. Blackburn (Ed.), *Biodegradable and sustainable fibres*. (Chapter 6). Woodhead Publishing Ltd.

The fibre years 2009/10; 2011 Report. http://www.oerlikon.com/ecomaXL/get_blob.php?name=The_Fibre_Year_2010_en_0607.pdf; http://www.oerlikontextile.com/desktopdefault.aspx/tabid-1763/.

Fibres, yarns & threads industry overview. http://www.teonline.com/fibers-yarns-threads/industry-overview.html.

Fourné, F. (1998). *Synthetic fibres: Machines and equipment, manufacture, properties*. Hanser Publishers (Section 9.3.25: Twist and intermingling tangling).

Fourné, F. (1999). *Synthetic fibres: Machines and equipment, manufacture, properties: Handbook for plant engineering, machine design and operation*. Hanser Publishers.

Fukuhara, M. (1993). Innovation in polyester fibres: from silk-like to new polyester. *Textile Research Journal, 63*, 387–391.

Galanty, P. G., & Bujtas, G. A. (1992). *Modern plastics encyclopedia* (pp. 23–30). McGraw-Hill.

Global Textiles and Clothing Industry. http://www.slideshare.net/333jack333/global-textiles-and-clothing-industry-by-hiresh-ahluwalia.

Hearle, J. W. S., Hollick, L., & Wilson, D. K. (2001). *Yarn texturing technology* (pp. 1–13). Woodhead Publishing Ltd.

Hongu, T., & Phillips, G. O. (1997a). Biomimetic chemistry and fibres: 4.1.3 morpho-structured fabrics imitate the insect morpho alae. In *New fibres* (pp. 81–85). Woodhead Publishing Ltd (Chapter 4).

Hongu, T., & Phillips, G. O. (1997b). High-touch fibres: 3.9 Iridescent textiles. In *New fibres* (pp. 66–67). Woodhead Publishing Ltd (Chapter 3).

Horrocks, A. R., & Anand, S. C. (2000). *Handbook of technical textiles*. Woodhead Publishing Ltd.

Lawrence, C. A. (2003). Yarn formation structure and properties. In *Fundamentals of spun yarn technology* (p. 309). CRC Press (Chapter 6).

Lewin, M., & Preston, J. (Eds.). (1985). *High technology fibers*. New York: Marcel Dekker.

Li, C., Liang, T., Lu, W., Tnag, C., Hu, X., Cao, M., et al. (2004). Improving the antistatic ability of polypropylene fibres by inner antistatic agent filled with carbon nanotubes. *Composites Science and Technology, 64*(13–14), 2089–2096.

Matsudaira, M. (1994). The mechanical properties and fabric handle of polyester-fibre shingosen fabrics. *Journal of the Textile Institute, 85*(2), 158–172.

McElhone, H. J. (2009). Fluorescent whitening agents. In *Kirk-Othmer encyclopedia of chemical technology*. John Wiley & Sons, Inc.

McIntyre, J. E., & Daniels, P.N. (Eds.), *Textiles terms & definitions*. Compiled by The Textile Institute Textile Terms and Definitions Committee.

McIntyre, J. E. (1998). Polyester fibres. In M. Lewin (Ed.), *Handbook of fibre chemistry*. Marcel Dekker (Chapter 1).

Meares, P. (1965). *Polymers: Structure and bulk properties*. D. Van Nostrand Company Ltd.

Morton, W. E., & Hearle, J. W. S. (1962). *Physical properties of textile fibres*. Butterworths & Co Ltd (Chapters 7–12).

M-type metallized yarns. http://www.indiamart.com/lakhotia-polyester-limited/metallic-yarn.html.

Miniknittingstuff. http://www.davytextiles.com/glossary_of_terms_and_definition.htm.

Mukhopadhyay, S. K. (Ed.). (1992). *Advances in fibre science*. The Textile Institute.

Natural Rubber. (2012). http://www.infoplease.com/ce6/sci/A0860822.html.

Overview of textile industry. http://www.legalpundits.com/Content_folder/THETEXTILEINDUSTRYREPORT290710.pdf.

Peters, R. H. (1963). Wet spinning. In *Textile chemistry: The chemistry of fibres*. Elsevier (Chapter 15).

Recent developments in the field of texturing technologies. http://www.expresstextile.com/20050630/technext01.shtml.

Reimschuessel, H. (1998). Polyamide fibres. In M. Lewin (Ed.), *Handbook of fibre chemistry* (p. 144). Marcel Dekker (Chapter 2).

Robinson, A. T. C., & Marks, R. (1993). *Yarns and their characteristics: Continuous-filament and spun yarns*. Woven Cloth Construction, The Textile Institute.

Rwimschuessel, H. K. (1998). Polyamide fibres. In M. Lewin (Ed.), *Handbook of fibre chemistry* (p. 105). Marcel Dekker (Chapter 2).

Salem, D. R. (2000). *Structure formation in polymeric fibres*. Hanser Publishers.

Schildknecht, C. E. (Ed.). (1956). *Polymer processes* (p. 841). Interscience.

Sjostrom, E. (1993). *Wood chemistry, fundamentals and applications*. New York: Academic Press.

Smith, R. L., Pieters, R., & Morrison, M. E. (1972). Fundamentals of false-twist texturing of thermoplastic continuous filament yarns. *Journal of Rheology*, *16*(3), 557–577.

Stackman, R. W. (1982). Phosphorus based additives for flame retardant polyester: polymeric phosphorus esters. *Industrial and Engineering Chemistry. Product Research Development*, *21*(2), 332–336.

Technicaltextile. (January 2012). http://www.technicaltextile.net/articles/.

Teonline. (January 2012). http://www.teonline.com/knowledge-centre/study-technical-textiles.html.

Textiles & apparel: Market opportunities, http://www.ibef.org/download/Textiles_Apparel_220708.pdf.

Textileglossary. (January 2012). http://www.textileglossary.com/terms/filament-yarn.html.

Trotman, E. R. (1975). *Dyeing and chemical technology of textile fibres*. Chales Griffin & Company Ltd.

Trougott Baumann, K. G., http://www.baumann-zwirne.de/en/services/twisting.html.

Ulrich, H. (1996). *The chemistry and technology of isocyanates*. New York: John Wiley & Sons.

Weil, E. D., & Levchik, S. (2008). Flame retardants in commercial use or development for textiles. *Journal of Fire Sciences*, *26*, 243–281.

CHAPTER

YARN TO FABRIC: WEAVING

11

S. Stankard
Universiti Teknologi MARA (UiTM), Selangor, Malaysia

LEARNING OBJECTIVES

At the end of this chapter, you should be able to:

- Identify different types of handlooms
- Dress a loom
- Understand draft plans and weave structures
- Document weaving processes

11.1 INTRODUCTION

Weaving is one of the oldest and most widely used methods of making a fabric. Simply put, weaving is the interlacement of two sets of threads; the warp threads run vertically through the length of the fabric and weft threads run horizontally across the width of the cloth. This chapter will explain the basic information to start weaving on a handloom. Initially, a term of definitions will provide explanations of the key terms used in weaving, followed by an outline of the types of looms that are used in hand weaving. Also detailed is the widely used process to dress a loom, that is, to make the warp, wind it onto the loom and make it ready for weaving to commence. An outline of warp drafting, and basic and derivative weave structures will illustrate how different patterns can be formed on a woven fabric. How to design for woven fabrics, including the use of colour in weaving, will be outlined. The standard documentation of weaving processes, which permits fabrics to be recreated, will be defined. Finally, a weave project case study will be documented, along with future trends, and sources of further information and advice will be presented.

11.2 LOOMS

There are several types of looms that can be used to weave by hand. All looms perform the same basic function, which is to hold the warp yarns taut and under tension, whilst weft yarns are inserted and

Textiles and Fashion. http://dx.doi.org/10.1016/B978-1-84569-931-4.00011-8

beaten into place to form the fabric. All looms have devices for making a shed, storing the cloth and producing patterns. The basic features common to all types of looms are:

- **Frame:** Secures components.
- **Beam:** Secures and stores the warp and is at the back of the loom. Some looms have more than one beam. If the loom has more than one back beam, it can be used for weaving two different types of warps for double cloth or warps at different tensions. The cloth beam at the front stores the finished weaving. The width of each beam determines the extent of the width of the cloth.
- **Shafts:** Consist of an upper and lower bar carrying the heddles. They control the rise and fall of the warp threads, thus forming the shed (there can be as few as 2 shafts or as many as 16 shafts on a table loom, or as many as 24 shafts on a floor loom).
- **Heddles:** Rest on the shafts of the loom. The warp yarns are threaded through the eye of the heddles.
- **Batten:** Pivoted frame holding the reed. It can hang from the top of the loom (overslung) or be pivoted at floor level (underslung).
- **Reed:** Used for spacing the warp and beating the weft. It can vary in sizes and ideally it should be made of stainless steel to prevent rusting.
- **Levers** (table loom) or peddles (floor loom): Raise and lower the shafts. Levers are placed at one or both sides of the frame. Peddles may be pivoted from the front or back.

11.2.1 RIGID HEDDLE LOOM

The rigid heddle loom is the easiest loom to work with, but is very limited as it can only be used for a plain-weave structure. The loom consists of a frame, one small beam at the back to hold the warp and one at the front to hold the woven cloth; there is also a rigid heddle and fixed reed. To vary the cloth produced, additional techniques can be used, such as grouped and spaced warps or the use of thick and thin yarns; also, threads can be doubled or trebled to give rib effects (see Figure 11.1).

11.2.2 TABLE LOOM

A table loom can measure up to approximately 24 inches (60 cm) in width and have between 2 and 16 shafts. There is one lever for each shaft, and shafts are raised and lowered by hand, using the levers at the side of the loom each time you insert a weft thread (pick) into the shed. Two-shaft looms are limited to

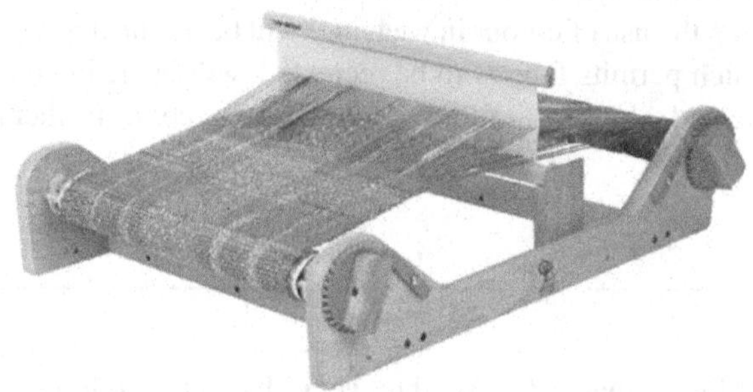

FIGURE 11.1

Rigid heddle loom.

plain weave, but looms with four or more shafts can produce many types of weave structures. Table looms are perfect for weaving samples or small items such as cushion covers and scarves (see Figure 11.2).

11.2.3 FLOOR/TREDDLE LOOM

Using a floor or treadle loom will enable the cloth to be woven more quickly than with a table loom. The use of peddles enables the weaver to throw the shuttle with the hands whilst forming the shed through the use of foot power. On a floor loom, the pedals can be connected to one, two, three or four shafts and will therefore raise them all simultaneously. They are not connected directly to the shafts, as this would make it impossible to lift the shafts evenly. The shafts are connected to lams or marches, which are supported slightly under the shaft; they centralise the lifting, and except on a counter-march loom, there are the same number of lams as shafts (see Figure 11.3).

FIGURE 11.2

Table loom.

FIGURE 11.3

Floor/treddle loom.

11.2.4 COUNTERBALANCED LOOM

A counterbalanced loom (Figure 11.4) is usually limited to four shafts, which are suspended by cords passing to horses, pulleys or rollers, which are attached to the frame of the loom from above. There is one lam or march to the bottom of each shaft, and the pedals are tied to a selection of these lams. Pressing a pedal therefore lowers the shafts to whose lams ties have been made, and through the system of pulleys, automatically raises all the other shafts – hence the name 'counterbalanced'. So to raise shafts one and three, a pedal has to be tied to the lams of shafts two and four. This loom's main disadvantage is that extra shafts cannot be added to the original four, but it is suitable for a weaver who is content with using four shafts (see Figure 11.4).

11.2.5 DOBBY LOOM

If a weaver wishes to use 16 or more shafts and regularly change the combinations of lifts, as when weaving samples, then a dobby loom will be speedier than a table loom. A dobby loom (Figure 11.5) has a device fitted to the top of the loom that is worked by a single floor pedal. The required lift of shafts for each pick are pre-programmed by a series of pegs that are manually placed in short wooden bars known as lags, which are chained together to form a loop. Pressing the pedal raises the shafts indicated by the pegs, and moves the whole chain around so that the next lag is presented to the dobby device. A disadvantage of the dobby loom is the heavy lifting of the shafts, which are held in the down position by weights or springs. Therefore, the weaver must work standing up to exert more pressure on the pedal. A further disadvantage is that a long pattern repeat requires many lags and can be strenuous to lift. A dobby loom is suitable for weave structures that require up to 16 shafts but have fairly small repeats (see Figure 11.5).

FIGURE 11.4

Counterbalanced loom.

11.2.6 COMPUTERISED LOOM

A computerised loom is basically a floor or treddle loom that has a computerised box attachment that controls the position of the warp yarns when the shafts are raised or lowered. By using computer software, the draft and the weave structures to be woven are inserted into a computerised panel attached to the loom. The loom usually has 24 shafts and 1 peddle. By pressing the floor peddle, the shafts, controlled by the software, are raised and lowered. This computer system is faster than manual methods of controlling and lifting shafts and allows for easy and quick pattern changes to the fabric (see Figure 11.6).

FIGURE 11.5

Dobby loom.

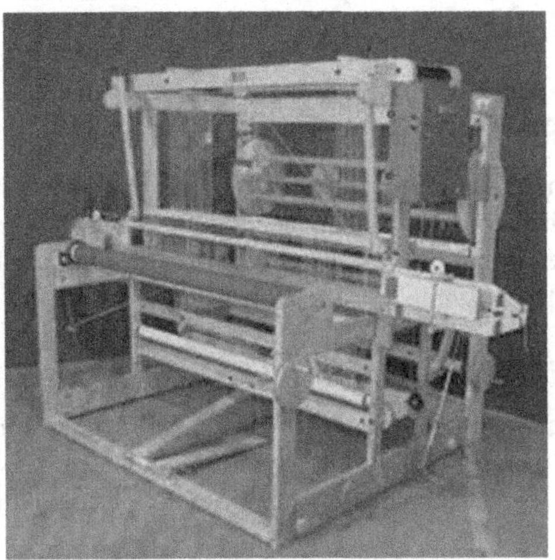

FIGURE 11.6

Computerised loom.

FIGURE 11.7

Jacquard loom.

11.2.7 JACQUARD LOOM

A Jacquard loom (see Figure 11.7) is used to produce fabrics that have intricate patterns and weave structures that use over 24 shafts. A Jacquard loom does not have shafts, it has individual heddles, and each heddle of the loom and corresponding warp yarn is individually controlled by the Jacquard mechanism. The lifting of each heddle was once controlled by a series of punched cards, but these systems have virtually all been replaced by microcomputer systems that control the lifting of heddles and corresponding warp yarns (see Figure 11.7).

11.3 MAKING A WARP AND DRESSING THE LOOM

When making the warp, all threads should be of the same length and tension and be able to be unwound evenly whilst weaving. Provision must also be made whilst making the warp for maintaining the order in which the threads are placed initially. **Dressing the loom** is the term used for placing the warp on the loom and preparing it ready for weaving.

11.3.1 SELECTING A WARP YARN

A warp yarn must be strong enough to be able to withstand the high tensions of the loom and the abrasions of weaving. Normally a warp yarn will be equal to, or finer than, the weft yarns to produce a good-quality cloth. Fancy or special-function yarns are usually placed in the weft.

11.3.2 CALCULATING THE WARP YARNS

The warp should be of the length of the cloth you wish to weave, plus an additional yard (1 m) for wastage. The width of the cloth should also be decided at this stage. To calculate the amount of warp ends

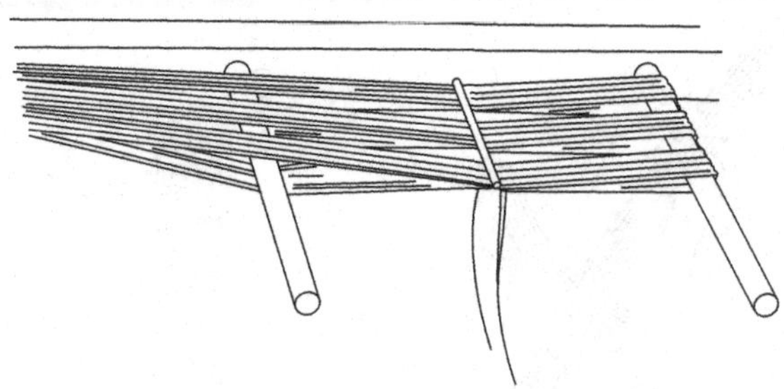

FIGURE 11.8

Lower cross.

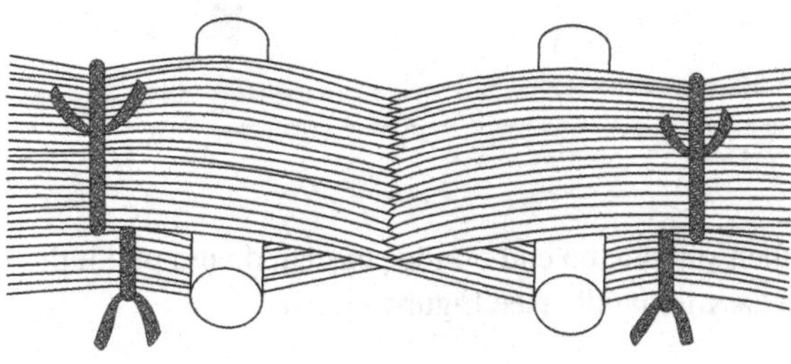

FIGURE 11.9

Four-tied sections of lower cross.

needed, wrap single warp threads around a piece of card so they are just touching, until it is covered by 1 inch (2.5 cm). Count the number of single warp threads that are within the 1 inch. This will provide the number of warp ends required per inch of cloth. Then multiply the number of warp ends per inch/cm by the width of the cloth, i.e. 48 epi × 12 inches = 576; therefore, 576 warp ends are required, multiplied by the required length of the warp. Note that additional warp ends should be included to provide a half-inch (1.25 cm) selvedge at each cloth edge.

11.3.3 MAKING THE WARP

A warping frame or mill is used to make the warp by winding the yarns at tension, including a cross at both ends of the frame or mill, until the number of warp yarns required has been wound. Warp yarns can be counted by inserting a yarn of a contrasting colour under and over the warp yarns at half-inch (1.25 cm) intervals at the lower cross (see Figure 11.8). Before taking the completed warp off the frame or mill, take threads of yarn in a contrasting colour and loosely tie around all four sections of each cross at the top and lower ends of the mill or frame. This will make the crosses easily identifiable (see Figure 11.9).

11.3.4 MAKING A CHAIN

Once all the ties have been made, slip the top end of the warp off the pegs, and slipping your wrist through the loop, grasp the warp and pull it through the loop. Continue this action to form a chain until

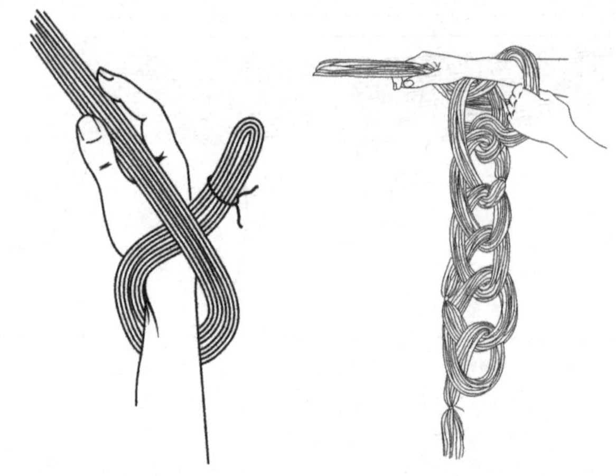

FIGURE 11.10

Warp tied in chain.

the lower end is reached, then slip the top end off the pegs but do not pull it through the loop. Producing this chain makes the warp easy to handle (see Figure 11.10).

11.3.5 DRESSING THE LOOM

To dress the loom, the weaver usually works from the back of the loom to the front. Firstly the warp is placed onto the back beam, and then threaded through the heddles on the shafts and through the reed, finally tying it onto the front beam. The process begins as follows:

1. **Inserting back sticks:** Take the warp and insert the back stick into the end loop of the lower cross of the warp; pass the cord attached to the back stick through the loop of the cross and tie it to the other end of the back stick. The cross of the warp is now secured and the ties that were earlier placed at this cross-section can be untied (see Figure 11.11).
2. **Raddling:** Securely tie the lower section of the raddle onto the back cross-bar of the loom. Locate the centre of the warp, and starting from the centre dent of the raddle, start to place the warp yarns into the raddle dents at half-inch (1.25 cm) intervals. Continue this process on both sides of the raddle until all the warp is placed into the raddle dents. Then place and securely tie the top section of the raddle onto the lower section (see Figure 11.12).
3. **Attaching the back stick to the loom:** Distribute the warp end loops evenly along the back stick and attach it to the back beam apron (see Figure 11.13).
4. **Beaming:** When the back stick and warp are firmly attached to the back of the loom, untie the chain of the warp. Then gradually wind on the warp, which is still spaced through the raddle, onto the back beam. Insert flat sticks into the first few revolutions of winding; this prevents warp yarns sticking together and forms an even tension through the width of the warp. After up to two revolutions using flat sticks, replace the sticks with sheets of thin paper; this will separate each revolution of warp yarns, preventing sticking. Continue winding the warp until most of the warp is

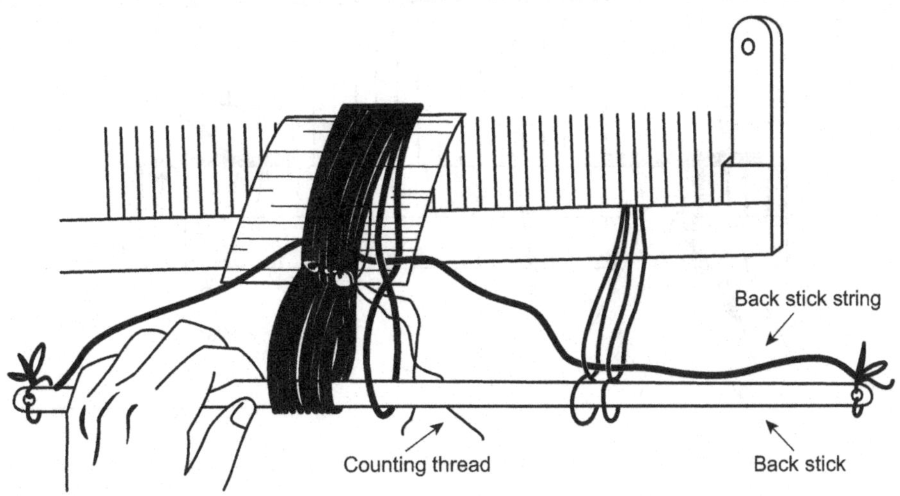

FIGURE 11.11

Warp ends tied to back stick.

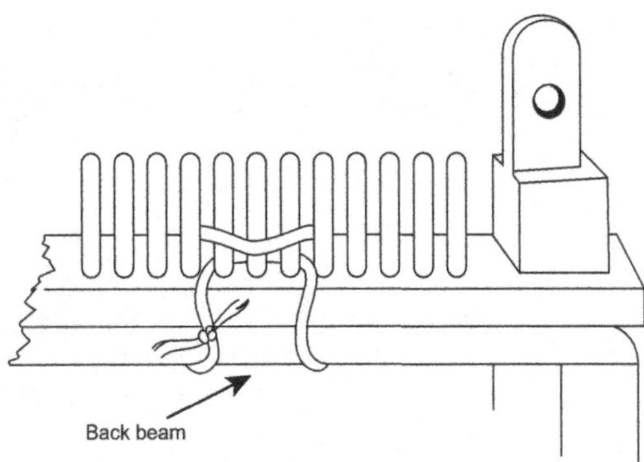

FIGURE 11.12

Raddle tied to back beam.

wound onto the beam, leaving around 18 inches (90 cm) unwound. This remaining warp will contain the second cross at the end of the warp (see Figure 11.14).

5. **Placing front cross-sticks:** Untie and remove the raddle, and place the remaining warp facing the front of the loom. Place a pair of cross-sticks through the remaining cross in the warp. Tie each end of the cross with cord and the front cross is now secured. Suspend the cross-sticks on cord behind the heddles on the shafts in preparation for threading through the heddles. From the front of the loom, gently push the cross-sticks through the warp yarns towards the back of the loom, so that around 11 inches (60 cm) of warp is left in front of the cross (see Figure 11.15).

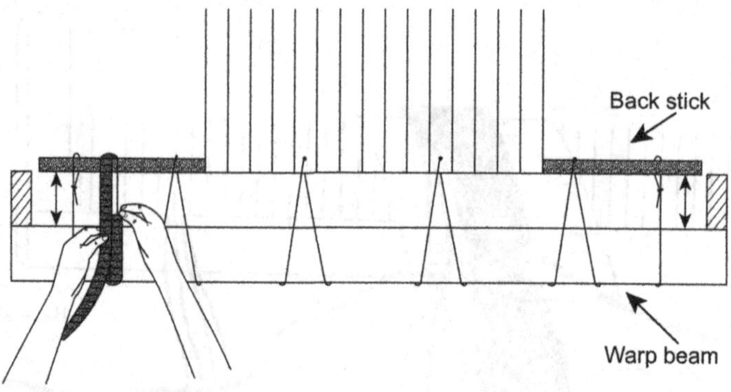

FIGURE 11.13

Tieback stick to back beam.

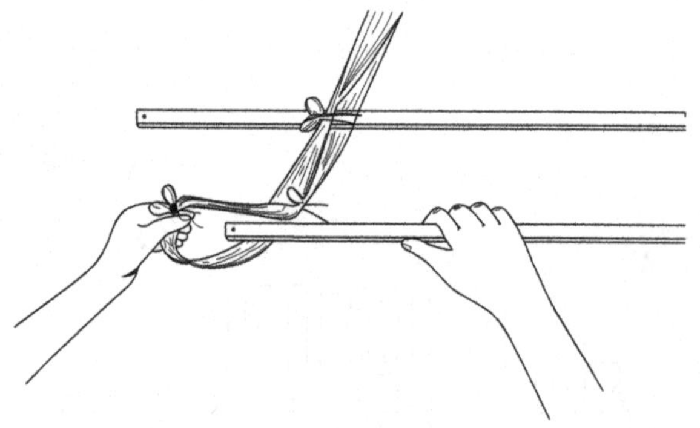

FIGURE 11.14

Insert cross-sticks through front cross.

6. **Threading heddles:** Firstly make sure there are enough heddles on each shaft for the number of warp ends allocated to it. In a plain-weave structure, this can be calculated by dividing the total number of ends by the number of shafts, i.e. 576 warp ends divided by 4 shafts, which equals 144 warp ends on each shaft. With scissors, cut the ends of the warp threads and divide the warp into two equal parts. Separate the heddles in to two equal halves and push the heddles to the right and left of each shaft. Bring the warp yarns through the centre of the shafts, and with cord, tie each half of warp ends to the front bar. The cross-sticks will permit the warp ends to lie in the correct order for threading, and this order should be followed by the weaver whilst threading the heddles. The heddles should be threaded with the warp ends, working from the centre outwards, using a threading hook. The order in which each heddle is threaded depends upon the draft (see Figure 11.16).
7. **Sleying the read:** Sleying is the placing of the warp ends through the reed. A reed hook is used to sley the reed, which is fine enough to be placed through the reed's dents. Starting at the centre of the reed and the centre of the warp yarns, the yarns are sleyed in the order that they were threaded through the heddles. The number of yarns sleyed into each dent should follow the previously

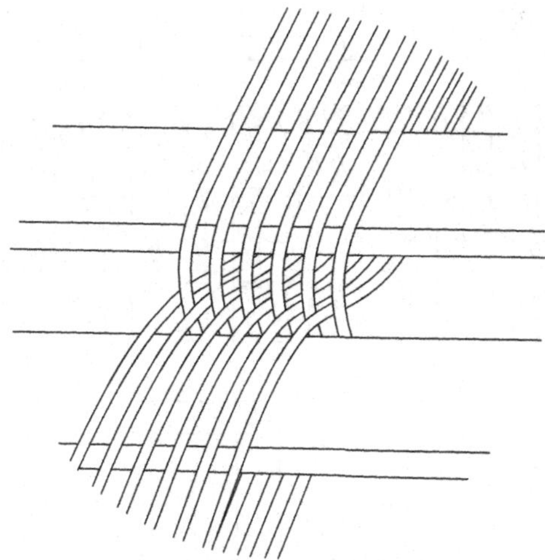

FIGURE 11.15

Gently pull cross-sticks through the warp yarns.

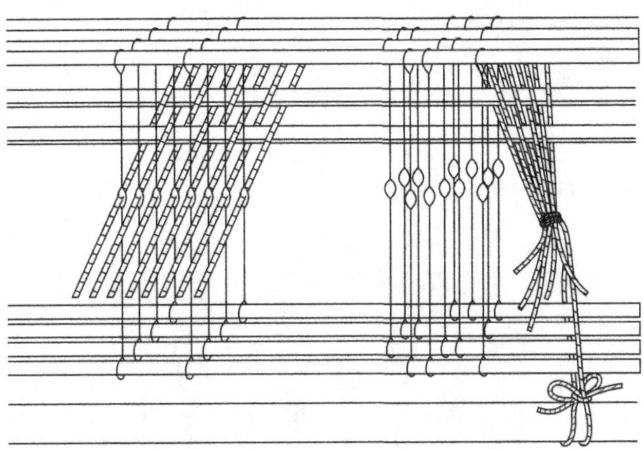

FIGURE 11.16

Thread warp yarns through heddles, one by one.

planned order – either two, three or four yarns. No more than four yarns are normally placed into a dent as this can cause **cramming** of the warp yarns and result in a line forming down the length of the cloth (see Figure 11.17).

8. **Tying on:** Once all warp yarns are sleyed through the reed, the warp should be tied on to the front beam. Starting at the centre of the warp, small bunches of warp yarns should be tied with one knot onto the stick that is attached to the apron of the front beam. When all the warp yarns have been attached, tighten each bunch of yarns with a second knot and ensure the tension is even along the width of the warp. The loom is now dressed and ready for weaving (see Figure 11.18).

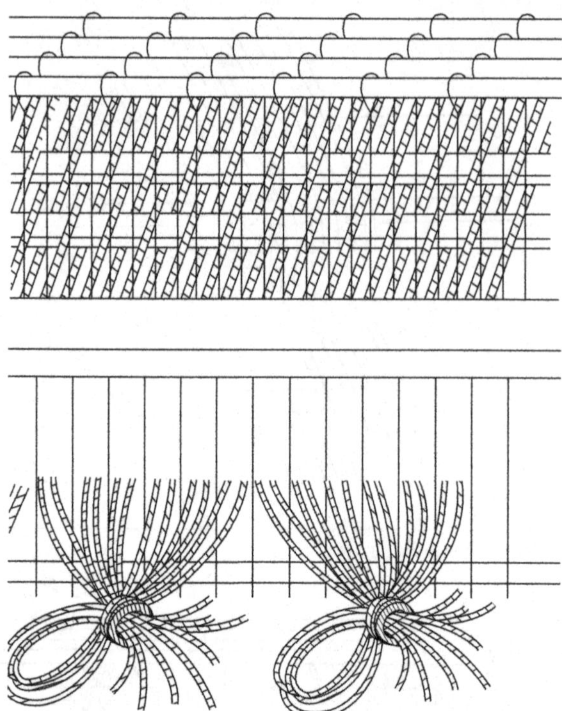

FIGURE 11.17

Thread warp yarns through dents in reed.

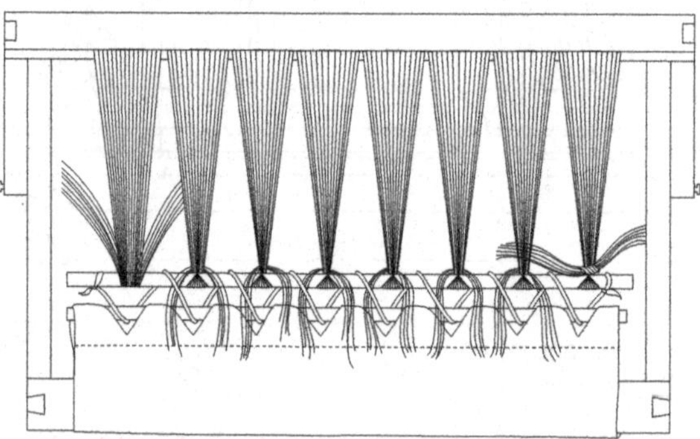

FIGURE 11.18

Loom is now dressed and ready for weaving.

11.4 DOCUMENTATION

Designing for woven textiles requires much pre-planning; the yarns, colours, woven structures and patterns all have to be decided prior to commencing weaving. This planning produces the documentation for the cloth and can used to recreate the cloth structure when required. This documentation includes the threading of heddles, lifting and lowering of shafts and sleying of the reed. Though the diagrams used in drafting may differ slightly from book to book, the principles are generally the same. Once the concepts of documentation have been learnt, it is relatively easy to follow most documentation of differing weave structures and patterns. Point paper is traditionally used for documentation, but there are also computer softwares that can be used to replace point paper when planning (Holyoke, 2013). The use of computer softwares can also provide documentation of the resulting patterns and structures that will be woven.

11.4.1 POINT PAPER

Point paper is similar to graph paper, with horizontal and vertical lines that form drawn squares. Within these squares each warp thread that is to uppermost when creating each shed is marked by shading in an opaque colour – usually black. The squares that remain without any shading, usually left white, depict the warp thread remaining down, therefore showing where the weft thread passes over the warp thread. Designs are marked on the paper starting from the lowest line and working upwards (see Figure 11.19).

11.4.2 THREADING PLAN

A **threading plan** is a diagram in which the order of threading the heddles is documented. The opaque squares placed within the threading diagram represent the shafts on the loom, and it is usual practice to number the shafts from front to back, i.e. lower to upper on the plan (see Figure 11.20).

11.4.3 LIFTING PLAN

The **lifting plan** is a diagram in which the lifting order of each shaft to produce particular weave structures and patterns is documented. The black squares represent the order in which the shafts are to be

FIGURE 11.19

Design on point paper.

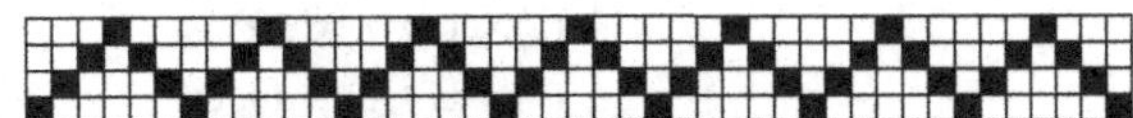

FIGURE 11.20

Threading plan.

lifted. The numbers indicating which shafts should be raised within the same shed are listed to the right-hand side of the weave plan (see Figure 11.21).

11.4.4 REED PLAN

The **reed plan** is a diagram in which the order and number of warp threads sleyed through the reed are documented. This is particularly useful if warp threads are to be grouped and spaced through the reed, or if differing numbers of warp threads are placed throughout the reed (see Figure 11.22).

11.5 PATTERN DRAFTING

The production of weave patterns depends upon how the warp yarns have been 'drawn in' to the heddles on the shafts/harness in the draft. The warp yarns are divided between the numbers of shafts available depending upon the pattern required to be produced. There are two methods of drafting the warp yarns – the first is 'front to back' the second is 'back to front'. Front to back is the most often used. There are many types of drafts available, but the most popular are straight, pointed, block (sometimes called grouped) and scattered (sometimes called satin).

11.5.1 STRAIGHT DRAFT

The **straight draft** is the basis for all other drafts and proceeds in one direction only. It can be used with any number of shafts. Each succeeding thread is drawn on a succeeding shaft, the first thread on the first shaft, the second thread on the second shaft, continuing in this regular order until the last shaft is reached, after which the operation is repeated, beginning with the first shaft (see Figure 11.23). With four shafts, the threads will be drawn from left to right as follows:

Shaft 1	Thread 1, 5, 9, 13, etc.
Shaft 2	Thread 2, 6, 10, 14, etc.
Shaft 3	Thread 3, 7, 11, 15, etc.
Shaft 4	Thread 4, 8, 12, 16, etc.

11.5.2 POINTED DRAFT

In a **pointed draft**, the warp yarns are drawn as in a straight draft, from front to back, and then from back to front, the shaft at the point of reversal receiving only one thread, and the other shafts each carrying two threads for the resulting double line. The thread at either point of reversal is called the point thread (see Figure 11.24). With four shafts, the threads will be drawn as follows:

Shaft 1	Threads 1, 7, 13, etc.
Shaft 2	Threads 2, 6, 8, 12, etc.
Shaft 3	Threads 3, 5, 9, 11, etc.
Shaft 4	Threads 4, 10, etc.

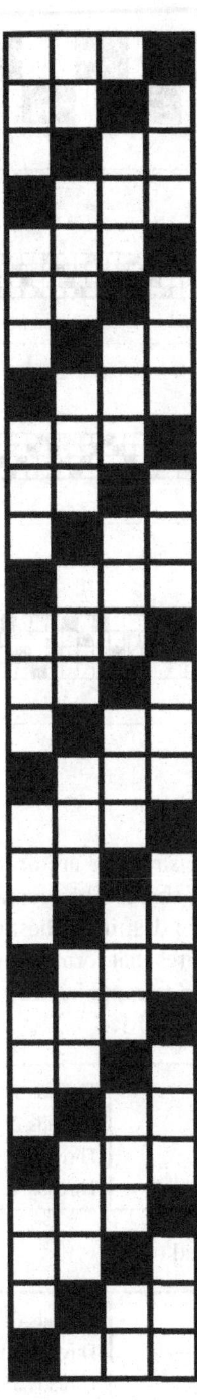

FIGURE 11.21

Lift plan.

FIGURE 11.22

Reed plan.

FIGURE 11.23

Straight draft.

FIGURE 11.24

Pointed draft.

FIGURE 11.25

Block draft.

11.5.3 BLOCK DRAFT

In a **block draft**, the threads of one weave structure are drawn in on one set of shafts, and those for another weave structure on a separate set of shafts. The separate blocks can incorporate different types of drafts – straight, pointed, etc. This type of drafting is best used on eight shafts or more, and is used to produce different blocks of weave structures that form alongside each other (see Figure 11.25). With eight shafts using a typical block draft, threads could be drawn as follows:

Block one straight draft, starting from thread 1.

Shaft 1	Threads 1, 5, 9, 13, etc.
Shaft 2	Threads 2, 6, 10, 14, etc.
Shaft 3	Threads 3, 7, 11, 15, etc.
Shaft 4	Threads 4, 8, 12, 16, etc.

Block two-pointed draft, starting from thread 61.

Shaft 5	Threads 61, 67, 73, etc.
Shaft 6	Threads 62, 66, 68, 72, etc.
Shaft 7	Threads 63, 65, 69, 71, etc.
Shaft 8	Threads 64, 67, etc.

FIGURE 11.26

Scattered draft.

11.5.4 SCATTERED DRAFT

In a **scattered draft**, the threads are scattered and not in a straight line; they resemble the draft for a satin weave. At least four shafts are required for this draft (see Figure 11.26). With four shafts, the threads will be drawn as follows:

Shaft 4	Thread 1, 5, 9, 13, etc.
Shaft 2	Thread 2, 6, 10, 14, etc.
Shaft 3	Thread 3, 7, 11, 15, etc.
Shaft 1	Thread 4, 8, 12, 16, etc.

11.6 WEAVE STRUCTURES

A **weave structure** is the order in which warp and weft threads are interlaced. There are many weave structures documented that are available to hand weavers; the basic ones are **plain, twill and satin**. These are described in the next sections. Basic weaves are those that are conducted on a loom without any modifications; 'basic' does not mean they are simple or for beginners, but that the weave is the foundation upon which derivative or modified structures are formed. Once learnt, they can be manipulated and other structures incorporated to design an individual cloth. The weave structures used in a cloth influence the finished fabric's properties, its strength or weaknesses, as well as the fabric's appearance.

11.6.1 BALANCED AND UNBALANCED WEAVE STRUCTURES

A weave structure is affected by its **sett**, that is, the ratio of warp to weft yarns. A balanced fabric or weave has one warp yarn to every weft yarn, a ratio of 1:1. An unbalanced fabric has significantly more of one set of yarns than the other. In a balanced weave, warp and weft yarns are usually of similar density or count. Unbalanced weaves usually have yarns that are different in density and count. Calculating the sett is important, as an inaccurate sett can render the woven cloth useless if it is sett too loosely or too tightly.

11.6.2 BALANCED PLAIN WEAVE

A **balanced plain weave** can be woven on two or more shafts. The weft thread (pick) passes over and under every other warp thread; the next weft thread then passes over and under the alternate warp thread. Each time a weft thread is placed through the shed it is then beaten down with the reed to hold the last weft thread in place (see Figure 11.27). A plain weave fabric has no distinguishable face or

FIGURE 11.27

Balanced plain weave.

back, and either side can be used. Interesting effects can be gained by using novelty or textured yarns, yarns of different sizes, high- or low-twist yarns and cramming and spacing of warp and weft yarns. Cloth woven with a plain weave has a finer appearance and harder feel, and is smoother, but possesses less elasticity than fabrics woven with other weave structures. This weave structure forms the maximum amount of interlacing and due to this produces fabric that is more prone to creasing.

11.6.3 UNBALANCED PLAIN WEAVE

An **unbalanced plain** weave can be woven on two or more shafts. In an unbalanced plain weave there are significantly more yarns in one direction than in the other. This can produce warp- and weft-faced fabrics and rib. Having more warp yarns than weft yarns will produce a warp-faced cloth. When less warp yarns are used and more weft yarns appear on the face on the cloth, this is known as a weft-faced cloth (see Figures 11.28 and 11.29). Warp-faced fabrics have a greater tensile strength than weft-faced fabrics.

11.6.4 BASKET WEAVE

A **basket weave** can be woven on two or more shafts. This structure is made with two or more adjacent warp yarns controlled by the same shaft, and two or more weft yarns passed through the same shed. The interlacing pattern is similar to a plain weave but showing two or more yarns in each warp or weft interlacement (see Figure 11.30). The most common basket weaves are of the ratio 2:2 or 4:4; however, they do not have to be of even ratio, they can be 2:1 or 4:2, etc. Basket-weave fabrics are more flexible and wrinkle resistant than plain weave as there are fewer interlacings per square inch.

11.6.5 TWILL WEAVES

The second basic weave, a **twill weave** is normally woven on four or more shafts. In a twill weave, each warp or weft yarn floats over two or more weft or warp yarns, with a progression of interlacing by one

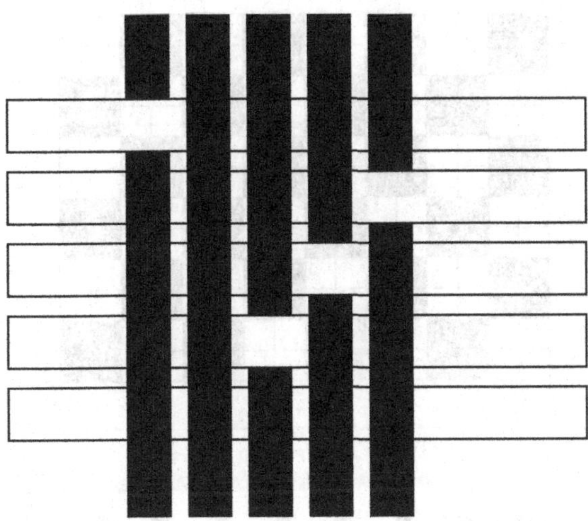

FIGURE 11.28

Warp-faced weave.

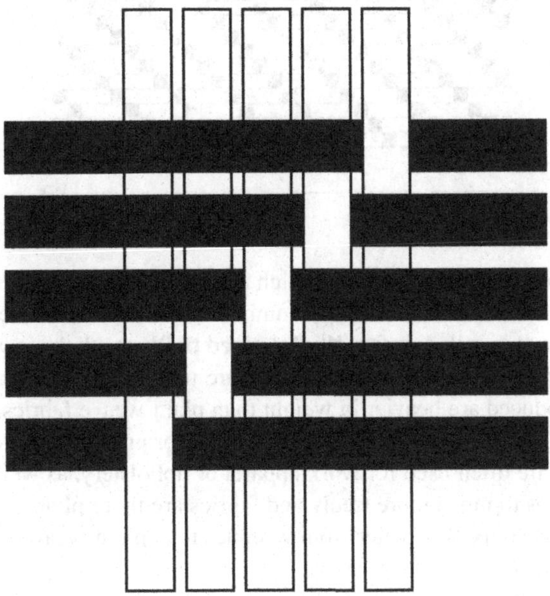

FIGURE 11.29

Weft-faced weave.

to the left or right, forming a distinct diagonal line (see Figure 11.31). Though a twill weave can be woven on three shafts, four or more shafts are normally used. The greater number of shafts available will enable more complex twills to be woven. Twill weaves have a technical face and back cloth; the face is the side that shows the most pronounced diagonal line. Twill weaves have two classifications,

FIGURE 11.30

Basket weave.

FIGURE 11.31

Twill weave.

uneven or balanced. **Uneven twills** are those in which the warp comes to the surface to either a greater or lesser extent than does the weft. If the warp predominates on the face, the weave is called warp twill. If the filling predominates, it is called weft twill. **Balanced twills** are those in which the warp and weft come to the surface to the same extent. Because there are more weft yarns than warp in twill-weave structures, the fabrics produced are heavier in weight than plain-weave fabrics. As twill-weave fabrics have textured and patterned surfaces they are not often used for printed fabrics. Due to the strength of twill-weave fabrics, they are often used for work apparel or upholstery, as well as denim jeans. Fewer interlacings allow the yarns to move more freely and fabrics are more pliable, lustrous and softer than plain-weave fabrics; they also recover better from wrinkles than plain-weave fabrics.

11.6.6 HERRINGBONE TWILL

A pointed draft is used to create a **herringbone twill weave structure**, causing the diagonal twill line to change direction intermittently, forming a zigzag line (see Figure 11.32). This structure is also known as a chevron. If the weave structure is carried out in a controlled reverse action, a diamond shape is formed. These structures are used in apparel and furnishings.

FIGURE 11.32

Herringbone weave.

FIGURE 11.33

Satin weave.

11.6.7 SATIN WEAVE

Satin weave is the third basic weave and requires at least five shafts to weave. In this structure, the weft yarns are predominant on the face of the cloth, and the warp yarns that bind the weft floats should be scattered as widely as possible. The farther they are removed from each other the more indistinct they become and the more attractive the cloth. No two interlacings are adjacent to each other, so no line forms as in twill weave; however, lines may appear in the back of the cloth (see Figure 11.33). All satin weave fabrics have a face and a back that look significantly different from each other. Satin-woven fabrics are strong due to the high number of yarns used, yet fewer interlacings provide pliability and resistance to wrinkling. Satin fabrics are almost always warp-faced and made of shiny filament yarns with very low twist to produce a lustrous finish. Satin-woven fabrics are used in clothing and apparel, particularly couture wear, wedding dresses, and drapery linings.

11.7 DERIVATIVE-WEAVE STRUCTURES

Derivative-weave structures are those that are developed from the basic weave structures. There are a number of Derivative-weaves, the main ones discussed here are mock leno, double weave, honeycombe and Jacquard.

11.7.1 MOCK LENO

A **mock leno** can be woven on a minimum of four shafts; it belongs to the open-weave group of structures and provides a cloth with a lace or gauze appearance (see Figure 11.34). A true leno structure requires the use of a doup, which provides intricate interlacing of warp and weft threads. A mock leno structure can provide a similar interlacing, but is simpler and quicker to weave than a true leno. In weaving a mock leno, three or more warp or weft threads are interlaced so that they group together in an opposite way to their adjacent group. This grouping together of threads forms openings in the cloth at this point (see Figure 11.35). The openings in the cloth can be enlarged by doubling the number of warp threads in each dent and leaving one or more dents empty in the reed between warp yarns. A mock leno structure can be woven alongside other weave structures, such as plain weave, and cause contrasting opaque and transparent effects in the same cloth. Finished fabrics with varying amounts of opacity and transparency can be used for apparel such as shirts, blouses and dresses, plus furnishings such as tablecloths, curtains and cushions.

11.7.2 DOUBLE WEAVE

A **double-weave fabric** has two layers of cloth, which are either joined at the selvedge or are interlaced between top and lower cloth. Reasons for weaving a double cloth are to form a tube, to produce a cloth of different colours on either side, to produce quilt-like effects, to provide a thick fabric or to trap objects inside

FIGURE 11.34

Mock leno weave.

the two layers of fabric. The interlacing of the two layers of fabric during weaving can be done by interlacing small amounts of yarns from the lower cloth with yarns from the top cloth, or by interlacing small amounts of yarns from the top cloth with those of the lower cloth (see Figure 11.36). Alternatively, equal amounts of yarns from both top and lower cloths can be interlaced, and can produce interesting colour 'interchange' if different-coloured yarns are used in the top and lower warp and weft yarns. The interlacing of top and lower cloths can also be conducted in such a way as to trap objects or 'stuffing' yarns into pockets or pleats. Interlacing top and lower cloths can also be used to form quilt-like patterns in the cloth. These are known as pique, and are best woven with the top and lower warp yarns set at different tensions on two back beams. The lower warp yarns should be at a higher tension than the top yarns, which will form a raised surface on the top cloth when woven. Double cloths are used for apparel such as jackets and coats that may be reversible (two different colours), and furnishings such as curtains and cushions.

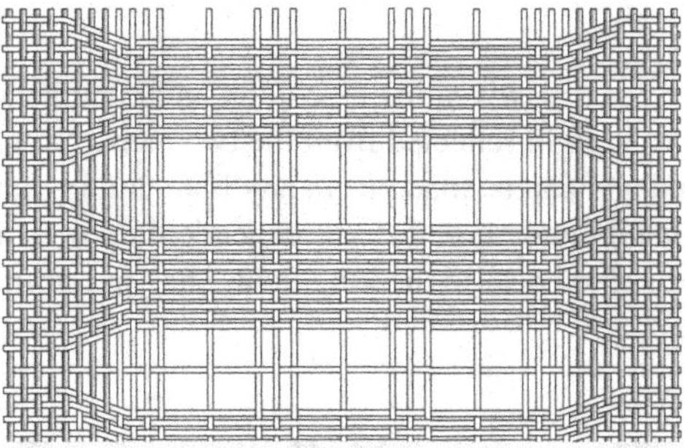

FIGURE 11.35

Openings in cloth formed through mock leno structure.

FIGURE 11.36

Double weave.

11.7.3 HONEYCOMB

A honeycomb weave structure permits the warp and weft yarns to form hollows and ridges that produce a cell-like appearance. A pointed draft is used for this structure, and both the warp and weft yarns float freely on both sides of the cloth (see Figure 11.37). Cotton yarn when used for this structure produces a very absorbent cloth, and is often used for the production of tea towels. Wool produces a very warm fabric as the air is trapped in the structure of the weave and also the fibres of the wool.

11.7.4 JACQUARD WEAVES

An advantage of Jacquard weave structures is that they can produce figured fabrics. Figured fabrics are those fabrics which have visual images on them rather than just patterns. Fabrics made on a Jacquard loom include damask, brocade, brocatelle, tapestry and others. They are used in apparel and clothing, furnishing and utility fabrics.

- **Damask:** These fabrics are usually woven in one colour, and patterns are subtle but visible because of slight differences in light reflected from the two areas. Damask structures have satin floats on a satin background and can be made from any fibre and in many different weights for apparel and furnishings. Quality and durability are dependent upon yarn count. Low-count (fine-yarn) damask is not durable because the long floats snag and shift during use.

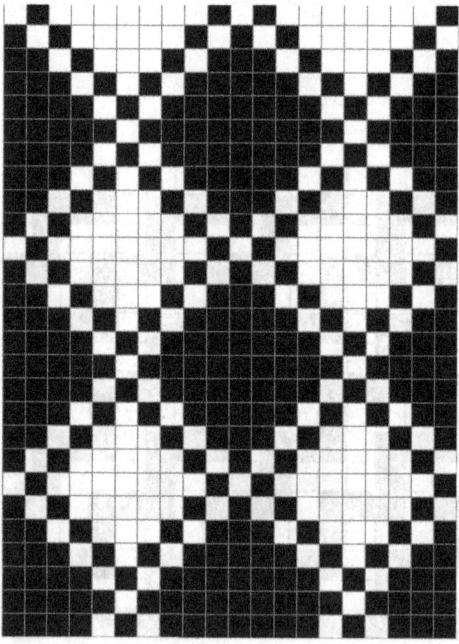

FIGURE 11.37

Honeycomb weave.

- **Brocade:** This fabric differs from damask in that the floats in the design are more varied in length and are often of several different colours. Brocade structures have satin or twill floats on a plain, ribbed, twill or satin background.
- **Brocatelle:** These fabrics are similar to brocade, except that they have a raised pattern. This fabric is frequently made with filament yarns, using a warp-faced pattern and weft-faced ground.
- **Tapestry:** Tapestry fabric woven on a Jacquard loom is mass-produced for upholstery and other uses. Jacquard-woven tapestry is a complicated structure consisting of two or more sets of warp and two or more set of weft interlaced so that the face warp is never woven into the back and the back filling does not show on the face. Upholstery fabric is durable if warp and weft yarns are comparable. With lower-quality fabrics, fine yarns are combined with coarse yarns, and the resulting fabric is not durable.

11.8 STARTING TO WEAVE

Shafts are raised by pressing levers and peddles according to the weave structure, and weft picks are placed through the shed created by the use of weaving sticks or shuttles. Once a pick has been inserted, the shafts are lowered and the weft pick beaten into position using the reed. The next shed is created by raising the next shafts in the series, and the next weft pick is inserted by taking the weft stick or shuttle through the shed in the opposite direction from the previous weft pick. This process is continued until the cloth is complete. The beat-up of the weft picks must be consistent to ensure an even quality in the cloth.

11.9 DESIGNING FOR WOVEN TEXTILES

Inspiration for a woven fabric can come from many sources – from nature, architecture, paintings, antique textiles, textiles from other countries and so on. These inspirations are translated into woven designs through the use of yarns, texture, stripes, checks and most importantly, colour (Holyoke, 2013; Shenton, 2014).

Texture can be created through the use of weave structures and yarns. Many weave structures can be used to create texture such as twill, pique and honeycomb. The use of two or more weave structures in a block draft can also produce textured effects. Even a simple two-shaft loom can be used to produce many interesting fabrics just by making use of differing yarn properties and loom manipulation techniques. Cramming and spacing techniques, where warp threads are spaced or crammed in the reed, and the weft threads are crammed and spaced whilst weaving, can produce textural or gauze-like effects. Yarns with reactive properties such as elastic, high-twist and those containing fibres such as wool provide exciting effects once taken off the loom and finished. Weaving by hand also permits the careful weaving of luxurious yarns such as cashmere, silk, lambswool and mohair. Confident weavers can also experiment with the use of monofilament, metal, paper, and plastic yarns, plus the combined use of thick and fine-density (low and high count) yarns.

Checks and stripes are a way of introducing contrast into a woven design. If the warp yarns are of differing colours and the weft yarns are all one colour, then warp stripes will occur. If the warp yarns are of the same colour and the weft yarns are of differing colours then weft stripes will occur. If both the warp and weft yarns are in multiples of different colours, then checks will occur. Checks and stripes do not have to be of equal size; varying the widths of warp or weft stripes will provide differing amounts of colour or checked squares and rectangles. The proportion, balance and rhythm of checks and stripes lead to a well-designed fabric, especially if it is to be repeated in the cloth. Sometimes a simple combination of single stripes or checks of colour is more effective than complicated designs and repetitions.

Colour is a way of being very creative in weave design; it can be applied to warp and weft yarns, prior, during or after the weaving process. Colour can be applied prior to weaving by dyeing or using yarns have been pre-dyed prior to purchase. Tie-dye or Ikat effects can also be woven by tying and dyeing warp and/or weft yarns prior to weaving. Colour can be applied during weaving by combining different coloured yarns in the interlacing process, forming new shades and tones. Colour can be applied after weaving and the cloth has been removed from the loom, by dyeing the whole woven fabric. There are several reasons why a weaver may choose to dye a fabric after it has been woven, a practice known as piece-dyeing. Fabrics that have been woven with different fibres may be dyed with only one type of dye, which could result in some of the yarns not taking the dye, or taking the dye at a different density, resulting in a different shade. This type of dye method is called cross-dyeing and can create multiple shades of colour in one fabric. Piece-dyeing is a very useful technique for yarns of a high-twist structure, which are difficult to unwind if pre-dyed in a hank, and for woollen yarns that are to be felted after weaving.

Colour and weave effects are patterns formed by using two or more contrasting colours for warp and weft yarns. These effects differ from stripes and checks as the patterns are formed through using twill and other weave structures. Colour and weave structures often change the appearance of the cloth completely, and the resulting patterns can produce very original patterns.

Though aesthetics are important in weave design, there are other considerations such as fabric function; is the fabric to be worn, hung, sat upon, walked upon or just looked at, how should it drape, should the cloth be warm or cool, should it be light or a heavier weight? These considerations require thought prior to weaving; it is the yarns chosen and the woven structures selected that will determine if the fabric is suitable for its proposed use. Therefore, it is important that sampling is conducted by the weaver prior to weaving larger fabric pieces.

11.10 DESIGNING FOR THE JACQUARD LOOM

Elaborate and intricate designs can be produced on a Jacquard loom, as each warp yarn can be lifted individually and a change in weave structure and pattern can be made at any time. Designing for the Jacquard loom includes the designer selecting his or her own design or repeat pattern, and then choosing which weave structures to incorporate into the design. Jacquard designs are usually created through a combination of a drawn or painted surface pattern design to which different weave structures are allocated. The painting or drawing is scanned into the computer (Holyoke, 2013) (on a computerised Jacquard loom) and then weave structures are inserted, thus eliminating the need for point paper. Editing can be done to simplify weave areas or to properly join weaves. The repeat size of the Jacquard loom is normally fixed and is dependent upon the maximum number of threads than can be lifted independently. As it is time-consuming to change the warp on a Jacquard loom; the warp lengths are usually very long and a neutral colour is used such as ecru or white. Colour is largely introduced into Jacquard fabric through the weft yarns. Also the different weave structures selected will produce different textures and a different proportion of warp to weft colour.

11.11 TAPESTRY WEAVING

A tapestry is a weft-faced woven pictorial image that can be realistic or abstract. The image to be woven is normally drawn first on point paper. The size of the image, known as the **cartoon**, is often the full size of the tapestry to be woven, especially if the image is very intricate. A frame loom is used for tapestry weaving, as it is particularly suitable for weft-faced fabrics. Tapestry looms vary in size; large upright tapestry looms have a rolling beam at both the top and bottom of the loom. The top beam is to hold the warp yarns and the lower beam is for winding the tapestry onto whilst weaving (see Figure 11.38). More popular, smaller looms may not have any rolling beams and are simply square or rectangular wooden frames, similar to a picture frame except without the glass or backing. Tapestry looms do not have any shafts or heddles.

The warp for this type of loom is made by winding the warp yarns across the length of the frame from one end to the other. If the frame does not have markers indicating ¼-, ½- or 1-inch intervals, these can be marked onto the loom by hand. Leave around 1 inch at the inner edge of the frame before starting to tie the warp yarn on, starting by tying the warp yarn onto the lower bar of the frame. Keep the tension tight whilst winding the warp yarns around the frame. The warp can be further tensioned after winding by inserting a stick or piece of doweling at one end of the frame holding the warp yarns. Finish tying on the warp by tying a knot onto the lower end of the frame. Ensure the warp threads are evenly spaced and in the correct position by spreading manually. Heading cords are then inserted into

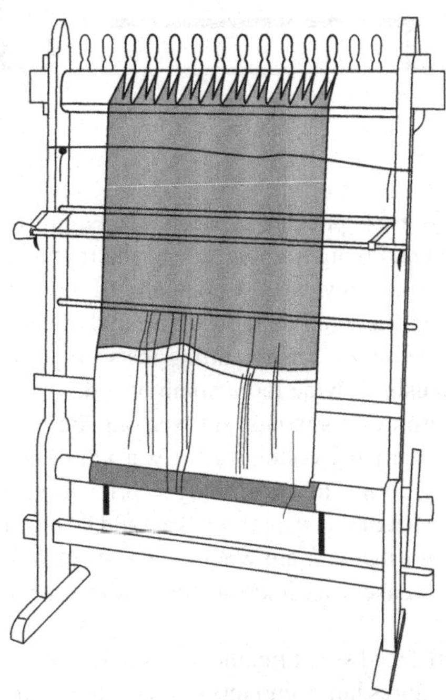

FIGURE 11.38

Upright tapestry loom.

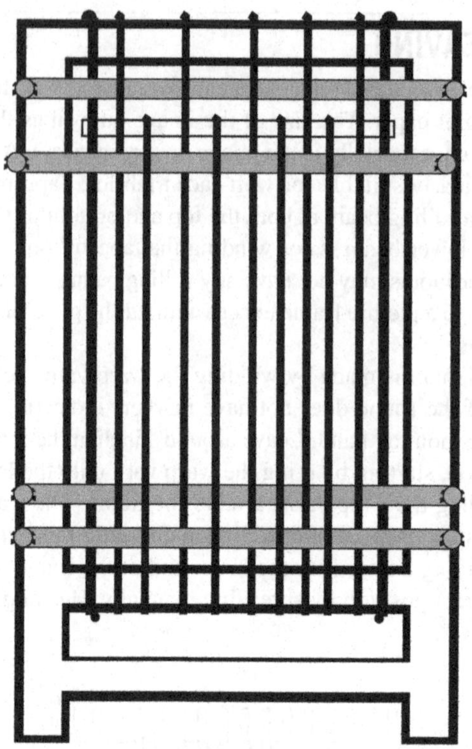

FIGURE 11.39

Warp yarns wrapped on tapestry loom.

the warp to ensure the warp yarns are kept evenly spaced (see Figure 11.39). Heading cords are made by taking a length of warp thread and tying it to the side of the frame at the lower end where weaving is to commence. Weave one pick in plain weave by manually lifting a shed in the warp yarns, take the yarn across the whole width of the warp and tie the yarn around the other side of the frame. Continue this in plain weave for two or three picks, finally cutting the weft yarn and tying onto the side of the frame. Beat the weft yarns down using a wide-tooth comb or fork.

Traditional yarns used for warp threads in tapestry weaving are wool, silk, linen and cotton; however, many tapestry weavers now only use cotton for the warp threads, due to its strength and its tendency not to stretch. Warp yarns do not need to be dyed prior to weaving as they will be covered completely by the weft yarns. The set of the warp threads should be between six and eight ends per inch (2.5 cm). Traditional yarns used for the weft are wool, due to its ability to cover the warp yarns well. But many contemporary tapestry weavers use linen and other textured yarns to differ the overall surface texture.

Though a tapestry is of a weft-faced structure, the weft does not travel all the way across the warp. Even if the image to be woven requires large amounts of one colour across the width of the warp, these are divided into several strands of threads across the same pick. This is conducted to keep an even tension between single and multicoloured areas. Tapestry is woven in plain weave with a weft-faced set.

Weaving a tapestry requires hand manipulation of weft yarns, and these yarns are inserted where and when the pattern demands, working in an upwards direction. Tapestries can be of any size, and different types of looms are used depending upon the size of tapestry to be woven. The tension required of the warp on a tapestry loom is higher than for ordinary weaving.

11.12 CASE STUDY: HONEYCOMB WOVEN STRUCTURES

The honeycomb weave structure derives its name from its resemblance to the hexagonal honeycomb cells in which bees store their honey. When woven, the honeycomb structure provides a very decorative and textured pattern in the cloth. With careful colour stripes in the warp and weft, the pattern can appear very intricate. A cloth woven with honeycomb structures can be used purely for aesthetic decoration, but it is generally used in furnishings and tableware, as the structure has absorbent properties. If the honeycomb structure is woven with cotton yarns then it becomes very absorbent of moisture and liquids. If it is made of wool, then it absorbs heat, especially body heat.

A weaver selecting wool yarns for a honeycomb-structure blanket or throw, for example, should be further aware of which type of wool to use. Shetland or mohair fibres may irritate the skin and cause itching. Cashmere or angora, though soft to the touch, would render the cloth very expensive to produce. Lambswool or merino wool, however, would provide comfort to the skin by its softness, retain body heat and provide warmth, and also have the ability to drape over the body.

The honeycomb structure has a series of horizontal and diagonal lines which start at the centre of the cell and work outwards. It is this cell-like woven intersection that traps air between its fibres, and that provides insulation properties to the woven cloth. Though there is flexibility within the size of the cells to be woven, care has to be taken with the stability of the cloth. A cloth to be used as a blanket or throw should have a structure that has warp and weft floats of sufficient length to avoid stiffness. Yet the floats should not be too long, as they will become snagged and loose with use.

If a further weave structure other than honeycomb is used in a single cloth, then it is good practice to sample each one separately to determine if they are mechanically compatible in such things as warp and weft take-up. The combined weave draft must be examined to ensure that warp and weft floats stay within the limits decided upon, especially in the areas of interchange between weaves.

11.13 FINISHING

When weaving is complete, the cloth is then taken off the loom. The condition of the cloth at this stage is known as 'loom-state'. To render the cloth in a finished condition and ready to use, 'finishing' processes are required. Differing finishing processes are required for different fibres, and the processes described here are for hand-finishing.

- **Cotton fabrics**: Wash in soap flakes or detergent in warm water, hot iron or stretch on a soft board and attach with pins.
- **Woollen fabrics**: Industrial processes include milling and fulling which felts the fibres together, making the fabric thicken and soften. Hand processes include using soap flakes and warm water and kneading the fibres using the knuckles. This will encourage the wool fibres to move towards

each other. This hand process requires at least half an hour of work; the longer the time spent, the more the cloth will felt and shrink, and the definition of the weave structure will soften. Ensure that all soap is rinsed out of the fabric when rinsing in cold water. The cloth should be dried under tension stretched out on a soft board and secured by pins. Once dry, the woollen cloth can be brushed; originally teasels would have been used, but a still brush will suffice. This raises the fibres on the cloth and gives it a softer handle.

- **Linen**: Wash in soap flakes and very hot water for about half an hour to clean and soften the fibres. Then rinse thoroughly in cold water and stretch dry securing with pins. Use a hot iron to remove creases when the fabric is almost dry.
- **Silk:** De-gummed silk (silk yarn with natural gum removed) fabrics can be washed in soap flakes and hot water for around half an hour to soften the fibres. If the yarn still contains gum, then the cloth will need to be boiled with soap flakes for around 1 hour, if gum still remains in cloth, then boil for a while longer until all gum is removed. Once the cloth has been washed, rinse thoroughly in cold water and stretch dry. Press with a hot iron once dry.

11.14 TIPS FOR WEAVING

It is worth keeping a record in notebooks of methods you use and their result. Include as many details as possible – information on yarns, sett, etc. This will allow you to build on your experience plus learn from mistakes. Include ideas on how the sample could be woven better or developed (Shenton, 2014). Collect samples of woven fabrics, and then deconstruct them in order to understand the techniques and weave structure used in their construction.

11.15 FUTURE TRENDS

It is computer systems that play a large part in the future of weaving equipment and design (Holyoke, 2013). Computer-aided design packages – such as APSO, AVL and Scotweave – eliminate the need for producing time-consuming design structures by hand on point paper. These software packages as well as others are able to produce weave structures, face fabric simulations and differing colour combinations just by the operator inputting basic draft information. The use of this software on the computerised looms has provided more accurate design drafting, speedier changes in design and fewer errors, making the weaver's task a little bit easier.

Hand-drawn designs can also be scanned into a computer and modified via a screen tool to incorporate weave structures and patterns. These scanning systems are normally utilised with Jacquard looms, which allow intricate patterning and the use of many different weave structures. Computer systems continue to be designed specifically for Jacquard weaving. Computers control looms that are instructed through electronic Jacquard heads. On a hand-operated Jacquard loom, this makes the task for the weaver much simpler.

Many Jacquard looms used to produce textiles today, even in some universities, are automated and not hand operated. And it is the speed of weaving through automation that has been further addressed by computer software. The computer controls the position of the warp yarns and the insertion of different weft yarns. An automated loom is fast, with weaving speeds of around 600 picks per minute. Furthermore, computers in Jacquard looms can detect incorrect weft insertions, remove the incorrect insertion, correct the problem and restart the weaving operation. All of this can be done without the assistance of the weaver.

Shuttleless looms: Because of the noise and slower speeds, shuttle looms continue to be replaced with faster, quieter and more versatile shuttleless looms in commercial operation. There are four types of shuttleless looms: air-jet, rapier, water-jet and projectile. In these looms, the weft yarns are measured, inserted and cut, leaving a fringe along the side rather than a woven selvedge. Many shuttleless looms can produce almost any weave pattern in various yarn types and sizes with multiple colours.

Air-jet and rapier are the most popularly used shuttleless looms and both can operate at speeds of up to 1000 picks per minute. The air-jet loom is suitable for spun weft yarns provided they are not too bulky or heavy, and is used in weaving sheeting and denim in industry. Smaller air-jet looms are available for use in colleges and universities. The rapier loom is suitable for use with spun, non-spun and filament yarns, and is often used in industry to produce cotton, woollen or worsted fabrics, as it is more versatile than the air-jet loom.

Triaxial looms: The triaxial loom weaves three sets of yarns at 60-degree angles to each other. Two of the yarns are warp and one is weft. The advantage of fabrics woven on the triaxial loom is that they are stable in horizontal, vertical and bias directions, unlike regular woven fabrics (biaxial woven), which are unstable on the bias. Triaxial fabrics are used for air structures such as hot-air balloons, sail cloth, truck covers and other technical fabrics.

Additions to loom technology include:

- Devices to weave intricate designs.
- Computers and electronic monitoring systems to increase speed, patterning capabilities and quality by repairing problems and keeping looms operating at top efficiency.
- Quicker and more efficient means of inserting filling yarns.
- Automatic devices to speed the take-up of woven cloth and let-off or release of warp.
- Devices that facilitate and speed up changing of warps.

11.16 SUMMARY

There are several types of handlooms that can be used to create a woven cloth, each of them presenting their own particular strengths and weaknesses to the weaver. All looms require much preparation before weaving can commence, such as making the warp, threading heddles and sleying the reed. The draft and structure of the woven cloth must also be selected prior to the commencement weaving. Only once all this has been completed can the weaver start to weave the cloth. The basic and derivative weave structures explained provide a starting point for the beginning weaver, who can then go on to explore more creative structures. Knowledge of the documentation of drafts and weave structures will permit the weaver to recreate a particular structure in a woven cloth. The case study of a woven blanket enables the weaver to understand that the selection of yarns and weave structures involves more than aesthetic considerations; in fact, their correct selection is essential to the cloth's chosen use.

11.17 REVISION QUESTIONS

1. What are the basic functions of all looms?
2. Name three basic features of all types of looms?
3. What is the minimum number of shafts a table loom can have?
4. What does 'dressing the loom' mean?
5. Why must warp yarns have strength?
6. What does 'sleying the reed' mean?
7. What is a 'straight draft'?
8. What is a 'weave structure'?
9. What is the 'sett' of a cloth?

11.18 SOURCES OF FURTHER INFORMATION AND ADVICE

George Weil Fibre Craft Supplies: www.georgeweil.com.
Hand-Weavers Studio: www.handweavers.co.uk.
Journal of Weavers, Spinners and Dyers: www.thejournalforwsd.org.uk.
Looms and Equipment: www.theloomexchange.co.uk.
The Weave Shed for Professional Weavers: www.theweaveshed.org.
Worshipful Company of Weavers: www.weavers.org.uk.

11.18.1 COLLECTIONS

Costume and Textiles Collection, Crafts Study Centre (University for the Creative Arts, Farnham, Surrey).
Costume and Textile Collection (University of Leeds International Textiles Archive).
Textile Collection, Whitworth Art Gallery, (University of Manchester).
Textile Collections, Victoria and Albert museum, London. The Constance Howard Textile Collection and Archive (Goldsmiths, University of London).

FURTHER READING

Fannin, A. (1979). *Handloom weaving technology*. Van Nostrand Reinhold Company.
Glasbrook, K. (2002). *Tapestry weaving*. Search Press Ltd.
Holyoke, J. (2013). *Digital Jacquard design*. London: Berg Publishers.
Joseph, M. L. (1976). *Essentials of textiles*. Holt, Rinehart and Winston.
Kirby, M. (1955). *Designing on the loom*. Select Books.
Oelsner, G. H. (1952). *A handbook of weaves*. New York: Dover Publications.
Shenton, J. (2014). *Woven textile design*. London: Laurence King Publishers.
Sutton, A. (1986). *The structure of weaving*. Batsford.
Sutton, A., Collingwood, P., & Hubbard (1982). *The craft of the weaver*. British Broadcasting Corporation.

Sutton, A., & Sheehan, D. (1989). *Ideas in weaving*. London: Batsford.
Tidball, H. (1961). *Two-harness textiles*. Shuttle Craft Books Inc.
Tovey, J. (1965). *The technique of weaving* (1st ed.). Batsford. Paperback edition 1983.
Yates, M. (1996). *A textile designers handbook*. W.W. Norton and Company.

CHAPTER

YARN TO FABRIC: KNITTING

12

E.J. Power
University of Huddersfield, Huddersfield, UK

LEARNING OBJECTIVES

At the end of this chapter, you should be able to:

- Identify a knitted structure
- Describe the process of mechanical loop formation
- Distinguish a warp knitted fabric from a weft-knitted fabric
- Describe the properties of a knitted fabric

12.1 INTRODUCTION

Knitting is one of the most versatile methods of producing a textile fabric. The structure is constructed from a series of intermeshing loops. Historically, there are many different ways of forming the loops, including knotting yarns together and using a spool with pins inserted. However, the technique that has become most associated with fashion knitwear (weft knitting) is derived from either hand or pin or needle knitting. Hand knitting usually involves two large needles or pins and a single end of yarn. This method can be traced back to the fifteenth century in the United Kingdom (UK), where the unique properties of the material obtained were utilised for the manufacture of stockings. Of course, as with all textile manufacturing, knitting is no longer a cottage industry. Hand knitting is considered a skilled craft, and most modern knitwear available on the high street is produced using sophisticated computer-controlled machinery. Since William Lee's pioneering invention of the stocking frame in 1589, there has been much technological innovation. There are now two distinct types of knitting technologies, which ultimately produce vastly different fabrics for specific applications (weft and warp). The mechanical knitting process has been perfected so that different weights of materials can be produced at extremely fast speeds to the highest quality, often combined with complex patterning and texture. The properties of knitted fabrics are vastly different from those of woven fabrics. Knitted structures are less stable, more flexible and generally have better drape than their woven counterparts. The recent trend towards casual dress has increased the popularity of knitted garments significantly. Today, the knitting industry holds a larger market share in clothing and fashion applications, with many retail companies employing specially trained knitwear designers and technical staff to create innovations season after season.

Textiles and Fashion. http://dx.doi.org/10.1016/B978-1-84569-931-4.00012-X

12.2 LOOP FORMATION

The mechanical process of forming loops differs from that of hand knitting significantly. In hand knitting, the movement of the wrists and fingers is performing a series of complex actions that would be nearly impossible to simulate mechanically. To simplify the mechanical process of knitting, each individual loop requires its own needle. The needle type may vary depending on the type of machine, but the process of knitting remains fundamentally the same (Figure 12.1). Assuming that the required number of loops (stitches) has been cast on (first row of knitting), six subsequent stages will follow to complete the loop formation process. **Stage 1** illustrates the standard position for the needle; this is the rest position to which the needle will return during the cycle of knitting. Initially, the needle will be moved forward to the knit position (**stage 2**); as the needle moves forward, the loop already formed (previously in the hook section) clears the latch (this is termed clearing). The new yarn is then inserted into the empty needle hook (**stage 3**). The needle then retracts backward causing the old loop to slip under the latch, forcing it to close (**stage 4**). The needle continues retracting into the needle bed, which enables the old loop to fall off the end of the needle – this is commonly referred to as knock over (**stage 5**). A downward force is applied (usually by a roller) to complete the knitting cycle – the force is referred to as the takedown force (**stage 6**), and the needle returns to the rest position identified in stage one. The yarns used to produce knitted fabrics usually have less twist than those used to produce woven structures – this is to ensure the yarn is flexible enough to allow the required deformation to occur in the formation of the loop shape.

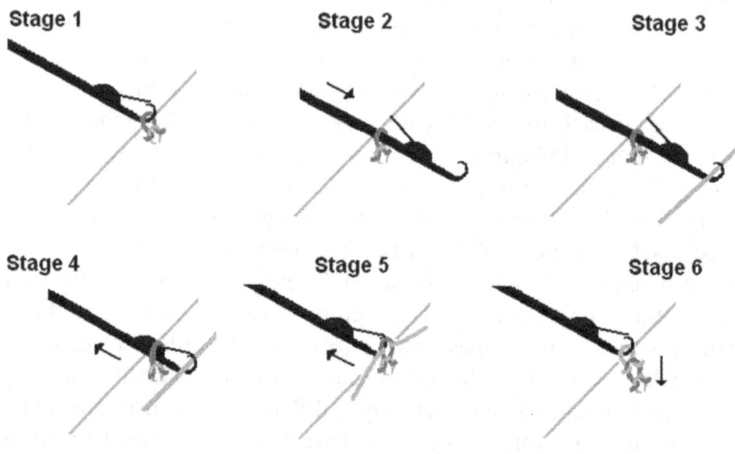

FIGURE 12.1

The process of loop formation.

12.3 KNITTING TERMINOLOGY

It has previously been identified that the intermeshing loops (more commonly referred to as stitches) are a key feature of knitted fabrics. To enable a fabric of specific dimensions to be produced the stitches (counted horizontally) and the number of rows (loops counted vertically) can be specified. **Stitches** and **rows** are a common feature in hand knitting patterns (instructions), but are rarely used in the commercial industry. The standard terminology is **wales for stitches** and **courses for rows**. The definition of a stitch in industrialised knitting refers to the type of stitch (knit, float or tuck), not the actual loop (this is discussed briefly in Section 12.4). Knitted fabrics vary significantly in weight, from ultra-lightweight (termed fine gauge) used in sporting and underwear applications, to heavier structures used in outerwear sweaters (termed chunky gauge). **Gauge** is a term often used to describe the fineness of the fabric, although it actually refers to the number of needles (per inch) within the needle bed (although there are notable exceptions: in some older weft-knitting machinery the gauge is expressed as number of needles per 1.5 inches, and in raschel warp knitting machines it is number of needles per 2 inches). But in general terms, 18, 14, 12 and 10 gauge machines produce lightweight fabrics; midweight fabrics are produced on 8 or 7 gauge; and 5, 3 and 2.5 gauges produce heavyweight fabrics. There are two distinct types of knitting technologies that ultimately produce vastly different fabrics for specific applications (**weft** and **warp**). The method that you will be most familiar with is that used in hand knitting, which is termed weft knitting – one continuous end of yarn, which feeds through each consecutive loop from the left selvage to the right selvage (Figure 12.2). Generally, weft-knitted fabrics are flexible and will extend in all directions, have good elastic recovery, superb formability and drape, provide excellent thermal insulation and are resistant to creases. However, they suffer from poor shape retention, are prone to pilling and ladder easily. In contrast warp knitted structures are more stable but lack drape properties. Warp knitted structures are produced using multiple yarn ends and the loops intermesh diagonally with the adjacent vertical columns (Figure 12.2). The resultant fabrics are ladder resistant and find end-uses in a variety of areas including, lace, openwork, net, underwear, sportswear and technical applications.

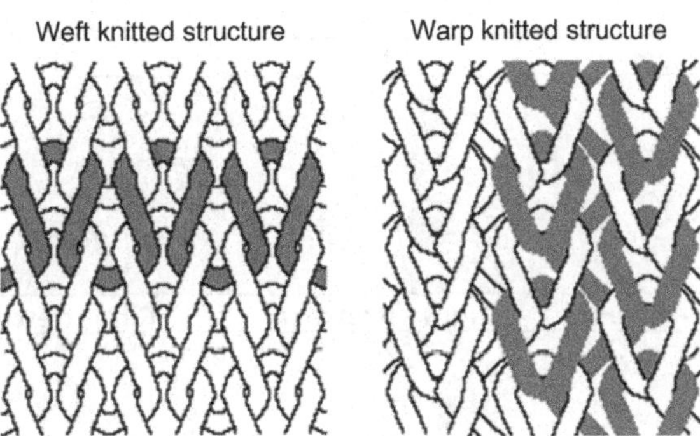

FIGURE 12.2

Weft and warp knitted structures.

12.4 WEFT-KNITTED STRUCTURES

The majority of knitted garments available on the high street are constructed from weft-knitted fabrics. This is by far the most versatile method of knitting, as the technology allows for a variety of structures to be produced that can combine extensive patterning in the form of texture and colour. This section identifies the three basic weft-knitted fashion structures (plain, rib and purl) illustrated in Figure 12.3. Plain structures are often referred to as **plain fabrics**, **single jersey** or **single bed** structures. This is the simplest structure since it is composed of a series of identical loops intermeshed together. On closer inspection, the fabric's visual appearance differs on each side of the fabric. The correct side (technical face) is the side where the loop legs can be seen clearly, while the reverse side (technical back) has a wave or rippled appearance. Plain structures are unstable at the edges (they curl backwards at the selvages, and forward at the top and bottom of the structure), and for this reason garments constructed from this structure are usually finished at the welt and cuff, by either incorporating a different structure such as rib or by hemming the fabric. One key factor of plain structure is that it will unravel from the top and bottom of the fabric and ladders easily. This structure generally has good elongation and elastic recovery and is the most drapable of the weft-knitted structures. It finds uses in a variety of applications including underwear, sportswear, casual wear (T-shirts, dresses, trousers and sweaters), fashion and classic knitwear.

Rib structures consist of **face loops** and **reverse loops** in the same course, which allows the fabric to collapse in the width direction and elongate slightly in the length (Figure 12.3). If the structure is a balanced rib (equal number of face and reverse loops repeating across the width of the structure) the structure will look identical on both sides of the fabric (hence there is no technical face). If the structure is unbalanced, it is usually designed for visual effect and it will be the designer's choice as to which is classified as the technical face. Traditionally this structure has been used extensively in cuffs, collars and welts due to its excellent elongation and recovery properties. It is a stable structure, so unlike plain structure it does not suffer from edge curling. Garments produced from this structure conform well to the body and are usually designed to be tight fitting. In a collapsed state, the structure is significantly

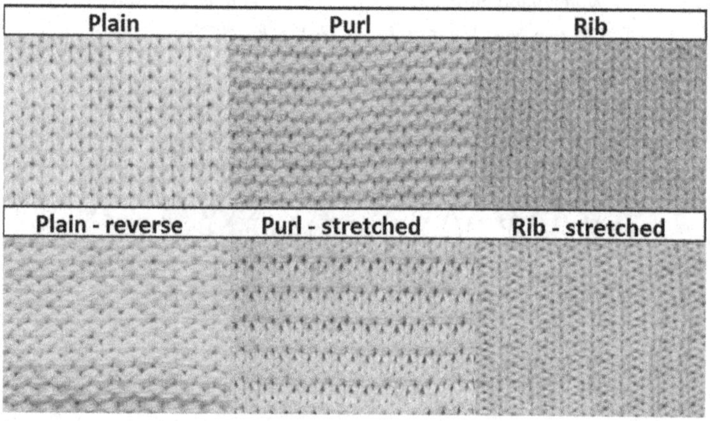

FIGURE 12.3

Weft-knitted structures.

thicker than plain structure and therefore has greater thermal insulation. In past years, anglers' outwear known as 'ganseys' would have been constructed utilising rib structures produced from chunky wools. The use of the Rib structure has become increasing important in fashion knitwear, especially during the late 1990s, and has enabled new garment shapes to be produced which fit like a second skin.

The final structure, purl, is used more as a decorative structure and tends to find uses in baby wear rather than fashion applications. Some knitting machine manufacturers refer to this structure as **links–links**. It is the opposite of rib: instead of every alternate wale being a face loop followed by a reverse loop, each course is constructed of identical stitches that alternate from front loops to back loops between the courses, producing a ripple effect (Figure 12.3) on both sides of the fabric (hence there is no technical face). Purl collapses in the length direction and elongates slightly in the width, which is a definite advantage in knitwear for infants and children, since they grow significantly in height during childhood. Although the structure collapses, it does not cling to the body like rib, and still retains reasonable drape properties. The structure is lofty, and when constructed from spun yarns it is incredibly soft. The major disadvantage of purl structures is the production cost; it is significantly more expensive to produce than rib or plain structures due to the continual transferring of wales between courses.

Unlike woven fabrics where there are many variations of structure types, the variations in weft-knitted fabrics occur when different stitches are introduced to change the surface texture or when colour is inserted to vary the visual effects. The options to change the texture of a knitted structure are limited to knit stitches, floats, tucks or movement of stitches to form Aran's and cable-type structures. Multiple coloured effects can be created by utilising two or more yarns in the same course to create a pattern or motif – coloured patterns in knitwear are described as stripes, Fair Isles, Jacquards or intarsias. David Spencer's (2001) book, *Knitwear Technology*, goes some way in explaining the functions of each.

12.4.1 WEFT-KNITTING MACHINES

Historically, there are three distinct types of weft-knitting machinery: circular, fully fashioned and flatbed. However, recent developments in technology have enabled a new type of flatbed machine to be manufactured specifically for the purpose of producing seamless garments; therefore, it could be argued that there are now four classifications. The most productive method of manufacturing weft-knitted fabrics is utilising the **circular knitting machine**. In this method, the knitting needles are arranged in a circular formation and can be fed from a variety of sources located around the circumference of the machine; hence more than one course is knitted in a single revolution. The fabrics obtained from this machine are generally continuous tubes that can be slit to produce an open width fabric. Single- and double-jersey fabrics utilised in T-shirts and sweatshirts are manufactured using this technology. One example of recent advances in circular technology include seamless Santoni technology. Santoni has produced small diameter machines that produce a tube of fabric to fit over the body contour. It is seamless in the sense that it has no side seams, but this machinery should not be confused with the flatbed complete garment process that can produce truly seamless garments (hence, a body with two integral sleeves).

Traditionally, high-class knitwear was produced on a special class of machinery termed the straight bar frame (or Cottons Patent Machines as they are more commonly referred to). The advantage of this machinery is that it can shape individual panels (front, back and sleeves). Traditionally, shaping was seen as essential for garments constructed from luxury yarns (cashmere, merino wool, lamb's wool and others) that were too expensive to produce using cut and sew manufacturing (Power, 2008).

The disadvantages of this machine type are that it knits at significantly slower speed than the circular knitting machine, and rib structures for cuffs and welts need to be produced on a separate machine (since it only contains one set of needles and not two).

The flatbed machine (or the rib machine) was developed to service the straight bar frame. However, as technology and computer software advanced this machine found a niche in the fashion markets. The modern flatbed machines are the most versatile of all weft-knitting technology. They are capable of producing ribs combined with complex structures, different patterning options, panel shaping and integral knitting (knitting pockets, collars and trims into the garment). The only disadvantage is speed, but this has improved significantly in recent years (the latest technology operates at 1.6 m/s). The majority of fashion knitwear (excluding T-shirts) on the high street is produced using flatbed machinery. The final knitting machine type could be classed as flatbed technology (since in reality it is) but its sophistication has earned it a classification of its own: complete garment knitting, or seamless. There are two major players in this area of innovation: the German company Stoll, with its Knit & Wear range of machinery, and the Japanese company Shima Seiki with its Wholegarment machinery range. These machines are designed to produce a complete garment; hence, the entire sweater or other garment produced by one of these machines is seamless and there are no post knitting operations to be completed after knitting (with the exception of labelling and sewing in the ends of yarn).

12.5 WARP KNITTED STRUCTURES

Warp knitted structures are constructed from intertwined loops, with the yarns that connect them crossing in a zigzag formation (Figure 12.2). A warp knitting machine consists of needles extending across the width of the machine like a weft machine, but each individual needle is fed from an independent yarn source. Hence, each needle is fed by its own yarn supply delivered by a guide that directs the yarn around the needle during the knitting action. It should be noted that all of the wales in one course (i.e. row) are formed simultaneously. At first glance, the appearance of a warp structure may appear similar to that of a weft, but on closer inspection one can observe that neighbouring loops of the same course are not created from the same end of yarn. In complex structures, it is quite common to have more than one guide per needle (the most common warp structure uses two sets). The yarns that feed through the guide bars are wrapped onto a warp beam (similar to the weaving process); a machine could consist of two to four beams depending on the fabric type to be obtained. The difference in the orientation of the yarn, fed from the weft to the warp direction, enables vastly different structures to be produced at extremely high speeds.

Warp knitted fabrics produced are continuous sheets of materials and usually produced from filament yarns which can be utilised in a variety of applications from industrial to fashion garments. The varieties of fabrics that can be produced are among the widest ranges of any textile manufacturing method. There is a wide range of machinery available in a variety of widths from small crochet and scarf making machines, to gigantic machinery (5 m wide) used to produce industrial fabrics. Whilst both warp and weft technologies are referred to as knitting, they have never really been in direct competition for market share, since the fabrics produced are so vastly different. Throughout history, warp knitting has remained the smaller sector, with specific niches, particularly in technical textiles. The mechanical properties of warp knitted fabrics are often similar to those of woven structures. However, the structure range is so diverse that a more cautious definition is that they combine the technological advantages of woven and weft-knitted fabrics. Fabrics can be produced that are extremely stable (like a woven structure) or very extensible like a weft-knitted structure – more often the range lies somewhere between the two. The two most common types of warp knitting machines are the tricot and raschel machines.

12.5.1 WARP KNITTING MACHINES

Tricot and raschel warp knitting machines have developed to find a niche in relation to the type of fabric that each produces. There are discrete differences in the configuration of the knitting elements that support distinctive structure types (see Spencer (2001) for more detail). The structures obtained from the tricot machine are generally of the plain type and find many applications, particularity in lingerie and sportswear; quite often, these incorporate elastane or Lycra to produce fabrics with two-way stretch. Raschel machines, by comparison, are useful for other structure types, since they can knit yarns in both filament and staple form into open works, laces, jacquards, fancywork (largest outlet) and pile fabrics. Structures produced from the raschel machine do not tend to stretch significantly and can be designed to be highly structural for technical applications. The most common warp knitted structures in clothing and fashion are those produced using the tricot type of machine. The simplest structure that is produced using one guide bar (Figure 12.2) is rarely used in any application because it is unstable (the loops incline to the direction of the yarn feed) when the knitting takedown is removed. The most common warp knitted structures to be produced are those of the plain tricot type knitted with two needle bars. Figure 12.4 shows one of the most common warp knitted structures, tricot or half jersey. However, in fashion applications locknit is by far the most popular option since it has good extensibility, cover, handle and excellent drape, and is flatter on the reverse of the fabric. In contrast, raschel machines can have one or two beds of needles (hence they can produce double structures). Many fabric types can be produced but the general structure categories can be divided into five types; openwork, inlaid yarns, double structures, pile structures and structures with spacers (there are exceptions to these structure types, but they tend to fall into the technical textile category that is outside the scope of this publication), Figure 12.5 illustrates the most popular fashion structures. The structures used in clothing applications tend to be split into three categories: those used in functional clothing such as bi-directional stretch fabrics; the supportive component in laminate structures and compression fabrics; and those which add aesthetic value such as laces, jacquards and fancy nets. The Italian design house Missoni continues to produce sophisticated warp knitwear designs (Black, 2002), and Karl Mayer, the German warp knitting manufacturer, remains a global leader in the development of warp knitting machines.

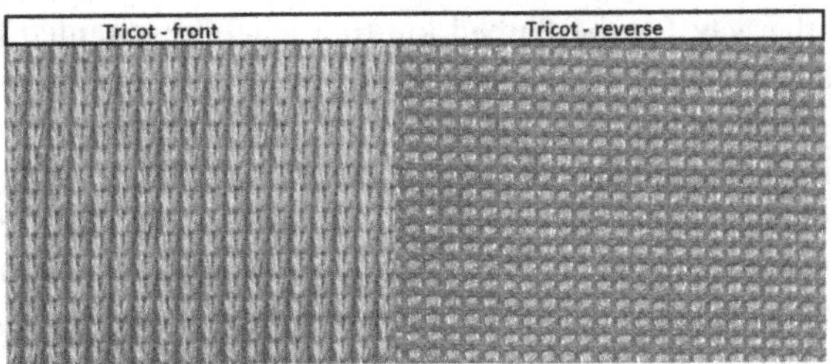

FIGURE 12.4

Tricot warp knitted structures (left) Tricot - front, (right) Tricot - reverse.

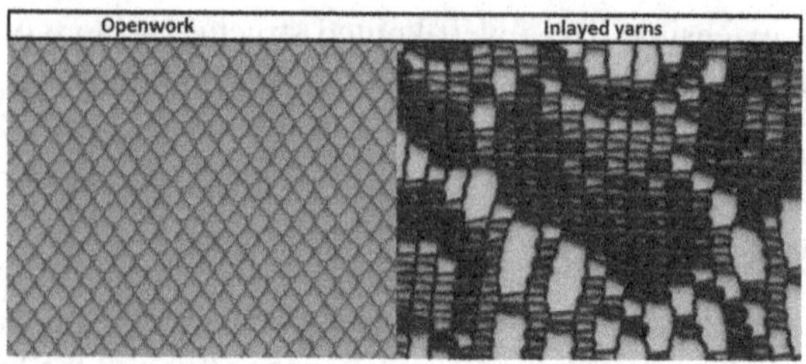

FIGURE 12.5

Raschel warp knitted structures (left) openwork, (right) inlayed yarns.

12.6 KNITTING DEVELOPMENTS

This chapter would not be complete without a brief mention of the key technological developments that enabled knitting to move from a cottage industry into the sophisticated global concern it has become. Since William Lee's pioneering invention of the weft-knitting machine in 1589, there has been much development in knitting technology. Initially, the mechanical knitting action was perfected so that machines were able to produce finer lightweight fabrics constructed from silk yarns. The next significant development came from Jedediah Strutt, who was responsible for the Derby Rib attachment (1759) more than 100 years after Lee's knitting frame was invented.

In the same century, alternative forms of constructing knitted structures were developed, notably the Crane and Porter warp knitting machine (1775) and Decroix's patent (1798) for the circular weft-knitting machine (Spencer, 2001). Mechanisation of the weft-knitting frame was achieved by Samuel Wise in 1769, and Dawson achieved the same for warp knitting in 1791. Once the techniques and technology options relating to knitting had been explored sufficiently, the quest was for productivity. During this era, there was significant change involving the transformation from cottage industry to factory production. The textile industry can be credited with providing the first examples of a factory system worldwide (Jones, 2002). During the nineteenth century, water and steam powered machinery was developed and productivity increased. UK hosiery and knitwear factories became the backdrop of the Midlands, just as spinning and weaving mills were associated with Lancashire.

The next major knitting development was in the modification of the actual knitting needle. Matthew Townsend patented (1849) a modification to the earlier developed latch needle (1806) that became extensively utilised in weft circular and warp knitting machines. This needle enabled each individual wale or loop to be produced sequentially, rather than all of the loops being formed simultaneously, which paved the way for the flatbed knitting machine patented by an American, Isaac Lamb (1865). Other notable developments around this time were the development of the straight bar frame (or Cottons Patent Machine), which had automatic panel shaping capability. It would be wrong to separate technical innovation from fashion, since changes in clothing trends from the latter part of the nineteenth century to the early twentieth century significantly changed the face of the knitwear industry and formed the basis of the modern knitwear and hosiery industries. Casual dressing became the vogue and knitwear was firmly established as a fashion item in the modern woman's wardrobe (Brackenbury, 1999). Post-1918 fashion gave birth to a new era for knitwear, with designers such as Chanel leading the way in cut-and-sew rayon jersey knitwear. After the war years, electronics emerged as the new force driving technology (Donofrio-Ferrezza & Hefferen, 2008).

Developments during the 1950s were in the synthetic fibres used to construct knitted fabrics and increasing the efficiency, quality and versatility of patterning. If only the magnitude of electronics innovation could have occurred in-line with that of the 1950s – who knows where knitted fashion would be by now (Power, 2007).

The 1970s–1980s were a key period for knitting technology. Protti launched the PDE flat knitting machine, which was the first machine that used electro/mechanical needle selection. Shima Seiki introduced the glove knitting machine around this period, and 1975 saw the first fully electronic flatbed machine, the Stoll ANV. The advantage of the flatbed electronic machine was that jacquard patterns previous produced using 'jacquard steels' were now produced using punched tape electronic control. The major factor that brought weft-knitted garments to the forefront of fashion was computer control of electronic needle selection on machines using flatbed technology. During this period, programming systems got simpler, stepper motors were introduced to control the amount of yarn in a loop and the cam boxes were refined. Large coloured motifs (jacquards) took advantage of this new individual needle control and were a popular feature in designer collections. Now that patterning was perfected on the flatbed machine, attention turned to the technique of shaping panels; up to now, although shaping could be achieved on some machines, it was not cost-effective due the cam box's inability to change direction halfway across a stroke. Stoll introduced the first CMS series in 1987, which had a variable stroke function (the cam box could change direction in the middle of a stroke). The machine also was capable of producing integral knitwear (knitting pockets and trims as part of the knitting process) and was able to knit 3D structures due to special holding-down sinkers. During the 1990s there was a significant change in knitwear design as a direct result of this technology: fully shaped panels were now being produced using less expensive yarns (acrylic) in a cost-effective way, and fashion knitwear now had the option of combining shaped panel features (reserved previously to the straight bar knitting technology) with extensive patterning and textural effects. This made flatbed weft-knitting technology extremely versatile and attractive to high street retailers and designers, who utilised the innovation in technology to develop new style lines that were not previously possible. During the mid-1990s, this resulted in the trend of figure-hugging stylish ribbed knitwear.

It was in 1995 when knitwear technology next came to the forefront of the textile world, as Shima Seiki introduced its Wholegarment range of machinery, and later Stoll introduced Knit & Wear. This technology was capable of knitting an entire sweater in one process, and hence there was no requirement to join the sleeves to the body, since this was completed as part of the knitting action. This technology was revolutionary at the time, but the industry was slow to realise the benefits. The earlier models of this machinery were considered too restrictive in terms of programming; however, in the last decade this technology has advanced significantly. There are examples of complete garments (seamless) right across the market, from designer and luxury ranges (John Smedley, Max Mara, DKNY Jeans, Versace, Burberry, and McQueen, to name a few) to high street and supermarkets (M&S, George, and Oasis). Production of complete garment knitting products is set to gain market share, particularly in sportswear and casual wear as the gauge range is extended to 18 gauge (needles per inch). Shima Seiki's most recent machine range is the Mach 2X and Mach 2S, which claims significantly higher productivity than earlier versions of Wholegarment, and is programmed through the SDS-ONE APEX design software. Various models of Stoll's CMS Knit & Wear machinery are available, and programmable using M1 Plus software. The advantages of producing using complete garment technology are elimination of waste, reduced labour (compared to other methods) and comfort (due to having no seams) in close-to-body-fitting garments.

12.7 THE IMPACT OF COMPUTERS IN DESIGN AND TECHNOLOGY

Knitwear dominated fashion in the late 1970s and early 1980s, with pioneering designers such as Kenzo, Rykiel, and Miyake (Black, 2002; Donofrio-Ferrezza & Hefferen, 2008) exploiting new technology and advanced programming capabilities to produce oversized motifs and asymmetrical designs that had not previously been possible in industrial knitting. Machine builders introduced dedicated programming languages for their range of machinery: Stoll devised its Sintral programming language, and soon after Shima Seiki introduced the Shimatronic design system.

Since the advent of electronics in the 1970s, knitting technology has advanced significantly, and needle selection is now controlled via computers running specialised software. Computer-aided design/computer-aided manufacturing (CAD/CAM) software packages are used to design complex patterns and stitch instructions and feed them to the knitting machine. Advancing technology resulted in the introduction of individual needle control, which changed the relationship between technology and fashion. In a modern electronic weft-knitting machine there can be as many as 1000 individual needles (depending on the needle gauge and the knitting machine width), which can be controlled individually using microchips.

The next major commercial development that significantly influenced knitwear design was the ability to knit short strokes, known as variable stroke or partial knitting. Although the understanding of this technique had been available for some time (Macqueen, 1960; Pfauti, 1960), computer control of individual needles, combined with better yarn control through stepper motors, enabled it to become a commercial reality. The first Stoll CMS knitting machines were introduced in 1987, followed in the early 1990s by the new SIRIX design package. During the 1990s, knitwear again took a dominant place in fashion, as the variable stroke function made it cost effective to produce shaped knitted panels that combined texture and colour in fashion yarns (acrylic). The fashion industry exploited this technology to produce skinny rib-fitted styles with pronounced fashioning marks around the armholes (Figure 12.6). This advancement made flatbed weft knitting the most versatile of all knitting technologies. Computer technology, certainly within flatbed knitting, has advanced significantly since the 1980s, and programming systems now offer a variety of user friendly features. However, there are two factors that have hampered the development and exploitation of the machinery: firstly the lack of

FIGURE 12.6

Skinny rib raglan shaped garment.

understanding by designers of machine capabilities, and secondly the need for skilled technicians to produce complex patterning techniques.

Since the turn of the millennium, there have been significant developments in computer graphics software that have resulted in increased design capabilities. Fast fashion has decreased the development time available to transform a 2D design into a 3D reality. The latest graphic software enables a design to be interpreted into a realistic view of a knitted pattern with accompanying technical information. Not only can software simulate complex stitches, but the most advanced design packages can express seamless garment designs in 3D, enabling design teams to see a true-to-life simulation of the garment prior to knitting. This virtual sampling provides a realistic photo simulation, which may provide many opportunities for design, marketing and buying teams worldwide.

12.8 QUALITY CONTROL

Before considering quality within the knitted structure, the reader must have some understanding of knitted fabric science. Pioneers in this area are Doyle (1953), Leaf (1958) and Munden (1959) – their work provided the underpinnings to later studies. It was acknowledged that a variety of factors influence loop dimension, such as the type of stitch, yarn construction, fibre type and friction during delivery. The one factor that could be used to determine and maintain quality (irrespective of all other variables) was the measurement of the loop length, or stitch length as it has come to be known. This measurement is obtained when a length of yarn used to form a knitted stitch is laid flat under a predetermined force. Of course, measuring the amount of yarn used in a single knit stitch is nearly impossible and open to a vast amount of human error. Therefore, in industry a set number of wales is specified relative to a larger standard measurement. Quite often, this figure is in line with the product the company will produce. An outerwear company producing adult sweaters will probably use 200 wales as the specified measurement, whereas a childrenswear manufacturer may stipulate 100 wales. Whatever length the manufacturer uses, the stitch length is still expressed as the length of yarn per knit stitch in either mm or inches. If more than one feeder is being used to deliver yarn (or in the case of modern weft knitting, more than one knitting system is used to knit), it is essential that all the yarns are tensioned appropriately, and that the cams are adjusted to ensure that each course, irrespective of the knitting machine, the carriage direction, the yarn feeder and the knitting system, is delivering a constant length of yarn per knit stitch. To ensure consistent quality, a sample of fabric must be obtained from the knitting machine, and each course should be unreeved and measured using the HATRA (Hosiery and Allied Trades Research Association) course length tester. This is a wall-mounted device with a series of wheels; at the start the yarn is secured with a spring clamp; the yarn then follows a defined path around the wheels; finally, a 2 g weight is hung at the end of the yarn, and a measurement is taken. Many modern weft-knitting machines claim to do this as part of the knitting process; however, it is prudent to cross-reference this with a manual measurement (BS EN 14970:2006). This is particularly important in weft flatbed knitting – in producing shaped panels, it is essential that yarn delivery is balanced (hence, the loop length is controlled throughout the panel).

The most important parameter after stitch length is stitch density. This is defined as the number of wales per cm or inch (w/cm or wpi) and the number of courses per cm or inch (c/cm or cpi); hence stitch density = wpi × cpi (or w/cm × c/cm). These two parameters enable the width and length of any garment panel to be calculated accurately, provided that the stitch densities are obtained from the fabric in its finished state. Other more complex knitting geometry relationships can be calculated from these figures, which provide acceptable values in terms of knitted fabric quality; however, they are becoming less frequently used in fashion knitwear due to the complexity of structures. Even so, the two quality factors that cannot be ignored are stitch length and stitch density.

12.9 CASE STUDY

Knitwear in recent years has gained a significant share of the global fashion market. This is accredited to the acceptability of casual dress (Jenkyn Jones, 2005; Power, 2008). If the properties that make knitted garments desirable are explored, many can be found:

- Good elastic recovery
- Excellent drape
- Good crease shedding
- Easy-care (dependent on yarn)
- Good thermal insulation
- Excellent formability
- Soft to touch
- Flexible
- Options for combining textures and patterns for visual or functional effect

When designing any product, including clothing intended for fashion applications, it is essential to consider user requirements in relation to functional and aesthetic values. If the user requirements listed are that a garment provides warmth, comfort and easy-care properties, knitted fabrics are a good choice. In particular, weft-knitted structures are desirable due to versatility in structure, good elastic recovery and drape. Of course, when considering a suitable fabric for a garment, careful selection of the raw material should be executed to ensure that the resultant material is fit for its intended purpose. Consideration should be given to fibre selection and the process by which the yarn is formed. If the knitted garment is intended to be a fashion garment with a short trend life, the cost of raw material should be significantly less. High-bulk acrylic would become a possible contender, since once the fibre is spun into a yarn with relatively low twist, it is easy to knit, and yarns are available in a wide spectrum of colours at a reasonable cost. Once the yarn is knitted, the resultant structure is soft, is flexible, has good stretch and recovery, is easy-care, sheds creases well and has a lofty fuzzy appearance similar to wool that will keep the wearer warm. If the fashion garment is to be tight fitting, a rib structure could be used in a fine gauge (perhaps 10 or 12 gauge); a skinny rib, such as a 2×2 or 2×1, would normally be the preferred option for this structure. However, if a rib appearance is desired without the clinginess, a mock rib would be used to prevent the structure from collapsing. When considering the technologies available to knit this structure from acrylic yarn, the selection would normally be flatbed technology, since this machine is versatile and offers the options of knitting both panels and fully shaped pieces, which has significant advantages in terms of aesthetic appeal.

Alternatively, if the end application were a modern sports base layer, the mid- and outer layers would need to be taken into consideration at the design stage. Since it is not desirable for moist clothing to be next to the skin, and muscles need to be sufficiently supported, fibres such as polyester or polypropylene would be desirable, due to their ability to move moisture and remain dry. Circular weft-knitted structures or warp knitted structures may be utilised, since they are available in ultrafine gauges, are particularly suited to the delivery of filament yarns and offer advanced properties of comfort, freedom of movement, breathability, easy-care, durability and supportive fit. In addition, they can be ultra-lightweight, and the structures can also be designed with texture to offer thermal insulation.

12.10 FUTURE TRENDS

Weft knitwear will continue to be prominent in fashion for the foreseeable future. The versatility of modern computer controlled flatbed machinery (including those able to produce complete knitted garments) remains unrivalled in terms of patterning and shaping capability. Globalisation has and will continue to have an impact on the knitwear industry, as the low-cost labour countries experience the technology revolution. Many luxury retailers and designers are looking towards advances in technology to provide new innovation and open-niche marketing opportunities. The most sophisticated knitwear being designed and developed today combines the benefits of advanced materials with the latest innovative technology on the market, leading to cutting edge designs. Opportunities within CAD/CAM are yet to be exploited to the same degree of utilisation as in the automotive and industrial design industries. But new developments in 3D graphic simulations are paving the way forward for innovative companies. These developments will assist in the transformation of the textile and clothing industries into high-tech, demand-driven and knowledge-based industries (Walter, Karsounis, & Carosio, 2009). The cycle of fashion has been shortened significantly from the traditional two seasons, and designers and retailers have many capsule collections with new products in stores being replenished every six weeks. Knitwear clothing and apparel is meeting positive consumer response, as technical advancements in raw materials and machinery are enabling manufacturers to respond quickly to consumer preferences in the production of cutting-edge fashions.

The retail environment of the future will look quite different from the establishments of today. Consumers will have been saturated by the fast fashion phenomenon and will be looking for customised alternatives. Individually designed garments that are tailored to specific consumer needs will be at the forefront of fashion. They are likely to contain 3D body scanning booths that will be linked to sophisticated 3D visualisation software. This will enable consumers to design and view garments on their own bodies prior to the products being produced. A knitwear manufacturing unit would be located on site that could produce the garment for collection on the same day. Already there has been research into this area with the Scan to Knit project (Dias, Cooke, Fernando, Jayawarna, Chaudhury, Mccollum, & Dix, 2002), which focused on custom-made compression garments, where a 3D scanner was used to capture the shape of the leg, which was then translated into a knitted program and fed directly to a fine gauge knitting machine. Other projects in this area include the 'Kit for fit' initiative that investigated customising a basic sweater to a specific individual's body shape, and Factory Boutique (Shima Seiki in Japan), a shop for on-demand production of customised whole garments, where customers co-designed garments in accordance with personal tastes (Black & Watkins, 2010; Peterson & Heikki, 2010). The preventing factors at the moment are the level of communication between different software and hardware providers, and the skilled personnel required to operate the systems, which are largely multidisciplinary in nature.

It can be predicted with reasonable confidence that the area of knitwear that will provide the highest growth during the next decade is sport and active apparel. Shishoo (2005) reported, from David Rigby Associates' 2002 market analysis, that the market in textile-related sports goods would increase from 841,000 tonnes in 1995 to 1,153,000 tonnes in 2005, and further growth was predicted. Certainly this was the case in the UK market – Mintel reported the market to be worth £5600 million in 2006 and predicted growth up to £5646 million by the close of 2010, despite the economic climate (Mintel, 2010).

The run-up to the Olympic Games in London 2012 stimulated growth in this sector as spectators and those that engage in leisure pursuits wanted a piece of the latest innovations. All of the foregoing factors provide significant opportunities for circular weft-knitted jersey products, including those obtained from the Santoni range of machinery; these include warp knitted fabrics used in compression applications, and flatbed technology where complete garment knitting will prevail as the leader due to the finer gauges that have become available. Polyester remains the most used fibre in knitted sports apparel, with new varieties of micro fibres and nano technologies offering garments having several desirable properties; they are ultra-lightweight, breathable, quick drying and supportive (i.e. they have even compression), and offer UV protection and excellent fit to body contours once knitted into a suitable structure.

The future described in the paragraphs earlier does not account for any changes in technology. What if there was a complete rethink regarding how loops were formed, and knitting needles became a technology of the past? Japanese engineers have developed rotors as a substitute for needles in forming loops, and this offers great potential for all knitting sectors. Currently it is being tried out on circular knitting machines; it is predicted that the resulting machines will be lighter and smaller, and use significantly less energy than even the most advanced technology currently on the market (Hirano, 2010). Recent T-shirt sales in the UK amounted to around £1248 million (Allwood, Laursen, Malvido de Rodríguez, & Bocken, 2006), so this technology is predicted to result in significant environmental savings. Knitted garments (T-shirts and pullovers) were reported to be among the top three best-selling clothing products in the UK in 2004 (Allwood et al., 2006). With the trend towards casual dress and sportswear continuing, knitted garments should be leading greener production, since there is no doubt that the future of clothing is energy-efficient reduced-waste production using ecofriendly self-cleaning biodegradable raw materials. The knitwear industry has proved throughout history that it is dynamic and eager to compete with other textile technologies. In the next century, knitwear will flourish, finding new innovative markets, and will be far removed from the industry as we see it today.

12.11 SUMMARY

- A knitted structure is made up of a series of intermeshing loops.
- The yarns used to construct knitted fabrics generally have less twist than those used in woven structures.
- There are two distinct types of industrial knitting technologies available, warp knitting and weft knitting.
- Weft knitting is the technique you will be most familiar with, since this is the process used by hand knitters.
- In commercial knitting a stitch is termed ‘a wale’ and a row is termed ‘a course’.
- Different knitted structures produce fabrics with dramatically different properties.
- The most versatile technology is that of the flatbed knitting machine, which is capable of producing a vast array of patterns and textures whilst simultaneously shaping panels.
- The latest flatbed weft-knitting technology, known as complete garment technology, should not be confused with the circular technology that is described as seamless. The former produces a garment in its entirety, whereas the latter produces a seamless tube (gussets, sleeves and straps need to be attached by auxiliary machinery).
- The advent of computer technology, in terms of hardware and software development, will continue to impact on design innovation of knitted products.

12.12 PROJECT IDEAS

Project 1 – Familiarising yourself with basic weft structures

Visit a department store and observe the knitwear collections. Identify the structure type – plain, rib or purl – and write a report of your findings, and compare with the trends for the season. Please refer to Section 12.4 for revision.

Project 2 – Identifying knitwear weights

Select five individual items of knitwear from your wardrobe. Take care to select samples from underwear, casual wear and outerwear. Categorise them in terms of the size of stitch, and place them under the headings of fine gauge, medium gauge or chunky gauge. Please refer to Section 12.3 for revision.

Project 3 – Calculating stitch density

Select an item of knitwear and count the number of wales in 10cm and the number of courses in 10cm. Calculate w/cm and c/cm, and use this to determine the stitch density. Please refer to Section 12.8 for revision.

Project 4 – Evaluating fitness for purpose

Select a knitted garment and identify the following:

1. what the primary use of the garment is,
2. the fibre composition (this will be displayed on the care label),
3. if it is warp or weft knitted, and
4. the structure type.

Discuss why the manufacturer selected the fibre, fabric and specific structure in relation to the garment's primary use. Evaluate whether the garment is fit for its intended purpose. Please refer to Section 12.9 for revision; you may also wish to revisit some of the other chapters in this publication for revision regarding fibre properties.

12.13 REVISION QUESTIONS

1. What is a knitted fabric?
2. List the properties of knitted fabrics.
3. Explain the basic principles of loop formation.
4. Briefly explain how you could identify a weft-knitted fabric from a warp-knitted fabric.
5. List some weft-knitted structures and suggest some possible end-uses.
6. On what types of industrial machinery is weft knitwear produced, and what are the advantages and disadvantages of each type?
7. List some warp-knitted structures and suggest possible end-uses.
8. Can you identify the two types of warp knitting machines on the market?
9. Detail the key developments in knitting technology and how they influenced fashion.
10. What are the latest innovations in CAD/CAM software for knitwear applications?

12.14 SOURCES OF FURTHER INFORMATION AND ADVICE

Most general textile publications cover some aspects of knitting technology. However, for a comprehensive read into detailed aspects of specific technology, David Spencer's publication *Knitting Technology*, currently in its third edition, is still the market leader. It provides detailed information regarding the fundamental principles of both weft and warp knitting and the complex aspects of knitting science. The breadth of this publication appeals to scholars and professionals alike, covering a wide range of applications in knitwear from clothing/fashion to advanced technical applications. In comparison, Brackenbury's (1992) publication focuses directly on knitwear. It provides a comprehensive overview of weft-knitted shaping technologies (with the exception of complete garment) and post-knitting construction methods. The reader will appreciate the many illustrations to assist with understanding what otherwise may be considered advanced processes. Power's (2008) contribution, *Advances in Apparel Production*, contextualized the basic weft-knitted structures into an easy reference chart. It also furthered Brackenbury's work by incorporating significant text relating directly to modern shaping techniques, including complete garment shaping.

Other notable publication that focuses on the design aspects of knitwear is Donofrio-Ferrezza and Hefferen's (2008) publication *Designing a Knitwear Collection.* This book provides an overview of key technological developments within the knitting industry, basic information relating to stitch geometry, details of industry specifications and a comprehensive overview of selected knitwear designers. Black's publications (2002, 2006) focus on knitwear's place as an item of contemporary fashion. The publications present knitwear as an art form – sculpture comparisons and many examples of contemporary designer knitwear are provided. Throughout the book, there are beautiful illustrations and specific reference to new technologies that have had an impact on fashion knitwear.

Jenkyn Jones' (2005) publication provides detailed information relating to employment within the knitwear industry. Despite focussing predominantly on the fashion industry, it is interesting that the text separates knitwear opportunities from those in mainstream womenswear, thus demonstrating that specific skills are required for designers intending to pursue a career in this sector of the fashion industry. Throughout the book, there are snippets of valuable information related specifically to knitwear, illustrating knitwear's growing significance in high fashion markets.

Publications written by historians should not be ignored in scholarly activity, since they provide valuable insights into the flexibility, versatility and resilience of the knitting sector as a whole. A few notable publications are: Wells (1972), Gulvin (1984), Chapman (2002) and Barty-King (2006). Finally, since technology is advancing at such an astronomical pace, journal and industrial publications offer the most up-to-date and comprehensive reads. Information regarding trend innovation, technological advancements and marketing briefings are readily published to keep the industry abreast of new developments. *Knitting International* magazine provides an archive of information related to the knitting technology evolution, including review of machinery (ITMA and IKME) and trend (Première Vision) exhibitions. In terms of marketing information, *Drapers* magazine provides a valuable source of knowledge and up-to-date information regarding current and predicted market trends in both the fashion and knitwear markets.

REFERENCES

Allwood, J. M., Laursen, S. L., Malvido de Rodríguez, S., & Bocken, N. M. P. (November 2006). *Well dressed?* Cambridge, UK: University of Cambridge.

Barty-King, H. (2006). *Pringle of Scotland and the Hawick story*. Norfolk: JJG.

Black, S. (2002). *Knitwear in fashion*. London: Thames and Hudson.

Black, S. (2006). *Fashioning fabrics*. London: Black Dog.

Black, S., & Watkins, P. (2010). *Considerate design for personalised knitwear – The knit for fit project*. The Textile Institute Centenary Conference. The Textile Institute. ISBN: 978-0-9566419-1-5.

Brackenbury, T. (1992). *Knitted clothing technology*. Oxford: Blackwell Science.

BS EN 14970: 2006. *Textiles – Knitted fabrics – Determination of stitch length and yarn linear density in weft knitting*. British Standards Institution.

Chapman, S. (2002). *Hosiery and knitwear: Four centuries of small-scale industry in Britain c. 1589–2000*. London, Oxford: Pasold Research Fund/Oxford University Press.

Dias, T., Cooke, W. D., Fernando, A., Jayawarna, D., Chaudhury, N. J., Mccollum, C. N., & Dix, F., 2002. Scan2Knit. In: World Congress - Knitting for the 21st Century, October 2002. UMIST Conference Centre Manchester UK, 7th – 8th Oct 2002.

Donofrio-Ferrezza, L., & Hefferen, M. (2008). *Designing a knitwear collection*. New York: Fairchild.

Doyle, P. J. (1953). Fundamental aspects of design of knitted fabrics. *Journal of the Textile Institute*, *44*(8), 561–578.

Gulvin, C. (1984). *The Scottish hosiery and knitwear industry*. Scotland: John Donald.

Hirano, H. (March 2010). Rotary knitting breakthrough. *Knitting International*, *116*(1377), 34–35.

Jenkyn Jones, S. (2005). *Fashion design* (2nd ed.). London: Laurence King.

Jones, R. M. (2002). *The apparel industry*. Blackwell: Oxford.

Leaf, G. A. V. (1958). A property of a buckled elastic rod. *British Journal of Applied Physics*, *9*(2), 71–72.

Macqueen K. G. (1960). Improvements in or relating to knitting processes and knitting machines, UK Patent Application 835270.

Mintel. (2010). *Sports goods retailing – UK – May 2010*. London, UK: Mintel.

Munden, D. L. (1959). The geometry and dimensional properties of plain-knit fabrics. *Journal of the Textile Institute*, *50*(7), 448–471.

Peterson, J., & Heikki, M. (2010). Mass customisation of knitted fashion garments, Factory Boutique Shima – a case study. *International Journal of Mass Customisation*, *3*(3), 247–258.

Pfauti E. (1960). Process of knitting, Swiss Patent Application 348499.

Power, J. (2007). Functional to fashionable: knitwear's evolution throughout the last century and into the millennium. *Journal of Textile Apparel Technology and Management*, *5*(4), 1–16.

Power, J. (2008). Developments in knitting technology. In C. Fairhurst (Ed.), *Advances in apparel production* (pp. 178–196). London: Woodhead.

Shishoo, R. (Ed.). (2005). *Textiles in sports*. London: Woodhead.

Spencer, D. J. (2001). *Knitting technology* (3rd ed.). London: Woodhead.

Walter, L., Karsounis, G., & Carosio, S. (Eds.). (2009). *Transforming clothing production into a demand-driven, knowledge-based, high-tech Industry*. London: Springer.

Wells, F. A. (1972). *The British hosiery and knitwear industry – Its history and organisation*. Newton Abbot, UK: David & Charles.

CHAPTER

FIBRE TO FABRIC: NONWOVEN FABRICS

13

N. Mao, S.J. Russell
University of Leeds, Leeds, UK

LEARNING OBJECTIVES

At the end of this chapter, you should be able to:

- Gain general knowledge of different nonwoven fabric formation processes and their main influences on fabric structure and properties
- Understand the characteristics of nonwoven fabric structure and properties and gain the ability to evaluate them
- Demonstrate understanding of the influence of the nonwoven fabric structure on the properties of nonwoven materials
- Understand the application of nonwoven fabric in various industrial sectors and their functional requirements

13.1 INTRODUCTION

Nonwoven fabrics are engineered fibrous assemblies that are essential to the functional performance of products used in diverse medical, consumer and industrial applications as part of daily life. Historically, the nonwovens industry has evolved from different sectors of the textile, paper and polymer industries, and today it has a separate and distinctive identity. The main market segments for nonwoven fabrics in terms of volume are hygiene, construction, wipes and filtration.

Although the word nonwoven is often used to describe fabrics that do not contain yarns, this is an over-simplification. The British Committee for Standards, International Organization for Standardization (ISO) and European Committee for Standardization (CEN) BS EN ISO 9092:2011 give the internationally accepted standard definition of nonwovens as follows:

> Nonwovens are structures of textile materials, such as fibres, continuous filaments, or chopped yarns of any nature or origin, that have been formed into webs by any means, and bonded together by any means, excluding the interlacing of yarns as in woven fabric, knitted fabric, laces, braided fabric or tufted fabric.

Textiles and Fashion. http://dx.doi.org/10.1016/B978-1-84569-931-4.00013-1

In 2010, EDANA (originally known as the European Disposables and Nonwovens Association) and INDA (The Association of the Nonwoven Fabrics Industry) proposed an alternative definition to the ISO as follows (http://www.edana.org/):

> A nonwoven is a sheet of fibres, continuous filaments or chopped yarns of any nature or origin, that have been formed into a web by any means, and bonded together by any means, with the exception of weaving or knitting. Felts obtained by wet milling are not nonwovens.
>
> Wetlaid webs are nonwovens provided they contain a minimum of 50% of man-made fibres or other fibres of nonvegetable origin with a length to diameter ratio equals or superior to 300, or a minimum of 30% of man-made fibres with a length to diameter ratio equals or superior to 600, and a maximum apparent density of 0.40 g/cm³.
>
> Composite structures are considered nonwovens provided their mass is constituted of at least 50% of nonwoven as per the above definitions, or if the nonwoven component plays a prevalent role.

A nonwoven fabric structure is different from some other textile structures in the following aspects:

1. It principally consists of individual fibres or layers of fibrous webs rather than yarns.
2. It is anisotropic both in terms of its structure and properties due to both fibre alignment (i.e. the fibre orientation distribution) and the arrangement of the bonding points in its structure.
3. It is usually not completely uniform in fabric weight, fabric thickness or both.
4. It is highly porous and permeable.

13.2 TECHNOLOGIES FOR THE FORMATION OF NONWOVEN FABRICS

Nonwoven fabrics are usually produced as sheet materials from fibres or filaments in two stages. Firstly, a web of fibres is formed (web formation), which is then treated to consolidate and increase its strength (bonding). Commercially, these are usually high-speed continuous processes delivering long lengths of fabric, may be many metres wide and are wound into rolls (roll goods). Subsequently, these roll goods are mechanically and/or chemically processed as they are 'converted' into many different industrial, consumer or medical products (converted goods).

A summary of the main nonwoven manufacturing processes for web formation and bonding is given in Tables 13.1 and 13.2. The web formation process, bonding methods and fabric finishing process have significant influence on the structure and properties of nonwoven fabrics.

Table 13.1 Web formation technologies

Drylaid	Spunmelt	Wetlaid	Others
• Carding (including carding and cross-lapping and perpendicular-laid webs, e.g. Struto, V-lap) • Airlaid	• Spunbond (S) • Meltblown (M) • S and M composites (e.g. SM, SMS, SMMS, SSMMS, SSMMSS) • Split and fibrillated films		• Electrospun • Centrifugally spun (including force spinning) • Self-assembled fibrous networks

Table 13.2 Web bonding technologies

Chemical bonding	Mechanical bonding	Thermal bonding	Others
• Saturation bonding • Spray bonding • Foam bonding • Print bonding	• Needlepunching • Hydroentangling (spunlace) • Stitchbonding	• Calender bonding • Oven bonding (through-air) • Ultrasonic bonding	• Combination bonding (involving two or more bonding processes) • Solvent bonding • Powder bonding

13.2.1 FIBROUS WEB FORMATION

Fibrous webs can be formed using various techniques including drylaid, wetlaid and spunmelt technologies.

Drylaid webs are usually formed from either staple fibres of 20–150 mm in length or short cut and pulp fibres of 1–10 mm in length. These materials may be of natural or synthetic polymer composition and can be processed alone or in blends. Carded webs are produced from either short staple fibres (20–60 mm) or long staple fibres (50–150 mm). Before carding, the fibres are mechanically disentangled to some extent and blended (if required). Carding produces a thin fibrous web, which is collected in open-width and deposited onto a continuous conveyor. This web may be transported directly to the bonding stage (parallel-laid), layered from side to side across its width (cross-laid) or corrugated such that the corrugations are oriented across the width of the web (perpendicular-laid). The terms in parentheses are used to describe, in very general terms, the nature of the preferential fibre orientation in the webs after each process. In contrast to carding processes, airlaid webs are produced by suspending relatively short fibres in air and then transporting this air–fibre mixture to a continuous air permeable conveyor, where the air is removed and the fibres are deposited to form a web.

Wetlaid webs are produced using technology that originated from the papermaking process, in that fibres are first suspended in water. There are four steps in this type of manufacturing process:

1. The fibres are dispersed in water to form a fibre suspension.
2. The fibre suspension is collected on a continuous screen to form a uniform web.
3. The web is drained and filtered to remove the water.
4. The web is heated to dry and bonded to form a sheet of wetlaid nonwovens (or paper).

Whereas paper is normally produced from short, fine fibres of cellulosic composition, the fibres in wetlaid nonwovens can be substantially longer and can be composed of many different natural, high-performance synthetic or inorganic materials.

Spunmelt (Butler, 1999) webs are formed from continuous filaments that are produced by extrusion processes derived from melt spinning. Commercially, most spunmelt nonwoven production is based on spunbond and meltblown technologies. Instead of being formed into tows or yarns as in conventional fibre spinning, fibrous webs are formed by the collection of the filaments on a conveyor belt. The spunbond production process (Rupp, 2008; Lim, 2010) is fully integrated, enabling polymer chip to be made into fabric in one continuous operation. The basic operational steps are presented in the following flowchart (the output of each stage is in parentheses):

Polymer melting (*Molten polymer*)→Filtering and extrusion (*Molten polymer*)→Quenching and drawing (*Filament*)→Laydown on forming screen (*Web*)→Bonding (*Nonwoven fabric*)→Roll-up (*Nonwoven fabric*)

The main difference between the spunbond and meltdown processes is that the meltblown process involves attenuation of the filaments using high-velocity hot air streams that impinge on the extruded filaments as they are emerged from the extrusion nozzles (Kevin McNally, 1998). The structure and properties of spunbond and meltblown fabrics are influenced by polymer properties, molecular weight and molecular weight distribution, as well as process settings such as production rate, die temperature, airflow and quenching conditions. There are distinct differences in the structure and properties of spunbond and meltblown webs, which result from the way in which each process is designed to operate. Meltblown fabrics consist of smaller diameter filaments, including submicron filaments and have superior filtration properties, while conventional spunbond fabrics contain coarser fibres and have much greater tensile strength. There is also a large variation in the diameter of filaments in meltblown webs as compared to spunbond.

In practice, spunbond and meltblown technologies are frequently used together in a continuous process in the production of meltblown–spunbond multilayer fabrics (e.g. Spunbond-Meltblown (SM), Sponbond-Meltblown-Spunbond (SMS). Such fabrics are important in applications that include protective clothing such as surgical gowns and hygiene products such as baby diapers. Bicomponent melt extrusion technology involves the production of continuous filaments consisting of more than one polymer type arranged in different configurations within the filament cross-section. Common configurations are side-by-side, core-sheath, segmented pie and islands in the sea. The last two configurations are utilised to produce microfibrous fabrics by splitting or fibrillating the filaments in the web after it has been extruded. This may be done as the web is being bonded or afterwards. Submicron filaments are released from the parent filaments thereby rendering the fabric softer, suede-like and more opaque. Alternatively, in the case of islands-in-the-sea bicomponent filaments, one of the polymer components in the fabric can be dissolved out to leave behind very fine sub-micron filaments.

13.2.1.1 Other web-forming technologies

Similar to the approach used for spunmelt nonwovens, a variety of extrusion processes are utilised to produce spunlaid webs based on, for example, wet spinning, gel spinning and flash spinning. Of these, electrospinning (electrostatic spinning) and centrifugal spinning technologies enable the production of webs containing submicron filaments. Fibres less than 100 nm in diameter achieve size-dependent properties and functions, or quantum effects (http://www.nano.gov/nanotech-101/special,accessed 30/09/2014). In addition, a nonwoven production line might include more than one method of web formation, such that a multilayer web is constructed. One example is spunbond–pulp–spunbond fabrics that are produced by first laying down a spunbond (S) web, followed by an airlaid web containing pulp (P), and finally another spunbond web (S). This SPS web is then bonded to produce a final, three-layer fabric. Other web combinations can easily be anticipated based on the properties that are required in the final fabric.

13.2.2 WEB BONDING TECHNOLOGIES

Bonding to produce a serviceable fabric follows web formation. The main methods may be summarised as follows:

1. **Mechanical bonding**: The fibres in the webs are physically bonded together by entangling, entwining and displacing fibres relative to each other (e.g. needlepunching, hydroentanglement) or by displacing and stitching the fibres or filaments (e.g. stitchbonding).
2. **Thermal bonding**: The fibres are bonded together by means of thermally fusing part of the fibres/filament surfaces (e.g. through-air bonding, calender bonding and ultrasonic bonding) or by the addition of heat-sensitive powders.

3. **Chemical bonding**: The fibres are bonded together by means of chemical adhesives applied to the web in the form of liquid dispersions, polymer solutions, powders and particles.

13.2.2.1 Mechanical bonding

Needlepunching

The fibres in most nonwoven webs produced by processes such as carding, carding and cross-lapping, and spunmelt processes are arranged in a planar fashion within the structure – that is very few are oriented in the thickness direction. In the needling process, barbed needles are repeatedly oscillated through the web to reorient groups of fibre (or more precisely segments of fibre) into the interior of the structure. As the needle penetrates the web, its ability to trap and transport fibres depends on factors such as the fibre diameter relative to the barb depth of the needles, frictional properties, the number of barbs on the needle (usually three barbs are available on each of the three sides of the needle) and the number of barbs that penetrate the web on each oscillation. The degree of bonding is strongly affected by the number of needles that penetrate the web per unit area.

The appearance of the needled fabric is affected by needle diameter and the fraction of fibres that are reoriented into the thickness of the fabric. It is also affected by needle selection and the design of the plate that is used to support the web during needling. For bonding purposes, a perforated metal bedplate is used to support the web as it is needled, which allows the needles to pass through as they are oscillated. 'Structured' needlepunched fabrics are produced by replacing the bedplate with either a brush surface or metal strips known as lamellae plates. When used in combination with needles that have one large barb located at or near to the needle tip, it is possible to produce a range of complex effects and patterns on the fabric surface such as velours (using the brush surface), ribbed, loop and pile effects (using the lamellae plate support surface). Structured needlepunched fabrics are commonly encountered as floorcoverings and mats in buildings as well as in automotive interiors.

Hydroentanglement

Hydroentanglement (also known as spunlace) refers to the process of bonding a fibrous web by entangling fibres using high-velocity water jets. During hydroentangling, the web is supported on a moving conveyor as the jets delivered from multiple injectors (manifolds) continually impinge upon it. Fibre segments in the web are entwined with others, displaced and reoriented during hydroentangling to increase frictional resistance to slippage and strength of the fabric. Therefore, to promote an energy efficient process, the fibres need to be sufficiently flexible when wet so that they can be easily deformed and rearranged into entanglements. As with all bonding processes, there is enormous scope to modify fabric structure and properties by altering process conditions. The design of the support surface is particularly important, as it facilitates the introduction of structural patterns, apertures, complex three-dimensional relief effects and combinations thereof. Additionally, hydroentangling can provide a convenient means of joining one or more webs without the need for thermal or chemical bonding.

Stitchbonding

Stitchbonding (Cotterill, 1975) is a method of forming nonwoven fabrics by stitching or knitting the fibres in fibrous webs together. In practice, it frequently involves warp knitting of the webs, either with or without additional yarns such that the fabric has the superficial appearance of a knitted fabric. Stitch-bonded fabrics can be formed using a variety of stitching methods and patterns such that flat, looped and pile structures can be formed on the surface. Examples of stitchbond fabrics include those produced by Maliwatt, Malivlies, Malimo, Malipol, Voltex, Kunit and Multiknit technologies.

13.2.2.2 Thermal bonding

Thermal bonding is only suitable for fibrous webs containing fibres (or particles) made from thermoplastic polymers such as polyethylene, polypropylene, polyester and polyamide – however, this does not mean that the entire web must comprise thermoplastic material. Web fibres are bonded by thermally fusing the thermoplastic components to the surrounding fibrous material. The thermal energy can be transferred to the fibre by various means to control the structure and properties of the thermally bonded fabric. Common approaches include hot air delivered via an oven system (through-air bonding) and direct contact via heated rollers (calender bonding). By engraving the rollers to create raised and recessed areas, it is possible to bond the web only at certain 'points' such that much of the fibrous structure is retained after bonding.

In ultrasonic bonding, fibre fusing is achieved by the conversion of acoustic energy into heat at ultrasonic wave frequencies higher than 20,000 Hz (Mao & Goswami, 1999), because of mechanical vibration of polymer molecules in the thermoplastic material. Other thermal bonding technologies include microwave, infrared and laser heating.

13.2.2.3 Chemical bonding

In chemical bonding, binders consisting of polymer latex adhesives in the form of emulsions, dispersions or solutions are deposited on to fibre surfaces in the web and then dried and cured to form a cross-linked film that bonds adjacent fibres together.

The binders (Williams, 1999) are a formulation containing the polymerised monomers and initiators stabilised in water by the use of surfactants. Binders must meet the specific requirements of the manufacturing process, so that they can be easily and economically applied, and of the final nonwoven product, since the binder can make up a large proportion of the final fabric. Important parameters include viscosity (influences ease of application in the manufacturing process), molecular weight (influences fabric toughness and handle), binder variety and versatility (influences important fabric properties such as hydrophobicity/hydrophilicity, softness, elasticity, flame-retardancy and antimicrobial performance) and binder cost (which influences product economics). The most widely used polymerised monomers in latex binders for nonwovens include various types of acrylic polymer, styrene acrylate, vinyl acetate, vinyl acrylic, ethylene vinyl acetate, styrene butadiene rubber, polyvinyl chloride and ethylene vinyl chloride. Additionally, some nonwovens are solvent or bonded, wherein fibres are treated with a solvent for the constituent polymer, frequently in gaseous form to produce autogenous bonding in the web. One example is solvent bonding of polyamide fibres using gaseous hydrogen chloride (http://www.cerex.com/PageView.asp?PageType=R&edit_id=172; Hoyle, 1989).

The fibrous web can be saturated with the binder, sprayed, printed or coated onto the web. In spray bonding, the binder usually stays on the surface of the web to form a nonwoven fabric with relatively high bulk but low tensile strength. In saturation bonding, a large amount of binder is introduced to the web, which forms an extensive adhesive matrix between fibres after the curing process, leading to relatively high fabric rigidity and stiffness.

13.2.3 NONWOVEN FABRIC FINISHING AND CONVERTING TECHNIQUES

Following nonwoven roll-good manufacture, and depending on the final product application, fabrics may undergo mechanical or chemical finishing treatments to modify properties. Such finishing processes may be applied immediately after roll-good manufacture or may form part of subsequent processes that 'convert' the roll-good fabric into final products. Such converting operations are often carried out by specialist companies that perform automated processes such as cutting, slitting, folding, layering and general mechanical manipulation of fabrics, as well as chemical impregnation, spraying, heat treatment, laminating, printing and many other operations, depending on the nature of the product. The converters produce final packaged products at very high speed that incorporate one or more nonwoven fabrics, as well as other components such as films and plastics in other forms. Examples of converted products include baby diapers, packs of wipes, filter modules, packaged wound dressings and other articles ready to be shipped or distributed.

Many of the functional performance characteristics of nonwoven fabrics, such as fabric softness, liquid absorbency, moisture transport, water/oil repellency and flame-retardancy, as well as aesthetic properties like fabric drape, conformability, appearance, colour, patterns and surface texture, are either enhanced or dependent upon chemical or biochemical (including enzymatic) treatment processes. As an alternative to wet chemistry, other means of modifying fabric properties include mechanical, thermal, acoustic, laser and radiation methods. Gas plasma treatment of fabrics is undertaken to modify the hydrophilicity, ultrahydrophobicity or oleophobicity of fibres as required. Additionally, plasma can assist in development of antimicrobial properties and is useful in surface sterilisation.

13.2.3.1 Dry finishing

Perforation and slitting of fabrics can be used to improve their drape. Another mechanical finishing process for nonwoven and paper fabrics involves microcorrugation of the structure to improve stretch and other properties. MICREX (http://www.micrex.com) is a microcreping process often used to improve the softness and surface appearance of fabrics. In this process, the nonwoven fabric is overfed, by a rotating conveyor roller with screw-shaped grooves in the surface, into compaction relaxation zones to form fine or coarse corrugations in the nonwoven fabric surface without significant impairment of fabric mechanical properties.

Calendering, pressing and hot embossing (Gunter & Perkins, 2007) are also frequently used to modify the surface characteristics of the fabrics by modifying the smoothness or by introducing patterns. Such processes usually operate under pressure.

13.2.3.2 Wet finishing

Finishing of nonwovens in the wet state involves treating fabrics with liquid formulations including aqueous solutions or dispersions. Wet processing includes (1) applying chemicals to the fabric; (2) fixing the chemicals to the fibres; (3) scouring and washing to remove unbonded chemicals; and (4) drying and curing (Wadsworth, Kamath, Dahiya, & Hegde, 1999).

In addition to substantial water consumption and the burden of wastewater treatment, most wet finishing treatments also involve thermal processes to dry and cross-link the applied chemistry on the fabric. For the majority of nonwoven fabric applications in the industry, there is currently no requirement for dyeing processes because of the nature of final products. Traditionally, some nonwoven fabrics such as shoe linings are dyed using methods similar to those used by the textile industry, but in many applications, fabric colouration relies on mass colouration and printing techniques including digital jet techniques.

Some of the basic wet finishing processes will now be considered.

Washing is probably the simplest wet finishing technique and is used to remove unwanted substances from fabrics and to soften them.

Chemical Finishing involves applying chemicals to the fabric that may be dissolved or suspended in a liquid medium such as water. Frequently used chemical finishing agents are listed below.

1. **Antistatic agents ('antistats')**: These chemicals can be either durable or nondurable once applied to the fabric. The selection is partly dependent on the intended life of the product. Examples of durable antistats include vapour deposited metals, conductive carbon or metallic particles applied in conjunction with binders, polyamines, polyethoxylated amine, ammonium salts and carboxylic salts. Examples of nondurable antistats include cationic (quaternary ammonium salts, imidazoles and fatty amides), anionic (phosphates, phosphate esters, sulfonates, sulfates and phosphonates) and nonionic antistats (glycols, ethoxylated fatty acids, ethoxylated fatty alcohols and sorbitan fatty acid esters).
2. **Antimicrobials**: The prevention and control of the growth of bacteria, fungi, algae and viruses on fabrics is important in many applications, particularly those intended for use in medical and hygiene products, where infection control is important. Antimicrobials may be durable or leachable. Industrially utilised antimicrobials include alcohols such as isopropanol or propylene glycol, halogens such as chlorine, hypochlorite, iodine, N-chloramine and hexachlorophene, polyhexamethylene biguanide, metals such as silver nitrate, mercuric chloride and tin chloride, various peroxides, phenols quaternary ammonium compounds, pine oil derivatives, aldehydes and phosphoric acid esters.
3. **Water repellents**: Chemical finishes can be applied to fabrics to increase the water contact angle on the fabric surface such that it is greater than 90° and to give the lowest possible critical surface tension. Water repellent agents usually include substances that have low surface tension such as waxes, wax dispersions, melamine wax extenders, chrome complexes, silicones and fluorochemicals.

Other finishes that are routinely applied to nonwoven fabrics for certain applications include UV absorbers, polymer stabilizers, flame retardants, softeners, re-wetters, fragrancies, adhesive binders, cross-linking agents, soil release chemicals, whiteners and fluorescent agents. Note that solid particles such as microcapsules, abrasives and fillers may also be applied to fabrics in conjunction with binders that adhere them to fibre surfaces.

13.2.4 COATING AND LAMINATING

Coating is a process to form a layer of material onto the fibre surfaces in a fabric, with various methods including padding, roller coating, knife coating, hot melt coating, slot die coating, foam coating, printing, spray coating, sol–gel coating, powder coating and vapour deposition.

Frequently used coating materials are polymers applied in the form of liquids that include solvent-based and water-based coatings. Common polymers include silicone rubber, polyvinyl Chloride (PVC), Poly (ethylene terephthalate) (PET), Polyterafluoroethylene (PTFE), Polyurethane (PUR or PU) and other elastomers.

Laminating is the process of permanently joining two or more layers of sheet materials with nonwoven fabrics to form composite fabrics. Composite nonwoven fabrics can be bonded together by adhesion, cohesion or friction. Commonly, chemical binders or thermoplastic binders are utilised in laminating processes.

13.3 CHARACTERISTICS OF NONWOVEN FABRIC STRUCTURE AND PROPERTIES

Although nonwoven fabrics share some compositional characteristics with textiles, paper and plastics, the fabric structures produced are particularly diverse and may be manipulated to obtain specific functionalities and performance characteristics. The structure and properties of a nonwoven fabric are determined by fibre properties, the type of bonding elements, the bonding interfaces between the fibres and binder elements (if present) and the fabric structural architecture. Examples of dimensional and structural parameters are as follows:

1. **Fibre dimensions and properties**: (a) dimensions – fibre diameter, diameter variation (e.g. in meltblown microfibre and electrospun nanofibre webs), morphology and cross-sectional shape, crimp, wave frequency and amplitude, length and density; and (b) fibre properties – mechanical properties (Young's modulus, elasticity, tenacity, bending and torsion rigidity, compression and friction coefficient), fibrillation propensity, surface chemistry and wetting angle
2. **Fibre alignment**: fibre orientation distribution
3. **Fabric dimensions and variation**: dimensions (length, width, thickness and weight per unit area), dimension stability, density and thickness uniformity
4. **Structural properties of bond points**: bonding type, shape, size, bonding area, bonding density, bond strength, bond point distribution, geometrical arrangement, the degree of liberty of fibre movement within the bonding points, interface properties between binder and fibre; surface properties of bond points
5. **Pore structural parameters**: fabric porosity, pore size, pore size distribution, pore shape and tortuosity

The physical, chemical and mechanical properties of nonwovens that govern their suitability for use depend on the properties of the composition and the fabric. The composition in this context refers to the fibres as well as any chemical binders, fillers and finishes present on, between or within the fabric's fibres.

This section focuses on the characterisation of nonwoven materials with reference to the influence of fabric structure on the physical and mechanical properties of fabrics. Some of the standard and recommended testing methods used to determine the structure and properties of nonwovens, as well as models designed to describe the relationship between a fabric's structure and some important nonwoven properties, are introduced.

13.3.1 CHARACTERISATION OF FABRIC BOND STRUCTURE

Nonwoven fabrics are held together by bond structures that can be characterised by type, shape, rigidity, size and density. Bond points can be grouped into two categories – rigid, solid bonds, and flexible, elastic joints– the prevalence of which depends on the choice of manufacturing process. The bond points in a mechanically bonded fabric (e.g. needlepunched and hydroentangled) are formed by the interlocking of individual fibres so that there is frictional resistance to separation. These bonds are flexible, and the component fibres are able to slip or move within the bonding points. By contrast, the bonds in thermally bonded and chemically bonded fabrics are formed by adhesion or cohesion between polymer surfaces, in which a small portion of the fibrous network is firmly bonded, and the fibres have little freedom to move within the bond points.

The bond points in thermoplastic spunbond and through-air bonded fabrics are formed by melting polymer surfaces to produce bonding at fibre crossover points, and the fibres associated with these bonds cannot move individually. In meltblown fabrics, the fibres are usually not as well bonded as in spunbonded fabrics, and in some applications, the large surface area is sufficient to give the web acceptable strength without the need for thermal, chemical or mechanical bonding. Stitch-bonded fabrics are stabilised by knitting through the web and the bonding points (fibres or yarns) are flexible but connected together.

The size of the bond points is influenced by fabric manufacturing parameters such as: the size of needle barb depth in relation to the fibre diameter, punch density and number of barbs that penetrate the batt on the downstroke (needlepunching); water jet diameter, specific energy and number of injectors (hydroentanglement); the land area and bond point area, pressure and the size of adhesive particles (thermal bonding); the method of binder application – for example, full saturation, spray or printing; and binder viscosity (chemical bonding).

The rigidity of solid bond points in most nonwoven materials can be physically characterised in terms of the measured tensile properties – for example, strength and elasticity – while the degree of bonding may be directly determined by microscopic analysis of the fabric cross-section. In mechanically bonded fabrics – specifically, needlepunched and hydroentangled fabric – the depth of bent fibre loops in the bonding points can be determined, and based on the depth of these fibre segment loops, a simple estimate of bonding intensity can be derived (Mao & Russell, 2006).

13.3.1.1 Needlepunched fabrics

Needlepunched fabrics have characteristic periodicities in their structural architecture that result from the interaction of fibres with needle barbs. Fibre segments are reoriented and migrated from the surface of the web towards the interior of the fabric, forming pillars of fibre oriented approximately perpendicular to the plane. On the fabric surface, needle marking is frequently visible and is due to the series of punch hole locations that may be joined by reoriented fibres in the fabric plane running in the machine direction (Hearle & Sultan, 1968) (see Figure 13.1(a) and (b)).

On a microstructural scale, needlepunched fabrics consist of at least two different regions. The first, between the impact areas associated with the needle marks, is not directly disturbed by the needles and retains a structure similar to the original unbonded web. The second region, the needle-marked area, contains fibre segments that are oriented out of the fabric plane. Some fibres are realigned in the machine direction. This rearrangement of fibre segments induced by the process effectively increases the structural anisotropy as compared with the original web, and therefore the structure of needlepunched fabrics is not homogeneous.

Both the number of needle marks and the depth of fibre penetration are related to the fabric bonding quality and fabric tensile strength. The shape and number of the holes depends mainly on the number of needles in the needle board, the size of the needles and the needle throat depth, fibre type and dimensions relative to barb dimensions, advance per stroke and punch density.

The depth of needle penetration, number of barbs that pass through the web and distance each barb and its attached fibres travel are important variables influencing the microstructure. Previously, the effects of changes in penetration depth and the number of barbs (Hearle & Sultan, 1967, 1968; Krcma, 1972) on the fabric structure have been investigated. These experiments have demonstrated that fabric strength is influenced by any changes in barb position as the needle passes through the web. Maximum fabric tenacity for a given web may be obtained with only three barbs per apex, if the depth of penetration is adjusted accordingly.

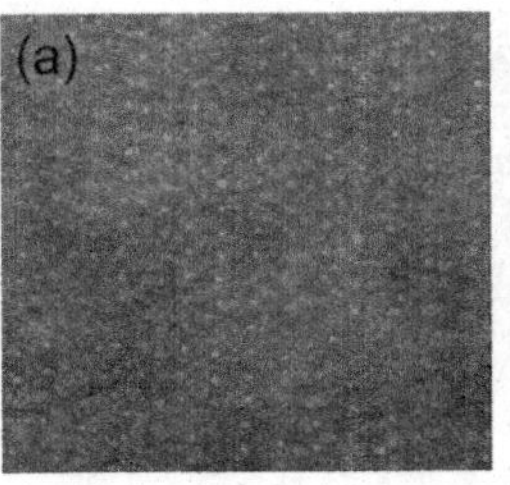

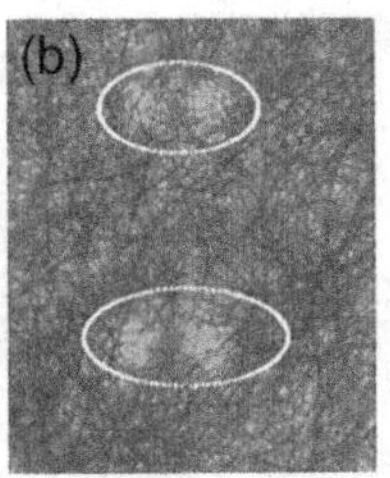

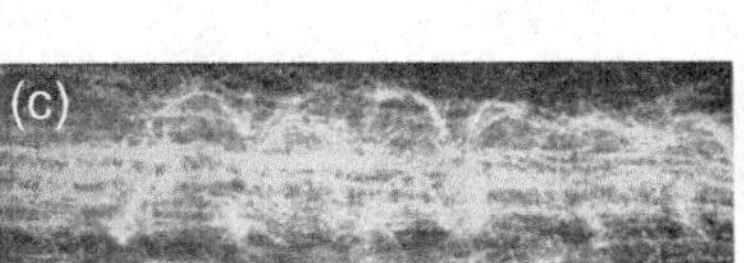

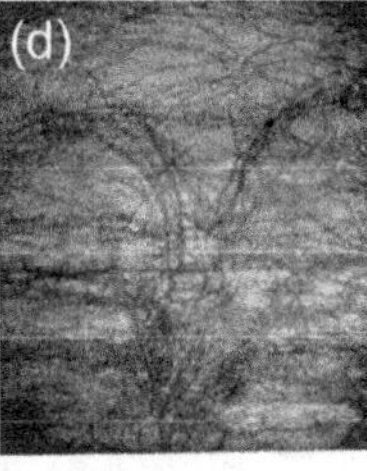

FIGURE 13.1

Structure of needlepunched fabrics: (a) appearance of needle marks in the surface of a needlepunched fabric; (b) two pairs of needle markings (holes) in the fabric surface; (c) cross-section of a needle mark (Hearle & Sultan, 1968); (d) pillar structure in the cross-section.

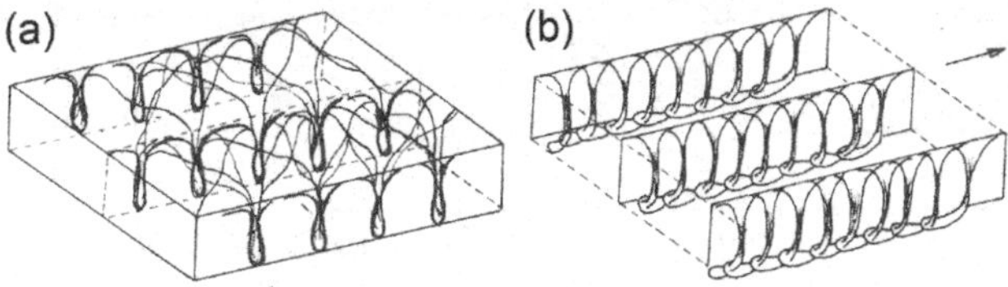

FIGURE 13.2

Needled fabric structures and the influence of process conditions (Hearle & Sultan, 1968): (a) low level of needling density and low needle penetration; (b) high level of needling density and high needle penetration.

Needlepunched fabrics have some fibre segments aligned in the transverse direction (Hearle & Sultan, 1967) (see Figure 13.1(c) and (d)), although the majority remain aligned in-plane and the fabrics have a greater porosity and a larger number of curved interconnected pore channels than woven fabrics. Hearle and Sultan (1968) observed that the punched loops of fibre do not protrude from the lower surface of the fabric when the needle penetration is small; the resulting fabric appearance and needle marks are illustrated in Figure 13.2(a). A pseudo-knitted appearance resulting from linked loops of fibre tufts produced by the needle barbs can be detected on the fabric surface when the needle penetration is large, as illustrated in Figure 13.2(b).

13.3.1.2 Hydroentangled fabrics

The microstructure of hydroentangled fabrics is quite different from that of needlepunched fabrics, in that the formation of discrete pillars of fibre in the fabric cross-section is absent. However, the incident high-velocity water jets locally migrate fibre segments, in both the transverse and the in-plane machine directions – some fibre segments exhibit substantial curvature after the process (see Figure 13.3). The number of fibre segments that is deformed in this way, and their maximum penetration depth in the fabric cross-section, can be linked to the specific energy that is consumed in the process (Mao & Russell, 2006). The fabric strength depends on the degree to which fibres are intertwined in the process.

Since fabrics are consolidated mainly in areas where the water jets impact, jet marks are formed on the fabric surface and appear as parallel 'lines' on the jet side of the fabric running in the machine direction (Figure 13.4). Jet marking becomes less pronounced as the number of injectors increases.

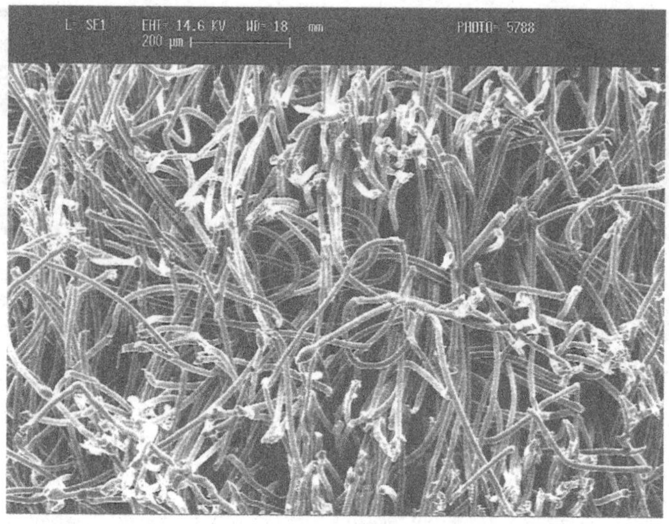

FIGURE 13.3

Example of curved 'U' shaped fibres in the cross-section of a hydroentangled fabric.

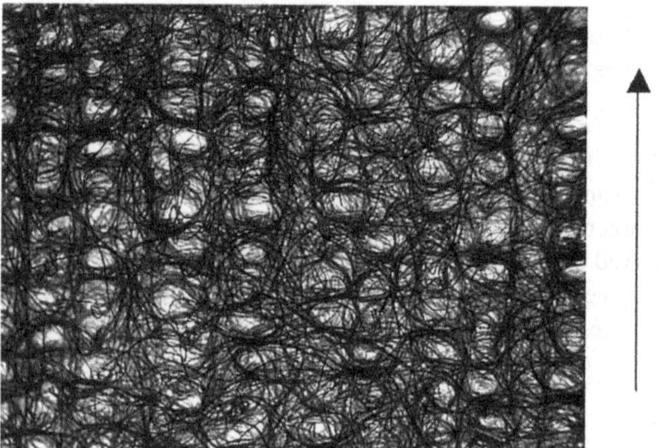

FIGURE 13.4

Jet marks on the surface of a thin hydroentangled fabric (30 g/m^2).

Hydroentangling produces local density variations or texture in the fabric that can influence tensile and fluid flow properties and introduce variations in local fibre segment orientation. Therefore, even if the original web is isotropic, structural anisotropy is introduced during hydroentanglement that may be of a periodic nature.

The structure of hydroentangled fabrics depends on process parameters and fibre properties. At low water-jet pressure, only a small portion of fibre segments on the surface of the web are entangled and intertwined. At high water-jet pressure, some fibre segments are reoriented towards the reverse side of the web, causing some fibre ends to project. Fibre rigidity and bending recovery influence the ability of the jet to

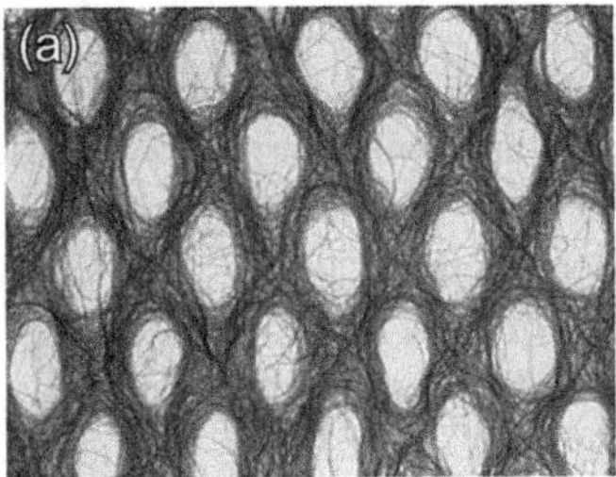

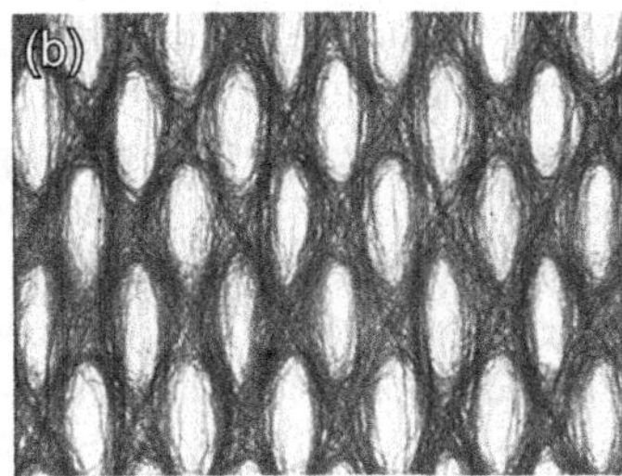

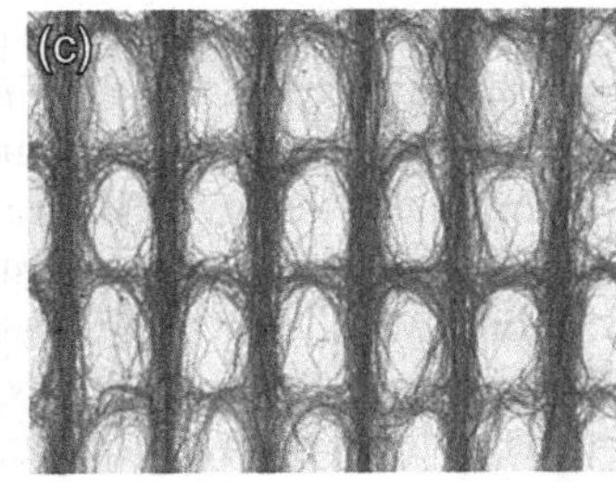

FIGURE 13.5

Examples of apertured hydroentangled fabrics: (a) elliptical holes; (b) extended-elliptical holes; (c) pseudo-rectangular holes.

produce fibre entanglements during hydroentanglement, and therefore the structural features of hydroentangled fabrics can differ according to fibre type. An example is fabrics made from polypropylene and viscose rayon using the same process conditions. The specific flexural rigidity of polypropylene fibre (0.51 mN mm/tex^2) is higher than viscose rayon (0.35 mN mm/tex^2), and polypropylene fibre has higher compression recovery, bending recovery and tensile recovery (Smolen, 1967) compared to viscose rayon fibre (Morton & Hearle, 1993). In a polypropylene-hydroentangled fabric produced at low specific energy, only the surface fibres are effectively bonded, while the fibres inside the fabric are poorly entangled. The surface is therefore more compact than the fabric core. In contrast, a viscose rayon fabric is more consistently bonded through the cross-section, and the compaction is greater than in the corresponding polypropylene fabric.

Where the support surface is three-dimensional, fibres can be displaced from the projections in the surface to form apertures, or other structural patterns can be introduced. The selection of support surface is therefore an important design tool, as it influences fabric appearance and texture. Various patterned structures including apertures can be introduced simply by changing the design of the support surface (see Figure 13.5). Adjusting the degree of pre-bonding in the web and water-jet operating conditions can influence the clarity and definition of these structural features.

13.3.1.3 Stitch-bonded fabrics

Stitch-bonded fabrics are produced by stitching a fibrous web together using either a system of additional yarns (filaments) or the fibres in the web by means of a warp knitting action. Because the formation of a stitch-bond fabric is a hybrid of warp knitting and sewing, it is reflected in the fabric structure. The fabric integrates stitched yarns and fibrous webs, the fabric structure exhibits a clear stitching pattern on at least one side of the fabric, and the stitches hold the fibres in the fibrous web together.

There are three basic types of stitch-bonded fabric structure: (1) fibres bonded with the constituent fibres in which the stitches are observed on one side of the fabric (Malivlies); (2) stitches of yarns on one surface and a projecting pile of pleated fibres on the reverse surface (Kunit); and (3) stitches of yarns on both surfaces (Multiknit, Maliwatt). The basic structural characteristics of these different types of stitch-bonded fabric structure may be summarised as follows (Raz, 1988):

- **Malivlies** fabrics are bonded by knitting fibres in the web rather than by additional yarns (filaments); therefore, the fabric consists of staple fibres. They have a warp knitted-loop stitch pattern on one side of the fabric, and the intensity of stitch bonding depends on the number of fibres carried in the needle hook, while carrying capacity depends on hook dimensions and fibre fineness.

- **Kunit fabric** is a three-dimensional pile structure made from 100% fibres stitched using the constituent fibres in the web. Fibres on one side of the fabric are formed into stitches, while the other side of the fabric has a pile loop structure with fibres arranged at an almost perpendicular orientation to the plane. The fabric has very good air permeability because of high-loft structure, and excellent compression elasticity because of the vertical pile loop structure.
- **Malimo Multiknit** constructions are formed from Kunit pile loop fabric, and both sides are formed into a closed surface by stitched loops of fibre. The two sides of the fabric are joined together by fibres orientated almost perpendicular to the plane. The fabric is stitched using the fibres in the original web rather than by additional yarns, and therefore a three-dimensional fabric composed of 100% staple fibres is formed.
- **Maliwatt** fabrics are fibrous webs stitched through with one or two stitch-forming yarns. Both sides of the fabric have a yarn stitch pattern and the fabric weight per unit area ranges from 15 to 3000 g/m^2 with a fabric thickness of up to 20 mm, and a stitching yarn linear density in the range 44 to 4400 dtex. Examples of Maliwatt bond structures are shown in Figures 13.6 and 13.7.

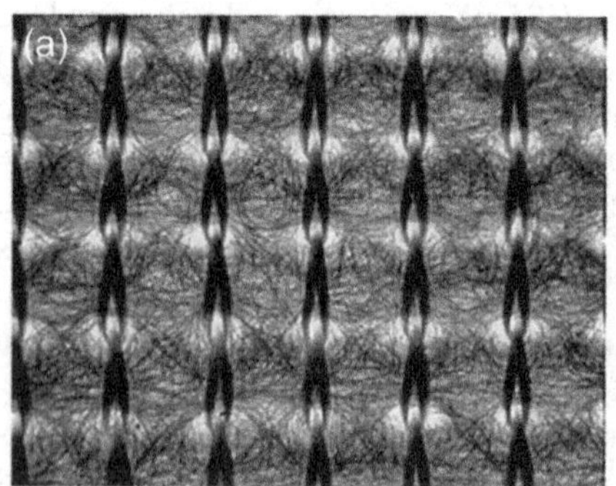

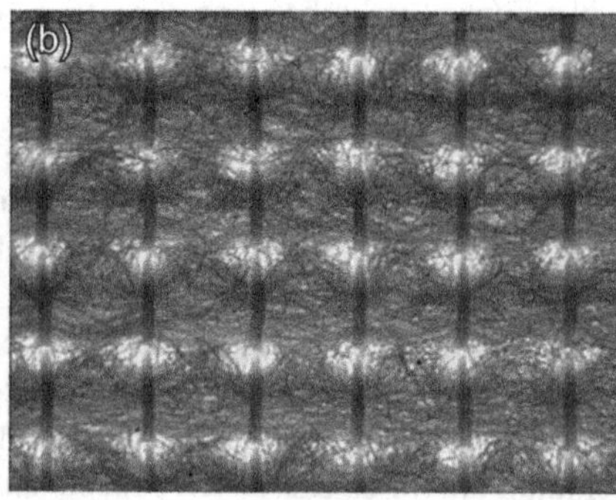

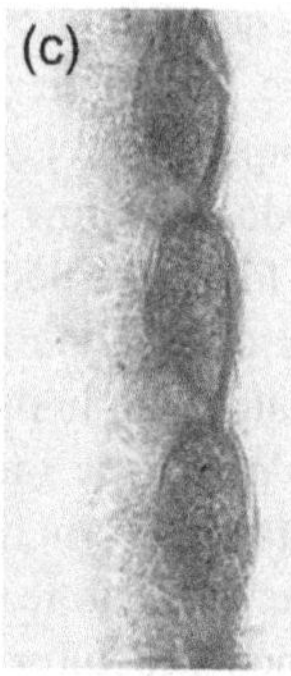

FIGURE 13.6

Example of the lengthwise bond structure of a Maliwatt fabric: (a) surface; (b) back; (c) cross-section.

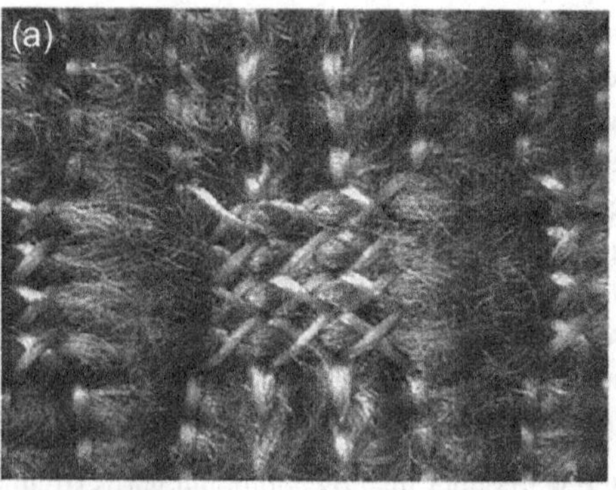

FIGURE 13.7

Example of the complex bond structure in a Maliwatt fabric: (a) surface; (b) back.

In stitch-bonded fabrics, yarn stitches are usually aligned in the fabric plane, while the fibre piles or the fibre pile loops are fixed by the stitches and are generally orientated perpendicular to the stitched fabric surface. Stitchbond fabric structure is determined by the warp knitting action applied by the machine, fibre properties and dimensions, web density and structure, stitching yarn structure, stitch density, machine gauge (number of needles per 25 mm), stitching yarn tension and stitch length. The stitch holes and the pile formed in the fabric surface are two unique structural characteristics of stitchbond fabrics. The number and size of the stitch holes depends on the properties of the stitching yarn, the properties of the fibrous web, machine gauge (number of needles per 25 mm), intermeshing intensity, interlacing and stitching yarn tension. The pile height, visible in certain stitch-bonded fabrics, ranges 2–20 mm and depends on how the oscillating element is set at the stitch-bonding position. Both the stitch holes and the piles formed in the fabric surface influence the fabric's properties. The warp knitting structure in stitch-bond nonwovens has an open fabric construction and short underlaps, and it is dimensionally extensible in the **cross direction (CD)** as well as in the **machine direction (MD)**. To increase the fabric tensile strength in the MD, a specific stitch construction is used (pillar stitch). To increase the widthwise stability, the underlaps are lengthened (e.g. satin stitch) and a three-dimensionally stable structure is achieved by combining these two types of stitch construction (e.g. pillar–satin).

13.3.1.4 Thermal-bonded fabrics

The types of bond structure formed in thermal-bonded fabrics depend on the method used to introduce heat to the fibres as well as the web structure and the type of binder fibre present. In calendered point-bonded fabrics (Figure 13.8), the fibres are compressed together and heat is introduced by conduction to only localised regions of the web. This produces deformation of the fibres and polymer flow around the bond points. Around the immediate vicinity of the bond points, the heating of surrounding fibres can introduce interfacial bonding at the crossover points of uncompressed fibres. This is known as secondary bonding and is particularly noticeable when bicomponent fibres are present as the binder component.

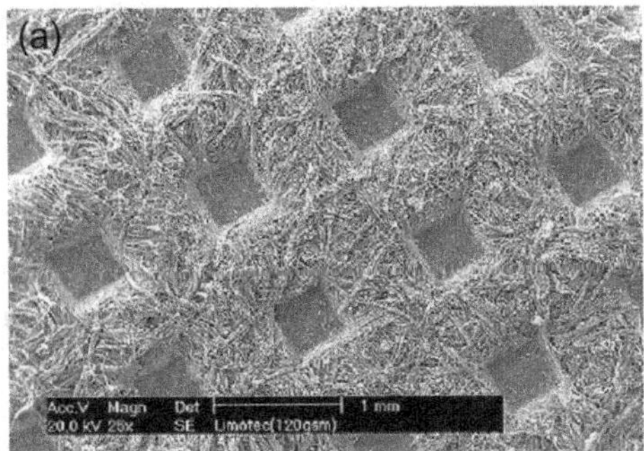

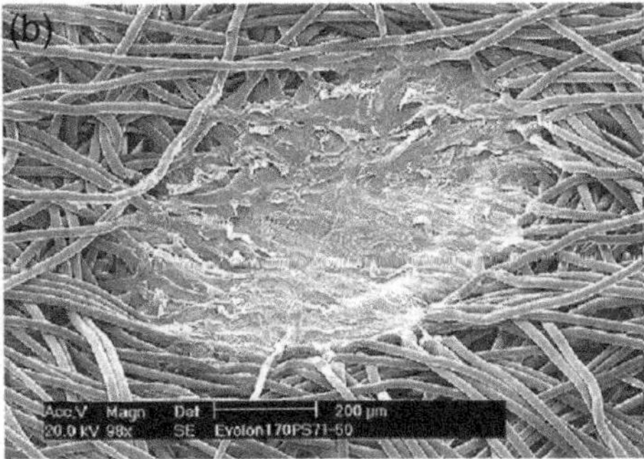

FIGURE 13.8

Examples of point-bonded fabrics: (a) thermal bonding points; (b) thermal bond points showing change in fibre morphology within the bond point.

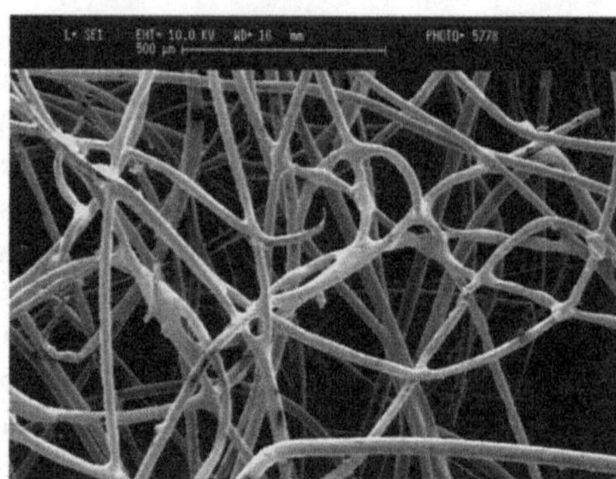

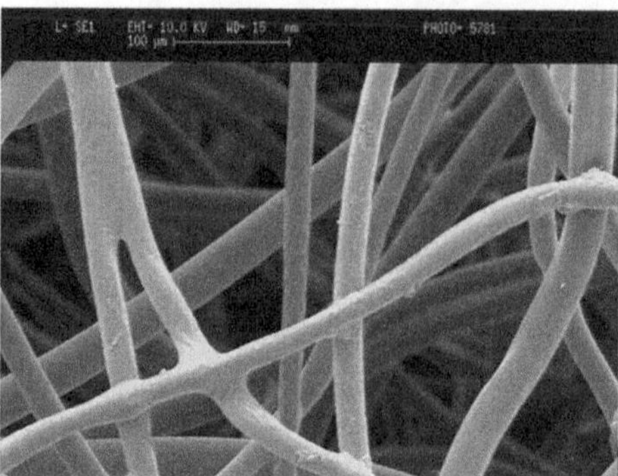

FIGURE 13.9

Bond points in through-air-thermal-bonded fabric.

In through-air bonded fabrics (Figure 13.9), core-sheath bicomponent fibres are commonly utilised, and convected heat introduced during the process produces bonding at the fibre crossover points as the polymer softens and flows. There is limited associated compression of the fibre assembly at these locations and therefore the resulting fabric density is lower as compared to a calendered thermally bonded fabric. Calendered thermal bond structures are found in a host of nonwovens both in monolithic and multilayer fabrics including SMS and other spunbond–meltblown web combinations.

13.3.1.5 Chemical-bonded fabrics

Chemically bonded fabrics are produced by the application of a resin emulsion (e.g. acrylic, polyvinyl acetate or other suitable chemical binder) to the web, and then the web is dried and cured. The distribution of the resin binder in the fabric is largely governed by its method of application to the web and the flow properties of the resin between fibres (Figure 13.10). Large numbers of fibres may be enveloped by a film binder connecting both fibre crossover points and interfibre spaces, and large segments of these film binders are visible in such structures that span adjacent fibres. Alternatively, the polymer may be concentrated at the fibre crossover points, producing localised bonding in these regions and either rigid or flexible bonds, depending on the polymer composition of the binder.

13.3.2 NONWOVEN FABRIC STRUCTURAL PARAMETERS

The structure and dimensions of nonwoven fabrics are frequently characterised in terms of fabric weight per unit area; thickness; density; fabric uniformity; fabric porosity; pore size and pore size distribution; fibre orientation distribution; and bonding segment structure and bonding segments distribution.

13.3.2.1 Fabric weight

Nonwoven fabric weight (or fabric mass) is defined as the mass per unit area of the fabric and is usually measured in g/m^2 (or gsm). Fabric thickness is defined as the distance between the two fabric surfaces under a specified applied pressure, which varies if the fabric is high-loft (or compressible). The fabric

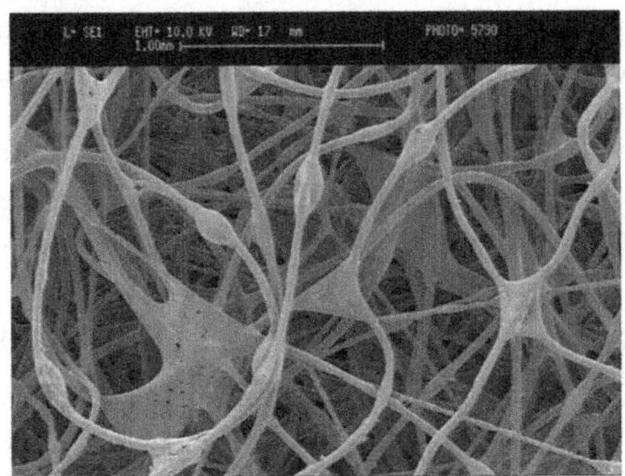
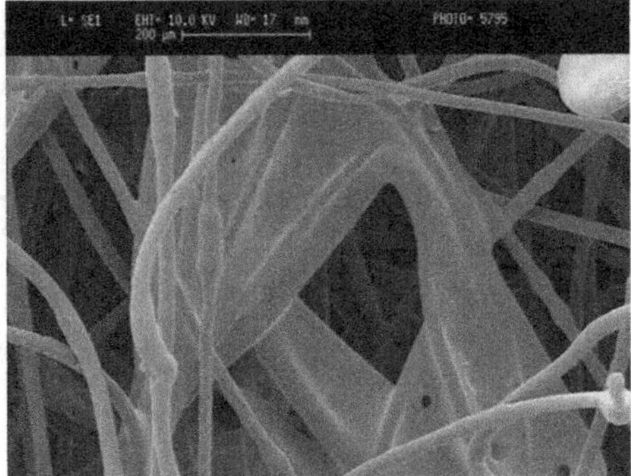

FIGURE 13.10

Examples of bond points in a chemically bonded fabric.

weight and thickness determine the fabric packing density, which influences the freedom of movement of the fibres and determines the porosity (the proportion of voids) in a nonwoven structure. The freedom of movement of the fibres plays an important role in nonwoven mechanical properties, and the proportion of voids determines the fabric porosity, pore sizes and permeability in a nonwoven structure. Fabric density, or bulk density, is the weight per unit volume of the nonwoven fabric (kg/m^3). It equals the measured weight per unit area (kg/m^2) divided by the measured thickness of the fabric (m). Fabric bulk density and fabric porosity are important, because together they influence how easily fluids, heat and sound transport through a fabric.

13.3.2.2 Weight uniformity of nonwoven fabrics

Fabric weight and thickness usually varies in different locations along and across a nonwoven fabric. Variations are frequently of a periodic nature with a recurring wavelength due to the mechanics of the web formation and/or bonding process. Persistent cross-machine variation in weight is commonly encountered, which is one reason for edge trimming. Variations in either thickness and/or weight per unit area determine variations of local fabric packing density and porosity as well as pore size distribution, and therefore influence the appearance; permeability; thermal and sound insulation; filtration; tensile, liquid barrier and penetration properties; energy absorption; light opacity and conversion behaviour of nonwoven products.

Fabric uniformity can be defined in terms of the fabric weight (or fabric density) variation measured directly by sampling different regions of the fabric. The magnitude of the variation depends on the specimen size; for example, the variation in fabric weight between smaller fabric samples (e.g. consecutive fabric samples of 1 m^2) will usually be much greater than the variation between bigger fabric samples (e.g. rolls of fabric of hundreds of metres). Commercially, to enable online determination of fabric weight variation, the fabric uniformity is measured in terms of the variation in the optical density of fabric images (Pound, 2001), the grey level–intensity of fabric images (Huang & Bresee, 1993) or the number of electromagnetic rays absorbed by the fabric (Aggarwal, Kennon, & Porat, 1992; Boeckerman, 1992), depending on the measurement techniques used. The basic statistical terms for expressing weight uniformity in the industry are the standard deviation (σ) and the coefficient of variation (CV) of measured parameters (e.g. fabric weight, fabric thickness, fabric density, optical levels, ray absorption amounts and grey level–intensity of images), as follows:

$$\text{Standard deviation: } \sigma^2 = \frac{\sum_{i=1}^{n}(w_i - \overline{w})^2}{n} \tag{13.1}$$

$$\text{Coefficient of variation: } \text{CV} = \frac{\sigma}{\overline{w}} \tag{13.2}$$

$$\text{Index of dispersion (Chhabra, 2003): } I_{\text{dispersion}} = \frac{\sigma^2}{\overline{w}} \tag{13.3}$$

where n is the number of testing samples, $\overline{w}$ is the average of the measured parameter and w_i is the local value of the measured parameter. Usually, the fabric uniformity is referred to the percentage coefficient of variation (CV).

Fabric uniformity in a nonwoven is normally anisotropic; that is the uniformity is different in different directions (MD and CD) in the fabric structure. The ratio of the index of dispersion has been used to represent the anisotropy of uniformity (Chhabra, 2003). The local anisotropy of mass uniformity in a nonwoven has also been defined by Scharcanski and Dodson (1996) in terms of the 'local dominant orientations of fabric weight'.

13.3.2.3 Fibre orientation distribution

The fibres in a nonwoven fabric are rarely completely randomly oriented, but rather, individual fibres are aligned in various directions, mostly in-plane. These fibre alignments are inherited from the web formation and bonding processes. The fibre segment orientations in a nonwoven fabric are in two and three dimensions and the orientation angle can be determined (Figure 13.11).

Although the fibre segment orientation in a nonwoven is potentially in any three-dimensional direction, the measurement of fibre alignment in three dimensions can be complex and time consuming (Gilmore, Davis, & Mi, 1993). In certain nonwoven structures, the fibres can be aligned in the fabric plane, with a small fraction oriented nearly vertical to the fabric plane. The structure of a needlepunched fabric is frequently simplified in this way. In this case, the structure of a

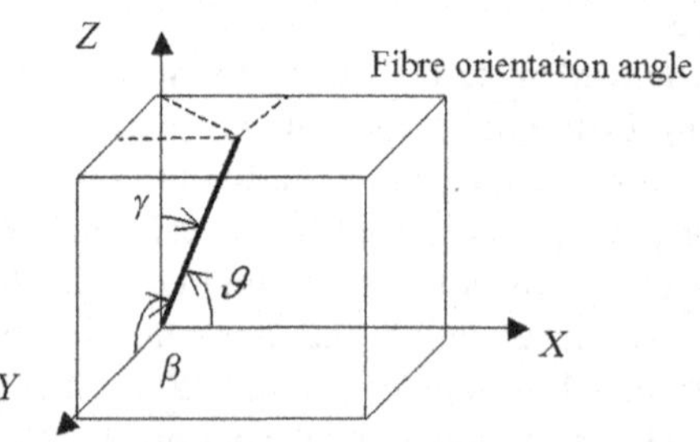

FIGURE 13.11

Fibre orientation angle in three-dimensional nonwoven fabrics.

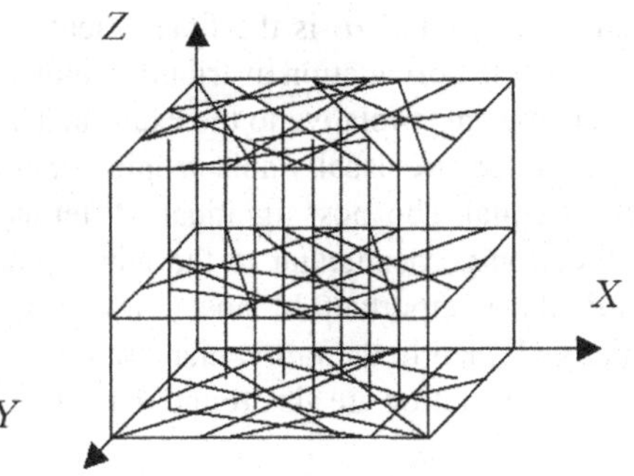

FIGURE 13.12

Highly simplified three-dimensional nonwoven structure.

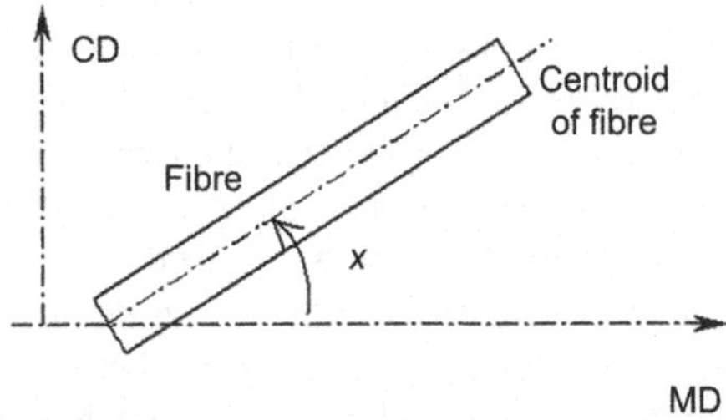

FIGURE 13.13

Fibre orientation and the orientation angle.

three-dimensional nonwoven may be simplified as a combination of two-dimensional layers connected by fibres orientated perpendicular to the plane (Figure 13.12). The fibre orientation in such a three-dimensional fabric can be described by measuring the fibre orientation in two dimensions in the fabric plane (Mao & Russell, 2003).

In the two-dimensional fabric plane, fibre orientation is measured by the fibre orientation angle, which is defined as the relative directional position of individual fibres in the structure relative to the machine direction as shown in Figure 13.13. The orientation angles of individual fibres or fibre segments can be determined by evaluating photomicrographs of the fabric or directly by means of microscopy, microcomputed tomography methods and image analysis.

The frequency distribution (or statistical function) of the fibre orientation angles in a nonwoven fabric is called fibre orientation distribution (FOD) or orientation distribution function (ODF). Frequency distributions are obtained by determining the fraction of the total number of fibres (fibre segments) falling within a series of predefined ranges of orientation angle. Discrete frequency distributions are used to estimate continuous probability density functions. The following general relationship is proposed for the fibre orientation distribution in a two-dimensional web or fabric (Krcma, 1972):

$$\int_0^{\pi} \Omega(\alpha)\, d\alpha = 1\ (\Omega(\alpha) \geq 0) \text{ or } \sum_{\alpha=0}^{\pi} \Omega(\alpha)\, \Delta\alpha = 1\ (\Omega(\alpha) \geq 0) \tag{13.4}$$

where, α is the fibre orientation angle, and $\Omega(\alpha)$ is the fibre orientation distribution function in the examined area. The numerical value of the orientation distribution indicates the number of observations that fall in the direction α, which is the angle relative to the examined area.

Fibre alignments in nonwoven fabrics are usually anisotropic – that is the number of fibres in each direction in nonwoven fabrics is not equal. The most significant difference between the fibre orientation in the fabric plane and in the direction perpendicular to the fabric plane (i.e. transverse direction or fabric thickness direction) is particularly important. In most nonwovens except some airlaid structures, most of the fibres are preferentially aligned in the fabric plane rather than in the fabric thickness. Significant in-plane differences in fibre orientation are also found in the machine direction and in the fabric cross direction in nonwovens.

Preferential fibre (either staple fibre or continuous filament) orientation in one or multiple directions is introduced during web formation and to some extent during mechanical bonding processes. A simplified example of an anisotropic nonwoven structure is a unidirectional fibrous bundle in which fibres are aligned in one direction only. Parallel-laid or cross-laid carded webs are usually anisotropic with a highly preferential direction of fibre orientation. Fibre orientation in airlaid structures is usually more isotropic than in other drylaid fabrics both in two and three dimensions. In perpendicular-laid webs, such as Struto or V-lap nonwovens, fibres are orientated in the direction of the fabric thickness. Spunlaid nonwovens composed of filaments are less anisotropic in the fabric plane than layered carded webs (Groitzsch), however, the anisotropy of continuous filament webs depends on the way in which the webs are collected and tensioned.

This structural anisotropy can be characterised in terms of the fibre orientation distribution functions. This anisotropy is important because of its influence on the anisotropy of fabric mechanical and physical properties including tensile, bending, thermal insulation, acoustic absorption, dielectric behaviour and permeability. The ratio of physical properties obtained in different directions in the fabric, usually the MD/CD, is a well established means of expressing the anisotropy. The MD/CD ratio of tensile strength is most commonly encountered, although the same approach may be used to express directional in-plane differences in elongation, liquid wicking distance, liquid transport rate, dielectric constant and permeability. However, these anisotropy terms use indirect experimental methods to characterise the nonwoven structure, and they are just ratios in two specific directions in the fabric plane, which can misrepresent the true anisotropy of a nonwoven structure.

13.3.2.4 Fabric porosity, pore size and pore size distribution

The pore structure in a nonwoven may be characterised in terms of the total pore volume (or porosity), the pore size, pore size distribution and the pore connectivity.

Porosity provides information on the overall pore volume of a porous material and is defined as the ratio of the nonsolid volume (voids) to the total volume of the nonwoven fabric. The volume fraction of solid material is defined as the ratio of solid fibre material to the total volume of the fabric. While the fibre density is the weight of a given volume of the solid component only (i.e. not containing other materials), the porosity can be calculated as follows using the fabric bulk density and the fibre density:

$$\phi(\%) = \frac{\rho_{\text{fabric}}}{\rho_{\text{fibre}}} \times 100\% \tag{13.5}$$

$$P(\%) = (1 - \phi) \times 100\% \tag{13.6}$$

where P is the fabric porosity (%), ϕ is the volume fraction of solid material (%), ρ_{fabric} (kg/m^3) is the fabric bulk density and ρ_{fibre} (kg/m^3) is the fibre density.

In a resin coated, impregnated or laminated nonwoven composite, a small proportion of the pores in the fabric is not accessible (i.e. they are not connected to the fabric surface). The definition of porosity herein refers to the so-called total porosity of the fabric. Thus, the open porosity (or effective porosity) is defined as the ratio of accessible pore volume to total fabric volume, which is a component part of the total fabric porosity. The majority of nonwoven fabrics have porosities >50% and usually above 80%.

A fabric with a porosity of 100% is a totally open fabric and there is no such fabric, while a fabric with a porosity of 0% is a solid polymer without any pore volume; there is no such fabric either. High-loft nonwoven fabrics usually have a low bulk density because they have more pore space than a heavily compacted nonwoven fabric does; the porosity of high-loft nonwovens can reach >98%.

Pore connectivity, which gives the geometric pathway between pores cannot be readily quantified and described. If the total pore area responsible for liquid transport across any distance along the direction of liquid transport is known, its magnitude and change in magnitude are believed to indicate the combined characteristics of the pore structure and connectivity.

13.4 PROPERTIES AND PERFORMANCE OF NONWOVEN FABRICS

Different nonwoven fabrics have different properties based on their structure. Examples of important nonwoven fabric properties and performance are:

1. **Mechanical properties**: tensile properties (Young's modulus, tenacity, strength and elasticity, elastic recovery and work of rupture), compression and compression recovery, bending and shear rigidity, tear resistance, burst strength, crease resistance, abrasion, frictional properties (smoothness, roughness and friction coefficient) and energy absorption
2. **Fluid handling properties**: permeability, liquid absorption (liquid absorbency, penetration time, wicking rate, rewet, bacteria/particle collection, repellency and barrier properties, run-off and strike time), water vapour transport and breathability
3. **Physical properties**: thermal and acoustic insulation and conductivity, electrostatic properties, dielectric constant and electrical conductivity, opacity and others
4. **Chemical properties**: surface wetting angle, oleophobicity and hydrophobicity, interface compatibility with binders and resins, chemical resistance and durability to wet treatments, flame resistance, dyeing capability, flammability and soiling resistance
5. **Application specific performance**: linting (particle generation), aesthetics and handle, filtration efficiency, performance in geotextiles, biocompatibility, sterilisation compatibility, biodegradability and health and safety status

The fabric properties rely on the constituent materials of the fabrics, fabric structural characteristics and the fabric surface properties. The performance of nonwoven products requires different properties of the fabric. The relationship of nonwoven fabric structure–property–performance can be either characterised in analytical and empirical models or simulated in numerical methods. The examples of some important models were developed in various research papers (Groitzsch, 2000; Mao & Russell, 2006) and summarised in some books (Mao, Russell, & Pourdehemy, 2006; Turbak, 1993).

13.5 METHODS FOR THE EVALUATION OF NONWOVEN FABRIC STRUCTURE, PROPERTIES AND PERFORMANCE

Various testing methods and techniques have been developed for the measurement of nonwoven fabric properties. These test methods can be grouped as follows:

- Standard test methods defined by standard authorities (e.g. ISO, CEN, BS (British Standard), ASTM (American Society for Testing Methods) and ANSI (American National Standard Institute)
- Test methods established by industrial associations (e.g. INDA, EDANA, AATCC (Association of American Textile Chemists and Colorists), IEC (International Electrotechnical Commission), CENELEC (European Committee for Electrotechnical Standardization), etc.) and individual companies
- Nonstandard test techniques designed for research purposes

The standard test methods, which are defined as orderly procedures in a reproducible environment, are designed to provide reliable measurements with certain precision for use in the trading of nonwovens and their products.

Industrial test methods are usually established for routine internal measurement concerned with the evaluation, benchmarking and quality control of semifinished or final products. In addition to these standard tests, numerous techniques are available to characterise nonwoven materials for either research purposes or the monitoring of the nonwoven production processes.

13.5.1 STANDARD TEST METHODS FOR THE EVALUATION OF THE STRUCTURE AND PROPERTIES OF NONWOVEN FABRICS

EDANA and INDA have worked together to produce a unified set of nonwoven test standards, Worldwide Strategic Partners (WSP), in 2005. The methods for the evaluation of nonwoven structure, properties and specific nonwoven products with reference to other major standards in Europe and North America (ISO, BS, EN with ERT, ASTM with ITS and AATCC) are summarised elsewhere (Mao et al., 2006). It includes: Fibre identification, Thickness, Weight, Tensile strength, Tear strength, Friction, Absorption, Abrasion resistance, Bursting strength, Electrostatic properties, Binder properties, Optical properties, Permeability, Repellency, Stiffness, Dry cleaning, Linting, Bacterial, Toxicity, Geotextiles, Degradable nonwoven fabrics, Superabsorbent materials, Absorbent hygiene products, Wiping efficiency, Interlinings.

The methods related to the evaluation of the specific nonwoven applications have also been defined in various standards, the typical examples of such technical products are nonwoven wound dressings and nonwoven filters.

13.5.2 STANDARDS FOR THE EVALUATION OF THE PERFORMANCE OF NONWOVEN PRODUCTS

Nonwoven products need to meet the standards related to both nonwoven fabrics and the specific industrial applications. For example, nonwoven wound dressings may need to comply with the standards for wound dressings in ISO, BP, ASTM and BS together with standards for nonwoven fabrics when few standards have been established for nonwoven wound dressings. The standard methods for the evaluations of the performance of two typical nonwoven products, nonwoven wound dressing and filter, are summarised below.

A detailed summary of the standards for the evaluation of the fabric structure, properties and performance of typical nonwoven products were also given in Chapter 9 of *Handbook of Nonwovens* (Mao et al., 2006).

13.6 NONWOVEN FABRICS AND THEIR APPLICATIONS

Because of the flexible and versatile formation processes, nonwoven fabrics have distinctive structural characteristics that are quite different to conventional textile materials. Furthermore, the structure and properties are readily modified during manufacture according to specific property requirements. They are engineered fabrics and are made to suit the needs of specific product applications.

Nonwovens are widely used in diverse consumer, medical and industrial products that are intended to be either disposal single use items or durable. The service life of nonwoven products can be measured in seconds, e.g. a wet wipe; minutes, e.g. a tea bag; hours, e.g. protective clothing; or years, e.g. car-interior trimmings.

A convenient classification of 12 areas can be used to categorise these applications: Agrotech, Buildtech, Clothtech, Geotech, Hometech, Indutech, Medtech, Mobiltech, Oekotech, Packtech, Protech and Sporttech (http://www.techtextil.messefrankfurt.com/frankfurt/en/besucher/messeprofil/anwendungsbereiche.html). A detailed list of nonwoven products and related statistics can be found in the websites of EDANA (http://www.edana.org/content/default.asp?PageID=37), and INDA (http://www.inda.org/enduses/enduses.html). Some examples of nonwoven applications are briefly summarised as follows:

Absorbent Hygiene products: A substantial proportion of global nonwoven fabric production is associated with hygiene products such as wipes, baby diapers, feminine-care and adult incontinence products. The fabrics are employed as functional elements such as top sheets, acquisition and distribution layers (ADLs), liquid retention layers and back sheets and thereby make a serious and widespread contribution to human health and wellbeing.

Agriculture and horticulture: Nonwovens are utilised as crop covers and for plant protection, seed blankets, and weed control fabrics either to reduce the use of pesticides and labour in land maintenance or to improve the quality and production rate of crops. They are also used as biodegradable plant pots, capillary matting and landscape fabrics to create a microclimates in which the heat and humidity are controlled against adverse weather conditions for optimal crop growth, at the same time to minimise the usage of water and fertiliser.

Functional clothing, footwear and baggage: Since the early days of the industry, nonwovens have been used as clothing components such as interlinings (fronts of overcoats, collars, facings, waistbands, etc.) and labels, shoe components (shoelace eyelet reinforcement, athletic shoe and sandal reinforcement, inner sole lining) and bag components due to their light weight and the ability to engineer desirable properties as shape retention, adaptation to the characteristics of other fabrics. They are also used to make entire clothing products such as disposable underwear, protective clothing and shoes, masks, clean room wears and hospital gowns and drapes.

Household products: Nonwovens are used in a multitude of household applications ranging from cleaning (abrasives, wipes and mops), adhesives, filters (vacuum cleaning bags, detergent pouches/fabric softener sheets), wall and floor coverings (blinds/curtains, carpet/carpet backings, wall coverings to furniture/upholstery, table linen) and bed linen (batting, pocket cloth for pocket springs/separation layer inside mattress, mattress covering, quilt backing, duvet coverings and pillow cases) to create comfortable and hygienic solutions for modern living. All these products can be delivered with additional functions such as dirt repellency, dust mite barrier properties, antimicrobial and anti-odour properties.

Nonwoven fabrics also extensively utilised in many industrial sectors where they are required for use in filtration, geosynthetics, thermal insulation, sound absorption and damp and fire protection in the civil engineering and building industries, automotive components, energy absorption and to maintain human comfort in protective products such as personal protective equipment. Safety and security applications include blast resistant curtains, burglarproof blinds. Nonwovens are also heavily used in food packaging as components in sport products.

13.7 NONWOVEN FABRICS IN FASHION

Nonwoven fabrics are extensively used in the manufacture of both single use (or disposable) and durable clothing, notable market segments being protective clothing, garment linings, interlinings, waddings, shoe linings and synthetic leather fabrics. However, in durable clothing, whilst nonwoven fabrics have long been utilised as garment components within clothing systems, they have had limited success in penetrating mainstream outerwear markets where woven and knitted fabrics still dominate.

During the 1960s, per capita consumption of products and packaging increased as disposable goods and a throwaway society developed (Radke, 1998). During this short fad paper clothing period, the 'Paper Caper' disposable dress was launched (Scott paper) to promote toilet and kitchen rolls and more than half a million dresses were sold (History of the Paper Dress, 2006). The fabric was a yarn reinforced tissue paper laminate to increase tensile strength (Palmer, 1991). Similar fabrics containing synthetic fibres with water resistance were also developed around this time by DuPont, prior to the industrial development of flash-spun nonwovens (Tyvek), a major market for which continues to be single-use disposable industrial-clothing fabrics (Welcome to DuPont Tyvek®, 2006). Owing to the inherent dimensional stability, abrasion resistance, tear strength, water resistance and adequate moisture vapour permeability of Tyvek® fabrics, other markets in clothing have also been explored (DuPont Shuts Neotis Studio as Nonwoven Apparel Fabric Venture Fails to Find Demand, 2002; Tilin, 2001). Apparel products such as sports shoes, including Nike's MayFly running shoes have utilised Tyvek. Other products include lightweight jackets, some of which were intended to be customisable by the user using permanent marker pens supplied with the garment at the point of sale (e.g. Nike Custom Windrunner) (Frings, 1996).

Intensive industrial development of nonwoven fabrics for clothing and domestic textile substitution was well underway in the 1960s and 1970s. Around this time, research was conducted on scrim-reinforced needlepunched wool blends, which after dyeing, finishing, milling and decatising produced nonwoven fabrics similar to melton cloths. In 1972, the shirtwaist dress containing Ultrasuede® was presented (Cole, 2002). Ultrasuede®, a needled synthetic suede fabric produced from bicomponent polyester/polyurethane fibre has been widely utilised by the fashion industry for over 30 years, including in outerwear (Knoll Textiles Partners with Toray Ultrasuede for Contract Market; Toray, 2004). Some of the most promising nonwoven fabrics for durable clothing have since emerged based on bicomponent fibre technology wherein microfibres are released from the original filaments in the fabric during or after initial bonding of the web. Spunbond fabrics containing splittable or fibrillating filaments produce extremely attractive aesthetics and handle whilst providing excellent strength, durability and abrasion resistance. If produced from polymers such as PA or PET they can also be dyed or printed. Leading producer of non-woven fabric, Freudenberg's Evolon fabrics are produced in an integrated process of bicomponent spunbond web formation followed by hydroentangling that simultaneously mechanically entangles and splits out the submicron filaments. The creation of submicron fibres by mechanical splitting or fibrillation of filaments in spunlaid webs during mechanical bonding processes such as hydroentanglement greatly influences fabric softness and drape. Textile-like aesthetics can also be promoted by introducing crimp in filaments during the quenching and stretching of side-by-side or eccentric sheath-core bicomponent filaments in extrusion (followed by thermal treatment). Appropriate mechanical and chemical finishing of such fabrics further enhances aesthetics and appearance.

By appropriate selection of the support conveyor, hydroentangled fabrics can be readily designed with many different surface textures, patterns and surface effects including pseudo-woven structural effects. This approach has been used to develop many interesting structural design effects in fabrics. One exemplar is Miratec® fabrics produced using Apex® technology (Polymer Group Inc.), which enabled dyed hydroentangled fabrics with the appearance of woven fabric to be produced for Levi's Engineered Jeans concept.

Development work by The Woolmark Company and by Canesis Network (AgResearch) to develop nonwoven fabrics suitable for wool garments (Anderson, 2005; Needlecraft Packs Punch, 2003) was mainly based on carding and mechanical bonding. A range of coloured, patterned and textured fabrics for apparel was directed at fashion applications (Continuing Innovation in Nonwovens, 2004). Durable sleeveless jackets produced from mechanically bonded nonwoven fabric consisting of 85% wool and 15% polyamide has been commercially available in corporate wear in Australia for many years.

Given the diverse range of nonwoven materials that are available together with the ease by which their properties can be modified, there is significant scope for creative use of such fabrics in fashion. Nonwoven fabrics have quite different appearances and property characteristics compared with woven and knitted fabrics, which provide opportunities to construct garments in different ways as well as produce articles of clothing that exhibit unusual aesthetic, appearance and performance characteristics. Some examples of nonwoven fabrics used in fashion are shown in Figure 13.14.

FIGURE 13.14

Fashion made from nonwoven fabrics in the School of Design, University of Leeds.

13.8 FUTURE TRENDS

The significant development of innovative materials, nanotechnology, energy and environmental technology, medicine and healthcare in recent years would impose great impact on the future of nonwoven technology and the markets of its products.

Nonwovens containing nanofibres produced by using technologies including meltblowing, electrospinning, melt electrospinning, meltblown electrospinning and centrifugal spinning are currently being developed. Unique filtration and barrier performance, as well as liquid absorption and wicking properties, will make nanofibre-based nonwoven webs irreplaceable in many applications.

Also, the functional treatment of nonwoven or its constituent fibres by using nanoparticles are another trend to obtain functional fabrics. Plasma (http://www.p2i.com/) treatment combining with nanoparticle treatment would have many special properties such antibacterial, antivirus, ultrahydrophobicity and ultrahydrophilicity.

The development of nonwoven products made from biomaterials, biodegradable and recyclable nonwoven materials are also in a great demand. The technologies for incorporating smart materials in nonwoven structures such as ink jet printing, digital printing also affects the future of nonwoven technologies and its products. All the new technologies emerged in various sectors including the examples mentioned above would have great impact on the resultant nonwoven products and their applications in medicine, tissue engineering, healthcare, military, energy and environment, personal protection equipment (PPE) and influence our daily life in the future.

13.9 PROJECT IDEAS

1. Nonwoven fabrics are widely used to apply lotions to skin. Could nonwoven clothing components be designed that are capable of conditioning our skin as we wear them?
2. Nonwoven fabrics typically have anisotropic properties, for example, mechanical properties and liquid transport properties. Could these anisotropic properties be exploited to produce new visual or structural effects that are difficult to replicate with knitted and woven substrates?
3. Nonwoven fabrics can be produced from a large variety of materials ranging from very short pulp fibres and recycled fibres to continuous filaments. Garments can also be constructed differently because nonwovens have unusual mechanical properties, are less prone to fraying and depending upon their polymer composition, are compatible with stitch-less joining technologies. Could we apply eco-design principles to construct new types of garment that could be assembled using fewer steps, could be homogeneous in terms of polymer composition, and which at the end of life could be rapidly disassembled to aid recycling or degraded with minimal environmental impact?

13.10 REVISION QUESTIONS

1. The structure of all nonwovens can be modified during processing, which influences the physical properties of the fabric. Give three to four examples of how a nonwoven fabric can be engineered to alter drape and softness.

2. Fibre diameter is an important parameter in the design of nonwoven fabric structure. Explain how fibre diameter influences fabric properties.
3. How does the fibre orientation distribution in a nonwoven fabric influence its suitability for use in different product applications?
4. What are the structural differences between spunbond and meltblown nonwoven fabrics and how do these influence fabric properties?
5. Why are wetlaid nonwoven fabrics stiff in comparison to needlepunched nonwovens?

13.11 SOURCES OF FURTHER INFORMATION

The more detailed information regarding nonwoven production, structure, property and performance can be obtained from the following books, journals and websites:

1. S. J. Russell (Ed.), (2006). *Handbook of nonwovens*. Woodhead Publishing Ltd.
2. Turbak, A. F. (1993). *Theory, process, performance, and testing*. TAPPI Press.
3. Hutten, I. M. (2007). *Handbook of nonwoven filter media*. Oxford, UK: Elsevier Ltd.
4. Horrocks, A. R., & Anand, S. C. (2000). *Handbook of technical textiles*. Woodhead Publishing Ltd.
5. *The Journal of Engineered Fibers and Fabrics (JEFF)*. http://www.jeffjournal.org.
6. *International Nonwovens Journal*. http://www.inda.org/INJ/index.html.
7. *TAPPI Journal (TJ)*. http://www.tappi.org/Bookstore/Technical-Papers/Journal-Articles/TAPPI-JOURNAL.
8. *Journal of the Textile Institute*. http://www.informaworld.com/smpp/title~db=all~content=t778164490.
9. *Textile Research Journal*. http://trj.sagepub.com.
10. *Journal of Industrial Textiles*. http://jit.sagepub.com.
11. *Textile Progress*. http://www.tandfonline.com/toc/ttpr20/current.
12. *AUTEX Journal*. http://www.autexrj.com/.
13. *Nonwovens World*. http://www.nonwovensworld.com.
14. *Textile Outlook*. http://www.textilesintelligence.com.
15. http://www.cottoninc.com/Cotton-Nonwoven-home/.
16. http://www.edana.org.
17. http://www.inda.org.
18. http://www.asianonwovens.org.
19. http://www.technicaltextiles.net.
20. http://www.textileworld.com.

REFERENCES

Aggarwal, R. K., Kennon, W. R., & Porat, I. (1992). A scanned-laser technique for monitoring fibrous webs and nonwoven fabrics. *Journal of Textile Institute*, *83*(3), 386–398.

Anderson, K. (2005). *Nonwoven fabrics in fashion apparel.* Available from World Wide Web: http://www.techexchange.com/library/Nonwoven%20Fabrics%20in%20Fashion%20Apparel.pdf, Accessed 30.09.14.

Boeckerman, P. A. (1992). Meeting the special requirements for on-line basis weight measurement of lightweight nonwoven fabrics. *TAPPI Journal*, *75*(12), 166–172.

BS EN ISO 9092:2011. Textiles. Nonwovens. Definition.

Butler, I. (1999). *Spunbond and melt blown technology handbook*. INDA, Cary, NC, USA.

Chhabra, R. (2003). Nonwoven uniformity—measurements using image analysis. *International Nonwovens Journal*, *12*(1), 43–50.

Cole, S. (2002). *Halston (Roy Halston Frowick) (1932–1990)*. Available from World Wide Web: http://www.glbtq.com/arts/halston.html [online]. Accessed 23.04.06.

Continuing innovation in nonwovens. (2004, June). Canesis Network. Available from World Wide Web: http://www.canesis.com/Canesis_News_Archive.shtm [online]. Accessed 23.04.06.

Cotterill, P. J. (1975). Production and properties of stitch bonded fabrics. *Textile Progress, The Textile Institute*, *7*(2), 101.

DuPont shuts Neotis studio as nonwoven apparel fabric venture fails to find demand. (2002). *Nonwoven Markets*, *17*(16), 20.

Frings, G. (1996). *Fashion: From concept to consumer* (5th ed.). London: Prentice-Hall International (UK) Limited.

Gilmore, T., Davis, H., Mi, Z. (September 1993). *Tomographic approaches to nonwovens structure definition*. National Textile Centre Annual Report, USA. http://www.ntcresearch.org/pdf-rpts/Bref1295/B95S9308.pdf.

Groitzsch, D. (2000). *Ultrafine microfiber spunbond for hygiene and medical application, August 16*, EDANA 2000 Nonwovens Symposium. Prague, 7–8 June 2000.

Gunter, S., & Perkins, B. F. (July 2007). *The basic mechanics of calendering and embossing nonwoven webs*. http://www.idspackaging.com/Common/Paper/Paper_320/calendering_wp.pdf.

Hearle, J. W. S., & Sultan, M. A. J. (1967). A study of needled fabrics. Part 1: Experimental methods and properties. *Journal of Texttile Institute*, *58*, 251–265.

Hearle, J. W. S., & Sultan, M. A. J. (1968). A study of needled fabrics. Part 2: Effect of needling process. *Journal of Textile Institute*, *59*, 103–116.

History of the paper dress, 2006. MPH Poster Dresses. Available from World Wide Web: http://www.mphdesign.net/page11.html [online]. Accessed 23.04.06.

Hoyle, A. G. (April 1989). *TAPPI Journal*, *72*, 109–112.

Huang, X., & Bresee, R. R. (1993). Characterizing nonwoven web structure using image analysis techniques. Part III: web uniformity analysis. *Journal of Nonwovens Research*, *5*(3), 28–38.

Kevin Mcnally, E. (1998). Melt blown technology innovations. *TAPPI Journal*, *81*(3), 193.

Knoll textiles partners with Toray Ultrasuede for contract market. (2006). Available from World Wide Web: http://www.knoll.com/news/hstory.jsp?story_id=3793&type=Press%20Releases&storyType=nf [online]. Accessed 23.04.06.

Krcma, R. (1972). *Manual of nonwoven textiles*. Manchester: Textile Trade Press.

Lim, H. (2010). A review of spunbond process. *Journal of Textile and Apparel, Technology and Management*, *6*(3).

Mao, Z., & Goswami, B. C. (1999). *Book of papers, INDA-TEC 99*. Cary, NC.

Mao, N., & Russell, S. J. (2003). Modelling of permeability in homogeneous three dimensional nonwoven fabrics. *Textile Research Journal*, *91*, 243–258.

Mao, N., & Russell, S. J. (2006). A framework for determining the bonding intensity in hydroentangled nonwoven fabrics. *Composite Science and Technology*, *66*(1), 66–81.

Mao, N., Russell, S. J., & Pourdehemy, B. (2006). Chapter 9. Characterisation and modelling of nonwoven fabrics. In S. J. Russell (Ed.), *Handbook of nonwovens*. Woodhead Publishing Ltd, 401–514.

Morton, W. E., & Hearle, J. W. S. (1993). *Physical properties of textile fibres*. London: The Textile Institute.

Needlecraft packs punch. (2003). Beyond the Bale, Australian Wool Innovations. Available from World Wide Web: http://www.wool.com.au/LivePage.aspx?PageId=555 [online]. Accessed 23.04.06.

Palmer, A. (1991). Paper clothes: Not just a fad. In P. Cunningham, & S. Lab (Eds.), *Dress in American culture* (Vol. 85). Bowling Green, OH: Bowling Green University Press.

Pound, W. H. (2001). Real world uniformity measurement in nonwoven coverstock. *International Nonwovens Journal*, *10*(1), 35–39.

Radke, L. (1998, June). Wisconsin's war on waste. *Wisconsin Natural Resources Magazine*. Available from World Wide Web: http://www.wnrmag.com/stories/1998/jun98/waste.htm [online]. Accessed 23.04.06.

Raz, S. (1988). *The Karl Mayer guide to technical textiles,* Obertshausen, Deutschland: Karl Mayer; Clifton, New Jersey: Mayer Textile Machine Corp.

Rupp, J. (2008, May/June). *Spunbond & meltblown nonwovens*. Textile world.

Scharcanski, J., & Dodson, C. T. (1996). Texture analysis for estimating spatial variability and anisotropy in planar stochastic structures. *Optical Engineering*, *35*(8), 2302–2309.

Smolen, A. (1967). *Polypropylene* (B.Sc. dissertation). Department of Textile Industries, University of Leeds.

Tilin, A. (2001). Slick as Teflon! Tough as Kevlar! Limber as Lycra!. *Wired Magazine*. Available from World Wide Web: http://www.wired.com/wired/archive/9.10/abfabs.html [online]. Accessed 23.04.06.

Toray. (2004). *The science of Ultrasuede®*. Available from World Wide Web: http://www.ultrasuede.com/about/science.html [online]. Accessed 24.04.06.

Turbak, A. F. (Ed.). (1993). *Nonwovens: Theory, process, performance, and testing*. Atlanta, GA: TAPPI Press.

Wadsworth, L. C., Kamath, M. G., Dahiya, A., & Hegde, R. R. (1999). *Finishing of nonwovens*. http://www.engr.utk.edu/mse/Textiles/Finishing of Nonwovens.htm.

Welcome to DuPont Tyvek®, 2006. Available from World Wide Web: http://www.tyvek.com/ [online]. Accessed 23.04.06.

Williams, M. M. (1999). Chemical binders for nonwovens fabrics. In: *Tappi Nonwovens conference proceedings*, March 15–17, 1999, Orlando, Florida.

CHAPTER

YARN TO FABRIC: SPECIALIST FABRIC STRUCTURES

14

R.H. Gong
University of Manchester, Manchester, UK

LEARNING OBJECTIVES

At the end of this chapter, you should be able to:

- Understand the difference between biaxial and triaxial fabrics, their properties and their applications
- Distinguish the methods for making different types of knotted fabrics
- Describe the methods used in manufacturing other multicomponent textiles
- Understand 3D structures, their applications and future development

14.1 INTRODUCTION

The textile industry is quite diverse, and there is a wide range of both raw material and manufacturing technology available. New technologies and raw materials are constantly being developed, leading to a virtually unlimited variety of final textile structures for an ever-expanding range of applications. This chapter provides a summary of only some of the most common specialist fabric structures. Among the fabric types covered are triaxial fabrics, pile fabrics, knotted fabrics, braided fabrics and three-dimensional (3D) fabrics. Detailed information may be found in the references list at the end of this chapter, especially the recent book edited by the author (Gong, 2011).

14.2 TRIAXIAL FABRICS

Conventional woven fabrics are biaxial structures composed of two orthogonally interlaced sets of yarns, warp and weft. Figure 14.1 shows a simple plain weave structure. It is obvious that when stress is applied in the bias direction, the two sets of yarns can rotate relative to each other at the interlacing points. Biaxial fabrics thus have relatively low shear modulus or low resistance to extension in the bias direction. This is desirable in most clothing applications, as the low shear modulus enhances drape and softness. However, in many technical applications such as structural textiles, sailcloth and composites, high material stability and greater isotropy are required.

Although there is evidence that triaxial fabrics have existed from ancient times (Tyler, 2011), modern triaxial woven fabrics were first developed by Dow (1969). As the name suggests, triaxial fabrics are

Textiles and Fashion. http://dx.doi.org/10.1016/B978-1-84569-931-4.00014-3

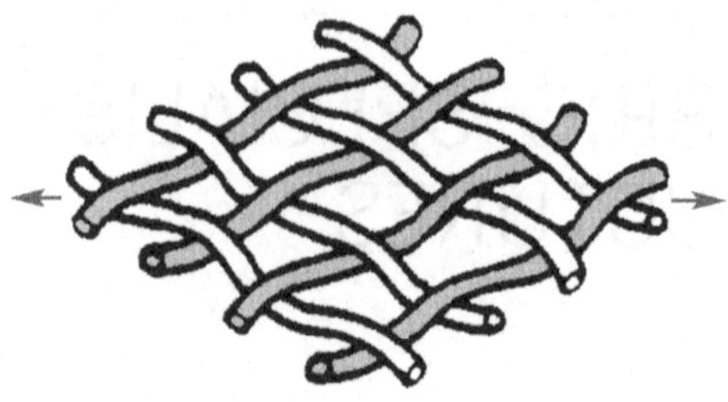

FIGURE 14.1

Plain weave structure.

FIGURE 14.2

Simple triaxial woven structure.

Source: Adapted from Dow (1969).

composed of three sets of yarns that typically intersect and interlace at 60° with one another. In a simple triaxial weave as shown in Figure 14.2, the three sets of yarns are interlocked, and all the yarns are compacted tightly at each intersection. The fabric has an area density about 50% that of conventional biaxial fabric having the same warp and weft composition and with the yarns similarly compacted tightly at all intersections. The structure shown in Figure 14.2 is highly porous with an apparent porosity of about 33%. These properties make triaxial structures very suitable for applications requiring light weight, high structural integrity and good ventilation, such as roofing structures or seat backing. Due to having three sets of yarns, triaxial fabrics provide relatively higher and more uniform resistance to extension, shear and burst deformation than comparable biaxial woven fabrics (Scardino & Ko, 1981). The greater isotropy (same properties when measured in different directions) of the triaxial structure makes these fabrics good reinforcements for composite materials (Pellegrino & Kueh, 2006; Rudo, 2007). Their open structure and anisotropy (different properties when measured in different directions) also make triaxial woven fabric composites a popular choice as the shell of antenna reflectors for communication satellites (Obst, Palermo, Ticci, & Santiago Prowald, 2005; Zhao & Hoa, 2005).

As in conventional weaving, the three sets of yarns can interlace with different patterns to produce fabrics with varied densities and properties. Figure 14.3 shows an example where the apparent porosity is zero and the fabric density is twice that of the simple triaxial structure (Dow, 1969).

FIGURE 14.3

Double-density triaxial woven structure.

Source: Adapted from Dow (1969).

Despite having some clear advantages over biaxial fabrics, triaxial fabrics are less frequently used. One difficulty is the manufacturing of triaxial fabrics, as the specialised and more complex machinery leads to higher costs. Biaxial fabrics are also more advantageous in traditional applications such as apparel, as the ease of yarn movement makes biaxial fabrics drape better. However, as technology advances, manufacturing difficulties will decrease and it is likely that triaxial fabrics will gain a greater market share in technical applications because of their better isotropy and dimensional stability.

14.3 PILE FABRICS

Pile fabrics are characterised by the tufts or loops of fibres or yarns that stand up from the base fabric. Pile fabrics exist in many forms such as velvet, terry towel, chenille and perhaps most commonly, pile carpets. They can be made by numerous processes including tufting, knitting, knotting, flocking and nonwovens. Velvet fabrics are believed to have existed as early as 2000 BC (Brandon, 2009). The most common type of velvet is a cut pile woven fabric used widely in apparel and home furnishing applications. The pile is produced by the extra set of warp yarns used during weaving. There are several production methods. One method is wire weaving by inserting wires in the shed formed by the warp yarns for the pile. The pile warp forms yarn loops on the fabric surface after the wires are withdrawn. The pile may be left as loop pile or cut to make cut pile. In terry fabrics, loops are formed on both sides of the fabric. Another method of producing pile fabrics is face-to-face weaving in which a double fabric, one on top of the other, is made with vertical links provided by the extra set of warp yarns. The linking warp yarn is then cut to produce velvet piles. This is illustrated in Figure 14.4. Clearly, only cut pile fabrics can be made this way. It is also possible to produce pile effects through knitting (Rock & Lohmueller, 2001; Starbuck & Shilton, 2010) or even from nonwoven fabrics (Erth & Wegner, 1999; Crawshaw, 2011). An example of a knitted double fabric is shown in Figure 14.5.

Another widely used pile fabric is tufted carpet. Tufting is done by forming yarn loops on the backing fabric by stitching with tufting needles that carry the pile yarn through the backing fabric. A looper

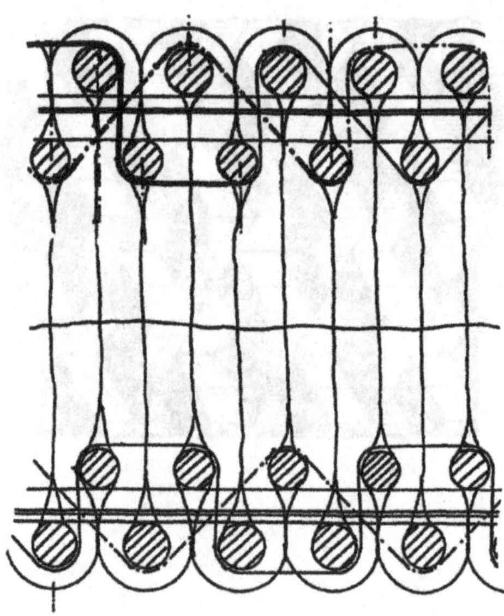

FIGURE 14.4

Woven double fabric (Gossl & Seidel, 2002).

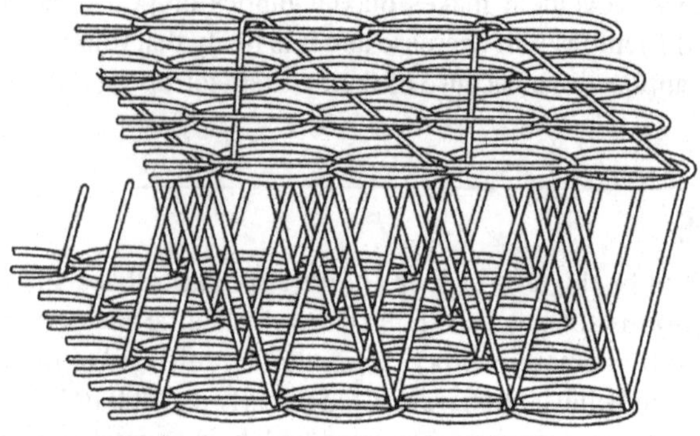

FIGURE 14.5

Warp-knitted double fabric (Rock & Lohmueller, 2001).

is then inserted between the yarn and the needle. The needle is retracted through the backing fabric to start the next cycle while the pile loop is formed over the looper. The loops formed on the backing fabric may be either cut or left as loops. The backing fabric is usually woven, but nonwovens may also be used. Pile yarn is traditionally a spun yarn, particularly from wool, but there is increasing use of bulked continuous filament yarns. Pile carpets may also be made through adhesive bonding and flocking (Crawshaw, 2011). In the flocking process, short fibres, cut from natural or synthetic fibres, are spread onto a fabric surface coated with adhesive; the fabric is then cured. The flock fibre spreading may be carried out mechanically, or more commonly, with electrostatic forces. In the electrostatic process, electrical field forces help to align the flock fibres perpendicularly to the fabric surface. The

FIGURE 14.6

Net structure.

flocking process is faster and cheaper than other pile fabric production processes, but the durability of the pile can be relatively low, as the anchorage of the pile fibres on the substrate fabric is limited and dependent on the bonding strength of the adhesive. Improvements in quality have led to increasing use of flocked fabrics in apparel applications as well as for technical uses such as insulation and high-impact composites (Kim, 2011).

14.4 KNOTTED FABRICS

There are many types of knotted fabrics. These include nets, macrame, lace and crochet (Philpott, 2011). Knotted structures have a wide range of applications including fishing nets, carpets, ornaments, accessories, and edging and decorative materials.

14.4.1 NETS

Net fabrics, as illustrated in Figure 14.6, are characterised by their low density, large open spaces and low resistance to deformation. Nets can be made by using either a single strand or a number of parallel strands interlinked by knots between neighbouring strands. The openings of the net can be a variety of shapes, such as triangles, quadrilaterals and hexagons. The most well-known uses are perhaps in fishing and mosquito nets. Nets are also used in many other applications such as tennis nets, veils and hammocks.

Netting may be done manually or mechanically. The hand process is fairly simple and consists of looping and knotting to produce an openwork fabric. Bobbinet lace machines and compound-needle tricot knitting machines are also used in net manufacturing.

14.4.2 MACRAME

Macrame is a heavily knotted fabric. It has found application in wall hangings, toys, shoes, belts, decorative fabrics, edging of rugs and carpets, plant hangers, place mats, and coasters and other

FIGURE 14.7

Knotted macrame structure.

tabletop coverings. Macrame has also been used widely in making jewellery such as earrings and bracelets.

It is essentially a handcraft technique and requires a minimum of equipment, although in complex work the yarns are wound onto bobbins to prevent entanglement. Figure 14.7 shows a macrame product that is part of a Chinese ornament.

14.4.3 LACE

Lace is a fine openwork fabric with a ground of mesh or net on which patterns may be introduced when the ground is formed or applied later. Laces are usually made using bobbins or needles, but can also be made using a crochet hook and knitting needles. Machine-made nets can be embroidered to give patterned laces (http://www.laceguild.demon.co.uk). Laces are used in lingerie, headscarves, as trim for dresses, gowns and skirts, and in the home as tablecloths, runners and curtains. Lace is generally divided into two categories according to construction, needlepoint lace and bobbin lace.

14.4.4 CROCHET

Crocheting usually involves only one loop at a time, and a single crochet hook is used instead of two or more knitting needles. The crochet hook is used to manipulate the yarn into interconnected knots. This is illustrated in Figure 14.8 (Fischer, 2009). The resulting crocheted structure may be expanded linearly or circularly to build up a larger piece. Crochet fabrics may be used for clothing, lace and other home accessories. The structure is more stable than knitted fabrics because the loops are interlocked in a complex pattern, while in knitted fabrics the loops are chained sequentially, and one broken loop may cause the whole structure to unravel.

14.4.5 KNOTTING

The knotting process is considered to produce the best-quality carpets. The piles are made from knotting cut lengths of yarns, as shown in Figure 14.9 (Breitschadel, 1934). Traditionally, knotted

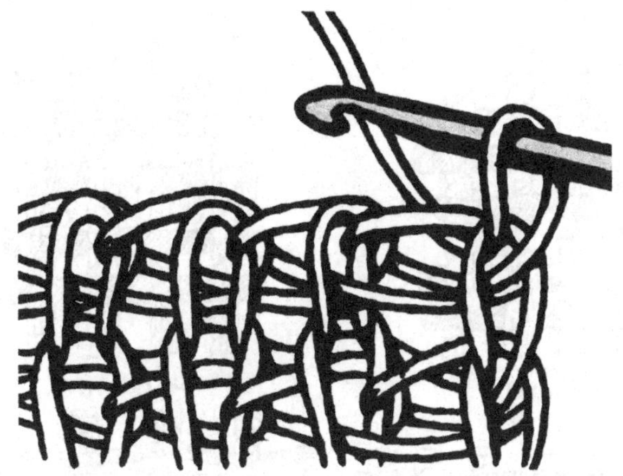

FIGURE 14.8

Crochet hook and structure.

Source: Adapted from Fischer (2009).

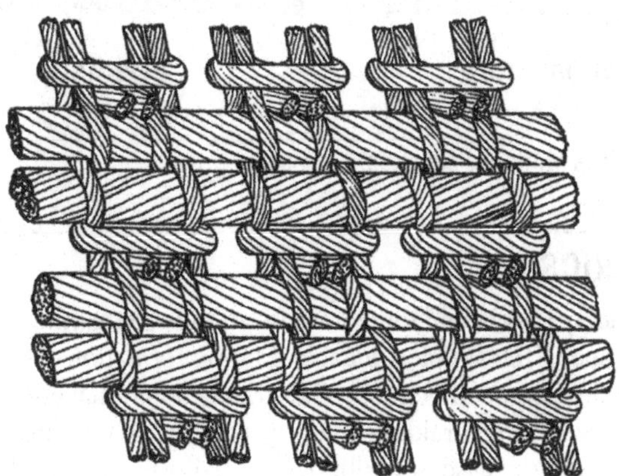

FIGURE 14.9

Knotted carpet.

Source: Adapted from Breitschadel (1934).

carpets are made by hand but may also be made with knotting machines. Several types of knots may be used: Turkish or symmetric, Persian or asymmetric, Jufti, and Spanish or single-warp symmetrical knots. Jufti knots can be either Turkish or Persian, but are tied around four warps instead of the customary two. These are illustrated in Figure 14.10 (http://arts.jrank.org/pages/9531/Carpet.html).

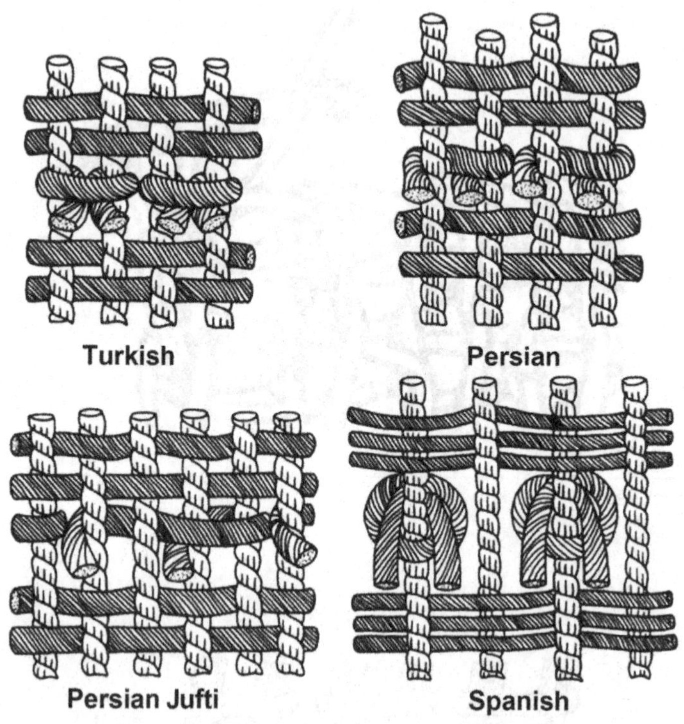

FIGURE 14.10

Types of carpet knots (http://arts.jrank.org/pages/9531/Carpet.html).

14.5 BRAIDED FABRICS

Braided structures are made by interlacing three or more strands so that they cross each other in diagonal formation. Braiding is one of the simplest means of fabric formation. The process of braiding does not require shedding, weft insertion and beat-up. The strands do not have to go through harnesses and reed, as is the case in weaving. During braiding all of the strands are interchanged in every cycle, while in weaving only one (or a few in the case of multiphase weaving) weft strand is interlaced with one set of warp strands. However, braided structures have a limited width, because there is a practical limit to the number of strands that can be accommodated in the braiding machine.

Circular braiding machines are the most widely used. There are two main types, maypole and high speed. In the maypole braiding machine (Potluri, 2011), the spool carriers themselves carry out the movements for interlacing. In the more widely used high-speed braiding machines, the two groups of spools carry out opposed circular movements, and the strands of one group are led alternately over and under the spools of the other group (Scherzinger, 1997).

Braiding can produce two-dimensional, flat or tubular, and solid products. The simplest form of braided structure is made of three strands as shown in Figure 14.11, which is well known as a style of hair plait. The pattern of strand interlacing can vary in a similar way to woven fabrics. The three most widely used interlacing patterns are diamond with 1/1 intersections, regular with 2/2 intersections and

FIGURE 14.11

Simple flat braid of three strands.

FIGURE 14.12

Triaxial braid.

Hercules with 3/3 intersections (Potluri, 2011). Braids are usually biaxial, which offers good extension and recovery parallel to the length and high conformability parallel to the width. In order to provide stability lengthways, a third set of strands may be used to produce triaxial braids, as shown in Figure 14.12.

Circular or tubular braids have a hollow centre. These can be used as cover for electric wires and hoses. Circular braids can be made directly over a mandrel to ensure tight fitting. Thicker braids may be made by braiding multiple layers on top of each other or by using a multi-ply braiding machine (Brown, 1987).

In two-dimensional braids, both flat and circular, strands are aligned diagonally; therefore, extension along the length can lead to large contraction and shape deformation along the width. In solid three-dimensional braids, the strands pass under and over each other in the thickness direction. Figure 14.13 shows a simple solid square braid with four strands. Solid braids have a higher resistance to shape deformation in the direction of thickness and are widely used in rope products. Complex shapes such as *I* or *T* can be formed for composite reinforcements.

Two-dimensional braids have very limited application in apparel, largely due to limited manufacturable width. One-dimensional products such as shoe laces, ropes and hoses are the most widely used applications. Braided structures offer good impact resistance, as all of the strands within a braided structure are continuous, are mechanically locked, take up the structural load and absorb energy. The interlocking structures also provide high structural integrity and resistance to delamination. For these reasons, there has been increased application of braided structures in composites, especially for 3D products such as reinforcements in sports equipment and aircraft fan blades (A & P Technology).

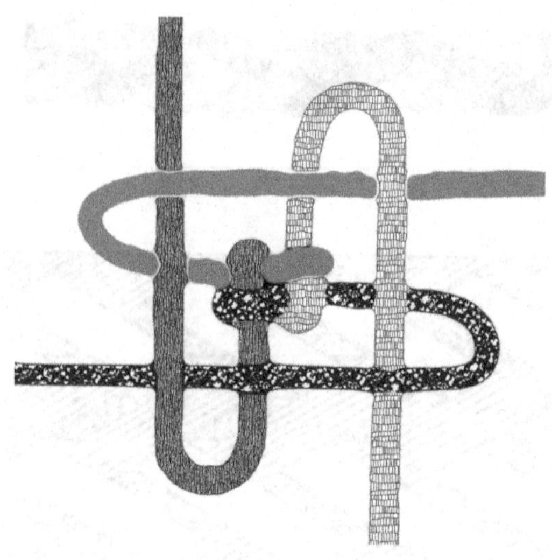

FIGURE 14.13

Solid square braid.

14.6 THREE-DIMENSIONAL FABRICS AND FUTURE DEVELOPMENTS

Most fabrics are produced as flat two-dimensional sheets. However, in some cases the ultimate shape of the end-use product is three-dimensional. Three-dimensional fabric structures have three broad categories – solid, hollow and shell – although more specific classifications may be used depending on structural details (Chen, 2011; Fukuta & Aoki, 1986; Khokar, 2001). Three-dimensional fabrics can be produced through most of the usual fabric making processes such as weaving, knitting, braiding and nonwoven. Three-dimensional woven fabrics can be made using standard weaving looms with some required modifications (Koppelman & Campman, 1963; Porat, Zhao, & Greenwood, 1996; Rheaume, 1970) or special weaving machines designed for 3D fabrics (Bryn, Nayfeh, Islam, Lowery, & Harries, 2004; Farley, 1995; Kang & Lee, 2010; Kimbara & Hayashida, 1993; Tsuzuki, 1994; Uchida et al., 1999).

14.6.1 3D SOLID STRUCTURES

Three-dimensional solid woven fabrics are essentially multilayer fabrics with integral stitching yarns. When compared with solid structures made by laminating flat fabrics on top of each other, 3D solid fabrics provide higher delamination resistance. Depending on the details of weaving, a variety of 3D solid structures can be produced to meet the needs of the end-use (Chen, 2011). An example of such a fabric structure is shown in Figure 14.14 (Van Schuylenburch, 1993). In this structure, a high degree of interlacing between yarns in the three principal axes provides more isotropic properties and additional straight yarns may also be added along all three principal axes to improve the modulus. Three-dimensional solid fabrics with various cross-sections, such as *I* and *T*, can also be produced for structural composite applications. For example, a process for producing a 3D fabric with an *I*-shaped cross-section, as shown in Figure 14.15, was described by Tsuzuki (1994). More complex structures that incorporate yarns in other directions, such as the bias, can also be made.

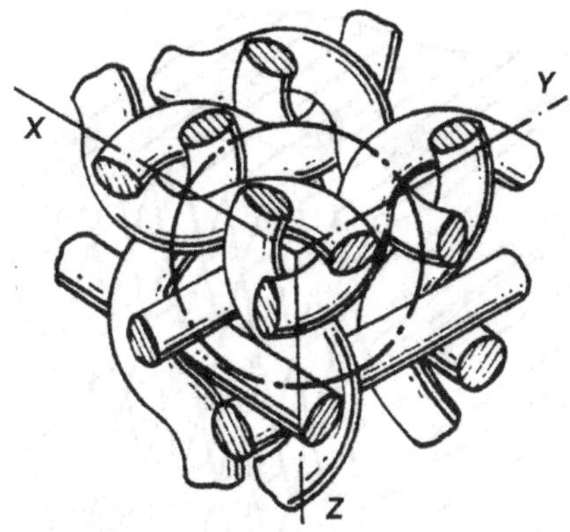

FIGURE 14.14

3D solid woven structure.

Source: Adapted from Van Schuylenburch (1993).

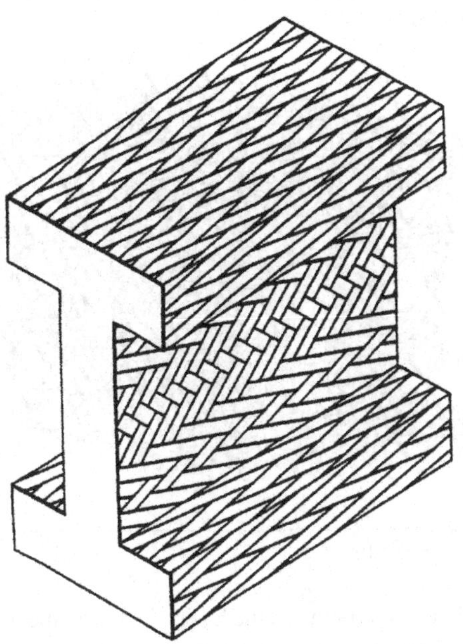

FIGURE 14.15

Woven I-beam (Tsuzuki, 1994).

14.6.2 HOLLOW STRUCTURES

Three-dimensional hollow structures contain various cavities within the structure. These may exist as continuous tunnels or isolated cells. Hollow structures can be used in many technical applications, such as reinforcement for lightweight structural composites, insulation materials and expansible materials for tents. In weaving, a flat fabric with interconnected multiple layers is often initially produced. The flat fabric is then

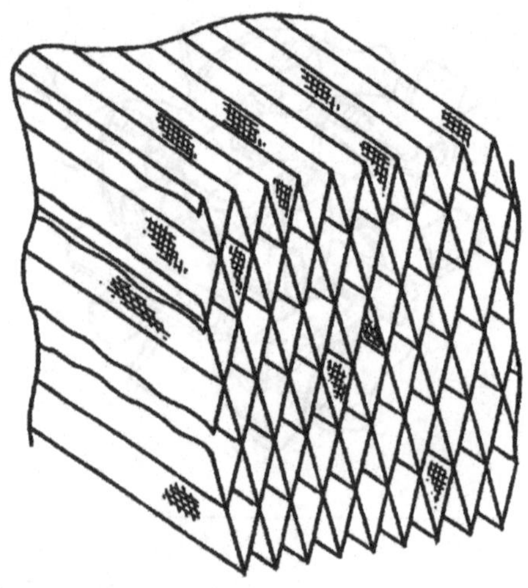

FIGURE 14.16

Woven tunnel structure (Yoshida, 2000).

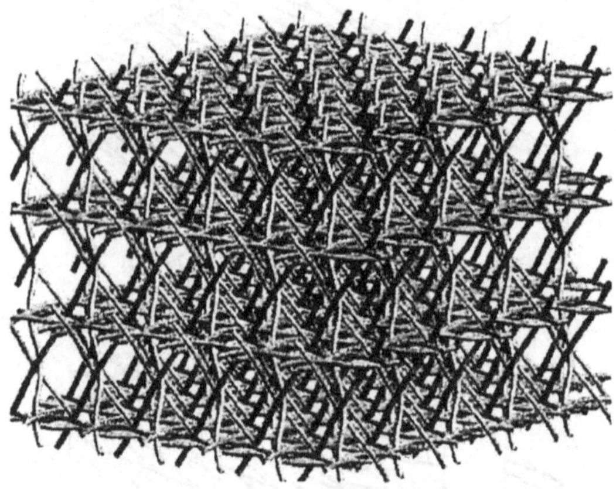

FIGURE 14.17

Woven cellular structure (Kang & Lee, 2010).

opened up after weaving. The connections between the layers then form the walls of the cavities (Koppelman & Campman, 1963; Porat et al., 1996; Rheaume, 1970). Figure 14.16 shows an example of 3D woven structure with tunnels (Yoshida, 2000) and Figure 14.17 shows a hollow structure with cells (Kang & Lee, 2010).

14.6.3 SHELL STRUCTURES

Three-dimensional shell structures are typically individual products with a relatively thin enclosure and one or more openings, e.g. as in a hat. Three-dimensional shell structures are often made from flat panels by moulding, stitching or fusing. However, forming 3D shell structures directly from fibres or yarns can result in more homogeneous properties and at lower cost. Three-dimensional shells produced

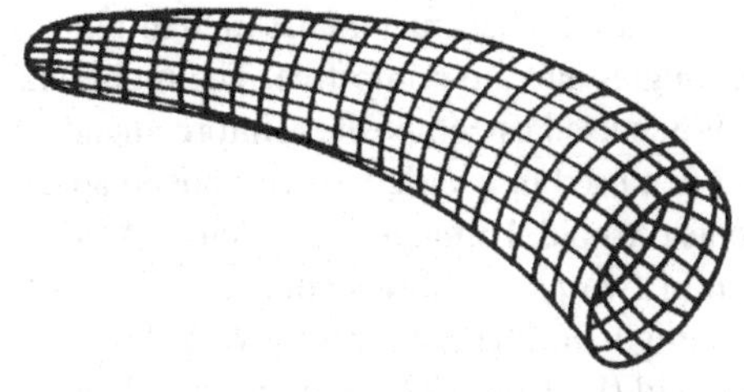

FIGURE 14.18

Woven shell structure (Horovitz, 2000).

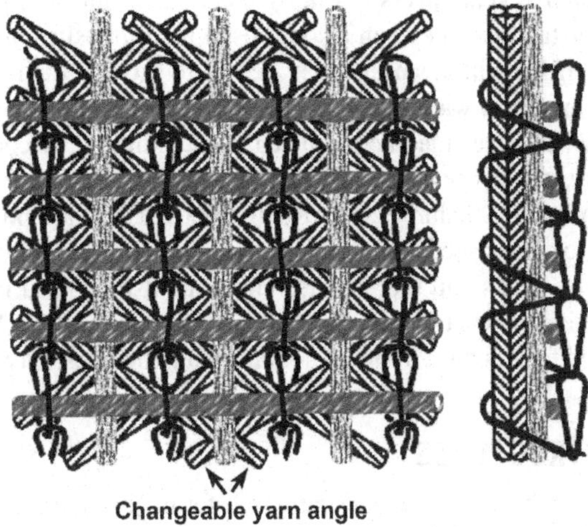

FIGURE 14.19

Multiaxial warp-knitted structure.

by weaving must be part of a continuous fabric because of the continuity of the warp yarns; the individual shells need to be cut out afterwards. Horovitz (2000) described an interesting method through which complex shell structures such as the one shown in Figure 14.18 can be woven for applications such as ballistic protection composites.

14.6.4 KNITTED STRUCTURES

Knitted fabrics generally have lower modulus and dimensional stability than woven fabrics due to the loop formation in the structure. However, knitted structures can conform to complex shapes more easily without developing creases, and knitting is a more flexible process in terms of forming preshaped structures. There are two basic types of knitting machines, warp knitting and weft knitting, defined by the loop yarn feeding direction. Warp-knitted structures have better structural stability and are more suitable as reinforcement material for composites. Knitted 3D fabrics may be divided into three varieties, multiaxial warp-knitted fabrics, spacer fabrics and fully-fashioned fabrics (Guo, 2011). Multiaxial warp-knitted fabrics are basically multiple yarn layers stitched together during knitting, as illustrated in Figure 14.19. In addition to the warp yarns along the length and the weft yarns across the width, yarn layers on a diagonal or along other angles can be used. The structure has high in-plane stability and modulus, because the yarns are aligned at multiple angles and are crimp-free. The multiaxial structure also has good conformability to complex shapes. When used as composite reinforcements, these structures are considered to be 3D solid fabrics and have been used in a variety of composite applications such as wind turbines, and car and aeroplane components (Arnold, Hufnagl, Stopp, Puffi, & Sfregola, 2004; Guo, 2011).

Spacer fabrics are composed of two-plane fabrics connected by pile fibres or yarns. These can be made by knitting (Ikenaga & Taniguchi, 2010; Shirasaki & Yukito, 2006) weaving (Bottger, 2000; Roell, 1996), or a nonwoven method (Le Roy, 1995; Poillet & Le Roy, 2003). They are essentially the same as the double fabrics described earlier in the section on pile fabrics, although the two linked fabrics are used as a single assembly instead of being separated afterwards. Spacer fabrics can be used on their own in applications such as seat covers to offer greater permeability and absorption performance than offered by single-layer fabrics. They can also be used in composites, and the space between the two surface fabrics may be filled with a variety of materials to provide the required properties for particular applications, e.g. as partition walls for improved noise and heat insulation.

Three-dimensional shell structures may be produced during knitting by controlling knitting parameters such as the stitch density, stitch type and number of working needles in one course (Kazuyoshi, 2008; Kobata & Nakai, 2001; Nobuo & Takahiro, 2010; Roell, 2000). These are produced using flatbed weft knitting machines with two or more needle beds. Three-dimensional knitted fabrics are used in seamless products such as hats, gloves, socks and seat covers. They can also be used in composite materials for applications such as jet engine vanes and T-shaped connectors (Guo, 2011). When compared with 3D woven structures, 3D-knitted fabrics have a lower modulus but better elasticity and shock absorbency.

14.6.5 NONWOVEN STRUCTURES

Nonwoven solid 3D fabrics can be produced in a single web-forming step using air-laying or by layering a number of webs on top of each other. High-bulk products with greater compression resistance and recovery can also be produced by folding a web vertically in the perpendicular direction of the web plane (Gong, 2011). These products are typically used for insulation applications. There have also been a number of developments for producing 3D shaped nonwovens directly. Most recently, a process based on air-laying was developed at the University of Manchester, UK (Gong, Dong, & Porat, 2003; Gong & Porat, 2001). Another based on melt-blowing was developed at North Carolina State University, USA (Farer, Seyam, Ghosh, Grant, & Batra, 2002; Farer et al., 2003; Velu, Ghosh, & Seyam, 2003; Velu, Seyam, & Ghosh, 2004).

The University of Manchester process is shown in Figure 14.20. Staple fibres are opened with an opening unit and then taken off the opening cylinder by high-velocity airflow. The airflow carries the fibres to the perforated 3D moulds to form 3D shaped webs that are then moved into a bonding chamber for consolidation. In order to produce a final product with the desired fibre distribution and other properties, the airflow in the web-forming and web-bonding chambers must be appropriately controlled. Using this process, it is possible to produce products with complex surface contours and large depths from different fibre types, a result which would be difficult to achieve using moulding techniques.

The process developed by North Carolina State University is shown in Figure 14.21. This process forms 3D shapes from molten polymers. The melt-blowing process uses high-velocity hot air to blow the molten polymer out of an extruder die mounted on a movable robotic arm, and draws the polymer flow into short fibres. The melt-blown fibres are sprayed onto a mould to form a 3D-shaped product. The mould itself can also be moved in order to produce complex-shaped products. To achieve the desired fibre distribution in the final product, the movements of the die and the mould must be

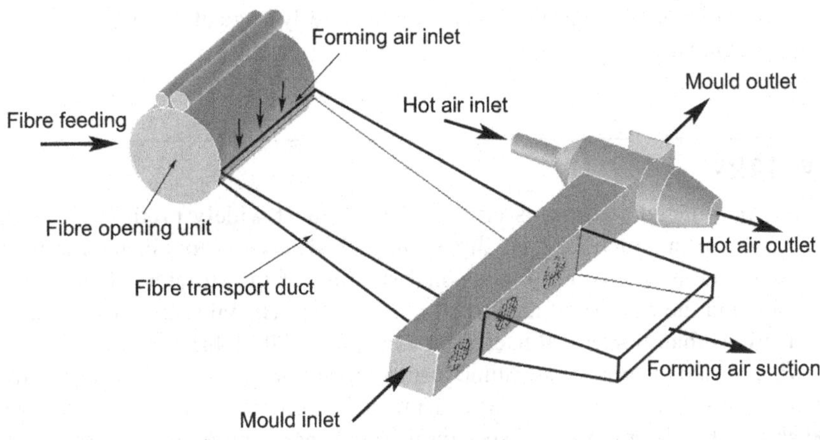

FIGURE 14.20

Air-laying 3D-shaped nonwoven process (Gong et al., 2003).

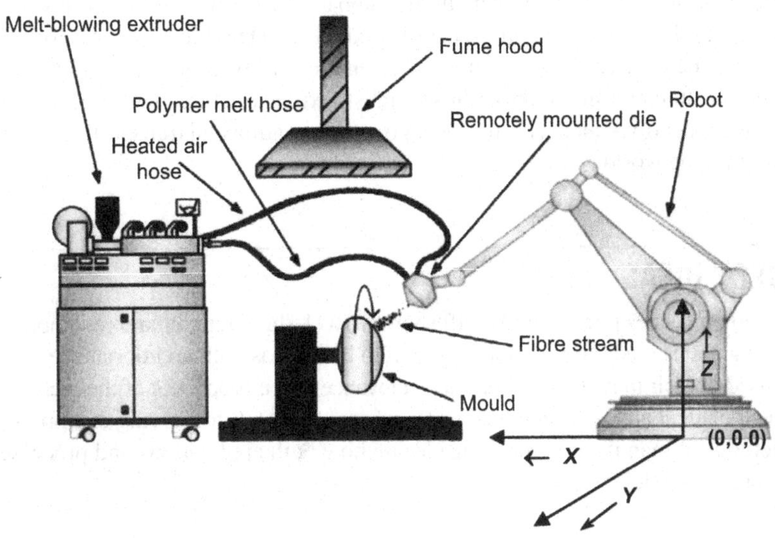

FIGURE 14.21

Melt-blown 3D nonwoven process.

Source: Adapted from Velu et al. (2003).

precisely controlled. The melt-blowing process involves numerous interacting variables and large numbers of fibres; the varying curvature of the 3D shape further increases the complexity. The properties of the final product will be affected by a range of parameters including the choice of polymer, flow conditions, fibre density distribution and fibre orientation distribution. Due to the uncontrolled nature of the drawing action by the hot air, the fibres tend to vary in length and thickness but are usually on the order of a few microns – the material is limited to synthetic thermoplastic fibres.

Three-dimensional shell nonwoven processes are still in an early stage of development and are limited to laboratory trials at present.

14.7 SUMMARY

This chapter has introduced and discussed some of the most widely produced specialist textile structures. Triaxial woven structures have three interlocked sets of yarns instead of the traditional two. This increases the isotropy and dimensional stability of the structure. However, the higher cost of production and the rigidity of the structure mean that triaxial structures are more suited to technical applications than traditional uses such as apparel. Pile fabrics provide soft surface textures and are widely used in apparel and home furnishing applications, while knotted structures are usually used as nets and decorative products. Braided structures are relatively narrow, and they provide good shock absorbency because structural strands are continuous. Braids are most widely used as one-dimensional products, but there is growing interest in using braids in composite applications.

Three-dimensional fabric structures can be made using almost any of the fabric production processes. Knitted 3D fabrics are already widely used in apparel, while woven 3D fabrics are used more in technical applications. The most difficult aspect of producing 3D textile structures for technical applications is obtaining the desired fibre density and orientation throughout the final product. However, there is no doubt that an increasing number of specialist textile structures will be developed for a wide range of technical applications such as aerospace, power generation and transport, as intensive research is being done around the world.

14.8 PROJECT IDEAS

1. Write an advertising/sales pitch for (a) a gift shop, (b) a clothing retailer and (c) a home furnishing retailer offering goods made from fabrics relevant to their areas with appropriate descriptions of the techniques used in their manufacture and their relevance to the properties of the merchandise.
2. Describe the technical and non-technical applications of 3D fabrics, giving examples of the different methods of manufacture and considering how both applications and processes may be expected to change in the future.

14.9 REVISION QUESTIONS

1. What are the distinctive properties of biaxial and triaxial fabrics?
2. Describe the end-uses of pile fabrics with reference to their manufacturing processes.
3. Describe the techniques of creating fabrics by knotting, and give examples of their end-uses.
4. Describe the process of braiding, and give reasons for the limitations on the use of these fabrics.
5. What are the main categories of 3D fabrics? Describe the latest developments in their manufacture.

REFERENCES AND SOURCES OF FURTHER INFORMATION

A & P Technology, Cincinnati, OH 45245-1055. http://www.braider.com/.

Arnold, R., Hufnagl, E., Stopp, J., Puffi, F., & Sfregola, P. (2004). *Method for producing multiaxial warp knit fabric*, US6711919.

Bottger, W. (2000). *Spacer fabric*, US6037035.

Brandon, K. (2009). *The luxury, the indulgence that is velvet*. Textile Fabric Consultants, Inc. www.textilefabric.com.

Breitschadel, F. (1934). *Process for the manufacture of knotted carpets on carpet knotting looms*, GB420651.

Brown, R. T. (1987). *Braiding apparatus*, GB2205861.

Bryn, L., Nayfeh, S. A., Islam, M. A., Lowery, W. L., Jr, & Harries, H. D., III (2004). *Loom and method of weaving three-dimensional woven forms with integral bias fibers*, US20040168738.

Chen, X. (2011). Interwoven fabrics and applications. In R. H. Gong (Ed.), *Specialist yarn and fabric structures*. Cambridge: Woodhead Publishing.

Crawshaw, G. H. (2011). Pile carpets. In R. H. Gong (Ed.), *Specialist yarn and fabric structures*. Cambridge: Woodhead Publishing.

Dow, N. F. (1969). *Triaxial fabric*, US3446251.

Erth, D., & Wegner, A. (1999). *Production of nonwoven fabrics with a velvet pile surface character*, DE19823272.

Farer, R., Seyam, A. M., Ghosh, T. K., Grant, E., & Batra, S. K. (2002). Meltblown structures formed by a robotic and meltblowing integrated system: impact of process parameters on fiber orientation and diameter distribution. *Textile Research Journal*, *72*, 1033–1040.

Farer, R., Seyam, A. M., Ghosh, T. K., Batra, S. K., Grant, E., & Lee, G. (2003). Forming shaped/molded structures by integrating of meltblowing and robotic technologies. *Textile Research Journal*, *73*, 15–21.

Farley, G. L. (1995). *Method and apparatus for weaving curved material preforms*, US5394906.

Fischer, U. V. (2009). *Crochet stitch/pattern*, US 2009/0061396.

Fukuta, K., & Aoki, E. (1986). 3D fabrics for structural composites. In *Proceedings of the 15th textile research symposium*. Philadelphia, PA: Textile Machinery Society of Japan.

Gong, R. H., & Porat, I. (2001). *Moulded fibre product*, GB2361891.

Gong, R. H., Dong, Z., & Porat, I. (2003). Novel technology for 3D nonwovens. *Textile Research Journal*, *73*(2), 120–123.

Gong, R. H. (Ed.). (2011). *Specialist yarn and fabric structures*. Cambridge: Woodhead Publishing.

Gossl, R., & Seidel, T. (2002). *Process for the manufacture of a double velvet fabric*, EP1180556.

Guo, Z. (2011). Developments in 3D knitted structures. In R. H. Gong (Ed.), *Specialist yarn and fabric structures*. Cambridge: Woodhead Publishing.

Horovitz, Z. (2000). *Two- and three-dimensional shaped woven materials*, US6086968.

Ikenaga, H., & Taniguchi, Y. (2010). *Stereoscopic knitwork*, US2010/0229606.

Kang, K. J., & Lee, Y. J. (2010). *Three-dimensional cellular light structures weaving by helical wires and the manufacturing method of the same*, US2010/0071300.

Kazuyoshi, O. (2008). *Three-dimensional knitting method, and three-dimensional article knitted by the method*, WO2008143172.

Khokar, N. (2001). 3D-weaving: theory and practice. *Journal of the Textile Institute*, *92*, 193–207.

Kim, Y. K. (2011). Flocked fabrics and structures. In R. H. Gong (Ed.), *Specialist yarn and fabric structures*. Cambridge: Woodhead Publishing.

Kimbara, M., & Hayashida, M. (1993). *Rod-type three-dimensional loom and continuous operating method*, US5273078.

Kobata, Y., & Nakai, S. (2001). *Method of knitting 3-D shape knit fabric*, US6318131.

Koppelman, E., & Campman, A. R. (1963). *Woven panel and method of making same*, US3090406.

Le Roy, G. (1995). *Method and device for producing composite laps and composites thereby obtained*, US5475904.

Nobuo, F., & Takahiro, Y. (2010). *Tube-shaped knitted fabric, and knitting method therefore*, WO2010098052.

Obst, A., Palermo, G., Ticci, L., & Santiago Prowald, J. (2005). Modeling of triaxial woven fabrics for antenna reflectors. In *Proceedings of the European conference on spacecraft structures, materials and mechanical testing 2005 (ESA SP-581)*. 10–12 May 2005. The Netherlands: Noordwijk.

Pellegrino, S., & Kueh, A. (2006). Thermo-elastic behaviour of single ply triaxial woven fabric composites. In *The 47th AIAA/ASME/ASCE/AHS/ASC conference on structures, structural dynamics, and materials*, May 2006. American Institute of Aeronautics and Astronautics.

Philpott, L. (2011). Knotted fabrics. In R. H. Gong (Ed.), *Specialist yarn and fabric structures*. Cambridge: Woodhead Publishing.

Poillet, P., & Le Roy, G. (2003). Needle punched 3D nonwoven structures with technical functions. *Asian Textile Journal*, *6*, 46–47.

Porat, I., Zhao, L., & Greenwood, K. (1996). *Weaving of preforms*, WO96/24712.

Potluri, P. (2011). Developments in braided fabrics. In R. H. Gong (Ed.), *Specialist yarn and fabric structures*. Cambridge: Woodhead Publishing.

Rheaume, J. A. (1970). *Three-dimensional woven fabric*, US3538957.

Rock, M., & Lohmueller, K. (2001). *Double face warp knit fabric with two-side effect*, US6199410.

Roell, F. (1996). *Textile spacer material, of variable thickness, production process and uses for it*, US5589245.

Roell, F. (2000). *Process for producing three-dimensional knitted fabrics and textile material thus produced* US6122937.

Rudo, D. N. (2007). *Triaxial weave for reinforcing dental resins*, US7186760.

Scardino, F. L., & Ko, K. K. (1981). Triaxial woven fabrics, part I: behavior under tensile, shear, and burst deformation. *Textile Research Journal*, *51*, 80–89.

Scherzinger, W. (1997). *Circular braiding machine*, GB2308389.

Shirasaki, F., & Yukito, K. (2006). *Warp knit fabric with steric structure*, US2006/0172646.

Starbuck, M., & Shilton, E. (2010). *Cut pile fabric and method of making same*, US7757515.

Tsuzuki, M. (1994). *Three·dimensional woven fabric with varied thread orientations*, US5348056.

Tyler, T. (2011). Developments in triaxial woven fabrics. In R. H. Gong (Ed.), *Specialist yarn and fabric structures*. Cambridge: Woodhead Publishing.

Uchida, H., Yamamoto, T., Takashima, H., Otoshima, H., Yamamoto, T., Nishiyama, S., & Shinya, M. (1999). *Three-dimensional weaving machine*, US6003563.

Van Schuylenburch, D. W. P. F. (1993). *Three-dimensional woven structure*, US5263516.

Velu, Y. K., Ghosh, T. K., & Seyam, A. M. (2003). Meltblown structures formed by a robotic and meltblowing integrated system: impact of process parameters on the pore size. *Textile Research Journal*, *73*, 971–979.

Velu, Y. K., Seyam, A. M., & Ghosh, T. K. (2004). Meltblown structures formed by robotic and meltblowing integrated system: the influence of the curvature of collector on the structural properties of meltblown fiberwebs. *International Nonwovens Journal*, *13*(3), 35–42.

Yoshida, S. (2000). *Three-dimensional woven fabric structural material and method of producing same*, US6010652.

Zhao, Qi, & Hoa, S. V. (2005). Finite element modeling of a membrane sector of a satellite reflector made of triaxial composites. *Journal of Composite Materials*, *3*, 581–600.

CHAPTER

YARN TO FABRIC: INTELLIGENT TEXTILES

15

H. Mattila
Tampere University of Technology, Tampere, Finland

LEARNING OBJECTIVES

At the end of this chapter, you should be able to:

- Understand the concepts of intelligent textiles, wearable technology and technical textiles
- Explain the main interactive properties of intelligent textiles
- Describe what kind of applications can be developed by using intelligent textiles
- Relate the importance of design issues to interactive properties of intelligent textiles

15.1 INTRODUCTION

Intelligent systems consist of three parts: a sensor, a processor and an actuator, all managed by controlling data, as presented in Figure 15.1. For example, body temperature monitored by the sensor is transferred to a processor, which computes a solution based on the received information and sends a command to the actuator for temperature regulation. This is a theoretical definition – an example of an intelligent textile system in practice is a jacket made using Outlast, a thermoregulating phase change fabric. When a wearer's body temperature rises above a certain point, the fabric cools the body by absorbing excess heat. At times when the temperature decreases below the threshold point, heat is returned and the wearer feels warmer.

According to Schwartz (2002), intelligent materials are capable of sensing and responding to their surrounding environment in a predictable manner. Intelligent textiles can sense and react to mechanical, thermal, chemical, magnetic or other kinds of environmental stimuli (Tao, 2001). Attempts by different experts to define the terminology:

- Intelligent textiles, also called smart textiles, interact with the environment. Based on information received, they perform predetermined actions repeatedly and often reversibly, such as phase change fabric Outlast given as an example above.
- Wearable technology products are textiles where electronic or mechanical components are attached to the textile material, and the textile part does not have any intelligent properties. By attaching solar panels and wires to a jacket, Ermenegildo Zegna has designed a product that can recharge your mobile phone battery – the textile parts of this jacket are made of conventional and nonintelligent fabric.

Textiles and Fashion. http://dx.doi.org/10.1016/B978-1-84569-931-4.00015-5

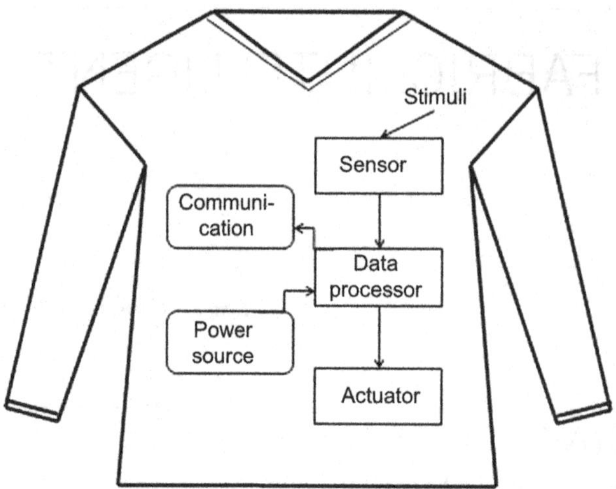

FIGURE 15.1

Three parts of an intelligent system with enabling technology.

- Technically advanced textiles are materials that have noninteractive special properties, and the material itself stays unchanged despite environmental change. Clothing made of Gore-Tex is water repellent and breathable, i.e. technically very advanced, but it is not considered intelligent, as its properties are always static and it does not react to environmental changes.

Shape memory fabric Diaplex is another good example of a smart textile. The molecular structure of the fibre changes with variations in temperature, increased or decreased ventilation, and moisture permeability. Musical jackets and similar products are also examples of wearable technology, although parts such as the user interface may be directly embedded in the textile itself. Fire retardant workwear fabrics are an example of technically advanced materials, but they cannot be regarded as intelligent because they do not interact with the environment in any away. Terminology varies from one author to another, but that is understandable because smart textile standards are only now being developed. This chapter focuses mainly on intelligent textile materials, but practical applications are also discussed.

15.2 WHAT ARE INTELLIGENT TEXTILES USED FOR?

Health and wellness is one of the main areas for smart textile applications. In the medical and health care sector, there is great hope for smart textile applications. Smart textile products are expected to genuinely transform the concept of health care in the future (Langenhove, 2007). Sensors and textile-based diagnostic systems are great tools for monitoring patients, managing risk and making diagnoses more accurate. VivoMetrics' LifeShirt is one of the best-known products for continuous monitoring of physiological data, including blood pressure, blood oxygen saturation, periodic leg movement, core body temperature, skin temperature and patterns of coughing. SmartShirt was originally developed by the Georgia Institute of Technology and later commercialized by Sensatex. It can transmit collected biometric data to a PC or other processors (Textile Intelligence, 2008). Smart textile applications may

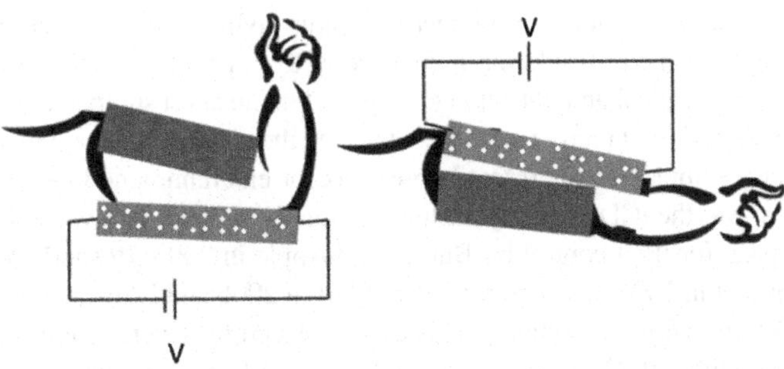

FIGURE 15.2

Electroactive polymer shirtsleeve expands and shrinks according to electric current. When voltage is applied, the polymer expands (patterned part), and when the voltage is removed, the sleeve returns back to its original length (block part).

help the aging population to stay home longer, and can be used for rehabilitation of injured or disabled patients. Infant monitoring, mobile health monitoring, drug-releasing textiles, surgical implants, wound care and even human spare parts are among current medical applications.

Sports and fitness is another primary area for smart textile applications, especially for consumer use. Monitoring of exercise performance, in terms of heart rate, body temperature, motion details, etc., interests sportspersons at all levels of athletic activity. Position and motion sensing can be used to monitor training and rehabilitation. Using sensors attached to the legs, it is possible to monitor a person's exact leg movements while running or walking on an exercise carpet. This could be useful for ensuring that the limb trajectory is correct – for example, when a person is being rehabilitated and learning to walk again after a major leg injury. Use of electroactive polymer actuators to produce artificial muscles could be applied for the enhancement of muscle performance. A shirtsleeve made of such material would expand and shrink according to electrical voltage, and muscular strength would increase and decrease accordingly as shown in Figure 15.2.

15.2.1 SMART TEXTILE APPLICATIONS

Smart textile applications for military and occupational safety include sensors incorporated into base layers of clothing for monitoring of the user's vital signs, as well as sensors attached to the outer layer for monitoring environmental conditions such as heat, gases, etc. The resulting information can then be transmitted wirelessly to the command centre. Monitoring can be done without interference, and the command centre can use the information to prevent injuries and reduce exhaustion and stress (Textronics, 2010). Applications are being developed for the military, firefighters, police, fishing professionals and others who work in hazardous and dangerous environments.

Both the US military and NATO (North Atlantic Treaty Organization) have actively developed wearable computing applications for battlefield soldiers. The Future Force Warrior research programme of the US military is developing a smart battle suit to be introduced by 2020 that will include GPS and network communications, and sensors to monitor physiological indicators such as heart rate, blood pressure and hydration. Liquid body armour and an exoskeleton are incorporated to enhance the soldier's protection and strength (Science.howstuffworks, 2009). By adding liquid to conventional armour materials such as Kevlar, a bullet's shock impact is spread faster over a larger area, thus giving soldiers better protection in combat. Exoskeletons are devices attached outside the human body to improve muscular strength. Traditionally, these devices have been mechanical and not very convenient, but the latest smart textile technology makes it possible to integrate such devices directly into garments using the electroactive polymers described earlier.

There are numerous examples of smart textile products for entertainment. One of the first to utilize a integrated music system was the ICD jacket by Philips and Levi's in 2000. This was followed by the Burton Amp jacket and backpack for iPod control by Burton and Apple in 2003, Rosner's MP3 player jacket in 2004, O'Neill's Hub jacket in 2005, CuteCircuit's Hug Shirt in 2006 and Ermenegildo Zegna's solar-powered jacket in 2007. Additional iPod clothing has been introduced by several companies including M&S, Bagir, Koyono, Kempo, JanSport, Quiksilver and Craghoppers. More recent introductions of smart textile products include O'Neill's NavJacket with GPS, Bogner-Osram's solar-powered LED jacket, Zanier's Heat-GX heated gloves, and the heat-transfer sock by Therm-ic and X-Technology Swiss (McCann & Bryson, 2009); however, none of these has been commercially successful. An illustrative example of a smart textile product is a dress by Enlighted Designs that makes use of LED devices, as shown in Figure 15.3.

FIGURE 15.3

A LED-lighted dress is an example of many entertainment and fun applications of smart textiles.

Source: Enlighted Designs.

15.2.2 RESEARCH AND DEVELOPMENT OF SMART TEXTILES

Research and development of smart textiles is often multidisciplinary – besides textile know-how, researchers need skills in areas such as electronics, telecommunications, biotechnology and medicine. Networking and knowledge exchange with academic institutes and businesses is often the only way to carry out successful smart textile research. The complexity and broadness of knowledge required makes smart textile research interesting but also challenging. According to Schwarz, Langenhove, Guermonprez, and Deguillemont (2010), a smart textile system can incorporate many functions: sensing, actuating, communicating, data storing and interconnecting to other systems. Smart textiles can be classified by their functionality (Tao, 2001):

- Passive smart textiles can sense the environment. A sensor is an example of these materials – it recognizes an impulse from the environment, but does not perform any activity.
- Active smart textiles sense a stimulus from the environment and perform an actuating function. Phase change material (PCM) absorbs and releases heat based on the temperature of the environment, and because PCMs can sense temperature and then perform a function based on that input, they are called active smart textiles.
- Very smart textiles adapt their behaviour according to environmental circumstances. Shape memory materials are often called very smart, as they can remember their original shape and return to it (Hu, 2007). Ultimately these materials are textiles that can learn and adapt to various stimuli. Such textiles have not yet been developed.

As defined earlier, a sensor is a passive but important part of a smart textile system – it is created by weaving or knitting electroconductive yarns into fabrics as signals are electrically transmitted. Sensors are used for monitoring heart rate, EKG, position, velocity, temperature, humidity and pressure, including pressure-sensitive sensors used in textile-based keyboards. Another important element is the actuator – actuators react to signals coming from sensors, with motion, sound or substance release. Thermal actuators can be used for warming or cooling. A drug-release actuator releases chemicals, for example for skin care. A power supply unit may be independent or integrated into the textile structure. Flexible solar panels, microfuel cells, flexible batteries and kinetic energy harvested from body motion are examples of integrated power sources. Solar panels are used to collect solar energy and convert it to electricity. They are normally made of rigid material, but in textiles flexible panels are needed. Batteries integrated into textile structures must be flexible. Microfuel cells are portable power sources that convert chemical energy into electricity. They can be attached to textiles using membranes having a fuel or oxidant catalyst layer. Kinetic energy harvesting is the conversion of human motion into electric power. Piezoelectric energy harvesting is the turning of pressure and vibration into electricity. Figure 15.4 illustrates a shoe sole that produces electricity from the pressure of each step.

There are many challenges that textile embedded power sources must cope with, such as flexibility, washability, heat from ironing, etc. Communication between a smart garment and its user may be arranged wirelessly. To facilitate this antennas and transmitters are attached to textiles. By means of fibre optics, flexible optical fibre displays can be attached to the textile surface and used for displaying messages and images. Philips Lumalive fabric is an example of photonics, which uses LED technology to display messages on textile surfaces. Sensors, actuators and data units must be connected in order to send and receive information, which can be achieved by embroidery, conductive Velcro tape, electroconductive yarns and snap buttons (see Figure 15.5).

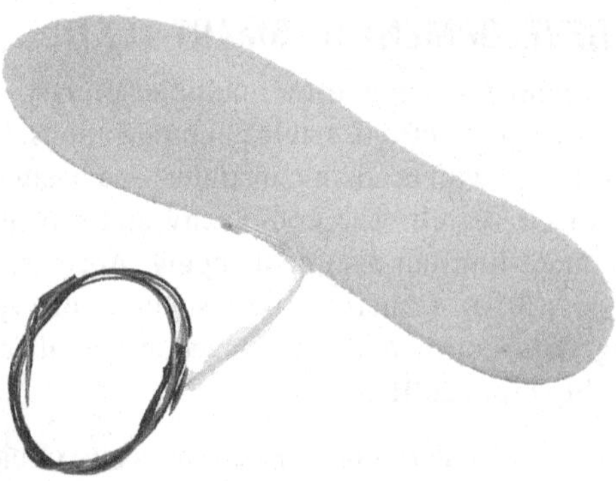

FIGURE 15.4

Smart Material Corporation has developed a shoe sole that can harvest piezoelectric energy by turning the pressure of each step into electric power.

Source: Smart Material Corporation.

FIGURE 15.5

Conductive hook and loop, or the so-called Velcro tape can be used for electric connection. Hooks and loops must be made of or coated by conductive material, for example silver.

Source: Hannah Perner-Wilson, PLUSEA.

15.2.3 PHASE CHANGE MATERIALS

Phase change materials are thermal storage materials used for regulating temperature fluctuations. PCM technology was originally created by NASA's space research programme in the early 1980s. Currently, phase change textiles are used in several areas, such as summer and winter clothings shoes, car seats and bed covers. Several military applications have also been developed, including a climate controlling antiballistic vest, combat uniforms, protective suits, helmets, boots, sleeping bags and underwear. Socks and underwear made of acrylic yarns, which in their fibres contain microcapsules filled with paraffin wax, are commercially available. Coated linings containing microcapsules are used for jackets. Midlayers between the shell fabric and the lining may also be coated with PCM. Outlast Technologies LLC is an American company and a pioneer in developing phase change technology for textiles. Comfortemp, an intelligent nonwoven textile into which PCM has been integrated, is a brand of

FIGURE 15.6

Schoeller-PCM fabric with microcapsules filled with phase change material (PCM).

German company Freudenberg, and is used for garments, shoes, gloves, hats and blankets. Schoeller is a Swiss textile company that has developed a phase change textile, called schoeller-PCM, that contains millions of tiny microcapsules filled with PCM for dynamic climate control, as shown in Figure 15.6.

The aforementioned materials control heat transfer by using chemical bonds to store and release heat as a material changes from solid state to liquid and back, and can thus cool or warm the user if attached to textile structures next to the skin (Mäkinen, 2006). The objective of all clothing is to help wearers to maintain a comfortable skin temperature, and this is achieved by keeping the microclimate temperature at a certain level. Microclimate is the thin layer of air between the skin and the piece of clothing next to the skin. Thermal insulation of textiles and garments can be enhanced with PCM treatment:

- surplus body heat is absorbed creating a cooling effect
- added insulation, as the textile structure creates a thermal barrier that prevents undesired emission of body heat
- the thermoregulating effect keeps the microclimate temperature within a desirable comfort zone

Heat is stored when the change from solid to liquid takes place, and released when the phase changes back to solid. PCM absorbs heat from the environment until it reaches its melting point. During the physical phase change from solid to liquid, temperature remains constant, and energy is absorbed to break down the bonding responsible for the solid structure. This so-called latent heat will be released into the environment when the material cools down as the phase change from liquid to solid takes place. Thermal energy storage materials must have high thermal conductivity and a large latent heat capability. In other words, the heat flow through the material must be efficient, and the material must be able to store heat well, but the melting temperature must be in a practical range when used in textiles. The PCM must also be nontoxic, chemically stable and low cost. During the past few years, PCM research has largely focused on hydrated salts, paraffin waxes and eutectics of organic and nonorganic compounds (Farid, Amar, Khudhair, Razack, & Al-Hallaj, 2004).

Paraffin waxes are the main PCMs used in textile applications. They are cheap and have a wide range of melting temperatures, enabling selection of those most practical for regulating human body temperature (see Table 15.1). They are also chemically stable and nontoxic. By selecting a paraffin wax with suitable crystallization and melting points, the preferred temperature, at which cooling or warming starts, can be selected, and the skin temperature stays within a comfortable range (see Figure 15.7). The main problem with current phase change textiles is a rather weak and short cooling and warming impact. This is due to the relatively low levels of PCM that can be embedded in the textile structure

Table 15.1 Phase change materials (PCMs)

PCM	Melting temperature (°C)	Crystallization temperature (°C)	Heat storage capacity (J/g)
Eicosane	36.1	30.6	247
Nonadecane	32.1	26.4	222
Octadecane	28.2	25.4	244
Heptadecane	22.5	21.5	213
Hexadecane	18.5	16.2	237

Mäkinen (2006).

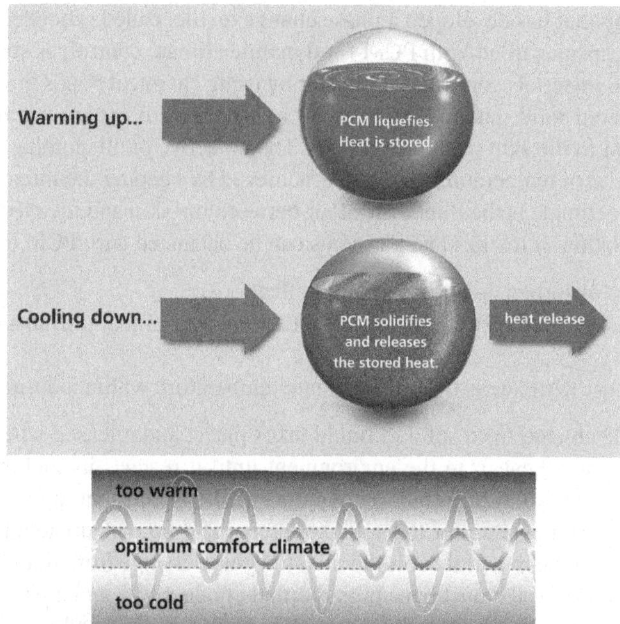

FIGURE 15.7

By storing and releasing heat, phase change material (PCM) keeps skin temperature within a comfortable range.

Source: Schoeller.

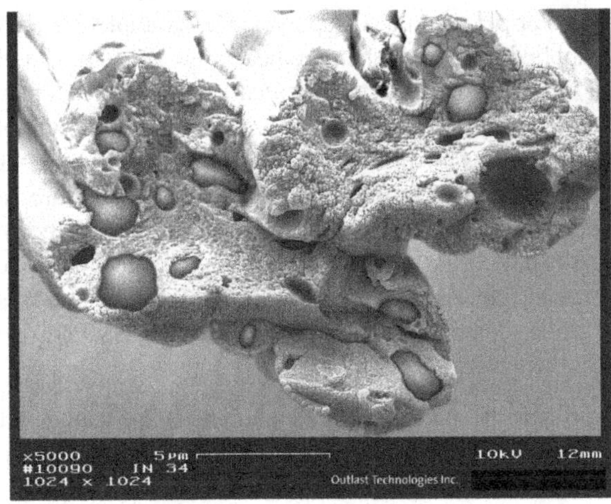

FIGURE 15.8

Melt-spinning, coating and laminating are the traditional ways of attaching phase change material microcapsules to textiles. The image illustrates capsules inside viscose fibres.

Source: Outlast.

using current technology. At present, microcapsules filled with paraffin wax are used to join temperature-regulating substances to textiles. The capsules are small, between 1 and 30 µm in diameter, or about half the diameter of a human hair. The capsules can be embedded in the spinning process, by coating or by laminating, as shown in Figure 15.8.

In order to increase the proportion of PCM in textiles, new technologies are being developed. Outlast introduced an award-winning new bicomponent PCM polyester fibre for enhanced thermoregulation at the Techtextil-Avantex exhibition in 2011. Another development is Novel Temperature Regulating Fibers and Garments (NOTEREFIGA),[1] a collaborative project with EU research and development funding. Its objective is to develop new type of fibres for enhanced temperature management, using two alternative methods. One uses bicomponent melt spinning where the fibre core is made of PCM, while the second incorporates PCM directly to cellulose fibre during the wet spinning process. The objective is to increase the proportion of PCM to 70–75% of total weight, and as a result gain a long-lasting and strong cooling and warming impact (Cordis, 2009).

15.2.4 SHAPE MEMORY MATERIALS

Shape memory materials can transition to a previously programmed shape in response to external stimulus, and often return back to their original shape when the stimulus is removed. According to Hu (2007), shape memory materials can be classified as very smart material, as they can remember their original shape. After deforming into a temporary shape, they return back to their original shape once

[1] Novel Temperature Regulating Fibers and Garments (NOTEREFIGA) is a European research and development project with 14 partners aiming at producing new type of phase change fibres. See http://extra.ivf.se/noterefiga/template.asp for further information.

the external chemical, thermal or pH stimulus disappears. For example, a polymer strap will bend and form a hook when heat is applied to it, and when the heat source is removed, the strap will automatically straighten out. Shape memory materials are also called stimuli-responsive materials. There are several types of shape memory materials, such as metals, ceramics, gels and polymers.

Shape memory textile applications are often based on shape memory polymers (SMPs), first developed in Japan in 1984. The stimuli causing a shape change can be temperature, pH, chemicals and light. Elasticity refers to the property of a substance to resume its original shape when distorting force is removed. For example, glass is not elastic and snaps when bent. A sheet of rubber can be bent back and forth, but returns to its original flat shape. Obviously, elastic polymers are useful when developing shape memory textiles. When designing thermally responsive SMPs, several temperatures are important:

- T_{trans} is the shape memory transition temperature where the shape change takes place
- T_m is the melting temperature and if $T_{trans} = T_m$ then the shape change takes place at the melting point
- T_g is the glass transition temperature above which the material is elastic and below which the shape is stable

SMPs with $T_{trans} = T_m$ include polyurethanes, polyurethanes with ionic or mesogenic components, block copolymers consisting of polyethylene terephthalate and polyethylene oxide, block copolymers containing polystyrene, polyvinyl ether and butyl acrylate, and ABA triblock copolymer made from polytetrahydrofuran (Hu, 2007).

Enhanced temperature regulation, vapour permeability, air permeability, volume expansivity and shrink-resistant finishes are achieved with shape memory materials in textiles, clothing, shoes and other apparel products. Vapour and air permeability, also called breathability, is one of the most important properties in the wearing comfort of apparel. If the perspiration cannot be transported through clothing the wearer feels wet and uncomfortable. Temperature dependent permeability is important especially in sportswear. The permeability increases once the body temperature of the wearer goes up and decreases again once the temperature goes down. In shape memory smart textiles, these properties vary significantly above and below the glass transition temperature T_g due to the difference in kinetic property of molecular chains. At glass transition temperature amorphous or noncrystalline materials perform a reversible transition from rigid state into elastic rubberlike state. Above T_g the moisture permeability increases and heat as well as vapour can be transported away from the skin of the sports person. In temperatures below T_g permeability is low and has an added warming impact on the wearer. Different technology can be used for building these properties into textiles. Manufacturing processes include finishing, coating, laminating and blending. One way is to laminate a thin film of shape memory material on normal outerwear fabric.

Diaplex is a shape memory fabric developed by Mitsubishi International. It is a solid film that is laminated on top of nylon fabric. Being nonporous, Diaplex is more water resistant and breathable than porous membranes. At elevated temperatures, kinetic energy increases and the molecular configuration changes, creating millions of tiny openings to allow excess thermal energy and moisture to escape. When the temperature drops, the molecular structure turns into a solid sheet, forming an insulator to prevent heat and moisture from escaping the body. Diaplex is manufactured using thermoregulating SMPs.

A smart suture that ties itself into a knot represents an SMP application in medical textiles. Other medical applications include polymer implants made of degradable polymers, such as orthodontic materials, bone screws, nails, plates, meshes and scaffolds for tissue engineering (Hu, 2007). Ascending and descending curtains are among the useful applications of shape memory textiles. When the sun shines directly on the window, curtains descend automatically preventing the room to heat up. When sunshine stops the curtains ascend. Several prototypes mainly for amusement purposes have also been developed, such as shirtsleeves that curve up when hot air is blown to them.

The Shape Memory Textile Laboratory of the Hong Kong Polytechnic University is one of the research centres concentrating on smart textile research. It has produced an impressive collection of research articles and books on shape memory textiles. Despite extensive research and development, only a few commercial SMP applications have been launched to the market so far. The advanced properties of SMPs are continuously being developed, however, and new commercially successful textile products are expected in the future with benefits in three areas (Hu, 2007):

- Waterproof, breathable and flame-retardant fabrics for apparel products
- Nonironing, wrinkle-free and shape-fixing garments
- Aesthetically active garments that can change shape or surface patterns

15.2.5 CHROMIC AND CONDUCTIVE MATERIALS

Chromic materials change, radiate or erase colour. Colour change is caused by external stimulus, and once the stimulus ceases to exist, the change is reversed. These materials are also called chameleon materials. Chromic materials can be classified according to the external stimulus that causes the colour change:

- photochromic – light stimulus
- thermochromic – heat stimulus
- electrochromic – electrical stimulus
- piezochromic – pressure stimulus
- solvatochromic – liquid stimulus
- carsolchromic – electron beam stimulus

Photochromic materials, which change colour due to change of light intensity, are the most common type of chromic material. Usually they are colourless in dim light or dark, but when brought to sunlight, their molecular structure changes and they exhibit the colours of underlying paints, inks or dyes. Colour change properties have primarily been used for fun applications, such as an embroidered logo or front print on a T-shirt that changes to bright colours when brought to sunlight, only to disappear again when brought inside. Chameleon-like camouflage fabrics are being developed by defence forces. Outfits made of such material will change colour according to the environment, with the goal of making the soldier invisible. SolarActive International supplies photochromic yarns, embroidery thread, T-shirts and toys. Super Textile Corporation of Taiwan produces photochromic fabrics that can be used for bags, toys and warning signs.

Thermochromic materials change colour at a predetermined thermochromic transition temperature. These materials can be made of liquid crystals, organic or inorganic compounds, polymers or sol–gels. The colour change is very noticeable, even dramatic. Organic colourants are normally used for colour-changing textile applications such as fabrics, yarns and printing inks, and Pilot Ink Co. Ltd of Japan has patented several techniques for producing thermochromic colourings, colour-memory compositions and microcapsule pigments. Commercial applications are similar to those of photochromic materials.

Electrochromic materials change colour in response to electric fields and return to their original colour once the field is reversed. Application areas include specialized displays, electronic books, paper-like displays, etc. Electronic inks can be used for printing images on textiles, which light up under an electric field. One of the current problems with chromic dyes and inks is that the colour change impact is not very stable, and it disappears quickly when the garment is repeatedly washed.

Photonics is a research approach that deals with the generation, emission, transmission, switching and sensing of light. Under this research topic, electrochromic materials, fibre optics, conductive yarns and LEDs may be combined – a textile surface can light up, displaying certain images or colours. One of the pioneers was France Telecom (now Orange France) with a fibreoptic display woven into the component panel of a garment – images and messages sent from a mobile phone would appear on the display. Philips Lumalive is a business venture that specializes in textile-based LED solutions – LEDs are used to display messages on the textile surface. CuteCircuit is a UK-based company that has created fantastic designs by combining conductive textile materials, photonics and wearable electronics. Thousands of LEDs have been embedded in garments that display videos and images in bright and changing colours.

15.2.5.1 Conductive materials

Conductive textile materials are, strictly speaking, not intelligent. They do not react to their environment, but they make many smart textile applications possible, especially those that monitor body functions. They are widely used in smart textile applications such as sensors, communication, heating textiles and electrostatic discharge clothing. Electroconductive materials are required in sensors, actuators and heating panels, and the best-suited materials are highly conductive metals such as copper, silver and steel. Stainless steel and copper yarns can be made flexible, soft and durable enough to be woven or knitted, and electroconductive plastics can be created by mixing conductive polymers, electroconductive fillers such as carbon black, and metal particles. Sensors, which can be either passive or active, can monitor heart rate, EKG and other vital life signals, as well as sense temperature, moisture and pressure. Passive sensors require an external power source, while active sensors use input energy to measure activity (Langenhove, 2007). Heart rate is monitored by measuring the electric impulses that the brain uses to control the heart muscle – this can be achieved using an electroconductive fabric that is placed close to the heart. A heart rate monitoring system for runners, by the Finnish company Polar, is based on conductive fabric on a strap around the runner's chest. The system has been on the market for several years and is selling well. A few years ago, Adidas and Polar joined forces and introduced a system called Fusion, which integrates heart rate monitoring into a T-shirt, bra or even running shoes. The system was not a huge commercial success, however, and it has been replaced by new developments, as Adidas acquired Textronics of the United States, a company whose mission is to seamlessly integrate microelectronics with textile structures (Textronics, 2010).

Eleksen, a division of British company Peratech, produces electroconductive pressure-sensitive fabrics for touch screen interfaces. Their technology, ElekTex, can locate the pressure position on a fabric – this XY positioning can be used in textile-based keyboards and similar user interfaces. The numbers for operating a mobile phone can be embroidered on a sleeve of a garment, as shown in Figure 15.9 – when the number 5 is pressed on the fabric, the system identifies it through its location. Conductive yarns can be used also in heating textiles, and Gorix was one of the first conductive fabrics used for heating – by using an external power source, the fabric can be heated like an electric blanket. Conductive yarns are used by International Fashion Machines to create textile-based light switches: ElectroPUFF lamp dimmer and POM POM wall dimmer.

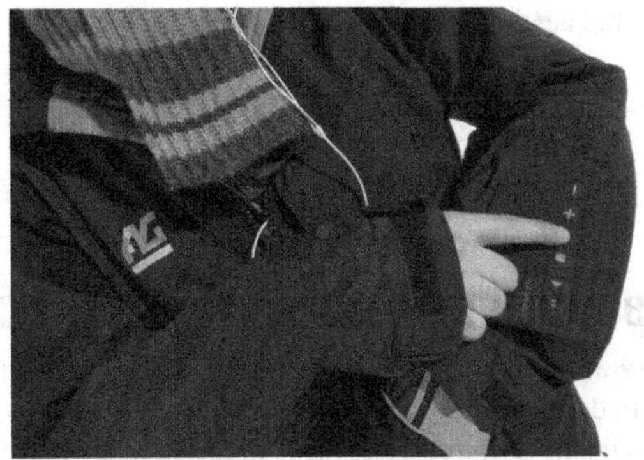

FIGURE 15.9

Textile-based user interface for operating a mobile phone.

Source: Centexbel.

15.2.6 STRESS-RESPONSIVE MATERIALS

Stress-responsive materials change their shape, flexibility or colour when stress is applied to them. The change is reversed once stress is removed. Normally, materials get thinner when stretched, but auxetic material swells when stretched. D3O Lab (http://d3olab.com) has developed specially engineered material for impact protection that normally is flexible, but hardens instantly upon being hit or on impact. Intelligent molecules flow with the motion of the user, but on shock, lock together and absorb the impact energy – this material can be used for protection in athletic and workwear. Beckman Institute at the University of Illinois has developed elastomers made with mechanophore-linked polymers that change colour when stretched. The amber-coloured elastomer turns progressively more orange as it is stretched, and finally turns red right before it snaps. This material could be used as a visual safety device in cords and ropes (Davis et al., 2009).

15.2.7 WEARABLE ELECTRONICS

As discussed earlier, intelligent textiles are defined as products where the intelligence is directly embedded in the textile structure. Wearable electronics means that electronic devices are attached directly to the body of the wearer or to clothing. In this way, the piece of clothing becomes a platform for carrying electronic devices, such as microprocessors, transmitters, cameras, etc. Most of the first attempts to commercialize intelligent textiles were actually garments with wearable electronics designed for entertainment, like the music ICD jacket by Philips and Levi's in 2000. None of the smart textile innovations between 2000 and 2010 was commercially successful. It seems that understanding consumer needs must be the starting point when designing wearable electronics products, and design needs to address the challenges of attaching electronic devices to clothing:

- Electronic devices cannot be bulky or rigid. The appearance and wearing comfort of the garment must be the same as it would be without electronics.
- Devices must be encapsulated in order to make them washable – requiring the removal of all devices before washing is not practical.

- Visible cable or wiring is not aesthetic and consumers may be concerned about electromagnetic radiation.
- Energy sources should be embedded in textile structures. Carrying batteries in pockets is not a robust solution.

15.3 CASE STUDY: BIOMIMETICS AND INTELLIGENT TEXTILES

Biomimetics and biomimicry are terms for research that, by examining models, elements, processes and systems in nature, aims to develop similar solutions to those obtained by conventional technology. Several applications of biologically inspired intelligent textiles have been developed and are already available commercially.

15.3.1 EXAMPLES OF BIOMIMETIC PRODUCTS

The US government's Sandia National Laboratories researches the ability of certain fish species to change colour and in this way blend with the environment. The power source for this colour changing ability is a basic cellular fuel called adenosine triphosphate (ATP), which releases energy as it breaks down. ATP is responsible for intracellular energy transfer in all living cells, and it transports chemical energy within cells for metabolism. Fifty percent of this energy is absorbed by motor proteins that aggregate and disperse skin pigment crystals in the cells, thus rearranging the colour display. Sandia believes that towards the end of the 2010s, they will have developed a fabric that automatically changes colour to fit different environments, using the same cellular fuel used by colour-changing fish (Sandia National Laboratories, 2009). The technology could be used for military purposes, as soldiers could make themselves less visible on the battlefield by changing colour with their surroundings. Other interesting biomimetic research topics at the moment are the strength of spider silk, the gecko's ability to walk on a ceiling, the shark's use of its skin to reduce water friction, adaptation of the butterfly wing for ultralight structures and the ability of polar bear fur to keep skin warm and dry while swimming in arctic water.

15.3.2 THE LOTUS EFFECT

The ability of the lotus leaf to stay clean even in muddy water has inspired the development of self-cleaning surfaces. The surface of the lotus leaf is also extremely hydrophobic, i.e. it cannot be wetted by water. Water droplets just roll off the surface taking all dirt, bacteria and spores with it preventing organic contamination, as shown in Figure 15.10. By scanning the lotus leaf surface with an electron microscope, Professor Barthlott and his team at the University of Bonn found out that the surface is nothing but smooth. It is covered by microscopic and bumpy cells, like nano-hair, that are covered by wax crystals around 1 millionth of a millimetre in diameter (Poole, 2007). Wax crystals are very hydrophobic. This combination creates a surface structure that minimizes the contact area of water and dirt particles. Drops of rainwater roll off the surface, removing all dirt and particles, and as result the leaf is always clean. After discovering the reasons for this self-cleaning ability, Professor Bartholtt named it the lotus effect.

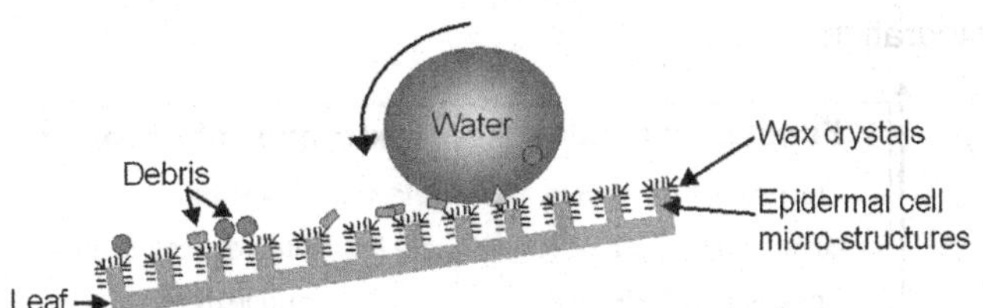

FIGURE 15.10

The lotus effect. Water droplets collect dirt while rolling on the tips of epidermal cell structures.

Source: Poole (2007).

The hydrophobic property of any surface depends on the contact angle, i.e. the angle between the surface and the tangent of the water drop. In order to be hydrophobic the contact angle must be over 120°. Further research has shown that the water contact angle on the lotus leaf is over 160° (Stegmaier, Arnim von, Scherrieble, & Planck, 2008). The minimum contact area between the water drop and the leaf increases the interfacial tension to air, and the drop gains very little adsorption energy. This keeps the water drops spherical, so instead of sliding down the leaf's surface, they roll.

The first applications of the lotus effect were architectural: roof tiles, self-cleaning windows and house paints (Lotusan). American company Nano-Tex has developed a textile finish that makes textile surfaces water and dirt repellent. It can be applied to garments and home textiles. Swiss firm Schoeller Textiles AG has named its lotus effect technology NanoSphere. The easy-clean clothes are becoming widely available. Development of new innovations in technical textiles is under way, and in the future we will see non-wetting self-cleaning marquees, awnings and sails (Forbes, 2008).

15.4 FUTURE TRENDS

Although commercially successful intelligent textile products are still few, the global science community, as well as companies, are investing in smart textile research. Schwarz et al. (2010) state that for smart textiles to achieve future success, several issues must be addressed: a coherent technical strategy; multidisciplinary approach and market applications. Further financing is needed in order to solve technical requirements and promote smart textiles.

15.4.1 FUTURE APPLICATIONS OF INTELLIGENT TEXTILES

As intelligent textiles is a relatively new research area, the scientific community has created novel terminology. The so-called textronics and fibretronics are the most challenging areas for wearable technology and smart textile research as illustrated in Figure 15.11 (there are also the companies Textronics Inc. and Fibertronics Inc., but they have no connection to this general terminology). Textronics refers to manufacturing of electronic components by textile manufacturing techniques – electronic components and textile structure are integrated and the product can be treated as textile. Fibretronics refers to the same techniques but for fibres rather than textiles. Electronic components are inserted into fibres, making them invisible. Catrysse, Pirotte, and Puers (2007) define the main challenges when integrating electronics in textiles as (1) how to attach electronic components to textile structures, (2) power

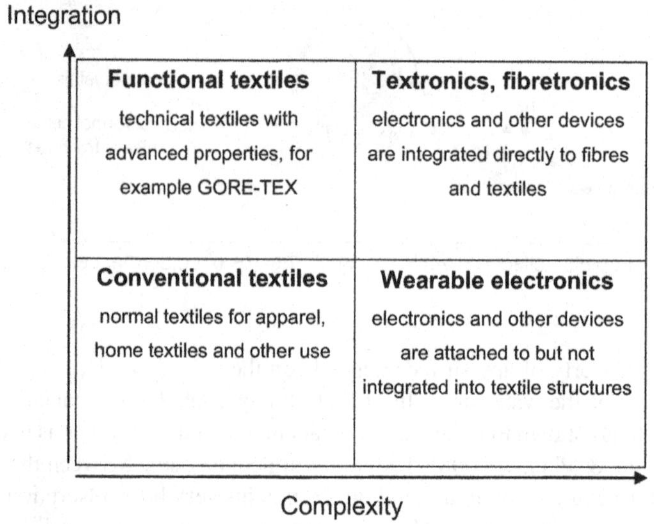

Integration ↑	Low complexity	High complexity
High	**Functional textiles** technical textiles with advanced properties, for example GORE-TEX	**Textronics, fibretronics** electronics and other devices are integrated directly to fibres and textiles
Low	**Conventional textiles** normal textiles for apparel, home textiles and other use	**Wearable electronics** electronics and other devices are attached to but not integrated into textile structures

Complexity →

FIGURE 15.11

Textronics and fibertronics are the most complex applications where intelligent properties are directly embedded to textile structures.

management and (3) washable, flexible and user-friendly packaging of electronic components. McCann and Bryson (2009) emphasize the importance of design, i.e. the end user should appreciate and find the product attractive and worth the money. Malmivaara (2009) argues that some of the main challenges as well as problems are related to the current sourcing–production–retailing value chain. Large sports-goods brands like Nike and Adidas are not direct manufacturers of intelligent textiles, and thus they have to rely on outsourced production, i.e. they are buying production capacity from subcontractors. The garment-manufacturing base in Europe is quite thin and many products, including smart garments, have to be made in East Europe and Asia, where it is difficult to carry out product development. Warranty, product life span and recycling issues may be different for the textile as compared with the electronic parts. Finally, personal electronics are sold in different stores than those that sell textiles and garments, and as a result it may sometimes be difficult to select an optimal distribution channel.

Challenges regarding technology relate often to the complexity of integrating flexible textile structures and rigid electronic components. Although flexible electronics already exist, there are still problems in embedding them in textiles. Current development of digital printing with conductive inks, printed electronics and miniaturization of electronics pave the way for improved textronics products. Inserting electronics inside fibres would be an ideal solution, as electronic components would become invisible and encapsulated. A transponder and antenna melt-spun inside a fibre would eliminate the need to hang or sew a radio-frequency identification (RFID) tag to the garment. RFID is a technology in which a microchip transponder is attached to an antenna on the surface of a tag. Data can be stored on the chip and read from a distance by means of a magnetic field. If this technology is miniaturized to a very small scale, the device could be inserted inside polyamide filament yarn during the melt-spinning process. Another solution is directly embedding such components in textile structures such as ElastoLite, the crushable, washable, mouldable and printable electroluminescent technology developed by Oryon Technologies.

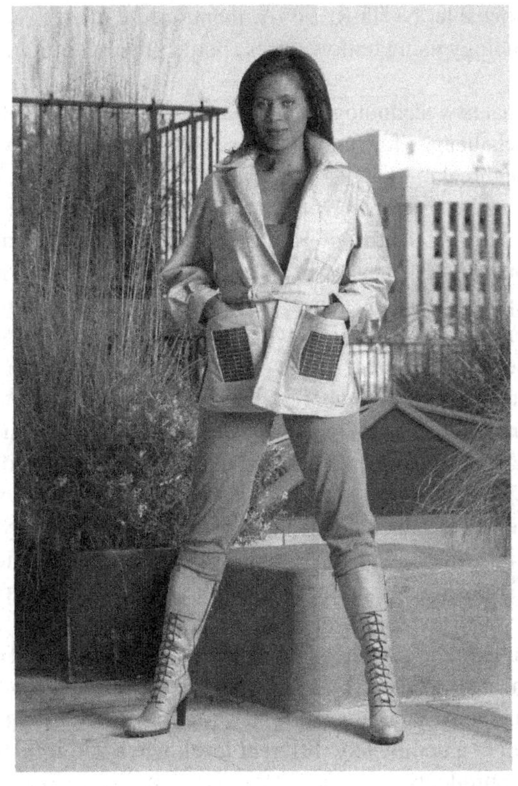

FIGURE 15.12

Solar-powered jacket by SilvrLining. Solar cells are attached to the pockets of the jacket.

Source: SilvrLining.

When electric power is required by the system, the source of power becomes a challenge, and carrying batteries in coat pockets is a primitive solution. Solar cells have been used in certain smart textile applications, for example in SilvrLining's GO solar power collection (Figure 15.12). Kinetic energy, perhaps in combination with solar energy, is an interesting option. Excess body heat could be recovered and used for powering embedded electronics by capitalizing on the Seebeck effect. The Seebeck effect was discovered by East Prussian scientist Thomas Seebeck in 1821. He noticed that a temperature difference between two metals in a circuit converts into thermoelectric current. Kinetic energy could also be harvested from a flexible piezoelectric-textile system, for example with a disk that generates voltage when deformed under pressure. Several research projects have focused on textile-based fuel cells, and flexible nanocomposite thin-film batteries are a new and interesting invention. These paper-like energy-storing devices could be laminated on textile structures (Pushparaj et al., 2007). Printing of organic high-efficiency solar cells on fabrics and light-activated flexible and lightweight power plastics are among the latest developments by Konarka Technologies. Organic solar cells are made of conductive organic polymers on a plastic film – being flexible, they are an ideal option for textiles. Improving thermoregulation of PCMs is another challenge: the cooling and warming impact of current PCMs is so weak as to be hardly noticeable. NOTEREFIGA, mentioned earlier, is an exciting new opening in this area, and this new technology using hollow fibres should greatly improve the warming and cooling impact of PCMs.

How can we make sure that new technological innovations can be turned into culturally accepted final products? One of the main challenges for future smart textile products is how to manage and guide the design process and understand real market demand. Prototyping is expensive, especially when we try to add properties that are not possible to achieve with current technology. Perhaps multimedia presentations can replace a prototype when testing the reactions of the potential users towards new products, like demonstrated by project MeMoGa (2003). This project focused on how to replace standard prototyping by 3D simulations and virtual prototyping. It seems that many of the final products brought to the market between 2000 and 2010 aroused a lot of attention, but at the end did not sell. How can the rapidly growing number of input and intermediate intelligent textile components be turned into commercially successful final products? Product managers and designers need to be aware of technical solutions, product positioning, price, promotion, branding, channel of distribution, cross-sector product testing, functionality of the product and the sustainable design issues in relation to the whole product life cycle. At the same time, the electronics industry is very technology driven, and wearable technology solutions may not always fulfil the aesthetic requirements consumers relate to garments (McCann & Bryson, 2009).

The electronic systems incorporated in smart textiles produce complications for sustainability and environmental protection. Disposing of a garment containing electronic components cannot be done in the same way as with non-smart textiles garments. Electronic components may be a problem during washing and cleaning of the garment as well as at the end of the life cycle. The retailers for smart textiles may have to set up schemes to take back used smart textile products and arrange safe disposal of the electronic components in them (Timmins, 2009). Requirements regarding sustainability and corporate responsibility will go up to a completely different level, when and if textiles with embedded electronics become mass-market products.

15.4.2 FUTURE MARKET DEVELOPMENT

There are various estimates for the value of smart textile market. Textile Intelligence classifies smart textiles into three categories:

1. Input components or enabling components for smart fabrics and interactive textiles (SFIT) include various devices such as electronics, conductive cables and power supplies, which must be built into a textile to produce intermediate SFIT components.
2. Intermediate SFIT components or SFIT-based modules are textiles obtained by incorporating input components such as electronics.
3. Finished SFIT-based textiles are garments and other textile products for sale to consumers and other end users.

The largest category is intermediate SFIT components, but the fastest growth is forecast for finished SFIT-based textiles.

Textile Intelligence estimated a total market size of US$ 329.2 million in 2006 and US$ 1129.5 million in 2010, as shown in Table 15.2. The Industrial Fabrics Association International estimated a US smart textiles market of US$ 193 million in 2008 (Technical-textiles net, 2009). Report Buyer, a UK-based market

Table 15.2 Global market for smart fabrics and interactive textiles (US$ million)

	2006	2007(f)	2008(f)	2009(f)	2010(f)	Annual change % 2006–2010
SFIT input components	121.7	137.9	157.7	182.1	213.8	+15.1
Intermediate SFIT components	221.5	258.3	318.4	431.7	678.1	+32.3
Finished SFIT–based textiles	26.0	44.6	77.4	133.5	237.5	+73.8
Total	**369.2**	**440.9**	**553.4**	**747.3**	**1 129.5**	**+32.3**

f = forecast.
Textile Intelligence.

Table 15.3 Global market for smart fabrics and interactive textiles by application (US$ million)

	2006	2007(f)	2008(f)	2009(f)	2010(f)	Annual change % 2006–20
Heat and energy management	155.5	170.5	190.2	217.2	258.4	+13.5
Sensing and monitoring	76.7	100.7	141.8	218.9	380.9	+49.3
Lighting	71.4	84.7	102.2	125.6	160.9	+22.5
Computing and communications	51.7	68.3	99.4	162.0	300.9	+55.6
Actuation and response	13.8	16.7	19.7	23.6	28.3	19.7
Location and position	0.1	0.1	0.1	0.1	0.1	0.0
Total	**369.2**	**440.9**	**553.4**	**747.3**	**1 129.5**	**+32.3**

f = forecast.
Textile Intelligence.

information company, more conservatively estimated a US smart and interactive textile market of US$ 193 million in 2012 – but also reports that growth has been rapid at more than 20% annually (Newswire, 2010). Forecasts by these three organizations cannot be compared directly, as definitions for smart textiles differ, and because Textile Intelligence estimates global demand while the others concentrate on the United States only. Table 15.3 shows Textile Intelligence's forecast by application. Sensing and monitoring, computing and communications, and heat and energy management, are estimated to be the main categories.

15.5 SUMMARY

Intelligent textiles are textiles that react to their environment. An intelligent system consists of a sensor, a processor and an actuator, all managed by the controlling data. Based on the received information, data are processed and a command is sent to the actuator to perform a preprogrammed action.

In broader terms, wearable electronics and functional fabrics are often referred to as intelligent textiles, with several classifications:

- PCMs are thermal storage materials for regulating temperature. Microcapsules filled with paraffin wax are attached to textiles by coating or in the spinning process. When the wax changes its phase from solid to liquid and back, heat is absorbed and released. By cooling and warming the wearer, phase change textiles maintain the microclimate temperature within the comfort zone.
- In response to external stimulus, shape memory materials change shape, and once stimulus is removed, return to their original shape. Thermally responsive SMPs are used in textiles for added temperature regulation and permeability. The shape change takes place at the transition temperature due to kinetic properties of molecular chains.
- Chromic materials change colour due to an external stimulus, which can be light, heat, electricity, pressure, liquid or electron beam. Inks and dyes are used for making chromic prints on textiles or for dyeing embroidery yarns.
- Photonics is a research topic for combining electrochromic materials, fibre optics and LEDs to textiles.
- Textiles made of conductive yarns are used for sensors, communication and heating.
- Stress-responsive materials change their shape when stress is applied to them. They may instantly turn rigid under a shock and return to a flexible state soon after. Such a property could be used in body protection.
- Biomimetics is a research agenda for mimicking biological reactions and properties in nature. The lotus leaf effect, chameleon-like colour change, the extreme strength of spider silk, the ability of a shark's skin's to reduce friction in water, and ultralight butterfly wings are some of the current research topics.

So far, different kinds of input and intermediate intelligent textile components have been developed in order to make ready-made consumer products possible, but commercially successful consumer products are yet to be created. Sensing, monitoring, computing, communications, and heat and energy management, are expected to be the main functions for the intelligent textile products of the future.

15.6 PROJECT IDEAS

Using conventional batteries for powering smart textile applications is not very convenient. What types of new technologies for integrating energy production directly into textile structures are currently developed?

What are the different concerns regarding sustainability, environmental protection and consumer safety with regard to intelligent textiles, and how should companies producing intelligent textile products handle these challenges?

Space agencies and defence forces carry out extensive research programmes, and they have greatly contributed to currently available intelligent textile applications. What kind of research and development projects are they now working on, and how can the possible outcomes from these projects be applied to consumer products?

15.7 REVISION QUESTIONS

1. Explain the differences among intelligent textiles, wearable technology and technical textiles.
2. How do phase change textiles function?
3. What kinds of technologies are shape memory materials based on?
4. What is the role of conductive textile materials in intelligent textile applications?
5. Give examples of stress-responsive materials and what are their advantages and disadvantages.
6. Explain the philosophy of biomimetics and how it is applied to intelligent textiles.
7. Describe future trends for intelligent textile design and its market.

15.8 SOURCES OF FURTHER INFORMATION AND ADVICE

http://www.amaterrace.com/en/product/dl_3.html.
http://www.biospace.com/company_profile.aspx?CompanyId=261204.
http://www.cutecircuit.com/products/.
http://electrochem.cwru.edu/encycl/art-p02-elact-pol.htm.
http://www.elektex.com/.
http://enlighted.com/index.html.
http://www.fashioningtech.com/.
www.goretex.com.
http://www.howstuffworks.com/exoskeleton.htm.
http://www.konarka.com/.
http://www.lotusan.de/.
http://www.lumalive.philips.com/.
http://www.nanotex.com/applications/apparel.html.
http://oryontech.com/.
http://www.osram.com/appscom/cgi-bin/press/archiv.pl?id=650.
http://www.schoeller-textiles.com/en/technologies/nanosphere.html.
http://science.howstuffworks.com/liquid-body-armor.htm.
http://silvrlining.com/index.htm.
http://www.smart-material.com/.
http://www.trendhunter.com/trends/zegna-ecotech-solar-jacket.
http://web.media.mit.edu/~leah/.

REFERENCES AND FURTHER READING

Catrysse, M., Pirotte, F., & Puers, R. (2007). The use of electronics in medical textiles. In L. van Langenhove (Ed.), *Smart textiles for medicine and healthcare*. Cambridge: Woodhead Publishing Limited.

Cordis. (2009). *Novel temperature regulating fibres and garments (NOTEREFIGA)*. Available from http://extra.ivf.se/noterefiga/template.asp.

Davis, D., Hamilton, A., Yan, J., Cremar, L., van Gough, D., Potisek, S., et al. (2009, May 7). Force-induce activation of covalent bonds in mechanoresponsive polymeric materials. *Nature*, *459*, 68–72.

Farid, M., Amar, M., Khudhair, A., Razack, S., & Al-Hallaj, S. (2004). A review on phase change energy storage: materials and applications. *Energy Conversion and Management, 45*, 1597–1615.

Forbes, P. (2008, August). Self-cleaning materials: lotus leaf-inspired nanotechnology. *Scientific American*, 88–95.

Hu, J. (2007). *Shape memory polymers and textiles*. Cambridge: Woodhead Publishing Limited.

van Langenhove, L. (2007). *Smart textiles for medicine and healthcare*. Cambridge: Woodhead Publishing Limited.

Malmivaara, M. (2009). The emergence of wearable computing. In J. McCann, & D. Bryson (Eds.), *Smart clothes and wearable technology*. Cambridge: Woodhead Publishing Limited.

McCann, J., & Bryson, D. (2009). *Smart clothes and wearable technology*. Cambridge: Woodhead Publishing Limited.

Mäkinen, M. (2006). Introduction to phase change materials. In H. Mattila (Ed.), *Intelligent textiles and clothing*. Cambridge: Woodhead Publishing Limited.

MeMoGa. (2003). Methods and models for intelligent garment design. Available from http://www.tut.fi/index.cfm?MainSel=14754&Sel=14773&Show=20828&Siteid=155. Accessed 20.07.11.

Newswire. (2010). *U.S. military and biomedical segments expected to push smart and interactive textile market to $ 193,3 million by 2010*. Newswiretoday.com. Available from http://www.newswiretoday.com/news/25577/ Accessed 04.04.10.

Poole, B. (2007). *Biomimetics: Borrowing from biology. The naked scientist*. University of Cambridge. Available from http://www.thenakedscientists.com/HTML/articles/article/biomimeticsborrowingfrombiology/. Accessed 18.11.10.

Pushparaj, V., Shaijumon, M., Kumar, A., Murugesam, S., Ci, L., Vataj, R., et al. (2007). Flexible energy storage devices based on nanocomposite paper. *Proceedings of the National Academy of Science of the United States of America, 104*(34), 13574–13577. http://dx.doi.org/10.1073/pnas.0706508104.

Sandia National Laboratories. (2009). *Sandia research points way toward chameleon-like camouflage*. Sandia National Laboratories. Available from http://www.sandia.gov/news/resources/news_releases/sandia-research-points-way-toward-chameleon-like-camouflage/. Accessed 31.03.10.

Science.howstuffworks. (2009). *How the future force worrier will work*. HowStuffWorks.com. Available from http://science.howstuffworks.com/ffw.htm. Accessed 30.03.10.

Schwartz, M. (2002). *Encyclopaedia of smart materials*. New York: John Wiley & Sons.

Schwarz, A., Langenhove, L., Guermonprez, P., & Deguillemont, D. (2010). A road map on smart textiles. *Textile Progress, 42*(2), 100–175.

Stegmaier, T., Arnim von, V., Scherrieble, A., & Planck, H. (2008). Self-cleaning textiles using the Lotus Effect. In *Biologically inspired textiles*. Cambridge: Woodhead Publishing Limited.

Tao, X. (2001). *Smart fibres, fabrics and clothing*. Cambridge: Woodhead Publishing Limited.

Technical textiles net. (2009). *Military driving US market: Smart textiles*. Technical-textiles.net. http://www.technical-textiles.net/htm/d20091120.619537.htm. Accessed 03.04.10.

Textile Intelligence. (2008). *Global markets for smart fabrics and interactive textiles*. Wilmslow: Textile Intelligence.

Textronics. (2010). *Energy-activated fabrics*. Textronics Inc. Available from http://www.textronicsinc.com/. Accessed 31.03.10.

Timmins, M. (2009). Environmental and waste issues concerning the production of smart clothes and wearable technology. In J. McCann, & D. Bryson (Eds.), *Smart clothes and wearable technology*. Cambridge: Woodhead Publishing Limited.

PART 3

FABRIC FINISHING AND APPLICATIONS

CHAPTER

FABRIC FINISHING: JOINING FABRICS USING STITCHED SEAMS

16

J. McLoughlin, A. Mitchell

Manchester Metropolitan University, Manchester, UK

LEARNING OBJECTIVES

At the end of this chapter, you should be able to:

- List and describe the principle stitches used to join fabrics
- List and describe the different types of seam
- List and describe the range of machines used to join fabrics and the key elements
- Describe the main seam defects and tactics to prevent them

16.1 INTRODUCTION

The performance of a garment or finished fabric product is affected not only by the quality of fabrics used in its manufacture but also by the technology of the manufacturing process, which can be lengthy and complex. Stitched seams are the most common methods for joining fabrics, with a variety of different options being used depending on the end use of the product.

When joining fabrics together with stitched seams, there are a number of factors that need to be taken into account. These include:

- Stitch type
- Seam type
- Machine choice and settings
- Needle size and point
- Thread properties
- Operator handling
- Fabric properties and characteristics

All of these factors are equally important in producing a quality seam. If one or more of them are incorrect, the result will be a poorly sewn finished garment. Manufacturers must therefore understand these factors and ensure that the manufacturing process takes them into account. Garment technologists, quality managers and clothing machine engineers can all contribute towards optimising the manufacturing process and minimising or eliminating production problems (McLoughlin, 1998, 1999, 2000). Efficient companies have assessment and quality assurance structures in place in order to minimise production downtime.

Textiles and Fashion. http://dx.doi.org/10.1016/B978-1-84569-931-4.00016-7

This chapter describes the stitches, seams and machinery used to join fabrics to produce garments, and focuses particularly on the lockstitch machine, giving a detailed explanation of the key elements of the machine and how choices and settings influence production quality. Problems that can occur while sewing garments and how they can be rectified are also discussed.

16.2 THE STITCH

The three methods of stitch formation are:

- Intralooping – the passing of a loop of thread through another loop of the SAME thread supply, e.g. 101 single-thread chain stitch.
- Interlooping – the passing of one loop of thread through a loop formed by a SEPARATE thread supply, e.g. 401 double-lock chain stitch.
- Interlacing – the passing of a thread around, or over, a separate thread supply or a loop of that supply, e.g. 301 lockstitch.

Stitches are defined by the way they are formed, and are classified into six different categories using three-digit numbering systems defined by the International Organisation for Standardisation (IOS, 1991). These classifications are:

- Class 100 – Single-thread chain stitches
- Class 200 – Hand stitches
- Class 300 – Lockstitches
- Class 400 – Multi-thread chain stitches
- Class 500 – Overedge/Overlock stitches
- Class 600 – Covering chain stitches

Specific stitch types are designated by the second and third digits of the number. For example, 301 is a single-needle lockstitch and is specifically identified by this number. Hundreds of different stitch types are used in the manufacture of textile products, but only the most commonly used will be discussed here. For information about other stitch types, refer to the further reading list at the end of this chapter.

16.2.1 CLASS 100 CHAIN STITCHES

Class 100 chain stitches are formed by the intra-looping of a needle thread supply on or around the fabric (Figure 16.1). Single-thread chain stitch seams are often used for temporary applications due to

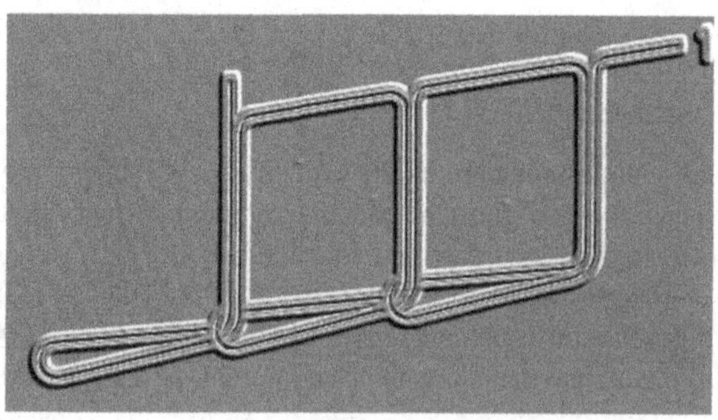

FIGURE 16.1

Single-thread chain stitch 101.

their ease of removal. This is because each successive loop is dependent upon the previous loop for security, so once a loop is cut, the stitching can be easily unravelled and removed.

Class 100 chain stitches are used for:

- Basting – holding fabric pieces together temporarily before final stitching.
- Closing industrial sacking – holds the sack closed, but is easy to remove.
- Cornelli embellishment – up to 65 needles can be employed to produce decorative seams with the chain on the face fabric.
- Buttonholes and attaching buttons.

It is important that whenever 100 chain stitches are used, the end of the seam is secured to prevent the stitches from unravelling.

16.2.2 CLASS 300 LOCKSTITCHES

The class 300 lockstitch is often referred to as a double lockstitch. It is formed by interlacing a single-needle thread supply with a bobbin thread supply from underneath (Figure 16.2), and has great strength and resilience if the correct types of thread, i.e. polyester/corespun, are used. Lockstitches are very secure, as a break in one stitch will not cause the seam to unravel, although it will compromise the overall seam performance. Lockstitch is the most widely used stitch in low-volume production. Lockstitches are used for:

- Comfort/stretch garments, because the stitch can extend by up to 30%.
- Fabrics where it is important to have the same appearance on both sides.
- Top-stitching collars, cuffs, etc., because this is the only stitch that sews reliably around 90° when the fabric is pivoted at the needle point.

16.2.3 CLASS 400 MULTI-THREAD CHAIN STITCHES

Multi-thread chain stitches are formed by inter-looping a needle thread with a separate looper thread on the underside of the fabric (Figure 16.3). This stitch is often referred to as a double-locked stitch because each needle thread loop is interconnected with two loops of the same single under-thread. The stitch therefore looks like a 301 lockstitch on the surface but has a double chain underneath.

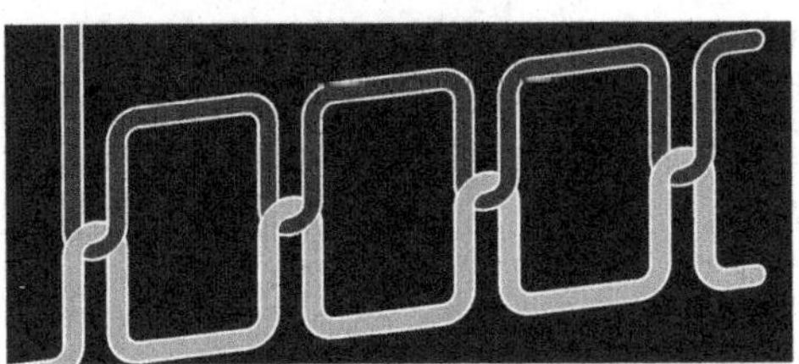

FIGURE 16.2

Single-needle lockstitch 301.

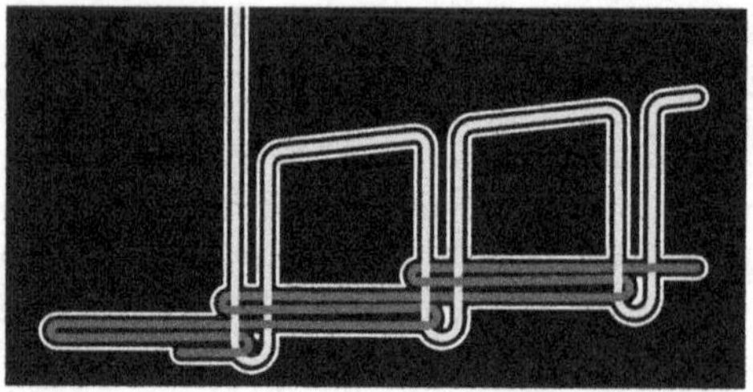

FIGURE 16.3

Two-thread chain stitch 401.

Multi-thread chain stitches are used for:

- Joins where good strength and extension/recovery properties are needed, because this stitch has lower static thread tension and inter-looped threads.
- Joins where preventing pucker is important, because these stitches are less prone to pucker (again due to lower static thread tension and inter-looped threads on the underside).
- Long seams, because of continuous thread supplies.

16.2.4 CLASS 500 OVEREDGE STITCHES

Overedge stitches are formed with at least one of the sewing threads passing around the fabric edge (Figure 16.4). There are many variations, and overedge stitching can use from one to four threads, of which only one is the needle thread.

The needle thread provides the seam strength while the looper threads provide seam durability and extensibility. The looper threads are often chosen for softness and appearance.

Overedge stitches are used for neatening the cut edges of fabric plies (pieces).

16.2.5 STITCH QUALITY

The top thread tension and bobbin tension on a lockstitch machine should be set as slack as possible in order to form a well-balanced, good-quality stitch (in fact, the tensions on any sewing machine should be set as loosely as possible). The bobbin thread should be wound with a low tension to enable the thread to be interlaced more easily with the top thread into the centre of the fabric.

Different stitch qualities are given in Figures 16.5–16.7 (Source: Union Special, 1988). The yellow thread is the top thread and the red thread is the underneath thread.

16.3 THE SEAM

Some form of seam joining is used in virtually all product manufacture that involves the stitching of materials using sewing threads (Table 16.1), and it is still by far the most important method used in the joining of textiles.

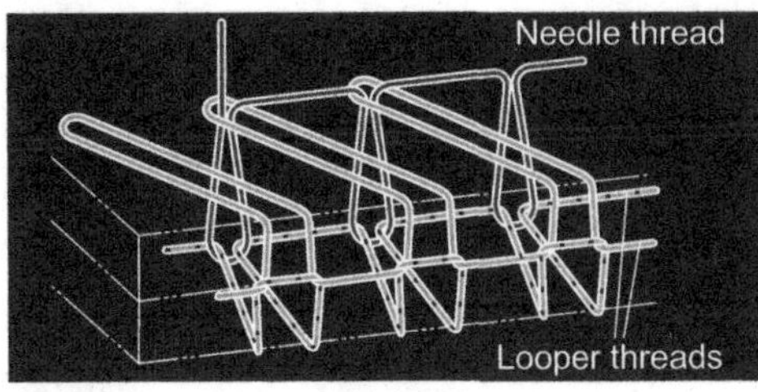

FIGURE 16.4

Overedge stitch 504.

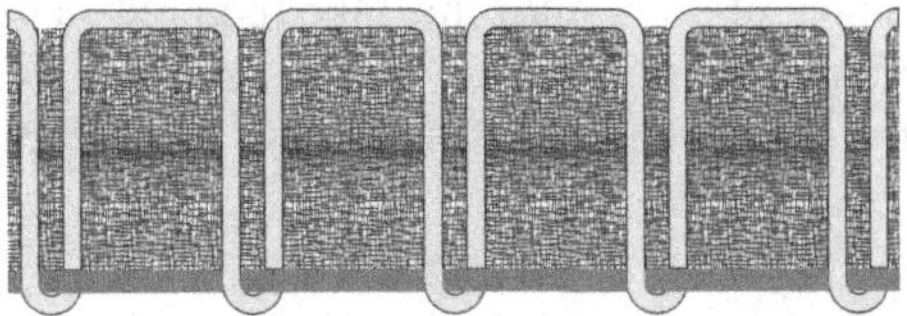

FIGURE 16.5

Unacceptable: stitch slack underneath, caused by top thread tension being too slack and bobbin thread tension too tight.

FIGURE 16.6

Unacceptable: stitch slack on top, caused by the bobbin tension being too slack and the top thread tension too tight.

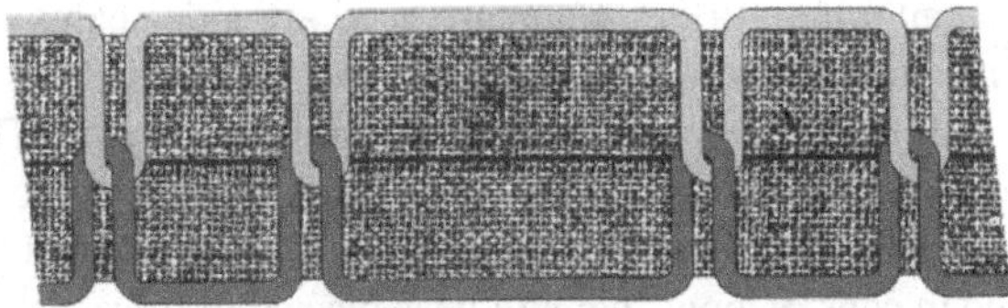

FIGURE 16.7

Acceptable: stitch formed correctly with both threads interlaced in the centre of the material.

Table 16.1 The British and United States standards for the main classes of seam

British Standard	U.S. Standard	Description	Example
Class 1	SS	Superimposed	SSa – 1
Class 2	LS	Lapped	LSc – 1
Class 3	BS	Bound	BSa – 1
Class 4	FS	Flat	Efa – 1

A seam should be suitable for the purpose for which it is intended, and the seam type depends on the product being sewn. The choice of seam must meet the required standards of appearance and performance, but it must also be economical if it is being used in a modern production environment, as cost can become an important factor. Thus when choosing a seam for a garment operation, the following factors should be considered:

- Aesthetic appeal
- Strength
- Durability
- Comfort in wear
- Ease of assembly
- Equipment availability
- Cost

A typical cost calculation would be as follows. A basic skirt contains 10 m of thread. What is the total cost of the thread in the skirt if the thread used costs £3.10 for 5000 m in a cone?

(Cost of thread/cone in m) × total thread used in seam.

i.e. $310/5000 \times 10 = 0.62$ pence.

The seams that are of major importance in garment manufacture are briefly described in the next sections.

16.3.1 CLASS 1: SUPERIMPOSED SEAMS

Superimposed seams (Figures 16.8 and 16.9) are the most basic and easiest to produce of all the seam types, and are frequently used in many areas of garment construction as well as for numerous other sewn products in the market.

Typical areas of a garment where these seams are used are side seams on a pair of jeans, side seams on blouses or shirts and trousers, and for attaching zips and trims etc. The fabric edges would usually be overlocked in order to prevent the material from fraying. The seam is durable and is easy to produce, as it only involves one fabric placed upon top of another.

A disadvantage of this seam is that its strength is limited by the strength and type of sewing thread. It should therefore not be used in constructions that will be subjected to force, for instance for a Judo suit, as the physical pulling and pushing of the fabric in use would cause the stitches to break and rupture the seam.

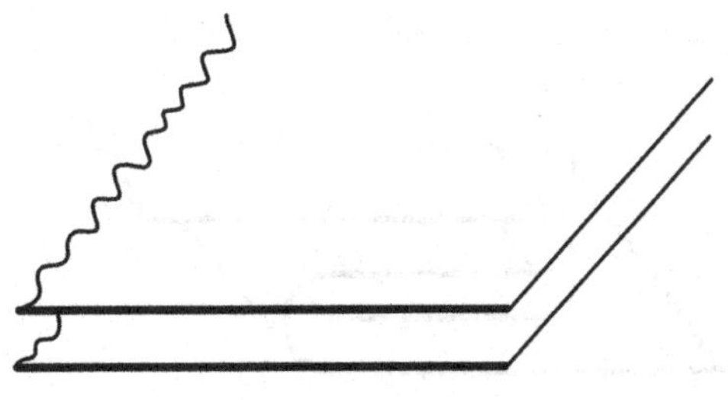

FIGURE 16.8

Superimposed seam.

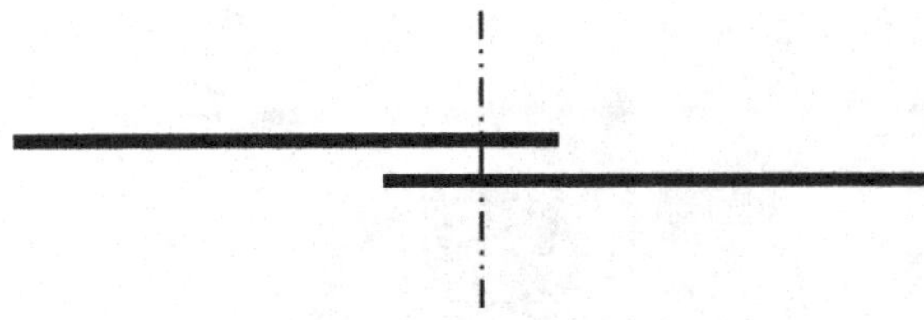

FIGURE 16.9

Superimposed seam cross-section.

16.3.2 CLASS 2: DOUBLE-LAP SEAMS

Double-lap seams (Figures 16.10 and 16.11) are among the strongest available and are used in many areas of sewn product manufacture where strength is required, particularly side seams on denim jeans, parachutes, tents and sports garments such judo suits.

16.3.3 CLASS 3: BOUND SEAMS

Bound seams (Figures 16.12–16.14) have many uses in apparel, such as hemming and binding operations for neatening fabric edges, and are also commonly used on luggage and inside tent joins.

16.3.4 CLASS 4: FLAT SEAMS

Flat seams (Figure 16.15) are produced with a minimum of two plies of fabric, with the raw edges butted together and usually stitched using a cover-seam machine. These seams are most commonly used in knitted products, particularly underwear. A major advantage in this seam type is its reduced bulk.

16.4 SEWING MACHINES

The first commercial sewing machines were made from wood, and sewed chain stitches. Chain stitch has remained a useful stitch, but since then, additional stitching methods have been developed, and commercial sewing machines today use the latest technologies in engineering, pneumatics, hydraulics and electronics to improve efficiency and increase production speeds.

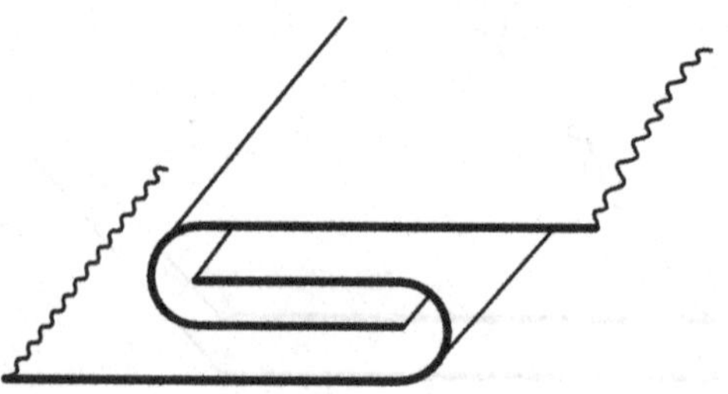

FIGURE 16.10

Double-lapped seam.

FIGURE 16.11

An outside leg of a pair of jeans using a double-lapped seam.

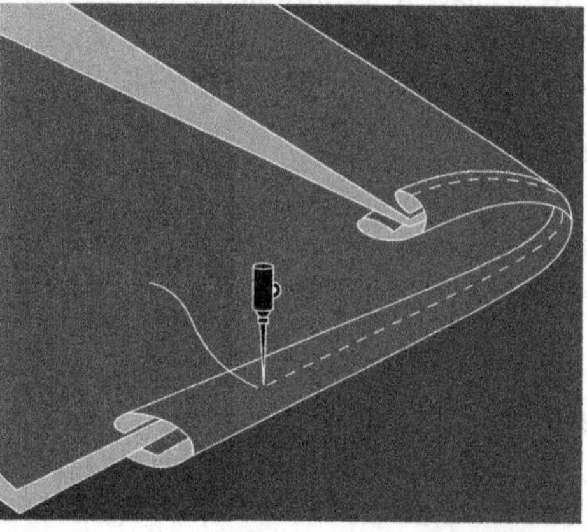

FIGURE 16.12

Bound seam.

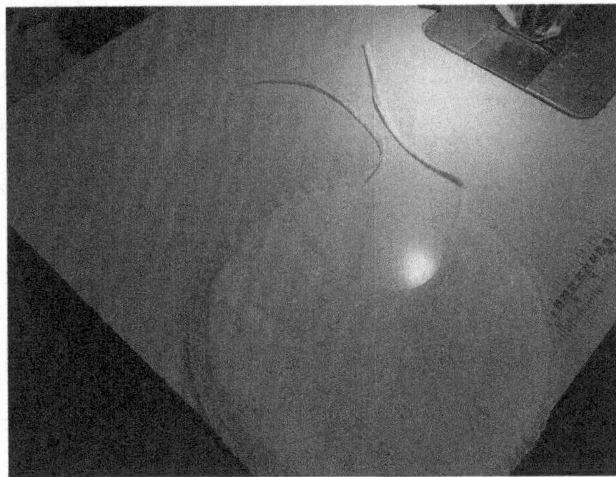

FIGURE 16.13

The binding of a necktie for a baby bib.

FIGURE 16.14

The binding attachment on the lockstitch machine.

Machines are usually categorised into four main types. These are:

- The basic sewing machine (Figure 16.16 and 16.17)
- The mechanised sewing machine (button sew, button hole machines)
- Semi-automatic machines (embroidery machines)
- Automatic transfer lines (used in curtain and towel manufacture)

It is important to use the correct machine for the type of operation on the garment. Hundreds of machines have been developed for many different and varied operations. These range from machines used in the production of apparel, including blouses, shirts and lingerie, to machines used in the production of tents, tarpaulins and bouncy castles.

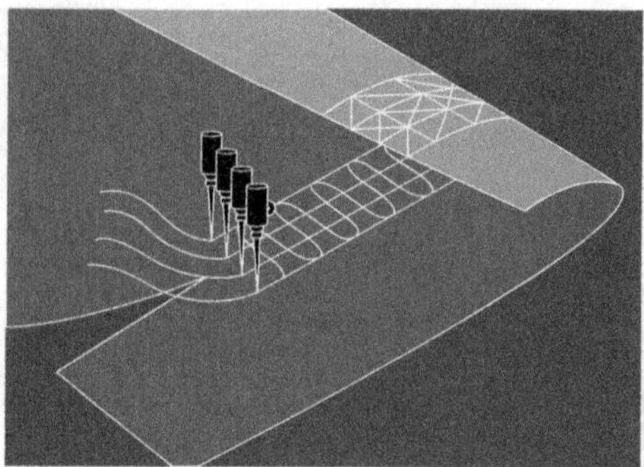

FIGURE 16.15

Flat seam.

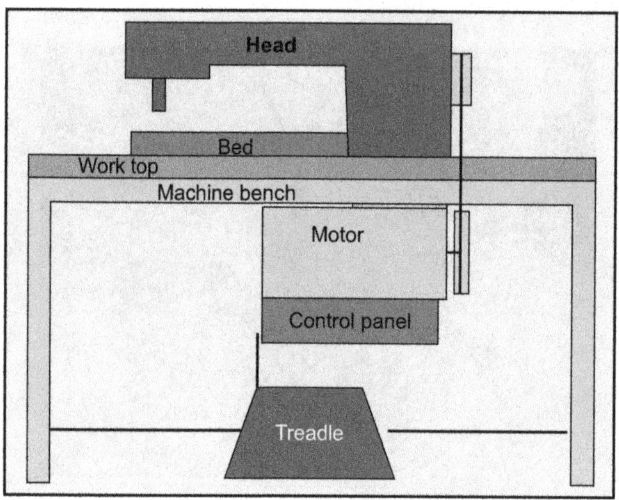

FIGURE 16.16

Schematic diagram of basic machine with all components.

The most commonly used machine in industry is the basic integrated sewing unit, or ISU, lockstitch sewing machine (Figure 16.18). Most companies manufacturing apparel and other sewn products will have this type of machine in their production lines. The ISU machine is a fully integrated unit that usually consists of:

- Automatic thread trimmers, enabling the threads to be cut without using scissors
- Automatic presser foot lift
- Automatic back tacking
- Stitch-counting and ply-sensing devices

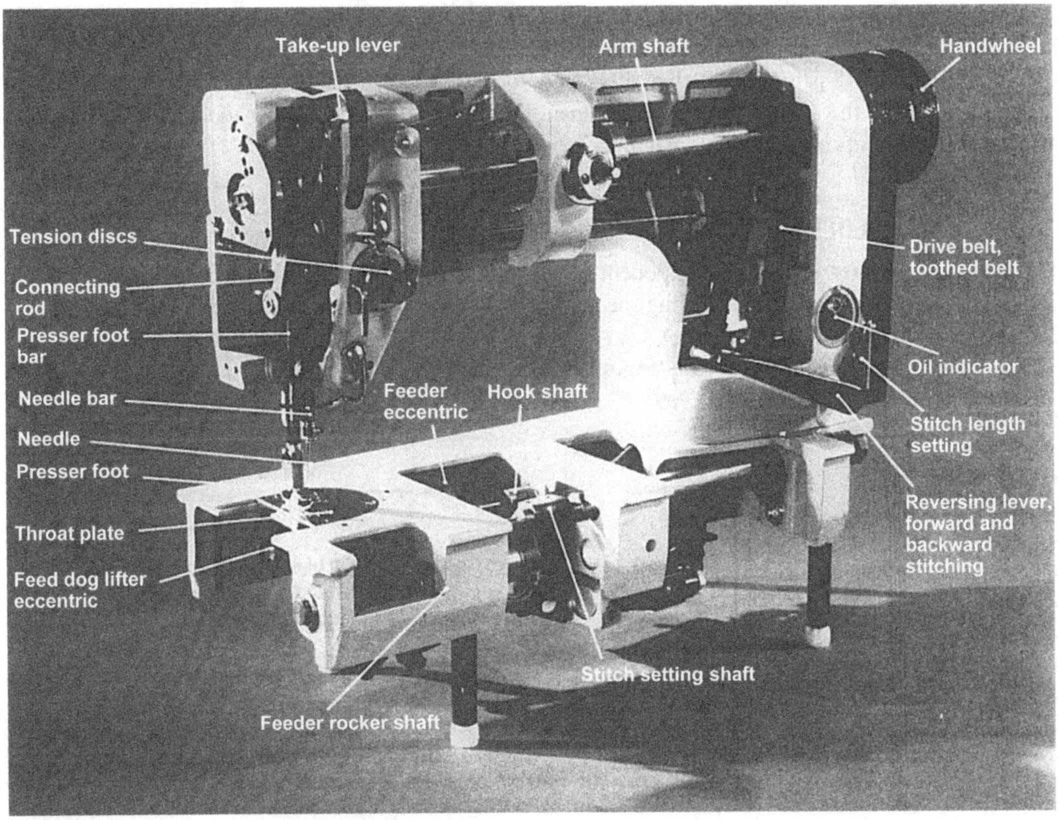

Take-up lever
Arm shaft
Handwheel
Tension discs
Connecting rod
Presser foot bar
Needle bar
Needle
Presser foot
Throat plate
Feed dog lifter eccentric
Feeder eccentric
Hook shaft
Drive belt, toothed belt
Oil indicator
Stitch length setting
Reversing lever, forward and backward stitching
Stitch setting shaft
Feeder rocker shaft

FIGURE 16.17

Diagram of a lockstitch sewing head.

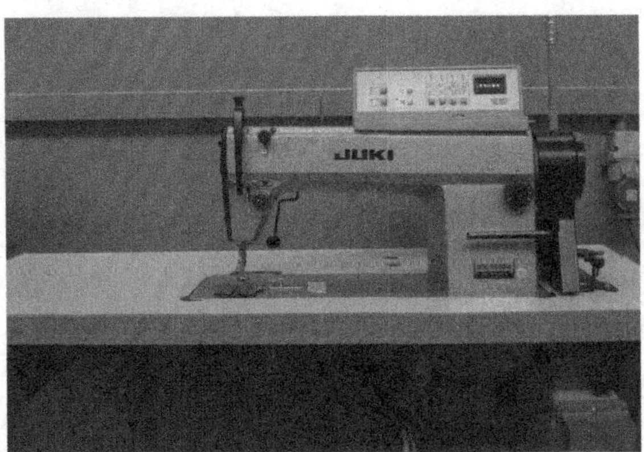

FIGURE 16.18

Juki single-needle ISU lockstitch sewing machine.

Stitch type 301, single-needle lockstitch, is the commonest of all the stitch types used in the clothing industry. The lockstitch machine is commonly called a flat machine in industry, and the example shown in Figure 16.18 is the Juki flatbed drop feed, lockstitch (301) machine.

The needle and the type of feeding mechanism that passes the fabric under the needle are the two most critical elements of the sewing machine.

16.4.1 THE NEEDLE

The needle is the most important component of the sewing machine due to the fact that it carries and delivers the sewing thread to the sewing mechanism and must penetrate the fabric whilst minimising damage to the material.

The component parts of the needle (Figures 16.19 and 16.20) are:

- Butt
- Shank
- Shoulder
- Blade
- Long groove
- Short groove
- Needle eye
- Scarf
- Needle point
- Needle tip

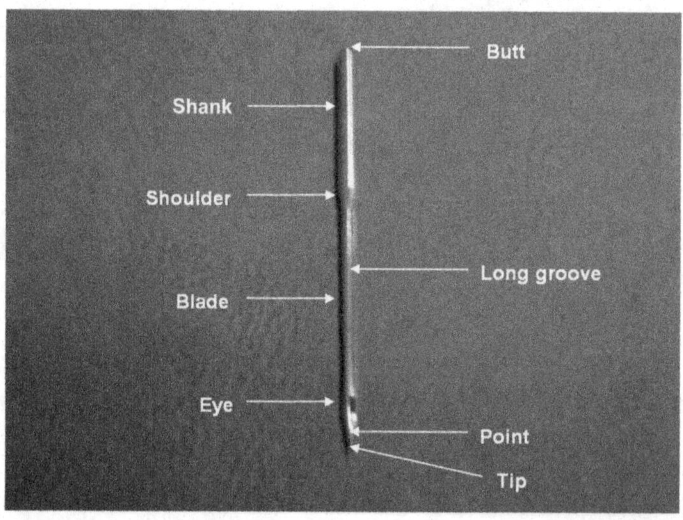

FIGURE 16.19

Needle component parts.

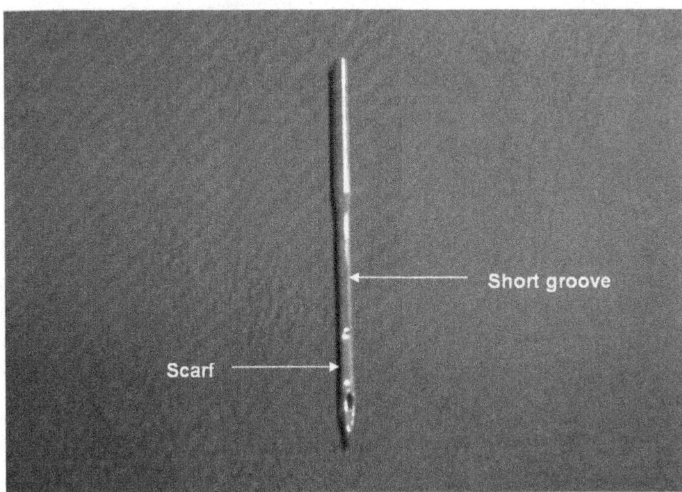

FIGURE 16.20

Needle component parts.

The point and the tip are the first point of contact with the fabric. The scarf (see Figure 16.20) is the flattened part of the needle, designed to enable the sewing mechanism (in the case of a lockstitch, the sewing hook) to pick up a loop of the sewing thread and thus form a stitch. The needle eye is threaded with the sewing thread, and some needle types have a groove that runs from the scarf of the needle up to the shoulder to serve as a protective channel for the sewing thread. The shank is the part that fits into the needle bar of the machine. Needle diameters can range in thickness from size 50s (0.50 mm) in excess of 150s (1.50 mm).

Hundreds of needle points have been developed for different fabrics, such as knitwear, woven materials and leather. The needle points most commonly used on fabrics are acute round-point needles and round-point needles (Figures 16.21 and 16.22).

More examples of needle points can be seen in Figure 16.23.

The factors to be considered when selecting a needle are:

- Fabric type
- Fabric density
- Fabric composition
- Seam thickness
- Type of machine

If the needle is not changed regularly, it can cause major quality problems. The needle is subject to the most use of all the machine parts as it penetrates the material at speeds of 5000–6000 times per minute for lockstitch and 8000–10,000 times per minute for chain stitch. The friction caused by the needle penetrating the fabric causes extreme heating, with needle temperatures in excess of 250 °C (Schmetz, 1998).

When fabrics are stitched together, the impact from the needle as it penetrates the fabric can cause buckling and distortion of the yarns and the fibres. A smaller-diameter needle reduces the mechanical

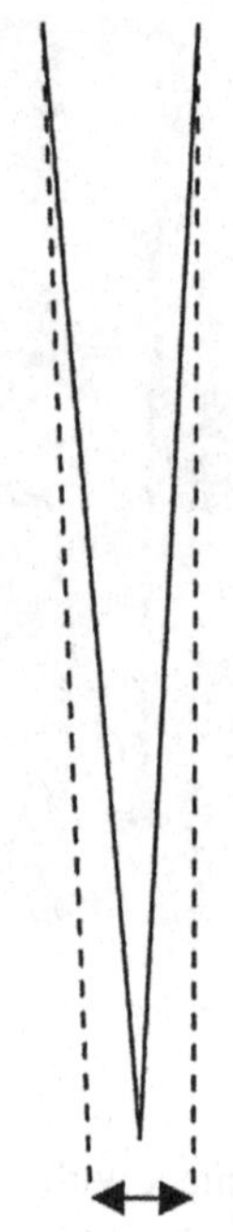

FIGURE 16.21

Acute round point.

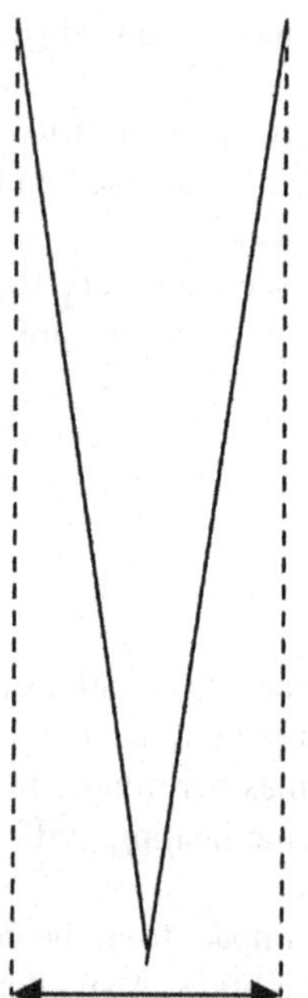

FIGURE 16.22

Normal round point.

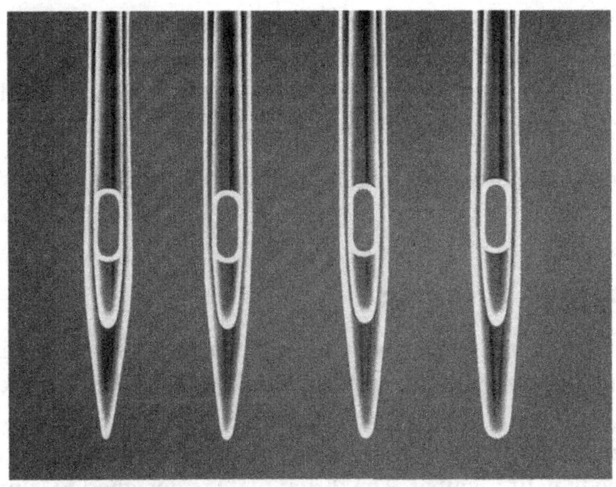

FIGURE 16.23

Examples of needle points. From right to left: acute round point, round point, light ballpoint, heavy ballpoint.

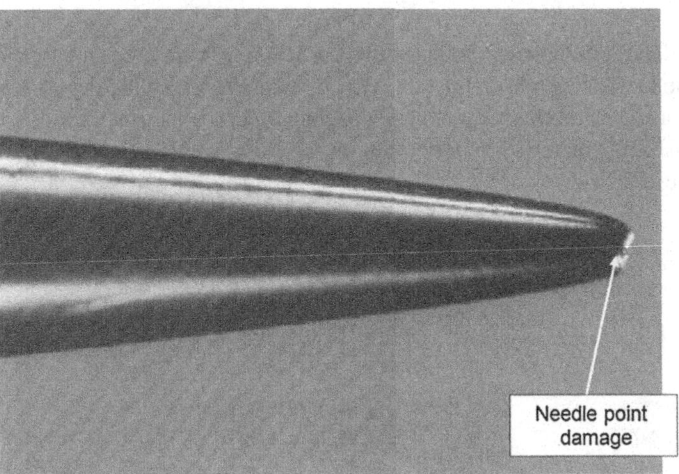

FIGURE 16.24

Damaged needle point.

Source: From Schmetz, 1988.

forces exerted on the yarns. The mechanical strain on the yarns increases if the needle is damaged (Figure 16.24), and can cause the fibres to rupture and hence reduce the seam strength significantly.

The following approaches can be used to help avoid problems caused by needles:

- Use a needle with the smallest possible diameter for the fabric and seam being sewn.
- Adapt the opening of the sewing plate to fit the needle size.

- Use a sewing thread with the correct diameter for the needle eye.
- Use the correct needle point for the type of fabric being sewn.
- Consider whether the type of seam being used to construct the garment could be changed, or use multiple seaming in order to divide the strain.

Diagrams for needle component parts are given in Figures 16.19 and 16.20.

16.4.2 MACHINE FEEDING SYSTEMS

The principle role of the feeding mechanism is to move the material from one stitch position to the next over the prescribed distance. In doing so, the material must be controlled precisely under the minimum amount of pressure.

The feed system is made up of the throat plate, presser foot and the feed dog. They are often called the fittings because they fit together. The throat plate is designed to support the material being stitched and allow the material to pass easily over its surface. The throat plate is manufactured with a needle hole to allow the needle to penetrate the fabric. The needle hole must be large enough for the needle and the sewing thread and there are slots in the throat plate that allow the feeder to rise and feed the material. The slots should match the width of the rows of teeth on the feed dog without allowing contact.

The metal feed dog feeds the material through the machine. There are many types available, and the choice of which one to use depends on the fabric to be sewn. When sewing denim for example, a heavy duty feed dog with a coarse tooth set should be used, but when sewing fine fabrics such as plain woven cotton or polyester shirting materials, a finer tooth set should be used. Examples of feeding components are shown in Figure 16.25.

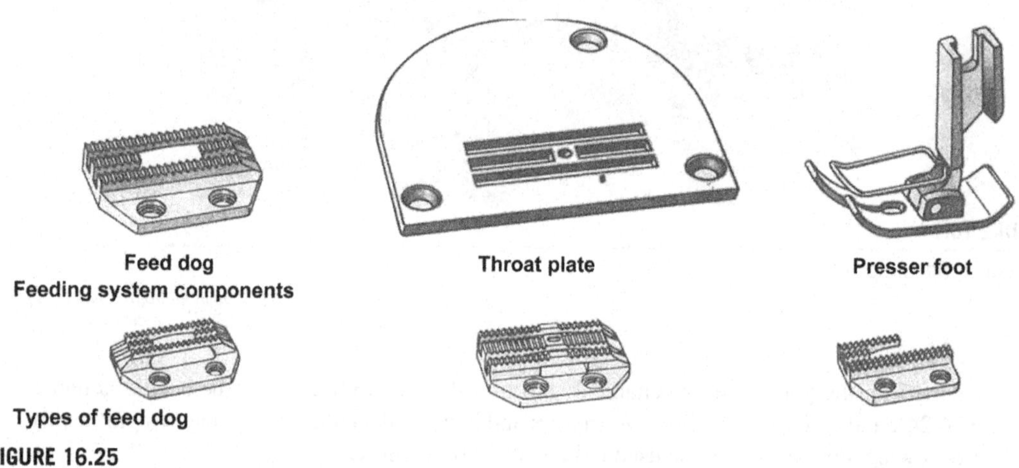

FIGURE 16.25

Examples of feeding components.

Many feeding systems have been invented for sewing textile products in order to accommodate the range of different types of operations and feed the material correctly as required for effective stitch formation. The different types of feed mechanisms include:

- Four-motion drop feed
- Differential drop feed
- Needle feed
- Compound feed
- Feeding foot
- Variable top and bottom feed
- Alternating foot
- Unison feed
- Puller feed
- Wheel feed
- Cup feed
- Manual feed

16.4.2.1 Four-motion drop feed

In the four-motion drop feed, the feeder engages the underside of the fabric ply intermittently and is set up (timed) to engage the material when the needle has risen clear from the top ply of the fabric (Figure 16.26). The feed is named 'four motion' as it has four movements:

- Motion 1 – Rising above the plate to contact the fabric.
- Motion 2 – Feeding the fabric the required stitch distance.
- Motion 3 – Descending beneath the plate to release contact with the fabric.
- Motion 4 – Travelling back underneath the plate the required distance to repeat the feeding process.

16.4.2.2 Differential drop feed

This system employs two feed dogs set in series that are driven in a similar manner to the single four-motion drop-feed dog (Figure 16.27). This feed system makes it possible to control the ratio of feed between the two feed dogs. By feeding more with the back feed dog than the front feed dog (in the same stitch cycle), it is possible to stretch the bottom ply of the fabric. Conversely, by feeding more with the front feed dog than the back feed dog, it is possible to introduce fullness into the bottom ply.

16.4.2.3 Needle feed

With the needle feed, the progression of the fabric is achieved by the longitudinal vibration of the needle bar alone. No feed dog is employed in feeding the material. This feed is rarely used in isolation today; it is mainly combined with a drop feed to give a compound feeding system.

16.4.2.4 Compound feed

The compound feed combines the needle feed and four-motion drop-feed systems (Figure 16.28). The feed timing is such that the feed dog rises and engages the bottom fabric ply at the same time as the

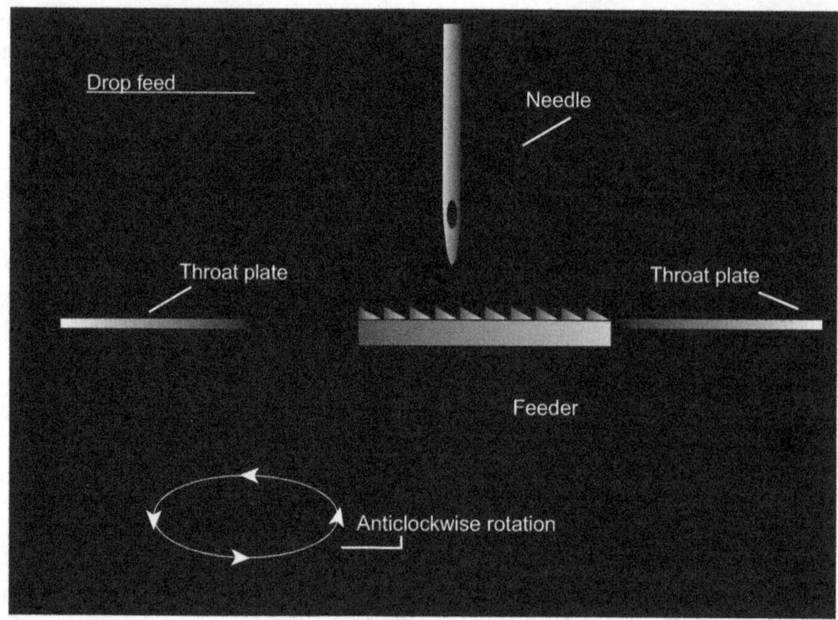

FIGURE 16.26

Four-motion drop feed.

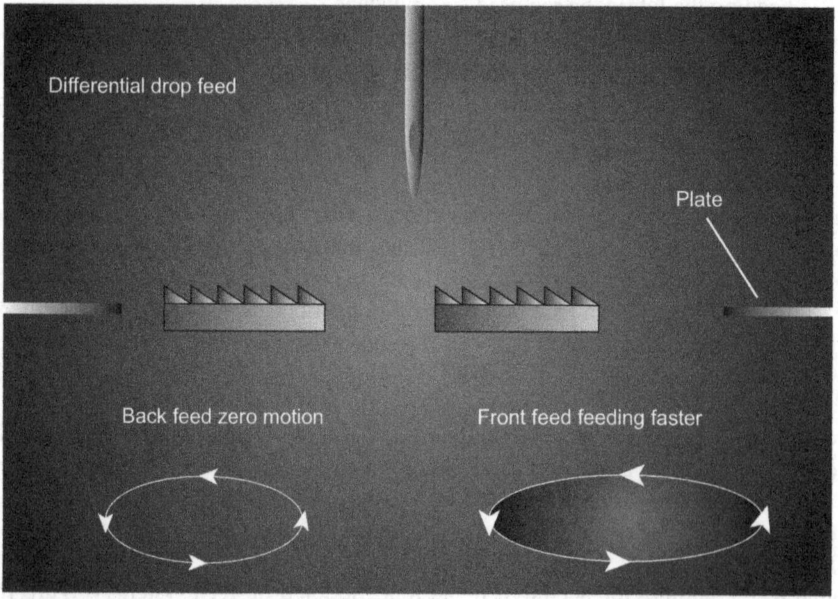

FIGURE 16.27

Differential drop feed.

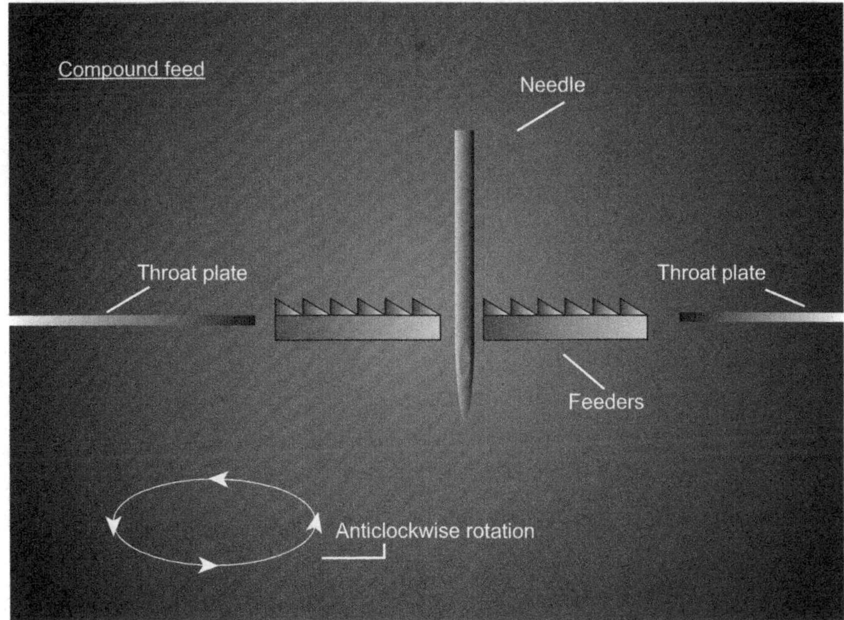

FIGURE 16.28

Compound feed.

needle descends into the fabric. Both the feed dog and needle move the fabric through the prescribed stitch length. This can help reduce feeding pucker and ply slippage during stitching because the fabric plies are pinned together during sewing.

16.4.2.5 Feeding foot

No feed dog is employed in the feeding foot system. The action is mimicked by the presser foot descending and engaging the fabric, moving it through the machine and then disengaging and rising before travelling forwards for the next stitch.

16.4.2.6 Variable top and bottom feed

This feed mechanism is a combination of a feeding foot synchronised with a bottom four-motion drop-feed system (Figure 16.29). These feeding mechanisms are often used for sewing high-friction materials such as simulated leather and composites, where the use of a static presser foot is unsuitable.

16.4.2.7 Alternating foot

The alternating foot mechanism consists of a feeding foot and a lifting foot in conjunction with a feed dog. The feed dog and the feeding foot transport the fabric while the lifting foot is clear of the work piece. The lifting foot descends to hold the fabric while the other two components return for the next stitch. These are excellent for seaming very bulky seams.

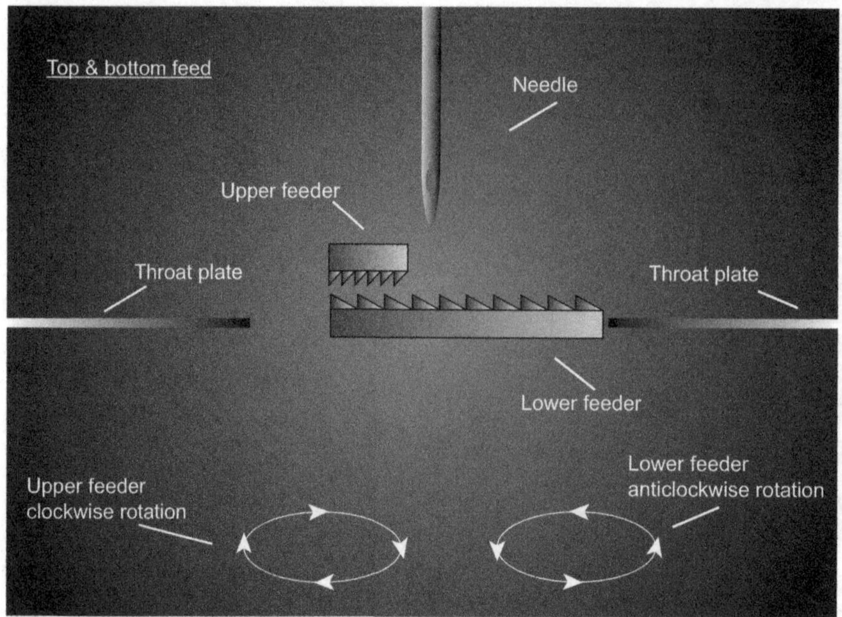

FIGURE 16.29

Variable top and bottom feed.

16.4.2.8 Unison feed

Unison feed is similar to alternating feed, but provides even more positive feed. Both sections of the two-part presser foot contact the fabric and transport it through the machine sequentially. This means that part of the presser foot is in contact with the fabric at all times.

16.4.2.9 Puller feed

The puller feed is an auxiliary feed (additional to the main feed system) that takes the form of a continuously or intermittently turning weighted roller positioned at the rear of the needle (Figure 16.30). It is mainly used in seaming heavy fabrics, or to maintain tension in lighter-weight materials.

16.4.2.10 Wheel feed

In the wheel feed system, a driven roller, wheel or belt is employed above, below or in both positions to move the fabric, instead of a feed dog and presser foot. The surface of the wheel is often knurled or coated to provide increased friction for feeding leather or plastics.

16.4.2.11 Cup feed

With the cup feed, two horizontally mounted feed wheels are driven simultaneously in order to move the fabric from right to left through machines with horizontal needles (Figure 16.31). The cup feed is most often used in the seaming of knitted goods and leather or suede. The positioning of the feed system allows greater visibility of both fabric plies.

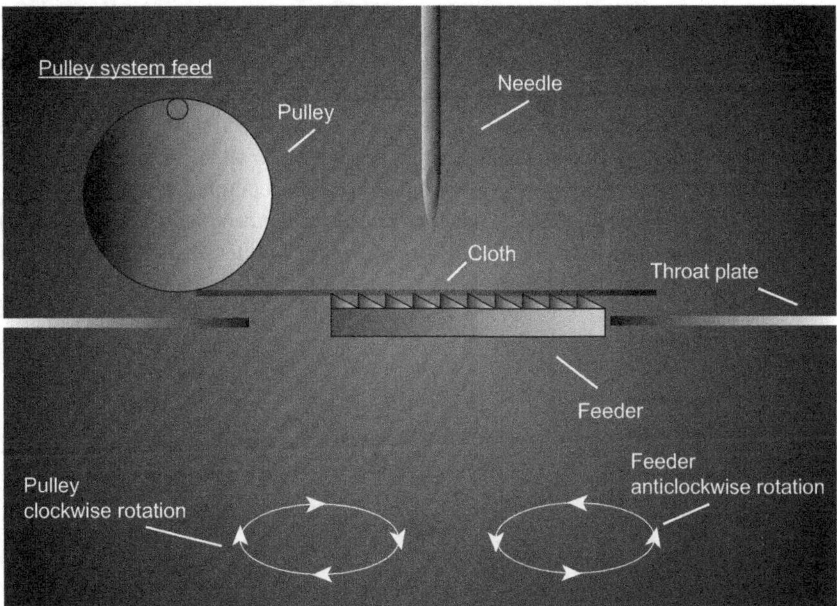

FIGURE 16.30

Puller feed system.

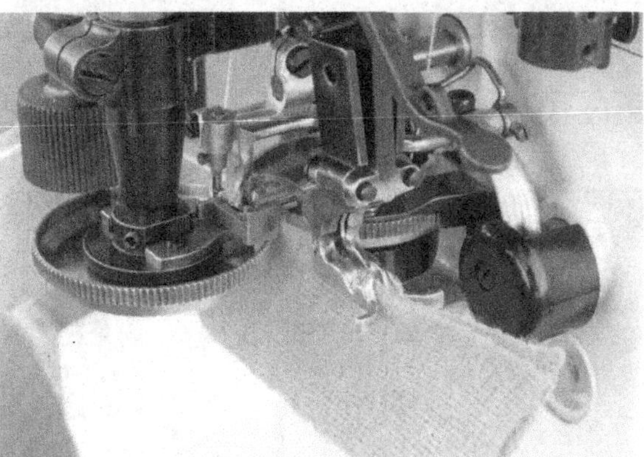

FIGURE 16.31

Cup seaming machine.

16.4.2.12 Manual feed

The operator moves the fabric beneath the needle. No feed dog is employed, thus allowing variable stitch lengths at the operator's discretion. This method is used principally for basting operations on 101 single-thread chain-stitch machines, or for decorative embroidery stitches. It is also employed for jigging collars for men's shirts.

16.4.3 MACHINES FOR DIFFERENT STITCHING OPERATIONS

Hundreds of machine types have been developed for sewing a varied and diverse product range. The machine bed casting, feed system and stitch formation can be adapted to many different types of sewn product operations, so, for example, a lockstitch machine can have numerous machine bed types, different feeding systems and different derivations of lockstitches associated with it depending on what it is required to do.

Choosing the correct machine for the operation must take into account the quantities to be produced and also the desired seam quality. Lockstitch machines used for the production of apparel can be compared to the same type of machine used for sewing larger pieces of material such as household, automotive and outdoor products (see Figures 16.32–16.34).

FIGURE 16.32

Long-arm cylinder arm lockstitch machine.

FIGURE 16.33

Long-arm lockstitch machine for curtains and bedding.

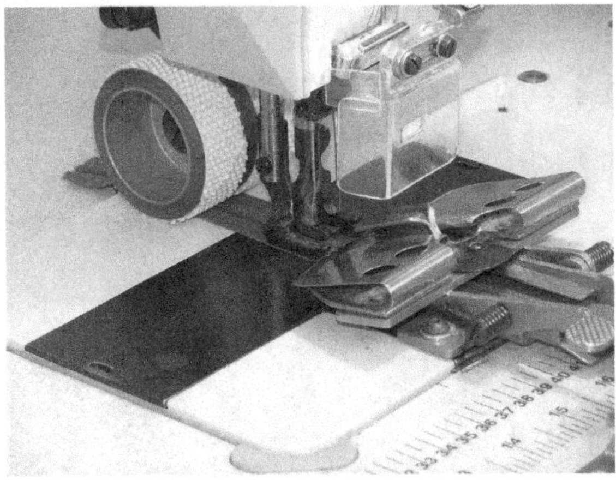

FIGURE 16.34

Lockstitch with puller and unison feed.

FIGURE 16.35

Overlocking machine with a puller feed.

Chain-stitch machines are also extensively used in industry for many types of products and seam operations (see Figures 16.35–16.38). Examples of these are:

- Side seams on garments jeans, shirts blouses, etc.
- Lingerie
- High-performance garments and tents
- Trousers, suits and skirts
- Hems on fleeces and knitwear

FIGURE 16.36

Cylinder arm cover-seam.

FIGURE 16.37

Overlocking a cushion cover closing using a bound seam.

16.5 SEAM QUALITY PROBLEMS

Seam quality problems can be time-consuming, frustrating and costly to a manufacturer. There are many types of problems associated with the quality of seams on a product, but only the most common problems are considered here.

16.5.1 PUCKER

One of the most common problems associated with lightweight fabrics is seam pucker, which is defined as an unequal crinkling or a gathering of the seam, and is typically due to the displacement of the fabric

FIGURE 16.38

Overhead hanging rail system on a production line.

caused by the stitching action, since lockstitch involves the formation of stitches within the material. Pucker is common in lightweight materials such as shirting, blouses and microfiber fabrics. These materials are less able to withstand the forces of the needle and sewing thread due to their more flexible nature. Seam pucker also presents problems in fabrics used in producing skirts, trousers and suits.

There are four main causes of seam pucker:

- Feed pucker
- Tension pucker
- Stitch density and fabric type
- Inherent pucker

16.5.1.1 Feed pucker

The shortening of one of the fabric layers (usually the bottom one) creates a wavy appearance on one side and results in what is known as feed pucker. This problem occurs when two plies of fabric to be joined are not fed uniformly through the sewing machine. The bottom ply is usually fed more positively by the feed dogs, while the top ply is only held and guided by the presser foot (Figure 16.39).

A great variety of feed mechanisms have been developed to try to improve feed pucker. Along with selecting the right feed system for the fabric and seam, tilting the feeder to provide the desired result on the seam may also help, but this adjustment can only be undertaken by a skilled sewing machine engineer.

FIGURE 16.39

Example of feed pucker on surf trousers.

FIGURE 16.40

Example of tension pucker.

16.5.1.2 Tension pucker

Tension pucker results from incorrect thread tension and/or incorrect needle choice (Figure 16.40).

The tension must be as slack as possible to produce a well-balanced stitch, and the smallest-diameter needle possible with the correct needle point type for the fabric should be used. The sewing threads must suit the seam position and the thread must be the minimum diameter possible to minimise disruption within the yarns of the fabric whilst still maintaining the strength of the seam.

Other factors that can affect tension pucker are the extension properties of the sewing thread and the possibility of shrinkage due to moisture and heat. The yarn twist of the sewing thread and frictional properties can both have a significant effect on the regularity of stitch inter-looping and seam appearance.

16.5.1.3 Stitch density and fabric type

Pucker induced by stitch density and fabric type should also be considered when studying tension pucker. It is directly linked with thread tension and the length of thread required by the seam. Increasing the thread consumption in the seam increases the seam strength. In a lockstitch, for example, a 3% increase in stitch consumption can give an almost 60% increase in seam strength.

However, a stitch is only complete after the fabric has moved past the needle. The fewer stitches per centimetre, the greater the distance the fabric must be moved for the next stitch insertion. Consequently a greater force is required to present the correct thread length for a perfect stitch. This can cause higher thread tension in the seam and induce puckering.

16.5.1.4 Inherent pucker

This type of pucker is the hardest to eliminate as it is caused by the displacement of the warp and weft yarns by the needle penetrating the fabric and inserting the thread. If sewn in the warp direction, the warp threads will be displaced laterally, causing an inevitable shortening of their length relative to the adjacent yarns. The fabric structure becomes jammed, resulting in swelling and puckering of the seam.

Factors that should be addressed in attempting to eliminate inherent pucker include:

- Choice of needle and thread – the finer the needles and threads, the less the risk of inherent pucker.
- Choice of needle plate – fine-holed needle plates are essential for reducing seam pucker.
- Choice of stitch type.
- Operator handling skills.
- Sewing direction.

The particular problem highlighted in Figure 16.40 is due to the fittings used on the machine, namely the feed dog and presser foot mechanism. With this type of garment, fine fittings must be used. A fine-toothed feeder and a foot would provide a smoother feeding process on the garment and reduce the amount of friction on the garment.

16.5.2 THREAD BREAKAGE

Thread breakage is one of the most frustrating problems to a machinist (Figures 16.41 and 16.42). If the thread breaks during the sewing of a seam, the thread has to be unpicked and the seam sewn again. This is because the joining of the stitch must not be seen on the face of the garment, as this would compromise the aesthetics of the garment.

16.6 FUTURE TRENDS

The sewing and joining of textiles has evolved over many hundreds of years, and this method of production looks set to continue for the foreseeable future. Research into seam quality issues is useful for developing techniques and methods to combat production problems.

Various approaches have sought to explain the factors involved in seam sewability (Stylios, 1983, 1997; Stylios & Lloyd, 1998). The problems identified included seam slippage, seam damage, seam grinning, seam cracking and seam pucker, all of which present difficulties when sewn.

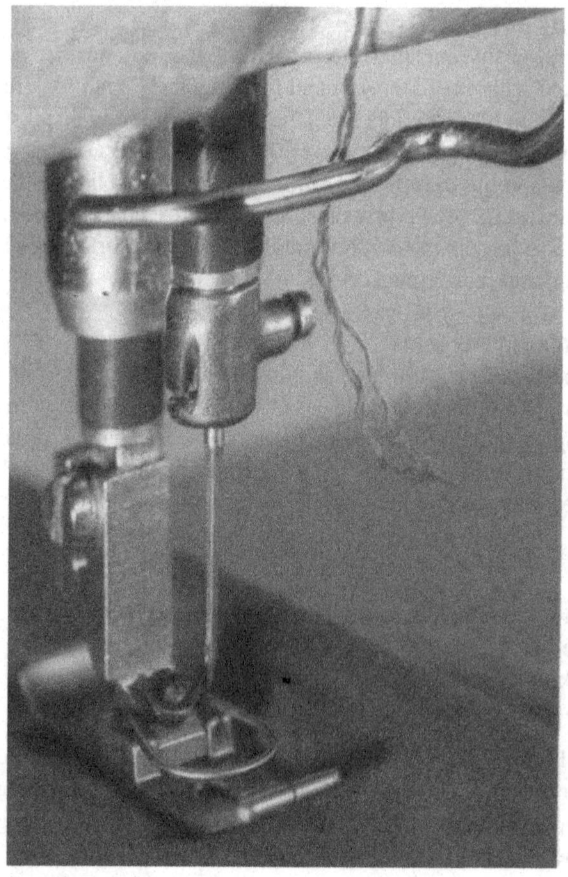

FIGURE 16.41

Example of sewing thread breakage.

McLaren Miller (1998) investigated lockstitch seam instability in the cross-grain construction of woven fabrics. Previous research had suggested that the difference between subjective quality and objective quality is that the former is perceived while the latter can be quantified in part from the mechanical properties. It was suggested that by linking these two aspects, it might be possible to improve control of irregularities arising in production. McLaren Miller quoted a production manager at a plant which supplied a UK retail chain:

> It is all very well having handling values for fabric, as numbers, to try to eliminate the pucker problem, but we have to work with what we're given and the turnover in fabric and design is high. We principally rely on the experience of the handlers and in-line assessment to eliminate the problems concerned

McLoughlin and Hayes (2007) developed a fabric sewability system that automatically analyses the results from Kawabata tests and, using this information, generates an automated textual report of the fabric properties. It also produces guidance as to the sewability of the material. The system was tested

FIGURE 16.42

Slipped stitches on a two thread stitch type 401.

by a team of industrial experts from the apparel industry, the Fabric Sewability Panel. Experts from industry were invited to analyse a number of fabrics and render a judgement on their prospective sewability. The level of agreement amongst the experts was measured using Kendall's coefficient of concordance and significance testing techniques. Comparisons were then made between the judgement of the experts and the results from the Fabric Sewability System. The results confirmed that there was a correlation between the judgements given by the experts and the Fabric Sewability System. During this exercise similar discussions took place between the researchers and the experts. As one expert stated:

> We are not dealing with sheet metal here! Fabrics are flexible structures that have external factors introduced into them during make up. Sewing threads, machine settings and operator handling are all important factors that influence the performance of a fabric during sewing

Many researchers have investigated the sewability and quality of garments using state-of-the art technologies (Lojen, 1998; Mallet & Du, 1999; McWaters & Clapp, 1994). Stylios and Sotomi (1996) have been at the forefront of establishing technologies for garment sewability and attempting to predict levels of seam pucker using laser and fuzzy logic techniques. Stylios and Sotomi investigated the possibility of developing thinking sewing machines for intelligent garment manufacture. They mention the fact that there is reasonable progress in relating fabric properties to sewing machine settings and stitching quality. There are, however, areas that have not been numerically defined because of the complexity of the dynamic interactions between needle, fabric and machine parameters.

Researchers have studied the behaviour of fabrics during apparel manufacture to develop systems that automatically adjust basic sewing machine settings. Ferriera (1997) developed an online control system to optimise seam production during the sewing process. He correlated sewing thread tension, presser foot forces and needle penetration forces with seam quality to generate quantitative information about seam production.

Theoretical models have been developed to explain interactions between the needle and bobbin sewing threads (Ferriera, Harlock, & Grosberg, 1994a; Ferriera, Harlock, & Grosberg, 1994b), and have identified that seam balance is a function of the stitch and formation cycles. Stylios and Sotomi (1996) developed methods for optimising sewing machine settings using fuzzy logic in a neural network. They claimed that optimum sewing machine settings were achieved under static and dynamic machine conditions. However, many adjustments can only be performed by hand (Needles Eye, 1996).

Gauges are necessary to ensure proper adjustment and optimisation of machine settings. Improper settings can result in sewing problems, poor stitch formation, seam deformation and a general decline in machine performance and seam sewability.

Fabric feeding can be a major cause of seam deformation. Kennon and Hayes (2000) have investigated fabric feed timing on lockstitch sewing machines and conclude that by retarding (causing to move more slowly) the feed timing by 25°, the tension in the stitch formation is reduced, therefore reducing seam pucker.

Zunic-Lojen and Gotih (2003) used computer simulations to model the kinematics of the needle bar and take-up lever mechanism on a lockstitch sewing machine in order to measure the thread tension forces in the sewing process.

Much of this state-of-the-art research, though important and necessary, does not seem to have an impact on the manufacturing shop floor. McLoughlin (2005) comments that:

> The gap between work practice methods and research methods could be seen to be large. Many companies do not have the resources to fund the purchase of an objective measurement system. In fact many do not know of the existence of such systems at all. It's apparent that there are major difficulties involved in joining fabrics together at the machine interface by the sewing process. (pp. 99–100)

A number of measures may be taken in order to help alleviate the problem of research not being implemented, including:

- Collating historical machine settings data for each style and fabric sewn.
- Establishing methods for dealing with seam pucker, understanding its causes and steps that may be taken to counter it.
- Giving technicians and production staff greater understanding of the properties associated with a fabric. This includes knowledge of fibres, yarns, yarn twist, frictional properties, shear forces, extensibility and bending rigidity.

- Extending the use of fabric objective measurement systems in fabric manufacturing companies in order to enable them to inform garment manufacturers about the fabric and its potential behaviour in different sewing systems.

Further research should be performed and a settings database may be created to determine optimum sewing conditions for each type of fabric sewn.

The use of low-cost instrumentation for machine optimisation should be promoted. Equipment for measuring thread tensions and strain gauges currently exists and is inexpensive to purchase.

16.7 SUMMARY

- Stitches are formed by thread(s) intra-looping, interloping or interlacing, and there are six classifications of stitch: 100 (single-thread chain stitch), 200 (hand stitch), 300 (lockstitch), 400 (multi-thread chain stitch), 500 (overedge or overlocking stitch) and 600 (covering chain stitch).
- Within each classification of stitch are numerous specific types, identified by the second and third digits of the classification number, e.g. 301 is single-needle lockstitch.
- Stitch quality is affected by the tension of the top and bottom threads.
- There are four classes of seam, Class 1 (superimposed), Class 2 (lapped), Class 3 (bound) and Class 4 (flat).
- Machines have been developed to perform a wide variety of sewing operations.
- The most important parts of a sewing machine are the needle and the feeder mechanism.
- Seam quality is influenced by a number of factors, including the choice of fabric, needle, thread, type of seam and type of machine.
- Pucker and thread breakage are the most significant seam quality problems.

16.8 CASE STUDY AND PROJECT IDEA

This case study is based on a real-life scenario at a production factory where flexibility and diversity are critical in an increasingly competitive and discerning market place.

Bodyware, a company that was established from humble beginnings in 1955, manufactures highly fashionable ladies knitwear for major High Street stores in the UK and other countries. Over its first three decades, the company grew considerably. Attention to quality and excellent customer service were deemed the major reasons for their meteoric success.

In 1983, the company had 15 production factories and employed a total of 1500 people. The factories produced other types of ladies clothing, including lingerie and dressing gowns, and Bodyware acquired a reputation for tackling difficult-to-manufacture products using any type of fabrics.

In 1985, the business saw an opportunity to go public, which would produce the capital needed for future investment and thus to ensure the continued success of the business.

However, in 1990, sales and demand for knitwear in the UK began to suffer due mainly to greater competition from overseas competitors. Leisurewear and sportswear were becoming more dominant and the board of directors felt that a different direction and product evaluation might be needed if the company was to remain competitive.

This decision was made for them later that year. The company was informed that orders in their knitwear division were being reduced by 60%. Cheaper imports from China and the African continent meant that the company could no longer compete on price. The search was on to find more innovative products.

Jim Smith, chairman of the board and co-founder of the business, stated that in order to guarantee the company's future, new innovative products would have to be developed that could give the company a competitive edge. The board therefore decided to engage a team of garment technologists to develop and recommend new products for manufacture.

Assume that you are part of a team employed by Bodyware to undertake the task of researching and recommending a suitable product to the board of directors. Within this assignment you may wish to consider:

- Your reasons for choosing the product.
- The types and suitability of the sewing machines and equipment you are going to use.
- Machine requisition and purchase.
- Machine setup, which includes types of needles, feed systems, stitch formations and machine bed types.
- Factors to be considered when sewing a quality product.

Your report should be in the form of a formal presentation to the managing director and the board of directors. Any material which is based on published sources must be references using the Harvard referencing system.

16.9 REVISION QUESTIONS

- List the six classes of stitches.
- Why is a chain stitch used in industrial sacking?
- For which three types of joins would you use Class 400 multi-thread chain stitches?
- Should thread tension on a sewing machine be tight or loose?
- What are the four types of seam?
- What type of seam would you use for a judo suit?
- List the 10 component parts of a sewing-machine needle.
- What type of feeding system on a sewing machine is used to stretch the bottom ply of fabric?
- What are the five factors that should be addressed to reduce inherent pucker?

REFERENCES AND FURTHER READING

Ferriera, F. B. N. (1997). Research and control on the seam process. *International Journal of Clothing Science and Technology*, *9*(6), 64–66.

Ferriera, F. B. N., Harlock, S., & Grosberg, P. (1994a). A study of thread tensions on a lockstitch sewing machine (part 1). *International Journal of Clothing Science and Technology*, *6*(1), 14–19.

Ferriera, F. B. N., Harlock, S., & Grosberg, P. (1994b). A study of thread tensions on a lockstitch sewing machine (part 2). *International Journal of Clothing Science and Technology*, *6*(5), 26–29.
International Organisation for Standardisation. (1991). *Stitch types – Classification and terminology.* Geneva, Switzerland.
Kennon, W. R., & Hayes, S. G. (2000). The effects of feed retardation on lockstitch sewing. *Journal of the Textile Institute*, *91*(Part 1), 509–522.
Lojen, D. Z. (1998). Simulation of sewing machine mechanisms using program package ADAMS. *10*(3/4), 219–225.
Mallet, E., & Du, R. (1999). Finite element analysis of sewing process. *International Journal of Clothing Science and Technology*, *11*(1), 19–36.
Mcloughlin, J. (1998). The expanding role of the clothing machine engineer. *World Clothing Manufacturer*, *79*(7), 37–41.
Mcloughlin, J. (1999). Implementation of a zero breakdown strategy. *World Clothing Manufacturer*, *80*(1), 12–16.
Mcloughlin, J. (2000). Time to value the production worker. *World Clothing Manufacturer*, *81*(3), 16–21.
McLoughlin, J. (2005). *Development of an automated reporting method for the analysis of sewability measurement* (Ph.D. thesis) (pp. 99–100). University of Manchester.
McLoughlin, J., & Hayes, S. G. (2007). Automating objective fabric reporting. In S. A. Ariadurai, & W. A. Wimalaweera (Eds.), *The Textile Institute 85th world conference*, 1st–3rd May 2007. Colombo (pp. 568–582).
McWaters, S. D., & Clapp, T. G. (1994). Computer simulation of fabric deformation for the design of equipment. *International Journal of Clothing Science and Technology*, *6*(5), 30–38.
McLaren Miller, J. (1998). *An analysis of lockstitch seam instability in the cross – grain construction of woven fabrics* (Ph.D. Thesis) (pp. 4, 15). University of Manchester.
Needles Eye (1996). The right adjustment needs the right gauge. (390), 24–28.
Schmetz the Needle Company. (1998). *Guide to sewing techniques* (3rd ed.). p. 139. http://www.Schmetz.com.
Stylios, G. (1983). *Seam pucker and structural jamming in woven textiles* (M.Sc. thesis). Leeds.
Stylios, G. (1997). Automation of sewing mazchine settings in difficult-to-see fabrics using objective measurement technologies. *International Journal of Clothing Science and Technology*, *9*(6), 7–9.
Stylios, G., & Lloyd, D. W. (1998). A technique for identification of seam pucker due to fabric structural jamming. *International Journal of Clothing Science and Technology*, *1*(2), 25–27.
Stylios, G., & Sotomi, J. O. (1996). Thinking sewing machines for intelligent garment manufacture. *International Journal of Clothing Science and Technology*, *8*(1/2), 44–55.
Union Special. (1988). *Stitch formation lockstitch type 301,* Technical Papers.
Zunic-Lojen, D., & Gotih, K. (2003). Computer simulation of needle and take-up lever mechanism using the ADAMS software package. *Fibres and Textiles in Eastern Europe*, *11*(4), 39–44.

CHAPTER

JOINING FABRICS: FASTENINGS

17

A. Mitchell, J. McLoughlin
Manchester Metropolitan University, Manchester, UK

LEARNING OBJECTIVES

At the end of this chapter, you should be able to:

- List the different types of fastenings used in the manufacture of textile products, and the key safety concerns of each one
- Select the best type of fastener for different uses

17.1 INTRODUCTION

Fastenings, whether functional or purely decorative, are now an integral part of apparel design. They are sophisticated in terms of the materials used and in the functions that they are required to perform, and include zips, buttons, hook-and-loop fasteners, press fasteners, cords, ties and belts, hook-and-eye fasteners, hook-and-bar fastenings, and buckles and adjustable fastenings.

A critical concern for manufacturers of textile products is safety, because loose or broken fastenings can lead to choking, particularly in small children, or to the failure of an essential piece of equipment. Government legislation and safety standards have been put in place, and failure to apply standards or comply with legislation on the part of manufacturers, suppliers and retailers can result in heavy fines and lawsuits if failure of the fasteners or their attachment causes injuries or even deaths.

17.2 ZIPS

The origins of the zip fastener or zipper date back to 1851, when Elias Howe patented a non-commercial automatic continuous closure using metal clasps (Figure 17.1). Various developments and new designs led to Sundback's hookless no. 2 zipper, introduced in 1913 (Figures 17.2 and 17.3). This was the first commercially successful zipper produced on a mass basis by automatic machinery, and became the foundation for the design of the modern zip fastener. Today, Japanese company YKK and Coats (under the Opti Brand) are the largest designers and manufacturers of high-quality zip fasteners.

Textiles and Fashion. http://dx.doi.org/10.1016/B978-1-84569-931-4.00017-9

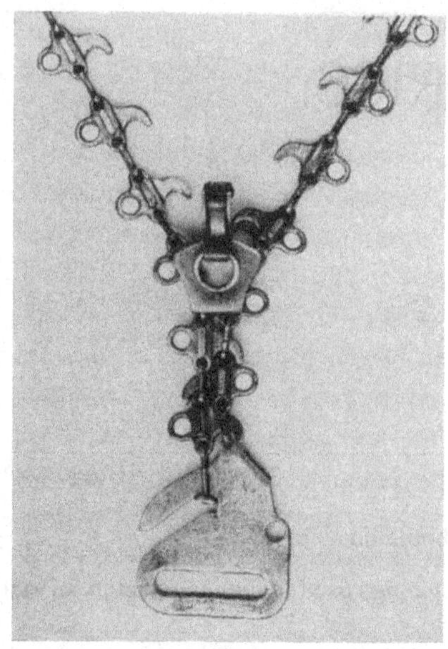

FIGURE 17.1

Whitcomb L. Judson's 'clasp locker', 1851.

Source: Talon Inc., in Friedel (1996).

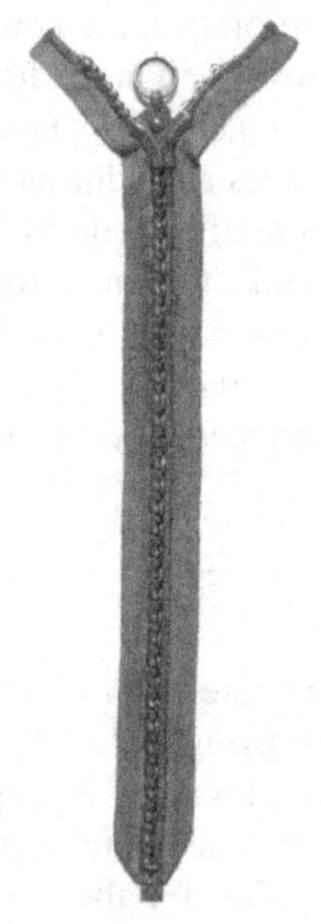

FIGURE 17.2

The Plako fastener, 1908.

Source: Courtesy of Talon International, vintage collection from Tom Allison.

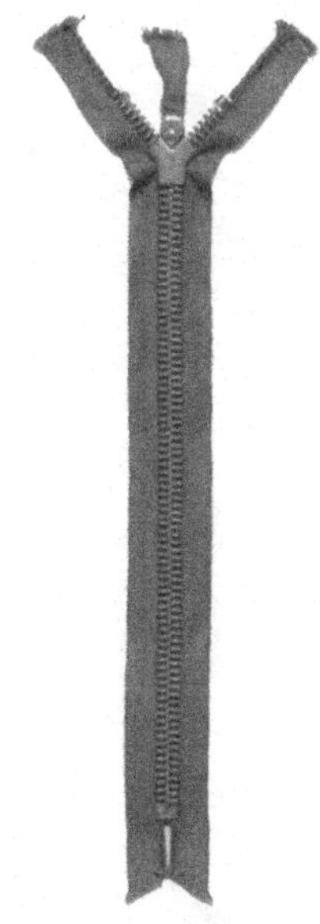

FIGURE 17.3

The hookless no. 2, 1913.

Source: Courtesy of Talon International, vintage collection from Tom Allison.

17.2.1 COMPONENTS OF A ZIP

A zip fastener is generally made up of five components (Figure 17.4):

- Zip tape
- Slider
- Elements, or teeth
- Top stop
- Bottom stop

The zip tape, depending on the end use, is normally made from polyester, nylon, cotton or vinyl tape. The latter is primarily used for the production of water-resistant or waterproof zip fasteners.

The slider joins or separates the elements. Various types of sliders are available, the choice of which to use depends on the function of the join. All sliders are equipped with a locking device that restricts free movement along the fastener length in the opening direction. The locking device may operate either automatically on release of the puller, or by manual pressure on the puller (BS 3084:2006). The main locking devices include autolock; non-lock, which tends to open independently and is not useful where modesty is required; semi-autolock; and reversible. There is also a pin lock, but this type has a sharp pin-type feature that stands proud on the underside of the slider, and is banned by most retailers and manufacturers for health and safety reasons.

The elements or teeth are applied to either side of the tape and are engaged or joined when the slider is passed along the teeth. The elements can be metal, moulded plastic or metal coil, and are normally available in different weights depending on the end use.

Top stops are found at the top of a zip and prevent the slider from going past the end of the zip.

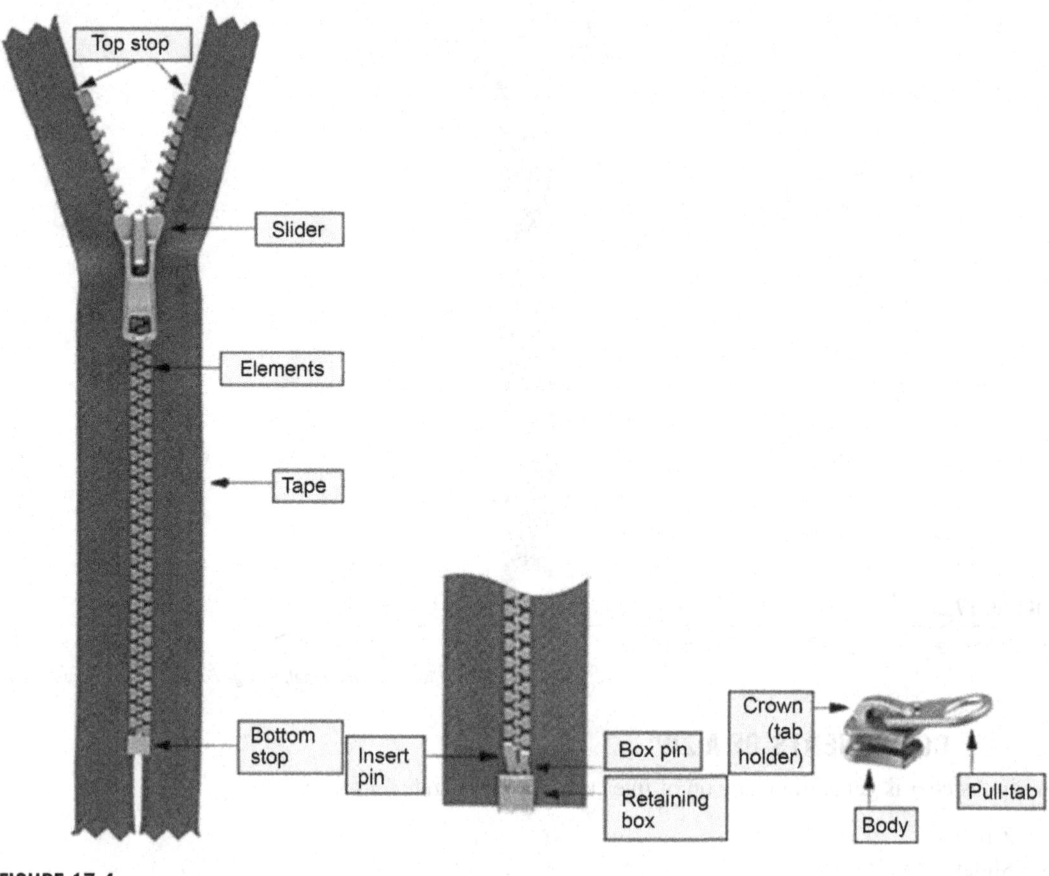

FIGURE 17.4

Components of a zip.

Source: Courtesy of YKK.

The bottom stop bridges the teeth at the bottom of the zip to prevent the slider from running off the end. Open-end zips do not have a bottom stop, as the retaining box replicates this function.

17.2.2 ZIP FUNCTIONS AND APPLICATIONS

Modern zip fasteners are designed to perform a variety of functions. Table 17.1 outlines the main varieties and their typical applications.

17.2.3 HOW TO MEASURE THE CORRECT LENGTH OF OPENING FOR A ZIP

Determining the correct length for a closed-end zipper is essential to prevent damage to the product during wear. If a zip is too short, the maximum opening circumference of the garment may be too small and the product may tear when it is being tried on. The length of the zipper when its slider is at its

Table 17.1 Zip function and their application

Function	Illustration	Typical end use
Closed end		Trouser fly, skirts, jeans, dresses, neck and sleeve openings, leg or ankle openings on trousers
Open-end separator	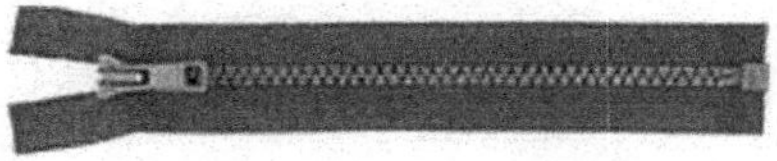	Jackets, coats, hooded tops
Closed end with double sliders		Items that require ventilation openings or items that can be fastened at either end for speed or convenience
Open end with reversible slider		Reversible garments and tents
Invisible zip closed end		Where a closed-end concealed opening is required. When inserted, only the puller is visible 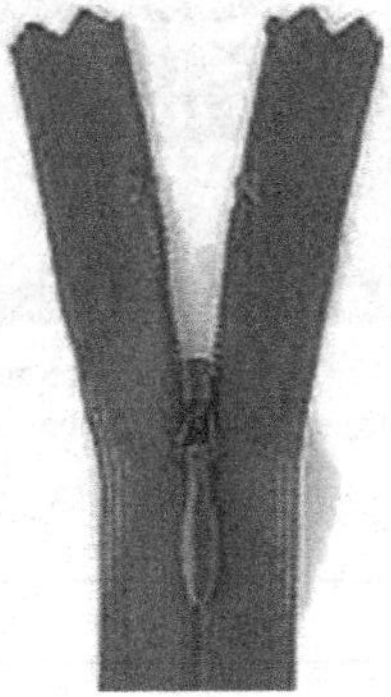

Continued

Table 17.1 Continued

Function	Illustration	Typical end use
Invisible zip open end	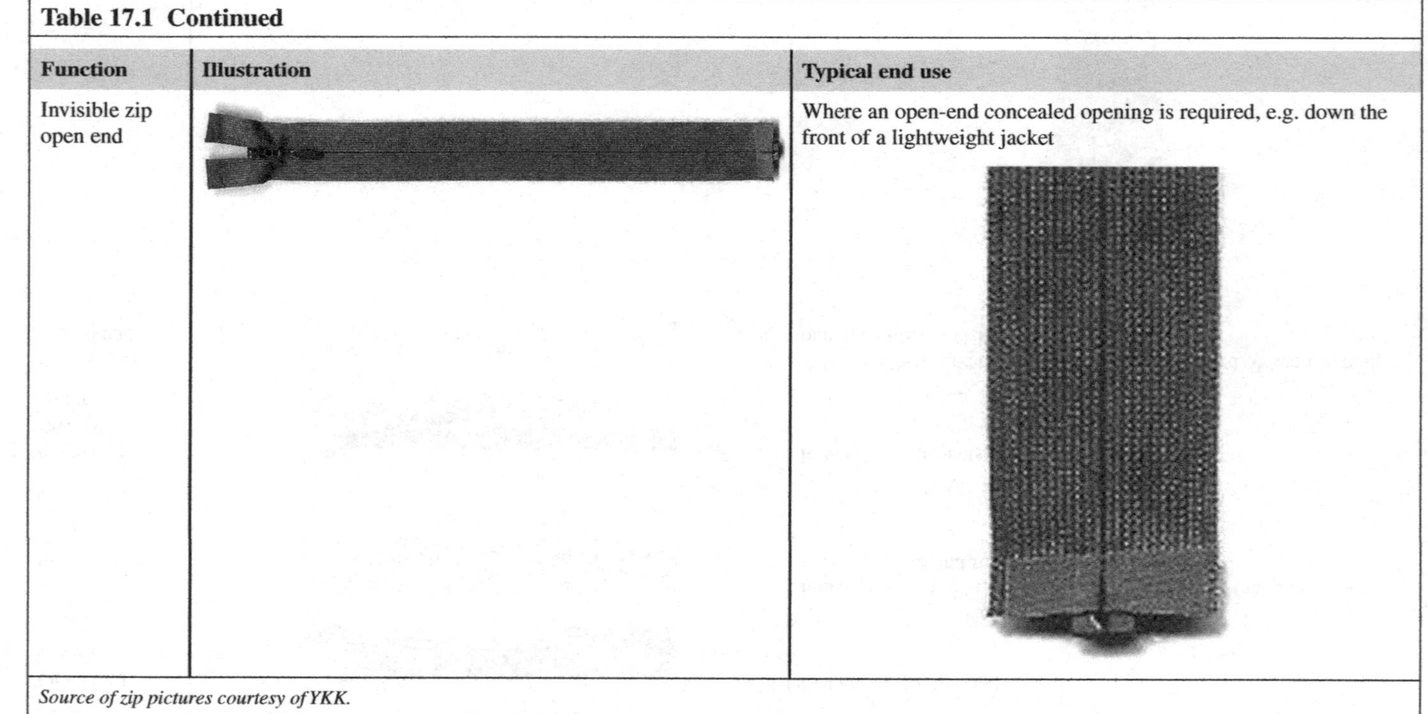	Where an open-end concealed opening is required, e.g. down the front of a lightweight jacket

Source of zip pictures courtesy of YKK.

lowest position should allow the garment to smoothly pass over the widest part of the body; this may be the head, shoulders or hips, or a combination.

17.2.3.1 Measuring a closed-end zip

As a general rule, closed-end zips should be measured from the top of the slider to the bottom of the bottom stop. However, if you wish to determine the overall opening length of the zipper, the slider should be lowered and the zip measured from the top of the top stop to the top of the slider. This accounts for the fact that the slider cannot pass beyond the bottom stop, and thus overall opening length is actually shorter, by roughly the length of the actual slider.

17.2.3.2 Measuring an open-end zip

Open-end zips should be measured from the top of the slider to the bottom of the box or zip tape.

17.2.3.3 Measuring an invisible zip

An invisible zip should be measured from the top of the top stop to the top of the slider when it is at its lowest position.

17.2.4 MACHINERY AND ATTACHMENTS USED TO APPLY ZIPS

Regardless of the zip type or function, zip fasteners are most commonly applied to a garment using a combination of 301 lockstitch and an appropriate zip foot. Table 17.2 summarises the foot or combination of feet used for some of the more common methods of zip insertion.

The exception to the above is the application of water-resistant or waterproof zips used for specialist outdoor or performance apparel, which can be applied using one of the following two methods:

- Bonding using thermoplastic adhesive; an example is given in Figure 17.5.
- Stitching and then sealing using hot melt tape.

These methods have been developed to either seal or eliminate needle holes, which would otherwise compromise the waterproof qualities of such items.

17.2.5 CONTINUOUS ZIPS

In some sectors of the apparel industry, zip fasteners are bought as a continuous roll and the sliders are bought separately. The required length of zip is cut when the fastener is being inserted during garment making up, and the slider is added afterwards. This can speed up application, as it eliminates the need to move the slider out of the way during insertion, an action that is not possible when using semi-automatic equipment such as a Pocket Jet machine that is equipped to insert a zip fastener simultaneously. A special metal zip jig (Figure 17.6) is used to mount the slider, allowing it to be easily fed onto the zip teeth, therefore speeding up the otherwise fiddly application of the slider. This system also means that a supply of zip tape can be kept in stock and new sliders can be ordered when style changes dictate.

17.2.6 SAFETY STANDARDS AND LEGISLATION FOR SELECTING AND APPLYING ZIPS

When selecting a zip fastener, both product and end-user safety are of paramount importance. Compliance with statutory legislation and key British and European standards is imperative to

Table 17.2 Attachments used to apply zip fasteners

Method	Zip function	Presser foot	Picture of presser foot
Invisible	Invisible fastening	Invisible zip foot, to apply the zip and a left-hand half foot to close the seam below the zip insert	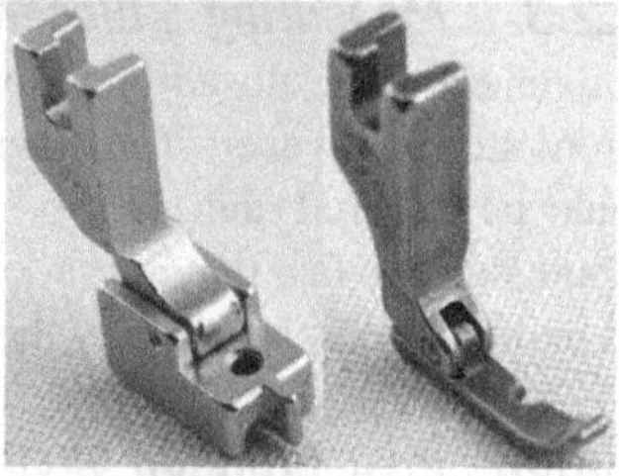
Channel	Open or closed end	Narrow-toe zip foot	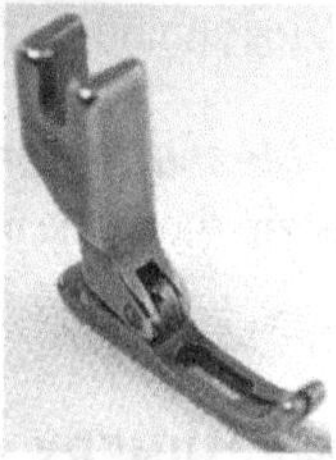
Open channel	Closed end	Narrow-toe zip foot	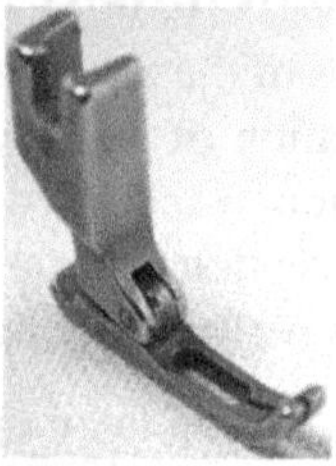
Concealed	Closed-end zip	Right-hand half-foot	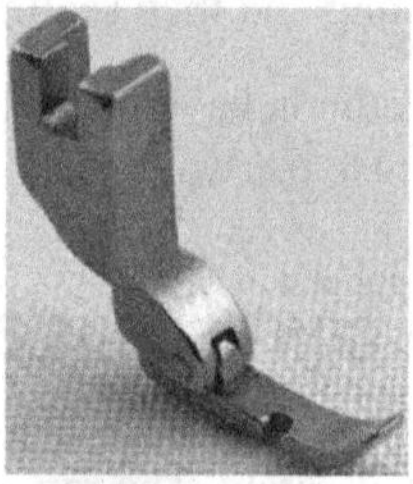
Fly front	Closed-end zip	Narrow-toe zip foot	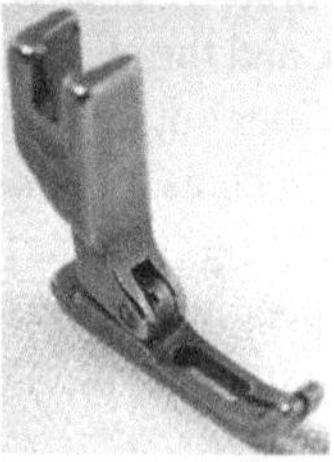

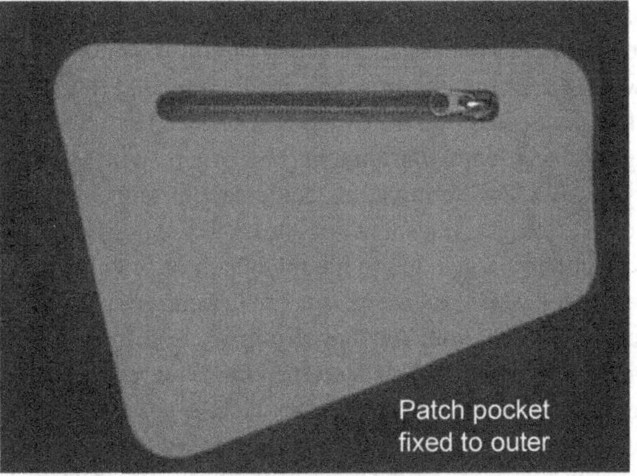

FIGURE 17.5

Bonded water-resistant zip fastener.

Source: Courtesy of Ardmel Automation Limited.

FIGURE 17.6

Zip slider jig used to apply the slider to the zip tape.

Source: Glover Bros, Taunton, Somerset.

protect the end-user and those involved in the supply chain of apparel that incorporates zip fasteners.

The following points should be considered when selecting a zip fastener:

- Will the garment undergo wet processing, such as enzyme wash, chemical finishing and dyeing? If so, the proposed zip should be sample-tested both prior to bulk production and at the commencement of bulk production to ensure that there is no corrosion or deterioration of the zip's components.
- If proposing a zip for children's wear, then a non-ferrous fastener should be selected to facilitate metal detection during and post manufacture.
- Pinlock sliders should never be used for children's wear and should generally be avoided due to the obvious health and safety risks associated with this type of locking device.

- Fly fronts that are fastened using a zip fastener should not typically be used in boys' wear for boys aged five years and under due to the associated risks of entrapment. BS7907:2007 should be the minimum standard when designing children's garments intended to include a zip fastener.
- For boys older than five years, a zip guard which is at least 20 mm wide should be incorporated to provide protection against accidental entrapment. Use of zips with plastic elements is preferable as they are less likely to cause severe injury if accidental entrapment does occur (BS7907:2007).
- Designers should avoid style/seam lines that cut across the seam where a zip is to be inserted, as this can result in costly quality issues due to the mismatch of seams when the zip is inserted. This problem is more pronounced where invisible zips are to be used, as this can result in partial jamming of the puller during fastening and can ultimately lead to zip failure.
- All metal zippers where prolonged contact with the skin is anticipated should comply to the statutory Nickel Safety requirement within regulation (EC) no. 1907/2006 (REACH).
- With reference to the dyeing and colouration of zip fastening tape, all zips should comply with the statutory requirement within regulation (EC) no. 1907/2006 (REACH), which prohibits the use of most azodye colourants.
- If zippers are to be cut from a continuous roll, the end should be securely boxed with a suitable material to prevent discomfort or injury from rough chain ends.
- The performance characteristics of zip fasteners should be tested to comply with BS3084:2006.

17.2.6.1 Statutory requirements

- Regulation (EC) no. 1907/2006. Registration, Evaluation, Authorisation and Restriction of Chemicals (REACH).
- Statutory Instruments 2005 Consumer Protection no. 1803. The General Product Safety Regulation 2005.

17.2.6.2 British standards

- British Standard BS 3084:2006 – Slide Fasteners (Zips) – Specification.
- British Standard BS EN 71-3:1995 – Safety of Toys – Part 3: Migration of Certain Elements.
- British Standard BS 7907:2007 – Code of Practice for the Design and Manufacture of Children's Clothing to Promote Mechanical Safety.
- British Standard BS EN 12472:2005+A1:2009 – Method for the Simulation of Wear and Corrosion for the Detection of Nickel Release from Coated Items.
- British Standard BS EN 1811:1998+A1:2008 – Reference Test Method for Release of Nickel from Products Intended to Come into Direct and Prolonged Contact with the Skin.

17.3 BUTTONS

Buttons have existed for thousands of years, and were originally used for decorative adornment. It was the Greeks and Romans who first used buttons to fasten their apparel, and this functional use of buttons spread to Europe from the thirteenth century. Buttons soon ranged from the simple to the highly decorative, and they remain one of the main methods of fastening apparel today. Materials and manufacturing equipment for buttons continue to be developed.

17.3.1 TYPES OF BUTTONS

17.3.1.1 Sew-through buttons

The vast majority of sew-through buttons are two-hole or four-hole; however, customised versions have included the three-hole button used on the famous American shirt brand Hathaway (which ceased production in 2002) (Figure 17.7).

Two-hole buttons are commonly used on lighter-weight clothing such as shirts and blouses, and are accompanied by a straight buttonhole.

Four-hole buttons are more commonly associated with heavier outerwear and tailored items, as the extra buttonholes offer a more durable attachment. They are normally accompanied by a keyhole buttonhole, which accommodates the additional thread shank when the button is attached and eliminates distortion of the buttonhole when fastened.

17.3.1.2 Shank buttons

There are a wide variety of shank buttons (Figure 17.8); however, most can be categorised as either self-shank, wire-shank or tunnel-shank. The more unusual cloth-shank button is usually formed when

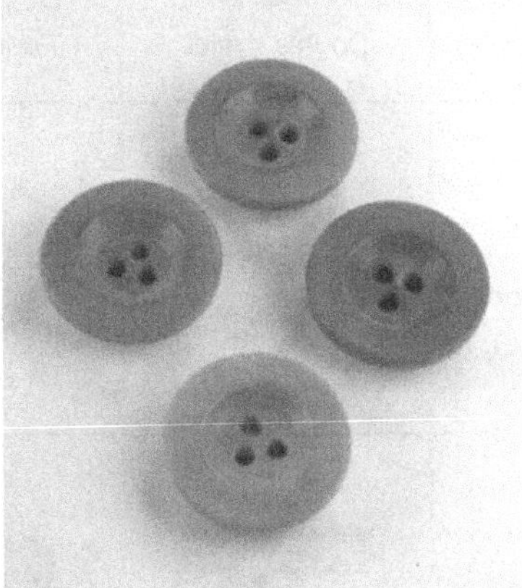

FIGURE 17.7

Sew-through three hole buttons.

Self shank	Wire shank	Tunnel shank	Cloth shank
		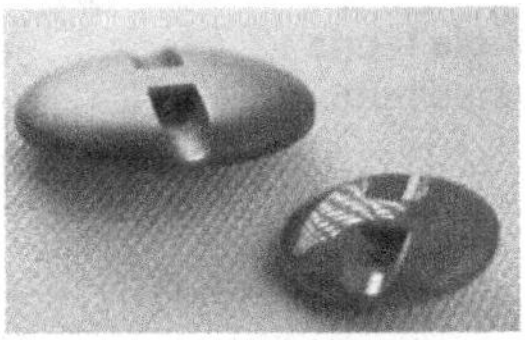	

FIGURE 17.8

Types of shank buttons.

constructing a special type of covered button. Shank buttons tend to be selected for their highly decorative appearance. Unlike the sew-through button, the sewing thread used to attach shank buttons is generally concealed.

17.3.1.3 Covered buttons

Covered buttons are fabric-covered plastic or metal forms with a separate backpiece that secures the fabric over the knob. Covered buttons are prone to changes in fashion; however, they remain the staple fastening for classical apparel such as bridal wear. There are many varieties of covered buttons; the main ones are shown in Figure 17.9.

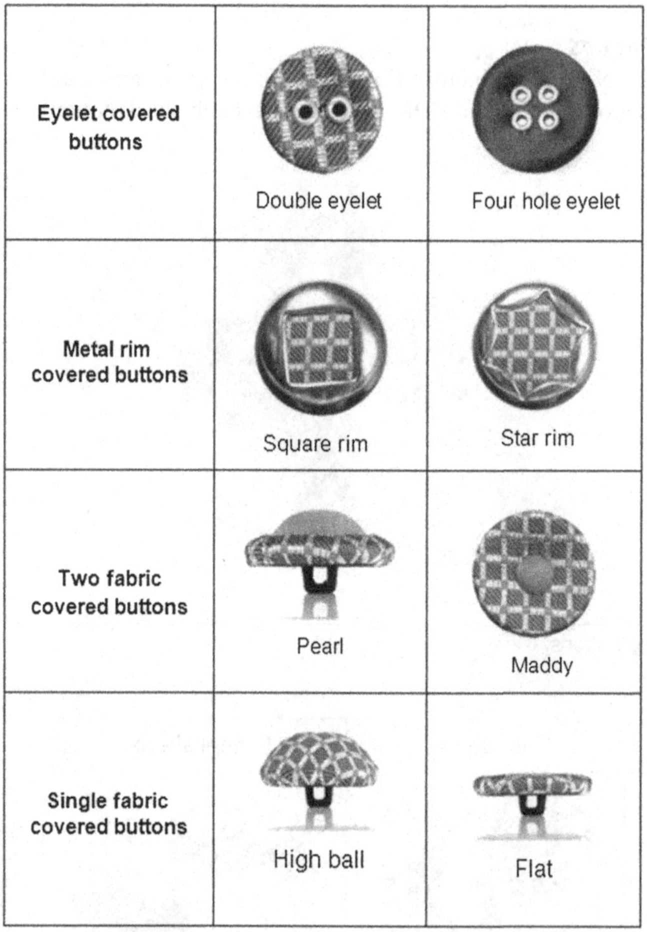

Eyelet covered buttons	Double eyelet	Four hole eyelet
Metal rim covered buttons	Square rim	Star rim
Two fabric covered buttons	Pearl	Maddy
Single fabric covered buttons	High ball	Flat

FIGURE 17.9

Types of covered buttons.

Source: Courtesy of Stock, N at Harlequin.

17.3.1.4 Multi-material buttons

Buttons are often fabricated using combinations of different materials; however, great care must be taken to ensure the components are compatible with the end use. This is particularly relevant where post-finishing, such as industrial wet processing, is to be undertaken. Figure 17.10 shows buttons made using a metal surround with a synthetic centre. If subjected to high temperatures, the synthetic centre could be damaged, while the metal rim might expand, making the construction loose. All such button designs should undergo thorough testing prior to final selection.

17.3.2 MATERIALS USED TO MAKE BUTTONS

Buttons can be made from natural and synthetic materials. Table 17.3 gives an overview of some of the more popular materials used to make buttons.

17.3.3 HOW TO MEASURE BUTTONS

Ligne is the internationally recognised measurement system used to record the size of a button. The origins of ligne are not clear; however, several authors suggest that ligne became the standard reference used by German button manufacturers in the eighteenth century. Figure 17.11 shows a typical brass gauge used to determine the ligne measurement of a button.

A 40-ligne button equates to 1 inch or 25.5 mm. The chart in Figure 17.12 can be used to convert millimetres to ligne.

17.3.4 MACHINERY AND ATTACHMENTS USED TO APPLY BUTTONS

British Standard BS 7907 states that the largest single cause of accidents and consumer complaints related to clothing involves buttons, as children are known to place such items in their mouth, nose and

FIGURE 17.10

Examples of multi-material buttons.

Table 17.3 Materials used to make buttons

Material	
Wood 	Like horn and bone, wood has been used throughout the ages to make buttons that have ranged from the plain to the highly elaborate having been decorated by burning, carving, painting or polishing. Most woods have been used, and during the eighteenth century hard woods such as cedar, maple, ash, and fruit tree wood such as rose and apple were used because of their interesting grains. Wooden buttons are not so common today, but became fashionable after the World War II due to material shortages.
Glass 	During the 1900s, black glass buttons were made for the masses in replica of Queen Victoria's black jet buttons; also referred to as mourning buttons following the death of Prince Albert. Jet is a soft black coal-like substance, known geologically as lignite; the best quality came from Whitby in Yorkshire, but it was also found in France and Spain. Glass buttons were made by pouring molten glass into patterned moulds, a process that was developed in America during the 1840s and then used all over the world. The delicate nature of glass was substituted with plastic in the twentieth century when cheaper, durable and safer products came into demand (Peacock, 1972).
Rubber 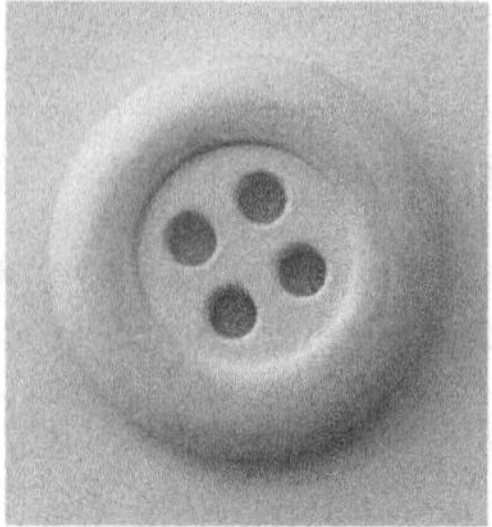	From the mid-1800s, rubber buttons were made from natural rubber, which was exported from Brazil, Peru and Java to America where it was vulcanised (hardened) and poured into button moulds. Today synthetic rubber materials are used to make rubber buttons, which are predominantly used for functional hard-wearing industrial workwear.
Horn 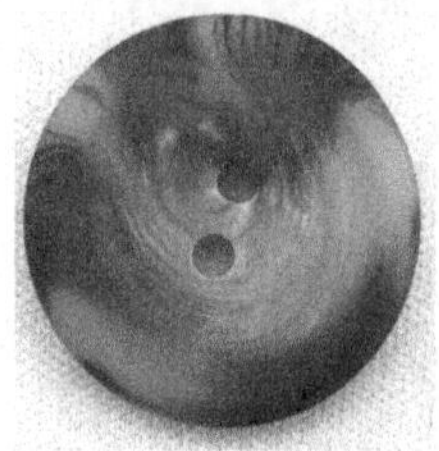	Horn is undoubtedly one of the oldest materials used to make buttons and has some limited use even today. Animal horns are collected as a bi-product of the meat trade, and once graded are used to produce exclusive buttons such as real-horn toggles used on duffle-style coats. Flat sew-on buttons can also be cut using tubular drills.
Bone 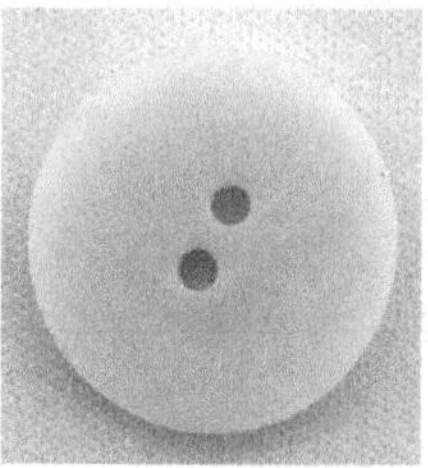	Bone is one of the oldest materials used for making buttons. During the nineteenth century they were made by cutting disks from the shinbone of animals, which were then punched or drilled with holes and polished. Bone buttons were discontinued after the introduction of vegetable ivory (Peacock, 1972).

Table 17.3 Continued

Material	
Tortoise shell 	Tortoise shell, which comes from the hawksbill turtle, has been internationally banned by the Convention on International Trade in Endangered Species (CITES) since 1973. Before this time, the hawksbill turtle shell was used in abundance for the production of buttons and other accessories, particularly during the 1960s. It has since been replaced by synthetic materials.
Vegetable ivory The corosa nut 	Vegetable ivory was the trade name for the kernel of the corosa nut. The corosa nut resembled a mini-coconut, and when the outside shell had been removed, it revealed the internal white kernel that resembled animal ivory. The nuts were sliced and kiln dried to prevent warping. Button blanks were then cut with tubular saws; they were then carved with patterns and drilled to facilitate attaching. Corosa nuts were exported from South America and Africa in the 1850s to button-making centres in Europe. The introduction of cheaper plastics after 1945 led to the gradual decline of its use (Peacock, 1972).
Metal 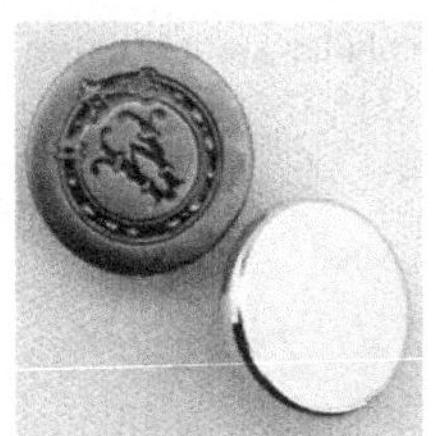	Many different metals have been used to make buttons, including brass, steel, lighter-weight aluminium and even silver. Britain has the longest and finest history of silverwork in the world, and silver buttons can be traced from the end of the seventeenth century, becoming highly popular in the early 1900s when heavily influenced by the Art Nouveau movement (Peacock, 1972).
ABS plastic (faux metal) 	Acrylonitrile butadiene styrene (ABS) is the name given to buttons made from a particular type of plastic that is electroplated or coated with a metal. The final product resembles metal but is much cheaper and lighter in weight. Consequently it is very common for metal style buttons to be applied to lightweight apparel.
Leather 	Some of the earliest leather buttons were fabricated using one strip of leather, which was braided and then compressed into shape, and were commonly found on sports jackets and trench coats. Today most leather buttons are machine made due to peaks and troughs in demand, as this material is likewise prone to fashion trends.

Continued

Table 17.3 Continued

Material	
Knotted thread 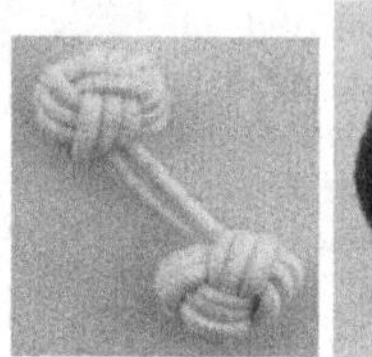 	The Chinese button knot is now more commonly used as soft cufflinks to fasten shirt cuffs. The single knot is commonly used alongside rouleau loops for fastening fashion items inspired by traditional dress styles.
Shell (mother of pearl)  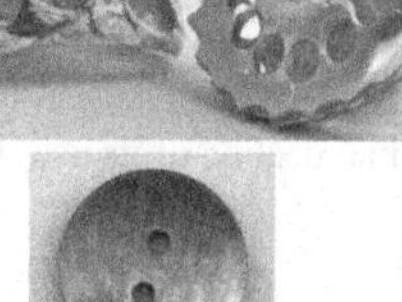	Mother of pearl buttons are often used in clothing either for functional and/or decorative purposes. The material is derived from a variety of seashells including mussel, agoya, abalone and trocas. Once harvested, shell buttons are cut using a tubular drill, and then graded to ensure uniformity of thickness. There is much waste involved in this form of button production, and today shells are more commonly farmed to prevent depletion or extinction.
Coconut 	Buttons made from coconut shell are cut using tubular saws, and then polished to give a natural woody appearance.
Magnetic buttons	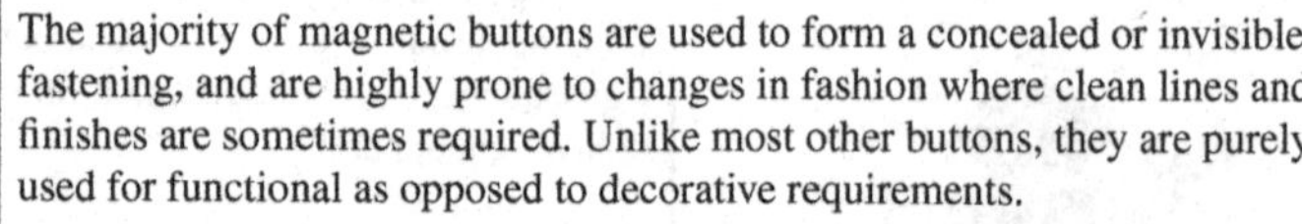 The majority of magnetic buttons are used to form a concealed or invisible fastening, and are highly prone to changes in fashion where clean lines and finishes are sometimes required. Unlike most other buttons, they are purely used for functional as opposed to decorative requirements.
Plastic: acrylic, nylon and polyester 	Since the 1940s most buttons have and continued to be made from plastic due to the fact that it is much more durable and cheaper than all of the other materials discussed so far. Throughout the twentieth century, button makers have experimented with a wide range of plastic materials from casein to Bakelite, and ultimately polyester has become the main material from which buttons are now manufactured. Sew-through polyester buttons are made from polyester resin, which is spread to form a sheet of resin that is then punched into button blanks. The blanks are then drilled to create the sew-through holes and then barreled to create a smooth finish. Polyester buttons are extremely durable, and are resistant to many chemicals and heat. The raw polyester material can be easily dyed prior to button punching, or the finished button blanks can be tinted afterwards, which allows quick response for garment manufacturers.

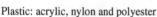

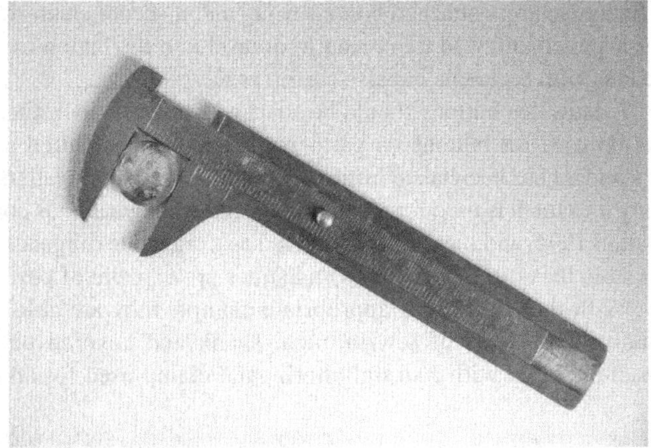

FIGURE 17.11

Example of a button ligne gauge.

Ligne	Millimetres (metric)	Inch (imperial)
14	9.2	0.362
16	10.5	0.413
18	11.6	0.457
20	12.5	0.492
22	14.2	0.559
24	15.0	0.590
28	17.8	0.701
30	19.0	0.748
32	20.5	0.807
34	21.5	0.864
36	22.9	0.902
40	25.5	1.00

FIGURE 17.12

Button ligne measurement conversion chart.

ears. For many years, manufacturers attached buttons using an un-secure chainstitch, which left a signature thread end that if pulled, allowed the thread to unravel and the button to fall off the garment, causing not only problems with accidents but also customer returns.

As a result, BS 7907 states that buttons should be attached using 301 lockstitch, and that class 100 chainstitch should not be used for buttons on garments intended for children aged three years and under. This minimum standard has been taken further by most high-quality retailers and manufacturers, who now insist that only lockstitch is used for all children's and adult wear, thus conforming to General Product Safety Regulation 1994, and therefore considered as a legal due diligence requirement.

Modern programmable lockstitch button-sew machines are capable of performing a wide variety of stitch patterns. With the addition of appropriate clamps, they are able to attach almost all button designs, including variations of sew-through, shank and covered buttons. Figure 17.13 shows a lockstitch machine fitted with a straight horizontal clamp used for attaching sew-through buttons.

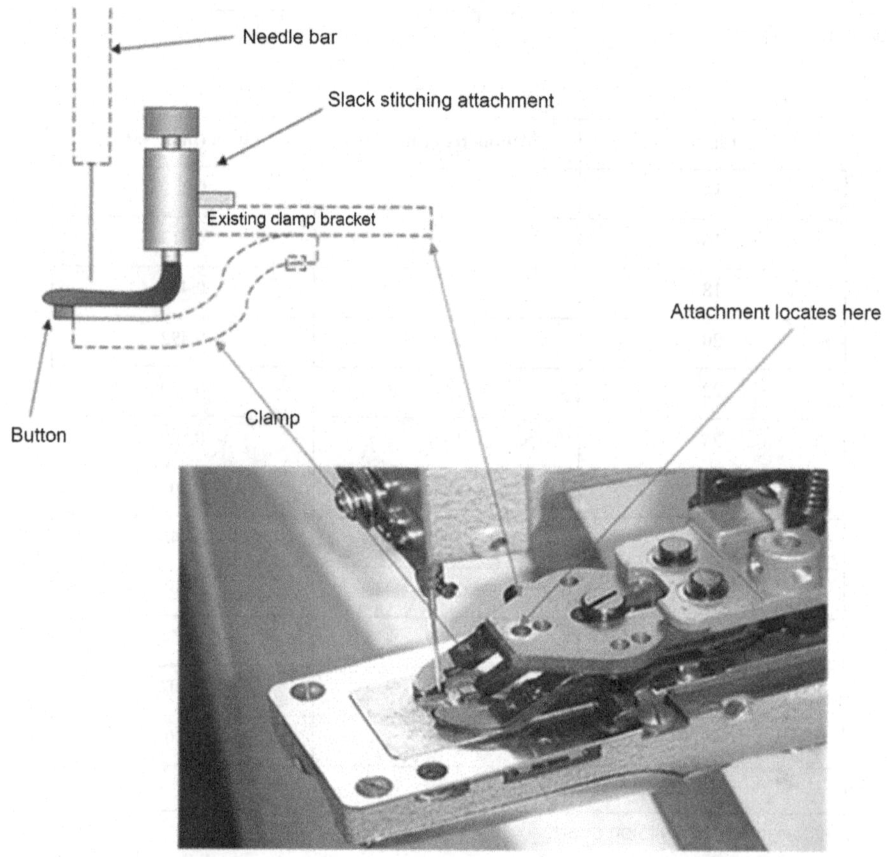

FIGURE 17.13

Button-sew machine clamp.

17.3.5 SAFETY STANDARDS AND LEGISLATION FOR SELECTING AND APPLYING BUTTONS

When selecting a button, end-user safety should be considered of paramount importance. Therefore compliance with statutory legislation and key British and European standards is imperative to protect the end-user and those involved in the supply chain of apparel that incorporates buttons.

The following factors should be considered in selecting buttons:

- Chainstitch should not be used for attaching buttons, as it is not secure and can cause fatal injury, especially important for children's wear. Lockstitch (class 301) should always be used.
- Buttons should be attached through at least two plies of fabric to prevent detachment.
- Only buttons that pass the mechanical tests outlined in BS 4162 should be used on children's clothing.
- Two-piece or multi-component buttons must be thoroughly tested to ensure there is no shrinkage or expansion during wet processing or laundering, as the centre can work loose from the rim. Such buttons should be avoided for products aimed at children aged 12 months to 3 years.
- Buttons containing metal, where prolonged contact with the skin is anticipated, must comply with the statutory Nickel Safety requirement within regulation (EC) no. 1907/2006 (REACH).
- Dyeing and colouration of buttons composed from textiles, such as covered buttons, must comply with the statutory requirement within regulation (EC) no. 1907/2006 (REACH), which prohibits the use of most azodye colourants.
- Shell buttons break easily and should not be used on children's wear, and all buttons should pass BS 1462:1983.
- Testing for colour-fastness is essential.
- Four-hole buttons rather than two-hole should be used for children's wear and babies' wear.
- Buttons that resemble food or sweets should not be used for children's wear.

17.3.5.1 Statutory requirements

- Statutory Instruments 1989 Consumer Protection no. 1291. The Food Imitations (Safety) Regulation.
- Regulation (EC) no. 1907/2006. Registration, Evaluation, Authorisation and Restriction of Chemicals (REACH).
- Statutory Instruments 2005 Consumer Protection no. 1803. The General Product Safety Regulation 2005.

17.3.5.2 British standards

- British Standard BS 4162:1983 – Methods of Test for Buttons.
- British Standard BS EN 71-3:1995 – Safety of Toys – Part 3: Migration of Certain Elements.
- British Standard BS 7907:2007 – Code of Practice for the Design and Manufacture of Children's Clothing to Promote Mechanical Safety.
- British Standard BS EN 12472:2005+A1:2009 – Method for the Simulation of Wear and Corrosion for the Detection of Nickel Release from Coated Items.
- British Standard BS EN 1811:1998+A1:2008 – Reference Test Method for Release of Nickel from Products Intended to Come into Direct and Prolonged Contact with the Skin.

17.4 HOOK-AND-LOOP FASTENERS

Hook-and-loop fastening was invented by a Swiss engineer called George de Mestral in 1941, after investigating how burrs from the Burdock thistle stuck to his clothing with thousands of tiny hooks (Figures 17.14 and 17.15).

George de Mestral patented Velcro in 1955 ('vel' from the French word velvet and 'cro' from the French word for hook, crochet), and subsequently sold his patent rights to Velcro SA, later to become

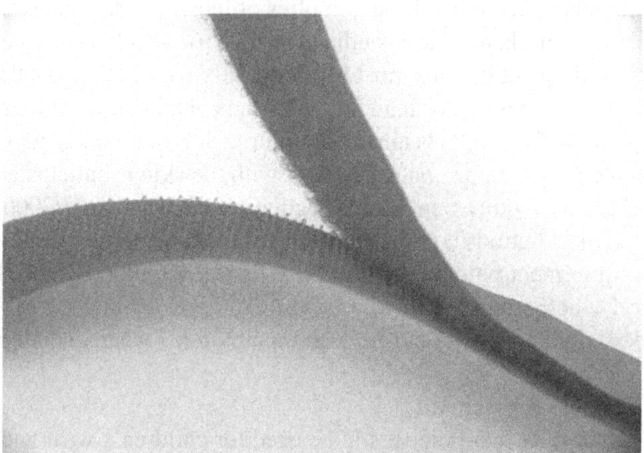

FIGURE 17.14

Hook-and-loop fastening tape.

FIGURE 17.15

Example of the magnified hooks on a thistle.

Velcro Industries NV (Strauss, 2002, Chap. 1, pp. 14–18). Over the decades the name Velcro has been used misleadingly to describe the wide variety of types and brands of hook-and-loop fastenings, which Velcro themselves now insist is actually the name of their organisation, and not the generic name for the hook-and-loop fastener that they produce (Velcro, 2011).

17.4.1 TYPES OF HOOK-AND-LOOP TAPES

There are three main types of hook-and-loop fastening tapes:

- Hook and loop
- Mushroom and velour
- Combined hook-and-loop tape (e.g. Omnitape™)

The principle of the fastening system requires the hook or mushroom of one tape to interlock with loops on another tape. Figure 17.16 shows a magnification of the hook-and-loop tape and mushroom-and-velour tape. The mushroom version can provide a very strong join and is commonly applied where a semi-permanent fastening is required, such as when mounting advertising or on displays, and is less commonly selected for apparel applications. The combined hook-and-loop tape is, as the name suggests, a combination of both the hook and the loop together on the same strip of fastening tape. This has an advantage over the other types as it eliminates the need to stock individual reels of hook-and-loop tape. It is also much less abrasive to the touch than the hook-only tape, and potential damage to more delicate fabric surfaces is minimized.

There are many other varieties of hook-and-loop tape, each with their own intended end use. For example, elastic hook-and-loop tape, woven from elastic Lycra®, is mainly used in the medical industry for fastening bandages, prostheses and orthopaedic footwear.

Most hook-and-loop tapes are manufactured in a range of weights and widths, the narrower of which, at around 10–25 mm wide, are the most commonly used in apparel products. The advantage of using this type of fastening relates to the fact that items can be easily opened and closed; this is particularly relevant for workwear, for speed and ease of donning and doffing, for children's wear and for

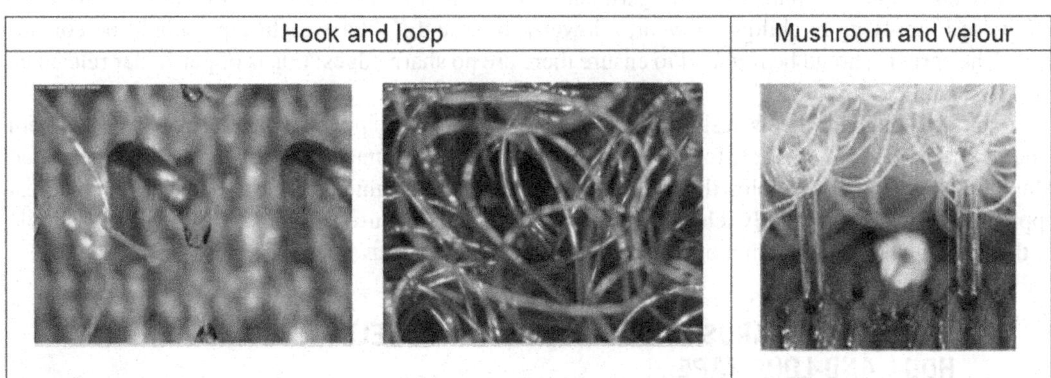

FIGURE 17.16

Hook-and-loop tape and mushroom and velour as viewed under a microscope.

clothing for the elderly or disabled. Lower profile and super soft light-weight versions are commonly available for children's wear, where a more delicate, non-scratchy finish is desired.

Depending on the security requirements of the fastening, the strength of hook-and-loop tape can be improved either by using a wider and longer piece or by selecting a version with a higher proportion of hooks and loops per unit area.

One of the disadvantages of hook-and-loop tape is the fact that over time the hook side of the tape tends to collect lint and fibres, which can reduce the effectiveness of the closure. Garments should always be laundered with the strip fastened to eliminate the collection of fibres and to help protect it from hot irons, as some tapes are not suitable for settings higher than two dots on a domestic iron.

17.4.2 MATERIALS USED TO MAKE HOOK-AND-LOOP TAPE

The very first sample of Velcro ever produced was manufactured from cotton (Strauss, 2002); however, this proved to be non-durable, and today the majority of hook-and-loop fastening tape is manufactured using synthetics such as polyamide, polyester, or a mix of polyester and propylene.

17.4.3 MACHINERY AND ATTACHMENTS USED TO APPLY HOOK-AND-LOOP TAPE

In terms of application, there are two main types of hook-and-loop fastening tapes:

- Sew-on
- No-sew adhesive

Sew-on is by far the most commonly used in the apparel industry, due to the fact that it can be applied using the standard lockstitch machine. No-sew adhesive versions generally make use of rubber or acrylic-based adhesives, which are usually activated by heat or a solvent. They are not widely used for apparel applications.

When specifying the method of attachment, careful consideration should be given to the placement and position of the hook-and-loop tape. The hook side of the tape tends to be extremely scratchy and should be placed on the garment panel, which is less likely to contact with the skin. Likewise if applying to children's wear, a lower-profile and lighter-weight tape should be considered. The corners should be rounded to ensure there are no sharp edges; this is of particular relevance to BS 7907:2007.

Hook-and-loop tape that is designed to be sewn on is normally applied using a lockstitch fitted with a narrow twin toe foot or a half foot. This allows the tape to be attached neatly around the perimeter. Manufacturers may also employ the use of a programmable machine to achieve a consistent and neat application. This is particularly relevant when rounded shapes are used. Figure 17.17 gives an example of the flag-stitched pattern commonly used on casual-wear products.

17.4.4 SAFETY STANDARDS AND LEGISLATION FOR SELECTING AND APPLYING HOOK-AND-LOOP TAPE

When selecting hook-and-loop tape, end-user safety is of paramount importance. Therefore compliance with statutory legislation and key British and European standards is imperative to protect the end-user and those involved in the supply chain of apparel that incorporate this type of fastening. This list

FIGURE 17.17

Hook-and-loop tape attached to a pocket flap using a programmable machine to produce a flag-style stitch formation.

is not exhaustive and in all cases a full risk assessment of the product should be undertaken to ensure the safety of the product and its end-user.

17.4.4.1 Statutory requirements

- Regulation (EC) no. 1907/2006 of the European Parliament and of the Council of 18 December 2006 concerning the Registration, Evaluation, Authorisation and Restriction of Chemicals (REACH).

  Note: With reference to the dyeing and colouration of hook-and-loop fastenings, there must be compliance with the statutory requirement within regulation (EC) no. 1907/2006 (REACH), which prohibits the use of most azodye colourants.
- Statutory Instruments 2005 Consumer Protection no. 1803. The General Product Safety Regulation 2005.

17.4.4.2 British standards

- British Standard BS EN 13780:2003 – Touch and Close Fasteners – Determination of Longitudinal Shear Strength.
- British Standard BS 7907:2007 – Code of Practice for the Design and Manufacture of Children's Clothing to Promote Mechanical Safety.

17.5 PRESS FASTENERS

Press fasteners are also commonly referred to as press studs, poppers or snaps. This type of fastening system comprises a male and female component that when forced together form a secure fastening device (Figure 17.18). The first patent for the press fastener was filed with the United States Patent Office by a German inventor called Heribert Bauer in 1885. During the 1930s, an American company

FIGURE 17.18

The 'S' spring press fastener.

called Scoville produced an improvement on the original version of the press fastener called the Gripper Snap®. This fastener used small teeth instead of a post, but it still operated on the same principal of an interlocking male and female component. The main advantage of the teeth or prong type means that the prongs can be forced through the material and do not cut a hole in the fabric, as is the case when the post/rivet type is used. Since holes can cause knitwear to ladder, the prong type offers better security for knitted products such as babies' wear (Figure 17.19).

The leading producers of press fasteners include Morito, Prym Fashions and YKK.

17.5.1 TYPES OF PRESS FASTENERS

A wide variety of press fastenings includes:

- Sew-on press fasteners (Figure 17.20), where each individual fastener component is hand-stitched onto garments.
- Popper tape (Figure 17.21), where the press fastener is intermittently welded onto a fabric strip, which is then stitched onto the garment.
- Press fasteners designed to be fixed on using pressure activated by a machine.

There are three main types of press fastener, primarily differentiated by the design of the female socket and known as:

- 'S' spring
- Ring spring
- Prong type

In the **'S' spring socket**, the socket of the female part incorporates a parallel spring mechanism that engages the stud (male component) (Figure 17.22). The spring is similar to the letter 'S'. It is commonly used on outerwear and also on leather goods, and most suppliers only recommend it for woven and **not** knitted materials.

FIGURE 17.19

Prong ring press fastener.

FIGURE 17.20

Example of a sew-on press fastener.

For a **ring spring socket**, the socket of the female part incorporates a single spring in the shape of a split ring, which fits around the inside diameter of the socket to engage the stud male component (Figure 17.23). This is generally only recommended for woven fabrics.

The **prong-type socket** is described as a 'petal like' female socket that engages the male prong (Figure 17.24).

17.5.2 MATERIALS USED TO MAKE PRESS FASTENERS

Press fasteners are made of metal or plastic.

Most high-quality fasteners are made from **brass**, which is usually electroplated or enamelled to allow a variety of finishes and colours. Stainless steel is also commonly used for particular end uses.

The most commonly used plastic material is **polyacetal**, a highly durable and resistant thermoplastic. Figure 17.25 shows the components of a polycetal press fastener.

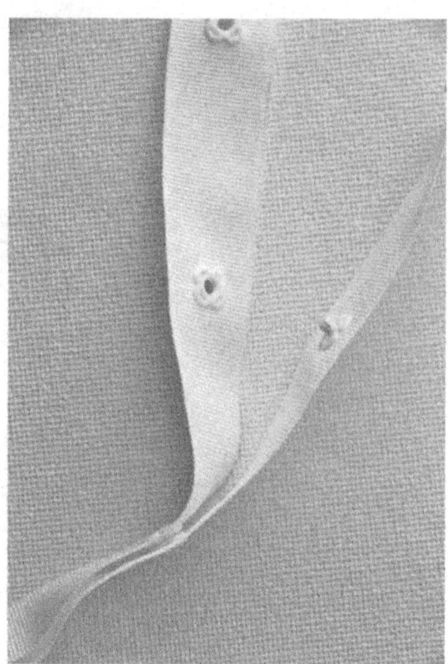

FIGURE 17.21

Examples of popper tape.

Male cap (A)	**'S' spring socket female (B)**	Male stud (C)	Post or prong (D)
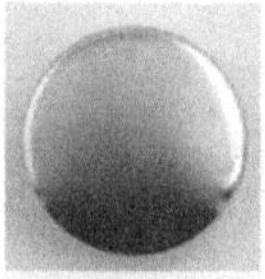			

FIGURE 17.22

Example of the 'S' spring press-fastener components.

Male cap (A)	**Ring spring socket female (B)**	Male stud (C)	Post or prong (D)
		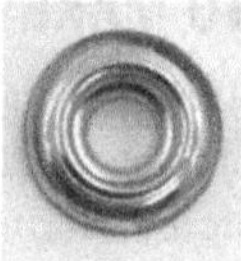	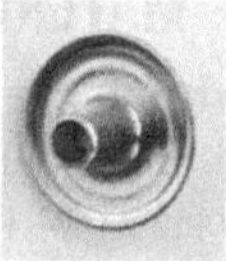

FIGURE 17.23

Example of the ring spring press-fastener components.

Male cap (A) open ring or pear cap	**Petal socket female (B)**	Male stud (C)	Post or prong (D)

FIGURE 17.24

Example of the prong ring press-fastener components.

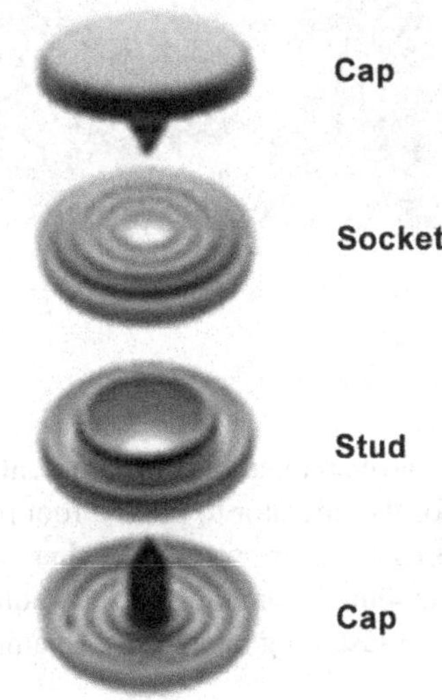

FIGURE 17.25

Example of a polycetal plastic press fastener.

17.5.3 MACHINERY AND ATTACHMENTS USED TO APPLY PRESS FASTENERS

Press fasteners are applied using pressure activated by a machine. Initially, fastenings were attached to garments by hand, which involved a high degree of accuracy and physical exertion. Tools were developed that could penetrate through the material, and these included hole punches and specially developed pliers that pressed down on the prongs of the fastener to rivet them together.

These gave way to manual devices (Figure 17.26), where an operator places the two pieces of the fastener (the male and female) into the top and bottom anvils of the device. The operator then presses a foot pedal down and the two parts of the fastener are clamped together. The use of these machines still created a certain amount of fatigue for the operator due to constant pressure on the lower limbs through pressing down on the pedal.

FIGURE 17.26

Manual fastening machine.

Advances in fastening technology have enabled the development of fastening machines that reduce this fatigue by eliminating the need for the operator to press a foot pedal. This is achieved by a combination of compressed air (pneumatics) and electronic devices that are operated by pressing a switch at the side of the machine. There are now semi-automatic and fully automatic machines available that are capable of undertaking a wide variety of fastening operations. Examples of these machines are given in Figures 17.27 and 17.28.

For the manual foot ratchet machine, the components are manually fed and activated using a foot ratchet press. Electric machines can be manual or hopper fed. The choice of machine is usually dependent on speed and volume of throughput required, but also access to equipment. For sample-making or low-volume requirements, the manual foot ratchet machine will normally suffice. Most suppliers of press fasteners offer a machine rental scheme, particularly when automatic equipment is required.

All machines whether manual or automatic, will require a corresponding set of dies, which are used to hold and attach the press stud components. Figure 17.29 shows a set of dies used to attach a particular configuration of press fastening components. Such dies are not universal and should be specified and supplied by the relevant press fastener company, who should also recommend the most appropriate press fastener as well as the optimum application, including the correct pinch setting.

1. The basic pinch setting refers to the measurement between the two surfaces of the components when attached to one another without any fabric.
2. The fabric to which the components are to be attached should be measured by a torque micrometre; this will measure the thickness to which the fabric will be compressed by an attaching machine.

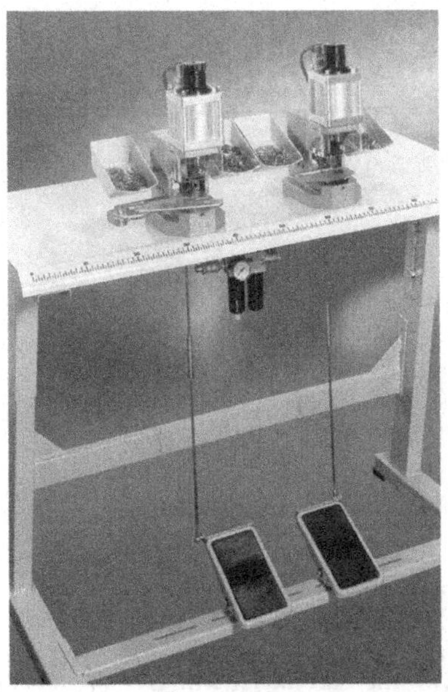

FIGURE 17.27

Semi-automatic example of both the male and female studding machines.

FIGURE 17.28

Automatic studding machine using a hopper to store the studs.

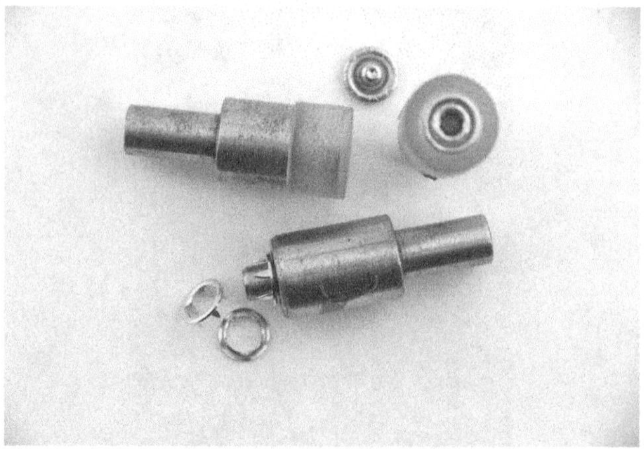

FIGURE 17.29

Example of dyes used to apply press fasteners.

> The combined effect of these two measurements will give the machine pinch setting to which the attaching machine should be adjusted for all applications of these relevant components on the appropriate fabric.
>
> **Joining Forces, ASBCI, 2011, p. 54**

17.5.4 OTHER NON-SNAP COMPONENTS

Several other components that are not press fasteners require the same safety procedures and machinery when being attached. They include the jeans button, also called the tack button, eyelets and rivets. They are all made from metal, usually brass, which has been electroplated to offer a variety of finishes, or a metal/plastic combination.

Jeans buttons have two components: the button (which can have a fixed or swivel top) and the prong, which allows the button to be fixed to the garment (Figure 17.30).

Eyelets also comprise two components, namely the eyelet and the washer (Figure 17.31). Careful testing and control of application using a safety data sheet is highly recommended for these components.

Rivets come in many different designs and finishes, and also have two components: the rivet, which is the part that you see when attached; and the tack, which is similar to the prong used to attach a jeans button (Figure 17.32).

17.5.5 SAFETY STANDARDS AND LEGISLATION FOR SELECTING AND APPLYING PRESS FASTENERS

The correct selection and application of press fasteners is of paramount importance, not only to ensure a high quality product but also to ensure the safety of the end-user. This is of particular relevance when proposing to use a press fastener on children's or baby wear; but the same standards should be applied across the whole spectrum of product sectors.

FIGURE 17.30

Example of a jeans button and prong.

Eyelet	Washer
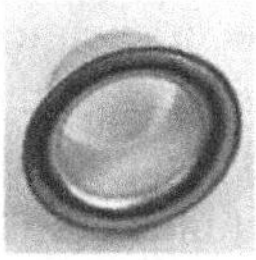	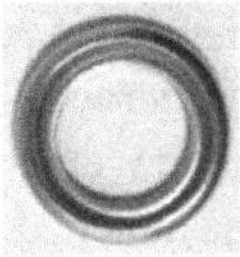

FIGURE 17.31

Example of an eyelet and washer.

FIGURE 17.32

Examples of a rivet and tack.

The supplier of the press fastener components should recommend the most appropriate press fastener for the proposed item of apparel. They are also responsible for specifying dies, appropriate machinery and the pinch setting.

The pinch setting and the press fastener components should be formally documented by the component supplier in the form of a Press Fastener Product Data Safety Chart, and this data should be carefully followed by the manufacturer. This includes the setting up of a quality control procedure where the machine settings and pinch settings are checked and recorded periodically throughout the working day. Regular daily visual inspection also forms an important part of ensuring high levels of safety and quality. Figure 17.33 shows an example of a typical Press Fastener Product Data Safety Chart.

The following quality and safety points are designed to assist in the initial selection and application of press fasteners; however, ultimate decisions should be informed both by the supplier of the press fastener components and compliance with relevant statutory legislation and key British and European standards.

- For press fastener components containing metal where prolonged contact with the skin is anticipated, there must be compliance with the statutory Nickel Safety requirement within regulation (EC) no. 1907/2006 (REACH).
- Fasteners should not be applied to a single ply of fabric; where necessary, a secondary fabric or interlining patch should be incorporated to stabilise the base material.
- Post fasteners should not be used for knitwear, because when applied they punch a hole in the fabric and can easily result in laddering. Prong fasteners are more appropriate for most knitted materials.
- Garments designed to undergo wet processing such as enzyme or chemical finishing should have the press fasteners applied afterwards to prevent detrimental effects on the fastening.
- There should be no gaps or movement between the fabric and component. For example, if you can fit your fingernail under the attached component, the pinch setting is too loose. Always refer back to the manufacturer's pinch setting.
- The component should not be applied too tightly, as this can cause the fabric to be cut or damaged. If the component is applied too tightly, the outer perimeter of the fabric will pucker. Likewise there should be no distortion of the component. Always refer back to the manufacturer's approved pinch setting.
- Fasteners should only be applied to a flat base, and should not be applied over seams or uneven surfaces, as loose components can be extremely sharp and therefore dangerous to the wearer.
- Where prong-type fastenings are applied, there should be no visible prongs sticking out around the prong ring. Always ensure the correct dye has been used and refer back to the press fastener data sheet for the correct dies and settings.
- Where possible, for the purpose of metal detection in children's wear, press fasteners that are made from high quality brass should be selected as opposed to those containing ferrous metals. Metal detectors should be calibrated at least every 4 h and records should be archived for at least one year.

17.5.5.1 Statutory requirements

- Regulation (EC) no. 1907/2006 of the European Parliament and of the Council of 18 December 2006 concerning the Registration, Evaluation, Authorisation and Restriction of Chemicals (REACH).
- Statutory Instruments 2005 Consumer Protection no. 1803. The General Product Safety Regulation 2005.

<table>
<tr><th colspan="5">Press fastener product safety data chart</th></tr>
<tr><td colspan="5">*This data sheet must be kept with all other product information relating to the order style reference number indicated below:</td></tr>
<tr><td>Factory</td><td></td><td>*Product style number</td><td colspan="2"></td></tr>
<tr><td>Supplier name</td><td></td><td>Product department</td><td colspan="2"></td></tr>
<tr><td colspan="2">Press fastener manufacturer</td><td colspan="3"></td></tr>
<tr><td colspan="2">Attaching machine type</td><td colspan="3"></td></tr>
<tr><td colspan="2">Machine setting tolerance</td><td colspan="3"></td></tr>
<tr><td>Garment thickness</td><td colspan="4">All areas where press fasteners are to be applied must be measured. No application allowed to uneven fabric foundation, all fabric foundations to be a constant thickness throughout</td></tr>
<tr><td rowspan="2">Position on garment</td><td rowspan="2">Compressed fabric thickness</td><td rowspan="2">Basic pinch</td><td colspan="2">Machine pinch setting</td></tr>
<tr><td>Seal</td><td>First bulk</td></tr>
<tr><td></td><td></td><td></td><td></td><td></td></tr>
<tr><td></td><td></td><td></td><td></td><td></td></tr>
<tr><td></td><td></td><td></td><td></td><td></td></tr>
<tr><td></td><td></td><td></td><td></td><td></td></tr>
<tr><td></td><td></td><td></td><td></td><td></td></tr>
<tr><td></td><td></td><td></td><td></td><td></td></tr>
<tr><td></td><td></td><td></td><td></td><td></td></tr>
<tr><td colspan="3" rowspan="2">Recommended fabric thickness</td><td>Minimum</td><td>Maximum</td></tr>
<tr><td></td><td></td></tr>
<tr><td colspan="5">Recommended press fastener reference numbers</td></tr>
<tr><td>Cap or prong</td><td></td><td>Socket</td><td colspan="2"></td></tr>
<tr><td>Stud</td><td></td><td>Stud eyelet/prong</td><td colspan="2"></td></tr>
<tr><td>Tack button</td><td></td><td>Nail</td><td colspan="2"></td></tr>
<tr><td>Eyelet</td><td></td><td>Washer</td><td colspan="2"></td></tr>
<tr><td>Rivet</td><td></td><td>Burr</td><td colspan="2"></td></tr>
<tr><td colspan="5">Comment:</td></tr>
<tr><td>Signatures</td><td>Sealing</td><td>Date</td><td>First bulk</td><td>Date</td></tr>
<tr><td>Press stud manufacturer</td><td></td><td></td><td></td><td></td></tr>
<tr><td>Factory</td><td></td><td></td><td></td><td></td></tr>
<tr><td>Sealing garment technologist</td><td></td><td></td><td></td><td></td></tr>
</table>

FIGURE 17.33

Example of a press fastener product safety data chart.

17.5.5.2 British standards

- British Standard BS EN 71-3:1995 – Safety of Toys – Part 3: Migration of Certain Elements.
- British Standard BS 7907:2007 – Code of Practice for the Design and Manufacture of Children's Clothing to Promote Mechanical Safety.

- British Standard BS EN 12472:2005+A1:2009 – Method for the Simulation of Wear and Corrosion for the Detection of Nickel Release from Coated Items.
- British Standard BS EN 1811:1998+A1:2008 – Reference Test Method for Release of Nickel from Products Intended to Come into Direct and Prolonged Contact with the Skin.

17.6 CORDS, TIES AND BELTS

Cords, ties and belts are probably one of the oldest known methods of fastening apparel and, more recently, have been used for drawstrings to pull in apparel, for instance at waists, ankles and in hoods (Figure 17.34). Ties are also used for fastening corsetry, lingerie and lace-up tops when combined with eyelets or rouleau loops used to fasten buttons, normally made from self fabric and pulled through using a rouleau hook (Figures 17.35 and 17.36).

17.6.1 MATERIALS USED TO MAKE CORDS, TIES AND BELT FASTENINGS

Cords, ties and belts are manufactured from almost any material, including natural materials such as cotton, wool, jute and leather, and synthetic materials such as polyester, nylon and associated blends. In more recent years, bungee cord has become a popular option, particularly for sports-style apparel.

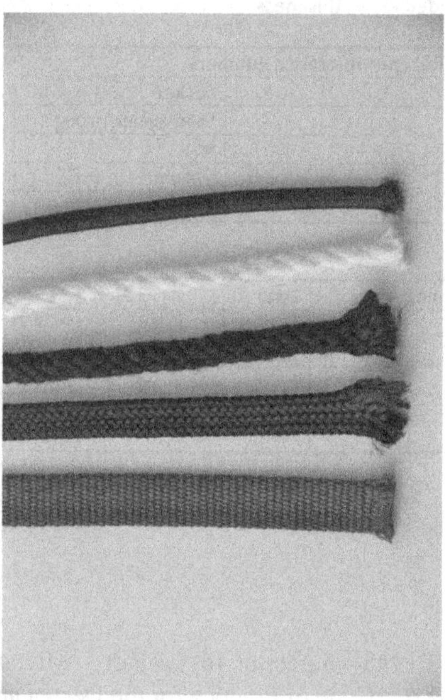

FIGURE 17.34

Examples of draw cords.

Bungee cord consists of a core of elastic strands that are covered in a nylon or cotton sheath, and is commonly supplied in a wide variety of colours and weights. The majority of natural and synthetic cords tend to be knitted, braided or woven and can incorporate elements of stretch depending on end use.

17.6.2 MACHINERY AND ATTACHMENTS USED TO APPLY CORDS, TIES AND BELTS

When applied to apparel to form an adjustable fastening, cords, ties and belts are usually encased within a channel during apparel construction, or threaded through afterwards using a large hook or

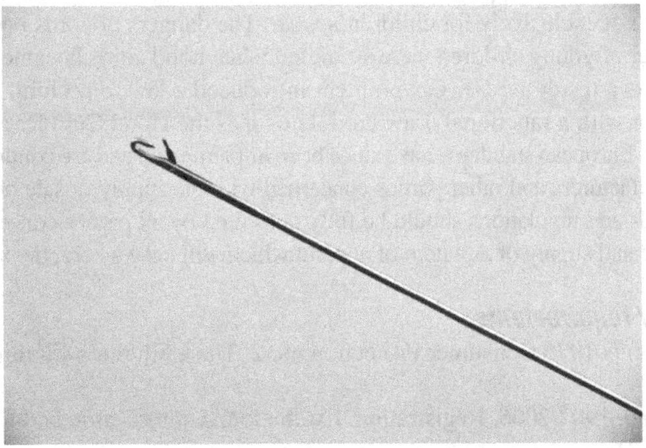

FIGURE 17.35

Example of a hook used to make rouleau loops.

FIGURE 17.36

Example of rouleau loops.

safety-style pin. Cords, ties and belts that are not encased within a channel are commonly held on using very fine knitted belt loops or narrow self-belt loops to prevent them from becoming detached from the garment either during transit to or within the retail outlet. For additional security they are usually kimballed to the belt loop using a small plastic tachit.

17.6.3 SAFETY STANDARDS AND LEGISLATION FOR SELECTING AND APPLYING CORDS, TIES AND BELTS

Although one of the oldest fastenings known, cords ties and belts are also regarded as one of the most dangerous fastenings used in the apparel industry today. The risks are strangulation, entrapment or tripping; particularly but not exclusively for children's wear. The dangers of cords became apparent in the 1970s, when a number of young children were strangled when hood cords became caught in play items such as park slides. As a result the UK Government introduced a law to prohibit the sale or supply of young children's wear with a functional draw cord, known as the Hood Cord Regulation 1976.

Additional UK and European standards have since been implemented and are continually updated to help guide designers, manufacturers and other parties concerned with the supply or sale of childrens' wear, and the following standards and regulations should be fully consulted by all parties concerned with the design, selection, manufacture and supply of any item of apparel which will have a cord, tie or belt fastening.

17.6.3.1 Statutory requirements

- Statutory Instruments 1976 Consumer Protection no. 2. The Children's Clothing (Hood Cords) Regulation 1976.
- Regulation (EC) no. 1907/2006. Registration, Evaluation, Authorisation and Restriction of Chemicals (REACH).

    Note: With reference to the dyeing and colouration of cords, ties and belts, there must be compliance with the statutory requirement within regulation (EC) no. 1907/2006 (REACH), which prohibits the use of most azodye colourants.

- Statutory Instruments 2005 Consumer Protection no. 1803. The General Product Safety Regulation 2005.

17.6.3.2 British standards

- British Standard BS EN 14682:2007 – Safety of Children's Clothing – Cords and Drawstrings on Children's Clothing – Specifications.
- British Standard BS 7907:2007 – Code of Practice for the Design and Manufacture of Children's Clothing to Promote Mechanical Safety.

17.7 HOOK-AND-EYE FASTENERS

The hook and eye was first patented in 1808 by Camus, for the production of hooks and eyes by machinery (Joining Forces, 2011). This method of apparel fastening became popular throughout the 1800s, particularly for fastening tight-fitting corset styles. Several improvements on the original

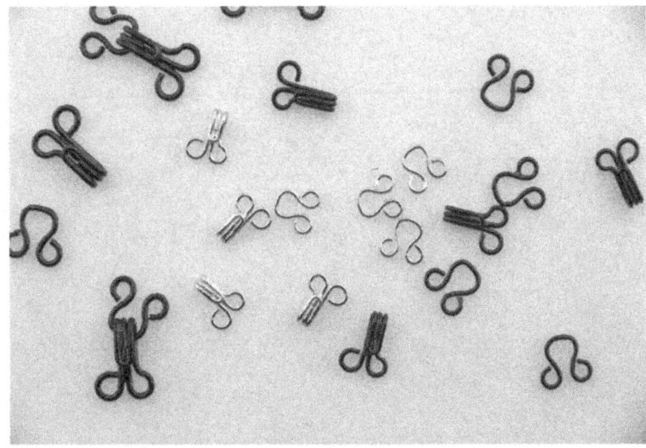

FIGURE 17.37

Examples of hooks and eyes.

patent design were submitted to the Unites States Patent Office between 1843 and 1964, most of which offered only minimum changes, such as flattening the metal/wire from which they were made (Figure 17.37).

17.7.1 TYPES OF HOOK-AND-EYE FASTENERS

There are three main types of hook-and-eye fasteners:

- Standard
- Fur
- Tape

Standard hooks and eyes are commonly classified from size 1 to 10, where size 1 is the smallest and size 10 is the largest. Size 1 hooks and eyes are sometimes attached to the top of a zip opening, as they hold the opening together and make it easier to fasten the zip. Standard hooks and eyes can also be used in place of a zip to fasten a complete opening or placket; however, hook-and-eye tape is more frequently selected for this type of opening.

Animal fur hooks and eyes for apparel are made from wire and then covered with a knitted sheath. An example is given in Figure 17.38.

Hook-and-eye tape is used where longer openings need to be fastened.

17.7.2 MATERIALS USED TO MAKE HOOKS AND EYES

Hooks and eyes are generally made from metal, usually brass wire that is electroplated or coated in plastic to give a variety of colours and finishes. Larger versions are also made from metal and are usually covered in a knitted sheath made from cotton, polyester or similar materials.

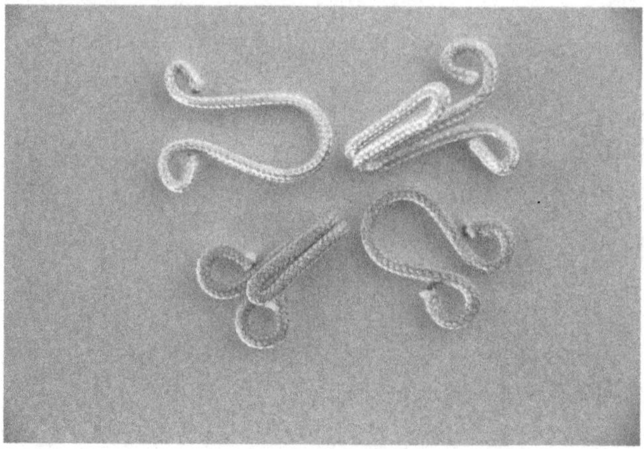

FIGURE 17.38

Examples of hooks and eyes used for apparel made from animal fur.

17.7.3 MACHINERY AND ATTACHMENTS USED TO APPLY HOOKS AND EYES

Hooks and eyes can be attached by hand sewing, but this method can be time-consuming and it is important that the sewing thread is securely locked off.

Alternatively, when high volume and speed of throughput time is important, hooks and eyes can be attached using a semi-automatic button-sew machine fitted with an appropriate hook-and-eye clamp.

Attachment of hook-and-eye tape is much quicker, and the standard lockstitch is most commonly used.

17.7.4 SAFETY STANDARDS AND LEGISLATION FOR SELECTING AND APPLYING HOOKS AND EYES

When selecting hooks and eyes, end-user safety should be considered of paramount importance. Compliance with statutory legislation and key British and European standards is imperative to protect the end-user and those involved in the supply chain of apparel that incorporate hooks and eyes. The following considerations and points are designed to help with selecting the appropriate hook-and-eye fastenings.

- If hooks and eyes contain metal and prolonged contact with the skin is anticipated, the application must comply with the statutory Nickel Safety requirement within regulation (EC) no. 1907/2006 (REACH).
- Dyeing and colouring of textile sheath coverings on hooks and eyes must comply with the statutory requirement within regulation (EC) no. 1907/2006 (REACH), which prohibits the use of most azodye colourants.
- Hooks and eyes should not be applied to apparel for children under three years of age.
- Lockstitch should be used for attaching hooks and eyes, not chainstitch button-sew, as it is more reliable and consistent.

- Hooks and eyes that do not contain ferrous metals should be selected so that metal detection is possible.
- Hooks and eyes should be used to correct zips that have been set too low, thus leaving a gap at the top. They should be used to help hold the opening together to allow ease of zip fastening.
- Only high-quality hooks and eyes should be selected to eliminate the possibility of sharp edges and burrs.

17.7.4.1 Statutory requirements

- Regulation (EC) no. 1907/2006. Registration, Evaluation, Authorisation and Restriction of Chemicals (REACH).
- Statutory Instruments 2005 Consumer Protection no. 1803. The General Product Safety Regulation 2005.

17.7.4.2 British standards

- British Standard BS EN 71-3:1995 – Safety of Toys – Part 3: Migration of Certain Elements.
- British Standard BS 7907:2007 – Code of Practice for the Design and Manufacture of Children's Clothing to Promote Mechanical Safety.
- British Standard BS EN 12472:2005+A1:2009 – Method for the Simulation of Wear and Corrosion for the Detection of Nickel Release from Coated Items.
- British Standard BS EN 1811:1998+A1:2008 – Reference Test Method for Release of Nickel from Products Intended to Come into Direct and Prolonged Contact with the Skin.

17.8 HOOK-AND-BAR FASTENERS

Numerous hook-and-bar fasteners have been patented, using a wide variety of terms to describe the hook and bar (Figure 17.39). One of the earliest patents was filed by American inventor John Ewig in 1889 (U.S. Patent no. 408,300), who referred to his hook-and-bar invention as a 'waistband fastener'. Countless versions have emerged since then; however, the main end-use of the hook and bar continues to be as originally conceived, to form a concealed fastening at the top of the fly on the waistband of trousers.

17.8.1 TYPES OF HOOK-AND-BAR FASTENERS

There are two main types of hook-and-bar fasteners, categorized by the method of application:

- Sew-on
- Machine applied

Sew-on hook-and-bar fasteners, as the name suggests, are designed to be attached either by hand sewing or by using a semi-automatic button-sew machine fitted with an appropriate clamp. An example of a sew-on hook and bar is given in Figure 17.39.

FIGURE 17.39

Example of a sew-on hook and bar.

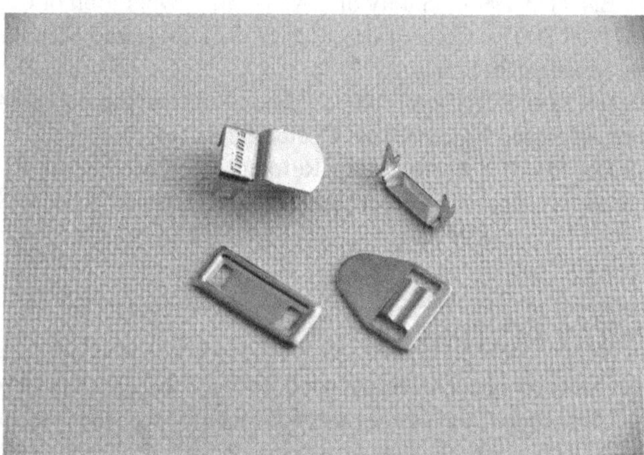

FIGURE 17.40

Example of prong-type hook-and-bar components.

Machine-applied hook-and-bar fasteners are applied using a machine similar to that used to apply press fasteners. There are two main designs:

- Prong type with a backing plate (Figure 17.40)
- Rivet/post type

17.8.2 MATERIALS USED TO MAKE HOOKS AND BARS

Hooks and bars are made from metal, usually brass, which is electroplated to give a variety of finishes resembling silver, gun metal, oxy brass, etc.

17.8.3 MACHINERY AND ATTACHMENTS USED TO APPLY HOOKS AND BAR/FASTENERS

Sew-on hooks and bars can be attached by hand sewing, but this method can be time consuming and it is important that the sewing thread is securely locked off.

Where high-volume and speed of throughput time are important, they can be attached using a semi-automatic button-sew machine fitted with an appropriate hook-and-eye clamp.

No-sew versions are applied using a special machine that is designed to apply the fastener with pressure; therefore a Safety Data Chart similar to that given in Figure 17.33 should be generated outlining the correct pinch setting, and the same rules for applying press fasteners should be followed.

17.8 4 SAFETY STANDARDS AND LEGISLATION FOR SELECTING AND APPLYING HOOKS AND BAR/FASTENERS

When selecting hooks and bars, end-user safety should be considered of paramount importance. Therefore compliance with statutory legislation and key British and European standards is imperative to protect the end-user and those involved in the supply chain of apparel that incorporate hooks and bars. The following considerations and points are designed to help with selecting the appropriate hooks and bars.

- Hooks and bars containing metal where prolonged contact with the skin is anticipated must comply with the statutory Nickel Safety requirement within regulation (EC) no. 1907/2006 (REACH).
- Hooks and bars should not be applied to apparel for children under three years of age.
- Lockstitch should be used for attaching sew-on hooks and bars, and not chainstitch button-sew, as it is more reliable and consistent.
- Only high-quality hooks and bars should be selected, to eliminate the possibility of sharp edges and burrs.
- Hooks and bars should only be applied to a flat, even foundation.
- Hooks and bars should not be applied to a single ply of fabric, where necessary, a secondary fabric or interlining patch should be incorporated to stabilise the base material.
- In no-sew versions, there should be no gaps or movement between the fabric and component. If you can fit your fingernail under the attached component, the pinch setting is too loose. Always refer to the manufacturer's pinch setting.
- The component should not be applied too tightly, as this can cause the fabric to be cut or damaged. If the component is applied too tightly, the outer perimeter of the fabric will pucker. Likewise there should be no distortion of the component. Always refer back to the manufacturer's approved pinch setting.
- For prong-type fastenings, there should be no visible prongs sticking out around the face of the applied component. Always ensure the correct dye has been used and refer back to the Hook-and-Bar Safety Data Chart for the correct dies and settings.
- For metal detection in children's wear, fasteners made from high-quality brass should be selected rather than those containing ferrous metals.

17.8.4.1 *Statutory requirements*

- Regulation (EC) no. 1907/2006 of the European Parliament and of the Council of 18 December 2006 Concerning the Registration, Evaluation, Authorisation and Restriction of Chemicals (REACH).
- Statutory Instruments 2005 Consumer Protection no. 1803. The General Product Safety Regulation 2005.

17.8.4.2 *British standards*

- British Standard BS 7907:2007 – Code of Practice for the Design and Manufacture of Children's Clothing to Promote Mechanical Safety.
- British Standard BS EN 71-3:1995 – Safety of Toys – Part 3: Migration of Certain Elements.
- British Standard BS EN 12472:2005+A1:2009 – Method for the Simulation of Wear and Corrosion for the Detection of Nickel Release from Coated Items.
- British Standard BS EN 1811:1998+A1:2008 – Reference Test Method for Release of Nickel from Products Intended to Come into Direct and Prolonged Contact with the Skin.

17.9 BUCKLES AND ADJUSTABLE FASTENERS

Visit any good museum with a collection of historical finds and no doubt you will come across an early example of a buckle or belt slide fastener. Many of these have survived because they were made from non-corrosive metals such as bronze or gold. Such metals are no longer used due to their cost and weight; however, many of the earliest adjustable fastenings designs have remained relatively unchanged.

17.9.1 TYPES OF BUCKLES AND ADJUSTABLE FASTENERS

Figure 17.41 shows some examples of buckles and adjustable fastenings commonly used on contemporary apparel to fasten or adjust the fit. Slide-release buckles, cord adjusters and strap adjusters tend to be used for casual wear or sportswear, while the ring and slider is predominantly used on lighter products such as lingerie, for adjusting narrow bra or camisole straps.

Some adjustable fastenings tend to be prone to changes in fashion, for example the brace clip is normally only seen on more staple items such as children's wear or workwear dungarees; likewise traditional shirt-sleeve arm bands rarely make an appearance unless fashion dictates.

17.9.2 MATERIALS USED TO MAKE BUCKLES AND ADJUSTABLE FASTENERS

The majority of adjustable fastenings are made from either metal (usually brass), plastic or a combination of the two. Most metal fasteners are electroplated to offer a range of finishes such as silver, gold, gun metal, etc., while others might be covered in real or faux leather. Where lighter-weight fastenings are required to resemble metal, ABS plastic versions are normally available.

The selection of an adjustable fastener or indeed any form of apparel fastening will be determined by the overall aesthetic appearance but also by its expected performance requirements, relating to

Slide release buckle	Cord adjuster	Belt slide fastener	Cam buckle	Strap adjuster
	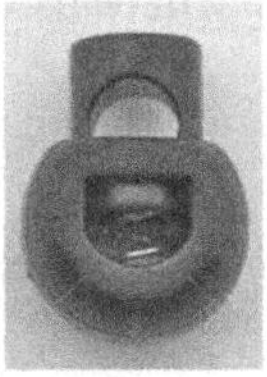	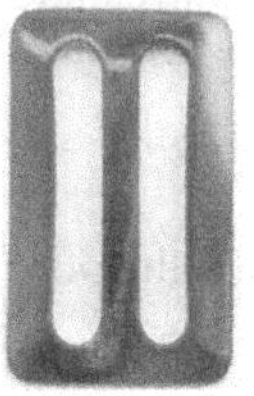		
D ring	Prong buckle	Ring and slider	Brace clip	Shirt sleeve arm bands
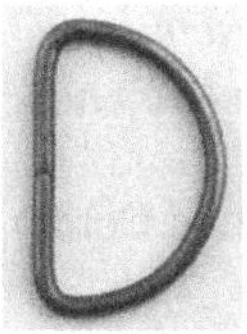	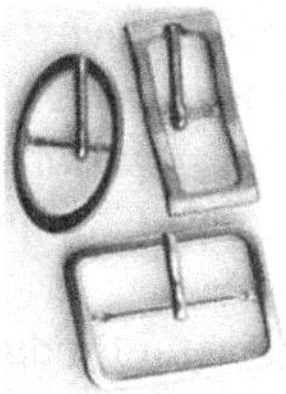	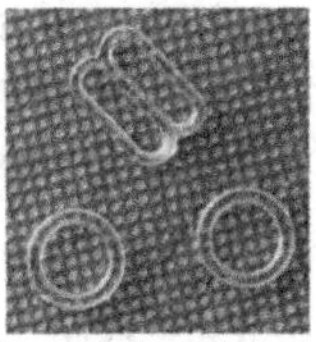		

FIGURE 17.41

Examples of buckles and adjustable fastenings.

strength, durability, laundering requirements and weight. For example, some plastics can become brittle when dry cleaned, and plastic D rings and slide-release buckles may not be robust enough for outdoor or performance apparel, but they may be useful for lighter-weight fashion wear. Therefore most fastening suppliers will offer almost identical products that are made from a variety of materials depending on the end-use performance requirements. Thorough risk assessment and testing are essential to ensure that the correct selection has been made.

17.9.3 MACHINERY AND ATTACHMENTS USED TO APPLY BUCKLES AND ADJUSTABLE FASTENERS

The majority of adjustable fastenings are designed to be applied to a cord, tie or belt. Most will be manually applied and then secured using a reinforcing stitch, knot or, in the case of some leather belts, can be secured in place using another fastening such as a rivet or stud.

17.9.4 SAFETY STANDARDS AND LEGISLATION FOR SELECTING AND APPLYING BUCKLES AND ADJUSTABLE FASTENERS

When selecting buckles or adjustable fastenings, end-user safety is of paramount importance. Compliance with statutory legislation and key British and European standards is imperative to protect the end-user and those involved in the supply chain of apparel that incorporates these types of fastening.

The following list of statutory requirements and standards should be referred to in selecting fastenings, but it is not exhaustive, and in all cases a full risk assessment of the product should be undertaken to ensure the safety of the product and its end-user, particularly as most will be used in conjunction with cords, ties and belts.

17.9.4.1 Statutory requirements

- Regulation (EC) no. 1907/2006. Registration, Evaluation, Authorisation and Restriction of Chemicals (REACH).
  Note:
  a. In relation to a buckle or adjustable fastening containing metal, where prolonged contact with the skin is anticipated, there must be compliance to the statutory Nickel Safety requirement within regulation (EC) no. 1907/2006 (REACH).
  b. *With reference to the dyeing and colouration of textile sheath coverings fastenings, there must be compliance with the statutory requirement within regulation (EC) no. 1907/2006 (REACH) that prohibits the use of most azodye colourants.*
- Statutory Instruments 1976 Consumer Protection no. 2. The Children's Clothing (Hood Cords) Regulation 1976.
- Statutory Instruments 2005 Consumer Protection no. 1803. The General Product Safety Regulation 2005.

17.9.4.2 British standards

- British Standard BS 7907:2007 – Code of Practice for the Design and Manufacture of Children's Clothing to Promote Mechanical Safety.
- British Standard BS EN 71-3:1995 – Safety of Toys – Part 3: Migration of Certain Elements.
- British Standard BS EN 12,472:2005+A1:2009 – Method for the Simulation of Wear and Corrosion for the Detection of Nickel Release from Coated Items.
- British Standard BS EN 1811:1998+A1:2008 – Reference Test Method for Release of Nickel from Products Intended to Come into Direct and Prolonged Contact with the Skin.

17.10 SUMMARY

- The main types of fastenings for joining fabrics are zips; buttons; hook-and-loop fasteners; press fasteners; cords, ties and belts; hook-and-eye fasteners; hook-and-bar fasteners; and buckles and adjustable fasteners.
- Choice of fastener is dictated by the end-use of the garment or textile product.
- Each type of fastener has advantages, disadvantages and its own methods of attachment.
- There are important safety standards and legislation concerning fastenings, and manufacturers should ensure that they know the implications of choosing a particular type of fastener and are able to follow guidelines and laws associated with that choice.

17.11 PROJECT IDEAS

- Design your own tent fabrics for camping at a pop festival.
- Design your own personalized bag using a choice of fabrics with your own choice of fasteners.
- Design a range of clothing for disabled people for their individual disability.

17.12 REVISION QUESTIONS

- What type(s) of fastener are appropriate for a baby-grow?
- What are the most important safety considerations when choosing to use a cord, tie or belt for a garment?
- How are zips attached during the production process?
- What stitch must *not* be used to attach buttons?
- What type of fastener is best for a concealed fastening at the top of the fly on the waistband of a pair of trousers?

17.13 SOURCES OF FURTHER INFORMATION

British Standards are available online at www.bsigroup.com.
Statutory Instruments are available online at www.legislation.gov.uk.
The REACH Regulation can be found online at http://www.hse.gov.uk/reach/legislation.htm.

REFERENCES AND FURTHER READING

ASBCI. (2011). *Joining forces: Fastenings*. Halifax, England: Association of Suppliers to the British Clothing Industry.

Bauer, H. (1885). Fastening for gloves and other articles, U.S. Patent No. 321940. Online http://www.google.co.uk/patents/about?id=R2BgAAAAEBAJ&dq=Bauer+Heribert+14th+July+1885. Accessed 21.04.11.

British Standards. (1983). *BS 4162:1983 – Methods of test for buttons*. London: British Standards Institution.

British Standards. (1995). *BS EN 71-3:1995 – Safety of toys – Part 3: Migration of certain elements*. London: British Standards Institution.

British Standards. (1998). *BS EN 1811:1998+A1:2008 – Reference test method for release of nickel from products intended to come into direct and prolonged contact with the skin*. London: British Standards Institution.

British Standards. (2003). *BS EN 13780:2003 – Touch and close fasteners – Determination of longitudinal shear strength*. London: British Standards Institution.

British Standards. (2006). *BS 3084:2006 – Slide fasteners (zips) – Specification*. London: British Standards Institution.

British Standards. (2007). *BS EN 14682:2007 – Safety of children's clothing – Cords and drawstrings on children's clothing – Specifications*. London: British Standards Institution.

British Standards. (2007). *BS 7907:2007 – Code of practice for the design and manufacture of children's clothing to promote mechanical safety*. London: British Standards Institution.

British Standards. (2009). *BS EN 12472:2005+A1:2009 – Method for the simulation of wear and corrosion for the detection of nickel release from coated items*. London: British Standards Institution.

Coats. (2011). *We're number two in the world and growing in zips*. Online http://www.coats.com/zips.htm. Accessed 21.04.11.

Ewig, J. (1889). Waistband fastener, U.S Patent No. 408,300. Online http://www.google.co.uk/patents/about?id=wnRjAAAAEBAJ&dq=waistband+fastener+John+Ewig. Accessed 21.04.11.

Friedel, R. (1996). *Zipper, an exploration in novelty*. New York: Norton.

Hse. http://www.hse.gov.uk/reach/legislation.htm.

Judson, W. L. (1893). Clasp locker or un-locker for shoes, U.S. Patent No. 504038. Online http://www.google.co.uk/patents/about?id=C9VHAAAAEBAJ&dq=Clasp+Locker+504038. Accessed 21.04.11.

Peacock, P. (1972). *Buttons for the collector*. Devon: David and Charles Publishers.
Prym Group. (2011). *A small fastener that's a big success*. Online http://www.prym.com/prym/proc/docs/press_fasteners_en.html?nav=0H04001dg. Accessed 21.04.11.
Statutory Instruments. (1976). *Consumer Protection No. 2: The Children's Clothing (Hood Cords) Regulation 1976*. Online http://www.legislation.gov.uk/uksi/1976/2/contents/made. Accessed 21.04.11.
Statutory Instruments. (1989). *Consumer Protection No. 1291: The Food Imitations (Safety) Regulation*. Online http://www.legislation.gov.uk/uksi/1989/1291/contents/made. Accessed 21.04.11.
Statutory Instruments. (2005). *Consumer Protection No. 1803: The General Product Safety Regulation 2005*. Online http://www.legislation.gov.uk/uksi/2005/1803/contents/made. Accessed 21.04.11.
Statutory Instruments. (2008). *Consumer Protection Environmental Protection Health and Safety No. 2852: The REACH Enforcement Regulations 2008*. Online http://www.hse.gov.uk/reach/legislation.htm. Accessed 21.04.11.
Strauss, S. D. (2002). *The big idea: How business innovators get great ideas to market, velcro*. Chigago, IL: Dearborne Publishing, Kaplan Professional.
Velcro. (2011). *Velcro history*. Online http://www.velcro.co.uk. Accessed 21.04.11.

CHAPTER

FABRIC FINISHING: PRETREATMENT/TEXTILE WET PROCESSING

18

P. Hauser

North Carolina State University, Raleigh, NC, USA

LEARNING OBJECTIVES

At the end of this chapter, you should be able to:

- Describe the major differences between batch and continuous processing
- List the different processing steps in fabric preparation
- Describe the fabric preparation processing steps
- Discuss alternative chemistries available for fabric preparation

18.1 INTRODUCTION

Fabric finishing is the set of different processing steps that textiles undergo before being made up into garments or home furnishings: preparation, dyeing and coloring, printing, physical/mechanical finishing, and chemical finishing. The processes involved for a particular fabric depend on a number of factors, including intended end use, making up procedures, and consumer demand.

Fabrics must be properly prepared before they are dyed and/or printed, in order to ensure that dye uptake is uniform, the desired colors are achieved, and the colors are fast. Preparation, the subject of this chapter, is a series of sequential processing steps that get the fabric ready for dyeing and subsequent finishing. Since water is key to many of them, the steps are collectively known as textile wet processing. The initial processes are also referred to as either fabric preparation or fabric pretreatment.

Few fabrics if any are presented for sale to the consumer in the "greige" or "grey" state (i.e., without any bleaching, dyeing, or finishing treatment). The overwhelming majority of fabrics go through a series of processes to prepare them for further processes such as dyeing and finishing. The goals of fabric preparation are threefold:

- to ensure uniform absorption of water across the width and length of the fabric;
- to ensure uniform removal of all impurities; and
- to ensure uniform whiteness values if the fabric is to be sold as white.

It is also important that, while achieving these goals, damage to the physical properties of the fabric is kept to an absolute minimum.

Textiles and Fashion. http://dx.doi.org/10.1016/B978-1-84569-931-4.00018-0

Uniform water absorption is important since all subsequent processing steps (coloring and finishing) involve treating the fabric with a water solution of a dye or chemical. If water absorption is not uniform throughout the fabric, then the dye or chemical may react differently in different parts of the fabric, and the result will be uneven.

Impurities must be removed because they can affect not only the appearance of the fabric but also the uniform application of dyes and chemicals. Natural impurities and chemicals, such as sizes and oil or wax lubricants added during previous processing, can impair the ability of the material to absorb water, and therefore restrict uniform access of the dye to the fiber. Raw wool can contain at least 50% by weight of impurities, including grease, suint (mainly solidified sweat), and vegetable matter. Among other natural impurities, raw cotton contains hydrophobic waxes, and raw silk has a coating of a gum, sericin. Synthetic and regenerated fibers present fewer problems to the dyer, but in all cases, lubricants such as oils, waxes, and sizes (based on starch or polyvinyl acetate)—which are added variously during the spinning, knitting, and weaving processes—must be removed as appropriate before dyeing commences.

For fabrics being sold as white, **uniform whiteness** is as important as uniform coloration for dyed fabrics.

All these processes must be carried out while minimizing damage to the physical properties of the fabric, so that the finished article will meet the minimum physical requirements for which it was designed and constructed.

Once a fabric has been properly prepared, it can go directly to a coloring (dyeing or printing) process. The usual steps in fabric preparation are:

- Desizing (woven fabrics only)
- Scouring (all fibers)
- Bleaching (typically only natural fibers)
- Mercerization (cotton fabrics only)
- Carbonization (wool fabrics only)
- Heat setting (synthetic fibers only)

18.2 PROCESSING METHODS

Textile wet processing is carried out either in batches, a continuous process, or a combination of the two. The choice of processing method depends on a number of factors, including the volume of fabric to be processed, the steps required, and the cost.

In **batch processing**, the fabric and calculated amounts of chemicals and water are put into a single vessel and stirred and heated as necessary. If there are several processes to be carried out on the same batch of fabric, the vessel may have to be emptied and cleaned between processes, making this method time-consuming. Batches of up to 1000 yards of fabric, and sometimes more, may be processed in this way.

Continuous processing uses a series of vessels, each of which represents one processing step. The combination of processing vessels is known as the range, and the total number of vessels in the range can be as many as 20 or more. The fabric is passed from one vessel to the next in the correct sequence for the processes being carried out. Continuous processing can reach speeds of

100 yards of fabric per minute. Because some fabric will be wasted when the range is first set up and adjusted to optimize performance, continuous processing is best for large yardages of fabric.

The preparation methods described in the following section are thus the first steps in any batch or continuous processing operation.

18.3 FABRIC PREPARATION PROCESSES

18.3.1 DESIZING

Prior to the weaving process, warp yarns are usually treated with a material designed to protect them from excessive abrasion during weaving, and as a result, increase weaving efficiency (see Chapter 12). These materials are referred to as **warp sizes** and are temporary treatments that must be removed before further processing can take place. The most common warp sizes are:

- modified starches;
- carboxymethyl cellulose; and
- polyvinyl alcohol.

Starches and **carboxymethyl cellulose** are used with cotton and cotton blends, while **polyvinyl alcohol** is used with polyester yarns. Since the warp size is intended as a single-use product (except for those that can be recycled), cost is an important consideration in the choice of warp size.

The desizing process is intended to remove warp sizes and must be tailored to the specific warp size used.

18.3.1.1 Desizing starch

Starch warp sizes must be solubilized in order to be removed. The traditional desizing method uses an α-amylase enzyme to catalyze the hydrolysis of the starch molecules into water-soluble pieces (Buschle-Diller, Radhakrishnaiah, Freeman, & Zeronian, 2002; Ibrahim, El-Hossamy, Morsy, & Eid, 2004; Pawar, Shah, & Andhorikar, 2002). Typical reaction conditions are given in Table 18.1. The desizing solution is maintained at pH 6–7.5, and a hot water rinse follows the desizing procedure. An alternative starch desizing procedure uses an oxidization reaction to solubilize the starch warp size (El-Rafie, Abdel Hafiz, El-Sisi, Helmy, & Hebeish, 1991; El Shafie, Fouda, & Hashem, 2009).

Since the starch molecules are fragmented during desizing, starch warp sizes cannot be recycled.

Table 18.1 Enzyme desizing conditions*

	g/L active enzyme	Temperature (°C)	Time (min)
Jet/Beck	0.25	70	60
J-Box	0.5	80	15
Steamer	1.0	95	2

**Note: 5 g/L of NaCl and 1 g/L of a nonionic wetting agent are often included.*

18.3.1.2 Desizing carboxymethyl cellulose

Carboxymethyl cellulose warp sizes can be removed by a simple warm water rinse at neutral pH (Cacho, 1980; Ibrahim, Abo Shosha, Fahmy, & Hebeish, 1997; Milner, 1997; Tomasino, Livengood, & Thorp, 1981). Theoretically, carboxymethyl cellulose warp sizes can be concentrated from the desizing solution and reused. However, cost considerations and the possibility of microbial growth in the recycled warp size have limited recycling of carboxymethyl cellulose to date.

18.3.1.3 Desizing polyvinyl alcohol

Polyvinyl alcohol, also referred to as PVA, can be easily removed with a hot water rinse. Recent research has shown that a plasma treatment prior to desizing can reduce both the time and temperature of the PVA desizing process (Cai, Qiu, Zhang, Hwang, & McCord, 2003; Ma, Wang, Xu, & Cao, 2009; Peng, Gao, Sun, Yao, & Qiu, 2009). Although polyvinyl alcohol is the most expensive warp size to purchase, it can be recycled and reused after ultrafiltration of the desizing bath (Achwal, 1995; Hoffman, 1981; Lin & Lan, 1995; Livengood, 1995; Robinson, 1993).

18.3.2 SCOURING

Scouring removes impurities that would otherwise interfere with the further processing of fabrics and can be carried out in both batch and continuous processes. Specific procedures (equipment and chemical auxiliaries) depend on the fiber type and the impurities to be removed. Natural fibers, such as cotton, wool, and silk, all have inherent impurities that can vary due to natural causes, while synthetic fibers have known processing aids as impurities.

18.3.2.1 Scouring cotton

Cotton fibers have a variety of impurities that must be removed. Table 18.2 gives a typical analysis of a greige cotton fiber. A proper scour of cotton will provide a fabric that is both absorbent and free of impurities that would adversely affect dyeing or printing.

Conventional scouring of cotton fabric involves a high-temperature treatment with a solution containing alkali, wetting agent, and detergent (Choudhury, 2006, p. 179). A chelating agent is often added to the scouring solution to complex any heavy metals present in the cotton. Typical cotton scouring conditions are given in Table 18.3.

Table 18.2 Typical composition of greige cotton fibers*

Component	Percentage
Cellulose	94%
Proteins	1.3%
Pectins	0.9%
Minerals	1.2%
Waxes	0.6%
Organic acids	0.8%
Sugars	0.3%
Other	0.9%

**Peters, 1967, p. 88.*

Table 18.3 Typical cotton scouring conditions*

Component	Jet/Beck	J-Box	Pressure steamer
Sodium hydroxide (50%)	0.5 g/L	3–6% owf[§]	6–10% owf
Detergent	2 g/L	0.5–1% owf	0.5–1% owf
Wetting agent	0.1–0.2	0.1–0.5% owf	0.1–0.5% owf
Temperature (°C)	95–98	100	130–140
Time (min)	60–120	30–120	1–2

**Choudhury, 2006, p. 179.*
[§]On weight of fabric.

18.3.2.2 Scouring wool

Wool fibers contain much higher levels of impurities than cotton fibers. Waxes, suint (perspiration salts), vegetable matter, and dirt can account for as much as 50% by weight of the wool fiber (Karmakar, 1999, p. 107). Therefore, wool fibers must be scoured prior to being spun into yarn. Raw wool is typically scoured at mild temperatures (around 50 °C) in tanks containing a mild alkali and nonionic detergent (Bateup & Warner, 1988; Wood, 1982a). Alternative scouring methods for raw wool include enzyme (Das & Ramaswamy, 2006) and solvent systems (Ke-Heng, Gray, Perry, & Bide, 1988; Wood, 1982b). These systems have the advantage of using less water and energy.

18.3.2.3 Scouring silk

Silk is a continuous filament fiber produced by silk worms. The raw fiber consists of two filaments glued together with a substance called sericin. Silk filaments are not sized before weaving, since the sericin protects the filaments during the weaving process (Peters, 1967, p. 323). After the silk fabric is woven, the sericin is removed in a process called degumming. Treatments with mild alkali, detergents, and enzymes are employed (Gowda, Padaki, & Sudhakar, 2004; Holds, 2007; Karmakar, 1999, p. 115).

18.3.2.4 Scouring synthetic fibers

Synthetic fibers present easier challenges in scouring, since any impurities were added during processing and therefore are of known composition. A scour bath with detergent and sodium carbonate at around 70 °C for 20–30 min is adequate for most synthetic fibers (Choudhury, 2006, p. 246).

18.3.3 BLEACHING

Bleaching is the process of removing naturally occurring color bodies from fibers. The object of bleaching is to obtain white fabrics without seriously damaging the fibers themselves. Only natural fibers should require bleaching, since synthetic fibers can be manufactured without added color. Optical brighteners are often added to bleaching formulas if the bleached fabric is to be sold as white rather than dyed.

18.3.3.1 Bleaching cotton

Cotton is typically bleached with oxidative chemistry, most often with hydrogen peroxide as the active bleaching agent (Evans, Boleslawski, & Boliek, 1997; Hickman, 1996; Juby, 1985; Shamey & Hussein, 2005a,b). Bleaching with hydrogen peroxide requires alkaline conditions to activate the bleaching

Table 18.4 Typical cotton bleaching conditions*

Component	Jet/Beck	J-Box	Pressure steamer
Sodium hydroxide (50%)	0.5 g/L	2 g/L	6–8 g/L
Sodium carbonate	1.8 g/L	–	
Sodium silicate	7 g/L	15 g/L	15–20 g/L
Hydrogen peroxide (35%)	1–2 g/L	2–4 g/L	40–60 g/L
Temperature (°C)	80–95	100	130–140
Time (min)	60–120	60–90	2–3

**Choudhury, 2006, pp. 274–278.*

agent and a stabilizer (often sodium silicate) to control the reaction rate, in order to avoid excessive fiber damage. A typical cotton-bleaching recipe for use with hydrogen peroxide is given in Table 18.4. Often, scouring and bleaching are combined for cotton fabrics processed in continuous equipment.

An alternative cotton bleaching procedure using reductive rather than oxidative chemistry to remove color is available (Karl & Freyberg, 1999).

18.3.3.2 Bleaching wool and silk

Wool fibers are sensitive to the alkaline conditions normally present when bleaching with hydrogen peroxide. As a result, special procedures must be followed when using this bleaching agent with wool (Cardamone, Yao, & Phillips, 2005; Cegarca, Gacén, & Caro, 1983; El-Khatib, 2002; Gacén, Cayuela, & Garcén, 2002; Karunditu et al., 1994; Palin, Teasdale, & Benisek, 1983; Reincke, 1998, p. 54). Table 18.5 gives typical reaction conditions for bleaching wool with hydrogen peroxide. Wool can also be successfully bleached with reductive chemistry (Cai, Evans, & Church, 2008; Gacen, Cayuela, & Gacen, 1999; Yilmazer & Kanik, 2009), a combination of oxidative and reductive chemistries (Karmakar, 1999, p. 176; Marmer, Cardamone, Arifoglu, & Brandt, 1994), or an enzyme assist (Levene, 1997).

Silk fibers are less sensitive to alkali than wool and can be bleached with hydrogen peroxide under more severe conditions. A recommended bleaching solution consists of 0.55% hydrogen peroxide, sodium silicate, and ammonia to pH 10 heated to 60–75 °C for 2–4 h (Peters, 1967, p. 325).

18.3.4 MERCERIZATION

In 1850, Englishman John Mercer patented a process for treating cotton with concentrated solution of sodium hydroxide. The process became so popular that it was named "mercerization" after its inventor. The benefits imparted to cotton by mercerization include a silk-like luster and increased strength, moisture absorbency, color yield after dyeing, and reactivity (Abrahams, 1994; Ghosh & Dilanni, 1994; Greif, 1996; Parikh et al., 2006; Sampath, 2001; Saravanan & Ramachandran, 2007; Shamey & Hussein, 2005b; Ströhle & Schramek, 2007). The concentrated alkali swells cotton fibers, producing irreversible physical and chemical changes to the internal fiber structure. Typical mercerizing steps are (Karmakar, 1999, p. 279):

- apply a 31–35% sodium hydroxide solution at a temperature of 15–18 °C;
- soak for 55 s while maintaining warp tension; and
- wash off the alkali solution while restoring the fabric width.

Table 18.5 Typical wool bleaching conditions*

Parameter	
Hydrogen peroxide (35%)	25–45 g/L
pH (with ammonia)	8–8.5
Temperature (°C)	50–60
Time (min)	45–190

**Karmakar, 1999, p. 176.*

Another approach to alkali treatment of cotton is treatment with liquid ammonia (Saravanan, 2005), with similar changes to fiber structure.

18.3.5 CARBONIZATION

After wool fibers have been scoured, spun into yarn, and woven into fabric, some cellulosic materials (leaf and seed fragments) may remain trapped in the yarn. The most common method for removing these impurities is to convert the cellulosic material to a brittle form with strong acids. This process is called carbonization (Edenborough, Nossar, & Chaikin, 1983; Mozes, 1988; Pailthorpe, 1991). After carbonization, the brittle material can be easily removed by mechanical means. A typical carbonization procedure involves applying 5% sulfuric acid by weight of fabric to wool fabric, drying the wool at around 55 °C, and heating the dry acid containing wool to 100 °C. The fabric is then passed through a crushing machine, washed, neutralized, and dried (Peters, 1967, p. 271).

18.3.6 HEAT SETTING

Heat setting is a process that provides dimensional stability, shape retention, and crease resistance to thermoplastic synthetic fibers such as polyester and nylon (Chattopadhyay & Renuka, 1994; Gacen, Bernal, & Maillo, 1988; Glawe, 1999, p. 83; Hearle, Wilding, Auyeung, & Ihmayed, 1990; Heller, 2001; Maynard, 1980; Sardag, Ozdemır, & Kara, 2007; Shenai & Saraf, 1986). These fibers will shrink when exposed to heat unless their internal structures have been stabilized by exposure to a temperature higher than they would encounter in subsequent processing or consumer use. Although heat setting is usually considered a finishing process, often specific fabrics are heat set during preparation to control shrinkage during processing. The fabrics are heated while held under tension in a tenter frame to maintain the desired fabric dimensions. Table 18.6 gives suggested heat setting times and temperatures for a variety of fibers and fiber blends.

18.3.7 DRYING

At the end of the preparation process, the fabric is dried using a weft straightener, a tentering (stentering) frame, and a drying oven. In the initial step of the drying phase, the fabric is straightened using the weft straightener. Then the edges of the fabric are stretched onto a tentering (stentering) frame. Pins on both edges of the frame pull the fabric and straighten the warp and weft in both directions during drying.

Table 18.6 Typical heat setting conditions*

Fiber type	Exposure time (s)	Setting temperature (°C)
Polyester	20–40	180–220
Nylon 6	15–20	190–193
Nylon 6,6	15–20	200–230
Acrylic	30	120
Polyester/cotton	30	180
Polyester/wool	30	170–190
Polyester/linen	30	180
Polyester/silk	30	190

**Karmakar, 1999, pp. 267–272.*

Table 18.7 Useful test methods for preparation process quality control*

Test method	Method title	Comments
AATCC test method 79	Absorbency of bleached textiles	Water drop absorbency
AATCC test method 81	pH of wet processed textiles	Measurement of fabric pH
AATCC test method 82	Fluidity of cellulose	Determines fiber damage from preparation
AATCC test method 89	Mercerization in cotton	Measures extent of mercerization
AATCC test method 97	Extractable content with enzyme, solvent, and water	Measure of extractable material
AATCC test method 110	Whiteness of textiles	Measure of bleaching efficiency
ASTM D1424	Tearing strength of fabrics	Determines woven fabric damage from preparation
ASTM D5035	Breaking strength of fabric	Determines woven fabric damage from preparation
ASTM D3786	Breaking strength of knitted fabrics	Determines knitted fabric damage from preparation

**AATCC (2010), ASTM (2010).*

18.4 QUALITY CONTROL IN FABRIC PREPARATION

A variety of quality-control methods have been developed for use in fabric preparation. The American Association of Textile Chemists and Colorists (AATCC) and the American Society for Testing and Materials (ASTM) publish test methods for evaluating fabric properties during the preparation process. AATCC provides performance test methods, while ASTM provides not only test methods but also suggested specifications for different product end uses. A listing of useful test methods for preparation processes is given in Table 18.7.

Although the absolute values of these tests are important in determining the quality of the prepared fabric, the uniformity of the results, side–center–side and lot-to-lot, is also an important factor in fabric preparation quality.

18.5 ENVIRONMENTAL IMPACT AND SUSTAINABILITY OF FABRIC PREPARATION

Preparation processes involve significant amounts of water, energy, and potentially hazardous chemicals. Desizing starch also puts a large biological oxygen demand into the effluent. A modern textile plant that carries out fabric-preparation processes will have a well-functioning waste treatment facility on site to treat wastewater before it is discharged into streams, rivers, or public sewers. Wastewater treatment is a significant expense, but must be carried out if the plant is to meet discharge regulations imposed by local, national, and possibly even international authorities.

18.6 RESEARCH AND FUTURE TRENDS

As the costs of providing well-prepared fabrics in an environmentally sustainable manner rise, new approaches to preparation processes are being explored. Recently, scouring systems for cotton have been developed that employ a mixture of enzymes (Csiszar, Szakacs, & Rusznak, 1998; Hartzell & Hsieh, 1998; Hsieh & Cram, 1999; Lin & Hsieh, 2001; Sahin & Gürsoy, 2005; Sangwatanaroj, Choonukulpong, & Ueda, 2003; Traore & Buschle-Diller, 1999; Tzanov, Calafell, Guebitz, & Cavaco-Paulo, 2001). These systems can provide well-scoured cotton without the use of harsh chemicals and high temperatures.

Another approach to scouring cotton that lowers the environmental impact is to expose the cotton fabric to a plasma treatment prior to scouring (Goto, Wakita, Nakanishi, & Ohta, 1992; Sun & Stylios, 2006). Fabric treated with plasma can be efficiently scoured with fewer chemicals and shorter process times and temperatures.

Enhanced hydrogen peroxide bleaching of cotton using specific enzymes as well as peroxide has been reported (Basto, Tzanov, & Cavaco-Paulo, 2007; Bouwhuis, Dorgelo, & Warmoeskerken, 2009; Pereira, Bastos, Tzanov, Cavaco-Paulo, & Guebitz, 2005; Shin, Hwang, & Ahn, 2004). Recent work has shown that the use of a peracid precursor leads to effective bleaching at lower temperatures and shorter times than conventional hydrogen peroxide bleaching (Gürsoy & Dayioglu, 2003; Gürsoy, El-Shafei, Hauser, & Hinks, 2004a; Gürsoy, Lim, Hinks, & Hauser, 2004b; Hebeish et al. 2009; Križman, Kovač, & Tavčer, 2005; Lavri, Kova, Tavčer, Hauser, & Hinks, 2007; Lim, Jung, Hinks, & Hauser, 2005; Malik & Das, 2006; Preša & Tavčer, 2008; El Shafie et al., 2009; Shao, Huang, Wang, & Liu, 2010; Tavčer, 2010; Topalovic et al., 2007). The precursor reacts with hydrogen peroxide to form a peracid that is more reactive at lower temperatures than peroxide alone.

Carbonizing wool with sulfuric acid can damage both the wool fiber and the environment. Recent work involves the use of a mixture of enzymes (cellulases, pectinases, and hemicellulases) to remove extraneous cellulosic material from wool with strong acids and high heat (Gouveia, Fiadeiro, & Queiroz, 2008).

The use of alternative chemistries such as enzyme mixtures for scouring and carbonizing, and peroxide activators for bleaching, will reduce operating costs and lower adverse environmental effects. Currently these alternative chemistries are more expensive than traditional chemicals on a per-pound basis, however, so a complete economic analysis of the entire preparation process is needed to determine if there are cost savings. Machinery innovations to reduce the amount of water needed to prepare fabrics properly are also being studied.

18.7 SUMMARY

- Fabric preparation is a sequential series of wet processes, the details of which can vary depending on the particular fiber and fabric construction.
- Desizing removes the warp size applied to warp yarns to facilitate weaving.
- Scouring is done to remove naturally occurring fiber impurities in natural fibers, and processing aids on synthetic fibers.
- The bleaching process provides the desired level of whiteness to natural fibers.
- Cotton fabrics can be mercerized to improve dye yield, add luster, and increase strength, while wool fabrics can be carbonized to remove any residual cellulosic material.
- All of these processes require considerable amounts of water and potentially harmful chemicals, but alternative chemistries are being developed to minimize adverse environmental effects.

18.8 CASE STUDY

A recent report (Carneiro et al., 2005) illustrates the value of examining alternative processes in fabric preparation. The authors compared the effectiveness of desizing, scouring, and bleaching of a plain-weave cotton fabric sized with starch, with and without a corona discharge treatment prior to processing. Corona discharge is an atmospheric pressure plasma process that exposes the fabric to high-energy electrons that can cause chemical and physical changes on the fabric surface. These changes make the surface more hydrophilic, and thus make the use of existing chemical treatments more efficient.

After exposing the greige cotton fabric to corona discharges of approximately 30 s prior to desizing, scouring, and bleaching, the following observations were made:

- The whiteness index of the corona-treated bleached fabric increased to the extent that bleaching could be carried out at a temperature 10 °C lower and achieve the same degree of whiteness as the non-corona-treated fabric.
- Optimal starch removal was seen for corona-treated fabric at a temperature 20 °C lower than for non-corona-treated fabric.
- Better mote removal was seen with the corona-treated fabric.
- The absorbency of the corona-treated bleached fabric was greater than non-corona-treated fabric.
- The uniformity of the bleaching was better with the corona-treated fabric.

The authors concluded that treating fabrics with corona discharge prior to preparation leads to better-prepared higher-quality fabrics at lower energy costs.

18.9 PROJECT IDEAS

- Review and report conclusions from recent publications and news articles that address the effects of textile wet processing (especially preparation) on global climate change.
- Interview a manager of a company that produces prepared textiles and report on that company's methods to meet environmental regulations.

18.10 REVISION QUESTIONS

- Explain why the preparation of the textile substrate is important for the dyeing process.
- What are the processing steps needed to properly prepare a cotton woven fabric?
- List the chemicals needed to effectively desize a fabric containing a starch warp size.
- Why is wool scoured in fiber form?
- What is the purpose of bleaching?
- Describe the mercerizing process, including the fiber to be treated, the chemicals needed, and the effects achieved by the process.
- List three quality control tests useful in fabric preparation.
- Describe some alternative chemistries and processes being developed to reduce the adverse environmental impact of fabric preparation.

18.11 SOURCES OF FURTHER INFORMATION

Choudhury, A. K. R. (2006). *Textile preparation and dyeing*. Enfield, New Hampshire: Science Publishers.

Karmakar, S. R. (1999). *Chemical technology in the pretreatment processes of textiles*. Amsterdam: Elsevier.

Peters, R. H. (1967). *Textile chemisty* (Vol. II). Amsterdam: Elsevier.

Textile chemical suppliers such as Huntsman (http://www.huntsman.com/textile_effects), BASF (http://www.performancechemicals.basf.com), and Clariant (http://www.textiles.clariant.com) can provide detailed information about the use of their products.

REFERENCES

AATCC. (2010). *AATCC technical manual*. Research Triangle Park, NC: American Association of Textile Chemists and Colorists.

Abrahams, D. H. (1994). Improving on mercerizing processing. *American Dyestuff Reporter*, *83*(9), 78,124.

Achwal, W. B. (1995). Recycling of sizes. *Colourage*, *42*(11), 41.

ASTM. (2010). *Annual book of ASTM standards*. West Conshohocken, PA: American Society for Testing and Materials.

Basto, C., Tzanov, T., & Cavaco-Paulo, A. (2007). Combined ultrasound-laccase assisted bleaching of cotton. *Ultrasonics Sonochemistry*, *14*(3), 350–354.

Bateup, B. O., & Warner, J. J. (1988). Selective scouring of dirt from greasy wool. Part III. Trials at commercial plants. *Textile Research Journal*, *58*(12), 707–714.

Bouwhuis, G. H., Dorgelo, B., & Warmoeskerken, M. M. C. G. (2009). Textile pretreatment process based on enzymes, catalyst and ultrasound. *Melliand International*, *15*(4), 150–151.

Buschle-Diller, G., Radhakrishnaiah, R., Freeman, H., & Zeronian, S. H. (2002). *Environmentally benign preparatory processes – introducing a closed-loop system*. Annual Report. National Textile Center.

Cacho, J. W. (1980). CMC warp sizing vs. polyvinyl alcohol. *Textile Chemist and Colorist*, *12*(4), 28–32.

Cai, J. Y., Evans, D. J., & Church, J. S. (2008). Amineborane: a unique reductive bleaching agent that protects cystine disulphide bonds in keratins. *Coloration Technology*, *124*(5), 318–323.

Cai, Z., Qiu, Y., Zhang, C., Hwang, Y.-J., & McCord, M. (2003). Effect of atmospheric plasma treatment of desizing of PVA on cotton. *Textile Research Journal*, *73*(8), 670.

Cardamone, J. M., Yao, J., & Phillips, J. G. (2005). Combined bleaching, shrinkage prevention, and biopolishing of wool fabrics. *Textile Research Journal*, *75*(2), 169–174.

Carneiro, N., Souto, A. P., Nogueira, C., Madureira, A., Krebs, C., & Cooper, S. (2005). Preparation of cotton materials using Corona discharge. *Journal of Natural Fibers*, *2*(4), 53–65.

Cegarca, J., Gacén, J., & Caro, M. (1983). 35 – the action of sodium laurylsulphate in the bleaching of wool with hydrogen peroxide in an acidic medium. *Journal of the Textile Institute*, *74*(6), 351–356.

Chattopadhyay, D. P., & Renuka (1994). Heat setting. *Man-Made Textiles in India*, *37*(9), 413–419.

Choudhury, A. K. R. (2006). *Textile preparation and dyeing*. Enfield, NH: Science Publishers.

Csiszar, E., Szakacs, G., & Rusznak, I. (1998). Combining traditional cotton scouring with cellulase enzymatic treatment. *Textile Research Journal*, *68*(3), 163–167.

Das, T., & Ramaswamy, G. N. (2006). Enzyme treatment of wool and specialty hair fibers. *Textile Research Journal*, *76*(2), 126–133.

Edenborough, B. W., Nossar, M. S., & Chaikin, M. (1983). Factors affecting the neutralizing of acidified wools with ammonia solutions. *Journal of the Textile Institute*, *74*(3), 131–137.

El-Khatib, E. M. (2002). Some characterization of wool bleached with activated hydrogen peroxide. *Colourage*, *49*(9), 39–44.

El-Rafie, M., Abdel Hafiz, S. A., El-Sisi, F., Helmy, M., & Hebeish, A. (1991). Fast desizing/scouring/bleaching system for cotton-based textiles. *American Dyestuff Reporter*, *80*(1), 45–48.

El Shafie, A., Fouda, M. M. G., & Hashem, M. (2009). One-step process for bio-scouring and peracetic acid bleaching of cotton fabric. *Carbohydrate Polymers*, *78*(2), 302–308.

Evans, B. A., Boleslawski, L., & Boliek, J. E. (1997). Cotton hosiery bleaching with hydrogen peroxide. *Textile Chemist and Colorist*, *29*(3), 28–34.

Gacen, J., Bernal, F., & Maillo, J. (1988). Detecting the source of irregularities in the heat setting of polyester fabrics. *Textile Chemist and Colorist*, *20*(2), 31–33.

Gacen, J., Cayuela, D., & Gacen, I. (1999). Rapid wool bleaching with reducing agents. *Melliand Textilberichte*, *80*(11/12), 949–950, E264–E265.

Gacén, J., Cayuela, D., & Garcén, I. (2002). Rapid bleaching of wool with hydrogen peroxide. *AATCC Review*, *2*(10), 28–31.

Ghosh, S., & Dilanni, D. (1994). Estimating the degree of mercerization using near-infrared spectroscopy. *Journal of the Textile Institute*, *85*(3), 308–315.

Glawe, A. (1999). Thermal analysis for determining the optimum conditions for heat-setting of synthetic fiber materials. *Melliand Textilberichte*, *80*(1/2) 83–85, E29–E30.

Goto, T., Wakita, T., Nakanishi, T., & Ohta, Y. (1992). Application of low-temperature plasma treatment to the scouring of gray cotton fabric. (English). *Sen'i Gakkaishi (Journal of the Society of Fiber Science and Technology, Japan)*, *48*(3), 133.

Gouveia, I. C., Fiadeiro, J. M., & Queiroz, J. A. (2008). Enzymatic wool treatment: preliminary evaluation of Trichoderma reesei cellulases and Aspergillus aculeatus pectinases and hemicellulases. *AATCC Review*, *8*(10), 38–44.

Gowda, K. N. N., Padaki, N. V., & Sudhakar, R. (2004). Enzymes in textile industrial applications. *Journal of the Textile Association*, *65*(1), 15–19.

Greif, S. (1996). Continuous mercerization – decision criteria for investment. *Melliand International*, 206(4), 209–212 .

Gürsoy, N. C., & Dayioglu, H. (2003). 2.2′ Bipyridine catalyzed peracetic acid bleaching of cotton. *Textile Research Journal*, *73*(4), 297.

Gürsoy, N. .Ç., El-Shafei, A., Hauser, P., & Hinks, D. (2004a). Cationic bleach activators for improving cotton bleaching. *AATCC Review*, *4*(8), 37–40.

Gürsoy, N. .Ç., Lim, S., Hinks, D., & Hauser, P. (2004b). Evaluating hydrogen peroxide bleaching with cationic bleach activators in a cold pad-batch process. *Textile Research Journal*, *74*(11), 970–976.

Hartzell, M. M., & Hsieh, Y. L. (1998). Enzymatic scouring to improve cotton fabric wettability. *Textile Research Journal*, *68*(4), 233–241.

Hearle, J. W. S., Wilding, M. A., Auyeung, C., & Ihmayed, R. (1990). A note on the heat-setting of nylon 6.6. *Journal of the Textile Institute*, *81*(2), 214–217.

Hebeish, A., Hashem, M., Shaker, N., Ramadan, M., El-Sadek, B., & Hady, M. A. (2009). New development for combined bioscouring and bleaching of cotton-based fabrics. *Carbohydrate Polymers*, *78*(4), 961–972.

Heller, J. (2001). High temperature heatsetting of tubular knitted fabrics containing spandex. *AATCC Review*, *1*(11), 32–34.

Hickman, W. S. (1996). Bleaching of cotton weft knitted fabrics. *Review of Progress in Coloration and Related Topics*, *26*, 29–46.

Hoffman, C. R. (1981). Recovering textile desizing effluent by ultrafiltration. *Journal of Coated Fabrics*, *10*, 178.

Holds, M. (2007). Use of enzymes in textile processing. *Colourage*, *54*(6) 64–66, 68–69.

Hsieh, Y.-L., & Cram, L. (1999). Proteases as scouring agents for cotton. *Textile Research Journal*, *69*(8), 590.

Ibrahim, N. A., Abo Shosha, M. H., Fahmy, H. M., & Hebeish, A. (1997). Effect of size formulation on sizability and desizability of some soluble sizes. *Polymer – Plastics Technology and Engineering*, *36*(1), 105–121.

Ibrahim, N. A., El-Hossamy, M., Morsy, M. S., & Eid, B. M. (2004). Optimization and modification of enzymatic desizing of starch-size. *Polymer – Plastics Technology and Engineering*, *43*(2), 519–538.

Juby, J. (1985). One-step bleaching. *Textile Chemist and Colorist*, *17*(5), 21–22.

Karl, U., & Freyberg, P. (1999). New reductive processes in textile finishing. *Melliand Textilberichte*, *80*(7/8), 616, 618–619, E161–E162.

Karmakar, S. R. (1999). *Chemical technology in the pretreatment processes of textiles*. Amsterdam: Elsevier.

Karunditu, A. W., Carr, C. M., Dodd, K., Mallinson, P., Fleet, I. A., & Tetler, L. W. (1994). Activated hydrogen peroxide bleaching of wool. *Textile Research Journal*, *64*(10), 570–572.

Ke-Heng, C., Gray, D. J., Perry, R. S., & Bide, M. (1988). The scouring of raw wool using freon TF solvent. *Textile Chemist and Colorist*, *20*(10), 19–24.

Križman, P., Kovač, F., & Tavčer, P. F. (2005). Bleaching of cotton fabric with peracetic acid in the presence of different activators. *Coloration Technology*, *121*(6), 304–309.

Lavri, P. K., Kova, F., Tavčer, P. F., Hauser, P., & Hinks, D. (2007). Enhanced PAA bleaching of cotton by incorporating a cationic bleach activator. *Coloration Technology*, *123*(4), 230–236.

Levene, R. (1997). Enzyme-enhanced bleaching of wool. *Journal of the Society of Dyers and Colourists*, *113*(7/8), 206–210.

Lim, S., Jung, J. L., Hinks, D., & Hauser, P. (2005). Bleaching of cotton with activated peroxide systems. *Coloration Technology*, *121*(2), 89–95.

Lin, C. H., & Hsieh, Y. L. (2001). Direct scouring of greige cotton fabrics with proteases. *Textile Research Journal*, *71*(5), 425–434.

Lin, S. H., & Lan, W. J. (1995). Polyvinyl-alcohol recovery by ultrafiltration – effects of membrane type and operating-conditions. *Separations Technology*, *5*(2), 97–103.

Livengood, C. D. (1995). *AATCC warp sizing handbook*, Research Triangle Park, American Association of Textile Chemists and Colorists.

Ma, P., Wang, X., Xu, W., & Cao, G. (2009). Application of corona discharge on desizing of polyvinyl alcohol on cotton fabrics. *Journal of Applied Polymer Science*, *114*(5), 2887–2892.

Malik, S. K., & Das, M. (2006). Low temperature bleaching of cotton using TAED activated peroxide bath. *Indian Journal of Fibre and Textile Research*, *31*(4), 588–590.

Marmer, W. N., Cardamone, J. M., Arifoglu, M., & Brandt, H. J. (1994). Optimizing process conditions in sequential oxidative/reductive bleaching of wool. *Textile Chemist and Colorist*, *26*(5), 19–24.

Maynard, W. M. (1980). Heatsetting of polyester fabrics. *American Dyestuff Reporter*, *69*(9) 34, 36, 38, 77.

Milner, A. J. (1997). Preserving the environment and protecting profits, realistic rationalization of pretreatment. *Book of papers – International conference and exhibition, AATCC* (pp. 496–506).

Mozes, T. E. (1988). Raw-wool carbonization. *Textile Progress*, *17*(3), 1–30.

Pailthorpe, M. T. (1991). Developments in wool carbonizing. *Review of Progress in Coloration and Related Topics*, *21*, 11–22.

Palin, M. J., Teasdale, D. C., & Benisek, L. (1983). Bleaching of wool with hydrogen peroxide and a silicate stabilizer. *Journal of the Society of Dyers and Colourists*, *99*(9), 261–266.

Parikh, D. V., Thibodeaux, D. P., Sachinvala, N. D., Moreau, J. P., Robert, K. Q., Sawhney, A. P. S., et al. (2006). Effect of cotton fiber mercerization on the absorption properties of cotton nonwovens. *AATCC Review*, *6*(4), 38–43.

Pawar, S. B., Shah, H. D., & Andhorikar, G. R. (2002). Enzymatic processing of cotton: a bio-technical approach. *Man-Made Textiles in India*, *45*(4), 133.

Peng, S., Gao, Z., Sun, J., Yao, L., & Qiu, Y. (2009). Influence of argon/oxygen atmospheric dielectric barrier discharge treatment on desizing and scouring of poly (vinyl alcohol) on cotton fabrics. *Applied Surface Science*, *255*(23), 9458–9462.

Pereira, L., Bastos, C., Tzanov, T., Cavaco-Paulo, A., & Guebitz, G. M. (2005). Environmentally friendly bleaching of cotton using laccases. *Environmental Chemistry Letters*, *3*(2), 66–69.

Peters, R. H. (1967). *Textile chemisty* (Vol. II). Amsterdam: Elsevier.

Preša, P., & Tavčer, P. F. (2008). Bioscouring and bleaching of cotton with pectinase enzyme and peracetic acid in one bath. *Coloration Technology*, *124*(1), 36–42.

Reincke, K. (1998). More economical and less environmentally harmful bleaching processes for wool. *Melliand Textilberichte*, *79*(1/2), 54–60, E13–E17.

Robinson, G. (1993). Technology of polyvinyl alcohol (PVA) warp sizing recovery. *Technical Textiles International*, *2*(1), 13.

Sahin, U. K., & Gürsoy, N. Ç. (2005). Low temperature acidic pectinase scouring for enhancing textile quality. *AATCC Review*, *5*(1), 27–30.

Sampath, M. R. (2001). Frequently encountered problems in textile wet processing and a diagnostic approach for prevention/solutions. *Colourage*, *48*(2), 23–26.

Sangwatanaroj, U., Choonukulpong, K., & Ueda, M. (2003). Cotton scouring with pectinase and Lipase/Protease/Cellulase. *AATCC Review*, *3*(5), 17–20.

Saravanan, D. (2005). Liquid ammonia treatment. *Journal of the Textile Association*, *66*(3), 133–138.

Saravanan, D., & Ramachandran, T. (2007). Forgotten fundamentals of mercerisation. *Asian Dyer*, *4*(5), 35–40.

Sardag, S., Ozdemır, O., & Kara, I. (2007). The effects of heat-setting on the properties of polyester/viscose blended yarns. *Fibres and Textiles in Eastern Europe*, *15*(4), 50–53.

Shamey, R., & Hussein, T. (2005a). 10. Problems in bleaching. *Textile Progress*, *37*(1), 19–22.

Shamey, R., & Hussein, T. (2005b). 11. Problems in mercerization. *Textile Progress*, *37*(1), 22–24.

Shao, J., Huang, Y., Wang, Z., & Liu, J. (2010). Cold pad–batch bleaching of cotton fabrics with a TAED/H_2O_2 activating system. *Coloration Technology*, *126*(2), 103–108.

Shenai, V. A., & Saraf, N. M. (1986). Some aspects of textile finishing. Part IV – setting processes. *Textile Dyer and Printer*, *19*(19) 21–24, 29.

Shin, Y., Hwang, S., & Ahn, I. (2004). Enzymatic bleaching of desized cotton fabrics with hydrogen peroxide produced by glucose oxidase. *Journal of Industrial and Engineering Chemistry (Seoul, Republic of Korea)*, *10*(4), 577–581.

Ströhle, J., & Schramek, G. (2007). Open width mercerizing of knitwear. *Melliand International*, *13*(2), 132–134.

Sun, D., & Stylios, G. K. (2006). Fabric surface properties affected by low temperature plasma treatment. *Journal of Materials Processing Technology*, *173*(2), 172–177.

Tavčer, P. F. (2010). Impregnation and exhaustion bleaching of cotton with peracetic acid. *Textile Research Journal*, *80*(1), 2–11.

Tomasino, C., Livengood, C. D., & Thorp, S. N. (1981). Cold water desizing for size recovery. *Textile Chemist and Colorist*, *13*(7), 19–22.

Topalovic, T., Nierstrasz, V., Bautista, L., Jocic, D., Navarro, A., & Warmoeskerken, M. (2007). Analysis of the effects of catalytic bleaching on cotton. *Cellulose*, *14*(4), 385–400.

Traore, M. K., & Buschle-Diller, G. (1999). Environmentally friendly scouring processes. *Book of papers – International conference and exhibition, AATCC* (pp. 183–189).

Tzanov, T., Calafell, M., Guebitz, G. M., & Cavaco-Paulo, A. (2001). Bio-preparation of cotton fabrics. *Enzyme and Microbial Technology*, *29*(6–7), 357–362.

Wood, G. F. (1982a). 4. Solvent-scouring. *Textile Progress*, *12*(1), 4–6.

Wood, G. F. (1982b). 5. Scouring with non-ionic detergents. *Textile Progress*, *12*(1), 7.

Yilmazer, D., & Kanik, M. (2009). Bleaching of wool with sodium borohydride. *Journal of Engineered Fabrics and Fibers (JEFF)*, *4*(3), 45–50.

CHAPTER

FABRIC FINISHING: DYEING AND COLOURING

19

P.R. Richards

Richtex Textile Consultancy, Newark, UK

This chapter has been adapted from Chapter 24, Methods of dye application, by P.R. Richards, originally published in J. Best (ed), *Colour Design: Theories and Applications*, Woodhead Publishing Ltd, ISBN 978-1-84569-972-7.

LEARNING OBJECTIVES

At the end of this chapter, you should be able to:

- Explain the basics of colour theory
- List the main classes of dye
- Relate dye classes to the dyeing of the main fibre groups
- Describe the processing routes for the dyeing of various material forms
- Describe the environmental challenges of dyeing

19.1 INTRODUCTION

Dyeing fabrics and the end result are a complex mix of chemistry and physics. The visual impression is created by light being reflected, and while highly subjective to the individual, can also be measured as a series of wavelengths and thus quantified. The dye itself adheres to the fabric as the result of one or more chemical reactions between the dye and the fibres, and there are many different types of dye that are suitable for different types of fibre and under different conditions. Thus the key issues in dyeing are the criteria for selecting dyes and the conditions under which they are applied, which in turn influence the choice of appropriate dyeing machinery.

19.2 COLOUR THEORY

Colour has a significant influence in the decoration of our environment, not least in the attraction of a potential customer to a textile article. But what is colour, and how do we perceive it? It is important to understand the basic physics behind colour, because this provides a way in which colours can be objectively measured and compared.

Textiles and Fashion. http://dx.doi.org/10.1016/B978-1-84569-931-4.00019-2

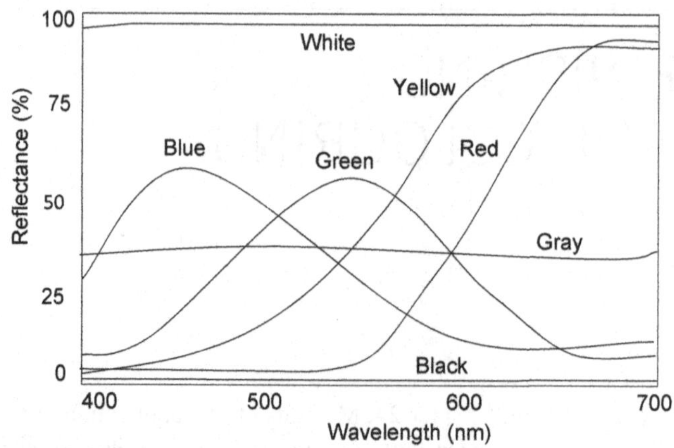

FIGURE 19.1

Examples of spectrophotometric curves.

Source: Reference: Park (1993).

19.2.1 LIGHT AND THE HUMAN EYE

Colour is described in the *Oxford English Dictionary* as 'the sensation produced on the eye by spectral resolution or selective surface reflection etc. of rays of light'. Colour is therefore a perception, and not an intrinsic part of an object.

Light is radiant energy capable of stimulating the eye and causing the sensation of vision (Society of Dyers and Colourists, 1988), and is in the form of wavelengths, which are measured in nanometres (nm), where one nm is equal to one billionth of a metre. The range of visible light is between 380 and 770 nm. Ultraviolet radiation is immediately below this range, while infrared radiation is immediately above.

When light strikes an object, some wavelengths are absorbed while some are reflected. When this reflected light reaches the photoreceptive rods and particularly the cones at the back of the eye, impulses are sent to the visual cortex of the brain, which interprets the impulses as colour (Broadbent, 2001, p. 427). Different wavelengths within the visible spectrum give rise to different colours.

Figure 19.1 shows examples of spectrophometric curves (Park, 1993). Table 19.1 shows the relationship between wavelength of absorbed light and hue (Ingamells, 1993).

19.2.2 COLOUR DESCRIPTION AND MEASUREMENT

When it is necessary to describe a colour, a definition more precise than 'red', 'green', 'blue', etc., is required. Colours may best be described in terms of their:

- **Hue**, or that attribute of colour whereby it is recognised as being predominantly red, green, blue, yellow, brown, violet, etc. (Society of Dyers and Colourists, 1988).
- **Strength**, or the colour yield of a given quantity of dye in relation to an arbitrarily chosen standard (Society of Dyers and Colourists, 1988).
- **Brightness**, or the converse of dullness (Society of Dyers and Colourists, 1988).

For the dyer, these verbal descriptions are not precise, as individual interpretations differ widely. A significant advance in standardising colour description was the development of a colour atlas by the

Table 19.1 Relationship between wavelength of absorbed light and hue

Absorbed wavelength	Hue of absorbed light	Perceived hue
400–440	Violet	Greenish-yellow
440–480	Blue	Yellow
480–510	Blue-green	Orange
510–540	Green	Red
540–570	Yellow-green	Purple
570–580	Yellow	Blue
580–610	Orange	Greenish-blue
610–700	Red	Blue-green

Reference: Ingamells (1993).

artist Albert Munsell early in the twentieth century. The Munsell system used **hue** (the actual colour), **value** (lightness or darkness) and **chroma** (strength or weakness) to define colour, and gave a reference to each nuance of shade thus produced. Further colour atlas systems have been developed, of which Pantone, for example, has gained universal acceptance.

19.2.3 INSTRUMENTAL COLOUR MATCH AND SHADE ASSESSMENT

While the logical arrangement of colour described was a significant advance in colour communication, it still relied on visual assessments and descriptions by individual observers. Instrumental systems have therefore been developed to overcome this difficulty.

In order to quantify the magnitude of the difference between two colours, mathematical colour difference equations have been developed. These complex formulae take into account problems such as the ability of the coloured substrate to absorb, scatter and reflect light, and the ability of the human eye to perceive colour differently across the spectrum. Kubelka and Munk created a theoretical model and formula in 1931, and refinements to this have continued to be made. Significant advances have been made in this field by the development of spectrophotometers capable of accurate measurement of reflected light.

The main colour difference equations are CMC, BFD (University of Bradford) and CIE-94 (Broadbent, 2001, p. 472). For many years Marks and Spencer used their own MS89 programme before changing to CMC 2:1.

A modified system can also be used to compare the spectral data of the standard with the stored reflectance values of dyestuffs, this information being used to predict a recipe to match the shade. Using instrumental systems to measure and compare colour is becoming an indispensible part of the dyer's working practice.

19.3 SELECTION OF DYES

There are 10 main classes of dyestuff (see Table 19.2), applicable to different fibre groups, with different dyeing properties and application methods, and with varying levels of hue and brightness within their ranges. The dyer has literally thousands of dyes available, and therefore must follow a selection procedure to ensure that the optimum dyes are chosen.

Table 19.2 Classes of dye and the fibres to which they are applicable commercially

	Cellulosic	Acetates	Protein	Polyamide	Polyester	Acrylic
Direct	✓					
Reactive	✓		✓	✓		
Vat	✓					
Sulphur	✓					
Azoic	✓					
Acid			✓			
Metal-complex			✓			
Mordant			✓			
Disperse		✓		✓	✓	✓
Cationic						✓

Table 19.2 shows the main dye classes and the fibres to which they are applied commercially.

19.3.1 ACHIEVING THE REQUIRED SHADE

The first step is to select dyes from a range applicable to the fibre, and check their suitability for achieving the desired shade. There are two main methods of colour matching:

- A traditional visual selection procedure.
- An instrumental colour match prediction procedure, which is the preferred method.

For the **visual selection procedure**, the dyer uses a previously formed shade library and dyestuff suppliers' shade cards representing different dyestuffs to predict the recipe needed. This purely visual procedure relies heavily on a succession of laboratory dyeings to fine-tune the recipe until the dyed sample gives a close visual match to the physical standard, and is very much a time-consuming 'trial-and-error' process.

Ideally, the dyer will have access to a **computerised colour match prediction** facility, with a database of the spectral data of dyes within the ranges considered suitable for dyeing the fibre in question. The dyer can then present either the physical shade standard or the numerical spectral data of the standard to the system. The physical shade standard could be in the form of fabric, yarn, paper, in fact anything which the customer has defined as portraying the desired colour to be matched.

The spectral data are the reflectance values (or the percentage of light reflected from the surface) of the standard at wavelengths between 400 and 700 nm at 20-nm intervals, so there are 16 points in total.

The system produces a reflectance curve from the data received by either route, and then attempts to match the shade by replicating this curve using combinations of the dyes available. In some cases, several dye recipes may be predicted, and the dyer needs to use further criteria to select the best recipe. Figure 19.1 shows examples of spectrophotometric curves for various colours (Park, 1993).

19.3.2 COMPATIBILITY OF DYES

A further consideration is that dyes within the same class may have different properties, for instance in rate of application to the fibre, and the dyer needs to ensure that the dyes within the chosen combination are compatible under the conditions in which dyeing will take place.

19.3.3 METAMERISM

Metamerism is a feature where the colour difference between the dyed material and the standard changes when the two are viewed side by side under changes of light source (commonly termed 'illuminant'). Typically, the dyer will be required to ensure a satisfactory match to the standard under the following illuminants:

- The customer's specified store light
- Artificial daylight
- Tungsten-filament light

Unless the standard and the dyed material are of the same fibre composition and physical construction, and dyed with the same combination of dyestuffs in the same proportions, their reflectance curves will be different. Therefore, although a good match may be possible under one illuminant, there is likely to be metamerism when the shade is viewed under other illuminants.

The instrumental match prediction system used to select the dye recipe can, by analysis of the reflectance curves, also give information on the degree of metamerism between the standard and each proposed dye combination.

If the dyer does not have access to an instrumental system, and uses a visual prediction procedure, a visual assessment of metamerism must be made by checking the laboratory dyeing against the physical standard under the specified illuminants.

19.3.4 COLOUR FASTNESS

Colour fastness requirements are dictated by the end use of the dyed material and by any further processing to which the material will be subjected. Test methods that are widely accepted have been devised by organisations such as the British Standards Institution, the International Organization for Standardization and the American Association of Textile Chemists and Colorists. In the retail sector, some retailers have developed their own variations of these tests to which their suppliers must comply.

The key tests common to most sectors of the industry are fastness to:

- Detergent washing
- Perspiration

- Water at 37 °C
- Wet and dry rubbing
- Light

Colour fastness is assessed generally by comparing any staining of specified adjacent fabrics during the test with a set of standard 'grey scales'. A numerical grading is given on a scale of 1–5, where 1 is very poor and 5 is excellent. Generally, a grade of 4 or above is deemed acceptable for commercial use.

Light fastness is graded on a system from 1 to 8, where 1 is very poor and 8 is excellent. For general apparel, a grade of 4 or 5 is acceptable, whereas for curtain or automotive fabric a grade of 7 or 8 is required.

The dyer must be aware of the customer's fastness requirements at the recipe prediction and dye-selection stage, and choose the best recipe accordingly. This will probably be by experience in the first place, confirmed by testing a dyed sample.

If dyeing is carried out at an early stage of the overall manufacturing process, the material may have to withstand further processing such as chemical shrink resisting, the application of easy-care resins, heat setting to provide fabric stability, etc. In addition to the end-use requirements, the dyer must select dyes to withstand the likely shade change encountered with these processes.

19.3.5 ENVIRONMENTAL CONSIDERATIONS

The dyer must consider the impact of dyes and associated application chemicals regarding residues that may potentially be retained by the dyed material, (e.g. formaldehyde, heavy metals, banned azo dyes), and also the content of the effluent following the dyeing process.

Dyes identified as carcinogenic, skin-sensitising or with a significant heavy metal content (e.g. chromium, copper) cannot be used.

A major area of concern is **azo dyes**. Azo dyes represent about 70% of all dyes used worldwide, but a very few of these dyes could potentially break down to create carcinogenic compounds. These compounds have been identified and are well known, and reputable dye manufacturers have eliminated any problem dyes from their ranges.

The application method and preparation processes or after-treatments related to the choice of dye must also be assessed. If the dyer is faced with potential problems in these areas, then either the choice of dye or the application method should be revised.

Many retailers have devised their own **Restricted Substances List** (RSL), which they circulate exclusively to their suppliers, and compliance with these requirements will help to address environmental concerns.

19.4 THE DYEING PROCESS

Fabrics and fibres are first prepared for dyeing as discussed in Chapter 18. These processes are critical to the success of the dyeing process and **must** be carried out effectively if dyeing is to be successful. There are then a number of considerations that the dyers must take into account in order to ensure effective uniform dyeing of the fibre or fabric.

19.4.1 DYEING CONDITIONS

Once the criteria of shade, metamerism and fastness are met, the dyer must ensure that the selected combination is technically suitable for successful application. This selection is based on the fibre involved and the form in which it is to be dyed (loose fibre, yarn, fabric or garment).

An important factor is gaining a balance between a dye's substantivity and its migration properties. **Substantivity** is defined as the attraction between a substrate and a dye or other substance whereby the latter is selectively extracted from the application medium by the substrate (Society of Dyers and Colourists, 1988). The degree of substantivity, usually rated as low, medium or high, has a bearing on the rate at which the dye is taken up by the fibre (**strike rate**), with dyes of lower substantivity having a slower strike rate. This is important in the selection of processing conditions, in that dyes of higher substantivity may need conditions where the strike rate is modified to ensure a dyeing with level results.

Migration is defined as the movement of a dye or pigment from one part of a material to another (Society of Dyers and Colourists, 1988). Thus, once the dye has penetrated the fibre, its migration properties have a bearing on the levelness, or evenness of application, of the resultant dyeing. Again, the processing conditions may be adjusted to ensure that optimum migration is allowed.

Processing conditions must be appropriate to the characteristics of the dyeing machinery used, based on:

- The machine's circulation properties
- The optimum liquor-to-goods ratio
- The degree of process control available

19.4.2 MACHINERY FOR DYEING

Textile materials are dyed in different physical forms. Machinery has been developed to process:

- Loose fibre
- Slubbing and tops
- Yarn
- Fabric
- Garments

A fundamental requirement for successful dyeing is the provision of movement of the goods through the dye liquor, or the movement of liquor through the goods. Depending on the form of the material to be dyed, the dyer needs to select the best machine for the purpose.

For example, the yarn dyer may need to decide whether to process the material in hank (Figure 19.2) or package (Figure 19.3) form (where the yarn is wound on to perforated spindles), the machinery having different configurations.

Similarly, the dyer of tubular knitted fabric may have the choice of jet (Figure 19.4) or winch (Figure 19.5) machinery, while the garment dyer may have to choose between paddle (Figure 19.6) or rotary (Figure 19.7) machines.

19.4.3 FURTHER TEXTILE COLOURATION

While the focus of this chapter is the dyeing of textiles in fibre, yarn, fabric or garment form, it should be pointed out that dyes are employed in other areas of the industry.

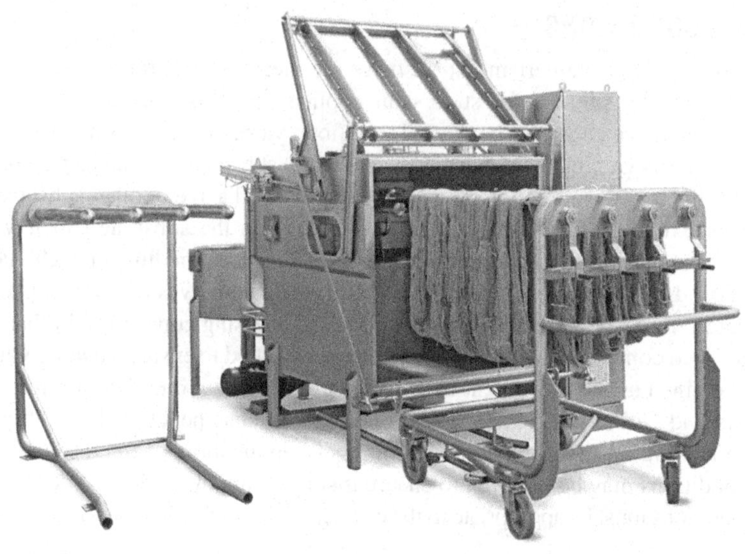

FIGURE 19.2

Thies GmbH & Co hank dyeing machine.

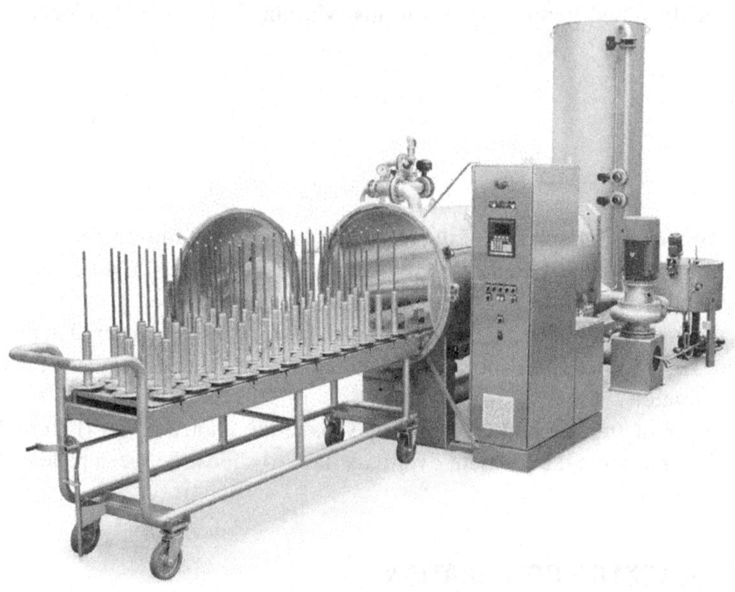

FIGURE 19.3

Thies GmbH & Co package dyeing machine.

FIGURE 19.4

Thies GmbH & Co jet dyeing machine.

FIGURE 19.5

Winch dyeing machine.

Source: Courtesy of Gagan Mechanical Works.

Dyes are used for **printing**, which may be carried out on knitted or woven fabrics and garments, generally with the same classes of dyes and pigments employed for dyeing the appropriate fibre. Fabric may be printed by continuous or semi-continuous procedures where the colourants, in a thickened paste medium, are applied via engraved rollers or patterned screens. Garments are normally printed either by screen printing, using the print paste medium and patterned screens, or by transfer printing and digital printing, where the

FIGURE 19.6

Flainox S.r.l side-paddle dyeing machine.

FIGURE 19.7

Flainox S.r.l rotary dyeing machine.

colourants are transferred by heat from previously impregnated paper. The methods of applying dyes in printing are discussed in Chapter 20.

Leather is dyed in hide form with anionic dyes similar to those used for dyeing protein fibres such as wool. Leather is processed by well-established preparation and after-treatment methods, the wet processing being carried out in large rotary machines.

In a process known as **dope dyeing**, it is possible to add dyes or pigments to a synthetic polymer solution before it is extruded into filaments. This gives excellent colour fastness, but needs an early commitment to a particular colour. Dope dyeing is valuable in the colouration of fibres such as polypropylene, which are hydrophobic and lack appropriate chemical groups to which the dye may attach, and are not dyeable by conventional methods.

19.5 CLASSES OF DYE FOR DIFFERENT FIBRE TYPES

The following outlines of classes of dyes for different fibres are based on commercial choices normally made by the dyer, while recognising that alternative application methods are possible.

19.5.1 CELLULOSIC FIBRES

This broad cellulosic category includes natural seed fibres such as cotton, coir and kapok, and the less pure bast fibres such as flax, jute, hemp, ramie and nettle, and leaf fibres such as sisal and abaca.

Regenerated rayon fibres, which include viscose and lyocell, have similar dyeing properties to natural cellulose fibres. However, the rate of dyeing and level of dye uptake are different to those of cotton, and vary across the range of rayon fibres. Rayon fibres are dyed with the same classes of dyestuff as those used for cotton.

Cellulose diacetate and triacetate fibres are also regenerated fibres, but are quite different in their dyeing properties to those fibres already mentioned. The chemical modifications made to the cellulose in their manufacture render them relatively hydrophobic (an inability to absorb water readily) and without the chemical groups to which dyes used for other cellulosic fibres may attach.

They are therefore dyed with disperse dyes, which are relatively insoluble in water and applied from a fine aqueous dispersion, where the dye molecules are suspended in water, rather than in solution. The dye enters the surface of the fibre when it becomes swollen by a rise in the temperature of the dyebath.

19.5.1.1 Direct dyes

Direct dyes have good substantivity for cellulosic fibres, and are relatively easy to apply. They are available in a wide range of hues, but generally lack brightness. Their colour fastness properties are not particularly good, especially in deeper shades. After-treatments are available to improve fastness to washing, but these can have an adverse effect on light fastness and can also change the shade of the original dyeing. Because of these shortcomings, direct dyes are usually confined to the production of cheaper goods where colour fastness is not paramount.

Direct dyes are generally applied to the material at 100 °C with the addition of an electrolyte such as sodium chloride or sodium sulphate to promote exhaustion of the dye to the fibre.

19.5.1.2 Reactive dyes

Reactive dyes are available in a wide range of colours, and are known for their brightness. They form bonds between their reactive groups and cellulose, which gives excellent colour fastness. The basic application procedure is carried out in three phases:

- Exhaustion, where the dye is transferred from the dyebath to the fibre
- Fixation, where the reaction takes place to fix the dye to the fibre
- Post-dye washing, where any excess dye is removed to give acceptable colour fastness

Reactive dyes have differing degrees of reactivity, and this has an influence on the details of the application method, particularly in the temperature of the exhaustion phase. In general, the higher the reactivity and therefore the strike rate of the dye (or the speed at which it exhausts onto the fibre), the lower its application temperature.

Once the dye has exhausted onto the fibre with the addition of an electrolyte such as sodium chloride or sodium sulphate, and migrated evenly, the fixation or reaction is promoted by the addition of an alkali. The alkali used is normally sodium carbonate, sodium hydroxide or a combination of the two.

While the alkaline phase promotes the reaction of the dye with the fibre, it has the side effect of encouraging reaction of the dye with water, or hydrolysis. This is undesirable in that hydrolysed dye is not available to react with the fibre, but still has some substantivity. Therefore not only is the hydrolysed dye wasted, but its presence on the fibre greatly reduces colour fastness.

There is always a degree of hydrolysis with reactive dyeing, and the dyer must minimise this by dyeing at as low a liquor-to-goods ratio as the machinery will allow, the theory being that the less water there is in the system, the lower will be the proportion of hydrolysed dye. The liquor-to-goods ratio is the ratio between the volume of the dyebath and the weight of goods. For example, a dyeing of 50 kg of goods in 1000 L of water has a liquor-to-goods ratio expressed as 20:1.

The final phase is to remove any hydrolysed dye from the fibre, and this is achieved by successive rinsing and soaping processes.

19.5.1.3 Vat dyes

Vat dyes have extremely good fastness properties on cellulose, and are used where fastness to washing and light are paramount, such as for awnings, upholstery, towels and shirting. However, their use is limited by their lack of good reds, their high cost and relative difficulty of application.

Vat dyes are derived from anthraquinone or indigo. Indigo is now manufactured synthetically. It tends to dye the surface of the fibre, lending itself to the typical distressed denim effects that are obtained by subsequent mechanical and chemical actions.

Vat dyes are insoluble pigments with little substantivity for cellulose, and are treated, in a process known as vatting, with a reducing agent under strongly alkaline conditions, to form what is known as the leuco compound. The leuco compound is both water-soluble and substantive. The reducing agent used is sodium hydrosulphite and the alkali used is sodium hydroxide.

Once dyeing is complete, the oxidation necessary for the development of the shade is achieved by washing in cold water or by treatment with an oxidising agent such as hydrogen peroxide. The final process is soaping with a detergent at 100 °C to remove any loose pigment, and is essential to give both the true shade and optimum colour fastness.

Solubilised vat dyes are available in the form of leuco sulphate esters, where essentially the vatting has already been carried out. Although the dyes have lower substantivity, they are useful in the production of pale shades in continuous dyeing where the vatting process is particularly difficult.

19.5.1.4 Sulphur dyes

Sulphur dyes are used widely for the production of dark, dull shades on fabrics such as drills, canvas, corduroys, uniforms, etc., where good wet fastness is required. They can be variable in their light fastness, and their chlorine fastness is poor. They are the dullest of dye classes, but are inexpensive and have worldwide importance.

Sulphur dyes are a type of vat dye, and are applied to the fibre from a substantive leuco compound in the same way as for vat dyes. The vatting is carried out using sodium sulphide rather than sodium hydrosulphite, which has too strong a reducing effect and can lower the colour yield. Sodium sulphide can also provide the alkaline conditions required for vatting, without the need for sodium hydroxide.

After dyeing, a rinsing process before oxidation is important not only to remove loose leuco compound but also to remove excess sulphur to minimise the possibility of later fabric damage. Under warm and humid storage conditions, sulphuric acid can be formed, particularly with black shades, and this reacts with the cellulose to degrade and weaken the fabric in a process known as tendering.

Oxidation to reform the pigment within the fibre is then achieved either through water or by the addition of an oxidising agent such as hydrogen peroxide or sodium perborate. Finally, soaping is carried out to optimise wet fastness by the removal of loose pigment, and to stabilise the shade.

There are soluble sulphur dyes available that have lower substantivity but are useful for paler shades on substrates where vatting and levelness of dyeing are difficult.

Environmental concerns, however, have been raised over the discharge of sodium sulphide into effluent systems, as this can lead to the formation of the foul-smelling and toxic gas, hydrogen sulphide. Steps should be taken to minimise this by oxidising exhausted dyebaths with hydrogen peroxide. Glucose can also be used as the reducing agent in vatting to replace or complement the sodium sulphide. There are low-sulphide ranges of dye available that also help with the problem.

19.5.1.5 Azoic dyes

Unlike other colourants described in this chapter, azoic dyes are actually insoluble pigments formed in the fibre by the reaction of two water-soluble components. Although they have been superseded largely by reactive dyes, azoic dyes have extremely good wet and light fastness, and are still important in the deep red shade area.

The pigment is precipitated in the fibre by a two-stage process involving first the application of a coupling component, which is substantive to cellulose, followed by the application of a diazonium salt. The reaction between these two chemicals produces the coloured pigment. There are several coupling components and diazonium salts available, and their use in various combinations can give a number of shades.

The coupling component is applied under carefully controlled conditions of temperature, liquor-to-goods ratio and sodium chloride concentration in order to regulate the eventual depth of shade.

The fabric is then rinsed to minimise excessive pigment formation on the fibre surface, and therefore to optimise the eventual wet fastness, followed by application of the diazonium salt at room temperature.

The final process is soaping at 100 °C with a detergent to remove surface pigment to optimise wet and rubbing fastness, and to develop the final shade.

19.5.2 PROTEIN FIBRES

In an industry where the volume of fabrics produced worldwide is dominated by cellulosic and synthetic fibres, animal protein fibres still have an important place in the higher quality section of the textile and fashion market.

Animal hair fibres, which include wool, cashmere, mohair and camel, llama, alpaca, vicuna and rabbit hair, are made up of complex proteins. These fibres have generally similar dyeing properties, and can be dyed with the same classes of dyestuff.

Silk is also a protein fibre, consisting of filaments of fibroin bound together with sericin. Although chemically less complex than the hair fibres, it is dyed with the same dye classes.

Regenerated protein fibres such as casein (from milk), and from sources such as soya beans and groundnuts can, with care, also be dyed with classes appropriate to natural protein fibres.

19.5.2.1 Acid dyes

Acid dyes, named for their application under acid conditions, are reasonably easy to apply, have a wide range of colours and, depending on dye selection, can have good colour fastness properties. The dyes are divided into three categories according to their levelling and fastness properties:

- Levelling (also known as equalising) dyes
- Milling dyes
- Super-milling dyes

Levelling-acid dyes have good levelling properties and are applied from a bath containing sulphuric acid to achieve exhaustion. Because of the ease of migration of dye molecules into and out of the fibre, levelling-acid dyes have poor fastness to washing, and are normally used for pale, bright shades where fastness is not paramount.

Milling-acid dyes have a greater substantivity for the fibre than levelling dyes, and therefore have poorer levelling properties. These dyes have better colour fastness properties than levelling-acid dyes, particularly under wet conditions. This is particularly important in the case of wool, where resistance of the shade to a subsequent fabric modification process known as alkaline milling may be required.

Super-milling acid, or neutral dyeing, dyes are applied in a similar way to milling-acid dyes, except that greater control over the strike rate of the dye is exercised. This is necessary to promote even application of the dye, and is achieved by the addition of levelling agents, the control of pH and the rate of rise of the dyebath temperature.

Super-milling dyes give very good fastness, and with an appropriate after-treatment can satisfy requirements for shades of medium depth, especially where reasonable brightness is needed.

There are considerable differences in the properties and application methods within the whole range of acid dyes available, and the dyer must ensure that the dyes chosen in combination are from the same group and have very similar properties.

19.5.2.2 Metal-complex dyes

Metal-complex, or pre-metallised, dyes are essentially acid dyes where a metal atom has been incorporated in the dye molecule during manufacture.

There are two classes of metal-complex dye, known as 1:1 and 1:2 dyes. In the case of 1:1 dyes, the ratio of metal atoms (usually chromium) to dye molecules is one to one, while for 1:2 dyes the ratio is one metal atom to two dye molecules.

For dyeing purposes, the application method for 1:1 dyes is the same as for levelling-acid dyes, whereas 1:2 dyes are applied in the same way as for super-milling acid dyes. With careful selection it is possible to mix acid and metal-complex dyes in the same recipe.

Both classes of metal-complex dye have very good fastness properties, usually superior to their nonmetal acid dye counterparts. However, they are less bright than acid dyes, and their use is limited to duller tones, although relatively deep shades are possible.

19.5.2.3 Mordant dyes

These dyes are acid dyes that are capable of reacting with chromium, resulting in a dyeing of excellent wet fastness, and are often referred to as chrome dyes. The shade range is dull, and their use is limited to colours such as black, navy and maroon, where the depth of shade achieved is unsurpassed and wet fastness is excellent.

There are three traditional methods of application:

- Prechrome, where the mordant is applied before dyeing
- Metachrome, where the dye and mordant are applied simultaneously
- Afterchrome, where the mordant is applied after the dye. Today, afterchroming is the process used almost exclusively.

Mordant dyes are applied as acid-milling dyes, and after migration and exhaustion of the dye, sodium or potassium dichromate is added to form the dye–metal complex. There is usually a complete change of colour from the original dyeing to the final shade following the addition of chromium.

There are serious environmental concerns over the introduction of residual toxic chromium from the dyeing process into the effluent. The dyer can take steps to minimise this, and any residual chromium on the dyed material, but there is pressure to use alternatives such as reactive dyes.

19.5.2.4 Reactive dyes

As their name suggests, these dyes contain chemical groups within the dye molecule capable of reacting with the fibre, and the resultant wet fastness is excellent. This range of dyes has superior brightness, especially in the red area. Although expensive, reactive dyes are often found necessary for superwash wool materials, where the shrink-resisting process lowers wet fastness and the nature of the product calls for exceptional dye fastness.

The application method for reactive dyes is similar to that for super-milling acid dyes. Although there is relatively little hydrolysis of the dye (as with cellulosic reactive dyeing, where the dye has a tendency to react with the water in the dyebath), dark shades in particular need an after-treatment to remove any unfixed dye to optimise wet fastness. This is achieved by an alkaline wash, usually using sodium carbonate. For particularly difficult shades, a further detergent wash may be necessary.

19.5.3 POLYAMIDE FIBRES

Synthetic polyamide fibres are chain polymers in which the recurring group forming the chain is an amide – hence their name. Although there are other fibres, the group is dominated by nylon 6 and nylon 6.6, and the methods of application outlined below relate to these.

Nylon fibres are susceptible to chemical and physical variations in their manufacture, which can lead to problems in dyeing. A significant problem is known as barré, where physical and chemical variations along the length of the filament give rise to colour variation on dyeing, which is visible as stripes in knitted and woven fabrics. Barré is a common problem which the dyer can minimise through dye and process selection.

Nylon fabrics often undergo a heat or steam setting process to add stability. For both nylon 6 and 6.6, setting before dyeing decreases the dye uptake, and setting after dye can cause yellowing of the shade. The dyer selects dyes and adjusts the shade accordingly.

The molecular structure of nylon 6.6 is more crystalline, or regularly aligned, than that of nylon 6, and therefore it is more difficult for the dye to penetrate the fibre. However, the colour fastness achieved is better than for the more amorphous nylon 6.

There are relatively few impurities on the fibre, and preparation for dyeing is normally limited to a detergent scour to remove spinning and knitting lubricants or weaving sizes.

Technically, nylon fibres can be dyed with direct, reactive, chrome, vat, sulphur and azoic dyes with various degrees of success. However, the commonest dye classes used are disperse, acid and metal-complex.

19.5.3.1 Disperse dyes

Insoluble disperse dyes are applied to the fibre from a dispersion, where the dye molecules are in suspension in the dyebath, rather than in solution. Their levelling properties are excellent, which is an advantage for overcoming the barré effect, but their fastness properties are poor. They are therefore confined to pale shades for articles such as ladies' hosiery and lingerie.

Disperse dyes differ widely in their molecular size and exhaustion rates. Dyeing is normally carried out at 100 °C. For those dyes of slower exhaustion, it is an advantage to dye at temperatures of 105–110 °C for nylon 6 or 115–120 °C for nylon 6.6 if the appropriate high-temperature machinery is available.

The different dyeing properties present within this class of dyes make it necessary for the dyer to choose combinations where the dyes are compatible.

19.5.3.2 Acid dyes

Acid dyes are widely used for nylon, and are noted for their good shade range and fastness properties, especially if a suitable after-treatment is applied.

Those dyes with low to medium substantivity give better barré coverage, but need relatively strong acid conditions for application, which could cause potential fibre damage.

Dyes of low and medium substantivity give poorer wet fastness, and an after-treatment with a proprietary agent known as a syntan (the term is derived from 'synthetic back tan') will probably be necessary.

Dyes of higher substantivity give better fastness, often without an after-treatment. Coverage of barré is poor, and levelling agents must be used to minimise this effect.

Dyes of low substantivity are typically applied by pretreating the material with an anionic levelling agent before dyeing. Anionic levelling agents have a negative ionic charge, as do acid dyes.

Dyes of higher substantivity, and where barré is expected, can be pretreated with an anionic levelling agent at 100 °C to exhaust it fully onto the fibre. The bath is then cooled and a cationic levelling agent added, followed by the dyeing process. Cationic levelling agents have a positive ionic charge, which is opposite to the anionic charge of the acid dye.

Anionic levelling agents compete with the dye for access to the fibre, thus reducing the rate of exhaustion. Cationic levelling agents react to form a chemical complex with the dye, the complex breaking down with increasing dyebath temperature to release the dye in a controlled manner.

The traditional after-treatment to improve wet fastness on nylon involves the application of tannic acid and tartar emetic, known as back tanning. Although very effective, it has the disadvantages of (1) being costly, (2) being environmentally questionable and (3) causing a significant change of shade.

Syntans are used widely to give good fastness results with minimal shade change. Products are also available to improve fastness to chlorine for material intended for swimwear.

Dyes must be chosen carefully to ensure their compatibility in combination.

19.5.3.3 Metal-complex dyes

Metal-complex dyes are duller in their shade range than acid dyes, but give relatively good colour fastness. However, in common with acid dyes, they exhibit problems with levelling and coverage of barré. The 1:1 metal-complex dyes give better coverage, but need acid conditions that can cause damage by hydrolysis to the fibre, where the amide is attacked and the fibre weakened. The 1:2 metal-complex dyes are more commonly used, and are applied as the more substantive acid dyes, as outlined above.

Again, dye selection for compatibility is imperative. For wool, with careful selection it is possible to mix acid and metal-complex dyes in the same recipe.

19.5.4 POLYESTER FIBRES

Polyester fibres are chain polymers where the linking group between the molecules is an ester – hence their name. There are several types of polyester fibre each with significantly different properties, and their processing cannot be included in a single description. Therefore the dye application methods for polyethylene terephthalate (PET) will be described, as it is the dominant fibre in this class.

Polyester materials in yarn or fabric form have a tendency to shrink unless stabilised, and heat setting at 200–225 °C is normally carried out to prevent unwanted shrinkage during the dyeing process.

PET is hydrophobic and crystalline and lacking in groups to which dyes may attach; hence disperse dyes are used almost exclusively. Although it is technically possible to use vat and azoic dyes, their use is limited.

Polyester fibres are relatively clean before wet processing, and a scour with detergent and sodium carbonate is usually sufficient to remove knitting and weaving lubricants if necessary.

19.5.4.1 Disperse dyes

Disperse dyes are the preferred practical option for dyeing PET, and have a good shade range. They have good wet fastness in pale to medium shades without after-treatment, but medium to dark shades usually require a process known as reduction clearing to optimise fastness.

PET's crystallinity makes it difficult for the dye to penetrate the fibre, and the rate of dyeing is very slow. If atmospheric dyeing machinery alone is available, pale shades are possible by dyeing at 100 °C, but the rate of dyeing usually makes this uneconomic and impractical.

Dyeing at 100 °C is aided by the use of agents known as carriers, which allow better penetration of the dye. Although effective in producing shades of good depth, carriers have disadvantages in their odour and toxicity, which has caused their use to decline, especially as machinery suitable for high-temperature dyeing has been developed.

Dyeing at temperatures up to 130 °C without a carrier has many advantages. The rate of dyeing is increased, and better penetration of the dye allows the use of larger-molecule dyes, which have better fastness but are difficult to apply at 100 °C. The better migration also gives improved coverage of variations in the material.

To optimise colour fastness where necessary, a process known as reduction clearing may be given after dyeing. This process uses sodium hydroxide and sodium hydrosulphite, and removes loose dye from the fibre.

19.5.5 ACRYLIC FIBRES

Acrylic fibres are based on polyacrilonitrile (PAN), the pure form of which is extremely crystalline and difficult to dye. Therefore other monomers are incorporated into the polymer to reduce the crystallinity. A fibre can be termed acrylic provided it contains at least 85% PAN.

Modified acrylic fibres (modacrylics) are produced to give specific properties (such as handle and dyeability), and must contain 35–85% PAN. Their dyeing properties are similar to those of acrylic fibres.

Acrylic fibres are almost exclusively dyed with basic (cationic) dyes, although disperse dyes may be used for very pale shades.

The fibre is prescoured with a neutrally charged nonionic detergent, as negatively charged anionic or positively charged cationic agents can affect adversely the basic dyeing process. (Basic dyes are cationic in nature, and the presence of anionic products causes undesirable chemical reaction with the dye, while cationic products compete with the dye and affect the rate of exhaustion of the dye on to the fibre.) Some acrylic fibres are inherently yellowish, and may need to be bleached before dyeing.

Acrylic fibres become plasticised in hot water, and since dyeing using both disperse and basic dyes is carried out at 100 °C, careful cooling of the dyebath is necessary to prevent irreversible creasing and distortion of the material.

19.5.5.1 Disperse dyes

Disperse dyes have limited build-up and wet fastness properties, and although they give level dyeing, are limited to pale shades.

Dyeing is carried out at 100 °C, after which the dyebath should be cooled slowly to prevent creasing and distortion, especially of knitted material.

19.5.5.2 Basic (cationic) dyes

Basic, or cationic, dyes have a comprehensive shade range and give excellent wet fastness on acrylic fibres.

Basic dyes have a particularly rapid strike rate over a short temperature range, and precautions must be taken to ensure level dyeing.

Acrylic fibres have a glass transition temperature of 80–90 °C, where the molecules within the fibre become less rigid and more mobile. Below this temperature the rate of dyeing is very slow, and above which it is rapid. Positively charged cationic retarding agents, which compete with the similarly charged cationic dyestuff for access to the fibre, are commonly used, along with careful temperature control over the critical range of rapid dye strike.

Sodium sulphate and acetic acid are also used to control the rate of dye uptake. Again, care must be taken with cooling the dyebath after dyeing to prevent distortion of the material.

Basic dyes give very good exhaustion. Therefore little loose dye remains on the fibre surface, and the resultant dyeing has excellent wet fastness without the need for after-treatments.

19.5.6 FIBRE BLENDS

Fibres are blended together for several purposes. For example, expensive and cheaper fibres can be blended for economy, while fragile and more robust fibres can be blended to give greater durability and performance.

Examples include the addition of:

- Acrylic to wool for economy.
- An elastomeric fibre to nylon for stretch in swimwear.
- Polyester to cotton for all-round performance.

The dyeing of blends can be carried out in fibre, yarn or fabric form. If a solid shade is required, the colour matching of the fibre components or yarns of single-fibre composition, before spinning or fabric construction, respectively, needs to be extremely accurate. This can be complicated by the fact that different dye-houses may be involved in the speciality dyeing of each component.

The dyeing of solid shades in multi-component yarns or fabrics presents a greater challenge. Dyes must be selected to give the appropriate shade on each component fibre with the minimum of cross-staining onto adjacent fibres and with the maximum colour fastness for each component. The processing conditions for each fibre must be addressed, and the effects on the most vulnerable fibre considered. The problem of degradation of one of the components can limit the use of certain delicate fibres in blends to be dyed. In their case, it is better to dye the components separately in fibre or yarn form and to construct the fabric from these materials.

The blending of fibres to be dyed in yarn and especially fabric form gives options to dye the fibres to different shades to produce various effects. These include:

- Dyeing one fibre while keeping the other fibre as white as possible.
- Dyeing the fibres to the same hue but to different depths.
- Dyeing the fibres to different shades to give a contrast effect.

In these cases, the fibres are chosen for their different dyeing properties, so that they may be dyed to achieve the desired effect.

Dye selection is critical for the dyeing of all blends. The dyer must be aware of the dyeing and processing properties of the fibres involved, and choose the dyes and their application methods to achieve the desired effect with the minimum of damage to the substrate and with optimum colour fastness.

19.5.7 FLUORESCENT BRIGHTENING AGENTS

The production of white shades is dependent on bleaching with a method appropriate to the fibre to reduce natural discolouration, and the application of a fluorescent brightening agent (FBA).

Most fibres have a naturally yellowish-green appearance, as they tend to absorb blue light, and have lower reflectance in the 400–500 nm wavelength region, which is the blue area of visible light. FBAs absorb ultraviolet radiation with wavelengths of 300–400 nm in the non-visible area and re-emit a proportion of this radiation at wavelengths of 400–500 nm in the visible violet-blue area. Therefore the fibre now has apparently more visible light being emitted than was absorbed, and appears brighter.

FBAs give excellent results with many fibres, notably with cotton, polyamide and polyester, but have limitations in effectiveness and light fastness with natural protein fibres such as wool and silk.

19.6 STRENGTHS AND WEAKNESSES OF NATURAL AND SYNTHETIC DYES

The emphasis today on sustainability and environmental awareness means that the question of why natural dyes play such a small part in the industry is often raised, with the suggestion being that it might be more environmentally friendly and sustainable to use natural dyes instead of synthetic dyes. However, there are a number of reasons why this is not a straightforward solution.

19.6.1 SAFETY

The term 'natural' and 'safe' are not synonymous; there are many naturally occurring substances, e.g. arsenic and asbestos, which are harmful. Natural dyes are derived from plant, animal or mineral sources, and while some dyes are quite safe to use, some are hazardous. For example, indigo and logwood are skin and respiratory irritants, and plants such as lily of the valley and bloodroot are toxic.

As with synthetic dyes, the safety of natural dyes must be considered within the criteria for banned and restricted substances. Extraction of dyes from natural sources can involve the use of hazardous chemicals such as sulphuric acid and sodium hydroxide. Many dyes need the use of a mordant to allow attachment to the fibre, and some mordants, e.g. tin, copper, chromium, etc., are also considered toxic.

Synthetic dyes are also under scrutiny for their safety, and the larger dyestuff manufacturers have removed contentious dyes from their ranges.

19.6.2 SHADE RANGE AND REPRODUCIBILITY

The shades available from natural dyes are intrinsically appealing, but the range is limited, even though different shades are available from the same dye, depending on the mordant and application methods employed.

The shade of any natural dye can be reproduced by combinations of synthetic dyes, but the reverse is not true. Also, synthetic fibres in general cannot be dyed satisfactorily with natural dyes. Therefore natural dyes cannot provide the complete range of colours expected by the consumer.

As with any natural product, the quality of natural dyes varies, and again the consumer's expectations in terms of shade variation may not be met. Synthetic dyes are produced to shade tolerances that lead to acceptable commercial results.

19.6.3 COLOUR FASTNESS

Colour fastness of natural dyes to washing and light is in general inferior to well-selected and applied synthetic dyes, and normally does not meet consumer demands. Some after-treatments applied to synthetic dyes can be environmentally unsound, and must be avoided or minimised to ensure safety of application and the final product.

19.6.4 AVAILABILITY OF NATURAL DYES

Natural dyes have limited availability. They are subject to the growing seasons of the plants or the life-cycles of the insects from which they are derived. Yields are low in relation to the ground used, and there may be competition with foodstuff production. There is just not enough natural dye production to support commercial demands on a large scale. Synthetic dyes, however, are produced continually on demand.

Natural dyes are used in small-scale craft industries, where the quantity of dyes required is low. Because of the limitations of producing these dyes on a commercial scale, regardless of any technical restraints, they will not replace synthetic dyes for the global mass production of dyed textiles.

19.6.5 THE WAY FORWARD

Although natural dyes have disadvantages in terms of large-scale textile production, they do have certain positive attributes. With careful dye selection and processing methods they are seen as contributing to an environmentally acceptable natural product. Although they take up land and other resources in growing and consume energy when being extracted, synthetic dyes have significant manufacturing demands of their own. It can also be argued that natural dyes have a degree of sustainability, as some may be obtained from renewable natural resources.

Natural dyes will not replace synthetic dyes for the foreseeable future but, with careful selection, they do have a place in the small-scale production of specialist articles where a marketing opportunity based on sustainability can be taken, and higher prices can be demanded.

19.7 ENSURING QUALITY AND EFFECTIVENESS OF DYEING

Dyers have responsibility for ensuring that the results of their processing meet certain minimum standards, e.g. shade, colour fastness and general substrate quality. Process selection and control is vital, but it is also incumbent on the dyer to carry out assessments of the customer's quality criteria for the particular product.

19.7.1 ASSESSMENT OF SHADE

The traditional method of shade assessment is to view a sample of the dyed material and the standard in a light cabinet under the relevant illuminants. The introduction of the light cabinet was a step forward in the standardisation of assessment in various locations and environments, but this type of visual appraisal has many flaws.

Variation in assessors' colour vision, fatigue, siting of the cabinet in relation to other light sources, and standards soiled from frequent handling are factors that combine to cast doubt on the results of visual judgement alone.

The preferred modern practice is to employ instrumental colour difference assessment, which, while certain procedures are necessary for standard operation, overcomes the problems of visual assessment.

The dyed material is assessed against the spectral data of the standard, and the colour difference is given according to the customer's preferred programme. Typically, the results are expressed numerically under store light, artificial daylight and tungsten. A pass or fail is recorded dependent on the differences between the standard and the sample in terms of hue, chroma and brightness and an overall ΔE (delta E) value, the tolerances having been determined by the customer.

Although there are variations based on the measurement system used and the customer's specification, in general the numerical tolerances are 0.6 for hue, 0.8 for chroma and brightness, and 1.0 for ΔE.

The use of instrumental assessment is a necessary part of the dyer's assurance system, especially with the supply base and the customer often being in widely different locations.

19.7.2 ASSESSMENT OF COLOUR FASTNESS

Colour fastness is the dyer's responsibility, and the dye selection and efficiency of the processing method must be assessed after dyeing. The customer will have determined the test methods and the acceptable grading results. There are worldwide standards for specific test requirements, and most retailers use these as the basis of their regime, while some have developed their own variations.

Ideally the dyer will have an accredited testing laboratory with personnel trained in the relevant tests. Failing this, an independent test house should be employed.

19.7.3 ASSESSMENT OF OVERALL SUBSTRATE QUALITY

The dyer must also assess the aesthetic attributes of the material, wherever it falls within the product chain. For those materials close to the finished article, assessments of the levelness of dyeing and the handle and surface appearance of the fabric are paramount. For those materials that require further processing, e.g. loose fibre and yarn, the lubrication requirements for the smooth running of the spinning or knitting and weaving processes must be considered.

Many dyeing processes are necessarily demanding of the substrate in terms of chemical or mechanical action, and the dyer must ensure that the selection of dyes and process conditions minimise any damage. The dyer must inspect and test the dyed material to ensure that the product's physical properties – for example, dimensional stability, strength, pilling resistance, etc. – are compliant with requirements.

Finally, it may be necessary to test the material for residual chemical content. Many retailers have Restricted Substances Lists with which the dyer must comply, and the complexity of this testing usually requires the use of a specialist laboratory.

19.8 ENVIRONMENTAL IMPACT OF DYEING

The dyeing industry has a justifiably poor reputation as a polluter of the environment and as a wasteful consumer of resources. In a welcome climate of tighter legislation, external pressure and internal awareness and intent, the situation is being addressed. There are steps that responsible dyers can take to minimise their impact on the environment, through process selection, modification and control.

The key environmental aspects that the dyer must consider are:

- The use of global resources such as water and fuel
- The quality of the air, water and land in the immediate surroundings of the operation
- The safety of the working environment within the operation
- The safety of the dyed product in terms of residual chemicals from the dyeing process

19.8.1 WATER CONSUMPTION

Water is both a precious natural resource and a necessary medium for most forms of dyeing. While accepting that water must be used, it is necessary for the dyer to implement ways of reducing water consumption.

Appropriate dyes of the highest substantivity should be chosen, as they can require less rinsing after dye. Processes can be modified to reduce the number of baths used for preparation, dyeing and after-treatment of the material, and machinery that allows the lowest workable liquor-to-goods ratio can be selected. The wet finishing process can be automated so that optimum fill levels are used throughout, and so that there is no wastage through human interference. Water employed for external cooling of machinery can be re-used, and water from final processing baths can be treated to remove impurities and recycled. Leaks in the water supply system should be repaired as they occur. Saving water is not only an environmental necessity; incoming water represents a cost to the dyer, and any reduction in consumption has an economic incentive.

19.8.2 ENERGY CONSUMPTION

Most industrial energy comes from fossil fuels, and this will be the case until alternative supplies such as solar, nuclear, wind, etc., are readily available at an attractive cost. Therefore it is incumbent on the industry to use its energy efficiently and thereby reduce the consumption of nonrenewable resources and the production of greenhouse gases.

The dyer uses energy mainly for raising steam to provide the heat necessary for processing, and power for running machinery, as well as lighting and heating of the premises. Efficient use of boilers for raising steam can be achieved by regular maintenance, and by the choice of fuel that gives the highest calorific value.

Re-use of heat from hot water used in processing, heat captured from the exhaust systems of operations involving high temperatures and from boiler chimneys can also assist in reducing the energy demand, as can insulating hot water and steam delivery systems.

Automating the dyeing and related processes can also help improve energy efficiency.

19.8.3 AIR EMISSIONS

Air emissions include oxides of sulphur and nitrogen from boilers and fumes from stenters (machines for the continuous heat treatment of fabrics) and dye application processes. Emissions must be quantified, the sources identified, and action taken to eliminate pollutants, or to reduce them to levels that have been set by local authorities. Maintenance of equipment, installation of adequate ducting and choice of chemicals are factors to be addressed.

19.8.4 EFFLUENT EMISSIONS

Depending on the location of the operation, the waste water produced by dyeing and finishing processes may be discharged to municipal facilities for treatment, or to watercourses such as streams and rivers, in which case the dyer needs to carry out treatment before release. The requirements covering the effluent content will vary with local authorities, but generally there will be focus on:

- Volume
- Biological oxygen demand (BOD)

- Chemical oxygen demand (COD)
- Suspended solids
- pH
- Heavy metals
- Pesticides
- Sulphates
- Ammoniacal nitrogen
- Colour

Each authority sets limits for these features, and usually has the power to monitor effluent and to impose fines for noncompliance, or to suspend or even close down the operation. Chemicals, dyes and process methods can be chosen to minimise critical effluent constituents, followed by complex treatment processes to ensure compliance before final discharge.

19.8.5 OCCUPATIONAL SAFETY

Dyeing is an inherently dangerous operation, involving hazards such as boiling water, handling of dyes and chemicals, and machinery with moving parts. Personal protection appropriate to the process must be provided by the dyer for the machine operators, and includes footwear, aprons, gloves and eye protection.

Dyes and chemicals may be hazardous in terms of inhalation and skin contact. Section 8 of the Material Safety Data Sheet (MSDS) for each product gives information on safe handling and the minimum protection required, and provides the basis for allocation of the necessary equipment. Figure 19.8 shows all 16 sections covered by the standard MSDS.

Generally, respirators or particle masks are necessary in addition to the equipment required by the machine operators. Automated weighing and dispensing systems are available that minimise the handling of dyes and chemicals, and therefore reduce the dangers of exposure.

In addition to ensuring a safe environment in terms of workplace layout, machine maintenance, dye and chemical storage, etc., the responsible dyer provides full and ongoing training in safe procedures for machine operation and materials handling.

19.8.6 SAFETY OF DYED PRODUCTS

The dyer must ensure that the final product falls within strict tolerances for chemical residues that may be harmful to the consumer. These residues may be prohibited by law, e.g. carcinogens; regulated by law, e.g. formaldehyde and heavy metals; or possibly harmful but not yet regulated, e.g. organo-tin.

Many retailers have introduced lists of harmful substances that are restricted on their goods, and the dyer must screen dyes and chemicals to comply with these requirements.

Major dye and auxiliary chemical manufacturers have addressed these concerns, and give excellent advice on the content of their products. The Oeko-Tex Standard 100 can be applied to dyed products, and provides independent and comprehensive analysis of residual substances, followed by certification. Details of the scheme can be found on the Oeko-Tex Website.

1. Identification – manufacturer details, chemical name
2. Composition/information on ingredients – list of any hazardous components
3. Hazard(s) identification – effects of exposure by eye, skin, inhalation and ingestion
4. First-aid measures – treatment for each route of exposure
5. Fire-fighting measures – flammability classification, extinguishing method, protection needed
6. Accidental release measures – containment and disposal of spillages
7. Handling and storage – handling practices, storage temperature, shelf life
8. Exposure controls/personal protection – ventilation required, personal equipment needed for each route of exposure
9. Physical and chemical properties – appearance, odour, normal physical state, boiling/melting/freezing points, specific gravity, solubility, pH
10. Stability and reactivity – stability of product, incompatibility with other chemicals
11. Toxicological information – toxicity data, epidemiology studies, carcinogenicity, neurological effects, genetic/reproductive effects
12. Ecological information – toxicity, environmental fate, physical/chemical data
13. Disposal considerations – waste disposal method
14. Transport information – shipping regulations, packaging group, hazard class, identification number
15. Regulatory information – local and international information on hazard ratings
16. Other information – supplier number, supplier release date

FIGURE 19.8

The 16 sections of the standard Material Safety Data Sheet, and the areas that they cover.

19.9 RESEARCH AND FUTURE TRENDS

The geographical shift of the traditional western supply base, mainly to the Far East, has led to reductions in retail prices for consumers, who at the same time are becoming more discerning over product quality, including colour fastness, ease of aftercare and product longevity. The consumer also shares, with legislative bodies, increasing concern over environmental issues. The dyer therefore has the challenge of maintaining quality within a given budget against a background of environmental requirements, including water and energy use, effluent disposal, occupational conditions and garment safety.

Future focus needs to be on the best use of available technology within the application process to ensure that the demands of product performance and safety are met, along with the implementation of appropriate treatments to ensure environmental compliance.

The best way forward is for all parties within the supply chain to work together closely, and input from retailers, designers, dyestuff and auxiliary manufacturers, environmental agencies, academic institutions and dyers is vital to achieve the optimum product in the safest manner.

Environmental considerations must include the best use of technology to reduce water and energy consumption through processing efficiency.

19.10 SUMMARY

- Colour is not an intrinsic part of an object, but a perception created in the brain.
- The criteria for selection of dyes are achievement of the desired shade, metamerism, colour fastness and environmental considerations.
- Dyes are specific to fibre groups, and are not applied universally.
- Instrumental match prediction and shade assessment are essential to modern dyeing.
- Machinery has been developed for the dyeing of all forms of textile substrate.
- Natural dyes are unlikely to challenge synthetic dyes on an industrial scale in the foreseeable future.
- Dyeing faces many challenges from environmental issues.

19.11 CASE STUDY: REACTIVE DYEING OF KNITTED COTTON GARMENTS

As discussed in this chapter, a key issue for the dyer is the selection of appropriate dyes and application methods for the substrate to be dyed. This case study explores in further depth the criteria and thought process involved.

The dyeing of garments in a practically complete form gives the advantage of leaving the choice of colour until late in the manufacturing process (which allows the retailer to assess market trends before making a decision on shade), but presents the dyer with specific challenges. The dyeing of garments in this form has been assisted by the development of knitting machinery that produces a garment with collars, plackets, etc., as an integral part of the garment, rather than having to be attached after knitting. Although a slow process, this does give a product ideal for garment dyeing.

As with all dyeing, the dyer must first address the nature of the material to be dyed. We will assume that the jumpers have been knitted from 100% ecru (natural colour) cotton yarn, with ribbed welts, cuffs and collars, and that all seaming and linking processes used to join the garment panels together have been carried out in 100% cotton. The shade required is a medium bright red, and the colour fastness and colour matching must meet High Street standards.

19.11.1 SELECTION OF DYES

The dyer first considers the classes of dye available for cotton dyeing.

- Direct dyes will not give the required fastness.
- Vat dyes are difficult to apply, and the range lacks good reds.

- Sulphur dyes are also weak in reds and will not give the required brightness.
- Azoic dyes, although strong in the red area, can be variable in their outcome.
- Reactive dyes will give the required shade with good colour fastness and shade reproducibility, and can be controlled to give a dyeing of the required overall quality, making them the best choice.

The next issue for the dyer to consider is that of the cuffs, collars, welts and seams of the garment, which are difficult areas for the dye to penetrate. Therefore the dyer must select dyes where the strike rate can be controlled, and which have good migration properties.

Reactive dyes differ in their degree of reactivity depending on their reactive groups. Dyes with highly reactive groups can be applied at temperatures of 30–40 °C, dyes with groups of medium reactivity are applied at 40–60 °C, while dyes with groups of low reactivity are applied at 80–95 °C, and are known as hot dyeing dyes.

To achieve satisfactory penetration of the critical areas of the garment, the extra diffusion of the dye into the fibre achieved at higher temperatures, combined with lower reactivity, make the hot dyeing dyes ideal for this particular purpose.

An appropriate dye recipe may then be formulated using an instrumental match prediction system.

19.11.2 SELECTION OF PROCESS METHOD

Ideal dyes for garment dyeing are from the Procion HEXL ranges from Dystar. These dyes will give the required shade and colour fastness, and the supplier assures their ecological safety in terms of dye manufacture, application and residues on the garment.

Figure 19.9 shows the profile of the method allowing the best migration of the dye, followed by alkali addition to effect the reaction between the dye and the fibre (courtesy of Dystar). Incidentally, dyers of all fibre types, employing all classes of dye, use profiles such as this schematic to plan and control their processing.

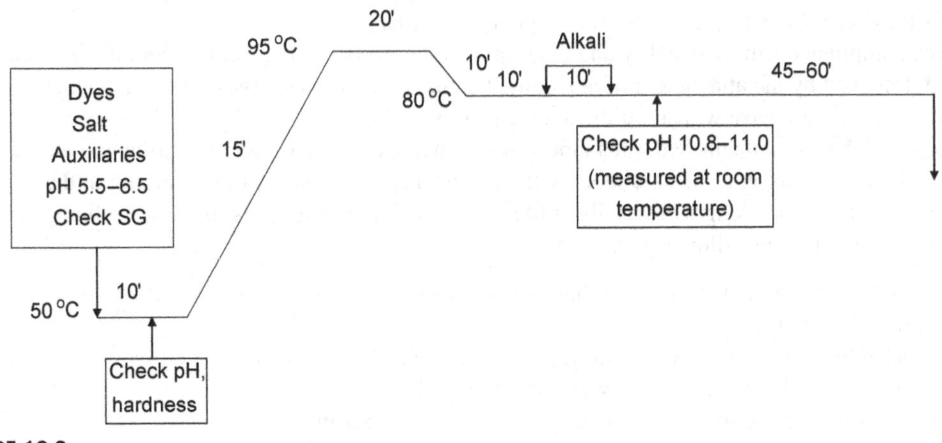

FIGURE 19.9

Special yarn migration process, bulk dyeing procedure.

Source: Reference: Procion (2006).

Prior to dyeing, the goods must be scoured to remove previously applied spinning and knitting lubricants, and bleached to give a base suitable for a bright shade. Hydrogen peroxide has superseded sodium hypochlorite as the preferred bleaching agent on environmental and product quality grounds.

Following the dyeing process, a series of rinses, including a 95 °C soaping bath, will be necessary to remove any hydrolysed dye to achieve the required colour fastness.

The final stage is to apply a softening agent as appropriate. Softeners range from basic handle modification agents to make the product attractive at point of sale, to complex silicones that have a degree of durability within the lifetime of the article. The type and amount of softener applied depends on the agreement with the retailer, bearing in mind that many consumers routinely add a softener to the domestic laundering cycle.

19.11.3 SELECTION OF MACHINERY

The garment dyer has two basic types of machines available. Traditional side-paddle and overhead-paddle machines operate at liquor-to-goods ratios of 30:1 to 40:1, while rotary drum machines can be operated at ratios of around 15:1. Bearing in mind the undesirable hydrolysis of reactive dyes in water during the alkali phase, the dyer will opt for the rotary drum machine to minimise this. This demonstrates the need for the choice of appropriate machinery for the article being dyed. Had the dyer been processing wool or acrylic garments, then the paddle machine would probably have been chosen for its gentler action.

19.11.4 POST-DYE OPERATIONS

On completion of dyeing, the garments are subjected to a hydro-extraction (spin drying) process to remove excess water, followed by tumble drying.

Quality checks are then carried out on shade, colour fastness, handle and garment dimensions, all according to the customer's specifications. Any checks required to confirm compliance with the customer's Restricted Substances List must also be made at this stage.

Once compliance with the quality and environmental criteria has been confirmed, the garments are pressed, followed by the attachment of any embellishments such as corsages, sequins, leather patches, etc., which would not have withstood the dyeing process.

The care label is then attached, plus labels indicating size, retailer's identification and any promotional material as required. The care label will have been agreed upon with the retailer at the garment development stage, and along with details of fibre composition, contains instructions for home laundering, usually covering the following aspects:

- Type of wash recommended – either hand or machine.
- Washing temperature.
- Dry cleaning – whether the article may, may not or must be dry cleaned.
- Bleaching – whether the article may or must not be bleached.
- Ironing – whether the article may be ironed, and at which setting.
- Tumble drying – whether the article may or must not be tumble dried.
- Other useful information such as 'dry flat', 'softener not recommended', 'wash separately', etc.

If the garments are to be packed in polythene bags, these must be free from butylated hydroxyl toluene (BHT). This is often used in the manufacture of polythene, but causes yellowing of pale shades during storage. Garments manufactured in, for example, China, for retail in Europe, may be stored for several weeks during the shipping and warehousing stages, and are particularly at risk from this problem.

Depending on the customer's requirements, the garments may then be passed through a metal detector to check for the presence of fragments of knitting or sewing needles that may have been broken during manufacture. For children's wear, this will be a definite requirement.

Finally the garments are either packed flat into metal-free cardboard cartons, or hung individually on rails, before being placed in a 20- or 40-foot long metal container for shipment. Although the flat-packing method makes better use of the space within the container, the garments are prone to creasing, and may need to be pressed in the country of destination before being sent to the retail outlet.

Air shipment is quicker, but should be avoided as it demands considerable cost in both monetary and environmental terms.

19.12 PROJECT IDEAS

These suggestions are intended to embrace the sections of the textile industry that have the most significant input into commercial dyeing and finishing. These are the dye and chemical manufacturers, the retailers and the dyers and finishers themselves.

For success in offering the consumer an innovative product of the required quality, at affordable cost and with acceptable environmental impact, these three sectors need to work closely together.

The suggestions for these projects involve contact with the technical departments of members of all three sectors, which will give the views of experienced textile professionals on technical and environmental issues related to dyeing.

The suggested started point, will be establishing contact with dye manufacturers, who will offer advice on contacts for the retailers and dyers with whom they work closely on product development and environmental issues. Although this list is by no means exhaustive, visit the Websites of Dystar, Archroma, Huntsman and CHT/Bezema for current contact details within their technical departments.

These investigations will not only provide very useful information on the nature of practices related to dyeing, but also an insight into the relationships between the different sectors of the industry.

19.12.1 PROCESS CONTROL TO REDUCE THE ENVIRONMENTAL IMPACT OF DYEING

The major dye manufacturers have invested in research to modify dye and chemical application processes to reduce water, energy and chemical consumption.

Contact at least two major dye manufacturers and obtain answers to the following questions. Collate the information gained to assess the seriousness with which the industry is viewing and addressing the environmental impact of dyeing.

- Which textile processes do you consider to be most wasteful in terms of water, energy and chemical resources?
- What steps have you taken to reduce the use of these resources through modification of the process profile and the introduction of new dyes and chemicals?

- How important are relationships between yourselves, retailers and processors in reducing the environmental impact of dyeing and finishing?
- What future plans do you have for either extending or reducing your ranges of dyes and chemicals based on their relevance to process control modification?

19.12.2 RESTRICTED SUBSTANCES LIST

Many retailers have issued their own Restricted Substances List (RSL) to which their suppliers must comply. Contact at least two major retailers to obtain answers to the following questions. Collate the information to determine the scope of a typical RSL, and the degree of response to its compliance from the processors in the supply chain.

- What criteria do you use to compile the RSL?
- On what legislation is the detail of the RSL based?
- Do you have direct contact with legislative bodies and nongovernmental organisations over the content?
- Do you confer with the major dye and chemical manufacturers in the compilation of the RSL?
- How often is the RSL revised?
- How is the RSL distributed to the supply base, and how is compliance monitored?
- How is noncompliance addressed?

19.12.3 CONTROLS WITHIN A DYEING AND FINISHING OPERATION

Major dyers and finishers have technical departments devoted to various aspects of their operation. Contact at least two major processors to obtain answers to the following questions. Analyse the information to ascertain the emphasis that a commercial operation places on internal quality procedures, product development and environmental issues.

- What is your policy on product quality management, and how is it implemented throughout the operation?
- What is your policy for product development, and what criteria do you use for deciding on which developments to pursue?
- What is your policy on worker safety, and how is it implemented throughout the operation?
- What is your policy on environmental management, and how is it implemented throughout the operation?
- How has the emphasis on environmental control changed over the last 10 years, and what are your plans to cope with expected future demands?

19.13 REVISION QUESTIONS

- List the main criteria for dye selection.
- Describe the key fundamental requirements of a dyeing machine.
- Describe the process by which colour is perceived.
- List the main classes of dye applicable to (1) protein fibres and (2) acrylic fibres.
- Explain the mechanism by which a fluorescent brightening agent is effective.

- Explain why natural dyes are unlikely to present an immediate challenge to synthetic dyes.
- Describe the criteria for assessing the quality of a dyed product.
- List the main areas of environmental concern for the dyer.
- Explain why the liquor-to-goods ratio is an important factor in dyeing.

19.14 SOURCES OF FURTHER INFORMATION

Broadbent, A. D. (2001). *Basic principles of textile coloration*. Bradford: Society of Dyers and Colourists.

Park, J. (1993). *Instrumental colour formulation*. Bradford: Society of Dyers and Colourists.

Ingamells, W. (1993). *Colour for textiles: A user's handbook*. Bradford: Society of Dyers and Colourists.

Shore, J. (1998). *Blends dyeing*. Bradford: Society of Dyers and Colourists.

Lewis, D. M. (Ed.), (1992). *Wool dyeing*. Bradford: Society of Dyers and Colourists.

Duff, D. G. & Sinclair Roy S. (1989). *Giles's laboratory course in dyeing* (4th ed.). Bradford: Society of Dyers and Colourists.

Johnson, A. (Ed.). (1989). *The theory of textile coloration* (2nd ed.). Bradford: Society of Dyers and Colourists.

(1988), *Colour terms and definitions*. Bradford: Society of Dyers and Colourists.

Duckworth, C. (Ed.). (1983). *Engineering in textile coloration*. Bradford: The Dyers Company Publications Trust.

(2002). *Textile terms and definitions*. The Textile Institute.

British Standards Institution: www.bsigroup.co.uk.

The Society of Dyers and Colourists: www.sdc.org.uk.

The American Association of Textile Chemists and Colorists: www.aatcc.org.

Oeko-Tex: www.oeko-tex.com.

The International Organisation for Standardisation: www.iso.org.

REFERENCES

Broadbent, A. D. (2001). *Basic principles of textile coloration*. Bradford: Society of Dyers and Colourists.

Society of Dyers and Colourists. (1988). *Colour terms and definitions*. Bradford: Society of Dyers and Colourists.

Ingamells, W. (1993). *Colour for textiles: A user's handbook*. Bradford: Society of Dyers and Colourists.

Park, J. (1993). *Instrumental colour formulation*. Bradford: Society of Dyers and Colourists.

(2006). *Procion H-EXL standard operating procedures*. DyStar.

CHAPTER

FABRIC FINISHING: PRINTING TEXTILES

20

H. Ujiie

Philadelphia University, Philadelphia, PA, USA

LEARNING OBJECTIVES

At the end of this chapter, you should be able to:

- List the different textile printing processes and their major characteristics
- Understand and describe the various mechanical methods and styles of textile printing technologies
- Describe the importance of CAD and the latest digital inkjet textile printing technologies

20.1 INTRODUCTION

The textile printing industries are typically categorized as two markets: industrial textile printing and soft signage printing. The industrial textile printing market includes apparel, home furnishing, and technical textiles; the soft signage textile printing market focuses on graphic advertisements printed on textile substrates such as banners, corporate flags, etc.

Industrial textile printing has shown annual growth rates of over 1% for some time, driven by the acceleration of fashion cycles and continuous growth of the world population, and today produces over 20 billion linear meters per year (Osiris, 2008). Apparel and home furnishing textiles account for over 54 and 38%, respectively, of the market share in printed textile production, with technical textiles making up the remainder (see Table 20.1). The textile printing industry is currently in a strong position with high demand, and it is predicted that this situation will continue (Osiris, 2008; Stork, 2002).

All fabrics to be printed need to be clean and free from impurities. Many printing problems result from improper fabric preparation. The processes that prepare fabric for printing are discussed in Chapter 18.

Although the contributions of traditional methods such as block printing and engraved copper printing to the textile design fields have been significant, they are barely used in the textile printing industry today, having been replaced by newer and more reliable technologies such as automatic flat-bed screen printing and rotary screen printing. According to a worldwide survey by Stork, more than 90% of printed textiles are produced by screen printing technologies, which include table, flat-bed, and rotary printing. Transfer printing accounts for 6% and digital inkjet printing for slightly over 1% (see Table 20.2) (Stork, 2002).

The choice of printing method must take into account several factors that are critical in today's competitive global textile printing market, including:

- short-run productions,
- sustainable printing conditions,
- quick response time,

Textiles and Fashion. http://dx.doi.org/10.1016/B978-1-84569-931-4.00020-9

Table 20.1 End use of industrial textile printing market

Annual production yardage	20 billion linear meters
Apparel textiles	54%
Home furnishing textiles	38%
Technical textiles	8%

Stork (2002) and Osiris (2008).

Table 20.2 Textile printing: mechanical methods

Screen printing	91%
Table screen printing	6%
Automatic flat-bed screen printing	27%
Rotary screen printing	58%
Transfer printing	6%
Digital inkjet printing	1%
Misc.	2%

Stork (2002).

- customized printing products, and
- new design possibilities.

In some cases, the choice of the printing method depends on the purchaser of the design, and in other cases it is up to the designer to develop the print designs with a particular printing technology in mind. Direct printing is the most common technique, but a number of special printing styles can be used to provide different effects, not only visual but also tactile, when desired.

20.2 DIRECT PRINTING

Direct printing is defined as a printing application in which colorants are applied to the cloth in a single operation (with the appropriate fixation and washing processes). All colorants prior to the mid-nineteenth century were derived from nature, and, since some of these colorants were not substantive (i.e., they did not adhere directly to the fibers) it was necessary to apply metal salts, in a process known as mordanting, to color the cloth. Using such compounds carries pollution and occupational safety risks. Direct printing does not involve mordanting, and most printed textiles today are produced by direct printing.

The images or patterns are formed by what is also referred to as "localized dyeing," where the colorants are restricted to certain areas of the fabric. In order to achieve this, it is necessary to formulate print pastes that will not bleed or wick through the fabric. Thus, in conventional textile printing, the print pastes consist of:

- the colorants (pigment or dyes),
- a thickening agent, and
- the fixing agent.

Table 20.3 Textile colorants and substrates

Pigment	All fibers including mixed fibers
Reactive dye	Cellulose fibers (cotton, linen, and rayon), protein fibers (Silk and wool) and some nylon It is mainly used for cellulose fiber
Disperse dye	Polyester, some nylon, acrylics, triacetates, and other synthetic fibers It is mainly used for polyester
Vat dye	Cellulose fibers and protein fibers It is mainly used for cellulose fibers
Acid dye	Protein fibers and some nylon

No single colorant is substantive to all classes of fibers; therefore, the formulations of print pastes are different and specific to each colorant and fiber type. Table 20.3 gives some of the formulations of major colorants.

20.2.1 PIGMENT PRINTING

Pigments are not substantive to any fiber, and are composed of colored substances in a solid state that are completely insoluble to solvents (including water). Pigment printing systems require a **binder** as a fixing agent, which helps the pigments to bind with the textile substrates. In the fixation process, the printed fabrics are baked in a curing oven. The heated dry air activates and allows the thermosetting binder to adhere the pigments onto the surface of the cloth.

Pigment printing pastes also contain a **thickening agent**. Most thickeners are synthetic compounds and can be categorized according their application system:

- emulsion, or
- aqueous.

The emulsion system is older and involves a solvent, whereas the aqueous system does not.

Unlike dye applications, fabrics printed with pigments do not require washing after fixations. Therefore, any thickening agents remain on the printed cloth and stiffen the printed areas.

The emulsion thickening system does not stiffen the printed areas as much as the aqueous system, and so it dominated the textile printing industry for many years. However, the emulsion system is associated with hydrocarbon emissions and has consequently been banned in many industrial countries due to environmental concerns. As a result, the aqueous thickening system has been improved significantly, and has now become the most popular method.

A disadvantage of pigment printing is that the polymer layers that enclose the pigments on the cloth surface can break easily, and so its crock (rubbing) fastness rating is poor. However, pigment printing dominates the textile printing industry, with 45% of the total colorant consumption worldwide (see Table 20.4) (Stork, 2002), mainly because pigment printing provides the following advantages:

- high fastness to light,
- vibrant colors,
- less expensive than dyestuffs, and
- it can be used on any fibers and blends.

Table 20.4 Colorants (pigment and dye)	
Pigment	45%
Reactive dye	30%
Disperse dye	18%
Vat dye	4%
Acid dye	1%
Misc.	2%
Stork (2002).	

20.2.2 REACTIVE DYE PRINTING

Reactive dyes have a wide color range and are mostly used for printing cellulosic fibers such as cotton and linen. In reactive dye printing, dye molecules are diffused into the fibers and establish chemical bonds with them after fixation occurs. Although pigments are the most popular colorants in textile printing, reactive dyes are the most frequently used out of all other dyes, accounting for 30% of the total colorants used in the printing industry (see Table 20.4) (Stork, 2002).

Unlike pigments, most dyes are substantive to fibers if the chemistry used in the coloration method is correct. Reactive dye printing chemistry requires an alkali, such as sodium bicarbonate, as the fixing agent to establish the molecular bonds between the dyes and the fibers. Printing with reactive dyes also requires a thickening agent. Sodium alginate thickener, a derivative of seaweed, is the most common. Other thickening agents, which are commonly used in other processes, are based on carbohydrates and would react with the dyes, creating poor printing results.

Fixation processes, which cure reactive dyes to the fabrics, generally include steaming the printed fabric with highly saturated steam followed by washing in hot water to remove the thickening agent and excess dye particles.

20.2.3 DISPERSE DYE PRINTING

Disperse dyes are used for sublimation transfer printing, but can also be printed directly to the cloth. Although disperse dyes are substantive to acrylics, triacetates, and other synthetic fibers, polyesters are the largest fiber group for this printing application.

Disperse dyes do not require any specific fixing agent because the dyes become substantive to the fibers through phase changes initiated by temperature (solid to vapor to solid). A wide range of thickening agents can be used, including crystal gum, modified starch, locust-beans, etc. However, in choosing the thickener, it is important to consider both proper adhesion of the thickener to the cloth and ease of removal from the cloth in the washing process. Fixation is typically performed with high temperature steaming and washing to remove excess dyes and thickeners.

20.2.4 VAT DYE PRINTING

Vat dyes have excellent wash and light fastness ratings and are used to print textiles for high-end furnishing and military uniforms. They are substantive to both cellulosic and protein fibers, but are mainly used for printing cellulosic fibers such as cotton and linen.

Vat dye printing employs reduction-oxidation chemistry, where the reducing agent becomes the fixing agent in the printing process. In the presence of a reducing agent, an insoluble vat dye becomes soluble in the print paste. The most popular reducing agent is sodium sulphoxylate-formaldehyde (commercial names are Formosul and Rongalite).

To fix the dyes after printing, the fabric is steamed, which initiates the chemical reduction process and diffuses the dye further into the fibers. Afterward, washing with an oxidation agent such as hydrogen peroxide converts the vat dye back to an insoluble state. Lastly, the printed cloth is treated in a hot water bath with detergent, which stabilizes the vat dye molecules and establishes maximum fastness and optimum shades.

20.2.5 ACID DYE PRINTING

Acid dyes are substantive to protein and some polyamide fibers, including silk, wool, and nylon 6.6. Acid dyes are characterized by clear and vivid colors and are used for printing fabrics such as those used for swimwear, high-end fashion dresses, and accessories. Depending on the selection of dye, the light fastness rating can become problematic for certain applications.

In the printing process, acid dye molecules need to establish ionic bonds with the fibers. Therefore, the print paste typically uses weak acid solutions, such as ammonium sulfate, ammonium tartrate or acetic acid, as a fixing agent to create the negatively charged dye site. The acid is mixed in the print paste together with a thickening agent made from guar gum, locust-bean, crystal gum, etc. The fixation processes are similar to those used for reactive dyes. After steaming with saturated steam, the printed fabrics are washed to remove excess dyes and thickening paste.

20.2.6 DIGITAL INKJET PRINTING

Inkjet inks do not contain any thickening or fixing agents, unlike conventional printing, which employs a paste with the colorant, thickener, and fixing agent. Instead, they are applied to the textile substrate as a pre-treatment. Generally speaking, inkjet inks consist of pure coloring matter and are formulated to achieve optimum printability through ensuring that the print head nozzles eject perfect droplets at the micro scale. These inks also have a longer shelf-life than conventional colorants, which tend to separate and sediment. Four classes of inkjet inks are currently most popular: reactive dye ink, acid dye ink, disperse dye ink, and pigment ink.

Prior to inkjet printing, the fabric needs to be treated with the appropriate thickening and fixing agents. Pre-treatments are required to establish crisp and optimal print qualities along with the proper chemical bonds for each coloration class. Fixation after inkjet printing is similar to conventional fixation for each different class of colorants. Cloth printed with reactive or acid dye inks is steamed with a saturated steam and washed. Fabrics printed directly with disperse dye inks are steamed at high temperature then washed, and fabrics printed with pigment inks are baked in an oven with dry air.

20.3 OTHER PRINTING TECHNIQUES

In current textile printing technology, special printing styles are defined as any printing style excluding direct printing. Although special printing styles are more time-consuming and expensive than direct printing, they create the kind of attractive and novel printed textiles that are in demand in niche markets. The three most common special printing styles are:

- resist,
- discharge, and
- burn-out (devoré).

20.3.1 RESIST PRINTING

Resist printing is achieved by printing a resist agent on to the cloth prior to coloring. Since the resist agent prevents the fixation of the colorant, the areas where it is applied will remain unaffected, and the uncolored areas create the pattern.

The resist agent can be physical or chemical. Physical resists include waxes, resins, thickeners, pigments, etc., which create physical barriers that prevent any interaction between the cloth and the coloring matter. Chemical resists are composed of chemical compounds that prevent the development of chemical bonds during the fixation process. Such compounds include acids, alkalis, salts, reducing agents, etc., and are chosen depending on the specific textile coloration chemistry.

20.3.2 DISCHARGE PRINTING

Generally, there are two types of discharge printing processes. One is white discharge (bleaching) and the other is color discharge.

In white discharge printing, the ground fabric is first dyed with a coloring matter that can be chemically bleached out by the discharge agent. The dyed fabrics are printed with the paste containing the discharge agents. During the fixation process, the discharge agent is chemically activated and bleaches out the ground color, creating the pattern.

Color discharge printing is similar to the white discharge process with pre-dyed fabrics, where the dyes can be chemically destroyed by the discharge agents. However, instead of printing with a paste containing only the discharge agent, additional colorants, called illuminating colors are applied. These illuminating colors are not chemically destroyed by the discharge agents, but introduce additional colors in the fixation process.

Discharge printing has the advantage over direct printing of being able to print bright and vivid colors on evenly dyed dark backgrounds. Moreover, because of the chemical reaction of discharging, the penetrations of the colorants are greater than direct printing. Discharge printed textiles therefore present visually appealing results and have been used for high-end fashion and home furnishing textiles.

20.3.3 BURN-OUT (DEVORÉ) PRINTING

In burn-out printing, the print paste contains a chemical compound that destroys or dissolves away one or more types of fiber in a blended fabric. After printing, the fixation process activates the chemical reaction that burns out the fibers. A typical application is to burn out the cellulosic fiber from a blend of cellulose and polyester, using aluminum sulfate or sodium hydrogen sulfate. After a dry heat fixation process, the cellulose fibers are dissolved away, leaving only polyester fibers.

20.4 TRADITIONAL PRINTING METHODS

Traditional printing methods include block printing and engraved copper printing. Although both technologies are almost obsolete, their contributions to the printed textile design fields are significant. For example, current textile design styles "traditional floral" and "toile" (see Figures 20.1 and 20.2) originated from these printing technologies. In traditional floral, three-dimensional tonal representations of the motifs are stylized by printing several layers of flat silhouette shapes separately. This style was a result of the mechanical constraints of wood block printing technology. Toile originated from engraved copper plate and roller printing technologies, and set the standard for the appearances of traditional textile designs. Today, both of these design styles are produced by modern printing technologies, but they retain their original names (Ujiie, 2001).

Block printing is considered one of the oldest textile printing technologies, though the exact origins are controversial and unclear. A block printed specimen found in China was dated at more than 2000 years old (Storey, 1974), while clay cylinder stamps were used for textile printing in Mesopotamia by 3000 BC (Wilson, 1979, pp. 96–103) (see Figure 20.3). The principal of block printing is based on relief printing, where raised pattern areas pick up the colorants and transfer them to the cloth. Block printing is typically executed manually and is one of the most time consuming methods of printing textiles.

Engraved copper printing technology originated from *intaglio* and *etching* on paper substrates. Lines and dots are cut into the copper surface to create edged recessed areas. Instead of the colorants being picked up by the raised areas, they are held in the recessed areas. The colorants are transferred to the cloth by applying massive pressure. Initially, this type of printing was done with copper plates, but in 1783, Thomas Bell developed and patented a roller printing system that replaced the copper plate with an engraved copper roller. Engraved roller printing technology became one of the main production methods from its invention in the late eighteenth century until the mid-1970s, responsible for half of the total worldwide printed production during this period (Miles, 2003, p. 6). Although it was the dominant printing technology and permitted high volume and low cost production for over 200 years, the

FIGURE 20.1

Traditional floral design style.

Source: ©2009 Heather Ujiie, photography Hitoshi Ujiie.

FIGURE 20.2

Toile design style.

Source: ©2009 Heather Ujiie, photography Hitoshi Ujiie.

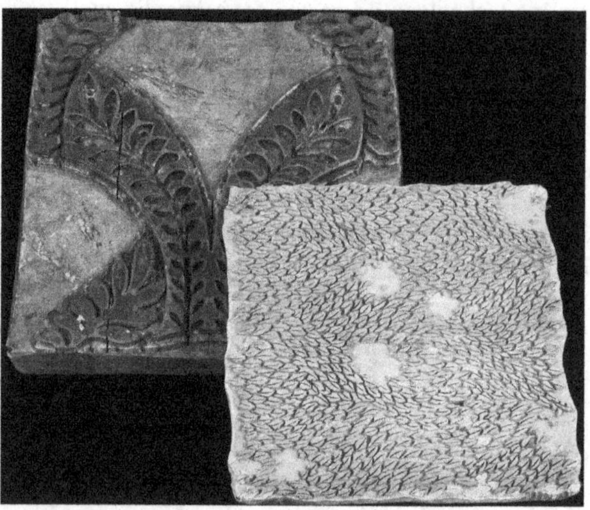

FIGURE 20.3

Wood blocks for block printing.

Source: Photography Hitoshi Ujiie.

technology had critical disadvantages, including the necessity of high production volumes because the cost of engraving the roller and wastage of cloth during preparation for printing were both high, the need for technical skills, and the danger to workers associated with the high pressure bowl.

Today, the textile printing market requires versatility and flexibility in design styles and production runs. Both block printing and engraved roller printing can no longer accommodate current market demands and have been replaced by newer and more flexible printing technologies.

20.5 SCREEN PRINTING

Screen printing is the most widely utilized technology in the textile printing industry. The origin of screen printing is not clear; however, the influence of traditional Japanese *katazome* papers, in which special papers are cut by hand for stencil dyeing processes, has been noted. Some *katazome* papers are constructed with fine meshes, which are used for reinforcement. These meshes are made of human hairs and fine silk threads and laid on top of open cut areas to adhere with the natural resins. In printing operations, *Katazome* pastes are applied through the meshes to the fabric. This concept is similar to screen printing technology, which uses a mesh (screen) and stencil. By the 1920s, hand screen printing was popular among early textile art and craft practitioners as they explored possibilities for new printed designs (Storey, 1974). Since then, screen printing has become one of the most important printing technologies.

Mechanically, screen printing is a form of stenciling with a print paste, which is forced through open areas of a screen by a squeegee blade to create printed impressions. Screen printing technology is divided into:

- table screen printing,
- automatic flat-bed screen printing (see Figure 20.4), and
- rotary screen printing (see Figure 20.5).

20.5.1 TABLE SCREEN PRINTING

Table screen printing technologies include hand screen printing, carriage screen printing, and turn-table printing. All of these printing mechanisms employ a mesh stretched over flat screens, and the process involves printing one color per screen at a time.

The cloth to be printed is fixed to the print table, which is normally 20–60 m long, and each screen prints every other repeat. Once the end of the fabric is reached, the printers go back to fill in the design between the first repeats. This is a slow process, but it can achieve the best printing results. By the time successive screen impressions are printed on the areas printed previously, those areas are dry. Therefore, the printing conditions become wet-on-dry, which facilitates the sharpest and most optimal printing quality.

The oldest method of table printing is hand screen printing. As the name implies, this process is operated manually by one or two printers, who physically handle the flat screens and print repeat by repeat with a squeegee. It is the most labor-intensive method, but it can achieve the highest quality.

Carriage screen printing is a semi-automated system. In this method, a metal carriage frame is equipped with a precisely secured flat screen, which can move along metal railings on both sides of the print table. The printing operation therefore requires only one person, so accurate screen registration can be achieved. Carriage and squeegee movements are automatic and controlled by computer systems that program all printing parameters, including repeat length, numbers of printing passages, squeegee angles, pressure, printing speed, etc. The printer operator monitors the carriage movements and refills the colorants when necessary, but does not perform any physical printing.

Turn-table printing is similar to the carriage system, and also a semi-automatic system, but here it is the fabric rather than the screen that moves. A flat screen is secured in a metal carriage frame and is arranged in a stationary position on the print table. The fabric is securely fixed to the rubber blanket of the conveyor belt on the table and moves repeat by repeat for printing. The system is controlled by computers that input the printing parameters, similar to the carriage printing system.

FIGURE 20.4

Automatic flat-bed screen printer.

Source: Courtesy Reggiani Macchine, Italy.

FIGURE 20.5

Rotary screen printer.

Source: Courtesy Reggiani Macchine, Italy.

Today, both carriage and turn-table printing are the most popular for high-end apparel, fashion accessories, and home furnishing industries due to their near-optimal printing qualities.

20.5.2 AUTOMATIC FLAT-BED SCREEN PRINTING

Automatic flat-bed screen printing (see Figure 20.4) was developed in the 1950s to increase the productivity of table printing. Instead of printing one screen at a time, all the flat screens used in the design print simultaneously, on a section of fabric that has just been printed with the previous screen. The

FIGURE 20.6

Rotary screen printer.

Source: Courtesy Stock Prints, Netherland.

average production speed for this technology is faster than any table printing technology, usually printing cloth at 10–15 yards per minute (9–13 m/min) (Ujiie, 2005).

In automatic flat-bed screen printing, the fabric is fixed to the moving rubber blanket of the conveyor belt and moves intermittently by a repeat length of the design. When the conveyor belt stops, all the screens lined up along the table print their impressions at the same time, thus the fabric furthest along the conveyor belt has had the most screens printed on it. At the end of the moving table, the complete design will have been printed, and the printed fabrics are automatically detached from the table for drying. The empty conveyor belt travels under the table and through washing and drying units, which clean its surface ready for the next batch of fabric. The print quality is slightly inferior to table printing due to it being wet-on-wet, but it is used for a wide range of printing applications, including fashion, interior, and bedding.

20.5.3 ROTARY SCREEN PRINTING

Currently, rotary screen printing (see Figure 20.6) dominates the textile printing industry. Introduced in the 1960s, it now accounts for about 58% of printed textiles (see Table 20.2) (Stork, 2002). Mechanically, the rotary screen printing system enables textiles to be printed continuously using rotating cylindrical screens, which contact precisely with the cloth while it is fixed to a moving blanket. Instead of the intermittent movement of automatic flat-bed printing, rotary screen production can print continuously, and can go as fast as 100 m/min, although average printing speeds range between 20 and 40 m/min.

In rotary screen printing, similar to the automatic flat-bed system, all screens are placed in a line on the moving conveyor belt table, and fabrics are printed continuously in a wet-on-wet manner. An automatic color supply system delivers the colorants to the inside of the cylindrical screens. Generally speaking, the printing quality is inferior to all other screen printing technologies because the engineering focus has been on speed, rather than the quality of printing. However, it is one of the most robust and fastest printing technologies, providing high volume and low cost production. Rotary screen printing is widely used for a variety of textile applications including printed apparel and printed interior fabrics.

20.5.4 SCREEN DESIGN AND PRODUCTION

Conventional textile printing technology requires one screen to be made for each color in a given design. This process is called screen engraving, and the number of colors used in a design represents the number of screens required for printing. Textile designers need to take account of the technical specifications for a particular printing process, including the required number of screens and repeat sizes. The repeat sizes are controlled by the actual dimensions of the screens, and also determined by the machine manufactures' specifications. Designers thus create textile patterns according to production specifications, including numbers of colors and sizes of the repeat, which are then sent to the engraver for color separation and creation of screens.

Screens used for printing are typically produced using a photochemical process. A special emulsion polymer is created from a mixture of polyvinyl alcohol and polyvinyl acetate with a dichromate salt as a photosensitizer, which is a light-absorbing substance that initiates a photochemical reaction. At first, this photosensitive polymer is applied to the surface of the screen mesh. After the coating is completely dry, a color separation film is placed on the screen and exposed to an ultraviolet light. After this process, the film is characterized by pure, opaque black patterns created on the surface of the clear film. The clear areas of the film are exposed to the ultraviolet light and the polymer layers on the mesh under the clear film become hardened and insoluble to water. The polymer layers under the black areas of the film remain unexposed. During the washing process, the unexposed polymer layers (under black on the film) are washed off and become the openings through which the print paste will pass.

In rotary screen production, there are generally two types of screens: lacquer and galavano. Lacquer screens are the most common, whereas galavano screens are used only under special conditions.

Lacquer screens are made of nickel mesh with uniform hexagonal openings. In traditional lacquer screen production, the photochemical process described above is used. However, today, laser engraving technology is the dominant method for creating lacquer screens. In this process, rotary screens are coated with a special polymer, which is different from those used for the photochemical process. After the polymer layers become hardened, laser beams burn only the polymer layers on the surface of screen mesh to create the openings that form patterns. The laser engraving machine is controlled by a computer system and the design (color separation) information in the computer directly controls the laser beams. This means it is a direct process with no need for color separation films. Since laser engraving does not employ a photochemical process, the polymer-coated screens can be stored for a long time prior to engraving without concerns over exposure to ultraviolet light.

Galavano screens are specialty screens that are durable and robust. They have the ability to create fine, tonal color gradients, and are produced by the electro-deposition of nickel. In this process, the pattern screens are generated in a solution of dissolved nickel. The dissolved nickel ions are attracted to the negatively charged electric current, and form a solid layer on the cylindrical nickel metal screen in a single process.

The non-pattern area is a solid nickel layer, and the pattern areas can have static or variable sized holes as openings. One of the advantages of using galvano screens is that they are highly durable, do not easily acquire pinholes, and can create optimal tonal print impressions with variable dot (hole) structures. In rotary screen printing operations, the choice of lacquer or galavano screens is typically determined at production meetings and depends on the quality and cost of printing requirements. Both types of screens can be used in the same printing operation to obtain the best printing results.

20.6 TRANSFER PRINTING

Transfer printing generally requires two processes. First, the design is printed on a thin, flat substrate, such as paper, film, or cloth, and then the image is transferred to the cloth. Historically, many textile printing techniques have utilized the concept of transfer printing. For example, one technique prints acid dye onto paper, which in turn is transferred onto silk; another uses pigment printed onto film, which is then transferred on to cloth.

Today, **sublimation transfer printing** is the most common technology and utilizes the sublimation properties of disperse dyes. In the printing step, designs are printed on paper with disperse dye inks, which contain the disperse dye molecules in the solid state. To transfer the designs onto the cloth, the printed papers are placed securely on the fabrics. By applying pressure at a temperature between 160 and 215 °C, the solid state disperse dye molecules are converted to the vapor state and migrate to the fibers. After cooling, the disperse dye molecules return to their solid state within the fibers.

Disperse dyes are only substantive on particular synthetic fibers, including the majority of polyester, acrylic, triacetate, and nylon. Historically, sublimation transfer printing was experimented with in the 1920s when disperse dye was developed, but it did not gain popularity because the available fibers were limited to cellulose triacetates, which do not function properly at the high temperatures needed for sublimation transfer printing to be successful. In the 1960s, Sublistic SA successfully commercialized a sublimation transfer printing system by using polyester fabrics and, since then, it has become one of the most important textile printing technologies (Miles, 2003; Storey, 1974).

20.6.1 GRAVURE PRINTING

The principals of the sublimation transfer printing process are identical to paper printing technology, and gravure printing is one of the main technologies used today. The basic principles of gravure printing are similar to engraved copper roller printing, so the initial step is to engrave the metal pattern cylinders. In modern printing operations, the surface of the metal cylinder is engraved by an automatic engraving machine equipped with diamond heads for etching the metal surface. The engraving machines are controlled by a computer system to ensure the accuracy of the engraving processes.

The printing unit consists of the pattern cylinders, color furnishing cylinders, and doctor blades. In the printing operation, each cylinder catches disperse dye inks from each color of the furnishing cylinders and all the excess inks on the surface are removed (scraped) by the doctor blades so the inks are only retained in the etched pattern areas. After pressure is applied to the pattern cylinders and papers, the inks are printed onto the papers. With current technology, printing speeds of over 100 m/min can be achieved and the systems can accommodate designs with as many as 10 colors. Disperse dye inks are required because of the high speed of operation (so that the paper dries quickly) and to prevent puckering and disintegration of the mark offs. Disperse inks for this process are typically formulated with volatile solvents such as petroleum distillates.

Gravure printing is a high volume printing system, mainly because printing volumes need to be balanced against the high set-up costs of the cylinders and paper. Nonetheless, there are numerous advantages for this technology, including:

- very fine detail and tonal shading effects in designs can be achieved,
- print mistakes can be found on paper prior to printing on cloth,

- can be used to print stretchy and unstable fabrics,
- it is a dry process so does not require steaming or washing, and
- equipment costs are low.

20.6.2 DIGITAL PAPER PRINTING

Since the late 1990s, large format digital inkjet plotters have been used for sublimation transfer printing. Using aqueous disperse dye inks (water-based disperse dyes formulated specifically for inkjet printing), many large format inkjet paper plotters have been converted to print disperse inks on transfer paper for sublimation. The printing operation in the soft signage industry was the first to use this process, to produce graphic banners, tradeshow fabric displays, and flags on synthetic fabrics.

20.6.3 HEAT TRANSFER PRESS

There are two types of heat presses for transferring the printed images to the cloth: web transfer heating machines, and flat-bed heating press.

Web transfer heating machines, or **heat calendar machines**, are used to produce continuous yardage. These machines consist of a heated roller and several supporting rollers, which control the continuous movements of the paper and the cloth. The printed papers and fabrics are fed together into the calendar machine, and sublimation takes place under heat and pressure when the printed paper and fabric make precise contact. Afterward, the printed fabric and used paper are wound onto separate rolls.

The flat-bed heating press is used for placement prints for garments and garment panels, and transfer printed images section by section. The garments and/or garment panels are placed on the flat-bed table and after the printed paper is laid on top of the fabric, the heating plate contacts the precise position to initiate sublimation.

20.7 DIGITAL INKJET PRINTING

Digital inkjet printing was first introduced in the carpet printing industry in the mid-1970s, when Milliken & Company in the United States developed the Millitron system, which utilized a computer injection dyeing system. Continuous streams of colorants were controlled by a mechanism using deflections caused by air. Due to the coarseness of the carpet substrate, a high resolution was not necessary and the printing system used only 10–20 jets per inch.

In the 1990s, digital printing technology became more widely used for industrial applications as technology improved and prices came down (Pond, 2000), so large format inkjet paper plotter printers became more affordable. Early pioneers of the textile printing industry began to adopt inkjet paper printing technology alongside their existing conventional textile operations.

Inkjet printing for textile production was introduced at ITMA (International Textile Machinery Association) in 2003, and, since then, innovations in the technology have led to far wider implementation in the textile printing industry (Ujiie, 2006). Today, many international textile print mills operate fully digital workflows from digital design to printing with production digital printers (Ujiie, 2002, pp. 254–257) (see Figure 20.7). Although the current average printing speed for digital inkjet textile

FIGURE 20.7

Industrial digital inkjet textile printer.

Source: Courtesy Reggiani Macchine, Italy.

printing is 60 m/h, the latest ISIS printer by Osiris of The Netherlands can print at over 20 m/min, which is comparable to automatic flat-bed screen printing speed (Osiris, 2008).

According to market research reports by IT Strategy and Web Consulting in 2005 and 2006, the current market volume of soft signage makes up half of the industrial textile market in terms of end user expenditure (see Table 20.5), and 45% of textile printing production in the soft signage print industry uses digital printing technology, with the majority of these using sublimation transfer printing. In contrast, digital inkjet printing technology accounts for only 1% of overall production in industrial textile printing (see Table 20.6). There are several reasons for this difference:

- The soft signage industry developed alongside the graphic design industry so both have a digital design workflow, whereas the workflow in the textile design industry is still transitioning to digital design. In a digital design workflow, graphic designs are developed in RGB color space, and sent directly to the printer for the final output. In the conventional textile design workflow, the patterns have to be designed with a countable number of colors, and the number of the colors corresponds to the number of screens used in the print production. After the designs are completed, they are sent to screen engravers to create pattern screens, and, after approval of test printing (called the strike-off process), the final printing is produced.
- The majority of printing jobs in the soft signage markets are short-runs, one-of-a-kind or customized products, such as banners, point of purchase displays, flags, etc.
- The business structure in the soft signage industry is based on individual projects, which includes printing, finishing, and installing the printed product. In contrast, the industrial textile printing market is based on volume of printed yardage. The level of customization means that soft signage production can have higher profit margins (Ujiie, 2010).

Today, printers and mills in the industrial textile printing industry use large format digital inkjet paper printers for sublimation paper printing, which enables them to provide short-run and customized

Table 20.5 Textile market (by end-user expenditures in dollar)	
Industrial textile printing (Apparel, home furnishing, and technical textiles)	67%
Soft signage (Graphic banners, tradeshow fabric displays, flags, etc.)	33%
Web consulting (2005) and I.T. Strategies (2006).	

Table 20.6 Inkjet printing penetration	
Industrial textile printing (Apparel, home furnishing, and technical textiles)	1+%
Soft signage (Graphic banners, tradeshow fabric displays, flags, etc.)	45+%
Web consulting (2005) and I.T. Strategies (2006).	

printing solutions, including engineered prints for high fashion and placement prints for high-performance sports wear and fashion accessories.

20.7.1 TECHNOLOGY AND CHARACTERISTICS

Digital inkjet printing (see Figure 20.7) is a form of non-impact printing, and involves streams of droplets of inks being ejected from the print heads onto the textile substrates. The size of the droplets ranges typically from 2 to 60 pL (a millionth of liter), and the size, speed, and placement of the drop formation and ejection are controlled electronically by the computer system. Printing resolutions range from 720 dpi for finely constructed textile substrates to 540–600 dpi for medium to coarse substrates (Ujiie, 2010).

While conventional textile printing technologies are spot color applications, in which each specific printing color in each design is mixed individually, inkjet textile printing systems employ the process color system, in which colors are preset. Cyan, magenta, yellow, and black are the fundamental colors in the process color system. However, more recent systems employ additional preset hi-fi colors (hi-fidelity colors), which increase the range of obtainable colors, typically including golden yellow, orange, red, blue, green, violet, etc.

Inkjet printing does not require the screens or other image transferring devices required for conventional textile printing. Design information is sent electronically from a design station to the printing system, which actualizes the image as a printed fabric on demand. Consequently, digital inkjet printing technology provides possibilities for small volume production and sustainable printing environments without the need for screen storage and without wasting colorants. Digital technology also allows designers to alter the design content quickly in response to changing market demands.

Until the early 2000s, digital inkjet printing technology was primarily used for samples rather than production. A **digital strike-off workflow** is more economical than the conventional strike-off process because it does not require the additional time and cost for screen engraving (Ujiie, 2002, pp. 254–257).

In the digital strike-off workflow, designs are presented to the potential buyers as digitally printed sample strike-offs. Depending on the potential buyers' responses, quick design alterations can be made. Upon approval, designs are sent to the printing mill and enter a conventional printing process.

Digital inkjet textile printing technology is still in its infancy and there are many disadvantages, such as:

- shallow penetration of colorants into fibers due to non-impact printing,
- higher costs of equipment and supplies than conventional textile printing,
- smaller color range due to the process color systems, and
- colorants can only be applied directly.

20.7.2 DESIGN APPLICATION

In conventional textile printing, all designs require repeat pattern screens, which are part of the engraving and printing process. Therefore, all designs need to be in a particular repeat size, since the size of the screen controls the dimension of the repeat. Each design is restricted by the number of colors, which corresponds to the individual screens. Some specialized textile software, such as Nedgraphics, Kaledo, and AVA software, can help in designing patterns in this reduced and limited color workflow.

Significantly, inkjet digital textile printing does not require this process. Just as with desktop paper printing systems, any graphic image created in a 24-bit color space with programs, such as Adobe Photoshop, and Illustrator, can be sent to a digital inkjet textile printer without going through color-separations or step-and-repeat requirements. Consequently, design pioneers have explored new creative possibilities using this technology, and new textile design styles have emerged, including:

- designs with millions of colors,
- very small images with extreme tonal and fine lines,
- photographic manipulation (see Figure 20.8),
- special digital effects with filters, and
- large, single-engineered images (Ujiie, 2006, p. 345) (see Figure 20.9).

Although digital inkjet textile printing currently accounts for only 1% of total textile printing production, it is predicted that it will gain popularity because of the new possibilities it provides that other conventional printing technologies cannot offer.

20.8 IMPACT OF CAD/CAM ON THE DESIGN OF PRINTED TEXTILES

The first computer system used in the textile printing industry was introduced by The International Business Machines Corporation (IBM) in New York City in 1967. This system, known as a **textile graphics system**, inputted the technical engraving data for engraved roller printing directly into a computer's memory. Throughout the 1970 and 1980s, computer technology was developed to support computer-aided manufacturing, **CAM**, by driving and managing mechanical functions in textile printing, including technical engraving and the operation of printers. Technical designers were able to trace and draw simple shapes (line drawings) directly onto a CRT monitor with a corded graphic tablet, and assign the color separations and engraving information. These technical color separations were then printed on clear film for engraving copper rollers (Lourie & Lorenzo, 1967; Textile World, 1967).

FIGURE 20.8

Photographic manipulation.

Source: ©2011 Hitoshi Ujiie Design.

FIGURE 20.9

Large single engineered image.

Source: ©2010 Heather Ujiie, photography Hitoshi Ujiie.

Computers were first introduced as a **creative tool** for printed textile designers in the late 1980s. Computer technology had begun to function as a metamedia (a medium that can dynamically simulate the details of any other medium) and computer-aided design, **CAD**, started to become an integral part of the design process (Kay, 1984). The CAD system consists of:

- design input by a flatbed scanner,
- design developments with a cordless stylus pen, and
- design output via inkjet printers (Ujiie, 2011).

With CAD systems, textile designers were able to edit and manipulate the original scanned designs on the display monitor with a stylus pen. However, although CAD systems became a creative tool for designers, the design and the production processes were still separate operations, and the functions of CAD and CAM systems remained as separate and different operations in the textile printing industry.

Today, the role of computer technology is changing. With the latest digital inkjet textile printing technology, there are no classical boundaries between CAD and CAM. In the digital inkjet textile printing workflow, designing processes and production processes are integrated seamlessly. Designers create textile designs on the computer and can send them immediately to be printed digitally on cloth on demand. Designers can have full control and responsibility for the entire textile printing production.

20.9 RESEARCH AND FUTURE TRENDS

Inkjet printing technology provides possibilities for new design styles and workflows, short production runs, sustainable printing environments, quick response time, and customization. One of the most significant contributions of this technology is the concept of the neo-cottage industry (Ujiie, 2005).

Initially, the mechanical functions of inkjet textile printers were similar to large format paper plotters, and when inkjet textile printing technology was still in its infancy, the only available textile printers were in fact adapted large format paper plotters. These modified plotters functioned identically to paper plotters with the exception of changing over media handling systems from paper to fabric. Using the appropriate inkjet inks for textile coloration, early inkjet printing pioneers working in textile mills operated multiple units of these modified textile plotters, and became able to produce significant yardage of inkjet printed cloth. Some mills in Italy produced over 10,000 meters of printed material with just a half a dozen of the modified textile plotters.

Today, many printing mills include inkjet textile printing in their production; however, the modified paper plotters, now called inkjet textile sample printers, are still available for short-run production and samples. In the neo-cottage industry model, individual textile designers can also become printers, manufacturers, and even brand owners using these inkjet textile sample printers. For short-run production, designers can design and produce their final product right in their studios. When a large volume is required, designers can outsource the print production by commissioning digital textile printing service bureaus. Unlike the past, when the textile printing industry consisted of a few large-scale printing mills, it is predicted that the future will include many small printing operations, including some using the neo-cottage industry model.

In considering the future of textile printing, inkjet textile printing coloration chemistry is crucial. Currently, it only has applications for direct printing, but inkjet printing applications for specialty printing styles are being explored. With the new concept of material deposition with micro (or nano) encapsulation technology, functional chemicals can be ejected from print heads onto textile substrates. There is much ongoing research on this subject; however, inkjet specialty printing systems have not yet been commercialized. Successfully developing the technology for inkjet printers to produce specialty prints would increase that technology's versatility tremendously.

20.10 SUMMARY

- Textile printing involves cloth preparation, printing and finishing operations, all of which are crucial for successful textile printing production.
- Textile printing technology is divided into two components: mechanical methods (printing mechanics) and styles (coloration chemistries). For successful printing production, it is necessary to address both components.
- Digital inkjet textile printing technology has begun to provide new design styles, workflows, and new business models.

20.11 CASE STUDY

Worldwide, approximately 60% of printed textiles are produced in Asia, with China producing about 30% and India another 30%. Western Europe and North America, which were once the capitals of textile printing, now have less than 10% market share each (see Table 20.7) (Stork, 2002; Xian, 2010).

Asian countries now produce printed textiles for international mass markets, based on large volume and cost-effective production, and rotary screen printing has become the dominant technology, providing printed textiles for both the apparel and home furnishing mass markets. Western European and

Table 20.7 Regional print production

Asia	60%
China	30%
India	30%
Western Europe	9%
North America	8%
Latin America	8%
Africa	7%
Middle East	5%
Eastern Europe	4%

Stork (2002) and Xian (2010).

Table 20.8 Worldwide installments of inkjet textile printer

Short-run sample printers	
Mimaki: (TX-1, TX-2) TX-3	2000+ units
Medium speed production printers	
Dupont: (3210) 2020	200+ units
Robustelli: Monna Lisa	100+ units
Konica/Minolta: Nassenger V	100+ units
High-speed production printers	
Reggiani/Huntsman/HP: DReAM	30+ units
Reggiani/Kyocera: ReNOIR	12+ units
Osiris ISIS	1+ units
Sublimation transfer printers	
Mimaki: (JV4)	4000+ units
Roland: (Hifi Pro)	1500+ units

The Center for Excellence of Digital Inkjet Printing of Textiles at Philadelphia University, March 2009 (Provost, 2010).

North American printing mills have thus been facing increasing cost pressure, and printers in these industrial countries have had to devise innovations in design and technology in order to remain competitive and keep their businesses viable. Digital inkjet textile printing has been one such innovation and has been employed by many companies since the late 1990s, and Western printing operations have begun to focus on producing digital textiles for high-end apparel and home furnishing markets, providing new competitive solutions including:

- creative and innovative designs,
- short-run productions,
- sustainable printing environments,
- quicker response time, and
- customized solutions.

According to a survey conducted by the Center for Excellence of Digital Inkjet Printing for Textiles at Philadelphia University, since the inception of digital inkjet textile printing technology in the late 1990s, more than 2500 inkjet textile printers (excluding those used for digital transfer printing) have been installed in printing mills internationally, with a majority of these units going to Western Europe, followed by North America (see Table 20.8) (Provost, 2010).

20.12 PROJECT IDEAS

- Worldwide inkjet textile printing currently shares only 1% of the total textile printing market, although many predict the share will go higher in future. Research the differences between digital printing and conventional printing in terms of design, technology, and business. Then, come up with your prediction for the future of the textile printing industries, by analyzing the design processes, technologies, and business workflows.

- Research the textile printing industry in your region/country and create solutions to make the printing industry in your region competitive with the worldwide market. You can discuss the technology, designing, workflow, and business aspects of this issue.
- Create textile designs for inkjet textile printing technology that cannot be visualized by any conventional textile printing method. The design can be for an interior or fashion application.

20.13 REVISION QUESTIONS

What are the three primary components of print paste?
How do the three components of print paste apply to digital inkjet printing technology?
What is the most dominant printing technology today?
Explain the advantages and disadvantages of the most dominant printing technology today.

20.14 SOURCES OF FURTHER INFORMATION AND ADVICE

The best sources for textile printing technology are journals and websites. Some of these are:

Textile World (www.textileworld.com)
Apparel Magazine (http://apparel.edgl.com)
Journals of Society and Dyers and Colourists (www.sdc.org.uk)
Journals of AATCC (American Association of Textile Chemists and Colorists) (www.aatcc.org)

Others are

Conference reports related digital textile printing including IMI conferences (http://imi.maine.com/) and NIP (Non Impact Printing) conferences (http://www.imaging.org/ist/Conferences/nip/index.cfm),
Websites of printing machine manufactures and dye manufactures including Huntsman Textile Effects (http://www.huntsman.com/textile_effects/), Mimaki Engineering (http://www.mimaki.co.jp/English), Reggiani Macchine (http://www.reggianimacchine.it/en/), Stork Prints (http://www.spgprints.com), Zimmer (http://www.zimmer-austria.com/en/),
Another helpful avenue is to research reports related to ITMA (http://www.itma.com) and related events
In terms of digital textile printing technologies,
Digital printing of textiles edited by Hitoshi Ujiie at Woodhead publishing is a good source for understanding the technology in general (Ujiie (2006), Digital Printing of Textiles, Cambridge UK, Woodhead Publishing)
Also the website www.techexchange.com has a variety of articles related to digital textile printing technologies.
The following books are also used to understand textile design: Bowles M and Isaac C (2009), *Digital Textile Design*, London UK, Laurence King Publishing Ltd; Clarke S (2011), *Textile Design*, London UK, Laurence King Publishing Ltd; Fogg M (2006), *Print in Fashion*, London UK, Batsford.

REFERENCES

Kay, A. (1984). Computer software. *Science American*, *25*(3), 52–59.

Lourie, J., & Lorenzo, J. (1967). Textile graphics applied to textile printing. In *AFIPS joint computer conference proceeding*, 14–15 November, 33–40.

Miles, L. (2003). *Textile printing*. West Yorkshire: The Society of Dyers and Colourists.

Osiris. (2008). Is digital production a reality? "ISIS" the velvet revolution in textile printing. In *AATCC/TC^2 symposium*, Durham, North Carolina.

Pond, S. (2000). *Inkjet technology*. Carlsbad, CA: Torrey Pines Research.

Provost, J. (2010). Chinese challenge. *Digital Textiles*, *1*(2010), 30–31.

Storey, J. (1974). *Manual of textile printing*. New York: Van Nostrand Reinhold.

Stork. (2002). *Developments in the textile printing industry*. Boxmear, The Netherlands: Stork Textile Printing Group.

Textile World. (January 1967). Now – it's fabric design by electronics. *Textile World*, 75–77.

Ujiie, H. (2001). *Interrelationship between textile design styles and production methods*. NIP-17. Fort Lauderdale, FL: IS&T.

Ujiie, H. (2002). *Textile education in digital inkjet fabric printing*. NIP-18. San Diego, CA: IS&T.

Ujiie, H. (2005). *Innovative product development in digital fabric printing* Presented at the Digital Textile, Berlin, Germany.

Ujiie, H. (2006). *Digital printing of textiles*. Cambridge: Woodhead Publishing.

Ujiie, H. (2010). *Inkjet textile printing status report 2010*. Presented at the digital inkjet textile seminar, Hangzhou. Available from http://www.hitoshiujiie.com/inkjetTechLibraryResearch.html/Honghua2010.pdf.

Ujiie, H. (2011). Computer technology from a textile designer's point of view. In J. Hu (Ed.), *Computer technology for textiles and apparel*. Cambridge: Woodhead Publishing.

Wilson, K. (1979). *A history of textiles*. Boulder, CO: Westview Press.

Xian, J. (2010). *17-year History in digital printing*. Presented at the digital inkjet textile seminar, Hangzhou.

CHAPTER

APPLICATIONS OF TEXTILE PRODUCTS

21

K. Canavan

University of Wales Institute Cardiff, Cardiff, UK

LEARNING OBJECTIVES

At the end of this chapter, you should be able to:

- Understand textile products as material culture and social messaging
- Understand the wide range of historical and contemporary textile applications and products
- Understand the roles of design, engineering, technology, and art
- Understand the practical, technical, and aesthetic uses of technical textiles in the clothing, interior, technical, and art industries
- Understand the different types of global clothing and textile industries and the effects of production on the environment
- Understand future textile uses and technological trends

21.1 INTRODUCTION

Since early times, human beings have been inspired to create and design textiles, and to manufacture and use textile products in virtually every aspect of their daily practical lives; for utilitarian purposes to protect or cover, but also in creative forms to fulfill a basic expressive demand for decoration and celebration, and spiritual and physical wellbeing.

In society, textile products clearly represent a message, for example, of wealth, status, or poverty. Textiles exhibited through dress or possessions at times of ritual celebration or festivals, and at important occasions throughout life, such as the birth of a child, marriage, or in respect of the dead, exude information as material culture, with products as symbols. Additionally, design motifs and decorative patterns on products create a silent, yet eloquent, language-like system messaging further information.

From the moment we are born, we are invariably touching, wearing, or using a textile product, and the visual language of textiles and the products we choose to own, wear, or use denote our own personal social requirements, practical demands, and instinctive aesthetic expressions. Historically, traditional crafts and technical developments, coupled with the wide diversity of textile design and products, can also be linked to religion, commerce, cultural exchange, and regional diversity across the world.

With the mechanisation of the textile industry in the early twentieth century, fast developing technological advances, and more recent computerised industrial production methods, have provided the

Textiles and Fashion. http://dx.doi.org/10.1016/B978-1-84569-931-4.00021-0

diverse opportunity of creatively fusing qualities of the traditional and the contemporary crafts with modern high technology and engineering.

Across the world, contemporary designers, engineers, and artists have been inspired to create and trade an extraordinary diversity of textile products for the fashion apparel, interior, mechanical, smart, and engineering markets. Design, engineering, technology, and art have merged into hybrid textile products, with a wider appreciation of sustainability and the environment, to create exciting, new, future possibilities.

21.2 APPAREL

The word 'apparel' means to equip, clothe, or adorn. Garments or fashion clothing products include fashion wear for men, women, and children as well as sportswear, hosiery, and casual and formal wear.

The ancient, fundamental social need for clothing products is to protect and to keep people warm or cool. These basic human needs are responsible for creating one of the oldest designs and manufacturing industries known throughout history. Clothing as protective items or fashion garments, and decorative products to embellish the body, indicate the individual's social status, personal aesthetic selection and self-expression, and acknowledges societies' standards of modesty and respectability. Fashion textiles and the apparel market are used to express a variety of different social status, and message aesthetic values within the echelons of human societies across the world. Expensive designer labels and haute couture compared to high street, ready-to-wear, mass production labels, and the contrast of the monarchy and high society wearing and owning expensively produced clothing and luxurious materials such as silks and metallic gold threads, with the lower classes using cheaper, plainer, or reproduction fabrics in smaller quantities, are two examples. The market and aesthetic value and price, and consequently the wearers' social standing, are reflected in the quality and make of materials used, whether with expensive or inexpensive embellished adornments, or produced by machine or hand finished.

Clothing and colours can inform and message important facts. For example, traditional clothing and colour, such as a priest's ecclesiastical purple and gold robes, or a Muslim women's black headscarf, express religious beliefs, whereas uniforms can exude a sense of authority and respectability, such as professional attire worn by academics and lawyers or socially recognised uniforms such as police, medical, or military garments.

Fashion designers and textile manufacturers have always created strong parallel links with innovative scientific and industrial research developments of the day. The use and availability of the sewing machine in the nineteenth century revolutionised and popularised the fast-changing clothing industry, and the invention and wide use of nylon in the twentieth century opened new and exciting possibilities for fashion designers and manufacturers to produce new, contemporary apparel.

Cultures from around the world have continued to customise and invent new frontiers of technological and fashionable developments, and creative innovation and the blending of design ideas and techniques have driven the global specialist apparel market into the twenty-first century. However, in the mid to latter parts of the twentieth century, specialist fashion and textile production radically declined, and with it many highly skilled craft techniques for the luxury products market. Mass-industrial production techniques for fashion clothing created manufacturers of mass-produced commodities, resulting in large conglomerates and standardised fashion products. Throughout parts of the world today, there are familiar high-street retail companies such as Marks and Spencer, Next, the Spanish retailer

Zara, and Swedish H&M outlets. All product ranges are inexpensive and highly popular with a fast turn over, but deliver similar fashion collections across the world, leading to social trends and pressure of social affiliation and group belonging.

Today, fashion textiles and apparel products are quickly changing, and reflect the times in which we live. Fashion trends continue to be used as a social indicator to display wealth and position. Seasonal trends are strongly linked to style, commerce, and social pressures, with affordable retail product lines enabling huge numbers of the global population to relay other messages of self-expression and create a throw-away-society, with the use of cheap employment and inexpensive production methods and materials.

The twenty-first century has embraced new technological discoveries and brought about new, future, smart ideas to redefine the commercial view of the design, production, and consumption of fashion apparel. This is driven not only by man's inherent ability to invent and improve, but also by demands from commerce and global industry, plus more recent social requirements to aspire to model-like, youthful looks for the consumer.

Apparel simply for adornment and to express decoration envelops a natural human instinct to assert creativity and individualism. However, attempts to conform to the idea of beauty can contradict the demands of comfortable products, such as between the sixteenth and twentieth centuries, when the traditional wearing of padded corsets to reshape the female body form was common place; or more recently, the popular wearing of uncomfortable and impractical high-healed shoes to give additional length and height to the body shape. Alluring evening wear and erotic fashion products, such as lingerie, cocktail, or evening dresses and accessories, can imply the sexual attractiveness of the wearer, and are made from soft, often silky materials that feel sensual to the touch and arouse sexuality.

Today, comfort, high-performance, durability, and ease of care and laundering are increasingly important considerations for modern-day apparel products, particularly in the leisure-clothing market. These garments have become visual indicators of fitness and health, as well as implying leisure status and wealth, and are worn not only by athletes, but a wide cross-section of society.

Personal choices and selection of textile products and individualised dress codes can be used to express psychological wellbeing and modernity. Current fashion and future trends, utilising futuristic fabrics, technical textiles and new technologies, indicate the interpretation of change and creativity of designers and consumers. New technologies and the development of smart textiles enable fashion designers to create innovative products in previously unimaginable and creative ways. New synthetics enable designers to create novel and exciting forms, with fashion clothing taking on experimental, new shapes and styles and looking towards the future. Heat-treated synthetics and digital CAD/CAM (computer-aided design/computer-aided manufacture) production eliminate hand processes, speed production methods, and reduce production costs leading to economic and ecological savings, leading to greater responsible production and environmentally aware consumption.

Japanese designers, such as Jun'ichi Arai, Issey Miyake, Yogi Yamamoto, and Reiko Sudo, experimented and led the way to create innovative new textiles and unique fashion garments. They combined the latest, sophisticated digital technology for manufacture with traditional craft forms, creating a synthesis for practical and wearable apparel. Interesting textures and mixes of natural and synthetic textiles enable the creation of sculptural fashion forms, which celebrate the body and invested in the latest technical materials and production processes.

Fashion garments and hosiery that embrace new technologies of chemical, biotechnical or electrical engineering can, for example, motivate designers to generate products that can emit light or change colour and shape according to the wearers' moods, temperature or time of day. More recently, fashion products have returned to retro styling and influences from the past, but with a modern twist using technical textiles. Designers of sportswear and practical clothing that shield the wearer from adverse conditions such as extreme cold and hot temperatures for athletes, or wet and dry conditions for mountain climbers and explorers, are forging new frontiers for hi-tech, specialist clothing products.

Fashion apparel for men and women, and more recently for children, is consequently a specialised form of practical body adornment, focusing on functions and performance, modesty, attraction, and decoration. Fashion products actively include symbolic messaging, social association, psychological wellbeing and enhancement, and modernism, while utilising quickly developing technical textiles and manufacturing processes.

21.3 FURNISHING OR INTERIOR TEXTILES, INCLUDING HOUSEHOLD PRODUCTS

In general terms, furnishing and interior textile products are primarily created for practical and functional purposes, and are used to cover and protect. Products for decoration, comfort, and wellbeing are also important, and exude messages of lifestyle and choice, comfort and wellbeing, with ancient woven textiles, printed furnishing fabrics, woven carpets, and, more recently, futuristic materials for interiors that are highly popular and commercial, being a part of everyday life. Today, traditional, natural materials are quickly being replaced by high-specification textile technology and man-made fibres, in order to create futuristic materials for contemporary interior settings and architecture.

Ancient tribal cultures used plain-woven linen cloth to wrap the internal organs of the deceased to preserve and protect the bodies, and to express respect for the person's spirit, in preparation for the after-life. Outer textile shrouds, decorated with important symbolic messages and motifs, were carefully prepared and covered the deceased, along with other practical requirements such as clothing, woven baskets and food supplies, in order to carry the spirit to the after-world and ensure life after death.

In vast regions of the world, nomadic Bedouin people traditionally created large woven tents and all of the interior textile furnishings required for nomadic living. Large tent dividers, floor coverings, bedding, storage bags, cushion items, and camel decorations were made from animal hair, to protect the family from extreme weather conditions and tribal warfare or invasion, and to create a comfortable yet practical living environment made entirely from textiles (see Section 21.7).

Today, textiles are used to surround us in our homes as household practical and decorative items and as interior textile membranes for modern architectural applications. Bedding, carpets, and curtains keep us warm and express our lifestyle choices and prosperity, and specialist fabrics provide particular requirements such as thermal insulation and translucent membranes to soften light emissions. Other interiors, such as public spaces and commercial properties, utilise textile products to create commercial recognition and status, plus practical comfort and atmosphere. Technical and specialist textile products such as innovative polyester-based fabrics, continue to advance and improve, enabling designers and engineers to create new interior art-forms and architectural outcomes.

Historically, peoples' perspective of aesthetic values led to colouring and decorating textiles products for the environment in which they lived. Early dyeing and printing from the Neolithic Age used colour extracts from minerals and plants that created temporary stains on fabrics. More permanent dyeing methods, such as indigo dyeing and patterning techniques, spread along the ancient trade routes to create beautiful shades of blue, whereas yellowish brown tones were obtained from a variety of barks and plant material such as roots and vegetable skins.

Early textile printing onto fibres and cloth for furnishing fabrics and clothing dates way back in mans' history. As evidenced in the ancient wall paintings of the Ajanta Caves in India, which date back to seventh century BC, sophisticated dyeing skills and complex patterning processes were known, and resultant textiles heralded high social status for royalty and the upper classes. Block printing and resist dyeing textiles were traded extensively throughout Asia and the Mediterranean region after Alexander's invasion of India, in 327 BC. This resulted in Indian textiles being traded and popularised in Europe for interior purposes, which later were copied and reproduced commercially. Arab traders in the second century AD, and exchanges with the British East India Company during the seventeenth century furthered the advance of design and production of domestic textile products. In the early eighteenth century, detailed prints for the interior became popular in Europe during the 'Rococo period', and the invention of the copper-plated printing process by Francis Nixon in Ireland provided improvements to the print process such as faster printing and the creation of larger repeat motifs and an improved quality of print.

Introduced to the West during the 1930s and mechanised after the Second World War, printing production became the most fashionable form of furnishing fabrics. Mass production of standard-controlled furnishing fabrics was available, which altered the design and availability for domestic and contract furnishing products such as upholstery, curtaining, floor covering, bedding, and table-wear.

In the 1960s and 1970s, the fast developing textile industry, with its high-speed, flexible, and affordable textiles production methods, enabled furnishing and interior textile products to reflect the rapid style changes, inline with fashion and contemporary design.

Recognised by consumers as more user-friendly than man-made fibres, natural materials provided optimum comfort and a positive influence on human wellbeing for healthy living and lifestyle, and are used for the high-end domestic and contract markets. Natural materials are used to produce upholstery fabrics, tapestries, curtaining, carpets, table-wear, and institutional fabrics, plus towels and mattresses, although more recently, developments of cheaper synthetics and technical textiles have become more acceptable and have replaced many natural material applications.

Upholstery products such as padding, filling, and barrier materials such as for seating and furniture are made from man-made fibres and have to adhere to strict health and safety requirements, such as fire and flame regulations, for homes, automobiles, airplanes, and boats. Interior designed fabrics are created not only for aesthetic purposes but also for fabric structure and composition for practical use and capability, such as light fastness and resistance to abrasion.

From the earliest times, wool and silk have been hand-woven into carpets on traditional hand-operated looms by craftsmen. Today, modern hi-tech operations by carpet manufacturers such as Wilton and Axminster Carpets produce luxury rugs and carpets in a wide range of colourful designs that assume a decorative focus to an interior. High-tech production manufacture and the inclusion of synthetic with natural fibres ensure products at affordable prices for the domestic and contract markets, while maintaining high standards of quality and health and safety requirements.

21.4 TECHNICAL TEXTILES

Technical textiles are used to produce functional products where aesthetic characteristics are not the main priority. Contemporary technical synthetic and microfibres coupled with performance textile technologies include flexible applications for industrial textiles, smart and medical textiles, wearable technology, and eco textiles.

Technical fibre engineering and science in the 1970s offered exciting new advances, which led to technology-rich production possibilities, and the merging of traditional crafts and technology for a variety of specialist outputs. Today, technology-rich textiles, designed with specialist characteristics to meet practical and economic solutions, and contemporary, high-performance demands, create new opportunities for designers and manufacturers to further research and develop new products.

21.4.1 INDUSTRIAL TEXTILES AND GEOTEXTILES

Unlike early synthetics, which consumers viewed as cheaper alternatives to natural fibres, modern sophisticated textile alternatives emerged with improved performance and acceptable and highly desirable aesthetic qualities. Lightweight synthetic woven textiles, prized for their versatility, strength, and non-corrosive properties, were developed for use in engineering, and for the manufacture of bridges, house building, and hybrid fabric structures.

Strong, lightweight composite materials become increasingly important when combined with non-textiles such as metal, glass, and ceramics, and were produced to be used in the manufacture of aircraft and spaceships. Cheaper, high speed extruded, nonwoven fabrics were developed and engineered to produce a wide variety of products, such as moulded linings in the automotive industry and baby's disposable nappies. Protective military clothing was manufactured to high specifications such as to withstand extreme temperatures and bullet penetration, whereas light-emitting fabric products were created by using optical fibres to illuminate discrete visual signals for military equipment and locate switches on planes, for example.

In the mid-twentieth century, textile composites and modern polymers enabled the production of performance geotextiles and specialist textile products for the civil engineering industry. Geotextiles now form a wide and diverse range of engineering membrane products that have advanced more recently from a combination of polymers and are used as products to improve soil conditions for road construction or to prevent coastal erosion. Woven and nonwoven composite products were combined with plastics to be used in drainage systems and soil reinforcement, whereas warp-knitted fabrics were designed to create net-like grid constructions, and produced for civil engineering ground construction.

21.4.2 SMART FABRICS AND INTELLIGENT TEXTILES

Smart fabrics can sense different environmental conditions and intelligent textiles or e-textiles can not only sense environmental changes, but can automatically respond to their surroundings or stimuli, such as thermal, chemical, or mechanical changes, as well.

Since its invention in the 1960s by the American firm DuPont, the synthetic fibre Lycra, or Spandex as it is known in America, is one of the most commonly used smart fabric and has been widely utilised, due to its exceptional elasticity, in the sport and fashion market. Lycra modernised traditional fabrics, when combined with natural fibres, enabling the creation of clothing, swimwear, underwear, hosiery, compressed surgical garments, and interior furnishings that were elastic, functional, and comfortable to wear.

The microporous Gore-Tex membrane, designed by Wilbert and Robert Gore in the late 1970s, and engineered by a similar technique to Teflon, was originally designed to enable the body to breath while being completely wind and waterproof, and thus was widely used for rainproof wear. Products for astronauts and practical clothing for use in extremes of cold temperatures were also developed, and more recently Gore-Tex fabrics have been used for medical implant products and in the prevention of the spread of bacteria on medical garments in hospitals.

Electronic or smart textiles, such as heat and light sensitive products, were initially developed as intelligent or smart clothing for the sportswear market, and created fabrics for products that were designed to embed computing or digital components. Advances in e-textiles, and emerging smart-textile electronics (see Chapter 15.4.2, for more information) and the microencapsulation technique enable thermal responsive clothing products, such as swimwear and evening wear to react to heat and change colour. Thermal responsive products in the interior markets became popular and interior textiles were developed that adapt to the temperature of the environment or exposure to sunlight and change colour accordingly.

21.4.3 MEDICAL TEXTILES

Due to innovative textile technology and advanced medical and surgical procedures, medical textiles products for the healthcare and hygiene sectors have grown rapidly in importance. Modern qualities of medical textile products used in medical operations and hospitals have developed at a rapid pace and have improved performance and become more human-friendly.

New medical fabric technologies and textile knitted and embroidered structures are now being regularly used as medical implants and healthcare products. Monofilament and multifilament yarns, woven and knitted and nonwoven structures have been engineered to provide the medical and surgical industry with vital applications that range from non-implantable materials, like dressings and bandages, to implantable materials, such as artificial ligaments, joints, and artificial organs.

Textiles as healthcare products are highly important and are enormously diverse. Examples of basic products such as surgical gowns and wipes compare to high performance, specialist fibres, which can be surgically used to replicate the elasticity and temperature fluctuations of human skin, reflect this importance and enormity.

More recently, advanced medical products such as bandages for wounds and burns have increased aesthetics values and comfort for the patient. These products have benefited and quickened recovery, and consequently survival rates in hospitals.

21.4.4 WEARABLE TEXTILES AND PROTECTIVE CLOTHING

Towards the end of the twentieth century, British designers, such as Nigel Atkinson and Bridget Bailey, and international fashion designers, such as Issey Miyake and Yogi Yamamoto, explored and experimented with exciting contemporary developments of technical textiles to create novel aesthetics for fashion garments. New processes and materials enabled inherent textile qualities of decorative texture and form to equal the importance of the garment design, and a new approach to fashionable wearable textiles and synthetic materials evolved.

Changes in the approach to the clothing market and more relaxed fashion trends led to an increased importance of ease-of-care and comfort products, without losing any of the aesthetic or performance qualities. Technology-rich textiles, such as Lycra and Gore-Tex, became highly fashionable and appeared on the haute couture fashion catwalks.

Advances of new fibres and engineered materials, with computer-driven enhancements, provided flexible solutions for wearable textiles. Seamless 3D shaped fabrics and one-piece styled garment products enabled high-performance yarns and external digital devices to be incorporated into the uncut fabric, while also reducing manufacturing costs.

Particular technical textiles enable the provision of high specification clothing and products suitable for athletes and specific protective clothing, vital for land, sea, and air defence military personnel, who work and live in extreme and hostile environments. High-performance clothing and equipment, with functional, practical, and cost-effective properties (also lightweight, compact, durable, and reliable), was designed. More specifically, military clothing and equipment products for active service are required to be lightweight and low bulk, durable, weather repellent for extremes of temperature, and produce low noise emission. Camouflage and deception requirements, with battlefield specifics of ballistics and chemical warfare, and heat and fire protection provide life protection and survival aids. Across the world, high levels of investment in research and development of military textile science and technical textile products continue.

21.4.5 ECO TEXTILES

It is important to consider the environmental impact of textile products and their applications. Fashion and functional clothing products, interior and architectural structures, and technical and engineered textile products provide fashionable demands and protection across the world. Degradation of textile products is inevitable, and resource depletion and pollution must be limited to protect the planet and reduce environmental harm. The depletion of world resources, due to high energy consumption in the manufacturing and distribution of textile products, requires vital attention to recycling, sustainability, and the production of environmentally friendly textile products.

Environmental awareness must be the responsibility of the textile designer, manufacturer, and consumer.

21.5 TEXTILE ART

Primitive people made simple textiles known as barkcloth. Later, ancient man knotted and wove simple cloth to construct a wide variety of utilitarian textile structures and products for tribal societies across the world. The importance of these textiles was multi-layered, and frequently expressed religious beliefs and association with spiritual life and the supernatural world. As possibly some of the earliest known textiles products, these cloths for everyday apparel and usage were decorated.

Aesthetic decoration fulfils mans' natural and instinctive creative abilities. Decorative textile techniques evolved, not only to impart special religious credence, or to impose a purpose beyond their social functions as ceremonial cloths, practical items of clothing, or modern commerce, but also to exude a feeling of pleasure and sensuous enjoyment to the observer. The decoration of textile products advanced, in some cases, as sophisticated art forms, which sustained aspects of a cultural legacy, or spiritual and mystical meaning up to the present day. The decline of strict religious beliefs and the introduction of outside influences and technological developments of the western world have impacted upon traditional production and aesthetics. Indigenous craftsmen continue to create intricate textile art forms for ceremonial and personal credence, but cheaper imitations, modern interpretations, and contemporary fashions are replacing and undermining traditional methods.

More recently, as opinions of industrial mass production began to change in favour for individual products, and the appreciation of quality and traditional artistry asserts itself, there has been renewed interest in the application of textile art. Designers and artists began to explore and create modern textiles utilising specific qualities derived from traditional and contemporary methods.

At the same time hand-made, studio-craft industries such as the traditional Scottish Fairisle hand knitters and the silk and cotton weavers of India and the Far East were creating textiles for the home and small-scale aesthetic and cultural market. These traditions continue today, and textile artisans create textile products as decoration and for aesthetic pleasure, for exhibition, or to one-off commission.

The aesthetic concerns of textile art products failed to equal the status of industrial textile products, due in part to historic prejudices against women as designers and practitioners, and the strong preconceptions of textiles being associated with the home. However, after the fibre arts movement in the mid-twentieth century, new and experimental interior textile hangings and tension fabric structures and installations predominated the market.

More recently, there has been an international revival and recognition of the value and status of craft and textile art. Today, contemporary textile artworks are being produced using the latest textile technologies, enabling artists to discover new and exciting methods of working, and free of the commercial demands of industrial production. The contemporary and flexible mix of traditional techniques, techno textiles, and digital processes have considered non-functional textile products to be important, with textile art becoming valuable and collectable.

This perception of contemporary textile art products as a hybrid mix of novel textile methods and material combinations creates new opportunities and broadens new horizons, leading to a narrative discourse between the maker and the public, which specialist international textile museums and galleries are increasingly recognising and responding to for an ever-widening audience.

Frequently, textile artists are environmentally aware and use environmentally friendly production methods, recycled materials, and non-pollutant chemicals.

21.6 TEXTILE INDUSTRY

In the eighteenth to nineteenth centuries, during the first industrial revolution, the main production of textiles was made from handspun wool and cotton yarns on basic handlooms. With the inventions of James Hargreaves' multi-spooled spinning jenny and Richard Arkwrights' water powered looms, mass production of clothing and textile products became mainstream.

The Englishman Thomas Saint is thought to have invented the first working sewing machine, and around 1840, the domestic sewing machine became popular, establishing greater efficiency and higher productivity standards and opportunities. Technical advances of the sewing machine were instrumental in the development and expansion of the textile industry and ready-to-wear clothing industry.

The early shuttle-operated power looms were replaced in the twentieth century with shuttleless looms and, more recently, new technical advances created specialist looms such as air-jet and water-jet looms, which greatly improved efficiency for particular textile production.

Today, with product development and textile manufacture being driven by product demand, modern fibres and materials, and new computerised technological advances have evolved into a flexible and highly sophisticated global industry. Fast-evolving research and technology has propelled the textile industry to become a computer-aided design and computer-aided manufacture (CAD/CAM) production industry.

Global cultural identities, traditions, and status dictate the textile design, interior, fashion, and technical industries, but function and performance of a textile product are vital, particularly in the specialist markets such as for smart, medical, or wearable clothing. Additionally, product promotion and quality of materials, manufacture, and handle are intrinsically linked with commerce, and all are important factors in product development, manufacturing specifications, and overall success.

The design, aesthetics, and use of a textile product are an important factor, and in the fashion and clothing market, the industry is governed by seasonal trends and styles, coupled with rapidly changing textile technology and production methods. The CAD/CAM textile industry continues to evolve into an adaptable and fast-responding production industry, capable of providing a number of different manufacturing systems. High-volume quantities of mass-produced materials for the apparel, interior, or technical markets can be produced, with batch production methods utilising quick, repeatable response methods of production using standard materials and fixed specifications with controlled monitoring systems to ensure quality and standards of production. Short-run manufacture of specialist fabrics and textile products for the higher end of the market is more specialised and consequently more expensive to produce, but nonetheless is highly commercial. More recently, the development of techno textiles and digital applications has enabled industrially produced individual designs to become more cost-effective for the elitist couture market and bespoke 'one-off' products.

The textile industry is responsible for high levels of environmental pollution. The textile chemical processing industries use huge volumes of water and generate effluent waste during the wet processing of textiles, which is costly and creates environmentally pollutant chemical substances. Approximately 20% of industrial water pollutants come from textile production, colouration, finishing, and distribution of textile products.

Recent programmes of pollution prevention and control are reducing water use and creating more efficient chemical processing. Cleaner production processes and pollution prevention measures provide environmental and economic benefits, while consumer awareness and green issues, such as recycling and upcycling, are more popular and socially acceptable.

21.7 CASE STUDY: TRADITIONAL BEDOUIN AL SADU HAND-WOVEN PRODUCTS AND CONTEMPORARY DIGITAL APPLICATIONS

Textiles are an essential element and integral part of material culture, representing an important human practical need and creative activity. Al Sadu or al Sedu is an ancient Bedouin tribal weaving craft, where women wove on simple ground looms, using hand spun, naturally dyed animal hair to create practical and decorative textile products.

Al Sadu weavings of Arabia convey the Bedouin's rich heritage and instinctive awareness of natural beauty, with patterns and designs messaging the traditional nomadic lifestyle, the desert environment, and the emphasis of symmetry and balance due to the making process. Traditional weaving is an expressive art form, and for Bedouin women, their textiles are testimony to their practical achievements, manual dexterity, and aesthetic values.

The resultant textiles provided all the necessary practical shelter and material provisions for nomadic tribal life. The most important woven items were the large traditional Bedouin tent itself and all the interior furnishings within, such as the rugs, storage bags, bedding, cushions, and a tent divider. The large tent divider protruded out from the tent in order to segregate the men and the women's quarters, and was the most decorated and impressive item produced, with an interesting array of motifs and symbols. It was regarded as an important part of a woman's possessions, or trousseau, and accompanied her into marriage. This was testimony of a tribe's wealth and prestige. With it's strong geometric and symbolic designs and deep rich colours, the creation and production of this extraordinary textile enabled the weaver to express her technical skill and her creative ingenuity. Symbolic patterns and motifs message the women's personal interpretation of nomadic lifestyle, her desert environment, beauty, and personal belongings.

With many of the regions textile traditions fast disappearing in the face of rapid cultural and economic change, the Bedouin nomadic lifestyle has radically diminished, and the number of traditional weavers in Arabia is in continual decline. No longer are the large traditional Bedouin tents and tent dividers required, and due to widespread illiteracy among the Bedouins, nothing is written down or recorded. These textiles still retain a role today, particularly with the older generations, at traditional ceremonies and on special occasions, but in the main, Bedouin weaving has lost its importance as a utilitarian and vital cultural craft form.

The aspiration is to preserve and reinforce the vital role that al Sadu textiles play in the regions cultural identity for the next generation, and to avoid the risk of losing important values and traditions. The challenge is to revive traditional crafts and skills, blending the traditional with modern technology, and education and training with commercial demand. This will inspire creative contemporary practitioners to produce new original designs and products, reinvigorating the lack of knowledge and apparent disinterest from the contemporary indigenous generation or local governments.

Having worked with al Sadu Bedouin weavers since 2004 throughout the Gulf region, I believe that the significance of preserving this cultural identity and the traditional nomadic textile products, with the associated knowledge and skills, is crucial if is not to be lost forever.

The young generation is familiar with computers and digitalised imagery with CAD/CAM production and is no longer interested in labour-intensive hand skills with little or no current demand for traditional products. To meet the requirements of traditional progress and young people's interest and aspiration, I initiated educational design projects and training with UK textile design students in collaboration with a Kuwaiti design-company. Influenced by Al Sadu textile patterns and traditional items, new and exciting digital designs for fast production emerged, for fashion, interior furnishings, household products, wallpapers, and paper products for the stationery market. Contemporary materials, using current textile technology and refined experimentation, produced sophisticated outcomes for modern applications with suitable performance qualities for everyday use.

The design projects evidence young peoples' interest in traditional crafts and trans-cultures, and meet the requirements of traditional progress and new product ranges, ensuring the survival of a modern al Sadu weaving tradition for the future.

21.8 FUTURE TRENDS

In today's highly evolving technological world, new and innovative technical developments have merged the fields of design, science, engineering, and the arts to explore related contemporary and novel digital possibilities, and to create the potential for the next generation of textiles.

Driven by innovation, new materials and technical processes, novel designs and demands have led to exciting, new or extended performance capabilities and products.

The fusion of interactive, historical skills and hand-crafted techniques with sophisticated, modern technology, and new digital applications methods and processes, inspired by computer-aided design, creates a contemporary vibrancy. The collaboration of artist/designer with textile engineer and cutting edge technology offers endless new computer design potential, resulting in the use of techno textiles and contemporary digital textile art. This satisfies novel aesthetic demands and provides advantageous commercial attributes for progressive textile manufacturers.

The awareness of the planet's ecology, and the energy flows of production and recycling associated with the textile industry, is changing consumers' attitudes to culture and technology, with new demands for environmentally friendly products. More recently, this customer demands and the 'market pull' has replaced the 'technological push', where customers dictate increasingly higher demands for technical solutions and cost-efficiency. In turn, the creation of a consumer industry has made radical changes on the textile industry and supply chain in order to meet the specifications of the customer.

Specialist textiles with particular qualities and characteristics for the functional and aesthetic requirements continue to evolve. For example, specialist medical textiles continue to exploit and create new ranges of simple and complex materials, using advancing scientific development of available fibres and fabric-forming textiles. Composite materials, producing higher performance characteristics with reduced costs and higher flexibility for surgery or healthcare, continue to be in demand and required for the medical industry.

Nanotechnology enables molecular manufacturing of functional systems at a molecular scale on textile substrates, with improved healthcare systems, protective clothing, and integrated electronics being some of the possible applications. Specialist self-cleaning fabrics, protective ultraviolet light, and fire resistance textiles in the smart and interactive market are incorporated into the aerospace, automatic, construction, and sports industries. Garments that sense their surroundings and interact with the wearer is of future interest, with smart clothing that could provide personalised healthcare systems and monitor health statistics.

Vital awareness and concerns regarding environmental sustainability and cultural sustainability are transforming what designers and manufacturers produce. Society and consumers have become more aware of fabric qualities and synthetic alternatives, with the promotion of eco-textiles and use of natural fibres and dyestuff. Ethical attitudes are growing fast, and biodegradable and recyclable or reusable materials are becoming more familiar and recognised as environmentally friendly, even though the results generally have a diminutive qualitative effect on products. The growing popularity of Fairtrade is an indication of consumers' interest in production processes, sustainability, and connections with producers in developing countries.

Currently, the dependency upon petroleum and oil for energy conflicts with ecological and environmental awareness, which determines future developments. Clean energy for manufacture and production of textile products, such as wind, tidal, or geothermal energy production processes, is being developed and used more widely. The reduction of waste effluent in the production of textile products is increasingly acknowledged as vital to reduce the damage to the planet and slow resource depletion.

Exciting new production techniques and wider applications of finishing methods and technical developments to make textiles handle and launder more efficiently are developing fast. Astonishing new smart properties of fabric are being developed and created across the globe, and new, advanced 'high-tech textiles' with specific high-performance characteristics are in growing demand, leading to flexible and creative future possibilities.

21.9 SUMMARY

- For thousands of years, and particularly since the industrial revolution, machinery manufactured textiles for the clothing, interior, or industrial market were produced more efficiently and at higher speeds of production, compared to earlier hand-crafted textiles.
- The importance of the textile industry and human's reliance upon textile products for protection, shelter, and decoration created a major merchandising exchange across the world.
- More recently, the computerised textile industry enables more flexible, fast-turn-around digital production runs for the mass market, or the more specialised, industrially produced 'one-off' market with new, sophisticated possibilities for future solutions.
- Today, as the world becomes more accessible with modern computerised links and new developing technologies, the global textile industry and application of textile products remain large and highly diverse.
- Technical advances and applications of textile products have developed into specialist industrial clothing. They have also developed into accessories such as chemical protection garments or outer protective sportswear, and mechanical parts for vehicles, aerospace, and building construction.
- Thermo-forming and 3D textiles have become widespread, and fabrics produced by nanotechnology for the healthcare and engineering industries are also available.
- Contemporary designers and artists are inspired by extraordinary modern technological developments and exchanges of historical and contemporary ideas, and produce a wide diversity of hybrid products and applications for the fashion apparel, interior, technical, and art market by fusing traditional craft skills with modern high technology.
- Future uses and applications of innovative interfaces between textiles technology, design, architecture, medicine, and wellbeing hold the capacity to transform the world. The changing role of textiles products and their applications, with new sustainable sources, will redefine the potency of textiles for the future.

21.10 REVISION QUESTIONS

- How do textile products evoke material culture?
- What are the main industrial applications of practical and technical textile products?
- Which market has merged aesthetics and contemporary technical textiles together?
- What are the three main textile production systems in the apparel industry?
- What is CAD/CAM?
- What are the most important future advances in the textile production industries?
- Describe future green issues related to the textile production industries.

21.11 SOURCES OF INFORMATION

Numerous trade publications, magazines, and journals
Interior Design Magazine
Websites provide useful information
For comprehensive information on fashion and colour trends, visit website World Global Style Network www.wgsn.com (subscription-free student website www.wsgn-edu.com/edu/).
For inspirational ideas, the organisation 'Textile Futures Research Group' promotes researches linking textile design with technology, culture, and ecology, website www.tfrg.org.uk.
Trend View and Selvedge publications are inspirational magazines for design and textile ideas.
Modern technology, including interior products and textiles and machine manufacturers – H. Stoll GmbH & Co KG, Stollweg 1, 72760 Reutlingen, Germany, TU Dresden.
Eco tradeshows
Sana: Bologne, Italy, www.sana.it.
BioFach: Nurnberg, Germany, www.biofach.de.
Trade shows
HEIMTEXTIL: Frankfurt, Germany, www.heimtextil.de.
INTERSTOFF: Frankfurt Germany, www.interstoff.messefrankfurt.com.
Magazines
NANO Magazine; Smart textiles and medicine. Issue 9 & Issue 17 2011.

FURTHER READING

Al Sabah, A. (2001). *Kuwait traditions. Creative expressions of a culture*. Kuwait: Al Sadu Weaving Cooperative Society.

Beith, M., Baulch, K., & Oppermann, K. (1997). *Textiles and technology*. Cambridge: University Press.

Braddock Clarke, S. E., & O'Mahony, M. (1998 & 2005). *Techno textiles 2: Revolutionary fabrics for fashion and design*. London: Thames and Hudson.

Canavan, K. (2003). *Dayak to digital: Traditional woven Ikat for contemporary knitted textiles*. Scotland: Heriot-Watt University.

Canavan, K., & Alnajadah, A. (2013). Material symbols of traditional Bedouin Al-Sadu Weavings of Kuwait. *Textile: The Journal of Cloth & Culture, 11*(2), 152–165.

Colchester, C. (1996). *The new textiles: Trends and traditions*. London: Thames and Hudson Ltd.

Colchester, C. (2004). *Textiles today: A global survey of trends and traditions*. London: Thames & Hudson Ltd.

Collier, B. J., Bide, M., & Tortora, P. G. (2009). *Understanding textiles* (7th ed.). Pearson International Edition.

Cresswell, L. (2009). *Textiles at the cutting edge*. Forbes Publications.

Crichton, A. R. (1989). *Al Sadu. The techniques of Bedouin weaving*. Kuwait: Al Sadu Pubs.

Dickson, H. R. P. (1983). *The Arab of the desert* (3rd ed.). London: George & Allen & Unwin Pubs. rev & abridged.

Eco, U. (1969). Function and sign: the semiotics of architecture. In G. Broadbent, et al. (Ed.), *Signs, symbols, and architechture* (pp. 11–70). Chichester: John Wiley and Sons (1980).

Eco, U. (1979). *A theory of semiotics*. Bloomington: Indiana University Press.

Hibbert, R. (2006). *Textile innovation: Interactive, contemporary and traditional materials* (2nd ed.). Linc Publications.

Horrocks, A. R., & Anand, S. C. (Eds.). (2000). *Handbook of technical textiles*. Woodhead Pubs Ltd.

Jensen, E. (2005). Straight down the line. *Selvedge* (6), 24–27.
Keohane, A. (1994). *Bedouin. Nomads of the desert.* London: Kyle Cathie Ltd.
McQuaid, M. (2005). *Textiles: Designing for high performance.* London: Thames & Hudson Ltd.
Newton, A. (1993). *Fabric manufacture: A handbook.* Intermediate Technology Publications.
O'Hara, G. (1989). *The encyclopaedia of fashion.* London: Thames & Hudson Ltd.
O'Mahony, M. (2002). *Sports tech: Revolutionary fabrics, fashion and design.*
O'Mahoney, M. (2007). Full recovery. *Selvedge* (15), 46–49.
Quinn, B. (2010). *Textile futures: Fashion, design & technology.* Berg Publications.
Rowe, T. (Ed.). (2009). *Interior textiles: Design and developments.* Woodhead Publishing in Textiles.
Rowley, S. (Ed.). (1999). *Reinventing textiles: Traditions and innovation* (Vol. 1).
Schoeser, M. (1995). *International textile design.* Laurence King Publishing.
Schoeser, M. (2003). *World textiles: A concise history.*
Schoeser, M. (2005). Learning curve: colleges bend to the wishes of students. *Selvedge* (6), 56–59.
Smith, B. (2010). Life's rich tapestry. *Selvedge* (35), 56–59.
Sparke, P. (1987). *Design in context.* London: Bloomsbury Pubs.
Wilson, J. (2001). *Handbook of textile design: Principles, processes and practice.* London: The Textile Institute, Woodhead Publishing Ltd.

CHAPTER

SUSTAINABLE TEXTILE PRODUCTION

22

M. Tomaney
University for the Creative Arts, Epsom, UK

LEARNING OBJECTIVES

At the end of this chapter, you should be able to:

- Understand the complexity and subjectivity of issues that govern definitions of sustainability in the production of textiles and apparel
- Attain a knowledge of the social impacts of textiles and apparel production
- Understand the implications of fast fashion on environmental and social responsibility
- Map the relationship between socio-economic development and textile production, both at craft and industry level.
- Understand the role of governance, governments, NGOs and other organisations working within the sector
- Understand the government of 'eco' labelling

22.1 INTRODUCTION

This chapter aims to help you map the complex and diverse issues that may be defined within the broader context of sustainability in the production and the consumption of fashion and textiles. In addition to outlining environmental aspects such as the overconsumption of textiles and direct environmental impacts in the processing of textiles, the chapter will discuss aspects of social responsibility. These aspects include the capital of fashion, textiles and other crafts as a driver for socio-economic development through social enterprise, campaigning, non-governmental organisation (NGO), and government-led initiatives that might be garnered by fashion, textiles and design professionals wishing to position their practices in the wider context of sustainability in this area.

Ethical and environmental debates are steering industry and government to address recent historical paradigms in textiles as in other sectors, and this is reflected within academia and in the depth of academic engagement with this debate. The current generation of design practitioners have been imbued from childhood with a sense of responsibility to the future of the planet, whilst the anarchic energy of youth is an opportunity to innovate and to reinvent systems of practice. This chapter aims to be enlightening and occasionally inspiring to a wide spectrum of fashion and related disciplines; creatives – designers, makers and those engaging in the creative promotion of fashion; students engaged with the

Textiles and Fashion. http://dx.doi.org/10.1016/B978-1-84569-931-4.00022-2

systemic processes of the fashion industry through management, marketing or buying; and those whose study is the dissemination or analysis of fashion, through journalism, theoretical engagement or social observation. Each will find information relevant to their practice.

The chapter supports individuals to research further and to confidently engage in a vital debate that has become increasingly relevant in the spheres of education, creative practice and business that touch fashion and textiles, a debate that continues to evolve whilst bridging an international community with inclusivity; The hope is that this chapter will arm its audience with the means to participate with an open and informed mind. Some of the issues that find place in this debate are extremely complex, reflecting the multi-faceted supply chain of the textile product outlined by others in this edition in greater detail. In some cases sets of issues appear in conflict, especially to the fledgling fashion practitioner, optimistic at the outset of a creative career, or the crafter wishing to create a product whose integrity is holistic. The chapter does not make promises to simplify the debate, rather to elucidate viewpoints and drivers that give it shape. I do not offer an A–Z of designers and other creatives whose work engages sustainability; in a digital age such information is profuse and easily obtained, as well as being widely defined. The aim here is that through better understanding what underpins ethical and sustainable fashion and textiles issues, the readers' contribution will be multi-dimensional and informed, with potential to innovate and influence.

22.2 KEY ISSUES IN SUSTAINABILITY

How do we define what is sustainable or ethical in fashion and textiles? The complexity of the systems employed in the production, the marketing, distribution and consumption of textiles offers a wide range of approaches to this question. Engagement with one aspect of sustainability in one part of the supply chain, from field to factory to showroom to consumer, might very well preclude a sustainable approach to another aspect of this chain. A designer or fashion buyer who wishes to offer a product that has been produced in a responsible way must first therefore understand the mechanisms for bringing a product from raw material to finished product, which for a fashion product, can include many layers of processing, not always completed in the same location. The manager of the product and its marketing (the designer, maker, brand or retailer) by the definition of fashion has an awareness of the consumption patterns in their own market, and they are responsible for the evolution of this in regard to their own designs at the point of sale.

A fractured supply chain makes traceability of processing (especially that of the textile component of the garment such as yarn, fibre and coatings) an enormous challenge, not only for big brands who have huge budgets to spend on factory inspections, but also for small entrepreneurial designers who would like to offer a personal guarantee to their customers. Independent designers are often dependent on the integrity of suppliers or 'middle men', who may themselves be relatively remote from the origins of the fabric production, dealing with commercial agents located close to sourcing bases.

For many ethical fashion companies, a kind of storytelling is employed in the communication of an ethical brand, enabling the detail of specific known elements of the supply chain to be disseminated directly to the consumer – for example, a relationship with a specific producer who is weaving cotton, or a particular natural dye or organic fibre that is used in the product; this is most successful where the design is strongly related to the story being told. The Body Shop's Community Trade Programme sources ingredients and accessories, in some cases organic or fairly traded, in some cases sourced from small community-based producers in the developing world, the branding effectively communicating the story of the product, often with an image of one of the actual producers on the label. Fairtrade cotton is communicated in a similar way, connecting the consumer directly with the producer in a very human way, in this case through Fairtrade campaigns and the associated 'Fairtrade Website'.

The majority of fashion consumers, however, are fairly remote from the concept of a complex supply chain. As an example, someone buying a pair of denim jeans might consider the cotton – is it organic? Is indigo dye an environmentally responsible process? How was the water used in the dyeing process disposed of? Or they may consider the production of the jeans – the factory in which they were stitched – having read somewhere in the press about labour rights abuses or child labour in developing countries. They might consider the sandblasting process that is commonly used in the finishing process of denim jeans, again something they may have heard about in the press, or the chemicals used in the final washes to give the faded quality that is universally demanded by consumers from Nairobi to Beijing to Bradford.

If a person is shopping for a pair of jeans, these concepts could be floating somewhere in their consciousness, but there is usually a very limited level of real information easily available. There is a limit to how far even the most ethically conscious consumer will bend his or her taste levels and shopping habits. As a society we tend to be brand loyal – I'm a Levi's guy, or a Wrangler girl, or I want a Top Shop cut or a Diesel finish. Fashion is not exclusive to the fashionable – our clothing is our identity, and the style or brand of jeans we choose to wear is an illustration of how subjective even basic fashion choices can be, reflecting the identity of the wearer in subtle ways. Thus, a designer who chooses to work in an environmentally or ethically responsible way takes on a huge responsibility that is not necessarily fully comprehended by the customer. Along with that responsibility come several layers of added costs that can be a barrier to communicating the environmental benefits to a modern consumer who is accustomed to buying a lot of 'stuff' at an 'affordable' price tag.

The choice and availability of sustainable fashion has mushroomed since 2000, yet an awareness of it remains largely niche. A real challenge to designers of sustainable fashion remains the aesthetics, the limited options in components – textiles, buttons, yarns, and so on, that can have a real impact on creative direction of the product, alongside the potentially higher costs that must be absorbed or passed on. An intelligent approach is to make the key sustainable feature the focal point of creative capital, as brands such as From Somewhere and Junky Styling; both have successfully created brand equity that *is* the socio/environmental equity their brands deliver. In both cases the key creative driver is unique methodology to upcycle used or surplus textiles, and in both cases this has become the story, and the point of engagement with press and consumers. As in any successful and long-lived fashion label, these brands are backed by determined, visionary designers with a market focused product – the special ingredient x that is difficult to quantify.

22.3 THE TEXTILE SUPPLY CHAIN

As has been noted, textile products typically pass through a complex life cycle beginning with the production of the raw materials required for manufacture through to manufacturing, distribution, use and, finally, disposal. This supply chain is characterised by a huge range of processes, people and places that are hard to track, even for major manufacturers or retailers. All of these stages have an impact on the environment. The impact of textile products is particularly significant because they are so widely used, whether as apparel or as technical textiles in such areas as medicine or civil engineering. A very simple diagram showing key stages in the life cycle of apparel is shown in Figure 22.1.

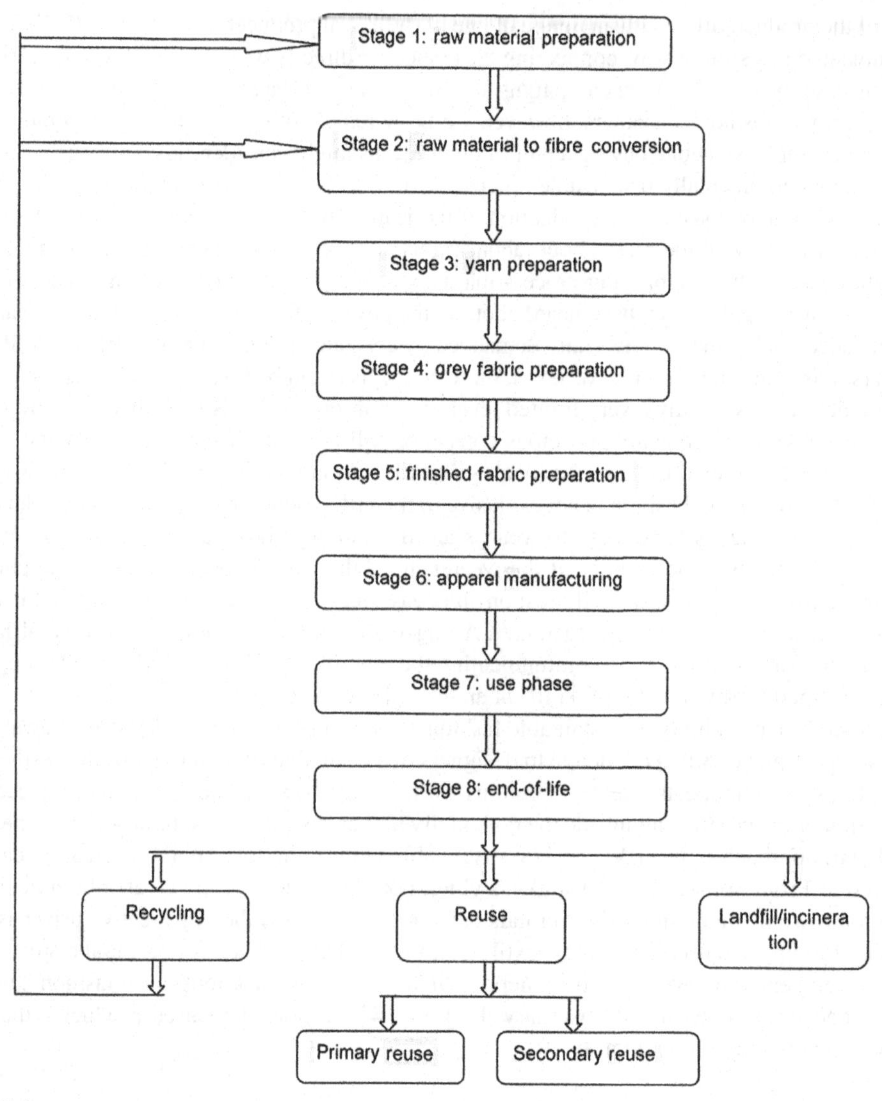

FIGURE 22.1

Key stages in the apparel life cycle.

22.3.1 SUPPLY CHAIN FOR FABRICS MADE FROM NATURAL FIBRES

The first step in a product's life cycle is production of the raw material. A basic distinction is between natural and synthetic fibres. Natural fibres can be classified into plant fibres (e.g. cotton) and animal fibres (e.g. wool) – see Chapters 1–3 for more on natural fibres. As an example, Figure 22.2 shows key production stages in the processing of cotton, the most widely used of the natural fibres. Although processes vary between fibres, cotton is a good example of the kind of environmental impact a natural fibre can have.

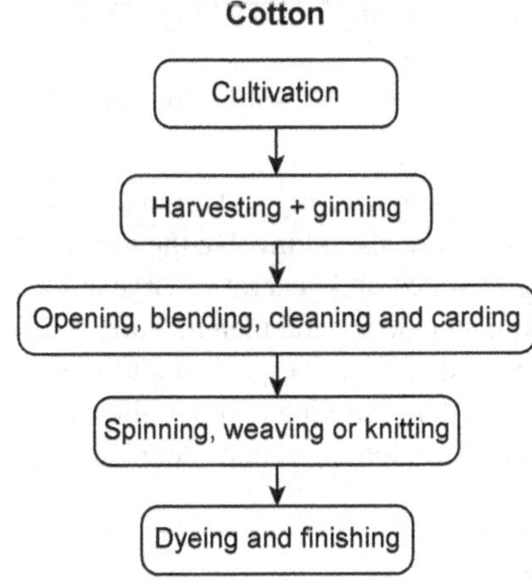

FIGURE 22.2

Key production stages in the processing of Cotton fibre.

Cultivation involves the preparation, planting and tending of the cotton plants. Resources involved include seed (which has its own resource requirements), water (for irrigation), pesticides and fertilisers (which again use resources in their production), machinery (e.g. for irrigation, spraying and harvesting), energy (e.g. to power the machinery), materials (e.g. packing materials for baling cotton) as well as labour. Other types of environmental impact include emissions and potential pollution (e.g. from the use of insecticides). It has been estimated, for example, that cotton uses 2.5% of the world's cultivated land and 16% of the world's pesticides (Muthu, 2014). This has led to increasing interest in growing organic cotton which avoids the use of synthetic fertilisers and pesticides, reducing overall resource and energy use, emissions and pollution as well as increasing biodiversity.

Once it has reached maturity, cotton must be harvested, the fibres separated from the seed by a process called ginning and packed together in bales to be transported to a processing plant. There the cotton bales must be opened, cotton fibres blended to achieve the desired quality, and then cleaned and carded to get them ready for spinning and other processes.

Spinning, weaving or knitting and other processes for fabric production require factory buildings (including humidification systems to create the right conditions for processing), spinning or other machinery, significant amounts of energy to run the machinery as well as materials such as chemicals to improve fibres for processing (e.g. use of sizing materials to strengthen yarns for weaving), lubricants and packaging materials. A process such as spinning also produces noise, dust, fibre and other types of waste. Whilst studies of the environmental impact of both weaving and knitting are less extensive than for spinning, it is generally assumed that knitting consumes less energy than either spinning or weaving.

Because of the chemicals and volumes of water required, dyeing has traditionally been a significant cause of pollution from disposal of waste water. Key finishing processes include:

- **Singeing:** passing fabrics over a flame that burns 'fuzz' off the surface to leave it looking smoother
- **Desizing:** removal of sizing materials (used to strengthen yarns for weaving)
- **Scouring:** application of chemicals to remove unwanted residues (e.g. waxes) from natural fibres

- **Bleaching:** application of chemicals to produce a whiter, more evenly coloured material for dyeing
- **Mercerising:** the use of sodium hydroxide to swell fibres and improve their strength and appearance

Many of these processes require subsequent washing and drying steps. Finishing processes (see Chapters 18 and 19 for more on textile finishing processes) involve the use of strong, potentially harmful chemicals as well as significant volumes of water, and generate a large amount of contaminated waste water.

Further stages in the supply chain include apparel manufacture, including garment cutting, sewing and finishing operations. Finished clothes must then be packed and transported to retail outlets for sale to consumers. Customer use involves washing, ironing and possible repair. It also involves issues of recyc ling or disposal once an item of apparel or clothing is considered to have reached the end of its useful life.

22.3.2 SYNTHETIC FIBRES

It has been estimated that over 60% of the world's apparel is made from synthetic fibres (see Chapters 4–7 for more information on synthetic fibres). The majority of synthetic fibres are polymer based. One of the most widely used is polyester which will be used here as an example. Figure 22.3 shows the key stages in the production of polyester. Polyester is made out of purified terephthalic acid, dimethyl terephthalate and mono-ethylene glycol. These raw materials are derived from crude oil. Extraction of crude oil requires very energy intensive processes (e.g. drilling) with significant emissions as well as pollution risks. Refining requires large amounts of heat as well as other resources and also generates significant emissions and waste. The production of the resins used in polyester manufacture also requires further energy and resource inputs.

The manufacture of polyester fibres starts by producing a solid fibre-forming material in the form of polymer chips. The chips are then converted to a spinning fluid (or dope) using heat or a solvent.

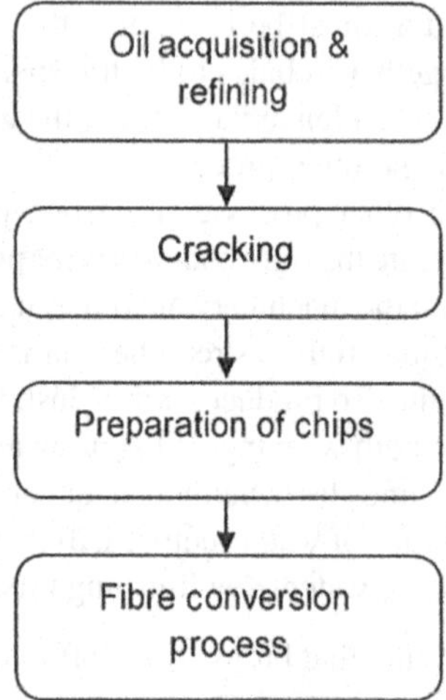

FIGURE 22.3

Key stages in the production of polyester.

Dope fluid is extruded, converting it into filaments that can then be solidified for spinning. Typical spinning techniques are melt, wet and dry spinning. All these processes require significant amounts of energy and extensive use of chemicals, and generate both emissions and waste.

It has been estimated that the energy required to produce polyester fibre can be as high as 125 MJ/kg of fibre. This compares to estimates for conventional cotton of 55 MJ/kg and as low as 15 MJ/kg for organic cotton (Muthu, 2014). Emissions of carbon dioxide are also much higher. Polyester production generates volatile organic compounds, which are both carcinogenic and particularly damaging to the ozone layer, potentially helping to accelerate global warming.

There are some areas where polyester may have less environmental impact than a natural fibre like cotton, for instance in the more limited requirement for water. Synthetic fibres may need fewer finishing treatments than natural fibres. However, fibres like polyester will not accept natural dyes, relying instead on more environmentally damaging synthetic dyes. Whilst synthetic fibres like polyester may be more durable than natural fibres, extending the life of apparel, polyesters are not completely biodegradable, making waste disposal a major problem.

22.4 ASSESSING THE ENVIRONMENTAL IMPACT OF THE TEXTILE SUPPLY CHAIN

As discussed in the preceding section, the production of apparel and other textile products involves a complex supply chain with a broad range of environmental effects, including:

- Use of scarce natural resources such as land and water
- The need for potentially harmful chemicals for raw material production and processing
- Energy use for processing
- Emissions and effluent from processing, transport and retail operations
- Large volumes of waste products requiring disposal

Defining and quantifying these impacts, identifying particular problem areas for improvement and making sensible comparisons between different fibres and processes is a huge challenge.

The best established methodology for assessing the environmental impact of textiles and other products is Life Cycle Assessment (LCA). LCA assesses the range of environmental impacts of a product 'from cradle to grave', i.e. from raw material production to disposal of the final product at the end of its useful life. The International Standards Organisation (ISO) defines LCA as the 'compilation and evaluation of the inputs, outputs and the potential environmental impacts of a product system throughout its life cycle'. LCA provides the basis for eco-labelling schemes designed to allow consumers to select more environmentally-friendly products.

The key steps in LCA methodology are:

- Goal and scope definition
- Inventory analysis
- Impact assessment
- Interpretation

Goal and scope definition involves setting objectives for the study. It requires definition of the product to be studied, the key production processes to be analysed and the type of environmental impacts to be measured (e.g. amount of energy or water used or the amount of carbon dioxide produced). A key aspect is the 'functional unit' which defines precisely what is being studied. An example might be 'a T-shirt made of 100% cotton to be worn 30 times in a 1 year life span'. In this case both the material and the assumed amount of use are clearly defined. A precise definition makes it much easier to define the scope of the study and to make useful comparisons with other products.

Inventory analysis involves constructing a flow chart of the overall production process and breaking it down into separate unit processes. Each unit process can then be analysed in terms of its inputs, e.g. energy and raw material requirements, and its outputs, e.g. emissions to air, water and soil, and generation of solid waste. This range of inputs and outputs takes in the product's range of impacts on the environment. This stage also requires a data collection plan showing where all the information for calculating different types of impact is coming from. Data collection typically involves direct data, i.e. data related directly to the production of the product, and indirect data, i.e. inputs and outputs related to those materials and processes that make production possible (e.g. fertilisers and pesticides for cultivating a crop like cotton).

Impact assessment defines the key categories of environmental impact, the key units of measurement and sets out calculations of the various inputs and outputs.

The interpretation stage sets out the key results and recommendations, e.g. those unit processes with the greatest environmental impact and what might need to be done to reduce that impact.

A good example of an LCA study is provided once commissioned by Levi-Strauss for one its jean products: the Levi's® 501® using a medium stonewash finish produced for the US market during 2006. The key results for carbon dioxide emissions (as a measure of contribution to climate change), energy and water use. The LCA indicated:

- Carbon dioxide: the consumer use phase accounted for the greatest impact, followed by fabric production and garment assembly
- Energy use: it was greatest in the consumer phase followed by fabric production and garment assembly
- Water use: cotton cultivation used the most water, followed by consumer use

Typically LCA studies of textile products conclude that the greatest environmental impact (75% or more) occurs during use by consumers in water and energy use for washing and drying clothes.

22.5 MINIMISING THE ENVIRONMENTAL IMPACT OF THE TEXTILE SUPPLY CHAIN

The use of methods such as LCA makes it easier to identify which stages in a product's life cycle have the greatest environmental impact. It also allows comparison between different processes such as conventional and organic cotton production. A designer must work within the brief set by the company developing a new product range. As has been noted, the product must meet the aesthetic and functional requirements of customers at a price they are willing to pay. It must be cost-effective and easy to manufacture. These requirements will determine the choice of fabrics, colours, patterns, method of manufacture, finishes and accessories such as fastenings. It may also determine where a product will be manufactured, which may further limit choice.

Within these constraints, designers need to take a life cycle approach to design, i.e. to think through the potential impact of their design from raw material production all the way through to the disposal of the product. A designer should be aware of the relative environmental impacts of different textile fibres in terms of their method of production, durability and recyclability. Durability can be important in that a product with a long life may have far less overall environmental impact than a series of more disposable items. Designers should also have an understanding of the environmental profile of different manufacturing processes. They should also be aware of how customers are likely to treat their product, e.g. washing and drying, which, as noted earlier, has a particularly significant environmental impact. Clothes that need less washing at lower temperatures, for example, mean substantially reduced emissions, energy and water use.

Within any design brief or project the ideal is to put sustainability at the heart of the design and the brand identity as companies such as From Somewhere and Junky Styling have done. Other options for more environmentally friendly design include:

- **Design for disassembly**, e.g. clothes using only biodegradable fabrics designed so that they can be easily dismantled for recycling
- **Design for durability**, e.g. modular clothing where worn elements (e.g. cuffs or collars) can be replaced whilst the core garment is retained
- **Design using recycled materials**, whether recycled textile fibres or non-standard materials

Designers can use their designs to stimulate greater environmental awareness amongst customers and create an eco-friendly dimension to an existing brand, which helps to differentiate it from competitors.

Manufacturers also need to be aware of LCA studies to identify which activities to focus attention on in reducing environmental impacts. A basic but significant contribution to reducing the environmental impact of manufacturing, whilst reducing costs and increasing overall quality, is to use 'best available technology' as well as implement good process control in conventional production (Tobler-Rohr, 2011):

- Selection of machinery with good energy efficiency, reliability and service life
- Good maintenance of process machinery
- Setting objectives for energy use, levels of waste and emissions
- Precise measurement and control of materials and energy consumption as well as emissions and waste
- Closed loop production systems e.g. for recycling and re-use of process water

Like designers, manufacturers need to be aware of customer expectations and behaviour in the use of clothing. Manufacturing durable clothes to a high quality of finish which retain dimensional stability and colour fastness after repeated washing and drying cycles, for example, will mean a longer product life and reduced waste. Clothes that can be washed at lower temperatures/reduced washing cycles will substantially reduce their overall environmental impact. Effective labelling and marketing by retailers can advertise these benefits to consumers in terms of cost, convenience and contribution to sustainability.

Within the constraints of cost, functionality and product positioning in the market place, both designers and manufacturers can also choose between different materials and technologies on the basis of their sustainability. It may be possible, for example, to select organically grown cotton or a biodegradable synthetic fibre such as poly(hydroxyalkanoates). Similarly, it may be possible to use a natural rather than synthetic dye, or select more environmentally friendly technologies such as biotechnologies (e.g. using enzymes) and plasma technologies for finishing (Blackburn, 2009).

22.6 CASE STUDY: CREATING SUSTAINABLE AND SOCIALLY RESPONSIBLE FASHION

In recent years, the fashion industry has been dogged with bad publicity and continued criticism from press and campaigning NGOs on ethical issues, particularly labour standards in fashion production chains where joined up international legislation has been less structured than in the environmental area. The consumer confounded by a lack of clearly available information at point of sale, and a confusing array of systems and labels (organic, Fairtrade or fairly traded, Better Cotton Initiative, etc.) that are used in marketing ethical and environmental interventions used in fashion.

The highly competitive fast fashion market that mushroomed at the beginning of the twenty-first century has its roots in the increase in number of fashion seasons – from the classic two seasons a year – autumn/winter and spring/summer, with a little seasonal injection at Christmas or high summer, to four, six or eight seasons a year, in the 1990s, to the current norm among mass market fashion retailers – that is, extremely short lifecycles, overlapping seasons that defy definition and very fast turnaround of new product at increasingly competitive price points.

Consumers at certain market levels demand new product on an almost weekly basis, with a direct effect on both social and environmental responsibilities in fast fashion sourcing locations. The high demand to produce and deliver orders very quickly that is commonly placed on suppliers in developing and emergent economies (largely but not exclusively China), when combined with cut-throat costing policies, effectively means that it is extremely challenging for a supplier factory to implement the responsible policies that are often requested by the buyer brand as a business requirement (this is especially true of some, certainly not all, of the bigger mainstream brands whose implementation of ethical sourcing codes can be a direct response to critical public relations). The costs and fast turnaround can preclude the level of intervention that is demanded by the buyer's company, with a rock-bottom wholesale price, often accompanied by the lack of long-term commitment that would support investment in supply chain ethics.

Fast fashion means fast changing product type – sequins and shiny satins one week, hand embroidered knits and cotton jerseys the next, so that brands and buyers often choose to keep their options open to switch suppliers or even sourcing locations, whenever they choose – this means that a supplier might delay costly infrastructure improvements in its factories, and may choose to work with migrant workers to avoid employment benefits and protected contracts, to deal with the short-term demands of an order that may not lead to a sustained business arrangement. So the workers lose out so that buyers can put newness and variability and perhaps innovation into European shopping centres and malls.

22.6.1 FUTURE THINKING IN SUSTAINABLE FUTURES

The last few years have seen some fledgling changes in fashion consumption, perhaps in direct response to the blandness and overconsumption of fast fashion. Vintage is now mainstream and widely available, and there is a resurgent interest in strategies to extend the life of textiles and garments such as creative repairs, re-inventing new from old, craft, customisation and swishing parties to exchange fashion and accessories among friends and colleagues. As noted earlier, brands such as Junky Styling have created successful business cases out of the concept of re-inventing 'timeless, deconstructed, re-cut and completely transformed' new clothes from old. The From Somewhere brand has successfully applied its model of taking textile surplus, manufacturing offcuts and mistakes to the Florence and Fred kidswear range for Tesco's, working directly with Tesco's suppliers in Sri Lanka to develop effective qualitative and economically viable production models – demonstrating that techniques starting out as low tech and craft based can be scalable in its application to mainstream business.

TRAID is a non-profit UK-based clothes recycling organisation founded in 2002. Clothing is donated by the UK public using TRAID's estimated 900 recycling banks across the UK. It is transported to a central warehouse where it is sorted by hand according to quality and style. A team of designers then redesigns and reconstructs clothes to create new one-off pieces which are sold under the award-winning TRAID remade fashion label in one of the organisation's charity shops. TRAID's activities offer an impressive illustration of a holistic sustainable fashion organisation; its activities consider ethics, socio-economic development, and longer term aspects of fashion consumption alongside a clear commitment to the environmental concerns for which it is primarily known. Profits are used to fund TRAID's other activities, as well as support local sustainable initiatives such as community 'learn to sew' workshops or overseas projects to alleviate poverty.

Many designers choose to work with developing countries for a variety of reasons that include access to unusual or beautiful textile craft skills and products as well as an intention to support poverty reduction through linking high-value markets in northern economies with poor artisans in the south. The vision of a designer is not usually something we quantify within the design community. However, to an artisan or producer who is working to build an enterprise that is extremely remote from the community it serves (i.e. the end user who will eventually purchase the product at retail or online), the designer's vision, knowledge of the market needs and the access it brings to new markets is a clear way for the artisan to add value by creating a market-focussed product with a defined market destination.

Both the designer and the supplier, which may be a small entrepreneurial co operative, or an NGO acting as a kind of benevolent middle man or marketing umbrella, need to have a clear idea of mutual expectations, as well as the systems that should be developed in order to deliver those expectations – concerns might include quality, colour matching, finishing and trims, to fundamental ability to deliver the order. Does the artisan organisation have the production and other capacity to fulfil the order on time? Is the raw material seasonal or crop dependant? How will the profits from business activity be invested, spent or shared and who will account for that?

Transparency is vital in small operations just as it is in mass market supply chains. It is important that designers are realistic about what they are able to deliver to the artisans they work with, and to understand the responsibility that is invested in the buyer by artisan communities with whom they are working, who will naturally have high expectations of the relationship from the outset, and may not understand fully the capricious and demanding nature of western markets – especially fashion & lifestyle markets!

Some of the most successful collaborations between designers and artisans in the context of international development happen when the designer or brand is able to slow down the consumption process by offering a product that has a potentially sustainable place in the market – which often means that the product is based much more directly on skills than the traditional fashion product, whose influences tend to come from the outside. A skill-based product still needs updating, often restyling to meet the changing market, but crucially, the same artisans would be able to work on the orders year on year, with seasonal training updates to adapt skills.

Pachacuti and People Tree both show very good examples of skill-based fashion. Pachacuti is a fair trade fashion brand who primarily work with artisans in Latin America. They specialise in Panama hats, which are adapted each season to extend the range with fashion colours and trims, modified shapes and fashion brand tie ins. The Ecuadorean hat cooperative they work with is able to harvest the grasses for production so there is no need to outsource – dying, weaving, finishing (often with bought in trims) and packing and the hats can all be done by the cooperative, so that almost the entire value chain is contained. This model is effective, keeping profits within the artisan organisation to facilitate training programmes, social benefits and environmental husbandry.

An effective small-scale supply route, like Pachacuti's, requires that systems be applied to what might be considered culturally to be everyday skills, so that they can be upscaled to meet the demands of overseas fashion and textiles markets, thus a very tightly guided systemic framework is essential to the success of delivery to the market, as in any business, big or small. Creating eco fashion within the framework of big business can be a challenge to mass market pioneers of ethical fashion, who have to negotiate the corporate machinery, but scalable systems and a good understanding of the way the retail world operates are essential to a small-scale eco designer trying to bring a handmade product to a worldly and demanding marketplace.

The creation of fashion and textiles artisan craft or ethical centres, can contain textiles skills that are passed from say mother to daughter, or those maintained through sewing circles, which exist all over the world. At the grass-roots level, the key elements of production – sewing, weaving, printing, embroidery, knitting – are activities that can be accessed with low-scale investment, making fashion or textiles a really useful tool for the progression of trading as a means of development, something that international agencies, governments and NGOs have accessed in the past few years.

Fashion and textiles primarily targets women, and thus offers an opportunity to build diverse sets of female stakeholders, the producers, craftswomen, designers and consumers of fashion, thus the international development community has accessed fashion and textiles as a part of gender programmes, where textile skills can be effectively harnessed to support female entrepreneurship. The International Trade Centre (ITC) is an agency of the UN, whose successful Ethical Fashion programme has garnered the support of Vivienne Westwood to create designs for one of their African projects aiming to build the capacity of women producers to export textiles. Westwood has incorporated cloth bags produced as part of the ITC programme into her mainstream collections, with a clear positive impact on the value and market for the bags.

Westwood's team is able to build their own product development and trading on the rigorous grounding of a capacity building exercise that represents several years work from the ITC's team of experts in the field – including experts in economics, enterprise development, international development and market access. The international high profile that the Westwood brand brings to the project adds value as well as providing a solid market – that is, a market for Vivienne Westwood's designs, that creates a brand equity for the African producers of the bag.

This is an excellent example of the fashion world and social science creating a coherent conversation, merging two worlds whose language is not always compatible. As more designers, especially young start-up brands, embrace the concept of design as a means to make social change, the communication between the government and campaigning sector and the creative community becomes more transparent with effective results, and the ITC project is one of many that works with designers to implement social and environmental change, whilst organisations around the world such as the UK-based Ethical Fashion Forum, Redress in Hong Kong and others offer guidance to enable the conversations.

22.7 SUMMARY AND PROJECT IDEAS

1. Project for management, product and design students.
   Design a sustainable fashion trajectory for a mainstream brand.
   Consider the market, product, communications and supply routes for the brand or business. Who is the customer, and the competition in the market? What defines the brand values?

How might the company successfully collaborate with an aspect of fashion that is more environmentally, and/or socially, sustainable, in a way that both fulfils the core brand values, and supports the sustainable objectives you have defined? Design a way for the company to build a sustainable fashion or textiles sub brand or brand extension into its current structure. It is important in this task that you consider the system and the market as well as the product.

2. Project or designer/makers
Using Websites such as World Fair Trade Organisation (WFTO) and Ethical Fashion Forum (EFF), research socially sustainable organisations in the developing world who are working with skills that complement your own. Consider the wider context of the organisations, and the artisans – for example, what political and economic factors impact on the lives and livelihoods of the community, and how might geography and economics impact on their ability to deliver their craft into export markets – for example, seasonality, or remoteness from a port or major supplier of fabric and threads.

Is there a bigger organisation supporting the community and the wider development plan?

Design a collaborative collection with one of the organisations – taking into account the capacity of the organisation to work with you, to understand your brief, produce and deliver your orders.

Consider where they will source materials, dyes, components – what is locally available, is it repeatable, and if not how can you creatively address this within your concept collection? Consider the available skill bases, and how you will manage them in production; does the product need finishing, for example, stitching or embroidery, that cannot be provided by the organisation you have selected to work with? How would you address that in real life?

Where, how and to whom would you sell the collection? How would you communicate the method of production?

22.8 SOURCES OF FURTHER INFORMATION AND ADVICE

Government & International Organisations:

ITC http://www.intracen.org/itc/projects/ethical-fashion/.

Defra Sustainable Clothing Roadmap https://www.gov.uk/government/publications/sustainable-clothing-roadmap-progress-report-2011.

Department for International Development https://www.gov.uk/government/organisations/department-for-international-development.

Campaigning and Research Organisations:

Pesticide Action Network http://www.pan-uk.org/.

Workers Rights Consortium http://www.workersrights.org/.

Better Cotton Initiative http://www.bettercotton.org/.

ETI http://www.ethicaltrade.org.

The Ethical Trading Initiative base code http://www.ethicaltrade.org/eti-base-code.

Labelling and Auditing Organisations:

Fairtrade Foundation www.fairtrade.net.

Soil Association http://www.soilassociation.org/whatisorganic/organictextiles.

The World Fair Trade Organisation http://www.wfto.com.

International Organization for Standardization (ISO 8000) http://www.iso.org/iso/catalogue_detail.htm?csnumber=50798.
REACH (the Registration, Evaluation, Authorisation and Restriction of Chemical substances – EU) http://echa.europa.eu/web/guest/regulations/reach.

Organisations That Are Focussed on Fashion & Textiles:

Redress http://www.redress.org/.
Ethical Fashion Forum http://www.ethicalfashionforum.com/.
Ecotextile News http://www.ecotextile.com.
Fashioning an Ethical Industry http://www.fashioninganethicalindustry.eu/.

REFERENCES

Blackburn, R. (2009). *Sustainable textiles: Life cycle and environmental impact*. Cambridge, UK: Woodhead Publishing Limited.

Muthu, S. (2014). *Assessing the environmental impact of textiles and the clothing supply chain*. Cambridge, UK: Woodhead Publishing Limited.

Tobler-Rohr, M. (2011). *Handbook of sustainable textile production*. Cambridge, UK: Woodhead Publishing Limited.

PART 4

DEVELOPING TEXTILE PRODUCTS: THE CASE OF APPAREL

CHAPTER

MATERIAL CULTURE: SOCIAL CHANGE, CULTURE, FASHION AND TEXTILES IN EUROPE

23

C. Ryder
Liverpool John Moores University, Liverpool, UK

LEARNING OBJECTIVES

At the end of this chapter, you should be able to:

- Understand and explain what is meant by the term 'culture'
- Understand and explain how a number of different influences can impact upon the culture of a specific region at a specific time
- Describe how the field of fashion and textiles relates to broader cultural contexts

23.1 INTRODUCTION

'Culture' is a term that describes the particular characteristics of a specific geographical region (e.g. British culture), civilisation (e.g. Mayan culture) or group (e.g. youth culture, drug culture), in a specific period of time. It can include the habits, attitudes, belief systems, values, artefacts and politics of a particular group of people at a particular point in history.

History has shown how shifts in society – political change, war, scientific developments and travel, leading to the discovery of new, foreign cultures – affect our own cultural practices, including those related to fashion and textile design. For example, during World War II, new clothing was difficult to come by, and a shortage of fashion fabrics meant that it was necessary to repeatedly wear and repair a family's existing clothes, resulting in a culture of 'make do and mend' (this is discussed in more detail in Section 23.4.3).

Changes in social thinking – reactions to the status quo and long-held beliefs – can also affect how we express ourselves through the arts that include:

- Music
- Film
- Painting
- Sculpture
- Ceramics
- Dance
- Theatre
- Dress

Textiles and Fashion. http://dx.doi.org/10.1016/B978-1-84569-931-4.00023-4

For example, in the 1960s, young people in Britain began to feel that the attitudes of their parents' generation – the wartime generation – no longer represented their own views. These young people had a positive vision of the future and expressed this positivity through changes in dress, as well as in music, theatre and art. The relationship between art and society is discussed in more detail in the following section, and the specific impact of culture on dress – fashion and textile design – is discussed further in Section 23.5.

23.2 ART AND SOCIETY

The art of a particular society is often described as its culture. Although the term 'culture' can be used as an umbrella term for a wide variety of characteristics related to a particular group of people, such as their behaviours and belief systems (for example, one could discuss the different religions that exist side by side in Indian culture), in modern society when one asks about the culture or cultural pursuits of a particular region, it is often assumed that one is talking about the arts.

Artistic endeavour has always been subject to societal influences. Throughout history, shifts in society such as technological advances, politics, war, travel and discovery have all helped to shape developments in the arts, including the performing arts (music, dance and theatre), literary arts (creative writing) and visual arts (painting, ceramics, sculpture, film and design – which includes fashion and textile design). Similarly, new developments in artistic endeavour have helped to shape society.

23.2.1 ADVANCES IN TECHNOLOGY: THE INDUSTRIAL REVOLUTION

Advances in technology regularly bring about changes in society's behaviour, and in turn these changes in behaviour influence the arts. One example of technological advances giving rise to a major upheaval in the arts is that of the Industrial Revolution. The term describes a period from around 1760 to the mid-1800s in Europe, during that time the gradual increase in mechanisation in a number of areas – including agriculture, mining, engineering and textile manufacture – led to a dramatic increase in production. The invention of mechanical processes of production meant that for the first time it was possible to quickly produce large quantities of a product. For example in textiles, innovations such as the Jacquard loom, the spinning jenny and steam-powered weaving looms speeded up the production of cloth from fibre, while in fashion, the design and development of the sewing machine meant that garment production moved from slow and laborious hand work to much faster machine production, giving birth to the mass-manufactured fashion that is so familiar today. Inevitably, the increase in speed of production was seized by industrialists as a means of increasing efficiency and profit.

The period that encompassed the Industrial Revolution also witnessed an increase in overseas travel, which meant that mass-produced goods could be transported more easily around the world to a dramatically increased customer base. At the same time, products from other countries – for example cotton, silk and tea – could be brought to Europe for trade. The East India Company, an English company founded to trade with the East Indies, became immensely powerful, and as its commercial function gradually became eclipsed by its military and administrative roles, the company eventually went on to rule large parts of India.

Against a backdrop of booming trade, Europe revelled in a climate of great wealth and power. The first line of the popular eighteenth-century British song: 'Rule, Britannia! Britannia, rule the waves!' embodies the mood of a euphoric nation, celebrating the prosperity and power brought about by its success in international trade. The second line of the anthem: 'Britons never, never, never shall be slaves', however, points to a deplorable practice that sadly contributed to the wealth being celebrated in the West. From the sixteenth to the nineteenth centuries, millions of people – mostly Africans, from central and western parts of the continent – were taken into captivity as slaves. Slaves were treated as property – without human rights – and were bought and sold as such. Bought by Portuguese, British, French, Spanish, Dutch and American slave traders from slave dealers who had taken them into captivity during 'slave raids', the slaves were transported to the America across the notorious 'middle passage' – a journey of six to eight weeks, sometimes longer, in horrendous, unsanitary conditions in which the human cargo was packed tightly together below deck, secured by leg irons. The mortality rate was sickeningly high:

> Surviving records suggest that until the 1750s one in five Africans on board ship died. Some European governments, such as the British and French, introduced laws to control conditions on board. They reduced the numbers of people allowed on board and required a surgeon to be carried. The principal reason for taking action was concern for the crew and not the captives. The surgeons, though often unqualified, were paid head-money to keep captives alive. By about 1800 records show that the number of Africans who died had declined to about one in eighteen.
>
> **The International Slavery Museum, Liverpool: http://www.liverpoolmuseums.org.uk**

On arrival in America, the slaves were sold by the traders and forced to work by their new masters, often on sugar, coffee, cotton and cocoa plantations, in industry, or as house servants.

As the Industrial Revolution progressed, and mass production increased, the design of goods became increasingly technology-led: that is, as machinery became available that could manufacture mass-produced goods more efficiently and profitably, items that could not easily be mass-manufactured were not considered for production at all, and the design of a product became secondary to its profitability. Plain or cheaply decorated machine-made products proliferated because manufacturers had little competition throughout the world, and minimising expenditure on design and decoration led to maximising profit.

By the early 1800s, artists, designers and patrons of the arts had become alarmed as the integrity of design was increasingly undermined, with the inevitable consequence of reduced employment for artists and designers. The architect Charles Cockerell (1788–1863), grandson of the diarist Samuel Pepys, commented: 'the attempt to supersede the work of the mind and the hand by mechanical process for the sake of economy will always have the effect of degrading and ultimately ruining art'.

A reaction to technology-led design eventually took hold, and in the second half of the nineteenth century, the Arts and Crafts movement, championed by William Morris (1834–1896) and his associates, celebrated the old ideals of beautiful craftsmanship, colour and artistic form. The movement led the world into a new appreciation of craft-based art and aimed to raise the status of the applied arts to that enjoyed by the fine arts. For a period of time, the concepts of design and craft – as distinct from the fine arts defined by the Royal Academy – enjoyed a new-found status at the height of fashion, and singly produced, hand-crafted artefacts were revered among the elite of society.

Textile design was a particularly important feature of the Arts and Crafts movement. The virtues of natural materials were emphasised, and design inspirations were also largely derived from nature. William Morris, the most famous of the Arts and Crafts pioneers, disliked chemical dyes and mass-production and responded by returning to traditional techniques such as vegetable-dyed hand-block printing and hand-loom Jacquard weaving (Figure 23.1).

FIGURE 23.1

William Morris, snakeshead printed textile, 1876.

However, a built-in problem existed within the very principles of the Arts and Crafts movement. The essence of the movement was backward-looking and elitist. It did not acknowledge the march of progress, and it ignored the benefits that the Industrial Revolution had brought about, such as the possibility for ordinary people to own inexpensive functional items for everyday use. The adulation of all things craft-based was not destined to last:

> In the twentieth century, particularly in the inter-war period, the arts and crafts movement was condemned on the grounds that its principles ran contrary to the interests of modernity, progress, and the importance of technology and industry. The emphasis on single items of furniture meant the neglect of thc importance of design in industrial production ... at the cost to British competitiveness.
>
> **McRobbie (1998, p. 21)**

Again, a social shift, this time political and economic, brought about a change of direction in the arts. The Arts and Crafts ethos became unfashionable, giving way to an emphasis on good industrial design. It is worth noting that even William Morris acknowledged the usefulness of modern technology; although he hated what he saw as the low quality of machine products, and is frequently regarded as 'anti-machine'. He *was* willing to use modern machinery as a means of producing his fabric designs and wallpapers more efficiently and at lower cost, although his company continued to produce its finer work by hand. This division in the production of different-quality goods – including fashion – continues today: while most fashion clothing is mass produced exclusively by machine, couture fashion (haute couture, literally meaning 'high sewing') is custom-fitted to a specific individual, often using time-consuming, hand-executed techniques, and its high cost means that it is available only to an exclusive, very wealthy clientele.

23.2.2 ADVANCES IN TECHNOLOGY: MODERN DEVELOPMENTS

In modern times, technological advances continue to affect art and design. New materials and production processes are being developed all the time, influencing what is possible in art and design, and providing artists and designers with inspiration for new methods of working. In textiles, advances in technology regularly give rise to new products. Throughout the twentieth century, the search for inexpensive alternatives to silk, cotton and wool led to the gradual introduction of synthetic fibres – most notably rayon, nylon, polyester and acrylic fibres – and these new materials revolutionised the production of fashion fabrics. In *Eco-Chic: The Fashion Paradox*, Sandy Black states: 'Manmade fibres now account for nearly 60 per cent of all textiles' (Black, 2008, p. 107).

The development of nanotechnologies in the 1980s has led to super-lightweight, durable fabrics and 'smart' or 'intelligent' textiles (materials and structures that sense, or react and respond to environmental conditions or stimuli). These include:

- Garments that change colour in response to external temperatures.
- 'Memory' fabrics, which return to shape after being stretched or distorted.
- Interactive electronic textiles, for example fabrics and garments with embedded music systems, telephones, computers and global positioning systems.
- Biotextiles, which are textiles used in specific biological situations, for example in medicine or sport.
- Microencapsulation (where microscopic capsules – microcapsules – trapped within textile structures are designed to release their contents under certain conditions for a particular purpose, for example to release fragrance, deodorant or antibacterial particles to protect the wearer from illness).

This enormous expansion in the range of available textiles has had a great impact on the way we live in the West, compared with 100 years ago (the first synthetic fibre, rayon, became available in 1904). Product designers, engineers and architects have been quick to utilise highly sophisticated fibres and fabrics with enhanced performance characteristics in the production of our modern, largely synthetic environment.

In the field of fashion, synthetic fabrics provide an inexpensive alternative to those made from natural fibres – which are now considered luxury fibres – and have been responsible for a dramatic change in the way fashion is consumed. Throughout the twentieth century, clothing became more and more affordable, and fashion – as distinct from clothing – was no longer a luxury, but something everyone could afford and increasingly took for granted. Most people in the West now have a diverse wardrobe as well as a range of regularly updated fashion items. Affordable technology provides us with specific clothing for specific activities, for example warm but lightweight clothing for sport and outdoor activities, and comfortable, crease-resistant easy-care clothing for work.

In fashion design, the relationship between people, clothing and computers is continually evolving. The use of computers has changed the way fashion design is rendered and communicated, and the demands of the design industry in a global marketplace – where separate components of a product might be designed and produced in different locations around the world – have meant that designing fashion or textiles has never been faster, or more efficient. Digital printing and knitting systems are now commonplace, as are computerised plotting systems, laser cutting machinery and automated production systems. The 3D printing technology, and 3D body-scanners integrated with knitting machinery, can create seamless fitted garments for a specific figure: seamless apparel now accounts for 20% of global underwear production.

Continuing research into the so-called 'techno-textiles', a term popularised by the book *Techno Textiles* (Braddock & O'Mahoney, 1998) and used to define those textiles that make use of technological advances, includes:

- Experimentations with electronic textiles that incorporate light-effects, sound, etc.
- Fibres that can be engineered from wood-pulp, seaweed or soya protein.
- Recycled plastics.
- Textiles that can be organically grown under certain conditions, for example 'spider-silk' – a protein fibre spun by spiders and having a tensile strength comparable to that of high-grade steel – can now be synthetically replicated. A version of spider silk called bio-steel can now be cultured synthetically in goats' milk.

The Biocouture fashion research project (www.biocouture.co.uk) is investigating the use of bacterial cellulose, grown in a laboratory, to produce clothing and simultaneously address the ecological and sustainability issues that surround fashion: 'Our ultimate goal is to literally grow a dress in a vat of liquid'. Innovations such as these will undoubtedly affect future developments in fashion and textiles design.

23.2.3 TRAVEL AND DISCOVERY

In the same way that technological advances have affected art and design throughout history, travel has traditionally yielded a rich source of inspiration for the arts. As improvements in transportation are made, for example during the Industrial Revolution, or with the advent of air travel in the twentieth century, an increase in the migration of people around the globe inevitably takes place with the result that cultural influences also migrate from one place to another.

The area of Spitalfields, a former parish of Tower Hamlets in the East End of London, situated near Liverpool Street Station and Brick Lane, illustrates the manner in which immigration can affect the culture of a region and is particularly relevant to the fashion and textile industries. Spitalfields' reputation for inexpensive living throughout the centuries has ensured that it has played host to a community built on waves of immigration from all over the world.

When the Catholic King Louis XIV of France declared Protestantism illegal by the Edict of Fontainebleau in 1685, he caused a mass exodus of between 40,000 and 50,000 French and Flemish protestants, known as Huguenots, who fled their own countries between 1670 and 1710 to seek refuge in

England, where King Charles II had offered them sanctuary. Historians estimate that around half of their number moved to London, with many settling in Spitalfields, attracted by cheaper food and housing.

The Huguenots who settled in London came from all walks of life – local records mention surgeons, jewellers, merchants and bakers – but the majority were involved in the textiles and clothing trade: silk weavers, tailors and lace-makers whose traditional skills had been honed in the Cévennes region of the South of France.

The silk-making expertise of the Huguenots had a huge impact on Spitalfields, in particular on its economy. There had always been a silk industry of sorts in the area, but with the diligence and skills of the Huguenots this industry flourished. For much of the eighteenth and nineteenth centuries, Spitalfields was recognised as a hub of high-quality garment and fabric manufacture, and Spitalfields silk became a world-famous export. Fashions for the British upper classes incorporated more of the readily available silk, including Queen Victoria's wedding dress in 1840. From the 1730s, Irish weavers also came to the area to find work in the silk trade after a decline in the Irish linen industry.

The silk trade died out in the 1930s, ironically in the face of French competition, and Spitalfields became better known for its fruit and vegetable market. The legacy of the Huguenots can still be seen, however, in the large windows of the weavers' lofts, and street names such as Fournier Street and Fleur de Lis Street, which reflect the area's French past.

From 1880 to 1970s, Spitalfields was overwhelmingly Jewish. Enterprising Jews from the Netherlands (known as Chuts) were followed by East European Jews. More than 2 million Jews left Eastern Europe between 1881 and 1941, fleeing from economic hardship and the anti-semitic pogroms (organised persecution or extermination of an ethnic group, especially of Jews) that swept across Russia and her neighbouring countries following the assassination of Tsar Alexander in 1881.

Large numbers settled in Spitalfields, again attracted by the inexpensive lifestyle offered by the area, and by the fact that it had been home to a Jewish population in previous centuries (Jewish families had first begun settling in Spitalfields in the seventeenth century).

Working in the clothing industry, the Jewish community thrived in Spitalfields for several decades. As they became wealthier, they gradually moved out to suburbs such as Golders Green and Hendon, a trend accelerated by the heavy bombing of the East End during the Second World War, the area being a prime target because of its docks.

Since 1970, the Tower Hamlets area in and around Brick Lane has been home to a thriving Bangladeshi community, mostly originating from the Sylhet region of Bangladesh. The Bangladeshi immigrants – known in Bangladesh as 'Londoni' – came to London to escape the political upheaval resulting from the Bangladesh Liberation War of 1971, when East Pakistan fought with India against West Pakistan, and the secession of East Pakistan gave rise to the independent nation of Bangladesh.

Following in the footsteps of their predecessors, many of the Bangladeshi community in Tower Hamlets work within the textiles industry, although many others have opened restaurants in the area, establishing Brick Lane – also dubbed Banglatown – as the curry capital of London.

The Tower Hamlets Londoni send money back to Bangladesh to support their families, with the result that Sylhet is now one of the wealthiest towns in Bangladesh.

23.2.4 ORIENTALISM

Another clear example of cultural innovation inspired by travel – this time travel from the West to other parts of the world – is that of 'Orientalism' or the 'Orientalist movement'.

During the Industrial Revolution, as improvements in transportation led to the discovery of new lands and previously undiscovered cultures, returning explorers brought home exotic goods and artefacts, which invariably inspired a fascination with distant lands and a fashion for products designed in a similar style.

Throughout the eighteenth and nineteenth centuries, increasing numbers of Europeans – colonialist explorers, missionaries or merchants – travelled to Asia, Africa and the Middle East, a broad geographical area referred to by Westerners of the time as 'the Orient'.

Of course, the nations and cultures encompassed by this broad term had distinct cultural traditions, but the Western mind grouped them into a unified whole: a colourful, sensual and mysterious place that represented exoticism, romance and adventure.

In fact, the Orient portrayed by Western artists was often an inaccurate, vague and fanciful place. Although many artists travelled East to witness different cultures for themselves, a great many more did not, and merely created fantastical images from their imagination. For example, the first great success of Eugène Delacroix (1798–1863), 'The Massacre at Chios' (1824), was painted before he visited either Greece or the East (Said, 2001).

Nevertheless, curiosity about these distant lands grew into an interest that radically influenced European taste. In design, fantastic creations proliferated as elements of the exotic pervaded the visual arts, literature and architecture.

Although the Orientalist art movement was predominantly a nineteenth-century phenomenon, from the seventeenth to twentieth centuries Orientalism influenced the fine arts, design, literature, theatre, architecture, music, poetry and philosophy of a number of different Western cultures, particularly European cultures such as the British, French, Russian and Dutch.

In design, the term 'Chinoiserie' describes the fashion in Western Europe for Chinese decorative themes that began in the late seventeenth century. In architecture, the use of motifs originating from the Indian subcontinent was known variously as Indo-Saracenic style, Hindoo style, Hindoostani Gothic, Indo-Gothic, Mughal-Gothic, Neo-Mughal or Hindu-Gothic. Examples include the façade of Guildhall, London (1788–1789) by George Dance, and the Royal Pavilion in Brighton (built in three stages between 1787 and 1822, with the designer John Nash responsible for the design of the present building).

After 1860, following the 'opening of Japan' described later in this chapter, shiploads of Japanese imports – fans, kimonos, bronzes, lacquers and silks – flooded into England and France. As a result, 'Japonism' or 'Japanese style' became an important influence in the Western arts, particularly for French Impressionist painters such as Monet, Degas, Manet, Pissaro and Cassatt. These artists drew inspiration from the flat areas of strong colour, compositional freedom, asymmetry and abstracted design of Japanese woodcut prints (Fukai et al., 1996).

In fashion, designs inspired by Orientalism were typified by luxurious, lavishly coloured fabrics, intricately patterned embellished surfaces, and dramatic silhouettes. Fur, a symbol of Orientalism, appeared on all manner of garments from outerwear to lingerie. Gowns inspired by the Orientalist movement were often accessorised with Oriental parasols and coolie-inspired 'lampshade' hats.

Imports from the East India Company during the first half of the seventeenth century made the cashmere Paisley shawl enormously popular (cashmere is the strong, soft, fine wool obtained from the cashmere goat; the word is a variation of the spelling of Kashmir, the Indian region where cashmere shawls had been produced since about the eleventh century). The Paisley pattern, characterised by the now-familiar bent-teardrop shape, originally hailed from India, Pakistan and Persia (now Iran), where

FIGURE 23.2

Paul Poiret Directoire-style designs.

Source: Paul Iribe illustration from 1908.

the pattern was known by a number of different names depending upon the region. 'Paisley', the Western name for the pattern, is the name of the central Scottish town that produced enormous numbers of the shawls in the nineteenth century, and whose weavers had a reputation for quality and intricacy of pattern, despite only using a quarter of the colours used in the original Kashmiri imports (Koda & Martin, 1995).

The creations of French designer Paul Poiret (1879–1944) were inspired by the Orientalist style. Poiret had become famous for designing garments with a columnar (straight, sheath-like) silhouette, described as 'Directoire' due to their similarity to the classical-inspired fashions of the French Directoire period 100 years earlier (1795–1799). At a time when S-bend corsets and bustles bent the female figure into a series of dramatic curves, Poiret's unrestrictive, un-corseted, straight silhouettes seemed revolutionary (Figure 23.2). His clients also wore Orientalist-inspired kaftans, turbans and draped Greek styles in rich brocades and velvets.

In 1903, Poiret designed a coat based on the shape of a Japanese kimono, and in 1911 he produced exotic designs for full pantaloons (or jupe-culottes) that were based on the Oriental harem pants worn under a tunic. The same tunic could also be worn over a skirt, in a style dubbed the Persian silhouette by the fashion press, in accordance with its supposed origins (Figure 23.3). The skirted variation was widely copied by other designers as it was more acceptable to society at the time.

Although it was some years since the idea of women wearing trousers had been introduced to Western society – the American Amelia Bloomer (1818–1894) had introduced her 'bloomers' costume in 1851 – at the time this mode of dress had only been adopted by fringe groups of Victorian dress reformers, who were subjected to widespread ridicule. By the time Poiret presented his pantaloon styles in 1911, the

FIGURE 23.3

Denise Poiret wearing Orientalist fashion by Paul Poiret, 1913.

fashion for cycling had instigated a gradual acceptance of divided skirts and bloomers-style trousers for cycling and more athletic pursuits, but the idea of wearing trouser styles as a fashion statement was still very unusual. It must be noted that in promoting trouser styles for fashionable women, Poiret was not attempting to make a political statement of any kind about liberating women from the constraints of their clothing: his intentions were purely aesthetic. Indeed, around the same time he produced designs that reflected the short-lived fad for the 'hobble' skirt (1911–1913), an extremely impractical skirt shape that was tight around the ankles and impeded a woman's ability to walk. Nevertheless, it is undeniable that the acceptance of his loose 'Oriental' designs as fashion statements by the elite of society did much to promote the idea of more comfortable clothing for Western women of all classes – a dramatic cultural shift triggered in the first instance by improved transportation and the increase in overseas travel.

23.2.5 THE SPACE RACE

Later in the twentieth century, another overwhelming design phenomenon was inspired by a combination of technology and travel. Post-war technological advances in the 1950s and 1960s meant that

FIGURE 23.4

Apollo bed-spread.

Source: Image courtesy of Peter Kleeman, www.peter-kleeman.com; www.spaceagemuseum.com.

travel to the moon had become a real possibility. The ensuing excitement about the 'space race' yielded a raft of futuristic designs for a variety of products including ceramics, furniture, fashion and textiles (Figures 23.4 and 23.5).

In fashion, designers such as Paco Rabanne, Pierre Cardin, André Courrèges and Yves Saint-Laurent pioneered the space-age look. Influenced by modern architecture, technology, modernism and futurism in art and design, their space-age creations featured geometric silhouettes that ignored the curves of the female figure, stark colour palettes (predominantly white, silver, black and primary colours), 'futuristic' materials such as plastic, metal and what were then new synthetic fabrics such as neoprene (1931), nylon (1938) and polyester (1960) (Figure 23.6).

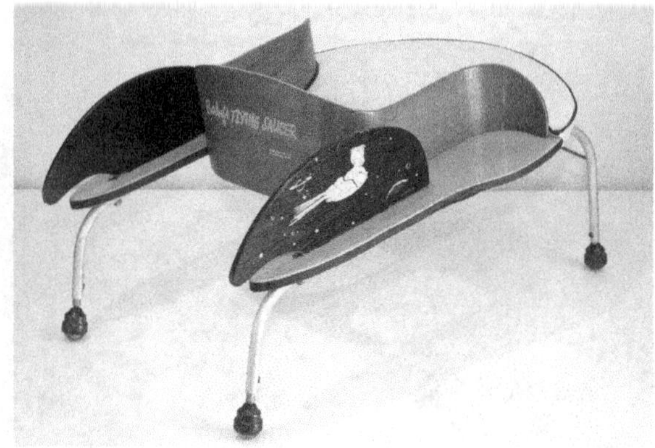

FIGURE 23.5

Space Age design.

Source: Image courtesy of Peter Kleeman, www.peter-kleeman.com; www.spaceagemuseum.com.

FIGURE 23.6

Pierre Cardin Cosmo Corps.

Source: Image courtesy of Pierre Cardin.

It is of interest that Paco Rabanne and Pierre Cardin had trained as architects and Courrèges as a civil engineer. The influence of their early career directions can be seen in their unconventional choice of materials such as plastics, metal and paper and their approach to fashion design: 'building' and 'engineering' garments rather than using traditional pattern-cutting routes.

Once again, the influence of travel – this time into space – prefaced a marked cultural shift where the old order with its associations of wartime deprivation, and where older generations held sway, was discarded. A more optimistic and thoroughly modern future was heralded by young people in particular, not least in their approach to fashion:

> French fashion designers explored futurism as a style that would appeal to young people, and 'Space Age' immediately became a metaphor for youth, charging fashion with optimism and vitality. In Paris the young clearly wanted to break away from couture; reportedly, when Chanel made a bid to dress Brigitte Bardot, the young star contemptuously replied. 'Couture is for Grannies'.
>
> **Quinn (2002, p. 7)**

23.2.6 MODERN DAY

In the twenty-first century, travel still provides designers – including fashion and textile designers – with a rich source of design ideas, including ideas for fashion and textile design. Improved communication and faster, more powerful transportation means that there are few communities of the world still to discover, but artists and designers in search of new ideas repeatedly rediscover and reinterpret the cultures of remote regions whose people retain the traditional skills and traditions of their heritage. For example, in his spring collection of 2007, following a tour of Japan, John Galliano borrowed elements from Japanese culture to create a couture collection for Christian Dior: a clear example of Orientalism continuing to flourish in contemporary fashion design the same way as it did 100 years ago (Figure 23.7).

23.3 POLITICS

Politics is another societal phenomenon that frequently has been shown to influence the arts. The so-called socio-political art is used to aid public understanding of a particular social or political issue, where artists present their interpretation of the social climate:

> The idea of the artist as an activist is not a new one. Whether fine artists, musicians, writers, architects or designers, artists have always used their 'art' as a means of expression. Art and design have expressed political, cultural and social movements in all cultures across the centuries.
>
> **Brown (2010, p. 6)**

23.3.1 POSTER ART

We are all familiar with the poster campaigns used by politicians to gain support during an election. Strong images and memorable slogans are used to bring them and their policies to the attention of the public in an easily digestible format with the intention of winning public support.

The idea of political poster art is not a new one. Since the fifteenth century (when every sheet was made painstakingly by hand), posters have been used to communicate and clarify social and political issues.

Immediately recognisable examples include the 1914 poster – used in both the First and Second World Wars – that shows Lord Kitchener proclaiming 'Your country needs YOU!' – designed to

FIGURE 23.7

John Galliano for Dior, spring 2007.

Source: Illustration by Carol Ryder, www.carolryder.com.

mobilise the British to support the war effort (Figure 23.8). Three years later, in 1917, it was re-designed by the American Army to show 'Uncle Sam' declaring: 'I want YOU for U.S. Army!' (Figure 23.9). Both British and American versions have since spawned numerous imitations, promoting anything from comics and shopping to beer and religion.

Similarly, the stylised image of Che Guevara (1928–1967) – a major figure in the Cuban Revolution of 1959 – has become an instantly recognised symbol of revolution, which inevitably reappears wherever there is conflict or dissatisfaction with the status quo (Figure 23.10).

23.3.2 SOVIET POSTERS

Another significant example of the power of political poster art can be found in the Soviet political posters of the 1920s, the golden age, of the Soviet poster. Following the Russian Revolution, the

FIGURE 23.8

WW I Lord Kitchener poster, 1914.

Bolsheviks seized power when the last Tsar of Russia (Nicholas II) abdicated in 1917. The Bolsheviks aimed to transform Russia into the world's first socialist society. However, this ambition relied upon the successful communication of the Bolshevik message to the Russian people at a time when the effectiveness of the written word was undermined by the high rate of illiteracy amongst the largely rural population. The communist government turned instead to the visual imagery of poster art. Graphic, dynamic and powerful, with bold colours and rousing slogans extolling the virtues of the workers of the world, the poster was a highly efficient medium of communication that reached a higher proportion of the population with a message that was easier to understand.

> Distinctively Bolshevik posters introduced the symbols of Communist revolution…including the hammer and sickle, the red star, heroic workers and peasants…In Soviet Russia, posters provided a façade of collective solidarity based on class, labour and the authority of the Party.
>
> **Aulich (2007, p. 136)**

An innovative style by the best of Russian artistic talent underlined the optimistic power of the message. Most Soviet artists of the time did not object to producing political propaganda; indeed, they found the prospect of speaking directly to thousands of people challenging and exciting. In the same way that modern advertising campaigns hope to change people's habits and encourage them to try new

FIGURE 23.9

Uncle Sam – 'I Want YOU for U.S. Army!' – 1917.

products, Communism's 'spin doctors' used motivational poster images to transform the mentality of an entire nation, encouraging people to feel optimistic about their future and to accept a new lifestyle along with new political ideas (Figure 23.11).

23.3.3 GUERILLA ART

In contrast to the giant mobilising power of governmental campaigns, political art can also be used by civilians as a form of protest, or to communicate dissatisfaction to politicians about issues of specific interest to the individual artist.

Unauthorised art that appears surreptitiously (and often suddenly) in a public place – particularly with the intention of making a political statement – has become known as 'guerilla art'. This is a form of street art with a political message, often of dissent, created by often anonymous artists.

Stencils, stickers and poster art are increasingly influential. In the United States, guerrilla poster artist Robbie Conal regularly uses Los Angeles as his personal gallery space:

> If I can make a surprise one-liner for people on their way to work in the morning – provide them with a little "infotainment" – and get them to think along with me about issues I think are important, I'm happy.
>
> **Robbie Conal's Art Attack: http://www.robbieconal.com/aboutrobbie.html**

FIGURE 23.10

Che Guevara.

Source: Photograph by Alberto Korda, 1960.

In the minds of many (particularly the police and local authorities) guerilla art is the same as graffiti. The word 'graffiti' is derived from the Italian word (itself derived from Latin) 'graffito', meaning 'a scribbling', a 'little scratch' (from the Italian verb 'graffiare' – 'to scribble'). Collins World Dictionary describes graffiti as 'drawings, messages, etc., often obscene, scribbled on the walls of public lavatories, advertising posters, etc'.

To their admirers, however, guerrilla artists have an important role in society, using stylish, considered, often witty means to communicate important political messages.

In the United Kingdom, the anonymous stencil-artist known as Banksy, often referred to as a graffiti artist, uses humour to communicate his political views to the general public (Banksy, 2005). These are mostly anti-war, anti-capitalist, anti-establishment or pro-freedom (Figure 23.12). Despite, or possibly because of, his reputation as a thorn in the side of authority, Banksy's use of humour makes his work immensely popular with the general public, especially those who share his dislike of heavy-handed authoritarianism:

> Even though artwork by the British graffiti artist Banksy is popular with celebrities such as Angelina Jolie and Christina Aguilera, some regard the artist's street works to be vandalism, pure and simple ... While Banksy made his name – or, rather, pseudonym – painting stencilled political and satirical images out-of-doors, in recent years his commercial pieces, including drawings, paintings and installations, have sold at auction for hundreds of thousands of dollars.
>
> **Logan (2008)**

FIGURE 23.11

Russian Revolution workers propaganda poster – 'Let's work, but rifle is ready!' (V.V. Lebedev, 1921.)

The result is that Banksy now commands a high commercial value, giving rise to a wide range of popular merchandise: posters, mugs, clothing and particularly T-shirts (Figure 23.13).

23.3.4 T-SHIRTS

In recent years, the unisex and ubiquitous T-shirt has become a popular and convenient means of conveying political opinion. Within a democratic society, the T-shirt can communicate the political viewpoint of the wearer in a direct but informal way that is more personal, and less anonymous, than poster art.

In the 1970s, Vivienne Westwood (b 1941 and now Dame Vivienne Westwood, DBE, RDI) and her partner Malcolm McLaren (1946–2010) were responsible for producing some of the earliest politically controversial T-shirts. In 1972, their shop at 430 Kings Road, Chelsea, previously called Let It Rock, had been renamed Too Fast to Live, Too Young to Die in a tribute to James Dean, as the pair turned their attention to designing and customising 'biker' style clothing. Let It Rock had already provided the clothes for the film 'That'll Be The Day' (directed by Claude Whatham, 1973), and the pair were fast gaining a reputation for subversive fashion amongst the anti-establishment youth of London, who were

FIGURE 23.12

Banksy rioter throwing a flower bouquet, 2010.

Source: From VanderWolf Images, Shutterstock.com.

looking for something more radical than the middle-class, ethnic and hippie clothes that populated the Kings Road at the time: an alternative fashion, that would help them to express their own ideas about society.

To add to the collection of anti-fashion 'biker' styles available at Too Fast to Live, Too Young to Die, Westwood began producing a series of controversial T-shirts 'collaged with feathers, nipple-revealing zippers, studs, chains, potato prints and found objects' (Wilcox, 2004, p. 12) – even using pornographic imagery and finely drilled chicken-bones to spell out words such as 'rock' and 'fuck'.

> Relatively affordable and with a provocative message, the T-shirts became the point at which cult fashion, sex and politics met, and were the perfect insignia of dissent for McLaren and Westwood. In 1975, they were prosecuted under the obscenity laws for 'exposing to public view an indecent exhibition' for a T-shirt showing two naked cowboys; others proved even more offensive, such as the 'Cambridge Rapist' and 'Paedophilia' T-shirts.
>
> **Wilcox (2004, p. 12)**

FIGURE 23.13

Banksy T-shirt: Shooting Panda.

Source: Image courtesy of Kapoww T-shirts, www.kapoww-t-shirts.

Speaking in 1981, in an interview with Jon Savage, an influential writer about the punk era (http://www.jonsavage.com), Vivienne Westwood restated her views on using fashion as a vehicle for political statement:

> My job is always to confront the establishment to try and find out where freedom lies and what you can do: the most obvious way I did that was through the porn T-shirt… I don't really want to talk that much about fashion. It's only interesting to me if it's subversive: that's the only reason I'm in fashion, to destroy the world 'conformity'. Nothing's interesting to me unless it's got that element.
>
> **Savage (1981, p. 25)**

Westwood, who is now an established and respected designer, twice-winner of British Fashion Council Designer of the Year (1990, 1991), and winner of the Outstanding Achievement Award in Fashion Design at the British Fashion Awards 2007, continues to use the T-shirt for political expression. In 2005, teamed with the civil-rights group Liberty, and protesting against the government's 'draconian' new anti-terror laws, Westwood produced a range of T-shirts proclaiming 'I am not a terrorist, please don't arrest me' (Figure 23.14). More recently, Westwood teamed with supermodel Naomi Campbell to help raise funds for the thousands of women and children left destitute by the Haiti earthquake disaster of 2010, unveiling her Fashion for Relief' T-shirt for the first time at the for Relief/Haiti fashion show, which opened London Fashion Week in February of that year.

Probably the best-known champion of the T-shirt's possibilities as both fashion statement and political loudhailer, Katharine Hamnett (b 1947) is a British fashion designer and well-known campaigner

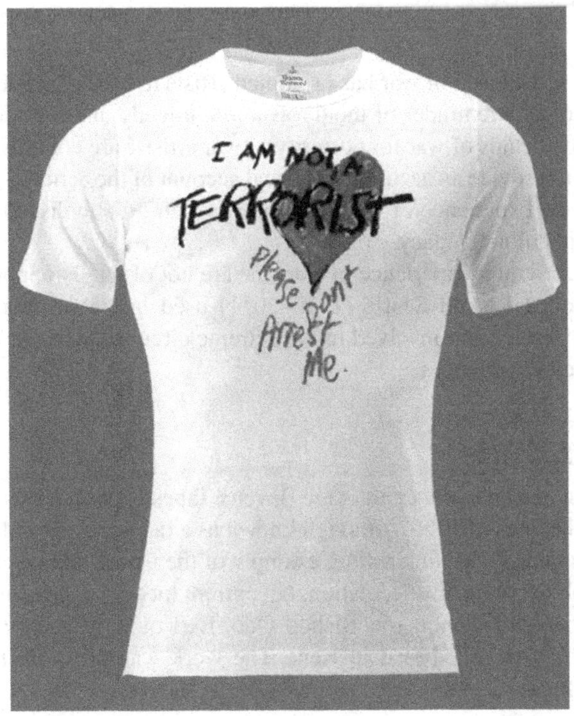

FIGURE 23.14

Vivienne Westwood 'I Am Not a Terrorist' T-shirt, 2005.

for ethical and environmental rights. In 1984, when Britain was living in fear of nuclear attack from the Soviet Union, Hamnett was named Designer of the Year by the British Fashion Council and Menswear Designer of the Year from the Bath Costume Museum. When invited to meet Margaret Thatcher (then British Prime Minister) at Downing Street, Hamnett wore an oversized anti-nuclear T-shirt proclaiming '58% DON'T WANT PERSHING' in an extremely public and controversial expression of disgust at the PM's complicity in the nuclear arms race.

In an interview with the *Sunday Times* in 2008, Hamnett explained: 'We had been invited to meet Margaret Thatcher but, to be honest, I didn't want to go. I thought she was dreadful. The way she was acting was appalling and what she had done to the country was deeply bad. As Jasper Conran said to me the day before our meeting, 'Why should we go for a glass of warm white wine with that murderess?' Then I thought: 'This is an obvious photo opportunity so I'll take advantage of it'. That is why I wore the anti-Pershing T-shirt. I kept it hidden under my jacket, but when I shook hands with her I got it out on display. She didn't notice it at first, but then she looked down and made a noise like a chicken, then quick as a fishwife she said: 'Oh well we haven't got Pershing here, so maybe you are at the wrong party', which I thought was rather rude as she had invited me. The Imperial War Museum wants the T-shirt, but I am hanging on to it' (Olins, 2008).

In a similar gesture during London Fashion Week in 2003, Hamnett sent her models down the catwalk wearing T-shirts emblazoned with the message 'NO WAR, BLAIR OUT'.

23.4 WAR

Throughout the centuries, the onset of war has prompted artists to express a variety of views, political, personal, triumphant – or as a reminder of the devastation, tragedy and loss that war creates. War art can document the specific details of war for posterity – 'war artists' are commissioned by governments to spend time under fire to provide an accurate historical account of the action – or else, to give a unique personal perspective of the experience of war according to their artistic discretion. In either case, war art becomes an important cultural legacy.

Some artists who record their experience of wartime are not officially appointed. For example, the British artist and cartoonist Ronald Searle (1920–2011) used his talent to record life in Japanese Prisoner of War camps. Servicemen involved in the extreme circumstances of a war may feel a strong compulsion to depict their experiences.

23.4.1 BAYEUX TAPESTRY

Probably the best-known example of war art is the Bayeux Tapestry, which was produced to record the events of the Battle of Hastings in 1066. Although known as a tapestry, it is in fact a large embroidery, 0.5 m high by 68.38 m in length – an interesting example of the use of textiles to tell a story of events. Its origins remain the subject of much speculation, but current theory suggests it was commissioned by the half-brother of William the Conqueror, Bishop Odo, Earl of Kent and Bishop of Bayeux (early 1030s–1097), who at the time was based in Kent. The work was therefore probably designed and carried out in England (Figure 23.15).

23.4.2 GUERNICA

Another famous example of war art is 'Guernica', a large mural-sized painting by Pablo Picasso (1881–1973) that powerfully depicts the pain, loss and tragedy of war, particularly that suffered by

FIGURE 23.15

Bayeux tapestry (detail), c. 1070.

FIGURE 23.16

'Guernica', Pablo Picasso, 1937.

innocent civilians. It is 3.5 m in height and 7.8 m wide, and was commissioned in 1937 by the Spanish Republican government for the Spanish display at the Paris Exposition Internationale des Arts et Techniques dans la Vie Moderne. A response to the bombing of the quiet Spanish market town of Guernica by Italian and German forces during the Spanish civil war, 'Guernica' remains a potent reminder of the tragedy of war, and a symbol of peace (Figure 23.16).

23.4.3 FASHION AND WORLD WAR II

Fashion design is also affected by war. The hardship and privations inflicted on a society during wartime inevitably mean that fashion – as distinct from 'clothing' – loses any real significance. For example, during World War II (1939–1945), restricted supplies of cloth and the rationing of goods meant that fabric for clothing was in short supply, and British people were encouraged to 'make do and mend' by re-using the wool from old knitwear, or cutting up old curtains to make skirts and dresses (Figure 23.17). Simplification of style was aligned with patriotism and garments became restrained and functional. This was known as 'utility' clothing (Figure 23.18). Fabric usage and garment details were strictly controlled by government regulations; for example, the length of men's shirts was specified, the number of jacket pockets limited, and turn-ups on men's trousers were banned in order to save cloth. Once the war ended, however, a reaction to wartime constraints gave rise to an entirely different fashion statement.

Convinced that the world was ready for a new style after the war, the French designer Christian Dior (1905–1957) launched the extravagant New Look in 1947. His first collection was defined by a voluptuous feminine silhouette, featuring extremely full, calf-length skirts, contrasted with nipped-in waists and rounded shoulders. Post-war, his indulgent use of fabric was shocking: 20 yards (approximately 18.5 m) of fabric in a single skirt was not uncommon (Figure 23.19). Nevertheless, the New Look stunned and delighted women of the time. The opulent celebration of femininity and glamour became symbolic of shrugging off the hardships of war and enjoying being feminine again. The success of the New Look ensured Dior's success as the most influential fashion designer of the late 1940s and 1950s and helped to restore Paris as the capital of fashion.

FIGURE 23.17

World War II – 'Make Do and Mend'.

23.4.4 TEXTILE DESIGN FOLLOWING WORLD WAR II

In the years following the war, Britain was badly in need of redevelopment, with much of London still in ruins. It was decided that a grand gesture was needed to restore optimism to the British people and pave the way towards recovery and progress. To this end, in May 1951 the Festival of Britain took place. Described by Labour Deputy Leader Herbert Morrison as 'a tonic for the nation', the Festival marked the centenary of the Great Exhibition of 1851, and promoted better-quality design in the rebuilding of British towns and cities. Designed by Sir Robert Matthew, Leslie Martin and Sir Hubert Bennett, the Royal Festival Hall was built for the occasion on the South bank of the River Thames.

The Festival Hall presented an ideal showcase for British designers. One of them, Robin Day (1915–2010), was a furniture designer who had trained at the Royal College of Art before the war, and had honed his craft during the war years. Noted for his highly functional, modern and stylish but low-cost designs, Day epitomised the progressive approach toward design embodied in the Festival of Britain. As a result, he was commissioned to design furniture for the Festival Hall and to produce two

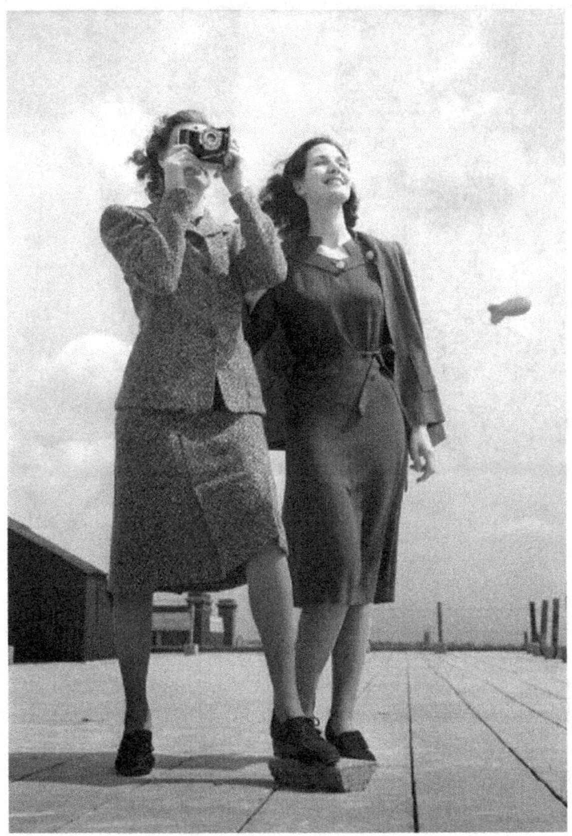

FIGURE 23.18

World War II utility clothing.

room settings for the House and Gardens Pavilion at the Festival – one 'budget', the other 'top of the range' – but both equipped with his latest storage furniture and chairs.

Day's wife, Lucienne (1917–2010), produced the furnishing fabrics for these displays. Having met at the Royal College of Art in 1940, Robin and Lucienne Day were drawn together by their passion for design, and married in 1942. A printed textile designer in her own right, Lucienne Day's designs were inspired by her love of modern art, particularly of contemporary abstract painters such as Wassily Kandinsky, Paul Klee and Joan Miró. In her work, she sought to create a similar energy through colourful, dynamic and vivacious compositions. In the preface to the first edition of Lesley Jackson's book *Robin and Lucienne Day: Pioneers of Contemporary Design*, Lucienne writes about her experience of the period after the war when she left the Royal College of Art to work in the field of textile design (Jackson, 2011).

Not only was the practice of design virtually unknown, but we were also faced with the restrictions imposed by war. Most manufacturers of textiles, for instance, had been forced to close or to transfer their production to materials needed specifically for the war – not an auspicious beginning … In the 1950s we both participated in the surge of modern design that was released by the end of the war and material

FIGURE 23.19

Eva Peron wearing Dior's New Look, 1950.

Source: Photographer unknown.

restrictions. There was a growing feeling of optimism and an anticipation of the emergence of a bright new world and we thought that progressive design could contribute to the quality of people's lives.

Lucienne produced the avant-garde design 'Calyx' for Robin's Festival of Britain room settings. Described by the Design Museum as 'an abstract pattern inspired by plant forms, composed of spindly lines and irregular cupped motifs in earthy and acid tones, Calyx was acclaimed both nationally and internationally.

As a result, the 1950s and 1960s were a time of feverish activity for Lucienne Day. As well as designing printed textiles, she was inundated by invitations to design carpets, wallpapers, tea towels, table linen and ceramics. Her principle client, the leading London furniture store Heal's, championed her contemporary style, and over the next 20 years Lucienne designed over 70 patterns for the store.

In Day's obituary in *The Guardian* (2010), Fiona MacCarthy paid tribute to the designer:

> Lucienne Day, who has died aged 93, was the foremost British textile designer of her period. Day's furnishing fabrics, of which the most famous was the Festival of Britain abstract pattern Calyx, hung in every "contemporary" living room in Britain. The reality of "art for the people", dreamed about by the Victorian William Morris, was finally achieved by a female designer in the middle of the 20th century.
>
> **MacCarthy (2010), http://www.guardian.co.uk/artanddesign/2010/feb/03/lucienne-day-obituary**

Although Robin and Lucienne Day worked separately on different projects for most of their lives, and in different areas of design, their distinct design styles always conveyed a synergy, underpinned as they were by identical principles of energy and dynamism, modernity and progress that were specifically conceived as an antidote to the misery and privations of war.

23.5 IMPACT OF CULTURE ON DESIGN, FASHION AND TEXTILES

> Far from being a hit-and-miss sideshow in the great circus of human thought, clothing fashions operate in such a way as to echo the larger mechanisms of aesthetic change....
>
> **Slade (2009, p. 1)**

Since fashion is defined by the endless search for the new, it is inevitable that any fundamental changes that take place in the wider context of the arts are seized upon to provide inspiration for both fashion and textile design.

23.5.1 CUBISM AND DELAUNAY

A notable example of how fashion can be influenced by wider movements in the arts is that of the Russian artist Sonia Delaunay (1885–1979), whose fashion and textile designs, typified by strong geometric patterns in bold, vibrant colours, were in part inspired by Cubism. As a painter, Delaunay took a revolutionary approach to fashion design, interpreting the body as a canvas, and in a lecture at the Sorbonne in 1927, articulated the influence of painting on fashion design. In effect, her garments were the result of applying contemporary aesthetic trends to the body, giving rise to simple, practical garments in geometric shapes that followed the lines of the natural figure rather than attempting to reshape or modify it (Figure 23.20).

Between 1920 and 1930, Delaunay produced a number of vibrant and original fabric designs, but returned exclusively to painting in 1930. Her decorated scarves are known to have influenced the work of artist Paul Klee, whereby textiles inspired by painting ultimately made a reciprocal mark on fine art.

23.5.2 SURREALISM AND SCHIAPARELLI

A decade later, in the 1930s, Italian designer Elsa Schiaparelli (1890–1973) also took inspiration from the world of fine art. Where Delaunay was influenced by Cubism, Schiaparelli's inspiration was provided by Surrealism, an art movement that depicted imagery from the unconscious mind without making any attempt to rationalise the outcome. Founded in 1924 by André Breton (1896–1966), Surrealism, in turn, was inspired by the psychoanalytical work of Freud and Jung. Introduced to the world of fashion by Paul Poiret, Schiaparelli was famous for her eccentricity and innovation. Closely connected to the artistic community, her infamous 'lobster dress' and 'shoe hat' resulted from collaborations with the most celebrated of Surrealist artists, Salvador Dali (1904–1989) (Figure 23.21).

23.5.3 POP ART

Later still, the radical new ideas and mood presented by the Pop Art movement served once again to revolutionise the world of fashion design. Pop Art – the term derived from 'popular art' – emerged in Britain in

FIGURE 23.20

Sonia Delaunay design, about 1930.

the 1950s as a response to the mass-consumerism of the post-war period and as a rejection of traditional views of art – particularly abstract art – which pop artists considered as pretentious and elitist. Pop art deliberately set out to subvert traditional highbrow views about art by using commonplace images from popular culture for inspiration – for example comic books, supermarket products, billboards and magazine advertisements. Labels and logos of familiar products feature prominently in the imagery selected by pop artists for immortalisation, for example Andy Warhol's 'Campbell's Soup Cans' (1962) (Figure 23.22).

Pop Art is also associated with the use of mechanical methods of rendering or reproduction, for example screen-printing. Again, overturning ideas about the 'preciousness' of art, pop artists argued that multiple images of familiar mass-produced products, reproduced in a mechanical way, could be aligned with the context of fine art. It could be said that Pop Art engendered the opposite view of that of the Arts and Crafts movement, as it placed great emphasis on the banal, rather than the precious, and celebrated mass-production rather than denouncing it.

Pop Art was 'art for the people', and although blurring the boundaries between high and low culture was seen as subversive, the absence of elusive intellectual concepts ensured its rapid, widespread appeal. Pop Art ultimately defined the look of the 1960s (Leslie, 1997).

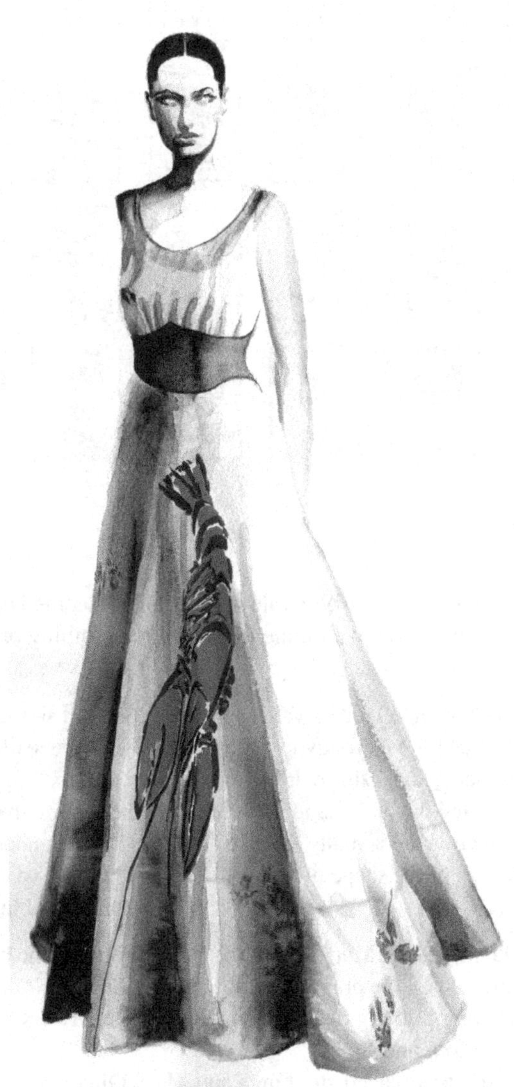

FIGURE 23.21

Elsa Schiaparelli Lobster Dress, 1937.

Source: Illustration by Carol Ryder, http:www.carolryder.com.

FIGURE 23.22

Andy Warhol Campbell's Soup Cans, 1962.

In fashion, restrictive femininity – so fashionable in the 1950s – seemed oppressive to a new generation of young women, who did not want to emulate the inhibiting clothing or behaviours of their mothers' generation:

> Fashion had its own establishment, a kind of Vatican, in the fifties and sixties and in this set-up they had dictators who set the lines for everybody to follow. The lines were set like edicts in the way of the old world... They were set by magazine editors for magazine readers. Vogue used to announce the colour of the season and up and down the land shops presented clothes in banana beige or coral red or whatever. In the fifties there were actually lines for fashion. Dictates about the shape a woman's clothes should be, irrespective of the shape of her....
>
> **York, quoted in Polhemus (1996, p. 28)**

In her autobiography (Quant, 1967), Mary Quant (b 1934), probably the best-known British fashion designer of the 1960s, commented on the prevailing mood: 'fashion wasn't designed for young people'. The result was a new generation of young designers, including Quant herself, who introduced to fashion an emphasis on youth, freedom and fun.

Skirts became shorter as tights replaced stockings and Mary Quant popularised the 'mini'. Simple, unrestrictive dresses became popular, and the invention of the contraceptive pill (USA, 1960) added to the sense of liberation felt by a generation of young women.

Shift dresses, designed with a minimum of fussy detail, provided the perfect canvas for Pop Art-inspired imagery. The invention of paper dresses underlined the 'throwaway' mood of Pop Art, including the celebrated 'Souper Dress', produced by Campbell's for an advertising campaign, which referenced Warhol's 'Campbell's Soup Cans': as art had imitated advertising, the world of advertising had, in turn, imitated art (Figure 23.23).

FIGURE 23.23

Campbell's 'Souper' dress, about 1966.

23.5.4 OP ART

Another 1960s art movement to make its mark on fashion was Op Art – the term derived from 'optical art'. Op Art followed Pop Art, but never quite established itself in the same way. Its style was typified by the use of optical illusion, often in black and white. Its most famous exponents were the English painter Bridget Riley (b 1931) and Hungarian-born Victor Vasarely (1906–1997).

Following an exhibition of Op Art called 'The Responsive Eye' (1965) at the Museum of Modern Art in New York, Op Art began to appear everywhere: in television advertising, in print and as a decorative motif for fashion and interior decoration. In fashion, the shift dresses and paper dresses that provided Pop Art with a mobile canvas presented the same opportunity for Op Art, and textile companies too began developing Op Art prints for furnishing fabrics.

Although Vasarely took the view that art should be for everyone, and collaborated with textile firms in the reproduction of his artwork, Bridget Riley took the opposite view and was dismayed

FIGURE 23.24

Op-Art headscarf, about 1965.

to see her original artwork co-opted for commercial purposes without her permission. In 1965, having witnessed row upon row of dresses in Madison Avenue shops bearing Op art designs copied from her paintings, she denounced the way her art was being 'vulgarised in the rag trade' (Figure 23.24).

23.5.5 POPULAR CULTURE/POP CULTURE

In the same way that the term 'Pop Art' was derived from 'popular art', the term 'pop culture', dating from the 1960s, is an abbreviation of 'popular culture': a term that had become established by the end of World War II.

Pop culture can be described as culture for mass-consumption. It is the sum of the ideas (e.g. political or religious), attitudes, behaviours, values, lifestyle (e.g. interior design, music, film and fashion) and products or items (including art) that are well known and generally accepted, considered mainstream and popular, and surround the everyday lives of people in a particular society. Popular culture can be heavily influenced by mass media. Advertising, for example, can bring a particular product into mainstream use and popularity. Television, journalism and Internet also regularly influence popular culture, for example the fascination with celebrity, which is currently popular in British society.

As pop culture reflects the preferences of the mainstream, and therefore represents big business, influences from pop culture invariably infiltrate every area of design, including fashion and textile design. Consequently, contemporary British fashion is heavily influenced by the dress codes of the celebrities who currently occupy such a massive portion of media coverage.

23.5.6 COUNTER-CULTURE

Counter-culture, on the other hand, is a culture that ignores, renounces or opposes the popular culture embraced by the mainstream. Counter-cultural groups include 'Goths', a music subculture with a dark 'Gothic' aesthetic hailing from the post-punk genre in the late 1970s; and the Japanese 'Ganguro', whose image, which includes dark brown make-up with white eyes, nose and lips, and whose musical preferences are chosen deliberately to run 'counter' to the mainstream.

Music is an important feature of any popular culture, or counter-culture, and like all aspects of culture, can exert its influence on other cultural sectors (Polhemus, 1994).

23.5.7 PUNK

In the 1970s, Punk Rock emerged as a reaction to the excesses – ostentatious, self-indulgent musical effects and technological exhibitionism – of mainstream 1970s 'stadium rock'. Fast, aggressive and energetic, punk songs were typically short, with stripped-down instrumentation and anti-authoritarian lyrics. By 1976, bands such as the Sex Pistols and The Clash in London and The Ramones in New York, were recognised as the vanguard of a new musical movement, which by the following year had spread around the world and went on to create a major cultural phenomenon in the United Kingdom.

The clothing associated with Punk grew out of a shop situated at 430 Kings Road, London, which had been opened by Malcolm McLaren, the founder and manager of the Sex Pistols, in 1971.

In 1974, after a number of incarnations, the shop was re-branded SEX, its name emblazoned in pink plastic letters above the doorway. The interior of the shop was covered in graffiti, while rubber curtains draped the walls. It was here that McLaren's partner, (now Dame) Vivienne Westwood, introduced London to her 'anti-fashions' – slashed, contentious T-shirts, fetish and bondage wear and garments peppered with safety pins.

McLaren, a street-smart, savvy entrepreneur who had managed 'Glam Rock' band the New York Dolls, wanted to create a band that would embody the spirit of his King's Road shop. Glen Matlock, a part-time employee at SEX, was already in a band with Steve Jones and Paul Cook, but they needed a singer and front man. Malcolm McLaren suggested that John Lydon (b 1956), a disreputable character who hung around the jukebox in SEX wearing an 'I Hate Pink Floyd' T-shirt, and who had never sung before, should audition. Famously, he got the job, changed his name to Johnny Rotten (because of his decaying teeth), and the 'Sex Pistols' were born.

Although Westwood's fashions played an enormous part in forging the image of the early UK punk movement, it was the fashion statements made by the fans and followers of the Sex Pistols that developed and popularised the look. Known as the Bromley Contingent, members included Billy Idol, Siouxsie Sioux and Jordan, a striking female figure also employed at SEX.

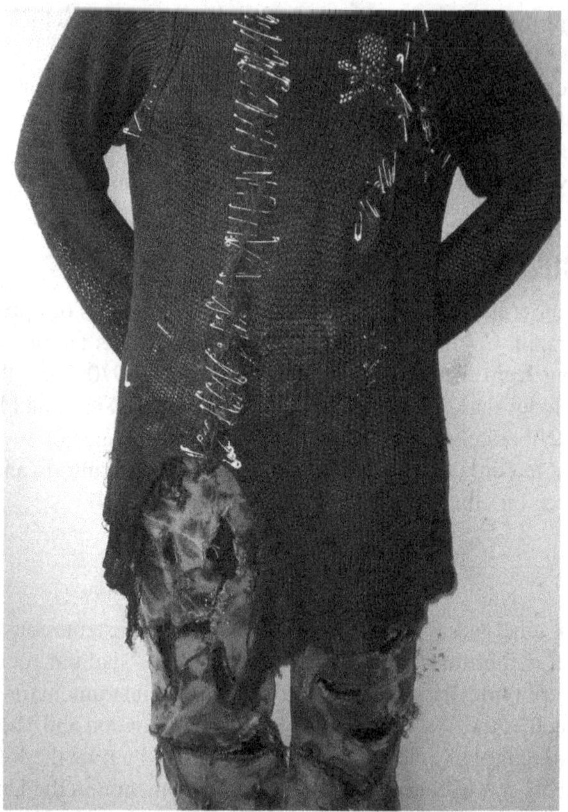

FIGURE 23.25

Punk anti-fashion.

In terms of fashion, as a founding influence at the inception of the movement, Vivienne Westwood is the name principally associated with the Punk movement. However, many other designers went on to be influenced by Punk. One of these was Zandra Rhodes (b 1940), trained as a textile designer and self-taught as a fashion designer, who had produced her first garment collection in 1969. By the 1970s, Rhodes had begun to move the focus of her collections away from printed textiles and decided that flowing, feminine and patterned dresses no longer seemed appropriate. Her designs began to take on a harder edge as she explored the possibilities of 'the beautiful qualities of a tear'.

Although inspired by the Punk movement, Rhodes' couture designs were very different from the slashed and pinned DIY anti-fashion worn by the Punks themselves (Figure 23.25). Instead, Rhodes' designs made a feature of 'safety pins straddling slashed holes in clinging jerseys that were beautifully stitched and jewelled' (Rhodes, 2005, p. 73). In this instance, the traditional fashion order had been overturned: instead of high fashion influencing the fashion of the High Street as was traditionally the case, music-inspired street-fashion had influenced haute couture (Figure 23.26).

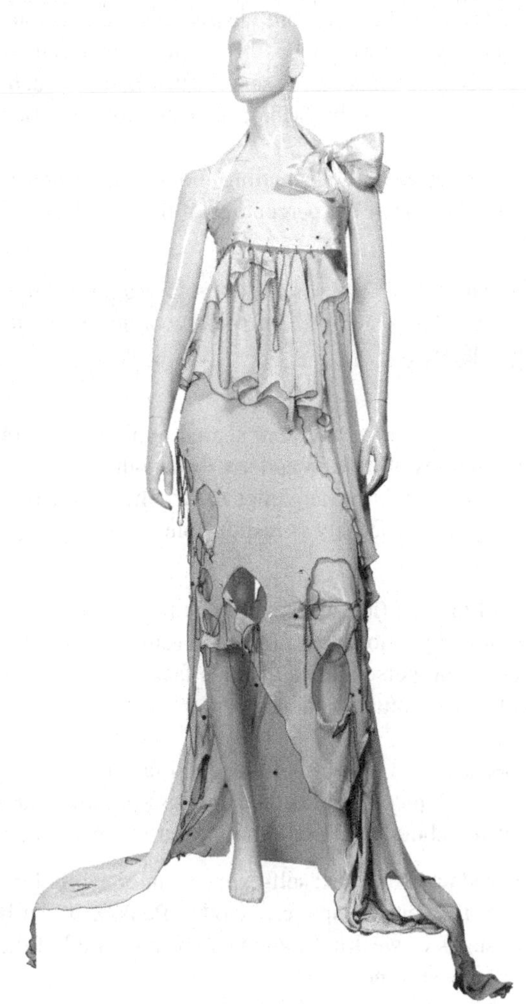

FIGURE 23.26

Punk couture – Zandra Rhodes 'Conceptual Chic' collection 1977–1978.

Source: Image courtesy of Zandra Rhodes.

23.6 DEFINITIONS OF TEXTILE CULTURE AND FASHION CULTURE: ARE THEY THE SAME?

Fashion and textiles are inextricably linked; jointly they form a stable relationship, one relying on the other not only for survival but also in order to prosper. Fashion adds value to textiles through the processing of raw materials, and textiles enable fashion to happen. Fashion develops continually and therefore demands new and exciting materials to work with from the textiles industry.

Gale and Kaur (2004, p. 61)

The twin disciplines of fashion and textiles are clearly related. Fashion would be unable to exist without fashion fabrics provided by the textile industry. While the textile industry (the industry responsible for producing fabrics of all kinds) undeniably serves other markets such as interior design and the construction industries, it has relied upon the fashion industry for centuries as a major outlet for its production.

Successful male fashion designers notwithstanding, both fashion and textile design are female-dominated disciplines, at least in terms of the percentage of the workforce, and as such have long had to contend with issues of sexism:

> Design history as originally taught in art and design colleges has tended to prioritise production in the professional 'masculine' sphere, re-enforcing notions of a subordinate 'feminine' area of interest into which fashion is generally relegated.
>
> **Breward (1995, p. 3)**

Despite their interdependence, however, and their shared feminine outlook, there exist a number of differences between the communities of fashion and textile design.

The term 'fashion', a widely used term that defies easy definition, can straddle both popular and high culture, and the wide range of its activity is responsible for an equally wide range of responses, which include:

- Derision for the triviality of fashion (its superficiality, impermanence, faddishness).
- Exasperation at the hauteur and preciousness of fashion (elitism, the extravagance and expense of haute couture, the 'prima donna' personalities of some fashion designers).
- Intellectual appreciation for conceptual fashion.
- Ridicule at extreme fashion.
- Childish delight at the spectacle and frivolity of theatrical fashion.
- Dismay at the wastefulness and questionable ethics of mass-produced fashion ('throwaway' culture, exploitation of cheap labour, exploitation of the earth's resources).

Fashion can be a functional product or fine art; self-expression, or symbol of conformity or belonging (e.g. uniforms, dress codes of different groups, e.g. Goths, Punks). It can be serious, casual, fun or formal. It can be a symbol of status or wealth. Fashion can be any or all of these things, irrespective of how contradictory they might appear to be.

In contrast, textile design has never provoked the same extreme responses – of shock, delight, derision or horror – that are inspired by fashion statements such as Punk, Dior's New Look, Poiret's Orientalist creations or the Japanese Avant Garde. Whenever textiles have provoked an extreme response, it has been when they are linked to either fine art or fashion – for example, the textile artwork of Tracey Emin, or the daring translucent muslin dresses fashionable in the Napoleonic era.

Whatever its incarnation, fashion is inevitably associated with temporariness. In accordance with the ephemeral nature of fashion collections, fashion shows are held in temporary marquees and fashion fairs in rapidly built environments that are there one day and gone the next.

In contrast, the parallels with age, history and tradition assumed by textiles are impossible to ignore. Textile design is generally regarded as a conservative, middle class occupation, while fashion, although admittedly sometimes stemming from the middle or even upper classes, just as often has its roots in working-class culture or anti-establishment rebellion. It could be argued that this is what gives fashion its 'edge'.

Is it inevitable then, that textile design will always be associated with the old, rather than the fresh, the young and the new? Is it too 'safe'? Not 'cutting-edge' enough? Quiet, rather than attention-seeking? Sedate, rather than frenetically fast-paced?

If so, this viewpoint hardly seems fair when, as described earlier, textile technology develops hand-in-hand with scientific advances. Designers of fashion fabrics work far in advance even of fashion designers, predicting which fabrics designers might want to use to create their garment collections. Burberry, Missoni and Pucci prove that a fashion company can be famed as much for its fabric as for its garments. Celebrated and revered fashion designers such as Issey Miyake, Basso and Brooke and Jonathan Saunders have built entire collections around textile design, where the design of the cloth provides the interest and garment shapes are secondary.

The reason for the less radical image of textiles may be embedded in the view that textiles have a close association with craft, and while fashion prides itself on representing the new and the cutting edge, 'craft' has connotations of history and unglamorous tradition from which fashion largely seeks to exempt itself. In the establishment of the Royal Academy (1768), a distinction was drawn between art and design. Design was perceived to be lowlier than art, primarily because of its connections with craft.

Despite its lowly beginnings in trade schools such as the Barrett Street Trade School (now the London College of Fashion) where dressmaking was taught, from the 1950s onwards fashion education fought hard to align itself with fine art – with lofty notions of higher purpose – and to distance itself from textile design (with its overt connotations of 'craft') in order to be accepted as a serious discipline in the art schools:

> The battle for fashion in the art schools was not easily won. Fashion retained a strongly feminine image in a male-dominated environment. And the further lingering associations of both craft and dressmaking skills meant that its passage into the status-conscious departments of the art schools was far from smooth.
>
> **McRobbie (1998, p. 22)**

Fashion's anxiety to shrug off its associations with craft seems to stem from the fact that since the demise of the Arts and Crafts movement, craft, in relation to art, simply became unfashionable. Even worse, the reason that craft had become so unfashionable was that it was guilty of being backward looking, nostalgically harking back to a simpler age, rather than embracing new ideas and facing the future with optimism. Perhaps it was at this point, given textiles' strong association with Arts and Crafts, that the twin cultures of fashion and textiles parted company. Perhaps fashion, ever eager to emphasise its identification with newness, to underline its fine-art aspirations, and discard its associations with trade, was equally keen to distance itself from textiles for its middle-class associations and ties to tradition and craft.

Worryingly, it is also possible – although astonishing, in this day and age – that despite fashion's still female-dominated workforce, the fact that successful male fashion designers are by no means uncommon may go some way to explain why fashion design has achieved greater success than textile design in being accepted as a serious discipline.

However, it is likely that fashion's dynamic identity is ultimately due to the animation, diversity and personality of the people who wear it. Fashion's close relationship with the human body ensures that fashion will always be more than just 'clothes'. Fashion carries with it connotations of image – supported by hairstyles, make-up, jewellery and accessories, as well as the personality of the wearer – and the vitality of the resulting characterisation is something that textiles alone cannot hope to achieve. Textiles – even fashion fabrics – only become animated when translated into fashion, and therefore always remain one step removed from the drama, endlessly playing a supporting role for the dynamic double-act of fashion and human beings.

23.7 PROJECT IDEAS

1. Since fashion is embodied by the endless search for the new, it is inevitable that any fundamental changes that take place in the wider context of the arts are seized upon to provide inspiration for both fashion and textile design. Bearing in mind historical examples set by Sonia Delaunay and Elsa Schiaparelli, who took inspiration for fashion design from art movements of their day, design a mini-collection comprising six pieces inspired by 'Craftivism', that are appropriate for the current High Street market and focused on an autumn/winter season.
2. Fashion and politics: Taking inspiration from artists such as Banksy and Robbie Conal, and designers such as Vivienne Westwood and Katharine Hamnett, carry out research into current political issues, and produce a range of graphic-based images that can be applied to T-shirts that express your feelings about one current political issue.
3. Orientalism continues to flourish in contemporary fashion design the same way as it did 100 years ago. Taking inspiration from the concepts of Japonism, Chinoiserie and Orientalism, research into traditional Japanese, Chinese and Indian motifs and design a collection chosen from one of the following disciplines:
   a. Printed textiles
   b. Knitted textiles
   c. Woven textiles

   and create a series of fabric designs appropriate for the current fashion or interiors textile market, for a season of your choice.

23.8 REVISION QUESTIONS

1. Explain what you understand by the term 'culture', giving examples of different types of cultures, and explaining what is meant by each.
2. Giving examples, explain how phenomena such as travel, politics and technological advances have affected fashion in the East and the West.
3. Explain the phenomenon of the Industrial Revolution, and how its influence affected fashion and textile design.
4. Describe the phenomenon of Orientalism, and how it affected European culture.
5. Explain how cultural shifts affect fashion and textile design, giving examples from nineteenth- and twentieth-century history.

23.9 FURTHER READING

As described earlier, 'culture' is an umbrella term that can relate to a wide range of societies – including those from the past as well as the present. Accordingly there are numerous sources of information to be found relating to different aspects of culture.

In particular, galleries and museums are excellent sources of cultural information. The V&A, Barbican, Tate Britain and Tate Modern in London, the Smithsonian Museum in Washington, the

Museum of Modern Art (MoMA) in New York, the Guggenheim Museums in New York and Bilbao, the Louvre and Musée des Arts Decoratifs in Paris, the Tokyo National Museum, the Berlin Museum Island, the Uffizi Gallery in Florence – to name but a few – provide lectures, gallery tours, interactive workshops and other activities as well as exhibitions to help students of fashion and culture gain as much as possible from their museum experience. In addition, most galleries and museums have excellent web sites that provide sources of information that can be accessed from home.

For current developments in art and culture, Internet is a valuable source of information. Portal web sites such as http://www.guardian.co.uk/artanddesign (hosted by UK newspaper *The Guardian*), and http://www.artandculture.com, provide a useful means of accessing current information on a wide variety of the arts.

A wealth of help and advice is also available at local libraries. As well as any books that they may have available on subjects of cultural interest, library staff will be able to organise inter-library loans that provide access to books from a number of other libraries and can give advice about alternative sources of information, such as journals, videos and DVDs.

For information specifically related to fashion, numerous fashion magazines are published at regular intervals – weekly, monthly or quarterly – and provide comprehensive information about current fashion trends. *Vogue*, *Dazed and Confused*, *AnOther Magazine*, *Love and Elle* are all excellent examples. As well as being available in hard copy, many of these magazines have online versions: see http://www.vogue.co.uk, http://www.dazeddigital.com, http://www.anothermag.com, http://thelovemagazine.co.uk and http://www.elle.com.

Dedicated fashion web sites and online fashion magazines such as Style.com (http://www.style.com), WGSN (Worth Global Style Network – http://www.wgsn.com), Fashion Telegraph (http://fashion.telegraph.co.uk), Fashion Insider (http://www.thefashioninsider.com), Net-a-Porter.com (http://www.net-a-porter.com) and Fashion.net (http://www.fashion.net) provide a continually updated source of information about designer collections, street style and fashion trends.

Television and radio programmes can also be useful in providing documentary information relevant to the fashion industry. The enlightening and often shocking BBC Three documentary series 'Blood, Sweat and T-Shirts' from May 2008 is available on DVD from Labour Behind the Label (http://www.labourbehindthelabel.org/join/item/760-blood-sweat-and-t-shirts). This series brought together a group of six young British fashion consumers to work in India alongside the people who make High Street fashion products, and experience first-hand the hardship and poverty they endure.

A defining feature of fashion is its continual evolution and reinventing of itself. It follows, therefore, that the best sources of information about contemporary fashion are those that are updated regularly, for example regularly published magazines, web sites and blogs.

Social networking sites such as Facebook (https://www.facebook.com) and Twitter (http://www.twitter.com) and a rising number of dedicated fashion blog sites such as TrendLand (http://trendland.net) and Style Bubble (http://stylebubble.typepad.com) are rapidly overtaking published magazines as the principle source of fashion information because they allow the user to keep abreast of fashion news on a minute-by-minute basis. In turn, fashion brands use social media as a means to humanise their online engagement with their customers, and to improve sales at minimal cost via regular updates about their product.

When considering future trends in fashion, it is particularly important to remember that fashion is shaped by a variety of cultural influences. Because of this, the fashion trends forecaster must keep abreast of news in a number of spheres – national and international events, technological advances, economic fluctuations and environmental changes – as well as debates about fashion itself. Newspapers, radio and television news programmes and their Internet-based counterparts are all invaluable in providing regular news updates.

For information about trends in sustainable and ethical fashion, books such as Sass Brown's *Eco Fashion* (Brown, 2010), or Sandy Black's *Eco-Chic* (Black, 2008) provide a useful overview of sustainable and ethical fashion trends in recent years. As the situation is continually changing, however, web sites such as the Centre for Sustainable Fashion (hosted by the London College of Fashion, http://www.sustainable-fashion.com), the Ethical Fashion Forum (http://www.ethicalfashionforum.com), Fashioning an Ethical Industry (http://fashioninganethicalindustry.org), Labour Behind the Label (http://www.labourbehindthelabel.org) and the Fairtrade Foundation (http://www.fairtrade.org.uk) are perhaps more useful, as they supply information about sustainable and ethical fashion trends that is updated regularly and therefore offers a more current picture in a rapidly changing field.

REFERENCES

Aulich, J. (2007). *War posters: Weapons of mass communication*. London: Thames and Hudson.

Banksy. (2005). *Wall and piece*. London: Century.

Black, S. (2008). *Eco-Chic: The fashion paradox*. London: Black Dog Publishing.

Braddock, S., & O'Mahoney, M. (1998). *Techno textiles: Revolutionary fabrics for fashion and design* (Vol. 1). London: Thames and Hudson.

Breward, C. (1995). *The culture of fashion: A new history of fashionable dress (Studies in design & material culture)*. Manchester: Manchester University Press.

Brown, S. (2010). *Eco fashion*. London: Laurence King.

Fukai, A. (1996). *Japonisme and fashion*. Tokyo: National Museum of Western Art and Kyoto Costume Institute.

Gale, C., & Kaur, J. (2004). *Fashion and textiles: An overview*. Oxford: Berg.

Jackson, L. (2011). *Robin and Lucienne Day: Pioneers of contemporary design*. London: Mitchell Beazley.

Koda, H., & Martin, R. (1995). *Orientalism: Visions of the east in western dress*. New York: Harry N. Abrams, Inc.

Leslie, R. (1997). *Pop art: A new generation of style*. New York: Todtri Productions.

Logan, L. (2008). Banksy defends his guerrilla graffiti art. *The New York Times* (online), October 28. Available at http://www.time.com/time/arts/article/0,8599,1854616,00.html#ixzz1MJwO5fWz. Accessed 24.01.11.

MacCarthy, F. (2010). Lucienne day obituary. *The Guardian*, February 3.

McRobbie, A. (1998). *British fashion design: Rag trade or image industry?* Oxford: Routledge.

Olins, A. (2008). My life in fashion: Katharine Hamnett. *The Sunday Times* (online), April 9. Available at http://women.timesonline.co.uk/tol/life_and_style/women/fashion/article3706444.ece. Accessed 24.01.11.

Polhemus, T. (1994). *Street style*. London: Thames and Hudson.

Polhemus, T. (1996). *Style surfing*. London: Thames and Hudson.

Quant, M. (1967). *Quant by quant*. London: Pan.

Quinn, B. (2002). *Techno fashion*. Oxford: Berg.

Rhodes, Z. (2005). *Zandra Rhodes: A lifelong love affair with textiles*. Surrey: Zandra Rhodes Publications Ltd.

Said, E. (2001). *Orientalism*. New York: Penguin.

Savage, J. (1981). Vivienne Westwood. *The Face*, January 9.

Slade, T. (2009). *Japanese fashion: A cultural history*. Oxford: Berg.

Wilcox, C. (2004). *Vivienne Westwood*. London: V & A Publications.

ELECTRONIC SOURCES

Art and Society

Art and Culture Portal, http://www.artandculture.com.
The Guardian – Art and Design Portal, http://www.guardian.co.uk/artanddesign.
The International Slavery Museum, Liverpool, http://www.liverpoolmuseums.org.uk.
Robbie Conal's Art Attack, http://www.robbieconal.com/aboutrobbie.html.

Technological Advancements

Biocouture, http://www.biocouture.co.uk/.
Interactive Custom Clothes Company, http://www.ic3d.com.
Nuno, http://nuno.com.

Case Study

Cultural Exchanges: Japan and the West
www.japanesestreets.com, http://www.japanesestreets.com.
meijishowa.com, http://www.meijishowa.com, http://www.meijishowa.com/?q=fashion&x=44&y=14.
oldphotosjapan.com, http://oldphotosjapan.com/en/category/women, http://oldphotosjapan.com/en/photos/323/bride-and-groom.

Sustainability and Ethical Fashion

Centre for Sustainable Fashion – London College of Fashion, http://www.sustainable-fashion.com/.
Ethical Fashion Forum, http://www.ethicalfashionforum.com/.
The Fairtrade Foundatiom, http://www.fairtrade.org.uk/.
Fashioning an Ethical Industry, www.fashioninganethicalindustry.org.
Labour Behind the Label, http://www.labourbehindthelabel.org/.
Marks and Spencer – 'Plan A', http://plana.marksandspencer.com/.
People Tree, http://www.peopletree.co.uk.
Practical Action, www.practicalaction.org.uk/education.
Sustainable Design Award, http://sda-uk.org/textiles.html.
UNICEF and child labour, http://www.unicef.org/pon95/chil0016.html.

CHAPTER 24

FASHION AND CULTURE: GLOBAL CULTURE AND FASHION

C. Ryder
Liverpool John Moores University, Liverpool, UK

LEARNING OBJECTIVES

At the end of this chapter, you should be able to:

- Describe the key features of a number of different cultures, in different geographical locations, at specific points in history
- Describe globalisation and how it might impact upon differing global cultures
- Describe a number of technological and societal developments that could shape fashion and textile design in the future

24.1 INTRODUCTION

'Global culture' can be defined as the whole that is created by every person on Earth; the sum total of who and what we are. In recent times, a dramatic increase in worldwide communication and travel has led to the breaking down of inter-cultural barriers (e.g. language, geographical distance, differences in behaviour) and has brought about a closer integration of the countries and people of the world. This is known as globalisation: an increasing connection between all the societies and cultures around the globe.

Globalisation has happened largely as a result of commerce, where entrepreneurial interest has been aroused by the possibilities of international business. Where there is economic globalisation, however, there is also 'cultural globalisation'. This refers to the idea that people around the world are conforming in their habits (e.g. watching the same television programmes, or eating the same food) and attitudes (e.g. converging beliefs about democracy and human rights). Depending upon the individual, this is a concept that can inspire both hope and anxiety.

The 'hopeful' view regards the idea of a 'one world' culture as a utopian dream, in which the mixing and blending of cultures enables different races to overcome national boundaries and embrace common causes, helping to create a more peaceful world.

In contrast, the 'fearful' view mourns the possibility that exciting diversity might be lost, and that the variety of different cultures currently in existence will become homogenised. Further, critics argue that globalisation mainly benefits the already powerful nations of the world, giving them leverage not only in trade with other countries of the world but also in influencing their politics and general lifestyle.

Textiles and Fashion. http://dx.doi.org/10.1016/B978-1-84569-931-4.00024-6

As Western (predominantly American) culture imposes itself around the world, for some, 'globalisation' equates to 'Americanisation' – a term from the early 1900s that described the process of turning new immigrants into 'Americans', through learning the English language and adjusting to American culture, customs and dress – whether or not they wanted to relinquish their traditional ways. Critics of globalisation are concerned that this has become a 'cloak' for the export of the American business model – an argument that gains credibility when considering that the most visible sign of globalisation is the spread of the American hamburger, via the McDonald's brand, and cola drinks, via Pepsi and Coca Cola, to almost every country on Earth.

24.2 IMPACT OF CULTURE IN EUROPEAN AND NON-EUROPEAN ARENAS

There is no such thing as a 'pure' culture. Culture is dynamic and changes over time. As new communication and transportation technologies allow for greater movement of people and ideas between cultures, external influences are assimilated and cultures adjust to changing environments. This has always been the case.

For example, the close proximity of countries within Europe has ensured that European countries have influenced each other's cultures for centuries. Notable examples of how invasion from other European countries has fundamentally influenced British culture include the Roman (Italian) invasion of Britain (c. 55 BC), the Viking (Scandinavian) invasion (c. AD 700) and the Norman (French) invasion in 1066. Each of these invasions left an indelible mark on the British way of life as the invading armies brought with them new technologies and other influences from their own culture (e.g. Roman art, architecture, education, language and religion).

The English language betrays cultural influences from numerous sources. Latin, for example, the language of the Romans, is responsible for a significant portion of English words. However, many other words in regular use have been plundered from a variety of sources, both from within Europe and further afield. Words such as 'husband' and 'mistake', for example, are of Icelandic origin. 'Pyjamas', 'bungalow', 'verandah', 'shampoo' and 'khaki' are all derived from Indian languages. 'Kayak' is derived from the Inuit 'qayaq', and 'jazz' originates from West African languages (Mandinka 'jasi', Temne 'yas').

In the fashion and textile industries, many words and phrases are borrowed from the French language. 'Haute couture' (or 'high sewing'), toile (meaning 'cloth'), lingerie, chic (meaning 'stylish'), art nouveau (literally 'new art'), art déco (an abbreviation of 'art décoratif', meaning 'decorative art'), crêpe de Chine ('Chinese crepe', a type of silk), chemise, décolletage or décolleté (meaning 'low neckline, lowered neckline') are all French words and phrases in common use in fashion and textile design.

English words have been similarly 'borrowed' and integrated into other languages. The French use 'le weekend', 'cool' and, surprisingly, 'un talkie-walkie'. In Italy, 'picnic', 'stress', 'shopping', 'manager' and the ubiquitous 'weekend' are all in common use.

Another clue to the influence of cultures from both inside and outside Europe can be found in supermarkets and restaurants. The type and variety of foods readily available to us clearly reflect influences from European culture – especially French and Italian – and from farther afield, for example, Japanese sushi, Indian curries, Thai, Chinese and Mexican foods – to the extent that the late Robin Cook, Foreign Secretary under Tony Blair (b.1953, Prime Minister of the United Kingdom from 1997 to 2007), suggested that 'chicken tikka masala' might be considered the 'national dish' of Britain.

Whilst the economic might of the United States has exercised a seminal influence on all cultures of the world, leading to concerns about globalisation (as described in Section 24.3), the United States itself is one of the most racially diverse countries in the world. 'Multicultural' American society is a blend of native Americans as well as a large number of immigrants from various countries, including Ireland, Germany, Poland, Italy, Latin America, Asia and Africa. Cultural influences from all these nations can be identified within modern American society, underlining the fluid nature of national identity as distinct cultural influences – from both inside and outside Europe – constantly interact and merge to create new cultural definitions.

24.3 CASE STUDY

24.3.1 CULTURAL EXCHANGES: JAPAN AND THE WEST

The case study of the cultural interaction between Japan and the West is particularly interesting because of Japan's isolation from the western world for over 250 years, during the seventeenth, eighteenth and nineteenth centuries.

Between 1603 and 1868, Japan was ruled by the shoguns (hereditary officials) of the Tokugawa family. The Tokugawa shogunate was established in Edo in 1603 by Tokugawa Ieyasu, the first shogun of the 'Edo' or 'Tokugawa period'. Edo was made capital of Japan by the Tokugawa shogunate, and was later re-named Tokyo, as we know it today (people born in Tokyo are still known as 'Edo-ko', or children of Edo). The Edo Period was a time of peace, and the richest period in the history of feudal Japan.

Ieyasu regarded European influence as a threat to the stability of his peaceful and prosperous nation and imposed a 'closed-door' policy on Japan, effectively isolating it politically, economically and culturally from the outside world. Those who ventured beyond Japanese borders were executed if they dared to return, to protect Japan from European 'contamination'. Only the Dutch and Chinese were permitted to trade with Japan, via a few closely-monitored ports. Because of this enforced isolation, the Edo Period became known as the *sakoku jidai* – the period of isolation.

The clothing of the Edo Period consisted of basic pieces such as the *Kimono* (traditional full-length robe, literally meaning 'thing to wear'), the *Obi* (sash) and *Hakama* (divided skirt, or trousers) – the style of dress that is still recognised as traditional Japanese costume (Figure 24.1).

Over the next 250 years Japanese culture developed almost completely without influence from the West until, in 1853, US Navy Commodore Matthew C. Perry travelled to Japan and negotiated hard for several months with Japanese officials for the right to trade with Japan. Perry eventually succeeded in forcing the Japanese government to open a limited number of ports for trade, following the signing of the 'Treaty of Peace and Amity between the United States and Japan', better known as the Treaty of Kanagawa, in 1854. This has become known as 'the opening of Japan'.

Later in 1854, Britain, eager to avail itself of newly-established commercial opportunities with Japan, used its fleet to force a treaty similar to the Kanagawa document on the Japanese. Arrangements with the Netherlands, France and Russia followed.

Japan's self-imposed isolation had finally been breached. The opening of its doors to the West meant that Westerners came into significant contact with Japanese culture for the first time and were spellbound by the richness and exoticism that they encountered. Western appreciation for Japanese art and design rapidly intensified, and by the end of the nineteenth century, the Japanese influence could be seen throughout fashion, textiles, interior design and art. This trend became known as 'Japonisme' (or Japonism in the United Kingdom), a term coined in 1872 by Philip Burty, a French art critic. In fashion,

FIGURE 24.1

Traditional Japanese costume.

the effect of 'Japonisme' was characterised by luxurious, lavishly coloured fabrics, intricately patterned embellished surfaces and dramatic silhouettes (Figure 24.2).

The cultural exchange between Japan and the West was not limited to one direction. As the 'opening of Japan' led to Japanese culture impacting upon the West, it was inevitable that Western cultural influences would soon travel in the opposite direction.

In 1868 the Edo period was supplanted by the Meiji (meaning 'enlightened rule') era. In order to persuade Western powers to revise unequal treaties imposed on Japan, Emperor Meiji and his government officials resolved to convert Japan into a modern nation state, demonstrating its high level of civilisation through an ambitious programme of military, political, social and economic reforms inspired by Western practices.

At the same time, the 'Industrial Revolution' in the United Kingdom had given rise to technological innovations such as steam-powered weaving looms and chemical dying techniques, which led to a revolution in the textile industry that soon spread to the rest of Europe, the United States and eventually Japan.

These cultural and technological innovations led to the adoption of Western-style dress (known as *yofuku*) in Japan, as a desirable symbol of progress and sophistication. When the Emperor and Empress, as public role models, adopted Western clothing and hairstyles at official events, government officials and the educated elite began wearing Western-style clothing in public (Hastings, 1993). Shortly afterwards, fashion-conscious women also began wear Western dresses in public, following the example of the Empress (Figure 24.3).

In 1912, the Meiji period gave way to the Taisho period (1912–1926), one of the shortest in Japanese history. During this period, traditional Japanese and Western dress became blended, giving rise to interesting combinations of East and West (Gomez, 2002).

The working women of Japan – nurses, elevator girls, bus conductors and typists – began wearing Western clothes in everyday life. Men's clothing also became gradually westernised, so that by

FIGURE 24.2

Japanese kimono inspired design – Paul Poiret, c. 1910.

Source: Image – Shakko Photos.

the time the Taisho period gave way to the Showa period (1926–1989), the business suit had become standard attire for most company employees. Throughout the 1920s and 1930s, fashionable Japanese women sported the short 'bobbed' hair and shorter hemlines of the Western 'jazz age' (Figures 24.4 to 24.6).

By the onset of World War II (1939–1945), the Japanese adoption of Western-style dress had more or less been established, and has remained normal outdoor attire for Japanese citizens ever since.

FIGURE 24.3

Japanese women in bustled nineteenth century Western dress, 1887.

Likewise, Western artists and designers have continued to plunder the East (or 'Orient') for inspiration into the twenty-first century, particularly in the field of fashion design, where the continuing cultural exchange between Japan and the West has remained a rich and popular source of design ideas.

For example, in the late twentieth century, the arrival of Japanese designers on the catwalks of Europe was responsible for redefining the character of international fashion design, hitherto dominated by European fashion designers.

Early pioneers Hanae Mori, the first Japanese fashion designer to show in the West (New York, 1965), Tokio Kumagai and Kenzo Takada (House of Kenzo) were followed in the early 1980s by a new generation of Japanese designers who shook the fashion establishment with collections that drew on a uniquely Japanese aesthetic.

FIGURE 24.4

Japanese actress Yukiko Tsukuba wearing Western fashions in 1925.

Issey Miyake, sometimes referred to as the founding father of the 'Japanese Avant-Garde', Rei Kawakubo (founder of the Comme des Garcons label) and Yohji Yamamoto challenged and overturned the prevailing European concept of fashion, which extolled the virtues of sexuality, glamour and status (Da Cruz, 2004). Conspicuous consumption and body conscious fashions – power suits and dramatic evening dresses by designers such as Georgio Armani, Ralph Lauren and Gianni Versace – dominated the European catwalks at the time (Figure 24.7). Instead, these Japanese designers ignored European stylistic trends and introduced garments that were monochromatic – often black – asymmetrical, sexless and unfitted:

> Radical and conceptual, challenging and uncompromising, functional and sometime incomprehensible, fashion from Japan demands attention. Since the 1970s and 1980s when Issey Miyake, Rei Kawakubo and Yohji Yamamoto established themselves as influential designers in international fashion, Japanese fashion has been acclaimed for its ability to challenge fashion conventions, embrace technology and point the way forward.
>
> **Fukai, English, and Mitchell (2005, p. 9)**

FIGURE 24.5

Crown Prince Hirohito, Princess Nagako and Princess Shigeko in Western-style clothing, 1925.

Not only did these garments overturn contemporary conventions of garment construction, they even challenged established notions of what could be considered 'beautiful'. A fundamental aesthetic concept of Japanese design is that of 'wabi-sabi', a compound word associated with Zen Buddhism that brings together the notion of 'a simple, austere beauty' (*wabi*), and 'a lonely beauty, such as the beauty of silence and old age' (*sabi*) (Figure 24.8) (Kawamura, 2004).

As well as imbuing their designs with a quiet contemplation, and a celebration of tradition, Japanese designers are also credited with embracing the new, revelling in cutting-edge technological developments in textile design. Japanese textile designers follow the same ideology, and although companies such as 'Nuno' (http://www.nuno.com) specialise in the creation of innovative textiles

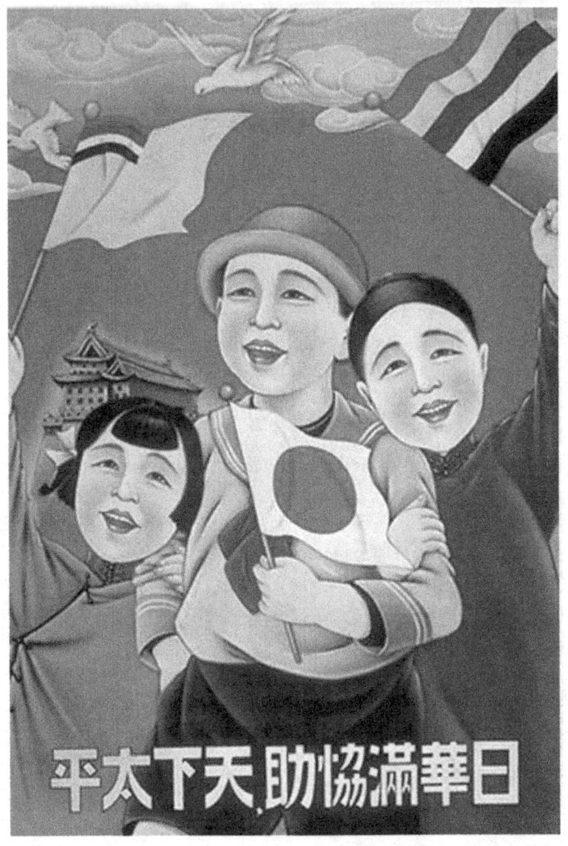

FIGURE 24.6

1935 Propaganda poster showing Manchurian, Japanese and Chinese people wearing a mixture of traditional and Western hairstyles and clothing.

using the latest technologies, their design philosophy is still influenced by traditional Japanese aesthetics:

> NUNO combines the best of past and present; drawing on traditional aesthetics and attention to creative processes to inspire today's fashions, while enlisting modern technologies to make Japan's 'lost art' more accessible to textile lovers worldwide.
>
> **Nuno Website (2011: http://www.nuno.com/Twist/Twist.html)**

In her essay: 'Fashion as Art: Postmodernist Japanese Fashion', Bonnie English describes the apparent paradox of Japanese design:

> …Japanese designers work within a post-modernist visual arts framework, appropriating aspects of their traditional culture, and embracing new technological development and methodologies in textile design. Yet, at the same time, they infuse their work with meaning and memory.
>
> **English, in Mitchell (2005, p. 29)**

FIGURE 24.7

Power dressing in 1980s – Anne-Marie Beretta, 1982.

Source: Image courtesy of The Woolmark Company.

Among the Japanese fashion designers, Issey Miyake in particular has become known for his experimental approach towards the use of materials and garment production.

In 1965, Miyake had travelled to Paris to study haute couture, where each season, fabric manufacturers would present their samples for fashion designers to select for their new collections. In Japan, however, fabric manufacturers were more concerned with textile innovation. Miyake wanted to be part of that innovation and to design and produce garments hitherto unimagined. In 1970 he returned to Japan to set up MDS – Miyake Design Studio – in Tokyo, and began a dialogue with manufacturers and traditional craftsmen that continues to this day.

In 1982 a Miyake model wearing a futuristic sculptural bodice in rattan and bamboo appeared on the cover of the American magazine Artforum – the first fashion ever to do so.

Miyake's 'Pleats Please' collection, launched in 1993, used an equally innovative method of garment production. Oversized garments in lightweight, stretch polyester fabric, two-and-a-half to eight times the size of the finished garments, were sewn together, then folded and sandwiched between layers

FIGURE 24.8

The Japanese Avant-Garde – Comme des Garcons, 1984.

Source: Illustration by Carol Ryder www.carolryder.com.

of paper and fed into an industrial heat press. The results were permanently pleated, architectural garments that were impossible to crease.

In 1997, Miyake launched 'A-POC', which enabled customers to cut different garments from a continuous roll of machine-knitted tubular cloth (A-POC, a play on the word 'epoch', stands for 'a piece of cloth'); then in 2010, when many assumed he had retired from design, Miyake introduced Miyake 132 5, a collection of garments made entirely from recycled materials in shapes created by a computer programme designed to generate complex three-dimensional shapes from a flat piece of paper.

The fabric used – originally produced by Japanese company Teijin – is made from a polyester fibre produced from recycled polyester offcuts and plastic bottles. Although the original Teijin fabric was too stiff to be comfortable; Miyake, working alongside his partners in the textile industry, produced a softer, more wearable version of the fabric.

In an interview for Telegraph Fashion (September 27, 2010), Sheryl Garratt (2010) describes Miyake's approach to fashion design:

> He (Miyake) is interested in the new, in invention, and there is never anything retro or nostalgic about his designs. 'Many people repeat the past,' he shrugs. 'I'm not interested. I prefer evolution.

Although largely derided when first presented to a bewildered European audience, the deconstructed and gender-neutral garments of the Japanese 'Avant-Garde' went on to influence Belgian designers such as Martin Margiela and the 'Antwerp 6' (Figure 24.9), and in modern times the radical, conceptual dynamic of the Japanese aesthetic is credited with overturning a tired and clichéd industry. The collision between Japanese and Western cultures had challenged the status quo and changed the face of fashion forever (Fukai, Vinken, Frankel, Kurino, & Nii, 2010).

FIGURE 24.9

The influence of the Japanese Avant-Garde – Martin Margiela, Spring 2014. Anton Oparin, Shutterstock.com.

In her influential book *Fashion Zeitgeist: Trends and Cycles in the Fashion System* (Berg 2004), Barbara Vinken (2004) reflects:

> With the 1970s, the fashion of a hundred years, the continuous line stretching from Worth to Saint-Laurent, comes to an end. …The Paris show of Comme des Garçons, in 1981, spectacularly marked the end of one era and the beginning of another.
>
> **Vinken (2004, p. 35)**

At the same time that pioneering Japanese fashion designers began to make their presence felt in Europe, American pop culture was making its mark among the young people of Japan. Since the 1950s, the Japanese press had recognised the concept of youth cultures, describing them as 'zoku' ('tribes'). In the late 1950s, Japanese motorcycle gangs called the 'Kaminarizoku' ('Thunder Tribe'), inspired by the films of James Dean and Marlon Brando and dressed similarly in blue jeans, black leather biker jackets and boots, caused unease in Japanese society.

The appearance of Japanese youth 'tribes' intensified in the 1960s and 1970s as rock 'n' roll, motorcycle culture, coffee-bar society, dance clubs and other American leisure pursuits flourished against a backdrop of rapid economic growth.

Despite undergoing numerous transformations, largely in line with Western trends, the devotees of numerous 'subcultures' among Japanese youth have established themselves as a permanent feature of Japanese cities – most notably in down-town areas of Tokyo such as Harajuku, Shibuya and Ikebukuro – identifying and expressing themselves through individual and often extreme dress codes (Aoki, 2001). 'Lolita', probably the most famous and recognisable of contemporary Japanese youth tribes (Figure 24.10), has itself spawned numerous off-shoots such as 'Classic Lolita', 'Sweet Lolita' ('Ama Rorita', or 'Amarori') (Figure 24.11), and 'Gothic Lolita' ('Goth-Loli') (Figure 24.12) – whose own variants include 'Elegant Gothic Lolita' and 'Elegant Gothic Aristocrat'. 'Kogals' (or 'Kogyaru') wear mini-skirted high school uniforms with baggy socks and platform boots; their style and conspicuous consumerism roughly approximate tanned Californian

FIGURE 24.10

Japanese street fashion: Lolita.

Source: Image courtesy of Eric Prideaux www.pripix.com

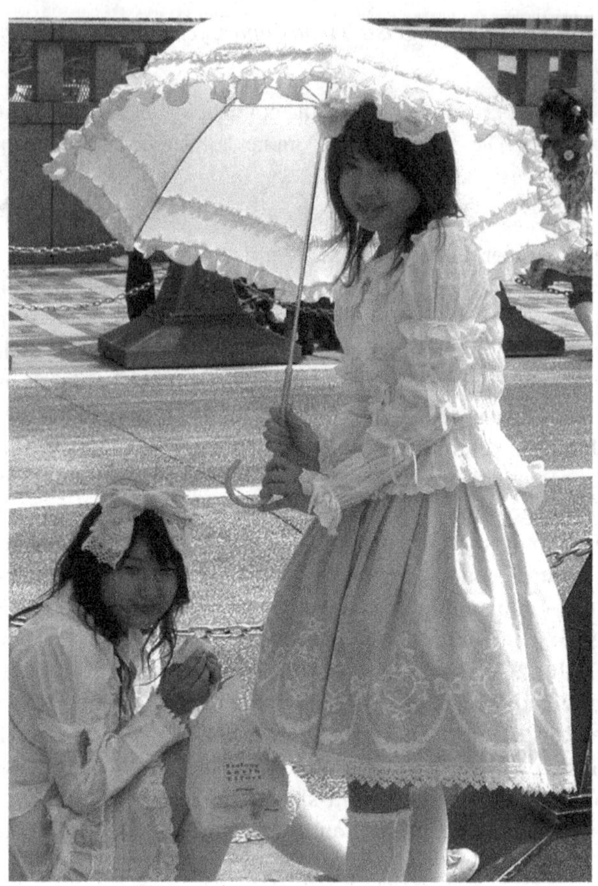

FIGURE 24.11

Japanese street fashion: Sweet Lolita.

Valley Girls (Figure 24.13). 'Ganguro' (literally 'black-face') wear dark brown or black make-up and paint their noses, eyes and lips white (Figure 24.14). Their male counterparts, 'Center Guys', are named after Center Street, a pedestrianised shopping street near Shibuya Station (Figure 24.15).

The creative energy and flair of these 'youth tribes' drive their feverish pursuit of fashion. There is an incredibly swift turnover and a bewildering variety of styles popular at any one time. They congregate in public urban spaces to socialise, see and be seen.

Although Western influences are still recognisable amongst the street style tribes of Japan – 'Gothic Lolita's' Victorian influence, for example, or the 'Kogals' reference to Californian Valley Girls – Japanese young people continually combine and augment these influences to create their own unique aesthetic, and new variants have evolved from the cultural mix, which in turn have been seized by Western youth as fresh, viable style statements (Steele, 2010).

Once again the cultural exchange between Japan and the West has come full circle.

FIGURE 24.12

Japanese street fashion: Gothic Lolita.

24.4 FUTURE TRENDS

Cultural change is brought about by a number of factors, including technological advances, politics and travel, as discussed in Sections 23.3.2 and 23.3.3. As part of a wider cultural context, developments in fashion and textiles can be informed by the same influences – as described in Section 23.6.

It follows, therefore, that future trends in fashion and textiles can be anticipated by considering the current climate in these areas. At the beginning of the twenty-first century, key global issues that are most likely to determine future trends in fashion and textile design are:

- Globalisation and the democratisation of fashion
- Sustainability in fashion and textiles: 'green' issues
- Ethical fashion

FIGURE 24.13

Japanese street fashion: Schoolgirl Kogals.

Source: Image courtesy of Sara Mari and 'Moments like Diamonds'. http://www.chouzuru.blogspot.com

24.4.1 GLOBALISATION AND THE DEMOCRATISATION OF FASHION

In the twenty-first century, fashion is more ubiquitous than ever. Its globalisation and widespread mass production has meant that, in one sense, fashion is more democratic than it has ever been, with low-cost, mass-produced fashion readily available in countries where 'fashion' is prevalent – that is, those countries where the consumption of clothing is based on desire rather than need.

Relative to income, clothing is far less expensive than it was a few decades ago.

Widespread availability and low cost have contributed to a shift in consumer behaviour. No longer considered a luxury, since the 1990s fashion has been regarded as a disposable commodity. Fashion-conscious consumers rarely continue to wear a garment to the end of its useful life; indeed it is common for a garment to be worn only a few times, sometimes only once, before being discarded and replaced – at which point the cycle begins again. This phenomenon is known as 'fast fashion'.

FIGURE 24.14

Japanese street fashion: Ganguro style, 1996.

> A key current trend in UK clothing demand is the growth in fashionable, low priced, 'disposable' clothing. H&M, Topshop and Inditex are examples of companies providing relatively low priced fashionable clothing through flexible, fast supply chains which allow clothing collections to be changed every two to three weeks. This encourages consumers to shop more often and the number of items bought annually is growing. In response to this trend, supermarkets like Asda and Tesco and clothing chains like Primark and Matalan, have developed high fashion brands at very low prices. In some cases these outlets are able to make rapid copies of famous designers' fashion items. In 2005, 19% of all clothing and footwear was bought in supermarkets: the combination of convenience and affordability is attractive.
>
> **Allwood, Laursen, De Rodriguez, and Bocken (2006, p. 38)**

In a cultural shift similar to the nineteenth century backlash against the low-quality mass-produced goods of the industrial revolution (described in Section 23.3.1), amidst the tide of low-cost, often low-quality fashion mass produced by 'value' fashion retailers such as Primark, Matalan and supermarket brands such as George (Asda), or Florence and Fred (Tesco), the signs of a backlash against fast fashion

FIGURE 24.15

Japanese street fashion: Center guy with Ganguro girl.

Source: Image courtesy of Eric Prideaux

have begun to emerge. In her article for Telegraph Fashion: 'Fast Fashion: is the party over?' (April 8, 2011), Belinda White reported:

> The flood of fast, throwaway fashion on the high street in recent years has inevitably led to questions about where the clothes are being made, who's making them and in what conditions, not to mention the environmental effects of mass production, and shocking waste.

> … it seems the tide may finally be turning…. in the final quarter of 2010 which saw profits fall 11 per cent, fashion comparison website Stylecompare.co.uk has today reported that year-on-year sales of 'low end' retailers fell by 21 per cent, as consumers flock to 'mid range' and eco brands for their fashion fixes.

> Julia Rebaudo from StyleCompare.co.uk comments: Our research has shown that the UK's shopping climate is set to change drastically… the notion of buying a dress just for the purposes of a Saturday night on the town seems fairly outdated. Quite simply, consumers are becoming more aware of the value of 'investment pieces', particularly at a time when being conscientious with your spending is a must.

> Eco brands are also set to make a big comeback, as the evidence is mounting that consumers associate eco with quality, and care more about where and when their clothes are manufactured.
>
> **(White, 2011)**

As the pendulum begins to swing away from fast fashion, mass production and cheap, low-quality goods, a new desire for individuality is gaining ground, giving rise to a growing trend for well-designed, high-quality handmade items that represent exciting novelty to a generation who have grown up with computer technologies and fast fashion but who have never learned to sew, knit or crochet. They are a generation for whom hand sewing and embroidery are lost skills that disappeared with their grandmothers' generation.

Replacing the much-derided 'knitting circles' of earlier generations, 'Stitch 'n' Bitch', or 'Knit 'n' Bitch' groups are springing up where women can revive these traditional skills. The term 'craftivism' – sometimes known as 'craftism' – combines the words 'craft' and 'activism', and was coined in 2003 by writer Betsy Greer to represent a more dynamic and politicised twenty-first century crafts movement that aims to reclaim hand crafts – knitting, sewing and other 'women's work' – using them to make political statements (mostly anti-war, anti-capitalist, or pro-freedom: similar to Banksy – see Section 23.4.3) and, by giving them significant meaning, elevating them to the status enjoyed by fine art. In the *Encyclopedia of Activism and Social Justice* (Sage Publications 2007), Greer defines Craftivism as: 'a way of looking at life where voicing opinions through creativity makes your voice stronger, your compassion deeper and your quest for justice more infinite' (Greer, 2008).

> Katie Bevan, curator of the UK Crafts Council exhibition, 'Knit 2 Together', commented on the resurgence in popularity of knitting as a response to global events, stating: 'There's a sort of zeitgeist: a make-do-and-mend spirit during this war on terror or whatever it is. Everyone wants to go home and knit socks.'
>
> **Turney (2009, p. 213)**

In 2006, in a gesture symbolic of 'craftivism', Danish artist Marianne Jørgensen covered an army tank with a giant pink blanket – complete with pink pom-pom on the gun barrel – in protest against Denmark's involvement in the Iraq war. As she explains on her web page: 'When it is covered in pink, it becomes completely unarmed and it loses its authority' (http://www.marianneart.dk).

In the same spirit, 'one-off' production is enjoying a renaissance in other areas: despite the speed and convenience of computer graphics, hand-rendering is once again at the 'cutting edge' of fashion illustration. Watercolour effects, characterful brush-strokes, collage and textured papers now appear new and exciting while computer-generated artwork, so progressive in the 1990s, seems commonplace, flat and boring.

Supported by the dystopian view of globalisation – the fear of a homogenised, faceless culture (see Section 24.3) – mass production is once more giving way to the desire for individuality and the unique. In fashion, 'one-off' has traditionally meant 'haute couture', with its associations of high prices. Customisation and low cost have always been mutually exclusive as mass production provides low cost uniformity, while the expense and exclusivity of customisation have made it the sole preserve of the rich.

Today, however, interactive technologies such as the internet allow customers to specify their unique requirements for a product which is then manufactured by automated systems. This process is known as 'mass customisation'.

Mass customisation has been defined as:

> producing goods and services to meet individual customer's needs with near mass production efficiency
>
> **Tseng and Jiao (2001, p. 685)**

In practice, specialist companies can now provide clients with one-off garments, including jeans, shirts, coats, jackets, skirts and shoes, with customised fit at a price far below couture level. In a New York Times article, columnist Michelle Slatalla describes a 'mass-customisation' company called the 'Interactive Custom Clothes Company' (IC3D):

> Shoppers can point and click to choose fabric, colour and cut, then type in their measurements. Beyond merely creating jeans that are the right size, IC3D offers customers the option of specifying every detail – from button colour to fabric weight to the number of pockets to how wide the ankle opening should be (the company's software can generate a pattern accurate to one-tenth of an inch).
>
> **Slatalla (2000, p. 64)**

As automated systems such as this become more commonplace, and body-scanning technology is no longer the preserve of science fiction, current difficulties in finding the 'perfect outfit', or 'perfect fit', especially for non-standard figure types, may soon become a thing of the past.

Thanks to mass-customising technology, the future seems to promise a more democratic fashion industry where individuality and freedom of choice are embraced and tastes and figure types as diverse as humanity itself are catered for, as standard, without inflated cost.

24.4.2 SUSTAINABILITY IN FASHION AND TEXTILES: 'GREEN' ISSUES

Also contributing to the backlash against 'fast fashion', the issue of sustainability has gained significant political impetus in recent years. Literally, sustainability means 'the ability to sustain'. In the twentieth and twenty-first centuries, the word has come to represent the idea of protecting the environment to support long-term ecological balance, and ensuring that the needs of current and future generations are provided for. A sustainable forest, for example, is one that is carefully managed so that as trees are felled to supply wood to communities or industry, seedlings are planted, ultimately replacing those felled.

> Sustainability is the conservation of life through ecological balance – human, animal, vegetable and planetary. A self-sustaining system is a system that does not take more from the environment than it gives back; it does not deplete resources, but sustains itself
>
> **Brown (2010, p. 6)**

The idea of sustainability has become increasingly important as the Earth's natural resources – for example oil and coal – become depleted, commonplace international travel increases the polluting impact of carbon emissions which results in climate change and globally sea and land are both contaminated by waste:

> There is universal recognition that time may be running out to find sustainable solutions for the future of the planet and its people. Whether these are to be based in nature or manufacture is by no means yet clear as debate continues.
>
> **Black (2008, p. 107)**

In Europe and the United States, land-fill sites where local authorities and industry take waste to be buried and compacted (also known as 'tips' or rubbish dumps) are overflowing. Toxic technological waste (also known as 'waste electrical and electronic equipment – 'WEEE', 'e-waste', or 'techno trash') is routinely exported, often illegally, to developing countries for cheap disposal. An area of the North Pacific gyre (a slowly rotating, ring-like system of ocean currents) contains over 6 million tonnes of non-biodegradable waste in an area of contamination estimated to be twice the size of Texas (Pope, 2009).

In terms of fashion and textile waste, synthetic fabrics represent the greatest problem. These are essentially plastic materials and may take up to 150 years to decompose in land-fill sites.

Because of this, it is commonly assumed that greater use of natural fibres would go some way to combating the amount of fashion and textile waste. As with many issues related to sustainability and ethical fashion, however, the problem is not that simple.

For example, the production of cotton – a natural fibre – also presents many environmental concerns. A key problem is that cotton is an exceptionally thirsty crop. It takes over 2000 L of water to produce just one cotton T-shirt. The Aral Sea in Uzbekistan, once the world's fourth largest inland lake with a thriving ecosystem, has shrunk to just 15% of its original size, mainly as a result of irrigation for the cotton industry. Its salinity has risen by almost 600%, and all native fish have disappeared from its waters, leading to infertile soil and huge areas of salty desert contaminated with pesticide residues that cannot sustain life.

Concerns about environmental issues have led to some fashion and textile design companies rethinking their current practices and considering more sustainable methods of production. For example, the use of replaceable, locally-sourced, recycled or organically-grown materials (such as organic cotton, which does not rely on heavy use of pesticides), exploring the concept of 'zero waste' (where garments are designed to eliminate any waste fabric in production), or 'closed-loop' production systems, which have no detrimental impact on the environment.

Research also continues into technological advances which support sustainability – for example, biologically-cultured textiles, recycled fibres, environmentally-friendly dyestuffs and 3D printing. Fashion produced using any of these methods is collectively known as 'sustainable' or 'green' fashion.

The ultimate reaction to the excesses of fast fashion is the burgeoning trend for 'slow fashion', where consumer behaviour is, predictably, exactly the opposite of that embodied by fast fashion. Slow fashion dictates that greater care is taken in the design and production of garments; the higher initial costs associated with better quality materials and careful manufacture are offset by a greater life expectancy from each garment, which ultimately represents better value. This is known as 'investment' dressing. As garments are required to last for a long time, classic styles that do not date are a key feature of slow fashion.

Another feature of slow fashion is the recycling of garments. Once a garment has reached the end of its useful life for a particular consumer, that individual might take the garment to a recycling centre to be recycled, to a charity shop, or sell it on an auction site such as 'Ebay' (www.ebay.com), to be purchased and worn by a new consumer.

Of course, the concept of slow fashion is not new; the Savile Row tailors in London, for example, have been producing classic men's suits in high-quality fabrics since the nineteenth century. Maintenance of garments by replacement or repair of worn areas is an important part of the service, reminiscent of the culture of 'make do and mend' during World War II, which also epitomised the ethos of slow fashion (see Section 23.5.3).

Another strategy designed to extend the life of a garment and delay its disposal still further, in line with the notion of slow fashion, is the idea of garments that from conception, have future lives 'built in' to the design. In *Sustainable Fashion and Textiles: Design Journeys*, Kate Fletcher describes one such project:

> One of the earliest examples of interconnected material cycles in fashion and textiles is a conceptual project developed in 1993… uncoloured virgin fibres in the first life produce a high quality fabric for use in high-end men's- and women's-wear. In subsequent lives, fibres are transformed into bulkier and lower quality fabrics (as quality deteriorates with each recirculation) – suitable for children's-wear. In the course of reprocessing, the fibres are overdyed, transforming them into brightly coloured pieces for the children's market.
>
> **Fletcher (2008, p. 109)**

Another concept related to slow fashion that addresses sustainability in general is that of 'cradle to cradle' design, as conceived by chemist Michael Braungart and architect William McDonough in 'Cradle to cradle: re-making the way we make things' (Braungart & McDonough, 2009).

Braungart and McDonough challenge the familiar model of 'cradle to grave' production, which is the idea that something is produced, used, and eventually discarded. 'Cradle to cradle' design takes its inspiration from nature, where dead and decaying plants provide fertiliser for the next generation – 'waste equals food' – and suggests that products, including fashion products, should be designed so that the end of one life cycle can represent the beginning of another, thus completely altering the approach to design and production:

> More control (being 'less bad') is not the same as being good. It is not protecting your child if you beat him three times instead of five, and it is not protecting the environment simply to use your car less often. When you do something wrong, don't try to improve upon it. …

> The long perspective is completely unlike the single reuse of popular 'recycling', when your plastic bottle becomes your parka … and in five years the parka goes to exactly the same dead-end cradle-to-grave where a few years earlier your bottle would have gone.
>
> **Braungart and McDonough (2009, pp. 4–6)**

'Cradle to cradle' philosophy demands a dramatic paradigm shift: that we view ourselves as equal participants in the Earth's ecology, rather than dominant over it – a traditional view taken by many tribal people around the world: 'What you people call your natural resources our people call our relatives' (Oren Lyons, faith keeper of the Onondaga, as quoted in Braungart & McDonough, 2009).

In the long term, instead of 'down-cycling' (where 'recycling' a product or material only results in an inferior product, as is the case with recycled paper) and 'minimising waste' (being 'less bad'), an approach needs to be taken where waste is 'designed out' of a product – including fashion – and the idea of 'saving' the planet by doing without should be replaced by a design model that encourages us to learn to live on it harmoniously and sustainably.

24.4.3 ETHICAL FASHION

'Ethical' fashion is the term used to describe fashion that is neither harmful nor exploitative of people, animals or the environment in its design, production or distribution.

The 'Ethical Fashion Forum' (EFF) describes ethical fashion as:

> … an approach to the design, sourcing and manufacture of clothing which maximises benefits to people and communities while minimising impact on the environment …

> For the EFF, the meaning of ethical goes beyond doing no harm, representing an approach which strives to take an active role in poverty reduction, sustainable livelihood creation, minimising and counteracting environmental concerns.
>
> **EFF Website: http://www.ethicalfashionforum.com/the-issues/ethical-fashion**

Ethical concerns related to the production of fashion include the large amount of fashion and textiles waste that is exported to other countries, which in turn stifles the textile trades overseas, or the hazardous pesticides used in the production of industrially-grown cotton – more insecticides are used on cotton than any other single crop – which harm people, wildlife and the environment. These pesticides can poison farm workers, drift into neighbouring communities, contaminate the ground and surface water and kill beneficial insects and micro-organisms in the soil. The result may be starvation and the migration of people who can no longer farm the ground to feed themselves.

As well as an environmental cost (described earlier in Section 24.6.2), fast fashion also has a human cost.

In order to satisfy the ever-increasing demand for cheaper clothes and rapidly-changing fashion trends, exploitation of cheap labour (often child labour) is commonplace in the fashion industry, especially in poor countries where people are willing to tolerate appalling working conditions in order to feed their families, and where governmental control on such labour is weak or insufficiently enforced:

> Millions of children work to help their families in ways that are neither harmful nor exploitative. But millions more are put to work in ways that drain childhood of all joy - and crush the right to normal physical and mental development …

> … By and large, it is the children of marginalised communities, their futures already threatened by inadequate diet and health care, who are at greatest risk from exploitation at work. In India, the majority of children in servitude are children from low castes or tribal minorities. In Latin America, the highest incidence of child labour is found among the indigenous people.

> Often such children are as young as six or seven years old. Often their hours of labour are 12 to 16 hours a day. Often their place of work is the sweatshop, the mine, the refuse heap, or the street. Often the work itself is dull, day-long, repetitive, low-paid or unpaid. Sometimes the child works under the threat of violence and intimidation, or is subject to sexual exploitation.

> In the 1990s, child labour has found a new niche in the rapidly expanding export industries of some developing countries. In one small carpet factory in Asia, children as young as five were found to work from six in the morning until seven at night for less than 20 cents a day. In another, they sat alongside adults for 12 to 14 hours in damp trenches, dug to accommodate the carpet looms on which they wove. In a garment factory, nine-year-olds worked around the clock sewing shirts for three days at a stretch, permitted only two one-hour breaks, during which they were forced to sleep next to their machines. Extracting such high human cost, child labour is nevertheless cheap. A shirt that sells in the United States for $60 can cost less than 10 cents in labour.
>
> **Source: UNICEF Website http://www.unicef.org/pon95/chil0016.html**

Although the practice of exploiting people, particularly children, for the purposes of cheap labour is often legislated against, the problem is not easily solved. For example, children often do not want to be taken away from the workplace as enforcing the ban on child labour can mean reducing or removing their family's income. Because of this, children often hide away from the people who are trying to help them.

A powerful illustration of the issues surrounding the exploitation of fashion industry workers in India can be found in 'Blood, Sweat and T-Shirts', a BBC Three television series shown in the United Kingdom from May 2008 that brought together six young British fashion consumers to experience working alongside the people who make the products sold on the high street.

As the issues surrounding sustainable and ethical fashion (often referred to as 'S & E' fashion, or 'Eco' fashion) gather momentum, a number of not-for-profit organisations – such as the EFF, the Centre for Sustainable Fashion (CSF – established by the London College of Fashion), Labour Behind the Label (LBL), and the Fairtrade Foundation – have grown up to inform, educate and support the fashion industry in its slow climb towards sustainability and ethical practices.

A registered charity established in 1992, the Fairtrade Foundation is an independent not-for-profit organisation that licenses use of the 'FAIRTRADE' mark on products in the United Kingdom in accordance with internationally agreed Fairtrade standards. The Fairtrade Foundation works with businesses, community groups and individuals, particularly in the South (Africa, Asia, Latin America and the Caribbean), to ensure sustainable livelihoods for farmers, workers and their communities, and improve the trading position of people and 'producer organisations' who produce goods such as chocolate, tea, fruit, flowers, coffee, gold and cotton. In buying products that carry the Fairtrade label, consumers can be assured that the farmers and workers who produce those goods have benefited from a fair price, fair wages and safe working conditions, do not work long hours and are able to afford food and healthcare.

Similarly, pioneering ethical fashion labels such as People Tree, launched in the United Kingdom in 2001 by Safia Minney, endeavour to uphold ethical practices in the production, distribution and retail of their products. People Tree is one of the longest-established ethical fashion labels. Since its inception, it has aimed to use only organic and Fairtrade cotton, uses strictly natural dyes, sources locally where possible and chooses recycled products over synthetics. The People Tree Website describes the company's design philosophy thus:

> For every beautiful garment People Tree makes, there's an equally beautiful change happening somewhere in the world.

> We like to think of it in terms of our little shoots-and-roots motif. We provide you with exclusive fashion–the shoots–while at the same time we work deep down through the roots, improving the lives and environment of the artisans and farmers in developing countries who work to produce it.
>
> **People Tree Website: http://www.peopletree.co.uk/content/26/about-us**

Other well-known ethical fashion brands include Ciel, founded by Sarah Ratty, a designer with long-established credentials in ethical practice, and Edun, founded by U2 frontman Bono and his wife Ali Hewson in 2005 to foster sustainable employment schemes in developing countries around the world and to encourage fair trade practice.

As sustainable green issues continue to gain public attention and become increasingly critical, and reports of ethical issues continue to impact upon social consciousness, the fashion industry has been forced to respond to mounting pressure to change its behaviour in line with more sustainable and socially acceptable practices. As a result, ethical and sustainable imperatives are gathering momentum in the fashion industry.

A small number of pioneering, conscience-led designers and companies such as those just described have led the way for large corporations in considering the wider impact of their familiar industrial practices. In the past decade 'corporate social responsibility' initiatives have become common amongst fashion brands and retailers. In response to the idea that businesses should be actively concerned with the welfare of society at large, the 'social responsibility' of a company can be described as its obligation to maximise its positive impact and minimise its negative impact on the society.

Corporate social responsibility involves a company voluntarily assuming responsibility for the social, ecological and economic consequences of its activities across its supply chain, reporting on these consequences and constructively engaging with its stakeholders, who include its customers, workforce and shareholders.

In 2007, Marks and Spencer, led by then Chief Executive Sir Stuart Rose (1949–), launched their 'Plan A' strategy for social and environmental improvements (named Plan A because 'there is no Plan B'). In this, they set out 100 commitments to achieve in five years: 'Through Plan A we are working with our customers and our suppliers to combat climate change, reduce waste, use sustainable raw materials, trade ethically, and help our customers to lead healthier lifestyles' (Marks and Spencer 'Plan A' Website: http://plana.marksandspencer.com). In 2010, they extended Plan A to 180 commitments to achieve by 2015, with the ultimate goal of becoming the world's most sustainable major retailer.

Similarly, other large fashion retailers have made recent commitments to improve their business models towards more sustainable and ethical practices. New Look, Asos, H&M, Top Shop and Gap have all produced 'eco-ethical' ranges as part of their retail collections, while Tesco and Sainsbury have pledged to increase their ranges of Fairtrade and organic clothing.

Despite these claims, it is clear that there is still a long way to go. Currently 'eco-ethical' ranges account for only a tiny percentage of the large retailers' output. Amidst accusations of tokenism and 'green-washing' – where claims of sustainable and ethical practices are exaggerated or misrepresented – press and consumers alike remain skeptical that enough is being done by big business to secure an optimistic future for the planet or its people (Mayhew, 2007).

As large corporations try to balance entrenched global production methods and the need to maintain healthy profit margins with increasing pressure to perform more sustainably and ethically, it is certain that for the foreseeable future at least, the imperative for sustainable and ethical fashion will gather momentum as politicians are put under pressure to implement initiatives that save lives, protect human rights across the globe and preserve the resources that future generations will need to live. These include laws to prevent illegal exportation and dumping of waste, increased taxation of air travel, reinforcement of laws that forbid child labour and codes of practice to ensure fair trade and safe working conditions.

24.5 SUMMARY POINTS

- 'Culture' is a term that describes the particular characteristics of a specific geographical region (e.g. British culture), civilisation (e.g. Mayan culture), or group (e.g. youth culture, drug culture), in a specific period of time. It can include, for example, the habits, attitudes, belief systems, values, artefacts and politics of a particular group of people at a particular point in history (e.g. the culture of ancient Rome).

- A great deal can be learned about the culture of a particular group of people by asking questions such as What did they do? What did they wear? What did they make? What did they believe? What did they eat? Even more can be understood by then asking Why?
- Influences such as travel, politics and war, societal shifts and technological advances can have a significant effect on the culture of a particular group of people.
- Cultural differences can be extreme. When two cultures encounter each other for the first time, the resulting interaction may create tension, fear and mistrust, but they can also give rise to new cultural phenomena as a result of positive collaboration and cultural exchange.
- Experiencing new cultures can lead to exciting developments in art and design, including fashion and textile design. For example, the culture of Japan has impacted upon Western fashions on a number of occasions throughout history, and this influence has been reciprocated as Japanese fashions have been similarly affected by the West.
- Future trends in fashion and textile design can be anticipated by considering current cultural phenomena as well as developments in technology, politics, society and the arts.

24.6 PROJECT IDEAS

1. *Sustainability and ethical fashion*: using the textile traditions of one African country, carry out focused research into materials, colours and processes, especially into those fabrics and production processes that are considered both 'sustainable' and 'ethical'. Use this research to inspire a capsule 'eco-collection' consisting of six pieces. These pieces should be appropriate to current trends, and for a specific season of your choice.
2. 'Elitist' versus 'democratic' fashion: carry out research into a culture of your choice. Taking this research as inspiration, design a collection of up to 12 outfits for the 'couture' market. Using the same research for inspiration, design another collection of up to 12 outfits appropriate for a specific High Street store. Ensure that you provide appropriate swatches of fabrics and trims for each collection. In addition you should provide costings for at least three comparable garments in each collection, and demonstrate your rationale for the market placing of each collection.

24.7 REVISION QUESTIONS

1. Explain how Japanese dress of the Edo period differed from that of the Taisho period, and how those differences came about. Illustrate your answer with relevant images of Japanese fashion.
2. Explain how Eastern culture has affected the West, and how Western culture has affected the East, in the nineteenth, twentieth and twenty-first centuries. Refer to social influences such as travel, politics and technological advances to support your answer.
3. In the nineteenth century, Western culture was heavily influenced by Japanese art and design, a phenomenon that became known as 'Japonisme', or 'Japonism'. Investigate the phenomenon of Japonism, and show how Japanese design ideals were integrated into Western design.

4. Investigate how Japanese fashions, both traditional and modern, have influenced Western fashion design in the twentieth and twenty-first centuries. Refer to specific designers – both Japanese and Western – to support your ideas, and illustrate your essay with relevant garment designs.
5. The fashion establishment was shaken in the late twentieth century by the arrival of Japanese designers on the catwalks of Europe. Explain what was so shocking about the work of these designers, using images of both European and Japanese fashion designers to support your answer.
6. Giving examples, describe how concerns about sustainability and related ethical issues might affect fashion and textile design in the future.
7. 'History has shown how shifts in society – political change, war, scientific developments, and travel, leading to the discovery of new, foreign cultures – affect our own cultural practices, including those related to fashion and textile design'. Carry out research into a culture of your choice, and explain how this culture has impacted on fashion and textile design.

24.8 FURTHER READING

As described, 'culture' is an umbrella term that relates to a wide range of societies – including those from the past as well as the present. Accordingly there are numerous sources of information to be found relating to different aspects of culture.

In particular, galleries and museums are excellent sources of cultural information. The V&A, Barbican, Tate Britain and Tate Modern in London, the Smithsonian Museum in Washington, the Museum of Modern Art in New York, the Guggenheim Museums in New York and Bilbao, the Louvre and Musée des Arts Decoratifs in Paris, Tokyo National Museum, the Berlin Museum Island, the Uffizi Gallery in Florence – to name but a few – provide lectures, gallery tours, interactive workshops and other activities as well as exhibitions to help students of fashion and culture gain as much as possible from their museum experience. In addition, most galleries and museums have excellent websites that provide sources of information that can be accessed from home.

For current developments in art and culture, the Internet is a valuable source of information. Portal Websites such as: http://www.guardian.co.uk/artanddesign (hosted by UK newspaper 'The Guardian') and http://www.artandculture.com provide a useful means of accessing current information on a wide variety of the arts.

A wealth of help and advice is also available at local libraries. As well as any books they may have available on subjects of cultural interest, library staff will be able to organise inter-library loans which provide access to books from a number of other libraries and can give advice about alternative sources of information, such as journals, videos and DVDs.

For information specifically related to fashion, numerous fashion magazines are published at regular intervals – weekly, monthly or quarterly – and provide comprehensive information about current fashion trends. 'Vogue', 'Dazed and Confused', 'AnOther Magazine', 'Love' and 'Elle' are all excellent examples. As well as being available in hard copy, many of these magazines have online versions: see http://www.vogue.co.uk, http://www.dazeddigital.com, http://www.anothermag.com, http://thelovemagazine.co.uk and http://www.elle.com.

Dedicated fashion websites and online fashion magazines such as 'Style.com' (http://www.style.com), 'WGSN' (Worth Global Style Network – http://www.wgsn.com), 'Fashion Telegraph' (http://fashion.telegraph.co.uk), 'Fashion Insider' (http://www.thefashioninsider.com), 'Net-a-Porter.com' (http://www.net-a-porter.com) and 'Fashion.net' (http://www.fashion.net) provide a continually-updated source of information about designer collections, street style and fashion trends.

Television and radio programmes can also be useful in providing documentary information relevant to the fashion industry. The enlightening and often shocking BBC Three documentary series 'Blood, Sweat and T-shirts' from May 2008 is available on DVD from 'Labour Behind the Label': http://www.labourbehindthelabel.org/join/item/760-blood-sweat-and-t-shirts. This series brought together a group of six young British fashion consumers to work in India alongside the people who make high-street fashion products, and experience first-hand the hardship and poverty they endure.

A defining feature of fashion is its continual evolution and reinventing of itself. It follows, therefore, that the best sources of information about contemporary fashion are those that are updated regularly, for example, regularly-published magazines, websites and blogs.

Social networking sites such as 'Facebook' (https://www.facebook.com) and 'Twitter' (http://www.twitter.com) and a rising number of dedicated fashion blog sites such as 'TrendLand' (http://trendland.net) and 'Style Bubble' (http://stylebubble.typepad.com) are rapidly overtaking published magazines as the principal source of fashion information because they allow the user to keep abreast of fashion news on a minute-by-minute basis. In turn, fashion brands use social media as a means to 'humanise' their online engagement with their customers, and to improve sales at minimal cost via regular updates about their product.

When considering 'future trends' in fashion, it is particularly important to remember that fashion is shaped by a variety of cultural influences. Because of this, the fashion trends forecaster must keep abreast of news in a number of spheres – national and international events, technological advances, economic fluctuations and environmental changes – as well as debates about fashion itself. Newspapers, radio and television news programmes and their internet-based counterparts are all invaluable in providing regular news updates.

For information about trends in sustainable and ethical fashion, books such as Sass Brown's 'Eco Fashion' (Brown, 2010) and Sandy Black's 'Eco-Chic' (Black, 2008) provide a useful overview of sustainable and ethical fashion trends in recent years. As the situation is continually changing, however, websites such as the 'Centre for Sustainable Fashion' (hosted by the London College of Fashion http://www.sustainable-fashion.com), the 'Ethical Fashion Forum' (http://www.ethicalfashionforum.com), Fashioning an Ethical Industry (http://fashioninganethicalindustry.org), Labour Behind the Label (http://www.labourbehindthelabel.org) and The Fairtrade Foundation (http://www.fairtrade.org.uk) are perhaps more useful as they supply information about sustainable and ethical fashion trends which is updated regularly and therefore offers a more current picture in a rapidly-changing field.

REFERENCES

Allwood, J., Laursen, S. E., De Rodriguez, C. M., & Bocken, N. M. P. (2006). *Well dressed? The present and future sustainability of clothing and textiles*. Cambridge: University of Cambridge Institute for Manufacturing.

Aoki, S. (2001). *Fruits*. London: Phaidon.

Black, S. (2008). *Eco-chic: The fashion paradox*. London: Black Dog Publishing.

Braungart, M., & McDonough, W. (2009). *Cradle to cradle: Re-making the way we make things*. London: Vintage.

Brown, S. (2010). *Eco fashion*. London: Laurence King.

Da Cruz, E. (2004). *Miyake, Kawakubo, and Yamamoto: Japanese fashion in the twentieth Century*. Heilbrunn Timeline of Art History (online). New York: The Metropolitan Museum of Art. Available at http://www.metmuseum.org/toah/hd/jafa/hd_jafa.htm. Accessed 15.01.11.

Fletcher, K. (2008). *Sustainable fashion and textiles design journeys*. London: Earthscan.
Fukai, A., English, B., & Mitchell, L. (Eds.). (2005). *The cutting edge: Fashion from Japan*. Sydney: Powerhouse.
Fukai, A., Vinken, B., Frankel, S., Kurino, H., & Nii, R. (2010). *Future beauty: 30 years of Japanese fashion*. London: Merrell.
Garratt, S. (2010). *Issey Miyake interview*. Telegraph.co.uk (online). Available at http://fashion.telegraph.co.uk/news-features/TMG8018669/Issey-Miyake-interview.html.
Gomez, E. G. (2002). When Japan tried on the modernist mantle. *The New York Times* (online), January 27. Available at http://www.nytimes.com/2002/01/27/arts/design/27GOME.html. Accessed 15.01.11.
Greer, B. (2008). *Knitting for good!: A guide to creating personal, social, and political change, stitch by stitch*. Boston: Trumpeter.
Hastings, A. (1993). The empress' new clothes and Japanese women, 1868–1912. *The Historian*, *55*(4), 677–692.
Kawamura, Y. (2004). *The Japanese revolution in Paris fashion*. Oxford: Berg.
Mayhew, L. (2007). How clear is the high street's conscience? *The Telegraph* (online), 18 July. Available at http://fashion.telegraph.co.uk/news-features/TMG3360626/How-clear-is-the-high-streets-conscience.html. Accessed 19.11.11.
Mitchell, L. (Ed.). (2005). *The cutting edge: Fashion from Japan*. Sydney: Powerhouse.
Pope, F. (2009). Mission to break up Pacific island of rubbish twice the size of Texas. *The Times*, May 2.
Slatalla, M. (2000). Circuits. *The New York Times* July 27.
Steele, V. (2010). *Japan fashion now*. London: Yale University Press.
Tseng, M. M., & Jiao, J. (Eds.). (2001). Mass customization. In G. Salvendy (Ed.), *Handbook of industrial engineering, technology and operation management* (3rd ed.). New York NY: Wiley.
Turney, J. (2009). *The culture of knitting*. New York: Berg.
Vinken, B. (2004). *Fashion zeitgeist: Trend and cycles in the fashion system*. Oxford: Berg.
White, B. (2011). *Fast fashion: Is the party over?* Telegraph.co.uk (online). Available at http://fashion.telegraph.co.uk/columns/belinda-white/TMG8438891/Fast-fashion-Is-the-party-over.html.

ELECTRONIC SOURCES

Art and Society

Art and Culture Portal, http://www.artandculture.com.
The Guardian – Art and Design Portal, http://www.guardian.co.uk/artanddesign.
The International Slavery Museum, Liverpool, http://www.liverpoolmuseums.org.uk.
Robbie Conal's Art Attack, http://www.robbieconal.com/aboutrobbie.html.

Technological Advancements

Biocouture, http://www.biocouture.co.uk/.
Interactive Custom Clothes Company, http://www.ic3d.com.
Nuno, http://nuno.com.

Case Study

Cultural Exchanges: Japan and the West

Brown, K H., Minichiello, S., & Mochinaga Brandon, R. (2002). *Taisho Chic: Japanese modernity, Nostalgia and Deco*. This exhibition catalogue can be found at http://agnsw.com/__data/page/11936/Taisho_edkit.pdf. Accessed 15.01.11.
Da Cruz, E. (2004). *Miyake, Kawakubo, and Yamamoto: Japanese fashion in the twentieth century*. This essay can be found at http://www.metmuseum.org/toah/hd/jafa/hd_jafa.htm#ixzz18ayQ0QM0. Accessed 15.01.11.
www.japanesestreets.com, http://www.japanesestreets.com.

meijishowa.com, http://www.meijishowa.com, http://www.meijishowa.com/?q=fashion&x=44&y=14.
oldphotosjapan.com, http://oldphotosjapan.com/en/category/women, http://oldphotosjapan.com/en/photos/323/bride-and-groom.

Sustainability and Ethical Fashion

Centre for Sustainable Fashion – London College of Fashion, http://www.sustainable-fashion.com/.
Ethical Fashion Forum, http://www.ethicalfashionforum.com/.
The Fairtrade Foundatiom, http://www.fairtrade.org.uk/.
Fashioning an Ethical Industry, www.fashioninganethicalindustry.org.
Labour Behind the Label, http://www.labourbehindthelabel.org/.
Marks and Spencer – 'Plan A', http://plana.marksandspencer.com/.
People Tree, http://www.peopletree.co.uk.
Practical Action, www.practicalaction.org.uk/education.
sda Sustainable Design Award, http://sda-uk.org/textiles.html.
UNICEF and child labour, http://www.unicef.org/pon95/chil0016.html.

CHAPTER

FASHION AND THE FASHION INDUSTRY

25

L. Drew[1], R. Sinclair[2]

[1]The Glasgow School of Art, Glasgow, UK; [2]Goldsmiths, University of London, London, UK

LEARNING OBJECTIVES

At the end of this chapter, you should be able to:

- Outline what influences changes in fashion
- Understand the processes behind the ideation and development of a fashion style
- Pinpoint factors influencing the relationship between consumers and fashion
- Recognise the differing roles played by retailers, the fashion industry and fashion designers in the creation and dissemination of new fashions
- Have an awareness of global sustainability and ethical issues related to fashion
- Understand the importance of digital technologies to the future of fashion design, production and distribution

25.1 INTRODUCTION

Fashion is 'the current popular custom or style esp. in dress or social conduct' (Oxford English Dictionary). This dictionary definition implies, by the very use of the word 'current', that fashion is constantly changing. However, it tells us nothing about the means by which something becomes popular or customary. Hopkins (2012) places fashion within the content of 'human phenomena and human behaviour' and also distinguishes it as distinct from dress or costume. Clark and Paulicelli (2009) extend this by identifying fashion as a process that integrates a system of consumer desire (Waddell, 2004) and as a mechanism for the fabrication of the self.

If something is to be acknowledged as 'fashionable', it must have been adopted by a significant number of people and given social approval (Hopkins, 2012; Shishoo, 2005). Fashion acceptance in wider society is perpetuated through a range of media such as film, photography, the Internet and social media (Clark & Paulicelli, 2009; Jones, 2011). Other traditional media, such as magazines, have grown around the concept of fashion systems and production. The personal nature of fashion is also significant and based on individual taste and choice. Healy (Hemmings, 2012) concurs that there are a range of 'philosophical, political, aesthetic and commercial values placed on fashion'.

Textiles and Fashion. http://dx.doi.org/10.1016/B978-1-84569-931-4.00025-8

Steele (2010) defines the fashion industry as consisting of symbiotic systems made up of four elements:

- Textile production
- Design and manufacturing
- Retail and advertising
- Ancillary services

Clark and Paulicelli (2009) concur, but also believe that fashion would not exist without the textile industry, the impact of the industrial revolution and mass production. Steele (2010) emphasises the importance of interconnected networks (Scoble, 2006; Weil, 2009) relying on each other and influenced by significant social and cultural influences. Fletcher (2008) argues that the constituent parts of the complex modern fashion system must be considered together with the means by which their interrelation creates a fashion system, taking account of global, environmental and ethical issues. Blossom (2011) argues for changes in design thinking and the place of social movements in changing the current fashion trajectory.

In the twenty-first century, new fashion futures are emerging (Diamond, 2002; Guerrero, 2009). The distance between the fashion and textile disciplines narrows as new materials and digital technologies (Martin, 2010; Quinn, 2012) dictate the nuances of change and interconnection with evolving textiles futures (O'Mahony, 2011). Studies of the impact of both the fabric and the end product, together with its ancillary processes and practices, seek to explore the material cultures of fashion (Kuchler & Miller, 2005), and use these practices to explore the 'barometer of change and social patterns of consumption' (Buckely & Fawcett, 2002). The contribution of the national, international and global constructs of the fashion industry cannot be underestimated and the definition of fashion and its uses is not always easy.

25.2 EMERGENCE, DEVELOPMENT AND CHANGE IN FASHION

How does a style initially emerge? A style may take months or years to reach full social integration with people in many age and social groups (Suoh, 2004) adopting the style. An understanding of fashion and how it evolves is an essential part of the designer's repertoire in creating and developing products. It is important to note that the fashion industry does not exist in a vacuum (Clark & Paulicelli, 2009). Today's fashion designer must understand the wider implications of the impact of fashion products on culture and fashion futures.

25.2.1 FASHION IS EVOLUTIONARY

Basic fashion shapes for women are also known as silhouettes. The silhouette is the shape that is instantly recognised from a distance. They originate from the desire to accentuate or conceal the natural shape of a woman's body. Some silhouettes are close-fitting, following the natural curves of the body, whilst others may be loose-fitting with a shape that is structurally independent of the form underneath. Most fall somewhere in between: close-fitting in some areas and loose in others. The fashion silhouette is influenced by a wide range of determinate factors, illustrating sometimes extreme change of shape or shock in the social world (Fan & Hunter, 2009; Jones, 2011), and its impact on clothing styles is visible.

25.2.2 WHAT IS FASHION STYLE?

A fashion style and fashion silhouette are sometimes confused. A fashion style is *a popular form of dress which may be emebedded in particular subcultures* (Polhemus, 2010); a fashion silhouette is *a visual representation of a particular silhouette, from a particular fashion era.* Fashion styles are based on silhouettes where variations in design details such as garment length, trimmings, colour, fabric, etc. are added. A fashion style is something that emerges and changes gradually over a period of time in response to wider influences. The style will eventually disappear, but not before it has become extreme, vulgar or simply unfashionable. Certain features within a style may have become overemphasised, for example, ultra-short skirt length, or tight trousers. When this style starts to be perceived as vulgar, it becomes unacceptable to many consumers. The style is then dropped in favour of one that by then will be emerging. However, experience shows that this is not a sequential ritual, but rather one of ever-emerging and waning styles that compete for dominance in the busy global fashion marketplace (Gerval, 2010). The advent of the Internet, and increased forms of written and visual communication (Bereton, 2009), mean that style and silhouette are constantly changing and merging (English, 2013; Jones, 2011). The number of possible silhouettes is obviously limited, as are the variations of styles based upon them. Styles therefore tend to be regenerated at intervals (Blanchard, 2004; Clarke, 2011), although the length of time between generation and regeneration changes due to fluctuation in consumer markets and other lifestyle influences such as the media and current affairs.

25.2.3 FASHION MOVES IN CYCLES

Marketing theory proposes the concept of a product life cycle (see, e.g., Kotler & Armstrong, 2010). Fashion product cycles can be identified in much the same way as those of other products but with some notable features. The idea that fashion is cyclical is a topic of much debate, as the word 'cycle' implies that a fashion will 'come in' and 'go out' in a regular way. However, such regularity is a false assumption. Five key stages are identified in the fashion cycle:

1. Fashion active: style leaders and early accepters
2. Rise
3. Maturity
4. Decline
5. Obsolescence

These stages can be described as follows.

- **Fashion Active:** The first stage of a cycle is where **fashion-active** consumers are first introduced to a new style. **Style leaders** and **early accepters** will pioneer the wearing of the style. Others will watch with interest (and some with incredulity!).
- **Rise:** By the second stage, the style has benefited from publicity. Style leaders have worn it, and it may have made headlines in the fashion press or in fashion blogs. **Fashion followers** will then pick up the style in versions interpreted by various retailers, brands, online retailers (or e-tailers) and other outlets. In these versions, it will be both less extreme and also less costly to the consumer due to adaptations in fabric, design and finish.
- **Maturity:** The peak of the maturity phase is known as the saturation point. This is where the majority of consumers in a particular market have accepted the style in an even more modified

format and at a price to suit various income brackets. These **fashion-average** consumers will shop in traditional retail outlets such as large multiple stores, variety chains or through e-tailer outlets. Fashion-average consumers will not worry about exclusivity at this stage, and the fashion-active consumers may well have dropped the style.

- **Decline:** By the time a style is deemed to be in decline, most average consumers will still be wearing it although it may be widely thought of as commonplace or even over-exposed in the fashion press. At this point, many stores or outlets will discount or mark down the style in sales, either at the end of a season or mid-season. During times of economic downturn, discounting may take place several times during a season, to prompt fashion consumers to buy flagging trends that are not selling well. The more price-sensitive or less style-sensitive consumer may adopt a style in decline. These consumers are known as **decline laggards** or **fashion reactive,** as they are slow to react to changes in fashion style.
- **Obsolescence:** Obsolescence can be viewed as the most embarrassing stage in the life cycle of a fashion style. Most consumers will express their distaste for the style and no amount of discounting will prompt the fashion-average consumer to buy any product at this stage. This is often heightened by poor-quality manufacture and use of inferior-grade textiles, often reducing the original features of the style to a parody or a poor style effect. It is unsurprising, therefore, that the style may stay in obsolescence for many years.

In fast-moving fashion outlets and online stores there is a degree of **planned obsolescence** (Fletcher & Grose, 2012; Rohr-Tobler, 2011). This consists of many styles that are intended to have a short selling period, particularly during the peak of the season. Fashion buyers are well aware that some styles, especially in accessories, have faddish elements that predictably will be destined for discounting in the same way as some fresh produce has a 'sell-by' date.

25.3 THE STANDARD FASHION-TREND CYCLE

25.3.1 STYLE REGENERATION

Trend cycles are complex (Jones, 2011; Raymond, 2010). If a whole style or trend is to be regenerated, it has to have become inappropriate for the generation that originally consumed and wore it. Examples include the myriad items you may have seen your mother wearing in the family photograph album (Gaimster, 2011). As there is an inevitable distance in terms of both identity and experience between the generations, it is often the case that the original wearers of a style will not re-adopt it because it has become unsuitable for their current lifestyle stage. It is therefore left to a new generation of consumers. This type of styling can also be regarded as retro, where whole elements of the style will be revived (Smith, 2004), including music, media and design artefacts. The more common form of style regeneration occurs where a consumer moves from one lifestyle stage to the next and a style they used to wear eventually returns to be worn by younger siblings.

Of course, a style will never return in exactly the same way, as technology (Fairhurst, 2008) and society will have moved on and re-framed the experience of the trend for new consumers. It may take as much as 5 to 15 years for a complete style regeneration. When this style is considered to be **on trend,** it may be interpreted again in many different versions, fabrics, colourways, trims, etc. (Hess & Pasztorek, 2010) until the style again reaches the saturation point.

25.3.2 THE CLASSIC

A classic cycle is really a misnomer, as all classic styles are considered to be timeless and not in need of reinterpretation. However, there are seasonal trends in elements of marketing fashion classics. An example is the highly successful brand resurgence of Burberry in the early part of this century with the designer Christopher Bailey retaining classic elements whilst incorporating trend features in the collections. Most classics were designed to fulfil practical and functional needs, so it is not surprising to find that many have their origins in military dress, for example, the trench coat, or work-wear such as denim jeans.

25.3.3 THE FAD

A fad cycle is also something of a misnomer. This is a style that is so transient that it often eludes description. Fads are often found in low-price or seasonal accessories. Essentially there is an element of humour or gimmickry to the real fad. It is cheerful and curious, but not necessarily designed for the keepsake drawer! It is this element that attracts the fashion consumer. Bright prints or trims including florals, fruits and fish have all been found in fad fashion recently. Like most jokes, however, the fun is short lived and the novelty wears off quickly. These fashion fads tend to be prevalent in the 'silly seasons' around the end of the calendar year and the high summer periods. Check out your own photos of parties and beach gatherings, especially those posted on social media sites such as Facebook at a moment's notice. Have you spotted some things you would rather edit out, even from last summer? Those leopard print hotpants or the dangling cherry earrings maybe? These are the times when fashion consumers are less inhibited and more likely to buy something 'just for the fun of it'. It is the large multiple stores and online fashion e-tailers that will take the risk and target the market for these high-turnover, low-price items.

Fashion cycles can be used as an aid to trend forecasting, but there are no hard and fast rules, so to understand how fashion trends are managed within the fashion industry, it is necessary to examine other aspects as part of the marketing mix.

25.4 WHY FASHION CHANGES?

The following case studies (Collins, 2008) illustrate how fashions develop by looking at the emergence of brands.

25.4.1 TOPMAN CASE STUDY

Topman's Dave Shepherd knows that product is key (he has been chief operating officer for trading at Arcadia since 2012), but the way the brand ensures its place in the market is through those with whom it associates. Despite being known in the fashion retail industry for his own sartorial style, as is fitting for the director of fashion retailer Topman, Dave Shepherd is something of an unsung hero. While sister Arcadia Group Topshop gets all the hype, Topman is actually the group's top-performing brand, according to its boss Philip Green. Topman often outperforms its siblings in Arcadia and the industry.

Passing by rail upon rail of clothing on the way into Dave Shepherd's Berner's Street office, just a short walk from the Oxford Circus flagship, it is very clear that for this business it is the product that is king. But that has not always been the case. Whilst Topman may be the crowning glory of today's Arcadia, it saw some pretty miserable times during the 1990s, when it lost its fashion sense and primacy in the hearts of young fashion-conscious men. Shopping in Topman had become unfashionable.

Enter Dave Shepherd, who in 1998 was given the top job at the chain, having already been with the business for five years. His route to the lead role was via buying and operations at Topman and Arcadia portfolio chains Principles and BHS. Shepherd started his buying career at Topman's High Street fashion rival River Island in its earlier incarnation, Chelsea Man. His task as the new head of the Topman was to refocus the brand on its core fashion values – and that has certainly been achieved. With 200+ branches, Topman is now as cool as its female counterpart Topshop and has as many marketing tricks up its sleeve as its big sister.

When it comes to articulating the essence of the brand, Shepherd says it is 'delivering fashion with authority'. 'If we believe that skinny jeans are right for our customers, then all of our customers in every branch will have a selection of skinny jeans. Of course there cannot be the depth of choice in the smaller branches, but they will be sold across the whole business'.

At the same time that Shepherd was appointed, both the marketing and design roles were given new impetus. Jason Griffiths was appointed to head marketing and Gordon Richardson to head design. Both are still with the business today, which means that the management team is very experienced in understanding the brand and its metamorphosis.

One of the key challenges for the team has been the launch of Topman.com, which became transactional in 2005, and its integration into the traditional bricks and mortar business. Not only have there been operational challenges, but the whole marketing universe has been shaken up to accommodate an integrated marketing strategy across the stores and Topman.com.

Shepherd says: 'Our philosophy is that the brand has to reflect the lifestyle of the customer. For our young male customers technology is already in their lifestyle. Online shopping makes perfect sense for them. And what we offer online is the best Topman, effectively equivalent to the Oxford Circus store. So now every customer gets the full selection, wherever they live, which is something we just cannot achieve through the stores'.

The launch of online trading has gone hand in hand with the need to move away from print-based advertising campaigns as audiences fracture across print, broadcast, online and mobile communication platforms. As a result, the £4 million marketing budget is now used in diverse arenas, including partnerships and sponsorship – but always with the aim of consolidating its fashion credentials.

'As a result of the downturn in print', Shepherd says, 'we've started to grow our sponsorship and other activities. We do fashion magazines such as GQ, FHM and Dazed, but only bi-annually. We also now use music magazines, and newer magazines like Men's Health. But the growing area for us is sponsorship, where it's all about awareness. We are involved in festivals, MTV, bands and student guides – it's about associating with brands that are right'. The business also has its own magazine Topmanzine, and has aligned itself with the testicular cancer charity Everyman since 2000.

The business has been ground-breaking in its associations. It is now in its fourth season as the first, and only, High Street menswear retailer to show during London Fashion Week (LFW). It partners Fashion East, which supports young designers, to produce the MAN event – a catwalk showcases of up-and-coming young menswear designers. Topman produces a catwalk of its own premium collection, which is one of the hottest tickets of LFW. If menswear gets a bigger profile at LFW, as is being discussed, Topman is well placed to be part of any increased presence.

Through its customer account base, provided by GE Capital, Topman already has a database of 750,000 people, and this is building further with incremental Internet customers. The retailer uses both the account list and online customer details to market both the stores and online businesses direct to customers using email – a much cheaper and more personal way of getting to customers than the traditional print routes.

A further upgrade of the system is being planned that will improve usage for customers as well as giving Topman better access to those customers. This includes the addition of Chetamail, an email marketing tool that enables online customer surveys. 'The upgrades are not just about usage, they are also about enabling us to build our relationships with our customers by improving our access to them. The new system will be managed in house – it doesn't need a third party involved'.

If technology is in the lifestyle of Topman customers, then so is music, and the business exploits this through its marketing partnerships and individual relationships. Bands at August's Leeds Festival performed on the Topman stage and the company has been negotiating with a music company to offer its customer pre-release downloads later in the autumn.

Some of the newer technologies are on the radar too. Topman has tried out a mobile phone marketing campaign in the Lakeside shopping centre in Thurrock, Essex, in which people were sent a text message about a Topman promotion. But Shepherd is not yet convinced. 'SMS marketing could be intrusive, and I'm not sure our customers want that. I'm not dismissing it – but I want to see how it pans out'.

Wireless technology is another area that Topman is keeping a close eye on – laptops, MP3s and premium mobiles are already part of its shoppers' lives. 'We're looking at the users to see if it works for them and for us. Is it in the lifestyle of our customer? Does it advantage our customer?'

Like all retailers who have jumped online, Topman finds that the web pulls customers into stores – so online is promoted in-store, and stores are promoted online.

When online was first launched, it offered separate promotions to those in-store, for example, price reductions, but as the business has developed all promotions were integrated and there is now no benefit in choosing one channel over the other. Occasional P&P-free promotions are used online. New lines are made available online as soon as they are in stores – sometimes sooner than a store delivery – and are flagged up for users just as they are through visual merchandising and graphics in-store. And the average transaction value is higher online than in stores, probably on the basis that customers want to cover the cost of P&P. Customers can return product bought online direct to stores.

In a further development of the marketing integration, Topman is to trial online terminals in-store so that customers can order any garments that are not carried in the store in which they are shopping. Despite this integration Shepherd says the best way to communicate with customers remains through the store branches themselves. 'The most powerful tool is still our shop windows and our in-store graphics. We use every online and direct mail communication tool, but the windows are still the place that has the most impact. So we use them to promote use of Topman.com. Our stores and online are completely in tune'. Topman.com is performing in the top five branches – and is expected to move to the number two spot after Oxford Circus.

All fashion retailers are subject to questioning about the supply chain and the quality of life of people who work in the overseas factories they use. Topman is addressing the problem through an independent audit system at the factories it uses. Shepherd says: 'Our customers are interested in ethical production and ethical trading. We have started to get questions from customers about it. In September we are launching a FairTrade cotton T-shirt collection branded Topman. We're also working with a small fashion brand called Ascension on a FairTrade collection that will be branded Ascension for Topman.

'We believe it is right, and that it's right for our customers. And the work we have done with factories has been pioneering within the Arcadia Group. We've introduced third party audits of factories and the programme is rolling out. And we'll use our windows and online to tell our customers what we are doing. There is a premium to pay on FairTrade product – but we will tell the story in our windows, on packaging and online, including a statement about our policy'.

Topshop trades internationally through 66 franchise stores in 18 countries. 'We control the brand very tightly. They have to follow our store fit, and any changes that we make, and we even give them product layout guidelines. Our international division has a team of merchandisers out in the market supporting the franchisees. We believe that the franchise is an extension of the brand and control it through the detail of the contract'.

Back in the United Kingdom, a major early autumn season promotion is being undertaken with a group of five young menswear designers all interpreting the classic white shirt. Like all Topman partnerships, Shepherd says: 'They carry to us as the right fashion associations. Our business partners have to fit with our brand DNA – it's about youth, about pioneering, being the market leader, showing authority in the field. It's all about like-minded partnerships' (Collins, 2008).

25.4.2 TOPMAN CASE STUDY: PROJECT IDEAS

1. What other menswear labels and brands can you identify with a strong online presence? Describe and illustrate the offer of two menswear brands – one that competes directly with Topman and one that differentiates its brand image by being more exclusive or niche in the range of menswear on sale.
2. Topman's Dave Shepherd describes an ethical position on garment manufacture and sourcing. What other ethical issues might a fashion retail brand consider in attracting a youth consumer and why? Use campaign literature and advertising as well as articles in magazines or online to illustrate at least one other ethical issue.
3. Visit a store outlet for Topman and a competitor (e.g. H&M). Draw or photograph their top trend styles (about 10 product lines in each store would suffice) and create a style or mood board for each store. What are the origins of each of the styles in store? Are they street, sportswear or designer influenced? How is the offering represented online? Use current web resources and magazines to make a complementary mood board for the influences of each store's current image.

25.4.3 ASOS CASE STUDY

'If you want to excite your customers then you have to invest in them – and online marketing is far more than a discount email'. So says Hash Ladha, marketing director at online fashion e-tailer Asos.com. The whole website is now branded as Asos.com and the editorial nature of the home page is reflected throughout the pages. Improved functionality enables shoppers to buy complete outfits as shown rather than search for individual items. A pure play online retailer, Asos.com launched in 2000, joined AIM within a year and has been growing like topsy since.

It started life as 'As Seen On Screen'. Its formula was to sell cheap, fast fashion inspired by celebrity looks. That remains central to the proposition today – indeed every page has a disclaimer to the effect that no celebrities have endorsed the product – but it has expanded the offer. Its own label collections now include a premium Asos Luxe line, plus mid-market and directional brands, premium and luxury brands, accessories including footwear, lingerie, beauty and menswear. It also has deals with retailers, including Mosaic Fashions' owned Coast and Oasis, to sell their collections.

If any disaster could be said to have a good outcome, the Bunsfield Fuel Depot explosion had one for Asos. The event put it out of business for five weeks in December 2005 – yet it still put on sales, and reaped the rewards of publicity. Charismatic chief executive Nick Robinson's image was shot around the world as he gave multiple interviews about the effect of the disaster on the adjacent Asos warehouse.

Today Asos.com has over 1.7 million registered users. It can attract six orders every minute. And the Website is consistently the number two online fashion Website according to research firm Hitwise. Category killer Next is yet to be knocked from its number one place. Topshop and River Island vie for third place.

Marketing has been pretty basic at Asos until now, according to Ladha, using emails to communicate with customers, pay per click, and a free fashion magazine, bagged with women's and fashion magazines and having a wide distribution to customers. But the focus of the new plan is to start managing customers – encouraging active shoppers to spend more, reactivating lapsed and dormant customers and, of course, to recruit new spenders. Ladha explains: 'With the Customer Relationship Management (CRM) tool the experience will be better from the moment a customer registers with us, from the way they are welcomed, how we explain what we're about, right through to the way we communicate with them.

'CRM will work for us because of the amount of information we have about our customers – their name, age, gender, email and postal address. We know what they bought, what size, when they bought it, and what they returned. We'll be able to cross sell to them'. Ladha cites Tesco as the pioneer in the use of this type of information, which is, as yet, underused by other fashion players.

'It was Asos.com that challenged the perceptions of how to sell fashion. Ten and even five years ago people said that you could not sell fashion online. And it has all but replaced mail order'. But now, of course, virtually every High Street retailer has an online store – competition that Asos did not have in its early years.

Ladha's background is in fashion retail marketing with names from New Look and Dorothy Perkins to Austin Reed, so he knows exactly how the High Street thinks. 'You always have to watch the competition and can never be complacent. Am I worried about High Street brands online? Yes. But they talk about their online offer as a traditional store, as another branch. That is immediately restricting. We approach the proposition from a different point of view'.

He goes on: 'What's interesting about the Asos business is that it is a fashion brand and a fashion retailer, but it is also a pure play online operation – that makes it unique. We do not have to work in the same way as fashion retailers because we are not limited by space or range planning [which allocates items to stores]. We have just one store representing all the product that we sell, so in those terms it's an easy challenge'.

The business is strongly aware that it does face challenges, however, and that they are changing – both because of the competition from other fashion retailers and the challenge of keeping the site fresh and compelling, with every page living and breathing the Asos ethos. 'We want to be the authoritative fashion brand that always has the right product, and has a wide appeal', Ladha says.

It may be counter-intuitive in this online world, but Asos is using its good old-fashioned customer magazine to spearhead its new marketing plan, which comes into play in April. The award-winning magazine produced by Seven Squared – which won the APA (Advertising Producers Award) for most effective consumer publication in November – was launched late in 2006.

From April, the magazine will be relaunched and sent only to active customers, as a benefit of shopping with Asos. Ladha says that contacting customers is not about discounting, although there may be discounts included. 'The magazine is about relevance. I believe the dwell time of a customer with the magazine is equivalent to 50 × 30 second ad slots. For me that makes it more effective than store windows or advertising'. He adds: 'The magazine builds the brand, allows interaction with our customers and drives sales. It is a retention tool, rewarding our most loyal customers – and that's not a bad thing. We have 5000 Stock Keeping Units (SKUs) online and launch 200 new products every week. It's all about getting our customers to come back. The objective is to keep them buying'. In addition, the business is setting up an incentive scheme for users to recruit their friends, which will have a discount attached to it.

Ladha is wary of contacting customers too frequently by email and is yet to be convinced that SMS text messaging is right for the UK market. 'We want to protect our customers from receiving too much. Asos emails are now just one of many communications. We want to talk often but with a high degree of relevance. An email may be a cheap and easy message, but what's the point if it's not the right message?' For non-customers, themed supplements of the magazine will be bagged with other magazines and ploys such as office drops will also be used. The strategy for men is quite different. A continuous advertising campaign of double-page spreads is planned, with two high-quality seasonal fashion guides, again to be bagged with men's interest magazines.

'Men behave differently to women in regard to fashion shopping, and men's usage of the Internet is not necessarily about fashion. We have to say to them "here is the image and the look – now you can go online and buy it"'. The marketing plan includes five major press events in the year, and the creation of 'more aspirational' press days – when journalists view the coming seasons' collections. Ladha says: 'We are investing in all areas of the business to make sure we deliver'. The PR team has been boosted to eight people since Ladha joined the business, and the marketing team is doubling to eight people. Customer care is also engaging more staff and a new warehouse in Hemel Hempstead – finally replacing the defunct Bunsfield site – was due to be fully operational by the end of February. The in-house studio team – models, photographer, stylists, grooming – are already creating some 80 images for the site each day.

'In the early years it was hard to keep up with the growth and there was just a basic level of marketing. Now we have a single vision – to encourage current customers to spend more and to get more customers. And we have to make sure that every other thing we do is right – we have to be right there on product, do more on customer care and keep marketing communications simple and engaging'. As the site itself and the marketing plan evolves, Ladha is confident that Asos.com will 'be the best online fashion store. As the pure play online specialist we have to own this sector' (Collins, 2008).

25.4.4 ASOS CASE STUDY: PROJECT IDEAS

1. Asos may not link its styling to celebrity identity and image, but you can! Make two mood boards of current Asos looks and celebrity images that you feel may have an inspirational or referential look to them.

2. Make a comparison of the styles and prices of Asos and a competitor e-tailer (e.g. Next, Very or Fashion Union). Compare five key styles: a trouser, a skirt, a dress, knitwear and outerwear. What are the key differences in your opinion?
3. Design a small collection (up to 10 pieces) for Asos next season. What will be the biggest influence on the trends for next season, and how will it build on a range currently selling with the e-tailer?

25.5 REVISION QUESTIONS

- What is fashion? How would you describe a style or a silhouette?
- What influences changes in fashion? Name at least two factors.
- How do consumers choose clothes? Give at least three examples.
- What is the role of retailers in disseminating new fashions? Is this different for online?
- Describe at least one issue each in relation to global issues, sustainability/ethics and the relationship with fashion products.

25.6 SUMMARY POINTS

- Fashion is constantly changing. This process must be placed in the context of social developments, sustainability and ethical considerations and their influences on consumer behaviour.
- Changes in fashion silhouettes and trends, along with fads, mean that 'fashion styles' are constantly changing, and are impacted on by changes in mass communication through the Internet, social media and online e-tailing.
- The emergence, growth and decline of a style is influenced by technological developments as well as by consumer reactions to social change.
- Case studies of successful fashion retailers indicate their awareness and identification of the lifestyles of their customer demographic and a readiness to adapt quickly to changing patterns of information technology.
- Online marketing of fashion has all but replaced mail order and is increasingly used to build a picture of a brand's typical consumer and their buying habits. Interactive advertising and shopping offer both the opportunity for the consumer to make purchases that are not constrained by their location and for the retailer to tailor their promotions to the desires of these consumers.

MAGAZINES/PERIODICALS

It's Nice That,
Elephantm,
i-D,
Dazed & Confused,
Another Magazine,
Grafik,
Pop,
Love,
Bon International.

REFERENCES AND FURTHER READING

Bereton, R. (2009). *Sketchbooks: The hidden art of designers, illustrators & creatives*. UK: Laurence King Publishing.

Blanchard, T. (2004). *Fashion and graphics*. UK: Laurence King Publishing.

Blossom, E. (2011). *Material change, design thinking and the social entrepreneurship movement*. USA: Metropolis Books.

Buckely, C., & Fawcett, H. (2002). *Fashioning the feminine, representation and women's fashion, from Fin de Siecle to the present*. London: I.B. Tauris Publishers.

Collins, J. (2008). Interviews with Dave Shepard (Topman) and Hash Ladha (ASOS), originally published in *Brand Management* and reproduced with permission.

Clark, H., & Paulicelli, E. (Eds.). (2009). *The fabric of cultures, fashion, identity and globalisation*. Oxon, UK: Routledge.

Clarke, S. (2011). *Textile design*. UK: Laurence King Publishing.

Diamond, E., & Diamond, J. (2002). *The world of fashion* (3rd ed.). USA: Fairchild Publications.

English, B. (2013). *A cultural history of fashion in the 20th and 21st centuries* (2nd ed.). UK: Bloomsbury Press.

Fairhurst, C. (Ed.). (2008). *Advances in apparel production*. UK: Woodhead Publishing.

Fan, J., & Hunter, L. (2009). *Engineering apparel fabrics and garments*. UK: Woodhead Publishing.

Fletcher, K. (2008). *Sustainable journeys*. London: Earthscan.

Fletcher, K., & Grose, L. (2012). *Fashion and sustainability. Design for change*. Laurence King Publishing.

Futuro Textiel. (2008). *Surprising textiles, design and art*. Sticting Kunstboek.

Gaimster, J. (2011). *Visual research methods in fashion*. UK: Berg Publishers.

Gerval, O. (2010). *Fashion from concept to catwalk*. London: Firefly Books.

Guerrero, J. A. (2009). *New fashion and design technologies*. London: A&C Black.

Hemmings, J. (Ed.). (2012). *The textile reader*. UK: Berg Publishers.

Hess, J., & Pasztorek, S. (2010). *Graphic design for fashion*. UK: Laurence King Publishing.

Hill, J. (2011). *The secret life of stuff*. London, UK: Vintage Publishing.

Hopkins, J. (2012). *Fashion design the complete guide*. Switzerland: AVA Publishers.

Hornung, D. (2005). *Colour; a workshop for artists and designers*. UK: Laurence King Publishing.

Hu, J. (Ed.). (2011). *Computer technology for textiles and apparel*. UK: Woodhead Publishing.

Jones, J. S. (2011). *Fashion design* (2nd ed.). UK: Laurence King Publishing.

Kotler, P., & Armstrong, G. (2010). *Principles of marketing*. USA: Prentice Hall.

Kuchler, S., & Miller, D. (Eds.). (2005). *Clothing as material culture*. UK: Berg Publishers.

O'Mahony, M. (2011). *Textiles and new technology*. USA: Artemis.

Marshall, L., & Meachem, L. (2010). *How to use images*. UK: Laurence King Publishing.

Martin, M.S. (2010). *Future fashion, innovative materials and technology*. Spain: PromoPress.

Polhemus, T. (2010). *Street style*. UK: PYMCA.

Quinn, B. (2012). *Fashion futures*. London: Merrel Publishers.

Raymond, M. (2010). *The trend forecasters handbook*. UK: Laurence King Publishing.

Roberts, L., & Wright, R. (2010). *Design diaries*. UK: Laurence King Publishing.

Rohr-Tobler, M. (2011). *Handbook of sustainable fashion production*. UK: Woodhead Publishing.

Scoble, R. (2006). *Naked conversations: How blogs are changing the way businesses talk with customers*. John Wiley & Sons.

Shishoo, R. (Ed.). (2005). *Textiles in sport*. UK: Woodhead Publishing.

Smith, P. (2004). *You can find inspiration in everything*. UK: Thames and Hudson.

Steele, V. (2010). *The Berg B fashion companion*. London: Berg Publishers.

Suoh, T. (Ed.). (2004). *The Kyoto Costume Institute, fashion from the 18th to the 20th century*. Koln: Taschen.
Waddell, G. (2004). *How fashion works*. Oxford, UK: Blackwell Publishing.
Weil, D. (2009). *The corporate blogging book: Absolutely everything you need to know to get it right*. Piatkus Books.

WEBSITES

www.newexhibitions.com.
www.davidreport.com.
www.wgsn.com.
www.style.com.
www.polyvore.com.
www.trendbible.co.uk.
http://showstudio.com.
www.thefashionspot.com.
www.thefuturelaboratory.com.
www.businessoffashion.com.
http://www.topman.com/?geoip=home.
http://www.asos.com/.
http://www.hm.com/gb/.

CHAPTER

VISUAL DESIGN TECHNIQUES FOR FASHION

26

J. Gaimster

London College of Fashion, London, UK

LEARNING OBJECTIVES

At the end of this chapter, you should be able to:

- Understand the design process
- Carry out basic research for design
- Recognise and use a range of design terminology
- Understand the basic garment design process from inspiration to final product

26.1 INTRODUCTION

The garments that we purchase in retail stores are part of a chain of supply that starts with the raw fibre and ends with a garment or textile product that appears in store. All of the garments that we buy have been through the design process. There are many different types of companies in the fashion industry; there are small designer labels that produce small runs for their own stores for boutiques or to sell online (see http://www.youngbritishdesigners.com/), and at the other end of the spectrum, there are companies like Dewhirst Group PLC who supply ready-to-wear garments to major retailers. There are companies who specialise in womenswear, menswear, lingerie, and sportswear or knitwear, and many other specialist areas such as accessories and footwear. There are also different levels of the market from designer goods to discount stores, but regardless of the market, the design and production process is fundamentally the same.

Every garment has to be researched, designed, and manufactured. The fabric and trimmings have to be selected and sourced, the price has to be appropriate for the end consumer, and the design has to be fashionable and appropriate for the market for which it is intended. This process is complicated and happens within a very tight timescale, sometimes in a matter of months, or in the case of fast fashion, within a few weeks.

26.1.1 RESEARCH FOR DESIGN

The designer will research into fashion styles and silhouettes or cuts (the overall shape of the garment and the way that it fits on the body). They will also look at the predictions for the season style and cut, colour, and fabric trends. They will also carry out inspirational research such as visiting museums, art

Textiles and Fashion. http://dx.doi.org/10.1016/B978-1-84569-931-4.00026-X

galleries, and watching people on the street. This inspirational research will be combined with market and trend research in order to decide upon a theme/story or concept for the collection. The concept helps the designer to create a look or story for their collection, which in turn helps them to explain the collection to the client, the press who are going to promote the collection, and then ultimately to the consumer.

Once they have selected a theme for the collection, the designer will gather fabrics, trimmings, and colour swatches and create mood boards (also referred to as concept boards) that will be shown to the buyer or the design director. From these initial ideas or concepts, the designer will then create sketches for the range or collection. These sketches will be edited and the best ideas selected for further development. The designer will then work out detailed drawings called specifications and think about the number of styles and pieces that are required within the collection. The designer may then work with a pattern cutter to create the patterns and send their ideas or patterns to an off-shore supplier, who will create the sample. Alternatively they may just send technical drawings from which the supplier will create the pattern. They may also work with a merchandiser to help decide which pieces to have in the collection and what price points they need to sell at. It is vital that the designer and the retailer understand what their customer wants to buy, how much they are prepared to pay for what they want, and at what price it can be produced, taking into account factors such as the price of raw materials, manufacturing, and transport costs. If the price of producing a garment is too expensive for the target consumer and the pricing structure of the brand, then adjustments may need to be made to the design to bring it into the price point. These adjustments may mean changing the fabric, simplifying the design, and removing details or sourcing cheaper trimmings. There are specialist suppliers of wholesale trimmings who show at the major industry trade shows, and there are also many directories on Internet that list suppliers, including http://www.fashiondex.com/ and http://www.apparelsearch.com/.

26.1.2 PLANNING THE COLLECTION

The collection has to be balanced so that it offers the consumer sufficient choice and allows them to create outfits from the individual pieces. Individual garments may be offered in a choice of colours or patterns to help the consumer to coordinate their outfits. The collection may consist of a number of smaller themes or stories that cater for different occasions, such as special occasion wear or office wear. It may also be divided into different stories that will enter the store at different times during the selling season. The traditional selling seasons are spring/summer and autumn/winter; however, many fashion companies now have several key points within these seasons where they introduce new stock. These key points are called drops and may coincide with special events such as Christmas or changes in climate, so there may be one drop for early summer and another for late summer. Fast fashion retailers may have more drops to ensure a constant supply of fresh ideas for the consumer.

Colour is a very important part of the design process. There are some colours that are never out of fashion: these are called classics or basic colours, like black, white, grey, brown, and neutrals such as beige. As well as these basic colours, designers will introduce new fashion colours (the colours that have been predicted as being on trend) each season. Colours can be in fashion for a short period or can develop into a trend over a longer period and stay in fashion for more than one season. Getting the colour right is important because if a colour is not in fashion or is not relevant to your market, then your consumers will not buy it. Designers also need to understand how to match and communicate colour and may work with colour specialists and technicians in order to achieve an accurate end result. This is an important process across all aspects of the industry from yarn suppliers and textile producers to accessory and garment manufacturers. Colour matching systems such as Pantone (http://www.pantone.co.uk/pages/pantone/index.aspx) ensure that the colour in the designer's original palette is the colour of the end product. Designers will usually request a lab dip (a sample of the dyed colour) on the cloth that the final garment will be produced in. The colour that a dye produces can vary depending on the cloth and fibre type being used, so this is done to ensure that the colour is correct before the garment goes into production.

26.1.3 DEVELOPING THE SAMPLES

Once the designs have been agreed upon, the patterns are cut, and prototypes called toiles are made, usually in a cheap fabric such as calico (a cheap, plain, woven cotton cloth). The toiles will be fitted on a model or dress stand and a first sample will be made. The sample may be made in the design studio or more often in the factory sample room. The designer will send the factory a drawing and a set/pack of information called a Specification or a Tech Pack (see further information in Chapter 27). You can see a good example of a detailed tech pack at http://www.techpackdesign.com/tp.htm.

This package gives the factory all the information that they need about the garment, including measurements, colour specifications, and trimmings. This information has to be very detailed and accurate or the factory will produce a sample that is wrong. Creating a new sample and shipping it is expensive and takes a lot of time, which can cause problems for the retailer, who has to keep a constant supply of stock in their stores.

The factory has to work out how best to produce the garment, taking into account the final cost and the required quality. Adaptations may need to be made to the design so that it can be produced effectively and at the right price. Once a final sample is agreed upon, this will be sealed and the final garments are checked for quality against this sample. The factory will then be given a deadline by which they have to supply the production to the retailer or supplier.

Often designs are created in one country, produced in another country where the production is cheaper, and sold in the original country or shipped elsewhere in the world; therefore, delivery times and costs also need to be taken into account. If a delivery is late, the clothes may have gone out of fashion and the retailer will not be able to sell them.

26.2 WHY CONSUMERS BUY NEW DESIGNS

We buy clothes for many reasons: to keep us warm, to maintain our modesty, to protect ourselves, to look good, and to create an identity. We buy clothing and products that reflect our personal interests and values. Clothing can be used to signal our interests and the groups (known as subculture) to which we wish to belong or to differentiate ourselves from other people (Berger, Heath, & Ho, 2008). These groups may define our occupation, our musical interests, or our social standing.

There are times when we need new clothing because we have grown or because something has worn out, or we have a special event to attend; however, fashion goes beyond these needs. Fashion is about wanting something new and different. We could wear the same thing everyday, but unless we are in a job where a uniform is required, most of use will use what we wear to express our personality and tastes.

According to (Solomon, 2009, p. 8) "fashion refers to a style that is accepted by a large group of people at a given time". Within fashion there are trends, which may be large and long term or small and short lived (in which case they are often referred to as fads). One example of a fad was the adoption of legwarmers, normally worn in the dance studio, which then became a fashion item for a while in the 1980s. Long-term global trends are often called megatrends. Ecology is an example of a megatrend that has been influencing fashion for a long time and is global in nature. There may be many trends that are in fashion at any given time, and these can vary across social groups and around the world. There are many different and very complex theories about how fashions develop and who influences that development. Very simply put, fashion is a reflection of the times we live in, sometimes referred to the "zeitgeist" or the "spirit" of the times.

Our purchasing decisions can be influenced by many things that are around us, including what other people are wearing, what we see in the movies on television, on the street, on the catwalk, and in magazines. We are also influenced by other people, including our parents, partners, and friends, as well as people we value as role models. This is why companies will often use a celebrity to endorse a product line. This happens with a lot of fashion related products, including perfumes, cosmetics, and clothes.

26.3 MARKET RESEARCH METHODS FOR IDENTIFYING EMERGING CONSUMER NEEDS

Fashion designers and retailers need to create garments that they know or think people will want to buy. In order to decide what they need to create, they will draw upon a wide range of sources of information. They will look at their sales history to see what has historically sold well for them and whether these sales are being maintained. They will consult professional trend forecasting companies and may employ market researchers or social scientists to look at what issues are influencing their core consumer. Every company will have a profile of the type of consumer who they want to attract. This is their core customer. The customer may be defined by several factors, including their age, gender, lifestyle, income, place of residence, interests and hobbies, whether or not they have children, the kind of job they have, etc.

Companies will use many different techniques to determine what their consumers needs are. These include using a mixture of quantitative and qualitative data, drawn from questionnaires, focus groups, and observation of consumer behaviour in the store. They will also look at demographics (statistical data on a particular population such as age, gender and income) and other data such as information about purchases made using loyalty cards or data from visits to websites. Online retailers use web analytics software to monitor how many unique visitors are coming to their site and what they are buying. They use this information in their marketing strategy to try and increase their sales.

People's attitudes and buying habits are influenced by a lot of different factors, including the season and weather, the economy, politics, lifestyle changes and environmental issues, social change and religion. A good designer will know what factors are important in influencing their consumer and what this means in terms of their fashion needs. For example, many consumers are becoming very concerned about how their clothes are produced. They want to buy clothing that has been produced without exploiting the workers who made them and without damaging the environment. This is often referred to as ethical or sustainable fashion (for more information, see Chapter 22).

The terms ethical and sustainable fashion are often used interchangeably. The Ethical Fashion Forum defines ethical fashion as "an approach to the design, sourcing and manufacture of clothing which maximises benefits to people and communities while minimising impact on the environment and they have adopted the definition of sustainability proposed by the Bruntland Commission in 1989 as the ability to "[to meet] the needs of the present without compromising the ability of future generations to meet their own needs" (http://www.ethicalfashionforum.com/the-issues/ethical-fashion).

Since the Rana Plaza disaster in April 2013, in which more than a 1000 factory workers died in Dhaka, Bangladesh, consumers have become much more aware of the conditions in which garments are produced and retailers have had to respond by improving their inspections of factories and responding more openly to issues of corporate social responsibility.

Fashion companies are responding to this by developing fair-trade policies and ensuring that their garments are clearly labelled as fair trade. Many companies are working to try and improve the conditions of their workers by using social compliance policies. This means that they pay their workers a living wage and ensure that they have appropriate, safe working conditions; they do not use child labour or force their workers to undertake overtime. You can find out more about social compliance in the fashion industry at http://www.ethicalfashionforum.com.

Other companies are trying to limit the impact that producing garments and textiles has on the environment by reducing the amount of chemicals and pesticides used in the production process or through recycling. You can find out more about eco fashion at http://www.ecofashionworld.com.

Today, people in the developed world are used to having a lot of choice as to where to buy their clothes, the number of styles available, and how much they pay for their fashion purchases. As this choice has increased driven by technology, consumers have become more demanding, wanting more for less, and they are also less loyal to particular brands. They will use Internet price comparison sites, discuss purchases with their friends, and use consumer sites for recommendations (Nielsen, 2008). There is a tension between this continual consumption and the need for ethical and sustainable fashion that is ongoing.

In the past consumers were used to buying new clothes for the summer and winter seasons but now retailers have educated them to expect a constant supply of new styles. Once such retailer, Zara, is able to turn around a new style within 15 days; they produce smaller quantities of a style so that the consumer knows that if they do not buy it at the time it might not be there when they return, they also offer a wide range of styles providing the customer with more choice and their stores are constantly refreshed with deliveries twice a week (Dutta, n.d.).

The speed of fashion and the turn around of trends has increased from two key seasons (summer and winter) and looks that defined them to a situation where there are many different trends that are in fashion at the same time and are adopted by different groups of consumers. These trends may last for a whole season or just a few weeks. This makes it increasingly important that the designer has a clear understanding of their consumer and which trends are influencing them.

26.4 FINDING INSPIRATION

Designers will constantly look for inspiration for new designs. They will visit exhibitions, museums, and music festivals. They will watch movies, go to fashionable areas of town to see what people are wearing on the street, and check what is happening online in fashion blogs. They will visit their competitors' stores to see what they are selling and designer stores to get inspiration. They will watch what is happening on the catwalk and what the celebrities are wearing on the red carpet. They may also be influenced by what is happening in architecture or interior design, and they often will use vintage clothing and fashion looks from other periods as starting points for their collections.

26.4.1 TRADE SHOWS

Trade shows are an important and excellent source of information for designers. There are shows that cater for every part of the supply chain: yarn shows, textiles shows, trimmings, and accessories, leather, knitwear lingerie, and other products. Many of these shows will employ trend forecasters to create mood boards or trend forums where they will show directional fabrics or colours. At the shows, the designers and buyers will order small sample lengths of fabrics or trimmings so they can test them before purchasing in bulk for their production.

The trade shows work ahead of the selling season. The fabric shows will be one year ahead and the garment shows six months ahead. A designer will visit a fabric show one autumn to buy fabrics for the autumn season of the next year. In September 2014, a major fabric show like Premier Vision (PV) in Paris would be showing fabrics for autumn/winter 2015/16. Meanwhile, at London Fashion Week, they would be showing designs in September 2014 for spring/summer 2015. The fibre manufacturers and textile manufactures have to work even further ahead of the selling season, as they have to develop their products ahead of the textile shows.

26.4.2 FASHION FORECASTING

Fashion designers and retailers get information about the latest fashion trends from fashion forecasters, such as Peclers, Nelly Rodi, and Promostyl, or from web-based services such as WGSN (www.wgsn.com). These companies employ lots of people to find and analyse trend information from all over the world. Companies use trend forecasters because it can be quicker and cheaper than trying to find and analyse the information. Forecast companies may have websites or produce trend books that their customers will purchase and use for inspiration and information. They will give information about colours, fabrics, styling, and also lifestyle trends (how people are living and what things are important to them).

26.4.3 COPYRIGHT

While the designer can look at all these things for inspiration, it is important that they do not copy. Directly copying a design is in breach of the copyright law, and it can cost a company a lot of money if the case is proven. For guidelines and information on copyright, how to avoid infringement, and how to protect your own designs, see the references at the end of this chapter.

26.4.4 CREATIVE THINKING TECHNIQUES

Many designers will use creative thinking techniques such as brainstorming and mind mapping to come up with new ways of thinking about an idea. Brainstorming is a process that usually happens in groups; everyone will put forward their ideas no matter how silly they may seem. These ideas will then be discussed and filtered to select those that have potential for further development.

In the mind mapping process (Buzan, 2010), the keyword or idea is written or visually represented in the centre of a sheet of paper. Related words or ideas are then added as branches to the initial idea. Each branch may have many other branches extending from it with other related ideas (see Figure 26.1). There may be unusual connections that can be formed between the various related ideas, or it can help the designer to reject ideas that may be too clichéd or inappropriate. You can find lots of information

FIGURE 26.1

Mind map for cruisewear.

Source: Photograph J Gaimster.

about creative thinking tools on Internet (see the references at the end of this chapter for some good resources).

26.5 AESTHETIC QUALITIES IN A GOOD DESIGN

Good design is not accidental. It is based upon an understanding of how many elements work together to create a result that is aesthetically pleasing. These elements are broken down into the key areas of

- Shape/Silhouette
- Proportion
- Colour
- Fabrics
- Trimmings
- Details
- Styling
- Prints
- Motifs.

26.5.1 SHAPE AND SILHOUETTE

The shape or silhouette of the garment is key to ensuring that it is appropriate for the current season. Throughout history, you can easily identify changing silhouettes in fashion. In the 1920s, dropped waists were popular. In the 1950s, full skirts with lots of petticoats were in fashion, whereas in the 1980s, there were lots of big shoulders. Different silhouettes will place an emphasis on varying areas of the body such as the bust, waist, or hips, or may create a shape that attempts to change the body shape completely. Sometimes a designer will deliberately introduce a new silhouette that will at first seem shocking or strange, such as Dior in 1947, or Vivienne Westwood's Mini-Crini (1985), but will then become acceptable as it works its way from the catwalk to the high street.

26.5.2 PROPORTION

Proportion in a garment is also important; understanding where to place the emphasis and draw the eye is very important. Having an outfit with a strong line at the waist will have a very different effect from one that has a line just under the bust. Use of colour or pattern or varying the length of the garment can also change the proportion of the figure.

Dark colours recede and make things look smaller, bold horizontal stripes make people look wider, while vertical stripes make them look slimmer. It is important to understand which styles of clothing and proportions look best on which type of figure and to be able to relate this to your core consumer.

26.5.3 COLOUR

Colour has to work in harmony with the shape and proportion of the garment. Often, very bright or strong colours will be used as a highlight or in smaller proportion to add emphasis to a look. It is a brave person who chooses a whole suit in acid green. If you look around you, you will see that people often choose classic or sombre colours for the main part of their outfit and add more vibrant colours as accessories. Of course there are exceptions to every rule, and if you are designing cruisewear or outfits for a tropical environment, then lots of vivid colours may be very appropriate and popular.

26.5.4 FABRICS AND TRIMMINGS

The fabric that the designer selects has to be appropriate for the silhouette and the function of the garment. If the designer wants a floating, soft effect, then chiffon will be more appropriate than heavier wool gabardine. However, if they were designing a tailored coat, then the wool gabardine would be the better option. Designers need to understand the properties and characteristics of fibres and fabrics so they can make an appropriate choice.

The same rules apply to trimmings. A very heavy zip does not usually work well with a delicate fabric; even the thread that you use can have an effect on the quality and finish of the garment.

26.5.5 PRINTS AND MOTIFS

The choice of print, motif, or embellishments such as embroidery, quilting, or beading can also add or detract from the value of the garment. If you were trying to create a very romantic mood, then a floral print would usually be more suitable than a geometric. A strong graphic can add value to a garment, whereas a poorly selected motif may dissuade your consumer from buying it.

26.5.6 DETAILS AND EMBELLISHMENTS

Details such as decorative stitching, seaming, pleating, and quilting can also be used to enhance a garment. These details should always be carefully worked out in the sample and tested before going into production. Too many details can make a garment look fussy and detract from its value rather than making it more saleable. However, there are designers whose signature is to use lots of details very successfully (Dolce & Gabanna, for example); it is about understanding what your consumer likes. Adding embellishments to a garment can add considerably to the price, and there are some processes and effects that can only be achieved by hand. If a garment is to be hand beaded, then there are issues around how much the worker is being paid to achieve this. There is always the danger that workers are being exploited, so it is important to ensure that the manufacturer is using ethical employment policies.

26.5.7 STYLING AND ACCESSORIES

A very simple outfit can be made to look exceptional with the addition of the right accessories and some strong styling. Styling is about the model used to display the garment, the hair and makeup, and the accessories, including jewellery, handbags, shoes, etc. Designers will often work with a stylist who selects the accessories and helps to create the desired look; when they are creating a fashion show or a photo shoot, the stylist makes sure that they are projecting the right image for their collection.

As well as these individual elements, the designer should also bear in mind the functional aspects of the garment and the intended consumer or market. What is required for an active sportswear garment? Snowboarding garments, for example, will be very different than the requirements for an evening gown. A garment intended for a high level designer boutique would have different criteria from a garment intended for a fast fashion retailer. A successful designer is able to pull all these elements together to create something unique, wearable, and commercial.

26.6 DESIGN TOOLS

The designer needs to be able to represent their ideas in both two and three dimensions. Traditionally, this has been done by sketching with a pen or pencil paints, marker, pens, etc., and/or by experimenting in three dimensional (3D) with calico or other cheap materials. The advent of the computer has changed the way that designers work and present their ideas. They may still use a pen and paper for their initial ideas, but increasingly these will then be transferred through scanning or other means of digitisation onto a computer where they can be quickly manipulated and processed into professional portfolio presentations.

There are several types of software that are useful to the designer. There are generic graphic applications such as Adobe Photoshop and Illustrator that help the designer to manipulate and present their drawings. There are specialist applications for fashion and textile design that help designers to produce different colourways for textile designs, create different repeat patterns, develop knitwear designs, and create technical drawings. There are computer applications that help the designer to develop and modify the pattern, to check how much fabric they will use by creating a layplan, and to working out a costing.

There are also 3D design packages where the pattern can be virtually stitched together and viewed on an electronic representation of a body called an avatar. In these applications you can emulate the type of fabric that the end design will use and see how it moves on a virtual catwalk, examples include Optitex, Modaris 3D by Lectra, and 3D ProPainter by SpeedStep. There are also other 3D applications like Rhino and Grasshopper for designing footwear, accessories, and jewellery. The advantage of these tools is that you can get a much better idea of the end product than with 2D drawing. You can also check for fitting issues before you make a sample. This can save a lot of time and expense. The more specialised tools are quite expensive and therefore are mainly used by bigger manufacturers that are able to afford the investment in the equipment and training their staff. However, as computer-aided design (CAD) packages become more affordable and the demands of the industry become more complex, the number of companies who are using these technologies is increasing. It is therefore important that fashion and textiles students understand CAD processes and are able to use the tools. These products can also help to reduce the need for travelling and creating multiple samples, thereby helping to reduce the amount of environmental impact in developing a new product. One company that puts a strong emphasis on the use of technology to improve sustainability is Adidas. They have initiatives that avid the use of oil-based plastic, which helps reduce carbon emissions. They have also developed thinner and lighter materials, which mean less waste and less embedded carbon (http://www.adidas-group.com/en/sustainability/products/sustainability-innovation/).

26.7 MOVING FROM SAMPLE TO PRODUCTION

It is very easy to make one beautiful sample, but when you are generating a design that is going to be produced in hundreds or thousands, there are a lot of things you need to consider. How are you going to ensure that the fabric is all dyed to exactly the same colour? How quickly can the factory produce the garment? Are there any details in the design that are making it too expensive for your customer? Will the fabric arrive in time? Can the factory source the trimmings that are needed? Is the production of consistent quality? Will the garment wash without shrinking or fading? How will the garments be shipped by sea (cheaper but takes longer) or by air (faster but very expensive and may not be affordable)?

If something goes wrong in the production process, the impact can be huge. The factory, or the retailer can lose a lot of money, the retailer can end up with no product in the store, or the product can arrive late and no one will want it. Companies use a lot of processes and checks to ensure a smooth flow of production, including software for tracking the progress of a collection (product data management and product life-cycle management) and sophisticated point of sale systems that tell them what is selling in store, so they know what they need to make in larger quantities. When a buyer or designer gets the design wrong, the garments will need to be sold off or marked down to try and retrieve some money and make room for new stock. The more garments they have to mark down, the less money the retailer will make.

26.8 FUTURE TRENDS: IMPACT OF NEW TECHNOLOGIES/PROCESSES

Internet has had a huge impact on the way that we shop, as well as how we consume and understand fashion. Sites are becoming more sophisticated, enabling us to customise designs to suit our own needs and tastes (NIKEiD, http://www.nike.com/gb/en_gb/c/nikeid). New technologies such as augmented reality and virtual mirrors (http://www.ray-ban.com/scotland/virtual-mirror) enable us to examine products in 3D and virtually try them on before we buy them.

Data from websites and the point of sale data that is collected by retailers enables them to more accurately assess what we are buying and what we may want to buy so that they can promote their products in a more targeted manner.

Technology has speeded up the length of time that it takes to create and realise a design and the time it takes for us to find it and have it delivered to our door. This means that trends in fashion are communicated and adopted more quickly and also change more often.

Technology can also help the designer to deal with some of the bigger global issues such as the environment, by helping to cut down on waste. This is done by reducing the amount of sampling that is required by enabling production to be developed according to need and customised to the consumer, and by developing new and more environmentally manufacturing processes and materials. Many companies are looking at ways of producing custom garments using mass production methods (mass customisation). These production processes allow the consumer to customise the product to meet their needs while still utilising the efficiencies of mass manufacturing. This customisation can include selection of different fabrics, colours, and trimmings, or enable clothing to be made to the customers' own measurements. These measurements can be captured by technologies such as 3D body scanners and then used to produce garments and accessories such as gloves and shoes.

Virtual sampling uses software applications such as OptiTex to improve communication between the designer and the manufacturer and reduce the need for producing real samples. Digital printing and rapid prototyping using special printers to produce 3D samples are techniques that also reduce waste and speed up the design process.

These technologies will become more important as more consumers become concerned about issues such as the environment. Designers in the future will need to be aware of new technologies and their impact on the design process and the environment. The fast fashion industry, where clothes are produced very quickly and cheaply is very wasteful, and there are organisations that are promoting a more sustainable approach. You can find out more by visiting the website of the Centre for Sustainable Fashion at http://centreforsustainablefashion.wordpress.com/.

26.9 CASE STUDY: THE DEVELOPMENT OF A GARMENT

This is the story of the development of a garment created by designer Tonia Bastyn for her own label collection for Spring/Summer 2011.

When she starts to create a new range, Tonia will look at previous key items that have sold well, she will think about her core consumer, trends that are emerging in terms of key items, and she will also consider the budget that she has to work to. Tonia describes the Bastyn woman as aged 35–45 with a career and possibly a family; she has a busy life but likes to be individual, and she knows her own style but is not a slave to fashion.

When she is designing the collection, Tonia will think about how many trousers, dresses, and jackets she needs, the colours that she will use, and the detailing. She will also consider the brand values that she is trying to communicate (beautiful, original, luxurious, versatile, modern, detailed). Tonia works closely with her merchandiser to ensure that the collection can be produced within budget.

Tonia tends to use classic and very muted colours because these fit with the lifestyle of her customer. The fabrics she uses are expensive and luxurious, such as 100% silk, wool, and cashmere, and she often uses stretch jersey fabrics, as these make the clothes easier and more comfortable to wear. Her key tailored pieces are interspersed with vintage (clothes from previous decades) inspired items and special occasion pieces such as draped dresses and silk crepe de chine tops. Tonia pays particular attention to detailing, as she feels that this is what makes her clothes special. Her garments sell in the store for between £80 and £395, placing her at the top end of the middle market.

Tonia sells her collection through concessions in two department stores: Brown Thomas in Ireland and House of Fraser in the United Kingdom. There are usually about 100 pieces in the total collection for each season, but they do not all go into the stores at the same time.

Usually there are four to five drops for each season. For example, the first items for Spring/Summer 2011 will appear in the stores in December. This may seem very early, but customers are ready for something different and a little lighter by this point. This early drop is followed by the main spring/summer collection that goes into the store at the end of January/the beginning of February. In February/March, she will introduce occasion pieces for weddings, and finally in April, the high summer pieces will complete the collection.

The autumn/winter range follows a similar pattern with the first pieces appearing in August, with the main collection in September and the Christmas stock in October/November.

The inspiration for the garment we are featuring in this case study was a vintage 1930s silk jacket that Tonia had bought about three years previously (Figure 26.2). She was inspired by the

FIGURE 26.2

Vintage jacket.

Source: Photograph J Gaimster.

embroidery on the jacket (Figure 26.3) and used this as a starting point to design a silk shift dress (Figure 26.4). Tonia took the original embroidery design, photocopied and manipulated it, and then created artwork to fit the dress pattern (Figure 26.5). A pattern cutter in the United Kingdom created the pattern for this piece, but sometimes a pattern cutter in the factory where the piece is to be produced will do this.

The artwork for the design, a specification sheet (Figure 26.6), the pattern, and a sample of a colour for the lab dip (Figure 26.7) are then sent to the factory that produces the first sample. The specification sheet is signed and dated as proof that the design is her copyright. This specification was accompanied by a pattern and therefore has no measurements; usually, a specification sheet will show the measurements as well and have care labelling instructions for how the garment should be laundered.

The factory producing this piece is based in India, and there they use traditional pot dyeing methods to match the colour. At the production stage, this colour will be matched using colour-matching technology by the mill that is going to create the fabric for the production.

The first sample is returned to the United Kingdom (this process usually takes around two weeks) and it is then fitted on a size 10 fit model. At the fit meeting, there will be Tonia, her merchandiser, the technical manager, and the production manager. They will discuss any problems with the garment and

FIGURE 26.3

Embroidery on vintage jacket.

Source: Photograph J Gaimster.

FIGURE 26.4

Silk shift dress by Tonia Bastyn.

Source: Photograph J Gaimster.

make comments to send back to the factory. At this point, a red-coloured plastic seal (see Figure 26.8) is attached to the sample. The factory will then create a second sample incorporating the comments and alterations. If approved, this second sample gets a black seal, indicating this is the approved sample; if the sample is still not right, a third one may need to be produced and this will be given a gold seal. The seal procedure ensures that the right version of the garment goes into production. Sometimes it is necessary to make more than three samples if the factory has not understood the nature of the alterations, so clear communication is very important.

It is also important to understand the strengths and weaknesses of the supplier, for example, Indian factories are usually very good at embroidery and decorative finishes, whereas Chinese factories are usually very good at technical production. To ensure that they get the best possible results, every designer needs to understand their supply chain. This requires research into the particular production methods and manufacturing capabilities that a country offers. Business to business websites, trade associations, and agents who work within a country can all be useful sources of information. Visiting the factory and building a relationship with the supplier is also very important, as this allows you to see the production capabilities and quality and also the working conditions.

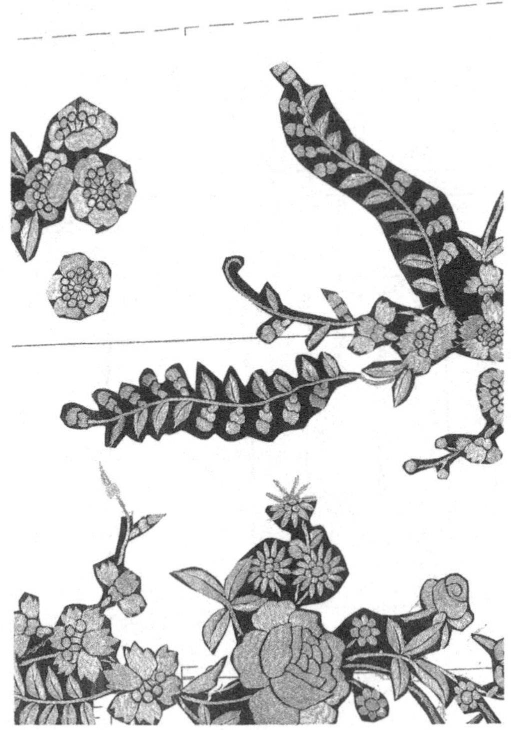

FIGURE 26.5

Artwork for embroidery Tonia Bastyn.

Once a correct sample is approved, there will then be a further two pre-production samples created. One is a size 10 and another in a size 14. This is to check that the design still works when it is graded (grading is the process by which a pattern is scaled to fit a range of different sizes).

Once everything is approved, the factory will produce the garment, in this case, a shift dress. They will produce around 200 pieces. The finished garments will be sent to the distribution centre, and from there they will be allocated to the stores.

As well as designing the collection, Tonia has to make sure that it is marketed and promoted. To do this, she uses several techniques. She will have a press launch to show the collection to the press in the hope that they will feature it in their magazines. She also produces a look book, (Figure 26.9) with very straightforward shots of the garments. This is used to show how the pieces in the collection work together. There are also mood shots; these are used for point of sale in the stores and the website and reflect the character of the brand.

Tonia also employs a public relations company to promote her collection. It is important to have a marketing strategy and to plan where she is going to spend her marketing budget, and also to consider how to maximise the coverage that does not have to be paid for through social networking sites, blogs, Twitter, and other media. Coverage that is not paid for is called editorial.

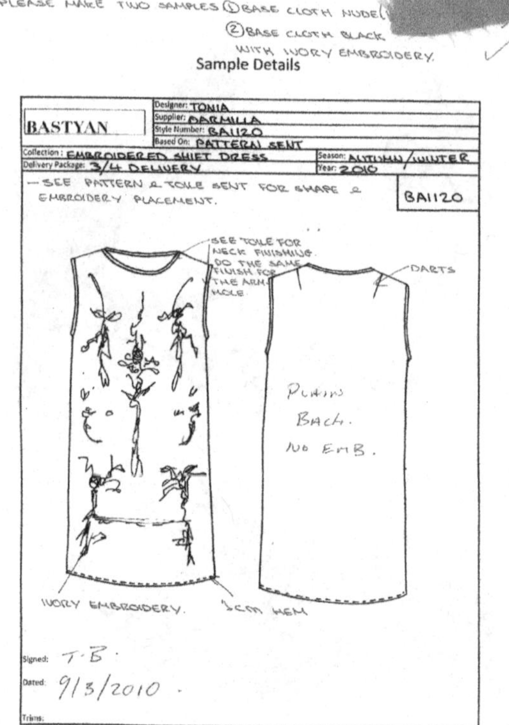
PLEASE MAKE TWO SAMPLES ① BASE CLOTH NUDE
② BASE CLOTH BLACK
WITH IVORY EMBROIDERY.

Sample Details

BASTYAN

Designer: TONIA
Supplier: DARMILLA
Style Number: BA1120
Based On: PATTERN SENT
Collection: EMBROIDERED SHIFT DRESS
Season: AUTUMN/WINTER
Delivery Package: 3/4 DELIVERY
Year: 2010

– SEE PATTERN & TOILE SENT FOR SHAPE & EMBROIDERY PLACEMENT.

BA1120

SEE TOILE FOR NECK FINISHING. DO THE SAME FINISH FOR THE ARM-HOLE

DARTS

PLAIN BACK. NO EMB.

IVORY EMBROIDERY.

1cm HEM

Signed: T.B.
Dated: 9/3/2010.
Trims:

FIGURE 26.6

Specification sheet for shift dress by Tonia Bastyn.

Tonia will also attend special events in the stores to promote the collection and meet her customers. She thinks it is very important to spend time on the shop floor so you understand how to sell and what customers want. Before setting up on her label, Tonia spent a lot of time working in retail, and she advises all students thinking of working in fashion to do the same.

Once the collection is in store, the sales are monitored on a weekly basis. The stores in which Tonia has her collections will take a percentage of the sales and they will set targets for these sales. This is because the stores have to make sure that the space is being used to create profit for them. Tonia also has her own sales targets for her staff. The weekly sales reports that she receives help her to know which stores require fresh stock, and they also help her with the planning of the collection for the following season.

If a garment is not selling, it may be removed for a while and then returned later in the season or held back until the end of season sale, when it will be marked down.

If a garment sells well, then Tonia may include a variation on it in her next collection; she may vary the colour or details while still retaining the essence of the garment. Most designers will have signature pieces that they reinvent and include in each collection; sometimes they will revisit and rework an idea from a season a few years previously if it feels right for the current trends.

Tonia will usually be working on two seasons simultaneously; for example, while she is working on S/S 2011, she was still finalising pieces for the late drop for her A/W 2010 collection. Tonia is

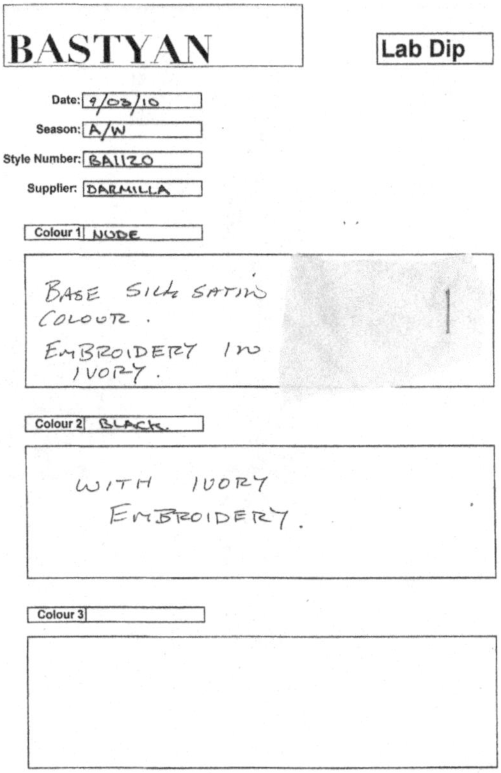

BASTYAN

Lab Dip

Date: 9/03/10

Season: A/W

Style Number: BA1120

Supplier: DARMILLA

Colour 1 NUDE

BASE SILK SATIN COLOUR.
EMBROIDERY IN IVORY.

Colour 2 BLACK.

WITH IVORY EMBROIDERY.

Colour 3

FIGURE 26.7

Example of a lab dip – Tonia Bastyn.

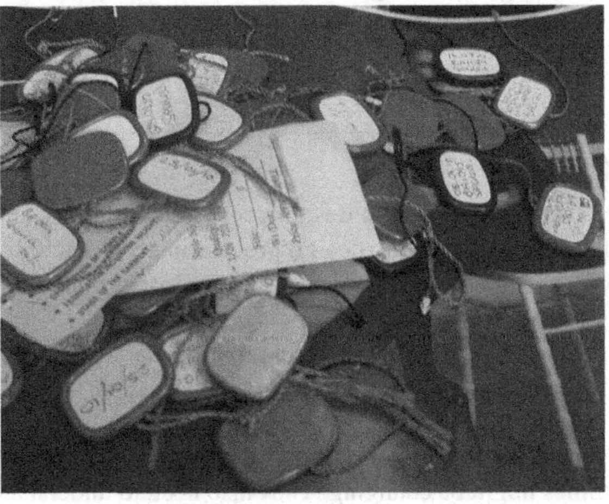

FIGURE 26.8

Selection of garment seals.

Source: Photograph J Gaimster.

FIGURE 26.9

Look book shot – garment by Tonia Bastyn.

Source: Photography Walter White model Clair at storm models.

passionate about her work, but says it is not glamorous; it is extremely hard work and she thinks that it is important that students going into the industry understand this. You can see more of her collection on her website www.bastyan.co.uk.

26.10 SUMMARY

In this chapter, we have discovered that the design process is complex and that it involves many people, not just the designer. In fashion, the design process is accelerated by the need to constantly provide the consumer with new ideas and to do so within seasonal boundaries and in line with fast-changing trends.

Understanding your consumer is the most important part of the fashion design process. Your ideas can be fabulously creative and trendy, but if they are too expensive or too extreme for your customer, they will not sell.

Designers need to be informed about a lot of things that are going on in the world so they can tune into the trends and how things might be changing. They also need to understand fabrics and colour and how garments are made. A good designer is not only creative but has a sound understanding of the technical and business aspects of the process.

26.11 PROJECT IDEAS

26.11.1 ANALYSING A COLLECTION

Visit a high-street store or boutique and look at the clothing for the current season (if you do not live near a shopping centre, you can look at a website). The store may have several stories or mini collections, so choose one. Can you work out what the inspiration for the collection was? How are the clothes coordinated? What are the main colours and fabrics in the collection? What are the key shapes, details, and prints or motifs? How many pieces are in the collection? How are they broken down into shirts, blouses, trousers, dresses, and skirts?

What is the price structure of the collection – the cheapest and most expensive items? Who do you think the target customer is?

26.11.2 FINDING TREND INFORMATION

To be a designer, you need to know what is happening with fashion trends. If you do not work for a large company, you may not be able to afford the services of a professional forecast company but there are still many sources of trend information on Internet. You can look at fashion blogs; the websites for trend companies may have some information on their home page or samples of the current season. You can also get some information from trade-show sites and companies that provide services for the fashion industry, like Pantone. Look at some of the web sites listed at the end of this chapter and see if you can identify the key trends that are being predicted for the coming season. Make a mood board to reflect your trend using your own images and photographs wherever possible. You can also look at sources such as Flicker www.flickr.com (check for images that are licensed under Creative Commons, a system that enables you to use images by attributing them appropriately).

26.11.3 ANALYSING A GARMENT

Select a garment from your wardrobe and analyse it. What is the fibre/fabric? Are there any trimmings? Where was it produced? How much did the garment cost and which level of the market do you think it belongs in? How should it be laundered, or should it be dry cleaned? What temperature should it be ironed at? What information would you have to give to a factory for them to be able to make it? Lay the garment on the floor and make a drawing of it. Think about the proportion and shape of the garment and draw the outline first, then fill in any details such as pockets or stitching.

26.12 REVISION QUESTIONS

1. How far ahead of the selling season do the textile trade shows such as Premier Vision work?
2. What methods do companies use to identify the needs of their consumer?
3. Why is colour an important aspect of the design process?
4. What are the different sources of information that a designer can use for trend information?
5. What impact has technology had on the design process?
6. What is a fad? Give some examples.
7. What are the key elements of a garment design?

26.13 SOURCES OF FURTHER INFORMATION AND ADVICE

http://creativecommons.org/ a non-profit organisation that increases sharing and improves collaboration.
http://www.apparelsearch.com/care_label.htm information on care labelling.
www.copyrightservice.co.uk/copyright/ Law, protection, registration.
http://www.londonfashionweek.co.uk/ London Fashion Week show information.
http://www.mindtools.com/pages/main/newMN_CT.htm creative thinking techniques.
http://www.own-it.org/ intellectual property advice for the creative sector.
http://www.premierevision.com/ fabric trade show.
www.fashion156.com fashion news and style information.
www.ftape.com fashion and catwalk trends.
www.infomat.com fashion industry information.
www.nellyrodi.com Nelly Rodi.
www.pantone.com pantone colour system and colour trend information.
www.peclersparis.com Peclers fashion forecasting.
www.promostyl.com Promostyl fashion forecasting.
www.style.com fashion news and style information.
www.texi.org the textile institute website.
www.trendhunter.com trend information.

REFERENCES

Berger, J., Heath, C., & Ho, B. Divergence in cultural practices: tastes as signals of identity. Manuscript in preparation. Available from http://www.chicagobooth.edu. Accessed 28.08.10.

Buzan, T. (2010). ThinkBuzan – Official mind mapping software by Tony Buzan [Internet]. Available from http://thinkbuzan.com/. Accessed 24.08.10.

Dutta, D. (2002). Retail @ the speed of fashion part 1. Available from http://thirdeyesight.in/articles/ImagesFashion_Zara_Part_I.pdf. Accessed 05.05.10.

Neilsen. (2008). *Trends in online shopping a global Nielsen consumer report*. The Nielsen Company. Available from www.nielsen.com.

Solomon, M. R. (2009). *Consumer behavior: In fashion* (2nd ed.). Upper Saddle River, NJ: Pearson/Prentice Hall.

FURTHER READING

Anstey, H., & Weston, T. (1997). *The Anstey Weston guide to textile terms*. London: Weston Publishing Limited.

Armstrong, J., Armstrong, W., & Ivas, L. (2005). *From pencil to pen tool: Understanding and creating the digital fashion image* (Illustrated ed.). New York: Fairchild Books.

Brannon, E. L. (2005). *Fashion forecasting/Evelyn L. Brannon* (2nd ed.). New York: Fairchild Publications.

Centner, M. (2007). *Fashion designer's handbook for Adobe illustrator/Marianne Centner and Frances Vereker*. Oxford: Blackwell.

Colussy, M. K. (2004). *Rendering fashion, fabric and prints with Adobe photoshop*. Upper Saddle River, NJ: Prentice Hall.

Cooklin, G. (1997). *Garment technology for fashion designers*. Oxford: Wiley Blackwell.

Davies, G. (2011). *Copyright law for artists, photographers and designers*. London: A & C Black Publishers Ltd.
Diane, T. (2004). *Colour forecasting/Tracy Diane & Tom Cassidy*. Oxford: Blackwell.
Feisner, E. A. (2001). *Colour: How to use colour in art and design*. London: Laurence King Publishing.
Gaimster, J. (2011). *Visual research methods in fashion*. Oxford: Berg Publishers.
Gerval, O. (2008). *Fashion: Concept to catwalk*. London: A & C Black.
Greenlees, K. (2006). *Creating sketchbooks for embroiderers and textile artists*. London: Batsford.
Hornung, D. (2004). *Colour: A workshop for artists and designers*. London: Laurence King Publishing.
Jackson, T., & Shaw, D. (2006). *The fashion handbook* (1st ed.). London: Routledge.
Jones, S. J. (2005). *Fashion design/Sue Jenkyn Jones* (2nd ed.). London: Laurence King.
Lowe, C. (2010). *Guide to copyright and intellectual property law* (A. revised ed.). London: Easyway Guide.
McKelvey, K. (2008). *Fashion forecasting/Kathryn McKelvey and Janine Munslow*. Oxford: Blackwell.
McKelvey, K., & Munslow, J. (2008). *Fashion design: Process, innovation and practice*. Oxford: Blackwell Science Ltd.
Renfrew, E. (2009). *Developing a collection/Elinor Renfrew, Colin Renfrew*. Lausanne; Worthing: AVA Academia.
Seivewright, S. (2007). *Research and design/Simon Seivewright*. Lausanne: AVA Academia.
Tate, S. L. (2004). *Inside fashion design* (5th ed.). Upper Saddle River, NJ: Pearson/Prentice Hall.
Udale, J. (2008). *Basics fashion design: Textiles and fashion*. Lausanne: AVA Publishing.
Underhill, P. (1999). *Why we buy: The science of shopping*. New York: Simon & Schuster.
Waddell, G. (2004). *How fashion works: Couture, ready-to-wear and mass production*. Oxford: Blackwell Science.

CHAPTER

27

COMPUTER-AIDED DESIGN (CAD) AND COMPUTER-AIDED MANUFACTURING (CAM) OF APPAREL AND OTHER TEXTILE PRODUCTS

S. Burke[1], R. Sinclair[2]

[1]Burke Publishing, London, UK; [2]Goldsmiths, University of London, London, UK

LEARNING OBJECTIVES

At the end of this chapter, you should be able to:

- Develop an understanding and appreciation of CAD and how it is used in the textile and fashion industry globally
- Develop an awareness and appreciation of digital design, presentation and communication, techniques, and skills
- Develop a basic understanding of the digital design process
- Understand the basic terminology used in the digital design process
- Understand why you need to develop digital design and presentation skills to be able to develop a career in the textile and fashion industry

27.1 INTRODUCTION

A technical revolution has been taking place in the world of textiles and fashion. Since the 1990s (Aldrich, 2008, p. 192; Jones-Jenkyn, 2011, p. 246), software for use in the fashion and textiles industry has become increasingly sophisticated. Diamond (2003, p. 18) identifies the way in which the fashion industry has been changed by the advent of new technologies and emphasises this 'technology has risen to new heights. Every segment of the fashion industry, from raw materials to the final distribution, to consumer, takes advantage of ever-improving technological discoveries'.

Guerrero (2009) and McCullough (1996) both emphasise that the immediacy of pen and paper as design tools will never be replaced, but that 'it is essential to note the increasingly important role that digital processes are playing in completing the representation of design' (Guerro, 2009, p. 38).

This period of rapid change coincided with the introduction of powerful and relatively inexpensive computers, systems, and graphics software such as Photoshop, Illustrator, CorelDraw, and the use of

Textiles and Fashion. http://dx.doi.org/10.1016/B978-1-84569-931-4.00027-1

software such as Excel for costings and production management has encouraged the textile and fashion industry, and the designers who work in these sectors, to use this versatile medium to help create and develop their designs, presentations and clothing ranges, and manage their workflow.

The latest graphics software offers a multitude of tools and techniques for sketching and design, image editing, page layout, and web design. These programs can be used to sketch a simple technical line drawing of a basic tank top or a highly detailed jacket, create the most amazing fashion illustrations, and develop dynamic presentations and page layouts for print and screen. The computer is an extension of one's hand, and an aid to creativity and visualisation.

Computer-aided design (CAD) was initially developed as an interactive computer design system for the textile industry, then introduced into apparel for pattern making and grading, and has been further developed for fashion and clothing design. More recently, powerful graphics software has been integrated into the fashion design process to help to create technical drawings of designs as flats, specification drawings, fashion illustrations and design presentations. This has speeded up the design process and presents a global standard for the visual communication of designs to the production, manufacturing, and marketing sectors within the industry.

Computer-aided manufacturing (CAM) was initially developed as a way of ensuring that the transition from 2D design process to 3D manufacture was more seamless, and also enable companies to benefit from quick responses (Diamond, 2002). The requirement to interlink the design and manufacturing processes can be seen specifically in the area such as knitting, in 3D knitting processes, weaving, and embroidery. For the textile designer, the ability to design, virtually check, and then send the design direct to the machine enables a more seamless process. Just as CAD enables design businesses to participate within a global field, so does the ability to integrate CAM, which enables companies to test physical prototypes (Jones-Jenkyn, 2011) and streamline production processes.

The advent and progression of the virtual world has united fashion with gaming and software technologies, such as the development of Avatars, online spaces such as Second Life, software such as Poser, online rendering, testing and virtual fit, and pattern sewing as seen in software such as Optitex and Marvellous Designer. Such technologies have and continue to be employed directly in fashion development and marketing, such as fashion company H&M's use of SIMS in its Fashionista program.

It is important for designers to recognise that the use of the computers in fashion and textiles is not just limited to the design of garments; such technology is used prevalently throughout the whole fashion and textile industry.

The CAD/CAM process can be viewed in the following ways for the development of a design from inception to manufacture:

1. Connectivity: The need for companies in design (buyers, designers, suppliers), production (manufacturers) and retailing (e-commerce/e-tailing) in the design and production chain to communicate in a global market.
2. Design creation/ideation: Through image/mood boards, designs, virtual fabrics/clothing, flats.
3. Production data management: The ability to use product data management (PDM) software to control the whole production cycle of a garment. This type of software allows tracking of the workflow and identification of the status of an individual garment at any one time.
4. Pattern design: This is divided into two areas, one being the specification drawings, the second at PDS (pattern design systems). It is in these systems that the pattern is generated for a garment.
5. Garment sampling: At this stage, the pattern created in the PDS can be modelled in both physical patterns and 3D virtual systems. In these systems, garments can be virtually

stitched together, tested for fit in a virtual environment, and then even visualised on virtual runways. It is also at this stage that patterns can be exported to digital textile printing machines, and prototype garments produced.

6. Sizing: The advent of body scanners had led to an increase in made-to-measure clothing, leading companies to develop mass customisation as a process for ensuring better customer fit, style, and options.
7. Pattern grading: Pattern grading systems now mean that the processing of patterns is leading to more streamlined process for pattern making. Pattern grading is done by inputting pattern data, creating the grading criteria, grading the pattern, and then sending this data to the production planner.
8. Production: Here the data created at the pattern grading stage is converted into a production pattern, known as production lay planning and marker making. Using the software the pattern can be specifically laid to create the best efficiency for cutting out, and lower costs especially related to material waste.
9. Plotting and cutting: Companies now utilise CAM systems to enable the accurate cutting out of garments and production pieces. These systems also enable companies to review waste management in terms of fabrics.
10. Product life cycle: Product lifecycle management (PLM) software that can be accessed on a global basis is database software enabling companies to view all aspects of a product from design to finish. It is usually integrated with PDM system software. This allows companies to monitor the styles to market, the workflow and the sales, from design inception to completion.

27.2 FASHION AND TEXTILE SOFTWARE PROGRAMS

There are a number of fashion and textile software programs designed specifically for the small business and freelance designers, but the larger apparel and textile companies are more likely to use the powerful CAD apparel and textile suites produced by Lectra and Gerber (see Table 27.1). These suites have been developed to integrate all areas of the apparel process from apparel and textile design, pattern making, grading, garment production through to merchandising, and data management. Consequently,

Table 27.1 Apparel and software fashion and textiles suites

Software manufacture	Apparel/textiles design suites	Platform: Mac or PC
Lectra	Kaledo	PC
Gerber	Fashion studio	PC
SpeedStep	Painter, sketch	*Mac or PC
Vetigraph	Stylgraph	PC
Adobe	Illustrator, Photoshop	*Mac and PC
Autodesk	Sketch	*Mac or PC and App
Textil studio	Textil studio	PC
Corel	Corel Draw	PC

**For some users, the Mac option requires the user to either have Virtual or Parellel windows software, and the latest Mac IOS operating system.*

the suites are expensive but enable a company to achieve economies of scale by reducing collection development costs and lead times, ensuring better quality standards while still maintaining an individual, distinctive style and a coherent brand image. Above all, it means greater productivity and profitability for the company. These suites can also be tailored to meet the company's or clients' specific needs within any sector of the textile and apparel business.

An added advantage is that the popular and more economical graphics software such as Illustrator and Photoshop are compatible with the Gerber and Lectra systems. This means that the drawing and image editing skills that designers use to create their designs in these packages are similar to the tools and techniques used by the more industry-specific fashion systems. For example, features such as 'drag-and-drop' can be used and/or files can be imported directly into the Gerber or Lectra programs.

CAD apparel and textile software specifically offers the designer a complete design software solution. It gives the designer the tools and techniques to digitally create designs from the initial concept to final presentation, to the creation of a pattern for a garment or product, to the grading (sizing) of a pattern.

This includes:

- Creating quick sketches and fashion illustrations
- Creating minibodies, croquis (figures), and line drawings for presentations/storyboards
- Creating and managing various colourways, colour palettes
- Using and adapting the latest style trends and seasonal colour palettes
- Creating and simulating print designs, knits, and wovens fabric swatches from libraries and from scratch
- Designing in repeat – any number of repeats can be made using different sequences and sizes
- Creating and managing colourways and colour matches for textile designs, fabric swatches, universal and seasonal palettes
- Working with style and trim libraries
- Editing images, scans, and photographs
- Creating individual portfolios
- Working with 3D images to create collections and virtual fashion shows
- Drafting 2D patterns and creating virtual 3D samples from those patterns
- Creating and viewing simulated garments on the body to evaluate fit
- Realistically simulating the drape of the fabric
- Sharing collections with buyers, suppliers, and retailers within the industry and globally

27.3 USING CAD TO DESIGN FASHION PRODUCTS

Fashion designers require a portfolio of computer drawing and design skills to sketch designs ranging from the simplest tank top through to the most highly styled creations. This includes:

- Drawing clothing as flats/working drawings/technical drawings
- Specification sheets (specs) including cost sheets
- Style sheets
- Illustrated designs drawn on a fashion figure body/croquis

Some of the key documents are discussed in the following sections.

27.3.1 FLATS/WORKING DRAWINGS

Flats, also referred to as Technicals, are explicit line drawings of garments, drawn to scale, using simple, clear lines, with no exaggeration of detail as you would find in a more stylised fashion illustration.

It is important in flats that all construction lines such as seams, darts, and styling details, such as pockets, buttons, and trims, are represented. Apparel companies use flats as their primary visual source to communicate and liaise with buyers, clients, pattern makers, and sample machinists – flats are the international fashion language. Increasingly flats are also added to trend reports, this allows the viewer to see key silhouettes and key design details. Flats form an essential part of the design process; digital drawings are the most efficient method to communicate designs from the fashion design studio to production, and to the buyers, merchandisers, and marketing teams.

27.3.2 SPECIFICATION SHEETS (SPECS)

In the fashion industry a specification sheet or a spec sheet, as it is known, is a document that contains an accurately drawn flat (line drawing of the design), and all relevant specifications (instructions and measurements). This information is needed to produce garments to the required standard and design. The spec sheet forms the basis of a binding contract between the design house or client, and the factory that produces the garment. With a large percentage of clothing manufacturing being outsourced offshore, it is important that this document be clear, precise, and self-explanatory. A spec sheet would be prepared by the fashion designer, fashion design assistant, and/or pattern maker and is of particular importance in the fashion industry in the design and production process. It is essential that it is checked for accuracy. Spec sheets can be created using the Excel spreadsheets software; there are other options available, both online and open source, the flat/line drawing is created using a vector program such as Adobe Illustrator©, Corel draw, open source option Inkscape, or specific apparel software.

Spec sheets would typically contain the following details:

- A technical drawing – front and back view and specific details to show the exact design and details such as the position of pockets, buttons, and labels.
- Specific construction details such as the types of seams required, hem, and topstitching details.
- Measurements of the garment/product, plus details of graded sizes required.
- Fabric details, a swatch of the fabric, and details of the trims and thread used.

The spec sheet would be adapted and updated depending on what stage the style/design was in the design and production process, for example:

- In the sample room where the pattern and sample garment/prototype is initially made.
- The production department where the garment will be costed.
- The factory floor where the garment orders are cut and constructed.
- It might also be adapted and used as a reference during the despatching and delivery of the garments.

Documents such as specs would be prepared by the fashion designer, fashion design assistant, pattern maker/team, and are of particular importance in the fashion industry in the design and production

Design style sheet/specification sheet			
Style no	Designer-Tina Fong	Customer/buyer	CMT/factory
Season: S/S Yr	Pattern maker	Department	Sample size
Commitment no	Machinist	Delivery	Created
		Color	Modified
		Units	Approved
			To grade

Fabric details		**Garment description**	
Fabric swatch	Description		
	Design	**General notes/trims**	**Notes cont'd**
	Type	Fusing info	
	Order no	Binding details	
	Composition	Zip	
	Quality	Seams	
	Weight	Seams	
	Width	Hems	
	Open/tubular	Wash	
	Sub sampling	Label position	
	Check repeat	Buttons (type, size, quantity)	
	Bulk del. due	Thread	
	Sample fabric	Swing ticket	
	Design	Wash	
		Rating	

Front design (or front and back)

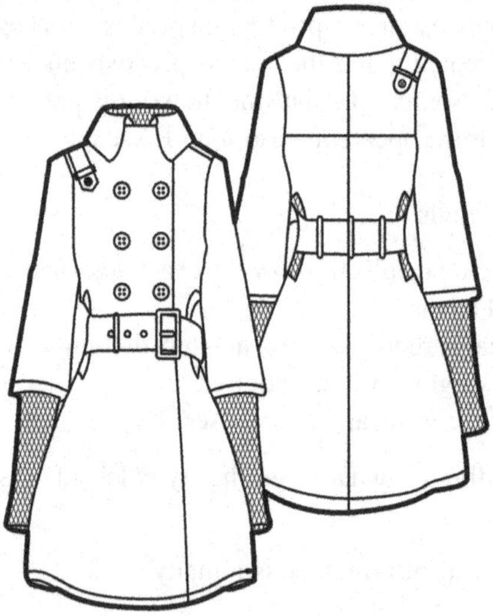

Pattern maker notes
(Specific measurements - lengths, widths etc.)

Cutter notes
(Specific cutting instructions)

Machinist notes
(Specific sewing instructions)

Back design (or specific details)

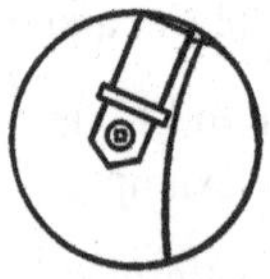

FIGURE 27.1

Spec sheet – presenting all the relevant information required to produce the garment to meet the required standard and design. Designer Tina Fong.

© Fashion Designer – Sandra Burke

process. Spec sheets are inherently now linked to the PDM and PLM data management and database software, and as such become integrated into the workflow program of a product. They present the key information about the garments/products to ensure that they are made to the required specifications and include some or all of the following information: the people responsible for the design, the technical drawing, fabrics and trim details, any specific measurements, etc. Figure 27.1 shows a spec sheet. Figures 27.2 and 27.3 show a spec sheet with measurement details whilst Figure 27.4 shows an extract of the style details required.

The costing sheets (typically in Excel or other spreadsheet software) list all the required information to calculate the total costs to manufacture the garments or products, and to establish the required selling prices (wholesale and/or retail) and include: the material/trims ratings and costs, processes, time costs, and all relevant information as shown in Figure 27.4.

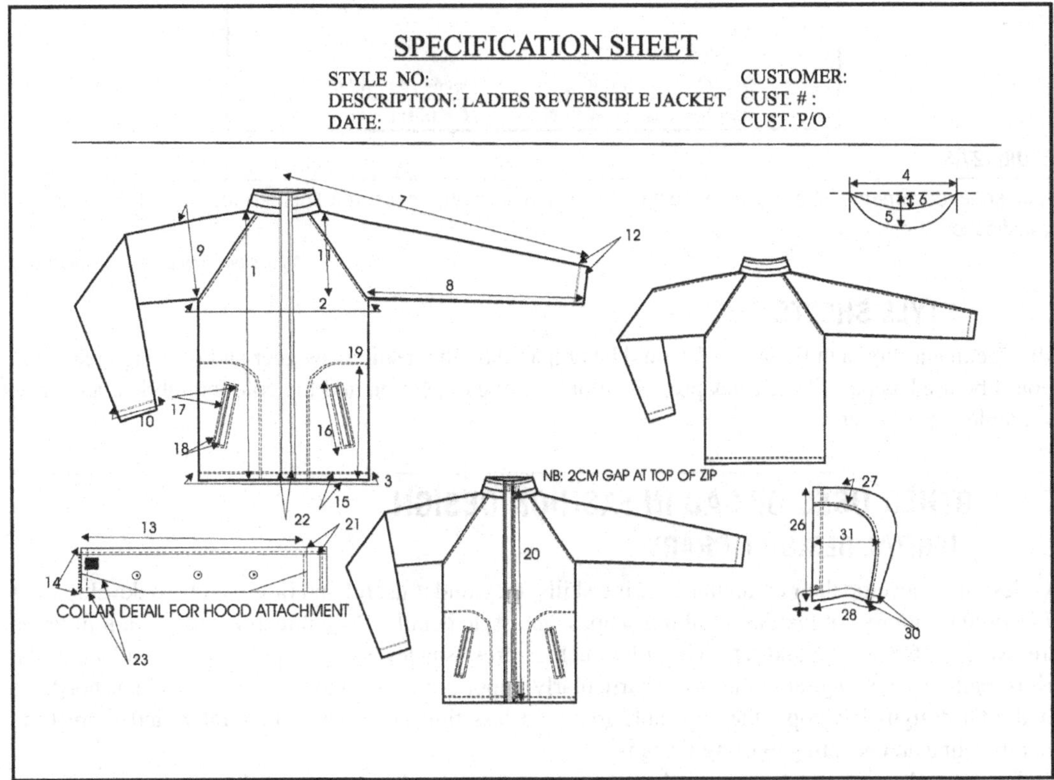

FIGURE 27.2

Specification sheet – with measurement details. The numbers refer to items in the spec sheet document.

© Fashion Computing – Sandra Burke

Finished Flat Measurements - size Medium
1. Total Length...............
2. Chest Circumference..............
3. Hem Circumference...............
4. Neck Width.............
Etc...............
Stitching: Single Needle: C.F. Zipper Edges, Raglan Seams, Hood Panels; Twin Needle: Hem, Sleeve Cuff, Pocket Bag Stitching
Labelling: Sew In Label - Woven Label for Sizes S-L, position.......... Care Label - As Per Garment Instructions
Accessories: Zip Slider: To Be Reversible Same Style as Per Original Sample Drawcord: Hood as Per Original Sample Velcro: 2cm Wide Velcro For CF
Fabric: Shell: 100% Nylon Taslon Lining: 100% Polyester Anti Pill Spun Polar Fleece - 280GM/M2
Color Combinations: Stone, Mink, Black with Contrast Lining

FIGURE 27.3

Spec sheet document – showing an extract of the style details and measurements required (see the previous specification sheet).

© Fashion Computing – Sandra Burke

27.3.3 STYLE SHEETS

The document shown in Figure 27.5 would be prepared by the fashion designer and/or design team and would be used as part of the sales process to present the collection and the colours, fabrics, and sizes available to the buyer.

27.4 OTHER USES OF CAD IN FASHION DESIGN

27.4.1 DIGITAL DESIGN LIBRARY

As designers develop their computer drawing skills, they find it useful and time saving to develop their own digital 'library' of textiles, clothing shapes, and style details. Digital textile and fashion libraries are excellent for storing and retrieving files such as dress shapes, skirt shapes, various collars, cuffs, trims, patterns, etc. Digital libraries are particularly important for designers because, as their portfolio of digital designs develops, they are able to spend less time drawing from scratch and more time manipulating and adapting existing designs.

For example, in a sales meeting, a buyer might order several styles from a designer's latest collection. In addition, the buyer might require another style such as a classic pencil skirt and perhaps in a slightly different fabric weave and these could then be instantly retrieved from the digital library folders, and immediately presented to the buyer (see Figure 27.6).

COSTING SHEET			
Style No	Buyer	Delivery Date	Units
Commit No	Dept	Style Description	

FABRICS								
Fabric Description	Color	Order No	Width	Rating	Rate + 7%	EST Cost	ACT Cost	Value
					0			0
					0			0
					0			0
					0			0
						FABRIC TOTAL		

TRIMS					RATING				
Supplier	Description	Size/Type	Order No	Reference No	Mtrs/Yds	Quantity	Price		Value
	Hanger	Dress 10							
	Brand Label	Woven				1			
	Fabric Label					1			
	Wash Care	Dry Clean				1			
	Swing Ticket	Embossed				1			
	Polybag								
	Cut Bias	25 mm				0.88			
	H/Tape	Branded							
							TRIMS TOTAL		

GARMENT SKETCH		
	Fabric + Trims	
	Waste	
	CMT	
	Grand Total	
	Mark Up	
	Selling Price	
	Gross Profit	
	Retail Selling Price	
	Sold to Buyer	
	Actual RSP	
	Actual Profit %	

FIGURE 27.4

Cost sheet – presenting all the relevant information for costing purposes.

STYLE: ANGEL SHIRT

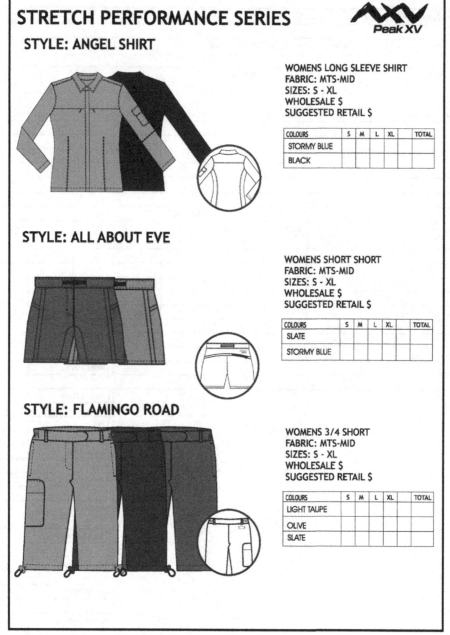

WOMENS LONG SLEEVE SHIRT
FABRIC: MTS-MID
SIZES: S - XL
WHOLESALE $
SUGGESTED RETAIL $

COLOURS	S	M	L	XL		TOTAL
STORMY BLUE						
BLACK						

STYLE: ALL ABOUT EVE

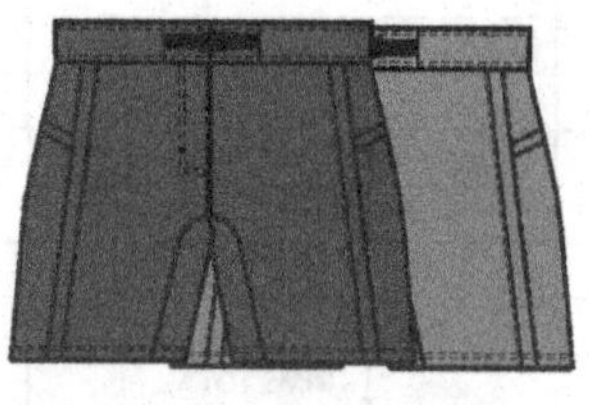

WOMENS SHORT SHORT
FABRIC: MTS-MID
SIZES: S - XL
WHOLESALE $
SUGGESTED RETAIL $

COLOURS	S	M	L	XL		TOTAL
SLATE						
STORMY BLUE						

STYLE: FLAMINGO ROAD

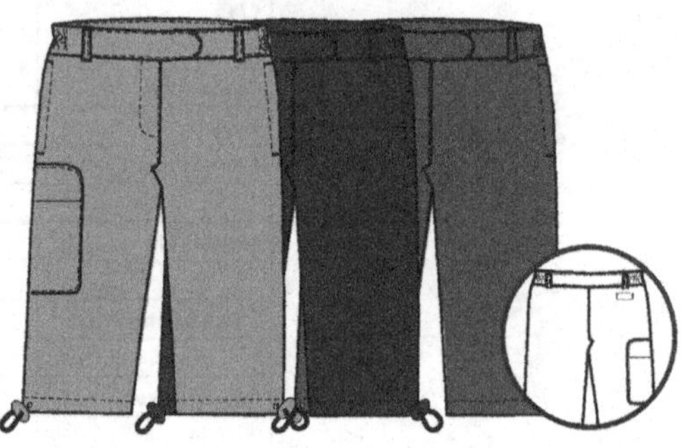

WOMENS 3/4 SHORT
FABRIC: MTS-MID
SIZES: S - XL
WHOLESALE $
SUGGESTED RETAIL $

COLOURS	S	M	L	XL		TOTAL
LIGHT TAUPE						
OLIVE						
SLATE						

FIGURE 27.5

Stretch performance series sheet – presentation of flats for retail buyers listing the available colours, sizes, cost, etc., within the collection.

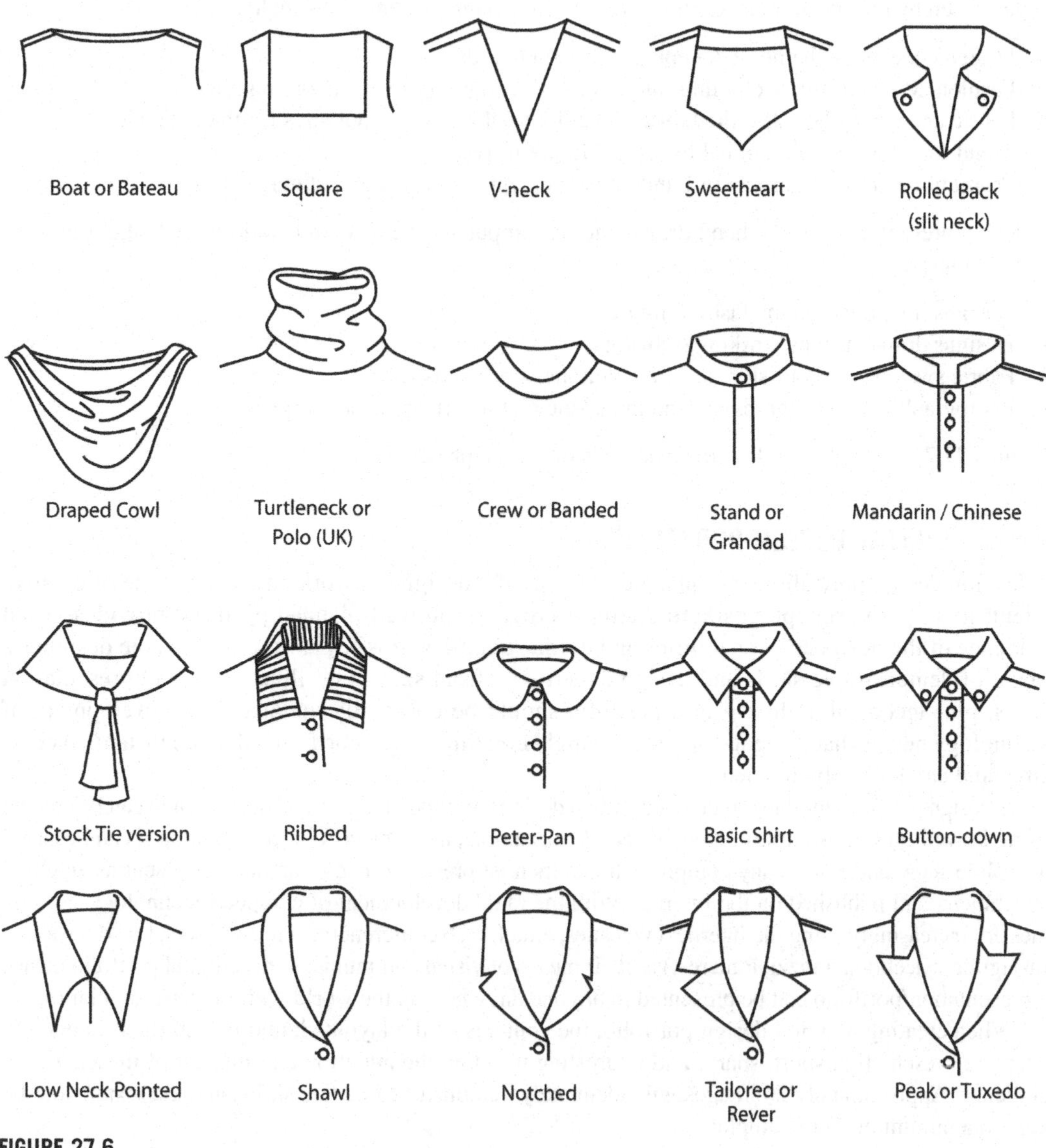

FIGURE 27.6

Digital design library of styles – necklines and collars – Penter Yip, Fashionary.

27.4.2 DESIGN PRESENTATIONS

Design presentations, also called storyboards (styling boards, range boards, etc.), are the designers' creative format to communicate textile designs, clothing designs and concepts to the design team, buyers, merchandisers, and marketing teams. These design presentations include:

- Design concepts – themes, concepts, and mood boards
- Fashion design boards – clothing ranges or collections for the forthcoming season
- Fabric and colour boards – the fabric and colour palettes or colourways for the collection
- Trend forecasting – directional looks and future trends
- Promotional artwork, brochures, and advertisements – for media and magazines

A design presentation can be hand drawn and/or computer generated and could include the following components:

- Designs illustrated on the fashion figure
- Designs drawn as flats/working drawings
- Fabric swatches, colour stories, trims (buttons, zips, studs, etc.)
- Photographic images, graphics, and magazine pictures (tear sheets/swipes)

Figures 27.7 and 27.8 provide some examples of design presentations.

27.4.3 DIGITAL DESIGN PORTFOLIO

A fashion design portfolio is not just a collection of a designer's work and an example of creative talent; it is also a principle marketing tool. A creative and well-planned portfolio provides visual evidence of the designer's capabilities and unique qualities. It is a means to express the designer's range of demonstrable skills and design expertise – from sketching, illustration and presentation skills, plus technical ability. And a portfolio should be constantly updated. The development of technology means that designers are increasingly creating online portfolios that are instantly accessible and can be easily updated.

A design portfolio might include hand-drawn designs, computer-generated designs, and can be created using PowerPoint, Illustrator, Photoshop, etc. For example, as a PowerPoint presentation it can be saved as a slide show and run on any computer. It can then be presented live to an audience, sent as an email attachment, and published on the Internet. With the rapid development of computer technology, companies are increasingly using the Internet (Websites, email, web conferencing, etc., and social media such as Instagram, Facebook, Prezzi, Pinterest) as their means of visual communication. A digital portfolio means a presentation/portfolio can be presented to anyone anywhere in the world with the push of a button.

When creating a digital design portfolio, the contents of the layout should be carefully planned to keep the presentation short, sharp, and interesting to retain the attention and interest of the viewer or audience. Approximately 20 images will adequately demonstrate creative skills, and the text should be kept to a minimum. For example:

- **Introduction**: Start with something about yourself; unique to you, perhaps a logo or graphic. Include your details (name, address, email, Website address, content).
- **Variety**: Display your creative talents and design skills by presenting a range of artwork to keep the portfolio exciting and informative.

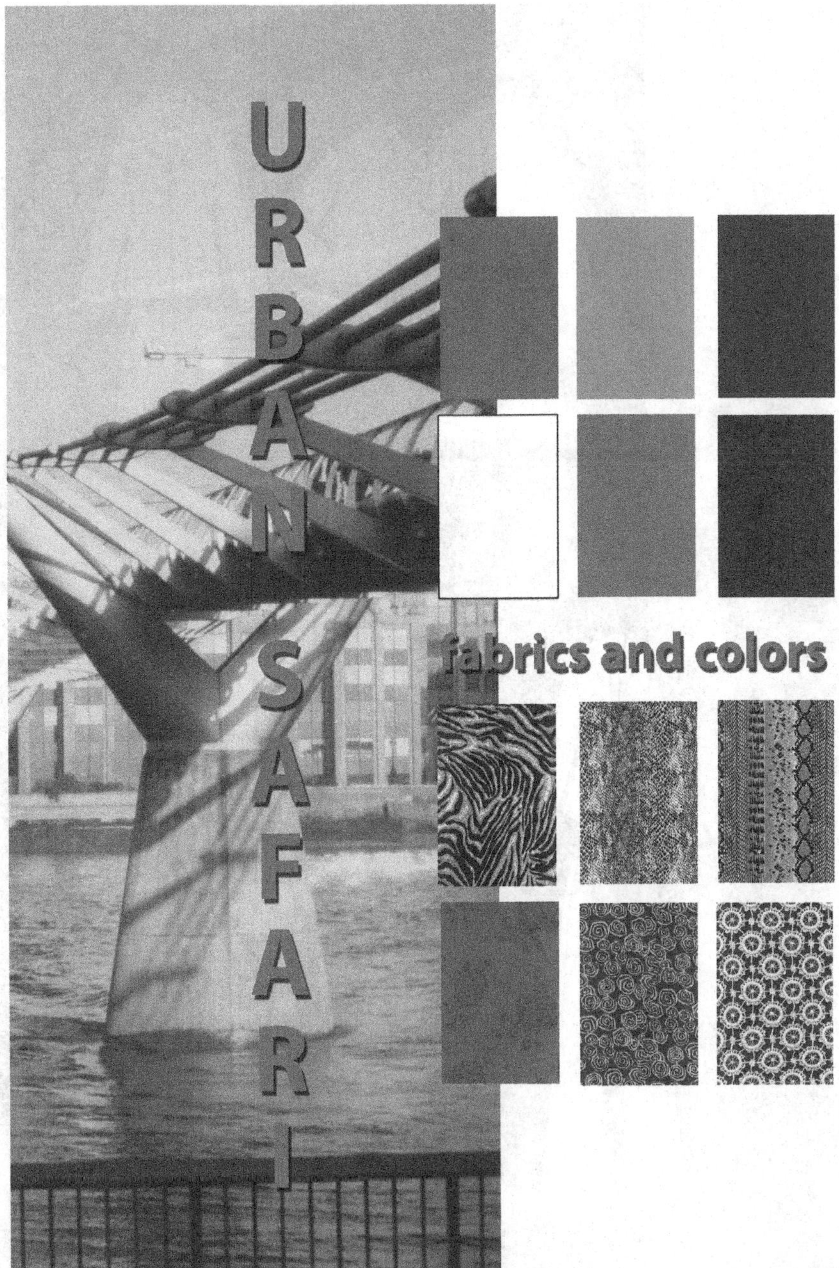

FIGURE 27.7

Fabric and colour presentation – Urban Safari (1): includes fabric swatches (scanned), colour story, background photograph, and digital type.

© Fashion Computing – Sandra Burke

FIGURE 27.8

Fashion design presentation, Flats – Urban Safari (2): includes digital flats, background photograph, and digital type.

- **Design Sense**: Keep your portfolio creative and graphically well presented, not necessarily wild on every page, but showing good design sense and strong themes.
- **Layout**: Organise your portfolio either by design or season (latest first) and finish with a dynamic finale (wedding, evening wear).

27.4.4 THE PLACE OF THE APP

As computers have changed the way in which designers utilise technology, so has the advent of the mobile phone and the tablet device. These devices run not on PC or Mac platforms, but on operating systems such as IOS or Android, or Blackberry. These new platforms allow small devices to run complex software, without the user installing large software programs. Many of the programs used on small devices operate from 'Cloud'-based servers, and utilise 'Wi-Fi' technology, meaning designers are no longer restricted to working from a desk, but can operate almost any where.

27.4.4.1 Firstly what is an app?

The term App is short for Application. Apps are basically software programs developed specifically for your mobile phone or portable tablet device. The programs range in size and complexity. Some come as standard on the mobile device you have. You can download additional apps, some are free and some you have to pay for, and some have additional in-app purchases that allow additional functionality, tools, or uses.

There are different operating platforms for apps just like you have for computers.

The main operating systems for apps are:

1. Apple's – IOS system – downloadable from the Apple Apps Store: https://itunes.apple.com/gb/genre/ios/id36?mt=8
2. Google's – Android system – downloadable from the Android Marketplace: https://play.google.com/store
3. Blackberry – downloadable from Blackberry app world: http://appworld.blackberry.com/webstore/product/1/?lang=en
4. Symbian – operate on phones such as the Nokia – http://store.ovi.com/
5. Windows surface but here is the latest link to the current windows apps store http://windows.microsoft.com/en-GB/windows-8/apps. Please note that you need Windows 8 to use the store and see the apps
6. The other competitors on the market also include E-Readers also known as E-Book readers, such as The Kobo, The Kindle, and Kindle Fire, The Nook, Sony E-reader, Archos, etc.
7. Just as open source software is used for mainstream computing, there is a growing field of demand for similar open source environs, which aims to keep overall cost of both software and hardware low, and more accessible

27.4.4.2 What kind of machines or computers do different apps work on?

Apps for the Apple IOS platform are for use on devices such as iPhone, iPad, Mini iPad, and iPod, and can also work on the Mac range of laptops and computers.

Apps for phones like HTC, Samsung, and mobile devices such as tablets, etc., utilise the Google Android operating system, and can operate on Desktop platforms, laptops, and tablets as well as smart phones. Blackberry uses the Blackberry platform.

27.4.4.3 *Why would you want an app?*

Apps allow you to customise your mobile device whether it be a phone or a tablet device to your specific set of wants and needs. The development of 'Cloud' based services where data is stored on virtual servers, rather than locally based servers, means that information is accessible any time we need it. As the mobile devices we use get smaller, thinner, and lighter, then the way we access the services we need to use are going to change; there are already a range of apps for both the fashion and textile designer, which provide a range of functionalities and options. A selection of some of the apps available are illustrated in Table 27.2.

27.5 CAM IN FASHION AND TEXTILES

27.5.1 3D DIGITAL AND VIRTUAL FABRICATION IN TEXTILES AND FASHION

CAM has become an integral part of the design and production systems that exist in fashion and textiles. Whilst there is growth in the development of online virtual systems, in textiles and fashion there is still the need to make physical prototypes. Although research in cloth modelling began in the 1930s, it would be the rapid development of computer graphics in the 1980s that would lead to developments in virtual cloth modelling, with the focus on modelling individual pieces of cloth, large clothing objects, as well how the cloth interacted with its environment (Breen et al., 2000, p. 20). Many of the early experiments would be later seen in gaming technology, and animated films, such as Disney's Fantasia 2000.

This early experimentation, would lead to further developments. As well as designers using specialist software, there was also growth using standard off-the-shelf software such as Maya and Poser, and today open source software such as Blender can be used to model fabrics. Other off-the-shelf such as Adobe's Photoshop can be used to do modelling using additional software plug-ins.

Inherent in any design system or software is the possibility to emulate the drape and fit of a garment on the body, and there is much development work in trying to ensure that the process becomes more seamless and real, with the focus on realism in virtual environments (Thalmann, 2010).

Designers and manufacturers already use virtual 3D prototype systems to visualise 2D patterns into 3D virtual prototyping as in the case of software such as Modaris 3D fit or Marvellous Designer. Other software such as Accumark V-stitcher and Optitex 3D runway, present the viewer with a 3D simulation, which seeks to demonstrate to the viewer, the fit of the garment and the drape of the fabric. Other software such as Optitex 3D runway and Marvellous designer, allow the designer to create virtual catwalks. This software is already been used outside the realms of fashion and textiles, and migrated into the realm of films and online video gaming, where realistic interpretation of apparel adds to the gaming experience.

Textiles designers are engaging in new methods of reimaging traditional design methods, for example, exploring the use of making printblocks, using CNC routers rather than traditional hand cutting block methods (Figures 27.9–27.11). This allows an altogether different mode of fabrication and design development, as well as experimentation, and the ability to work across disciplines; Gamister (2011) refers to this as thinking laterally.

Table 27.2 A selection of design apps and their operating platforms

	Adobe collage	My pantone	iPen note	Compliance intertek	Moodboard lite	Paper	Fashion dictionary	Style studio
What does it do	A presentation and drawing program. Can be interlinked with software on your own PC/Mac	Take a photo or use an existing image, stored in your picture gallery and extrapolate the colours. Can be emailed and used in composing designs and making decisions about colour. Great way to start design development	The iPen can be connected directly to an iPod/iPad or write or draw. You can also use the iPen with just plain paper. Gives you the option to develop ideas on paper and digitally, allowing working on a device and paper then back to computer	A guide to the terms related to restricted substances that can be used in textiles and fashion. Great for technical terminology	An interactive presentation app that allows you to get images from a range of sources and arrange them for inspirational starting points. Can be integrated directly into E-portfolios	Create your own collection of sketch books, great ideas for developing sketch books for different projects	A fashion dictionary in 5 languages-created for understanding textile and fashion terms. Fantastic starter tool for use in a multilingual classroom	A fabric and fashion design app, covering male and female fashion garments, accessories and prints. Great introduction tool
Cost	Y	Y	N	N	N	N	N	Y
IOS (iPhone or iPad)	Y	Y	Y	Y	Y	Y	Y	Y
Android	Y	Y	X	X	Y	X	Y	
Blackberry	X	X	X	X	X	X	X	X
Symbian	X	X	X	X	X	X	X	X

Availability of app on different platforms Y = Yes, N = No, X = N/A.

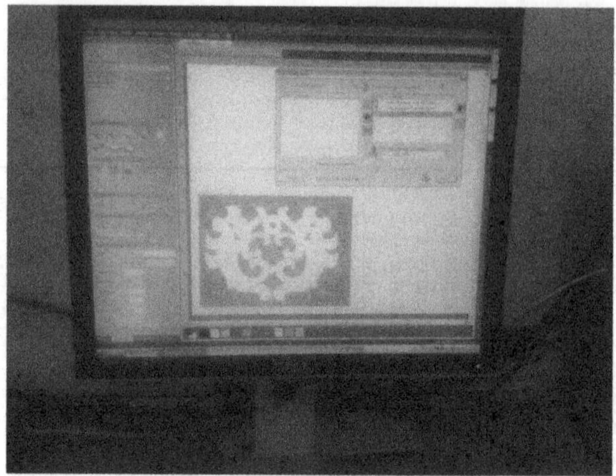

FIGURE 27.9

Developing design on CNC software and calibration process.

FIGURE 27.10

CNC router begins milling process for print block – prototyping in blue foam.

27.5.2 NEW 3D PRINTING AND FABRICATION IN TEXTILES AND FASHION

3D printing is not a new process; it has been in use in the manufacturing industry since the 1980s (Barnatt, 2013). What has made the technology take hold, like all new technologies, is the rapid change in the hardware, technology, and software.

The impact of new technologies in other spheres of design are migrating to fashion and textiles through a form of printing known as 3D printing. Originally in the fashion sphere, the technology was utilised in developing fashion accessories such as jewellery using software such as RhinoGold and

FIGURE 27.11

Completed prototype block.

3Design. Companies such as Nike and Materialise have used 3D printing technology in creating prototype boots and shoes. Materialise has developed shoes specifically for the fashion catwalk. The nature of 3D printing and the complexity of the software has led to fashion and textiles designers co-designing and co-creating to develop interesting products.

27.5.2.1 What is 3D printing?

3D printing is a computer-controlled process of building up a 3D form by adding or extruding material. Other terms for 3D printing may include additive manufacture, rapid prototyping, and layer manufacturing.

In order to get an actual 3D printed object, the design can be created in 3D vector modelling software and is then further processed using complex programs, which convert the design so it can then be printed on the printer in a series of layers. After printing, which can take several hours depending on size of object and type of printer, further finishing colour, etc., can be applied.

Alternatively, bureau services have sprung up that allow designing and configuration online, using a range of viewers and proprietary software allowing some design input. Once the design is created it is paid for, printed and despatched in the post. This process puts design input into the hands of the consumer, whilst at the same time enabling new forms of cross collaboration between design disciplines, as can be seen in the work of fashion designer Iris van Herpen and 3D print specialist company Materialise.

27.5.2.2 Materials that can be printed on a 3D printer

All digital technologies rely on materials in some form, and 3D printing is no exception.

Some of the current substrates used for digital printing include Polylactic acid (PLA), Nylon, and Polycaprolactone (PCL). All the substrates require different speeds or printing, and different finishes can be applied to aid the finishing process.

FIGURE 27.12

FOC for: Andreia Chaves. The first commercial studio series by designer Andreia Chaves was launched at Mercedes-Benz Fashion Week, 2011.

However, the desire to have flexible materials for fashion and textiles products has led to companies such as Belgium-based Materialise to develop flexible materials such as Thermoplastic Polyurethane, which provides a soft material and was used in the Iris Van Herpen's catwalk collection Hybrid Holism in Paris Fashion Week in July 2013.

Companies such as UK-based Freedom of Creation (see Figures 27.12–27.15) and New York–based Continuum have used a range of materials to create fashion-based garments. Continuum designers Jenna Fizel and Mary Huang created a 3D printed bikini. Freedom of Creation has created 3D bags and jackets that emulate the knitted stitch. Dutch designer Iris Van Herpen's designs explore the sculptural aspects of the technology in her sculptural fashion. Others designers such as Catherine Wales exhibited the way in which 3D printing can be incorporated into fashion products at the Design Museum exhibition in London, The Future is Here in 2013.

Current 3D technology, as it exists, does not currently enable a whole garment to be 3D printed, they still need to be printed in parts then 'bolted together'. Meanwhile designers are looking at how the technology can be used to create components such as fasteners and zips.

Whilst this technology is still at early stages in the world of fashion and textiles, it will no doubt enable designers to rethink fashion futures.

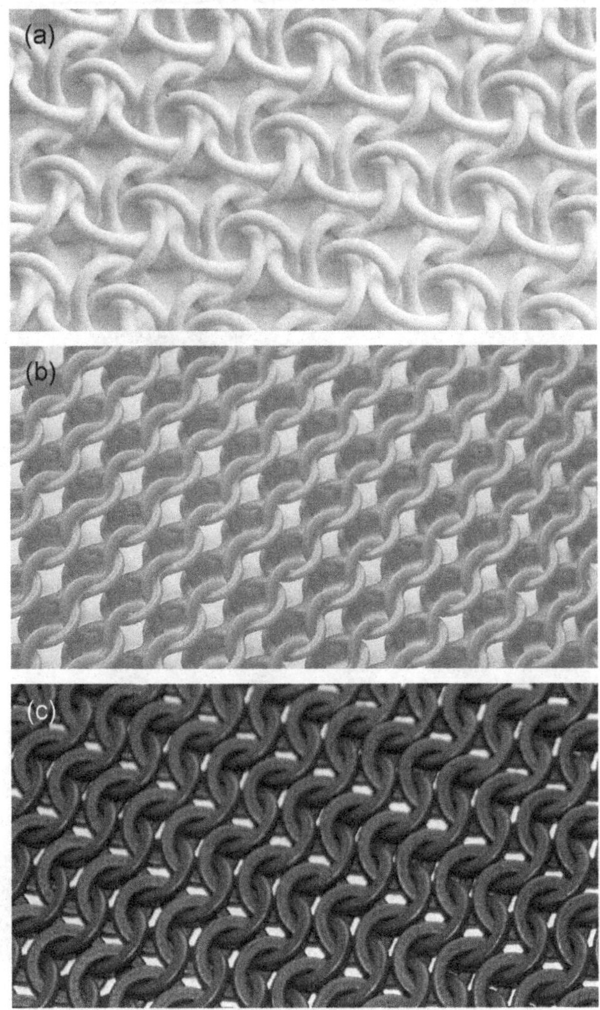

FIGURE 27.13

FOC textile mobius (a) FOC textile 8 (b) FOC textile four in 1 (c).

27.6 CASE STUDIES: FASHION DESIGNERS INTERVIEWED BY SANDRA BURKE

27.6.1 LAURA KRUSEMARK

Laura Krusemark is co-founder and designer for International Citizen Design House, LLC (www.ictzn.com).

1. As a fashion designer, how would you describe your progression from hand drawing to computer? Or have you always worked with computers to create your designs?

FIGURE 27.14

FOC: 3D printed shoe for Brazilian designer Andreia Chaves, as part of an exhibition entitled 'The Invisible Shoe'.

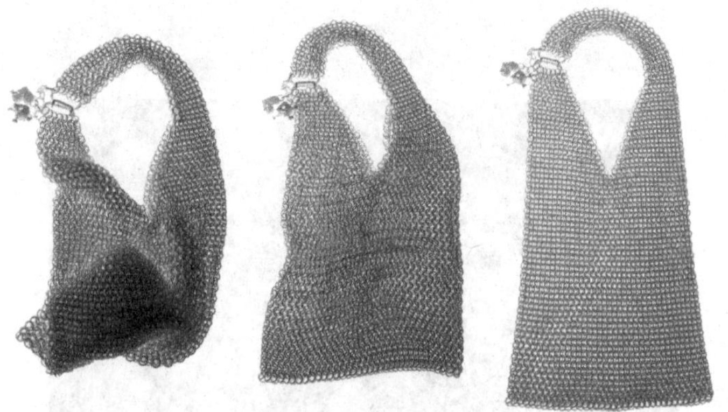

FIGURE 27.15

FOC V bag, based on the concept by Jiri Evenhuis of Rapid Manufactured (RM) Textiles 1999. The pattern used for the design is inspired by armours, knights used to wear in medieval times.

I have always loved hand rendering for my fashion designs, but in the industry it's very useful and efficient to have computer-generated illustrations for quick and easy changes. When I studied fashion design at the Illinois Institute of Art, Chicago, I learned Adobe Illustrator and Photoshop, which helped me to bring my hand-rendered style into a more technical way of illustration.

2. What design software do you use and why?

I mainly use Adobe Photoshop and Illustrator for rendering my designs as these programs are quite user friendly and universal. Illustrator is used in my design company for the technical illustrations, which accompany specification sheets given to manufactures, contractors, pattern makers, etc.

Photoshop is used more for creating a visual prototype, including fabric type, colour, and details. This program helps me as a designer to visualise what the garment will look like before it's been created as a sample.

3. Do you use, or have ever used Gerber and Lectra design suites or specific fashion design software to create your designs?

I have used Lectra designs suites, particularly U4ia and Kaledo, for the creation of prints, weaves, knits, repeats, and colourways. They are very useful programs for creating samples of weave and knit types to show to buyers, manufacturers, sales representatives, or even to have printed out digitally.

4. How important is it to use CAD/computer software for your work? Do you create all your designs on computer or do you mix hand drawing with digital – use a scanner, digital photography, etc.?

CAD software is extremely important in the process of creation with our designs. I create all of our technical illustrations in Adobe Illustrator and our visuals of the garments in Photoshop. Photoshop is also useful for our marketing materials. My partner and co-founder, Sheena Gao, uses the program for re-touching photos, cropping images from the photo-shoots, line plan sheets, look books, creation of the Website, and other promotional materials (see Figures 27.16–27.18).

5. How do you approach designing a new collection and how does the use of CAD fit into your design process?

I typically start out hand rendering the idea or design and then create a technical sketch in Illustrator. I then bring my Illustrator file into Photoshop and design the details and trims placement as well as colour ideas. Many times I go back into Illustrator and refine the technical illustrations and add measurements for manufacturers as well. I scan images to use as reference on the Photoshop prototypes as well as resourcing from the Internet.

6. What other software and digital technology do you use for your work excel, etc., for marketing, Facebook, Twitter, YouTube, etc.?

Software is used constantly in our designs, marketing, and other parts of the business. We use Microsoft Excel spreadsheet software to keep track of expenses, mission statements, vision, references for buyers, manufacturers, important contacts, and so on. We also use InDesign for creating look books and other advertising materials such as promotional and business cards. Sheena handles the social media sites such as Facebook, Myspace, Twitter, YouTube, and Polyvore to promote brand awareness and help market our designs and company image to the public.

27.6.2 ALISSA STYTSENKO

I grew up in a family of artists and was encouraged to explore my creativity from a very young age. I studied fashion and received a Bachelor of Art and Design (Fashion Design) and then gained excellent experience working in the fashion industry. Throughout my career, I leaned towards Fashion Graphics – illustration, branding, textiles. In 2006, in collaboration with my husband Mark, I started a design studio which is entirely dedicated to Fashion – called VDNA Fashion Design 'N' Art Studio (www.vdna.com.au). VDNA studio is 'tailored' to provide complete graphic services to fashion brands – from logo design and brand ID material to apparel range design, storyboarding/illustration, textile print design, and production specifications.

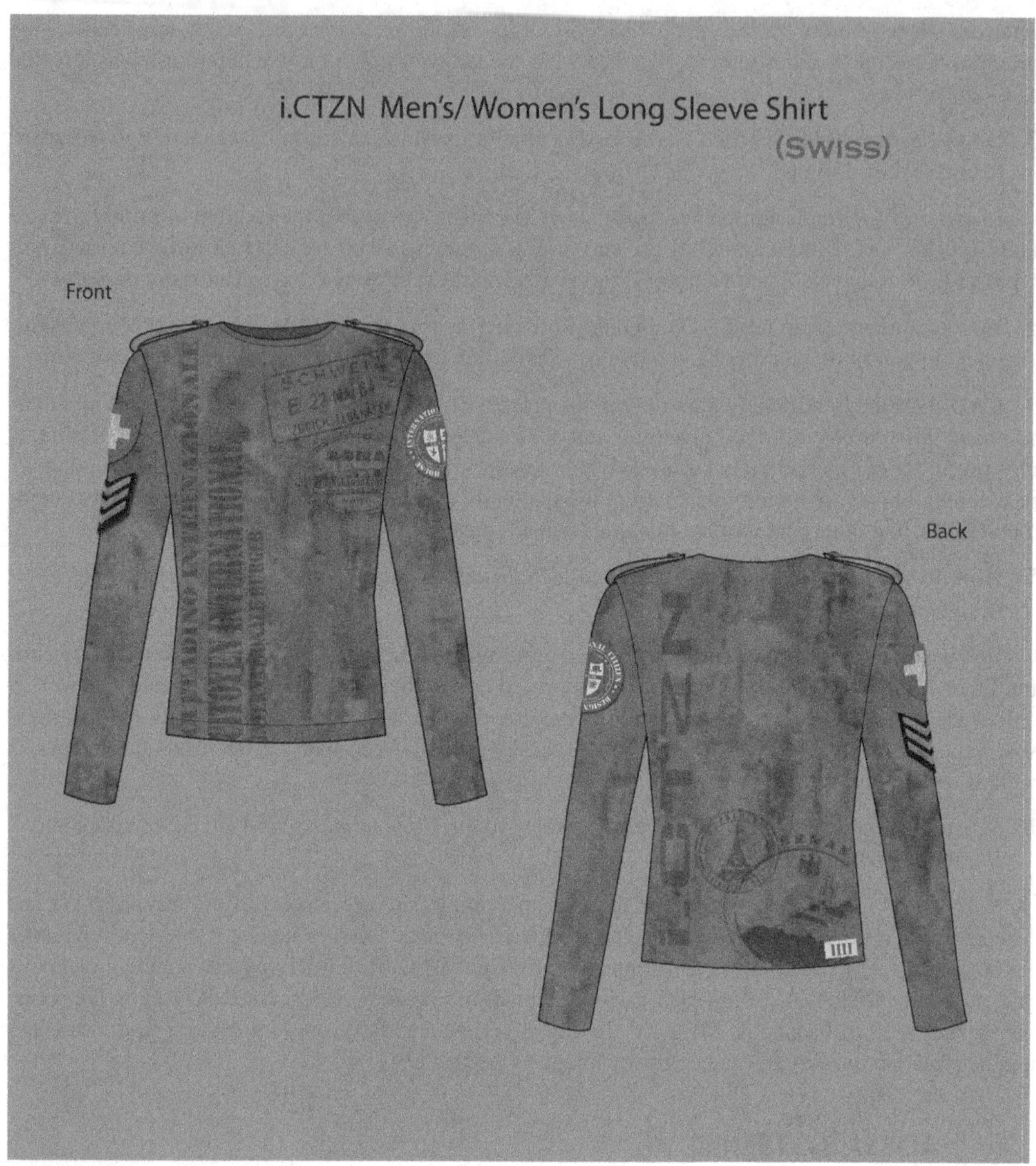

FIGURE 27.16

i.CTZN men's/women's long sleeved shirt: created using Illustrator and Photoshop. Designer Laura Krusemark.

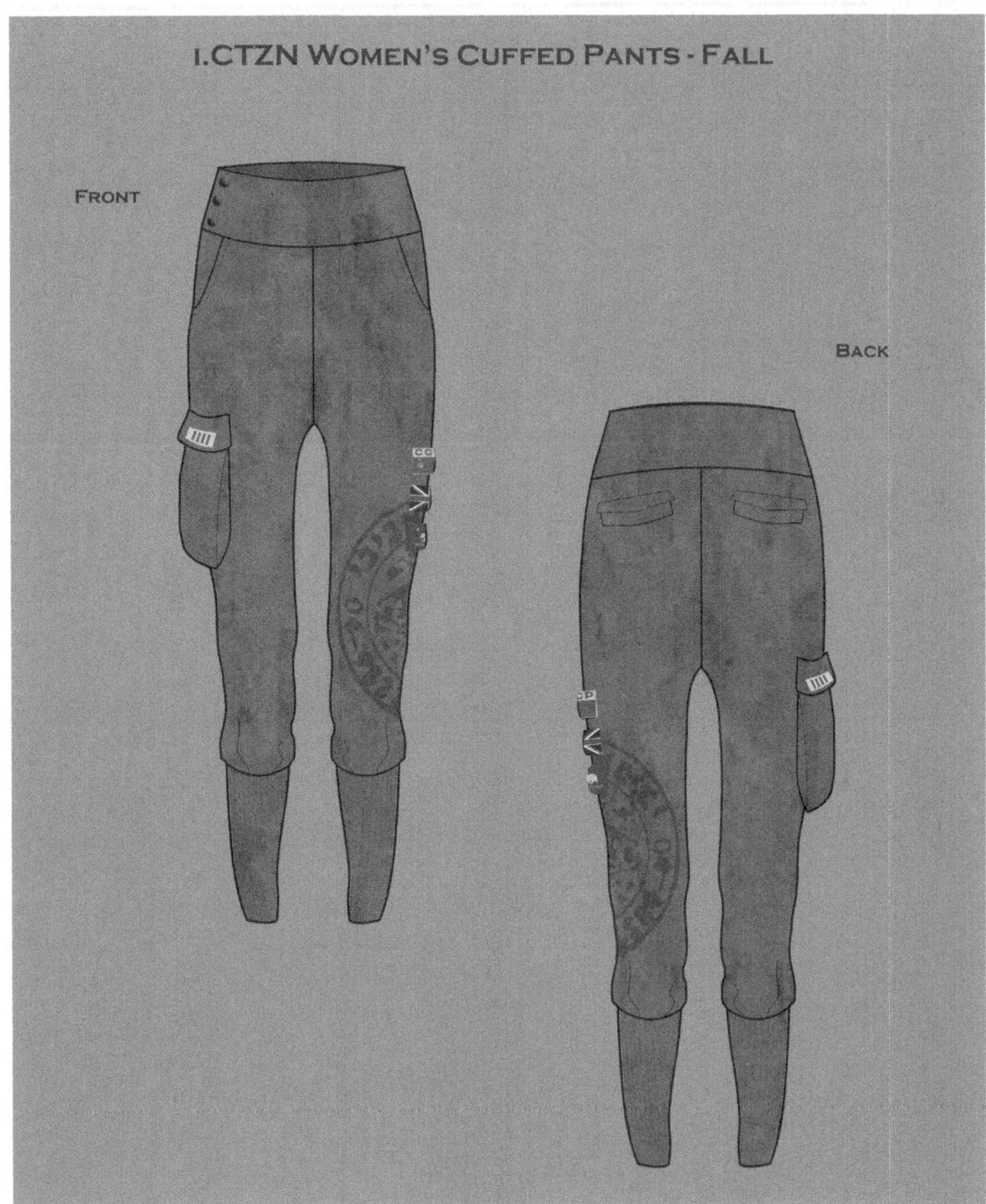

FIGURE 27.17

i.CTZN women's cuffed pant: created using Illustrator and Photoshop. Designer Laura Krusemark.

© Fashion Computing – Sandra Burke

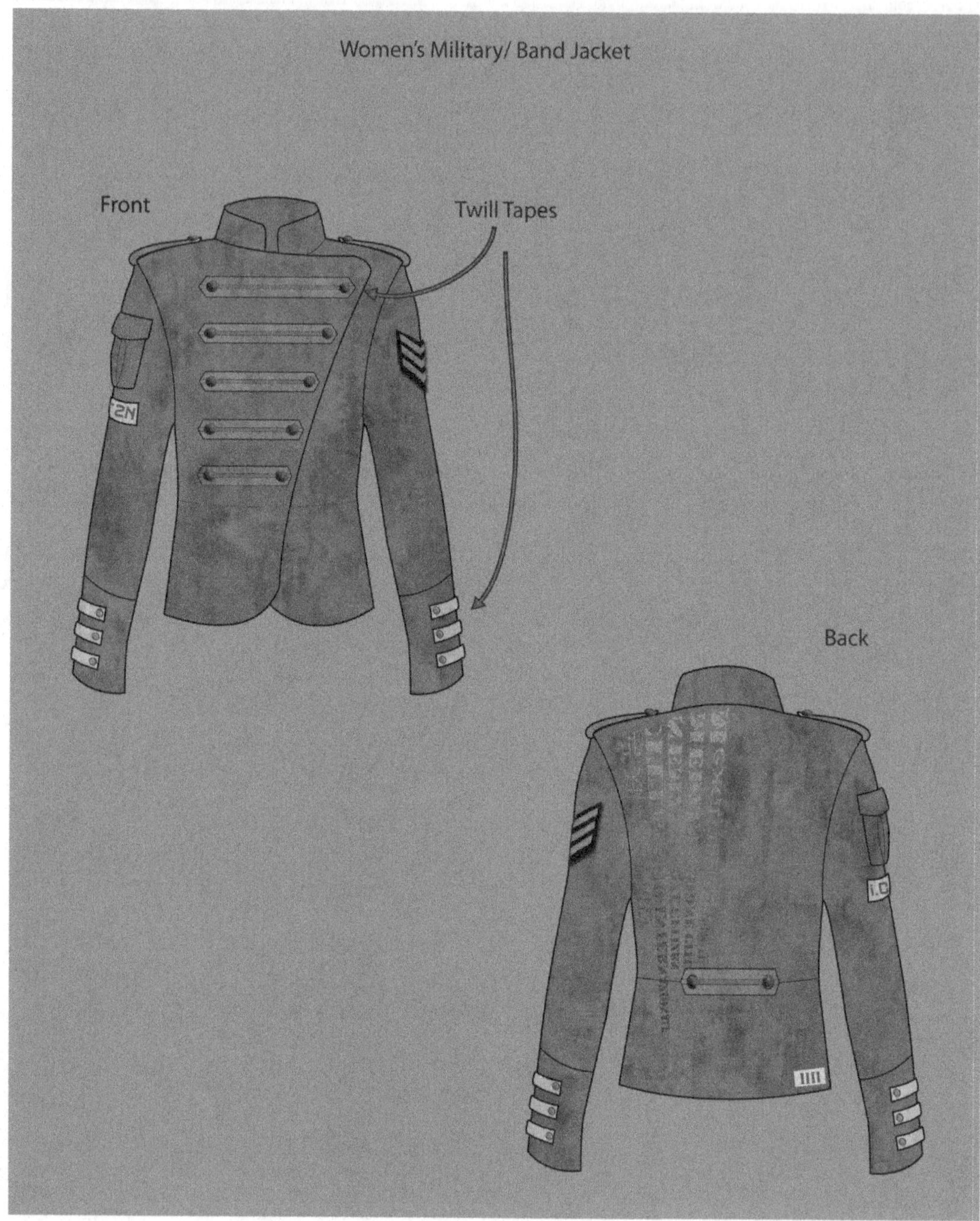

FIGURE 27.18

i.CTZN women's military jacket: created using Illustrator and Photoshop. Designer Laura Krusemark.

© Fashion Computing – Sandra Burke

1. As a textile designer, how would you describe your progression from hand drawing to computer? Or have you always worked with computers to create your designs?

When I started designing I did all my fashion drawings and everything by hand. This demanded strong concentration as each mistake meant starting the work all over again. I started textile design when CorelDraw and Photoshop were already widely used and popular. Later, I started to use Illustrator for all my graphics and never looked back. It is extremely complex and clever software and is suitable for pretty much any kind of graphics, but I still use Photoshop for freehand drawing and digital painting.

I used to scan my hand-drawn sketches to use them as a guide in Illustrator. I still do sketch from time to time especially if I am researching and sketching outdoors or during 'field-trips'. But mostly I use Photoshop and my trusty Wacom tablet for all freehand sketching and planning. I can't imagine computer-free textile design! Repeats and pre-production preparation are so less time consuming using the current software.

2. What design software do you use and why?

As mentioned, I mostly use Adobe Illustrator and Photoshop for all my graphics. Sometimes I turn to InDesign (part of Adobe creative suite) for storyboarding and the compilation of advertising material (brochures, booklets, catalogues, etc.). For website compilation, I prefer Dreamweaver and I also have a basic knowledge of Flash, which I use for animation.

3. Do you use or have you ever used Gerber and Lectra design suites or specific fashion design software to create your designs?

I'm familiar with the Gerber design suite as we learnt to use it during my fashion studies. While working in the fashion industry I have also used Lectra Modaris for pattern-making. I was impressed with both software packages; they are easy to use and include lots of useful technical tools for fashion businesses. However, my work is more on the creative side in apparel design and I am not so much involved in the pattern-making, production, costing and preparation side of the industry, which is why I am more of a specialist in software that has a lot more complex tools for creative designing – like Illustrator and Photoshop.

4. How important is it to use CAD/computer software for your work?

Extremely important! I work on a computer most of my working hours. Computer software saves time and allows precision. Mistakes can be easily fixed (without repeating the same work again and again); work can be edited and re-arranged numerous times; production sketches can be specified and drawn up to present even the smallest stitch detail, not to mention the ease of setting up a yardage repeat in Illustrator or similar programs (see Figures 27.19 and 27.20).

But every designer knows that the computer is merely a tool, and a tool is only useful in the hands of an expert. Hand-drawing techniques and a thorough understanding of artistic rules like composition, proportion, and colour science is the basis for every good designer.

5. Do you create all your designs on computer or do you mix hand drawing with digital – use a scanner, digital photography, etc.?

I definitely mix hand-drawing and computer drawing. I try to sketch from nature by hand from time to time as practice. Hand-sketching is also useful for capturing visual information for research. Inspiration comes from different sources, anytime and in any form, so it's always handy to be able to make a quick sketch of an idea that can be scanned for future use.

FIGURE 27.19

VDNA print design: digitally created using Illustrator and Photoshop. Designer Alissa Stytsenko.

I'm also learning digital painting at the moment – the technology is developing fast and there are many wonderful tools available for artists. Digital painting is a mystical experience – it allows fantastic freedom of expression! Digital photography is also useful for research and collection of information. I rarely apply photography to artistic work but it is used a lot for advertising and web-designing.

6. How do you approach designing a new collection and what is your design process using CAD?

It all starts with a brief. The brief needs to be precise and cover as much information about the project as possible, for example, inspiration, styling, mood, colours, elements to be used and elements to avoid, scaling, quantities, production techniques and peculiarities, deadlines, etc. The brief has to be clear and written down.

The next step is to collect the information, do the research, make the initial sketches, and do the planning. The elements of the design are then drawn in Illustrator if the artwork is graphical in styling, or in Photoshop if the artwork is more free-flowing or natural. Then the design elements are arranged into a composition (usually in Illustrator). It sometimes takes several variations to get the design right.

When the design is done and compiled I usually try several colour options to figure out the best one for the theme. This is where computer software is extremely useful. Once the design is complete and

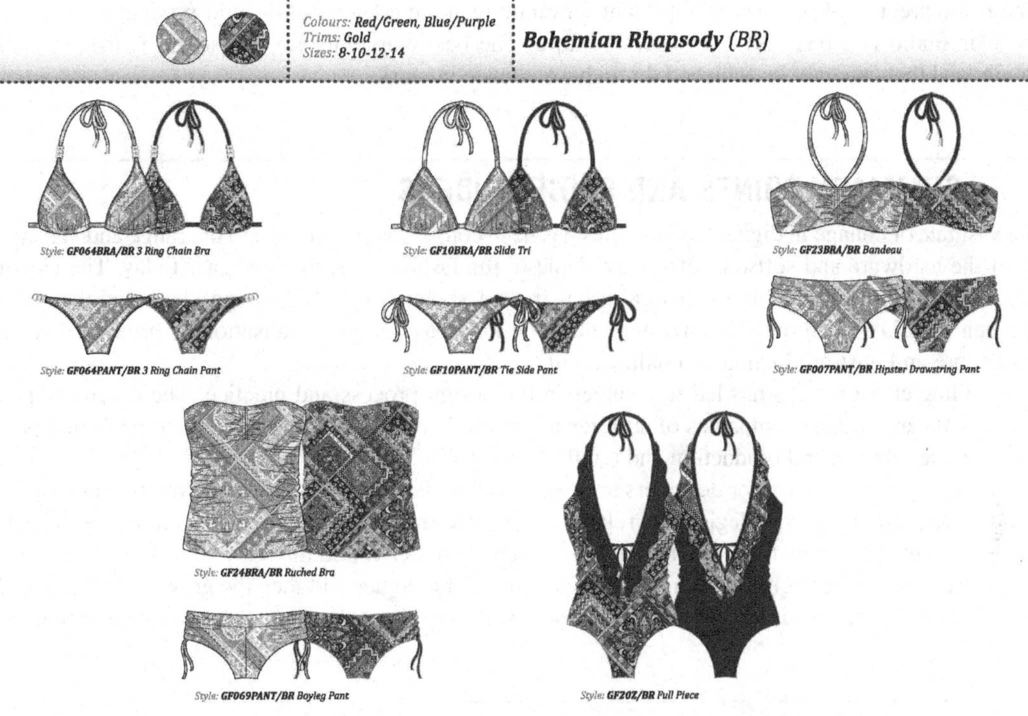

FIGURE 27.20

VDNA storyboard/styling board: digitally created using Illustrator and Photoshop. Designer Alissa Stytsenko.

© Fashion Computing – Sandra Burke

has been approved by the customer I prepare it for production. For textile design this means checking scaling, setting up a repeat, flattening the colours and effects, separating colours for screens or optimising for digital-printing, and saving files in various formats for manufacture. Sometimes artwork might need a few alterations after the first strike-off by the manufacturer.

7. What other software and digital technology do you use for your work – Excel, etc. for marketing, Facebook, Twitter, YouTube, etc.?

I can't imagine working without my Wacom tablet. It takes time to get used to a tablet, but it's definitely worth it. It's much more precise then a mouse and easier to manipulate when free-handing. I use Microsoft applications often (excel, word, outlook, etc.) and most of my communications are through emails. I also find my iPhone useful for snap photographing and noting. I have an app in my iPhone for time sheets on the go and budgeting. Within the company we sync information and emails through Google Apps. This was set up for us by our technology company who looks after all our technology and hardware. We have also installed VOIP phones (Internet phones) for all our designers; this has proved to be very effective for international telecommunications and extremely cost-effective.

I try to use Facebook for sharing news and information about the studio, but it is not such a strong marketing tool for our type of business. It is also time-consuming, and in the fashion design industry time is too precious. Facebook is important for retailing and marketing of the end-product to consumers. Our studio is a business-to-business venture. The best marketing for our studio is the word of mouth, and that can only be achieved through excellence in work.

27.7 SUMMARY POINTS AND PROJECT IDEAS

The vast rate of change in digital technologies has been coupled with a change in the range and versatility of the hardware and software that is available to the fashion or textile designer today. The use of digital technologies has led to fresh approaches from designers to present design ideas, and the move between the 2D virtual and 3D environments. The move to online visualisation of product both in design, pre-and post-production and online retail.

e-Tailing environments has led to changes in the design process and practice. The range of open source software enables companies of all sizes to engage in developing and implementing digital processes in their design and production and retail arenas.

There is a growing need for designers to be involved in the whole design and development process of a product, and the growing legislation related to sustainable issues in textiles and fashion is also leading the way in which new technologies can be utilised from ideation to finish.

The advent of both digital printing and 3D printing technologies and then the growth of App-based technologies are leading designers to look at new ways of making and engaging in new design futures.

27.8 REVISION QUESTIONS

1. How has the use of CAD changed the way in which designers work since the 1990s?
2. How have new digital technologies impacted on the workflow process from design ideation to manufacture?
3. How are individual designers changing their workflow systems because of the use of CAD?
4. Choose a fashion or textiles company that has an online presence. Examine this Website and how it utilises technologies to enhance the customer experience.
5. New 3D printing technologies are being acknowledged as creating the third Industrial Revolution. Using this statement as a starting point, explore what will be the new textiles and fashion futures and how you see this being combined with current technologies and utilised by designers of the future.

27.9 SOURCES OF FURTHER INFORMATION AND ADVICE

27.9.1 BOOKS

Adobe illustrator for fashion design + Myfashionkit. (2012) Pearson, Prentice Hall.

Aldrich, A. (2008). Metric pattern cutting for women's wear. 5th ed. UK: Blackwell Publishing.

Armstrong, J., Lorrie, I., & Armstrong, W. (2006). From pencil to pen tool. New York: Fairchild Pubs.

Barnatt, C. (2013). 3D printing-the next industrial revolution. Explaining the Future.com

Bowles, M., & Isaac, C. (2009). Digital textile design. London: Laurence King.
Braddock, C., O'Mahoney, S., & Harris, J. (2012). Digital visions for fashion. London: Thames & Hudson.
Briggs-Goode, A., & Townsend, K. (2011). Textile design, vol. 112. Oxford UK: Woodhead Pub. [in association with] The Textile Institute.
Burke, S. (2006). Fashion computing, vol. 3. London: Burke Publishing.
Burke, S. (2011). Fashion designer – Concept to collection, Burke Publishing.
Centner, M., & Vereker, F. (2011). Fashion designer's handbook for Adobe illustrator, 2nd ed. USA: Wiley Chichester.
Chipkin, F. L. (2007). Adobe photoshop for textile design. Kew Gardens, NY: Origin Inc.
Colussy, M. K., & Greenberg, S. (2004). Rendering fashion, fabric, and prints with adobe photoshop. Upper Saddle River, NJ: Pearson Prentice Hall.
Drudi, E., & Paci, T. (2011). Figure drawing for men's fashion. Amsterdam: Pepin Press.
Gardiner, M., & Hewat, K. (2008). Everything origami. Heatherton, Victoria, Australia: Hinkler Books.
Gaimster, J. (2011). Visual research methods in fashion. Oxford, UK: Berg Publishers.
Fernandez, A. (2009). Fashion print design, from idea to final print. London: A&C Black.
France, A. K. (2013). Make: 3D printing, the essential guide to 3D printers. CA: Make Media.
Guerrero, J. A. (2010). New fashion and design technologies. English language ed. London: A&C Black.
House, D., & Breen, D. E. (2000). Cloth modeling and animation. Mass: A K Peters Natick.
Kight, K. (2011). A field guide to fabric design. CA: C&T Pub Lafayette.
Lazear, S. (2010). Adobe photoshop for fashion design. Upper Saddle River, NJ.: Prentice Hall.
Lee, J., & Steen, C. (2010). Technical sourcebook for designers. New York: Fairchild Books.
Lipson, H., & Kurman, M. (2013). Fabricated. Indiana, USA: Wiley Press.
Magnenat-Thalmann, N. (2010). Modeling and simulating bodies and garments. London: Springer.
McCullough, M. (1996). Abstracting craft. Mass: MIT Press Cambridge.
Myers-McDevitt, P. J. (2004). Complete guide to size specification and technical design. New York: Fairchild Publications.
Myers-McDevitt, P. J. (2011). Apparel production management and the technical package. New York: Fairchild Books.
Riegelman, N. (2003). 9 Heads, 2nd ed. Los Angeles, CA: Nine Heads Media.
Schlein, A., & Ziek, B. (2006). The woven pixel. Greenville, SC: Bridgewater Press.
Schneider, R. Adobe for fashion. UK: Lulu.Com.
Shillito, A. M. (2013). Digital crafts, UK: Bloomsbury Press.
Szkutnicka, B. (2012). Flats. London, UK: Laurence King.
Tallon, K. (2013). Creative fashion design with illustrator. UK: Anova Books.
Tallon, K. (2011). Digital fashion illustration with photoshop and illustrator. London: Batsford.

27.9.2 WEBSITES

Fashion/textiles specific software

http://www.browzwear.com/
http://www.gerbertechnology.com/

http://www.lectra.com/en/index.html
http://www.marvelousdesigner.com/
http://www.speedstep.de/
http://www.vetigraph.com/
http://www.optitex.com/

Generic open design source software

http://www.blender.org/
http://www.gimp.org/
http://www.inkscape.org/en/

Digital printing

http://printpattern.blogspot.co.uk/
http://www.spoonflower.com/welcome

Off the shelf software

http://poser.smithmicro.com/
http://www.autodesk.co.uk/products/autodesk-maya/overview

3D printing

http://www.materialise.com/
http://www.irisvanherpen.com/
Project DNA – catherine wales: http://www.dezeen.com/2013/06/27/project-dna-3d-printed-accessories-by-catherine-wales/
Continnum fashion: http://www.continuumfashion.com/
Freedom of creation: http://www.freedomofcreation.com/

27.9.3 WEB RESOURCES

Crafts Council
  Crafts and the Digital world (11 May 2011): http://www.craftscouncil.org.uk/about-us/press-room/view/2011/craft-the-digital-world/
  Crafts Council – Labcraft Project (2011): http://www.labcraft.org.uk/
Design Museum 2013 The future is here: http://designmuseum.org/exhibitions/2013/the-future-is-here
Not Just a Label: http://www.notjustalabel.com/editorial/open_source_now

27.9.4 OPEN SOURCE SOFTWARE

Inkscape: http://www.inkscape.org/en/
Gimp: http://www.gimp.org/
Sophiesew: http://sophiesew.com/SS2/index.php
Burda style: http://www.burdastyle.com/
Burda style (UK): http://www.burdastyle.co.uk/

REFERENCES AND FURTHER READING

Jones-Jenkyn, S. C., Briggs-Goode, A., & Townsend, K. (2011). *Textile design*. (vol. 112). Oxford: Woodhead Pub. [in association with] The Textile Institute, 232–262.

Future lab. http://www.futurelab.org.uk/resources.

Educase. http://www.educause.edu/eli http://www.educause.edu/eli/publications?filters=sm_cck_field_super_facet%3A%22EDUCAUSE%20Library%20Items%22%20tid%3A33152%20tid%3A33438.

Horizon report – K12 edition. http://www.nmc.org/publications/2012-horizon-report-k12.

Horizon report – Higher education edition. (2013). http://www.nmc.org/publications/2013-horizon-report-higher-ed.

CHAPTER

ADDING FUNCTIONALITY TO GARMENTS

28

L. Hunter[1], J. Fan[2]

[1]*CSIR and Nelson Mandela Metropolitan University, Port Elizabeth, South Africa;* [2]*Cornell University, New York, NY, USA*

LEARNING OBJECTIVES

At the end of this chapter, you should be able to:

- List the six ways in which the functionality of fabrics and garments may be compromised
- List the reasons behind each of the six functionality problems
- Describe how functionality problems can be minimized

28.1 INTRODUCTION

Consumers increasingly expect their garments to have functional properties that match the specific end-use conditions under which the garments are to be worn, and further that these properties will be engineered into the fabrics and garments themselves. Therefore, everyday garments are no longer expected to merely satisfy the basic and traditional requirements of everyday wear such as basic comfort, adornment, modesty protection and fashion, but also to satisfy other performance requirements and functionalities including antiwrinkling, breathability, waterproofing, physiological and psychological comfort, UV protection, etc. Different types and constructions of fibres, fabrics and garments require different approaches and treatments to achieve the desired level of functionality.

However, in many cases it is necessary to compromise and optimise when engineering such garment performance. There are frequently trade-offs to be made because improving one particular property or functionality (e.g. soft handle) may come at the expense of another property (e.g. pilling and abrasion resistance). Furthermore, cost constraints, and sometimes health and environmental aspects, can limit the functionality and performance that can be engineered into a fabric or garment.

Clothing manufacturers require their fabrics to have good tailorability, i.e. to be easy to make up, to pass through the garment manufacturing process easily and without undue problems, and for the finished garment to have a good appearance and the required wear performance. The fabric manufacturer has to meet the technical and other expectations of the garment manufacturer within an acceptable cost budget and time frame, while the garment manufacturer needs to meet similar requirements from the retailer, such as design and fashion, that ultimately reflect the desires of consumers.

Garment functionality can be added or improved at the fibre, yarn, fabric or garment stage, or by using a combination of those stages, but it most frequently takes place at the fibre or fabric stage. However, this depends on the specific functionality required and what other performance and aesthetic requirements the

Textiles and Fashion. http://dx.doi.org/10.1016/B978-1-84569-931-4.00028-3

garment has to satisfy. You are referred to the source used for this chapter, Fan & Hunter, 2009, for a detailed and in-depth treatment of the topics covered, and a full list of references. Functionalities not covered here include UV resistance, antimicrobial protection, odour resistance.

28.2 FACTORS AFFECTING GARMENT FUNCTION

Most garments are highly complex products, resulting from intricate fibre, yarn, fabric and garment structures combined with mechanical and chemical finishing procedures to provide a garment that meets all the requirements of the wearer and wear conditions, whether aesthetic, design, fit or performance. The various factors interact in an intricate manner to ultimately give the garment its specific characteristics.

If high strength, abrasion resistance and durability are required, with comfort of secondary importance, a high-performance synthetic fibre such as polyester would be selected, together with an appropriate fabric structure – fairly tightly woven or possibly knitted – and weight. On the other hand, if softness and comfort are most important, the fibre choice would most likely be a soft and fine natural fibre such as cotton (particularly for against-the-skin comfort), linen or wool, or possibly even cashmere if garment price is not a major consideration. Similarly, if a close-fitting (body-hugging) garment allowing for sufficient body movement is required, a knitted fabric structure would be the most obvious choice, possibly incorporating a stretch yarn, with a woven fabric containing a stretch yarn also a possibility. Extreme heat and sun, such as the conditions prevailing in a desert, would require a loose-fitting garment, covering virtually the entire body but with minimal contact between the garment and body, and constructed with vents and openings that produce a billowing action during walking and other physical activities in order to allow air movement and circulation.

The above are but a few examples of how fibre, fabric and garment characteristics can be selected and designed to produce a garment that will meet the properties and performance required for a specific end use.

28.3 IMPROVING FABRIC HANDLE AND TAILORABILITY

This Section is from Hunter & Hunter (2009). Fabric handle and making-up performance (tailorability) are interrelated, and they represent key quality parameters for clothing manufacturers and consumers. Handle (or hand) is defined as 'the subjective assessment of a textile material obtained from the sense of touch' (*Textile Terms and Definitions*, 11th edition, The Textile Institute, Manchester, 2009). Although good making-up and wear performance is a basic requirement, different types of 'handle' are often required, e.g. crispness rather than softness.

Traditionally, the quality of fabrics and their 'fitness for purpose', including their performance both during making-up and while wearing the garment, were assessed subjectively in terms of fabric handle by highly skilled clothing industry experts. In assessing the fabric, they used sensory characteristics such as surface friction, bending stiffness, compression, thickness and small-scale extension and shear, all of which play a role in determining handle and garment making-up and appearance during wear. They assessed fabrics using their hands to perform certain physical

Table 28.1 Fabric properties that are related to tailoring performance, wear appearance and handle

Property	Test	Tailoring performance	Wear appearance	Handle
Physical	Thickness	–	–	+
	Mass per unit area	+	+	+
Dimensional	Relaxation	+	+	–
	Shrinkage			
	Hygral expansion	+	+	–
Mechanical	Extensibility	+	+	+
	Bending properties	+	+	+
	Shear properties	+	+	+
	Compression properties	–	–	+
Surface	Friction	–	–	+
	Surface irregularity	–	–	+
Optical	Lustre	–	+	–
Thermal	Conductivity	–	–	+
Performance	Pilling	–	+	–
	Wrinkling	–	+	–
	Surface abrasion	–	+	–

+ *Important.*
– *Less important.*
Source: De Boos (1997).

actions, such as rubbing, compressing, bending, shearing and stretching. They expressed what they felt in terms of subjective sensations, e.g. stiffness, limpness, hardness, softness, fullness, smoothness and roughness, which formed the basis for the fabric selection (Potluri, Porat, & Atkinson, 1995). Because of the way the fabric was assessed, i.e. by tactile/touch/feel, and the terminology used, i.e. 'fabric handle or hand', it is sometimes incorrectly assumed that the assessment was purely aimed at arriving at a subjective measure of the fabric's tactile-related properties (i.e. handle). In fact, the fabric handle, when so assessed, provided a composite measure of the overall garment-related quality of the fabric, including garment making-up, comfort, aesthetics, appearance, performance and other functional characteristics as shown in Table 28.1 (De Boos, 1997).

Figure 28.1 (Niwa, 2001) summarises the approach involved in developing an 'ideal' fabric having good handle with good appearance and comfort in the final garment.

28.3.1 FIBRE PROPERTIES

Fabric quality (handle) and tailorability (making-up), and the subsequent appearance and performance of garments, is related to five basic fabric mechanical properties (Hunter & Hunter, 2009): diameter, cross-sectional shape, modulus, length and crimp, with the first two being particularly

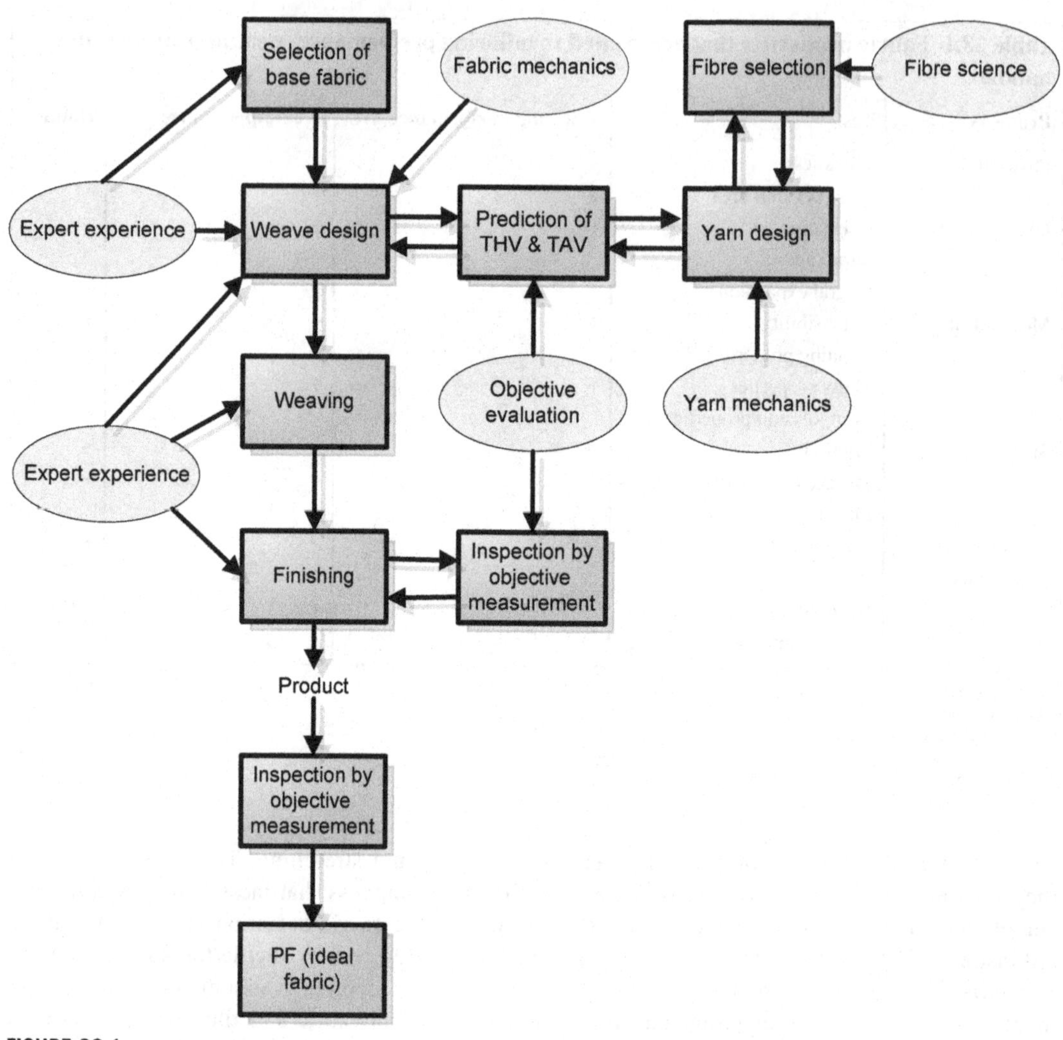

FIGURE 28.1

Summary of the approach involved in developing an 'ideal' fabric, having good fabric handle, good garment appearance and garment comfort.

Source: Niwa (2001).

important. The mechanical properties of the fibre, notably bending and tensile strength, are related to the corresponding mechanical properties of the fabric and garment, but the relationship is affected by the structural parameters of the yarn and fabric, notably those that affect the freedom of movement and alignment of the fibres within the yarns, and the freedom of movement of the yarn within the fabric. Fibre bending stiffness (flexural rigidity) has a major effect on fabric and garment handle, it being a function of fibre bending modulus, diameter (to the fourth power) and cross-sectional shape, as illustrated in the following equation (Morton & Hearle, 2008):

$$\text{Flexural rigidity} = \frac{\eta E T^2}{4\pi\rho} \tag{28.1}$$

Morton and Hearle (2008) provide tables of values (see their Tables 17.1 and 17.2) for η and E for different fibres.

Where:

E = specific modulus in N/kg m
T = fibre linear density in kg/m
η = fibre shape factor (=1 for a circular fibre)
ρ = fibre density in kg/m^3

In practical units, the equation becomes:

$$\text{Flexural rigidity} = \frac{\eta E T^2}{4\pi\rho} \times 10^{-3}\ \text{N mm}^2 \tag{28.2}$$

where

E is in N/tex, T is in tex and ρ is in g/cm^3.

For a fibre of unit tex (i.e. eliminating fibre fineness), this equation becomes:

$$R = \text{Specific fibre flexural rigidity} = \frac{\eta E}{4\pi\rho} \times 10^{-3}\ \text{N mm}^2/\text{tex}^2 \tag{28.3}$$

where

R can also be taken as the 'intrinsic stiffness', which is independent of fineness (i.e. per unit of fineness or tex).

The shape factor η, and consequently rigidity, increases as the distance of the material from the fibre centre increases, it being greatest for a hollow cylinder. For the same type of fibre, T is a function of diameter squared, and therefore flexural rigidity is a function of diameter to the fourth power. Clearly, fibre diameter has by far the major effect on fibre stiffness, as well as fabric stiffness, softness and handle. The greater the fibre diameter (i.e. the coarser the fibre), the stiffer, harsher and crisper (harder) the handle becomes, and the rougher (harsher) the fabric surface. A good example is the softness associated with cashmere garments, which is almost entirely due to its low fibre diameter (≈15 μm). Increasing fibre diameter also increases the sensation of scratchiness and prickliness; wool fibre diameters (i.e. coarseness) of more than 30 μm are mainly responsible for the sensation of prickliness and scratchiness. Microfibre fabrics will tend to be very soft and smooth, even limp. Noncircular fibre cross-sections will tend to produce less stiff (i.e. softer and more flexible) fabrics than circular fibres of the same linear density.

Fibre bending modulus is mainly dependent upon the type of fibre and to a lesser extent on chemical (softening) treatments applied to the fibre and fabric. Chemical treatments can change Young's modulus, and through that the handle and making-up properties. The fibre torsional modulus is also related to the diameter to the fourth power. In essence, fibre-diameter changes will have the greatest effect on fabric stiffness, and therefore on all handle and related making-up properties, followed in importance by the fibre's Young's modulus (E) and shape (η).

Longer fibres will produce less hairy yarns and fabrics, making them sleek and smooth, while increasing the fibre crimp will improve fabric bulk, fullness and compressible properties.

28.3.2 YARN PROPERTIES

The effect of yarn properties on fabric handle and making-up is mainly a function of the constituent fibre properties, with yarn structure and twist playing some role. The effect of yarn structural parameters on fabric properties, such as handle, depends upon fabric structure, particularly in the way it affects yarn mobility. The yarn tightness, which is largely a function of twist factor, affects fabric handle and making-up, as it affects the fibres' freedom of movement and compresses them within the yarn structure.

In addition to yarn twist factor, yarn structure, due to different spinning systems, also affects fibre configuration and freedom of movement within the yarn. For example, rotor-spun (open-end) yarns, having a high number of wrapping fibres, tend to be stiffer and rougher than ring-spun yarns, thereby producing stiffer (i.e. harsher) fabrics and garments. Higher yarn twist factors reduce fibre mobility and yarn diameter, surface roughness, friction and compressibility, but increase yarn stiffness, harshness, interfibre frictional forces (cohesion) and smoothness. At a constant twist factor, finer yarns tend to produce smoother and softer fabrics.

28.3.3 FABRIC PROPERTIES

The handle of a fabric is a function of bending, shear stiffness, bulk compressibility and surface frictional properties, while its scratchiness and prickliness are mainly a function of the surface fibre end density and buckling force, as already discussed. Fabric structure has a major effect on fabric handle and making-up properties, inasmuch as it affects yarn mobility within the structure and the compression properties of the fabric. More tightly constructed fabrics, containing a greater number of yarn interlacings per unit area, generally have a stiffer handle and poorer making-up performance than looser fabrics with fewer interlacings. For example, a twill-weave fabric is preferable to a plain weave fabric, and a coarser yarn to a finer one, other factors such as fabric weight per unit area being constant. Similar considerations apply to knitted fabrics.

28.3.4 DYEING AND FINISHING

Tomasino (2005) reviewed the effects of dyeing and finishing on fabric handle and making-up properties. Fabric dyeing and finishing, particularly the latter, have a major effect on fabric handle and making-up performance, generally reducing the internal stresses and energy in the fabric. This results in large reductions in fabric bending, shear rigidity and hysteresis, with associated improvements in fabric handle and making-up. Generally, dyeing and finishing affect the stiffness of tightly constructed fabrics more than they do loosely constructed structures. Table 28.2 (De Boos & Tester, 1994) illustrates the effect of finishing operations on wool fabric. Any fabric or garment dyeing and finishing procedure that reduces fibre-to-fibre and yarn-to yarn friction, bends and flexes (mechanically manipulates) the fabric, induces relaxation, and increases fabric surface hairiness (i.e. brushed effects), will improve fabric handle, notably softness. At the same time, any chemical finish that increases interfibre friction and/or bonding (spot welding) will increase fabric stiffness. Handle is improved by finishes that reduce the energy needed to deform the fabric, while increasing fabric smoothness.

Table 28.2 Effect of finishing operations on the properties of wool fabric

	Fabric property					
Operation	**Relaxation shrinkage**	**Hygral expansion**	**Extension**	**Bending**	**Shear**	**Compression**
Wet setting	X	X		X	X	X
Scouring	M	M	M	M	M	M
Milling	X	X	M-X	X	X	X
Dyeing	X	X	X	X	X	M-X
Drying	X		X			
Cropping	M		M			M
Singeing						M
Damping						M
Relaxing	X		X			X
Pressing	M-X		M-X	M	M	X
Decatising	X	X	X	X	X	X
Sponging	X		X			M

X, large effect; M, small but significant effect; M-X, effect is normally small, but can be large under the appropriate conditions.
Source: De Boos and Tester (1994).

For conventional fabric (piece) dyeing, the major impact on fabric handle comes from the way the fabric is handled during dyeing (i.e. dyeing conditions), rather than from the type and depth of dye and dyeing auxiliaries used (Tomasino, 2005). This is attributed to relaxation of internal stresses and increasing the ease with which the yarns can move relative to each other within the fabric structure. Thus, Beck dyeing of polyester gives a softer handle than Thermosol dyeing, because of differences in the dyeing process and conditions.

Hand modifiers, such as softeners and hand builders incorporated in the final finish bath, have a pronounced effect on fabric handle (Tomasino, 2005). Generally, hand builders are divided into two main groups, the one adding fullness (increased bulk) and the other adding stiffness.

Mechanical (dry) finishing generally involves subjecting the fabric or garment to a mechanical or physical action (manipulation) such as surface finishing (brushing, raising, napping, sueding, sanding or shearing/cropping), compaction (e.g. Sanforising), calendaring (schreinerising), setting or pressing. Mechanical finishing nearly always produces a softer fabric handle and improved making-up performance.

28.3.5 MEASUREMENT OF FABRIC HANDLE AND MAKING-UP

As already mentioned, fabric handle and tailorability were originally evaluated by experts who used their hands to manipulate the fabrics. Although the experts were highly skilled and their judgements sensitive and reliable, the end result was still subjective and qualitative by nature and suffered from the weaknesses inherent in all subjective assessments, including dependence upon the skills, training and background (cultural and other) of the evaluator. Hence, there was a need to develop an objective (i.e. instrument based) measurement system for assessing fabric quality, and fabric objective measurement (FOM) is one such integrated system of measurement.

FOM instruments were designed to measure mainly the low deformation forces encountered when the fabric is manipulated by hand and also during the garment making-up process, and they remove much of the guesswork from garment manufacturing. In certain cases, dimensional stability and crease-related parameters are included in FOM. At present, and as used here, FOM refers to the instrumental measurement of those fabric properties that affect the tactile, making-up/tailorability and appearance-related properties of fabrics in garment applications, and generally involves small-scale deformation characteristics (bending, shear, compression and extension) as well as dimensional stability–related characteristics such as hygral expansion and relaxation shrinkage (see Table 28.1). Various research workers, notably Peirce, Shishoo, Kawabata, Niwa, Lindberg and Postle, carried out pioneering work in this respect (Hunter & Hunter, 2009).

Although a number of instruments have been developed over the years, including the Wool Handle-Meter recently developed for knitted fabrics, by the CSIRO, Australia (Wang, Mahar, & Postle, 2013) only two systems, namely Kawabata and FAST, have found wide application and acceptance. The Kawabata Evaluation System for Fabrics (KES-F, later to become KES-FB) laid a solid foundation for the accurate and routine measurement of those fabric properties that determine fabric handle and garment making-up and appearance. Along similar but greatly simplified lines, Council for Scientific and Industrial Research Organisation (CSIRO) of Australia developed the FAST system many years later for measuring the main fabric properties that affect garment making.

The Kawabata and FAST systems measure similar low-stress fabric mechanical properties (compression, bending, extension and shearing), although they differ somewhat in the measurement principles that they use. There are good correlations between similar parameters measured on the two systems, and also as measured by other systems. The results obtained on the two systems are plotted on control charts, sometimes called 'fingerprints', and comparisons between fabrics, as well as diagnosis of tailoring problems, can be made more easily when information is presented in this way.

The following fabric properties are measured by FOM systems:

- Compression, essentially providing a measure of fabric softness or fullness and the stability of the fabric finish.
- Dimensional stability:
  - Relaxation shrinkage
  - Hygral expansion
  - Thermal shrinkage
- Tensile and shear, including hysteresis (energy loss), measured under low deformation forces, providing an indication of deformability and moulding and seam pucker problems.
- Surface friction and roughness related to fabric handle.
- Bending rigidity related to fabric handle and making-up; the latter, for example, is measured in terms of fabric distortion during cutting, and seam pucker during sewing.

28.4 REDUCING WRINKLING

This Section is from Hunter (2009a). Wrinkling may be defined as the unwanted residual deformation, largely random in nature, that occurs during wear and does not disappear spontaneously, resulting in wrinkles or creases that can make the fabric appear unsightly (Denby, 1982).

By their very nature, all textile apparel fabrics bend and fold with remarkable ease during wear and laundering, often to relatively high curvatures. This pliability, so important in wear, comfort, aesthetics,

Table 28.3 General details of wrinkles that occur in wear

Type of wrinkle	Wrinkle forming mechanism	Nature of wrinkles	Location and orientation of wrinkle w.r.t. fabric direction
Compressional wrinkles	Fabric is compressed between body and another surface or between two different parts of the body. Pressure relatively high	Sharp, high frequency	Back of blouse – random trouser seat – bias and across warp skirt seat – trouser crotch – skirt tail – random trouser behind the knee – across warp
Movement wrinkles	Repeated flexing	Rounded, low frequency	Generally found in coat sleeves, skirt laps, trouser knees and trouser and trouser lap (front) Generally horizontal and perpendicular to axis of bending and therefore across warp yarns

Source: Smuts (1989).

and garment construction and shape, often leads to wrinkling during use. The tendency for a fabric or garment to wrinkle or crease when subjected to sharp folds (bends, creases or wrinkles) under pressure (load) during wear or laundering also has a bearing on ease-of-care related properties (durable-press, easy-care, minimum-iron, after-wash appearance, etc.). In effect, it is not so much the ability of the fabric to withstand creasing or wrinkling that is important in practice, but rather its ability to regain its original shape and smooth appearance, i.e. its wrinkle or crease recovery.

Two types of wrinkles occur in garments during wear, viz. pressure (sharp) wrinkles and movement (rounded) wrinkles (Smuts, 1989) (Table 28.3), both of which are normally confined to small areas and specific locations in the garment. Pressure wrinkles are typically produced when the wearer sits and creases the fabric, and are sharp and numerous (Denby, 1982). Movement wrinkles are produced when the wearer bends, and are frequently accompanied by bagging. These wrinkles are formed in those areas where the fabric is compressed between two parts of the body as a result of movement. Sharp wrinkles are probably the most bothersome, and are the ones usually simulated in laboratory tests.

28.4.1 FACTORS AFFECTING WRINKLING AND WRINKLE RECOVERY DURING WEAR

Wrinkling occurs because the deformation forces imposed during wear generally far exceed the resistance of the fabric to bending, and the subsequent recovery forces are generally inadequate to overcome the forces opposing complete recovery (i.e. a return to the undeformed state). The most important factors that influence the wrinkling (including type of wrinkles formed) of fabrics or garments during wear are given below, the importance and magnitude depending upon various other factors, notably the fibre type (Smuts, 1989; Wilkinson & Hoffman, 1959):

- The temperature and moisture content of the fabric in contact with the body
- The pressure that deforms the fabric, and the sharpness of the wrinkle or crease inserted by the applied pressure
- The number of times the fabric is deformed (bent or creased) in the same locations

- The length of time the fabric is deformed, and the recovery time and conditions, wrinkling being worse when subsequent recovery is at a lower RH and temperature
- The length of time the fabric is in a particular atmosphere before wrinkling commences
- The changes in atmospheric conditions (or in the fibre regain and temperature) and associated de-ageing effects, as well as the aged state of the fabric relative to the conditions prevailing during wrinkling. These are important considerations, more particularly for wool and other animal fibres
- The conditions during wear – this pertains to environmental conditions before, during and after creasing. The ageing/de-ageing phenomena can also be included, particularly for fabrics containing animal fibres such as wool and mohair
- Factors relating to the wearer, e.g. the activity, size and shape of the wearer
- Wearer and garment interactions, e.g. garment fit, cut, style and activity of the wearer

Table 28.4 (Leeder, 1976, 1977) lists the various fibre, yarn and fabric parameters that play a role in wrinkling and wrinkle recovery. When fibres in folded or bent fabric are subjected to strain during wear or laundering, particularly in the presence of moisture and heat for prolonged periods, they invariably retain a certain degree of residual deformation after removal of the deforming forces. The residual deformation largely depends upon the fibre's viscoelastic properties and to a lesser extent on the frictional couple or coercive couple (M), the latter being related to interfibre and inter-yarn friction and pressure. In addition, it is greatly dependent upon the degree and duration of the original deformation, as well as the conditions of temperature and moisture prevailing during and after actual deformation. The most severe form of wrinkling generally occurs under conditions of sharp deformation that last for a prolonged time and occur under conditions of high humidity and temperature.

Fibres cannot be regarded as perfectly elastic, but are viscoelastic: during bending or stretching, they undergo stress relaxation; when allowed to recover, they show an instantaneous partial recovery followed by a slow time-dependent recovery, the latter being characteristic of viscoelastic materials.

Table 28.4 Fibre, yarn and fabric parameters that influence wrinkling

Fibre	Yarn	Fabric
Type	Type	
Viscoelastic bending parameters	Ply and twist level	Weave structure and crimp
Viscoelastic torsional parameters	Twist direction	Mass per unit area
Cross-sectional shape	Crimp	Thickness
Crimp	Cross-sectional shape	Surface hairiness
Diameter	Linear density	State of relaxation (i.e. energy state)
Diameter distribution	Diameter	Sett
Friction	Lateral pressures	Compactness
Chemical or physical treatment	Construction (e.g. woollen-spun, worsted-spun, friction-spun, etc.)	Mechanical finishing

Source: Leeder (1976, 1977).

The greater the stress relaxation, the poorer the wrinkle recovery. The wrinkle recovery properties of a garment can be improved in two main ways (Smuts, 1989):

- Micro level: By changing the fibre at molecular level, e.g. by changing fibre type and chemical/amorphous structure, and by chemical treatment (e.g. resin treatment of cellulosic fibres); this can also be achieved by annealing, in the case of fibres such as wool, thereby improving the elasticity and viscoelastic properties of the fibre.
- Macro level: By constructing the yarn, fabric and garment in such a way that either the strains imposed during bending (use) or the interactions between fibres and yarns, or both, are minimised. The yarn and fabric construction largely affects the frictional couple, and to some extent the nature and severity of the fibre deformations when the fabric is deformed during use.

28.4.2 FIBRE PROPERTIES

Certain apparel fibres, such as polyester and wool, generally have good wrinkle recovery properties, whereas others, notably natural cellulosic fibres such as cotton and flax (linen), do not (see Figure 28.2) (Leeder, 1976). Therefore, selecting a fibre with good wrinkle-recovery properties is one of the best means of achieving good wrinkle recovery in a garment. It should be noted, however, that

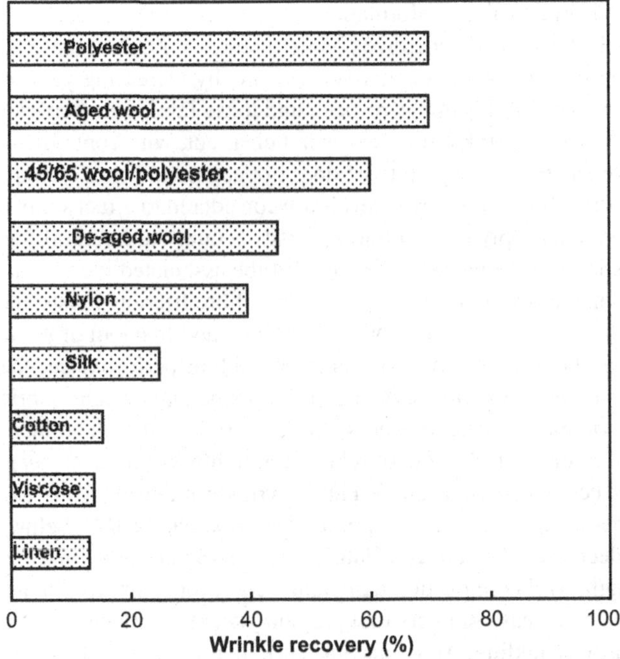

FIGURE 28.2

Relative wrinkle recovery of fabrics made of different fibre types.

Source: Leeder (1976).

conditions prevailing prior to, during and after wrinkling, such as temperature and humidity, particularly in the immediate vicinity of the fabric where it is being deformed (creased) and of the fabric itself, play a decisive role in wrinkling performance, and the relative wrinkling performance of different fibre types.

The viscoelastic properties of the fibres tend to have the greatest effect on the wrinkling behaviour of the fabric, and various fibre types exhibit a wide range of relative performance in terms of their viscoelasticity (Hunter, 2009a). Furthermore, the viscoelastic component combines with the frictional component to produce the coercive couple. For a well-relaxed fabric comprising highly wrinkle-resistant fibres, the frictional couple is small and does not play a role nearly as important in wrinkling as do the fibres' viscoelastic properties. This is not the case for fibres with low resilience (recovery). The glass-transition temperature (Tg) of a fibre plays an important role in wrinkling, since it is at this temperature that the behaviour of an amorphous polymer changes from glasslike to rubberlike, and changes quite drastically when subjected to bending and other deformations. The higher the Tg, under conditions of high humidity or moisture content, the better in terms of wrinkling performance (Hunter, 2009a).

28.4.3 YARN AND FABRIC PARAMETERS

Yarn and fabric structural parameters affect wrinkling inasmuch as they affect (1) the freedom of fibre movement, (2) interfibre frictional forces (frictional coercive couple), both within the yarn and within the fabric and (3) the degree of fibre deformation during wrinkling. These are related to the factors affecting fabric bending stiffness and coercive couple (Hunter, 2009a); therefore, fabric wrinkling can be reduced by reducing the frictional (coercive) couple, i.e. by increasing yarn mobility. Thus, knitted fabrics are superior to woven fabrics in this respect, and loose fabrics are superior to tight fabrics.

The effect of yarn twist on wrinkle recovery is not clear cut, with contradictory results reported in the literature; therefore, changing the yarn twist does not appear to be an effective or practical way of reducing wrinkling, although very high twist levels are considered to effect some improvement in wrinkling. Similar considerations apply to yarn linear density (count), although if fabric mass is kept constant, coarse yarn should improve wrinkling because of the associated increase in fabric thickness and decrease in fabric tightness, as well as inter-yarn and interfibre frictional resistance.

The state of relaxation (i.e. stress relaxation) of the fabric and freedom of yarn movement (mobility) are the two main fabric parameters affecting fabric wrinkling performance, their effects depending upon the recovery properties of the fibres. All other factors being constant, fabric wrinkling improves with an increase in fabric mass, fabric thickness and relaxation, and with an increase in yarn mobility, the latter being related to the fabric's state of relaxation, tightness (cover factor) and yarn float length (number of yarn intersections per unit area). Fabric wrinkle recovery tends to increase as sett, and therefore fabric tightness, decreases, up to a point of approximately 10% below maximum sett, after which it levels off or decreases. The effect of fabric weave structure on wrinkling depends upon fabric tightness and float length, twill being better than plain weave, for example. An increase in fabric weave crimp, accompanied by a decrease in interstitial pressure, decreases shear modulus, resulting in a concomitant improvement in wrinkling. At a constant fabric mass, the use of coarser yarns together with the required lower sett would provide better wrinkling performance than the use of finer yarns and higher sett. Furthermore, the use of a weave structure with fewer intersections (longer float lengths) is preferred in terms of wrinkle recovery performance.

Although variations in yarn and fabric structural variables are unlikely to effect large improvements in wrinkle recovery, it is possible to effect significant improvements by optimising them:

- Decrease cover factor (tightness), up to a certain point.
- Increase float length.

28.4.4 MECHANICAL AND CHEMICAL FINISHING TO REDUCE WRINKLING

Essentially, any treatment (e.g. cross-linking and resin) that improves fabric relaxation, reduces the frictional couple (F) and improves fibre and the fabric elastic recovery (resilience) will have a beneficial effect on fabric wrinkling performance. For example, autoclave decatising of wool fabrics improves their wrinkling performance, and so does heat-setting of fabrics containing thermoplastic fibres, such as polyester and nylon. Ultimately, to effect an improvement, the treatment (usually chemical) needs to increase the ability of the fibre to return to its undeformed state, i.e. the state prior to the wrinkling deformation. Furthermore, any treatment that makes the fibres less sensitive to the effects of heat and moisture, e.g. increasing the glass-transition temperature (Tg), will have a beneficial effect on wrinkling performance.

Mechanical and chemical treatments (e.g. silicone) that reduce interfibre and inter-yarn frictional resistance (i.e. frictional couple) will also improve wrinkling performance. Care must be taken, however, that no treatment adversely affects the desirable properties of wool, notably comfort, durability and handle. The application of external polymers, such as polyurethane and silicone elastomers, can improve fabric wrinkling performance. Polymers that substantially improve wrinkle recovery generally fall into the class of elastomers, which have high elastic recovery, high glass-transition temperatures and low permanent set. Such polymers can improve wrinkling performance by reducing interfibre frictional forces, and by the introduction of elastic polymer bonds between fibres that do not stress relax, thereby assisting recovery after deformation. Such surface polymers can also impede the flow of moisture in and out of the fibre and add an elastic (resilient) component to the fabric (e.g. elastic cover/sheath to the fibre and/or elastic interfibre bonds). As an example, the wrinkle recovery of untreated cellulosic fabrics is poor because moisture disrupts inter- and intrachain hydrogen bonds (Kernaghan, Stuart, McCall, & Sharma, 2007), which re-form to stabilise the wrinkled (creased) state. Improved wrinkle resistance in these fabrics is generally achieved by cross-linking adjoining cellulose polymer chains, which increases flexibility and resilience (Bhat, 2007).

In practice, fabric wrinkling is largely dependent upon the viscoelastic properties (V) of the component fibres, and to a lesser extent upon the frictional couple (F), the former being largely a function of the fibre type and chemical treatment (e.g. resin) applied to it, and the latter, i.e. F, being largely a function of the state of relaxation of the fabric and interfibre frictional forces and elastic bonds. The effect of F is less for highly elastic (resilient) fibres. Table 28.5 (Hunter, 2009a) presents a summary of the factors that affect wrinkling, as well as ways of improving wrinkling performance.

28.4.5 MEASUREMENT OF WRINKLE AND CREASE RECOVERY

In most cases, wrinkle or crease recovery is measured rather than wrinkle resistance. There are various test methods for measuring fabric wrinkle and crease recovery (Fan, Hunter, & Liu, 2004). These can be divided into two broad categories, those that involve the insertion of a single sharp crease (fixed deformation) and those that insert a family of largely random creases or wrinkles (random deformation) in the fabric. In both cases, the conditions of fabric deformation, as well as of recovery, are critically important and need to be carefully controlled and consistent. Also important are the atmospheric conditions, relative humidity in particular, and the fibre moisture content during both creasing and recovery. Wear trials represent a third, though highly subjective, category of assessing wrinkling performance. It is important to mention that the results generated by different test methods are often not very highly correlated.

A popular method used by industry to assess fabrics is American Association of Textile Chemists and Colorists (AATCC) Test Method 128 'Wrinkle Recovery of Fabrics: Appearance Method', in which largely random wrinkles are induced in the fabric under standard atmospheric conditions using a standard wrinkling device under a predetermined load for a prescribed period of time. The specimen is then reconditioned and rated for appearance by comparing it with three-dimensional (3D) reference standards (AATCC Wrinkle Recovery Replicas). Considerable research has led to the development of objective assessment techniques, including computer vision (photometric stereo technology combined with ANFIS – adaptive neural-fuzzy inference system), for evaluating the fabric wrinkle grade.

In the case of crease recovery testing (e.g. using a Shirley crease recovery tester), the fabric specimen (either wet or dry) is creased and compressed under specified load and atmospheric conditions for a

Table 28.5 Ways of improving fabric wrinkling performance

Parameter	
Fibre	
Type	Use resilient fibres (e.g. polyester, nylon, wool) that are insensitive to temperature and moisture.
Fineness	Coarser fibres preferable (small effect).
Glass-transition temp (Tg)	Higher Tg, also under high moisture conditions, advantageous.
Viscoelasticity (V)	Good viscoelastic properties are critical.
Interfibre friction	Reduce interfibre friction, particularly for fibres with relatively poor viscoelastic properties.
Yarn	
Construction	Select constructional parameters that reduce interfibre frictional forces (small beneficial effect).
Linear density	Coarser yarns slightly better.
Twist	Effect of twist unclear, very high levels sometimes considered beneficial.
Fabric	
Construction	Looser structures (lower tightness within certain limits), longer float lengths and thicker fabrics preferable.
Blend and fibre type	Sensible selection of blend components. Fibre composition critically important.
Finishing	
Frictional	Reduce interfibre frictional forces, by reducing interfibre friction and/or replacing the interfibre frictional forces by elastic interfibre bonds.
Viscoelasticity	Internal resin and polymer treatments that, by cross-linking, etc., improve the elastic recovery from deformation of the fibres. External (surface) resin and polymer treatments that coat the fibres with an elastic sheath.

Source: Hunter (2009a).

predetermined period (e.g. 5 min) After this, the load is removed and the specimen is allowed to recover, once again under specified conditions and times (e.g. 5 min), and the recovery angle (crease recovery angle) is measured. Test methods include AATCC 66, BS EN 22,313 and ISP 2313. This test is frequently used to assess durable-press- and easy-care-related properties of treated cotton fabrics.

AATCC Test Method 124 (ISO 7768) is used for evaluating the appearance, in terms of smoothness, of flat fabric specimens after repeated home laundering. This provides a measure of the durable-press, easy-care or minimum-iron properties of the fabric. The test procedures and evaluation methods are almost the same as in the two methods mentioned above, except for differences in specimen preparation and standard replicas.

28.5 REDUCING PILLING

This Section is from Hunter (2009b). Pilling is defined as 'the entangling of fibres during washing, dry-cleaning, testing or in wear to form balls or pills which stand proud of the surface of a fabric and which are of such density that light will not pass through them (so that they cast a shadow)' (*Textile Terms and Definitions*, 11th Edition, The Textile Institute, Manchester, 2009). A pill is therefore a cluster or entanglement (bundle) of fibres in the form of a ball, attached to the fabric surface by one or more anchor fibres (see Figure 28.3; Anonymous, 1972), sometimes incorporating lint or foreign matter.

Pilling, when it occurs, represents a serious quality problem, particularly in apparel and upholstery fabrics, and can be one of the reasons why a product becomes 'unsightly' and no longer acceptable for further use. Although pilling is an unacceptable fabric fault or weakness, it is often a necessary trade-off when a high degree of softness is required, which frequently comes associated with increased pilling propensity. Furthermore, when easy-care is required, strong synthetic fibres with a long flex life are often blended with weaker natural fibres, such as wool and cotton, with the synthetic fibre anchoring pills to the fabric and preventing them from wearing off.

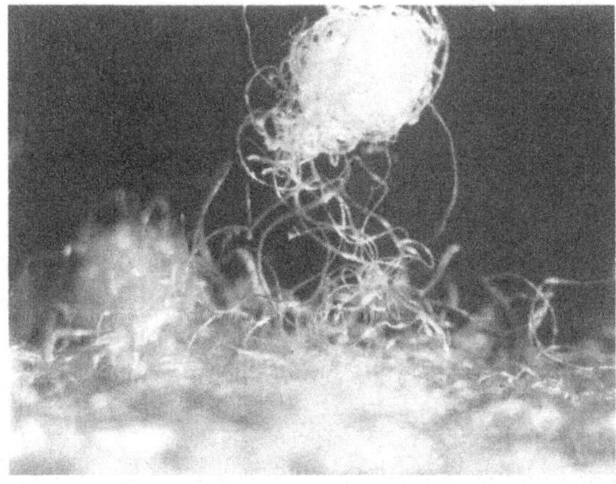

FIGURE 28.3

A typical pill in knitted fabric.

Source: Anonymous (1972).

28.5.1 HOW PILLS ARE FORMED

Broadly speaking, there are three different stages involved in pilling – fuzz formation, pill formation and pill wear-off – and the rate at which each occurs depends on fibre, yarn and fabric properties and their interactions. In order for pills to form, a sufficient density and length of fibre ends (hairs) is required on the fabric surface, so that they can entangle during mechanical action such as rubbing the fabric surface. Therefore, pilling cannot occur where the fabric contains continuous filament yarns, unless the fabric is subjected to a treatment that breaks the filaments on its surface. Mechanical action applied to the fabric surface, for example, the rubbing action during wear or cleaning, or even during dyeing and finishing, can cause surface fibres to break and become entangled into clusters or balls (i.e. pills). The mechanical action can also cause fibres to migrate to the fabric surface (i.e. to be pulled out from the body of the fabric) and to extend further from the fabric surface until the density and length of surface fibres (fuzz or hairs) are such that further mechanical action and frictional forces cause them to become entangled into pills that remain attached to the fabric by one or more fibres, the latter often being the stronger fibres in a blend. Loose fibres, or even contaminants, on the fabric surface often form an integral part of the pill-formation process and pill.

Continued mechanical action can cause the pills to wear off. Generally, pills are removed during wear as a result of the anchoring fibres being pulled out of the fabric or breaking; flex life is a critical factor in the latter case, and fibre length and binding in the former.

There are many factors that affect the degree of pilling that occurs during use, including the following:

- End use
- Wear conditions
- Washing and dry-cleaning conditions
- Fibre properties
- Yarn properties
- Fabric properties
- Fabric finishing
- Relative humidity

28.5.2 FIBRE PROPERTIES

Fibre composition and properties play a major role in the formation and wear-off of pills.

28.5.2.1 Length

Longer fibres produce fewer fibre ends (i.e. lower hairiness) and also resist migration better than shorter fibres, since they are more securely bound within the yarn structure. Therefore, longer fibres should lead to lower fuzz and pill formation, but could reduce pill wear-off due to better anchoring of the pills. On balance, however, longer fibres are generally preferable in terms of reduced pilling propensity.

28.5.2.2 Diameter and stiffness

All other factors being constant, a change in fibre diameter (or fineness) is associated with a corresponding change in fibre stiffness, which is proportional to diameter to the fourth power or linear density (fineness) squared. An increase in diameter (or linear density) and fibre stiffness, all other factors being constant, will lead to a reduction in fibre surface hairs, fuzz and pill formation; stiffer fibres also resist entanglement into pills better than do more flexible fibres.

28.5.2.3 Friction

An increase in interfibre friction will generally reduce pilling, since it reduces the tendency for fibres to migrate to the fabric surface. Fibre surface treatments, such as plasma or chlorination, can affect fibre friction by increasing the surface roughness, thereby decreasing pilling propensity.

28.5.2.4 Cross-section

It is generally held that a noncircular fibre cross-section – for example, trilobal, elliptical, irregular or flat (ribbon) – is preferable from the standpoint of reduced pilling, as the fibre cross-section affects both fibre friction and bending.

28.5.2.5 Strength

Stronger and more elongated fibres generally lead to greater pilling, whereas weaker and more brittle (lower elongation) fibres reduce pilling, since stronger fibres anchor pills to the fabric, thereby resisting pill wear-off. The strength (and flex life) of synthetic fibres such as polyester may be reduced by means such as chemical damage or a reduction of average molecular weight in order to reduce pilling propensity, particularly when used in blends with natural fibres. Such fibres are referred to as 'low-pilling'.

28.5.2.6 Flex life

The ability of a fibre to withstand flexing, bending and twisting, notably the first of these, plays a crucial role in the pilling propensity of a fabric. Fibres with higher flex life generally produce fabrics with greater pilling propensity, because such fibres anchor the pills and do not allow them to wear off. A combination of fibres with low and high flex life (or low and high strength) is particularly bad from a pilling perspective.

28.5.2.7 Crimp

An increase in fibre crimp is generally associated with a reduction in fabric pilling, possibly because it reduces yarn hairiness and increases interfibre cohesion and frictional resistance.

28.5.2.8 Composition

Fibre composition, in terms of fibre blend, plays a major role in fabric pilling, the actual effect depending greatly upon the characteristics of the blend components as well as the proportion of each blend component. It is probably safe to say that blending fibres with properties that differ widely will usually lead to an increase in pilling propensity. This is particularly the case where fibres with relatively high strength and flex life (e.g. polyester and nylon) are blended with fibres of a relatively low strength and flex life (e.g. cotton) – the weak fibres break and form the pill, while the stronger, high flex-life fibres anchor pills to the fabric and reduce pill wear-off.

28.5.2.9 Electrostatic properties

Fibres having a greater propensity to electrostatic charge (static) will be more inclined to collect loose fibres or other contaminants that can aggravate pilling.

28.5.3 YARN PROPERTIES

The effect of yarn properties on pilling can be ascribed largely to yarn hairiness and the ease with which fibres can migrate to the surface of the yarn and fabric, which in turn is related to yarn structure, construction and hairiness. Finer, higher twist-factor yarns, as well as less hairy (e.g. compact) and plied yarns, tend to reduce pilling. Wrapper fibres also reduce pilling. It follows that yarn singeing, which reduces hairiness, will reduce pilling.

28.5.4 FABRIC PROPERTIES

Fabric structure and properties affect pilling mainly because they influence the number of fibres on the fabric surface and the ease with which fibres can move (migrate) from the body of the fabric to its surface. The fewer and shorter the surface fibres, and the more difficult it is for fibres to migrate (i.e. the more securely they are bound), the lower the pilling propensity. The latter factor implies that pilling propensity should decrease as yarn crossover points per unit area increase, yarn float lengths become shorter, fabric surfaces become flatter, and fabrics become more compact/dense (i.e. have a higher tightness factor). Therefore, the tighter and more compact the fabric, in both knitted and woven fabrics, the lower the pilling propensity generally. An increase in woven fabric sett or knitted fabric stitch density will reduce pilling, with knitted fabrics more prone to pilling than are woven fabrics. Nevertheless, a more tightly constructed fabric could reduce the ease with which anchoring fibres detach from fabric, thereby reducing the tendency for pills to wear off and cause increased pilling.

28.5.5 DYEING AND FINISHING

Various chemical and mechanical finishing treatments for fabrics are effective in reducing pilling, mostly by reducing the surface hairs and/or the ease by which fibres migrate to the fabric surface. Any treatment that decreases fibre strength and/or flex life, i.e. embrittles the fibres, will reduce pilling propensity. Care needs to be exercised, however, that any treatment applied to the fabric does not cause an unacceptable deterioration in other desirable fabric properties such as handle and durability. Treatments that can reduce pilling include:

- Singeing
- Shearing or cropping, preferably preceded by sanding and brushing
- Heat-setting and steaming
- Cross-linking (e.g. acrylic)
- Any treatment, e.g. resin, polymer or adhesive, that increases interfibre friction and cohesion or binds the fibres more securely within the fabric
- Biopolishing and enzyme treatments
- Caustic solution treatment of polyester and viscose fabrics

Table 28.6 (Hunter, 2009b) summarises the factors that affect pilling.

28.5.6 MEASUREMENT OF PILLING

A large number of pilling test instruments and associated methods (ISO EN BS12495) have been developed, and these have been covered in detail elsewhere (Fan et al., 2004; Hunter, 2009b). In brief, pilling testers utilise either a generally flat rubbing/abrading action or a tumbling action to develop pills.

Table 28.6 Fibre, yarn, fabric and finishing parameters that* can reduce fabric pilling propensity

Parameter
Fibre
Increased length
Decreased flex life/increased brittleness
Decreased strength/tenacity
Increased stiffness/modulus
Increased diameter
Increased noncircularity
Increased fibre friction
Yarn
Decreased hairiness
Increased fibre binding
Increased twist factor
Decreased yarn linear density
Increased yarn evenness
Plying/folding
Yarn structure (e.g. compact, open-end and wrapped yarns)
Fabric
Increased compactness/tightness
Decreased yarn float length
Decreased hairiness
Increased fibre binding
Flatter surface (reduced raised effects)
Increased weight
Finishing
Removal of surface hairs (singeing, cropping, shearing)
Increasing fibre binding/cohesion (latex, milling, resin, etc.)
Increasing fibre friction (as above, plasma, chlorination)

**Some of which are interdependent. Not necessarily in order of importance.*
Source: Hunter (2009b).

Examples of the former include:

- Martindale abrasion tester
- Stoll Quartermaster universal wear tester (using an elastomeric friction pad)
- Brush and sponge pilling tester

Examples of pilling testers based upon a tumbling action are:

- Random tumble pilling tester
- ICI pilling box

Traditionally, the pilling induced on the fabric sample during laboratory testing has been assessed (i.e. quantified) subjectively by a rating system, often with the aid of photographic standards, and additional objective and sophisticated techniques have been developed in recent years.

Generally, tumbling pilling testers are used mostly for knitted fabrics, and abrasion pilling testers for woven fabrics. Correlations vary greatly when comparing the results of different pill testers, pill testers with actual wear, and corresponding rankings of fabrics, and depend upon the fabric type and construction, as well as the testing and wear conditions involved.

28.6 REDUCING BAGGING

This Section is from Hunter (2009c). During wear, a garment conforms to body movement by slippage over the skin, space allowance between body and clothing, and fabric deformation (Kirk & Ibrahim, 1966), the latter leading to the phenomenon known as bagging. Bagging may be defined as the residual (i.e. unrecoverable) 3D dome-shaped deformation (distortion or stretch) in fabrics (notably knitted garments) caused during wear that often leads to the garment becoming unsightly and unacceptable in appearance.

Ideally, the fabric and garment should change shape (i.e. deform) easily to accommodate body shape and movement and provide 'dynamic comfort', but then recover the original shape once deforming forces are removed, and this should be so for the entire wear life of the garment. Bagging is largely the consequence of the fabric's inability to fully recover from such deformation (Zhang, Yeung, Miao, & Yao, 2000). This 3D deformation and bagging most often occurs at the elbows of jerseys and jackets and at the knees and seats of pants and trousers, as a result of forces exerted on the garment during movement and during sitting, bending, etc., causing the fabric to deform into a 'dome' and bulge.

Bagging-related deformation is dependent upon many fibre, fabric, garment and wear parameters:

- Elastic and viscoelastic properties of the fibres and fabric
- Interfibre and inter-yarn friction
- Fabric and garment construction
- Type of garment and garment fit, notably tightness or snugness
- Wearer size and activities
- Level and nature of deforming forces
- Duration and number of deformation and recovery cycles

In essence, the bagging behaviour of a fabric is determined by its ability to resist 3D forces (bagging resistance), together with its ability to recover from any deformation resulting from such forces (bagging recovery), notably from repetitive deformation cycles, with the latter being more important. The fibre, yarn and fabric properties required to produce good bagging performance (i.e. low bagging propensity) are most similar to those required to produce good wrinkling performance (see Section 28.4).

28.6.1 FIBRE PROPERTIES

Elastic, viscoelastic and fibre-to-fibre frictional properties, particularly the first two, play the main roles in fabric bagging. If the fabric constructional parameters and wear conditions are constant, the bagging propensity of the fabric will mainly depend upon the fibre viscoelastic (elastic and stress-relaxation) properties, and, to a lesser extent, on fibre frictional properties.

All other factors being constant, the fibre initial modulus will determine the fabric resistance to bagging deformation, resistance increasing as the fibre's initial modulus increases. Recovery from bagging deformation will depend mainly upon the elastic recovery and stress-relaxation properties of the fibres (i.e. elastic and viscoelastic properties) and to a much lesser extent on the fibre-to-fibre frictional properties. Any factor that affects the fibre viscoelastic properties, such as temperature and fibre moisture content, will therefore affect bagging.

28.6.2 YARN PROPERTIES

Yarn properties, such as increased yarn twist, can reduce interfibre slippage within the yarn and enable the fibres to withstand increased strain during bagging deformation, thus improving bagging performance. Fine high-twist smooth yarns with low inter-yarn friction have better bagging performance. The use of stretch yarns such as Spandex and elastane, which have good recovery properties, also improves bagging performance.

28.6.3 FABRIC PROPERTIES

Woven and knitted fabrics should be discussed separately, because their responses are quite different. The following factors are common to both woven and knitted fabrics:

- Tightness: all other factors being constant, tighter fabrics will have better bagging resistance and usually bagging recovery as well
- Thickness: the thicker the fabrics, the better
- Weight: heavier fabrics should generally resist bagging better and recover better.

28.6.4 GARMENT CONSTRUCTION

A garment constructed so that it accommodates body contours and allows easy movement for various body parts, with the fabric sliding over rather than sticking to the skin, should provide the best bagging performance and smoothest appearance (Crowther, 1985).

28.6.5 FINISHING

Finishing can reduce bagging by increasing the fibre's elastic properties and decreasing its viscoelastic properties; interfibre elastic bonding (e.g. resin treatment) and/or reducing interfibre friction (e.g. by silicone or softener applications); and reducing the fabric's internal energy by such means as fabric relaxation and heat-setting.

28.6.6 MEASUREMENT OF BAGGING

Bagging involves deformation and elastic recovery from that deformation. Testing methods attempt to simulate and measure this, either by monaxial or biaxial deformation and recovery. Three approaches are used:

- Strip or uniaxial
- Biaxial
- Simulated 'arm' (DIN 53860)

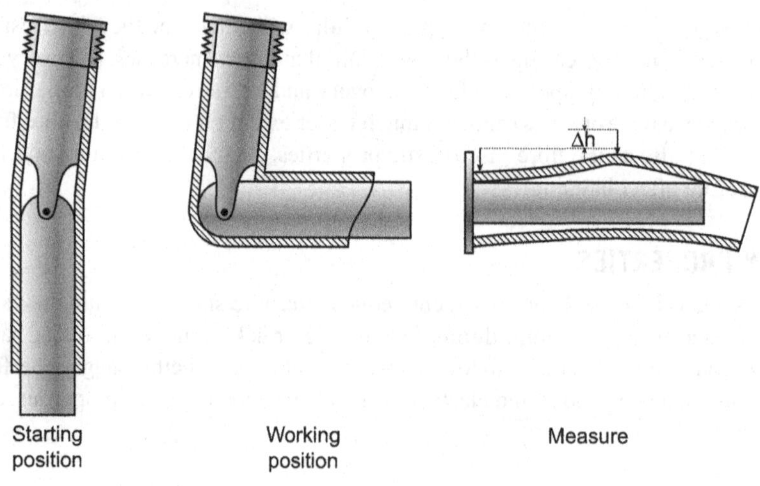

FIGURE 28.4

Schematic diagram of 'artificial-arm' bagging tester and bagging height (Δh) measurement.

Source: Grünewald (1977).

All three tests involve static and dynamic elements. Most of the techniques and test methods used to measure bagging involve securely clamping the fabric sample, usually circular in shape (e.g. Celanese test), around its perimeter and then subjecting the central unclamped section to deformation, generally cyclic, by means of a dome-shaped 3D deforming sphere or force. The amount of residual deformation after the deforming force is removed is then measured and used as a measure of bagging propensity.

An example of a bagging test is that illustrated in Figure 28.4 (Grünewald, 1977), also referred to as the Zweigle-type bagging tester, in which a fabric tube is sewn and then drawn over an 'artificial arm' with an 'elbow joint', and held in position under tension. The 'arm' is flexed and held bent for a number of hours to simulate the action that occurs in practice at the elbow (or knee). It is then straightened for 10 minutes, after which the fabric tube is carefully fitted over a vertically suspended tube and then tilted to a horizontal position, and finally illuminated by parallel light to cast a shadow on a surface placed behind the tube. The height of the 'bagging shadow' (i.e. bulge) is measured directly with a millimetre-graduated scale.

28.7 IMPROVING FABRIC AND GARMENT DRAPE

This Section is from Hunter, Fan, Chau (2009). A highly desirable and critical characteristic of a textile material is its ability to undergo large, recoverable 3D draping (bending and shear) deformation by buckling gracefully into rounded folds of single and double curvature (Stump & Fraser, 1996). This characteristic plays a critical role in the fit, body conformation, comfort and aesthetic appearance (visual beauty) of garments, and when translating 3D body shapes into 2D patterns and vice versa.

According to *Textile Terms and Definitions (11th edition)* of The Textile Institute, drape is defined as 'the ability of a fabric to hang limply in graceful folds', while Cusick (1965) defined drape as 'a deformation of the fabric produced by gravity when only part of the fabric is directly supported'.

During draping, a fabric undergoes large deflections but very small strains due to its high flexibility, with the largest deflections coming from bending, and only a small component being due to in-plane extension and shear deformation. This, together with the effect of seams, interlinings and linings (e.g. bonded), determines the way that a garment fits and moulds itself to the shape of the body.

Drape appearance depends not only on the way the fabric hangs in folds, but also upon the visual effects of light and shade, and fabric lustre, colour, design and surface decoration. A fabric is considered to have good draping qualities when it adjusts into folds or pleats under the action of gravity in a manner that is graceful and pleasing to the eye. In practice, drape is usually assessed visually and subjectively, and the assessment greatly depends upon often-changing factors such as fashion, personal preference and human perception. Frequently, there is also an element of movement – for example, the swirling movement of a skirt or dress – and therefore dynamic properties are also involved. As a result, in recent years a distinction has been made between static and dynamic drape.

Drape is a critically important parameter in the application of body scanning, mass customisation, CAD–CAM and automatic pattern making in clothing design and manufacturing. Drape modelling, in particular 3D visualisation of designed garments in draped form, has become one of the key technologies in computer-aided garment design and internet apparel and clothing systems, since it enables designers to assess the design, fabric suitability and accuracy of garment patterns in a computer environment. It is also essential for effective online trading and retailing, because without it buyers and consumers are not able to assess garment style, appearance, fit and suitability onscreen.

Drape can be summarized as a complex combination of a fabric's mechanical, optical, seam and interlining properties, in addition to subjectively assessed properties and factors. All of these factors need to be taken into consideration when engineering drape.

If the required drape is known, then the drape can to a large extent be engineered or predetermined to meet the specific end-use requirements. Fabric drape is determined largely by the fabric bending and shear stiffness properties, with fabric thickness, low deformation tensile properties and weight also of importance. To predetermine the fabric drape accurately, the precise relationship between fabric drape and fabric bending and shear stiffness must be known, along with the effects of fibre, yarn and fabric physical and structural variables on bending and shear stiffness, as illustrated in Figure 28.5 (Anonymous, 1981). For the sake of simplicity, changing fabric and garment drape will be considered here from the point of view of the commonly measured drape coefficient (DC) only.

28.7.1 FIBRE PROPERTIES

Fibre bending stiffness plays a major role in determining fabric bending stiffness and therefore fabric drape, the higher the fibre and fabric bending stiffness, the higher will be the fabric drape coefficient. The fibre bending modulus, fibre diameter (cross-sectional area), fibre cross-sectional shape and area variability all play a role in fibre bending stiffness (see eqns 28.1–28.3 in Section 28.3). If a softer or more flexible drape is required, then a fibre with either a lower bending modulus or a finer fibre, or both, should be selected. In practice, the drape coefficient can be changed

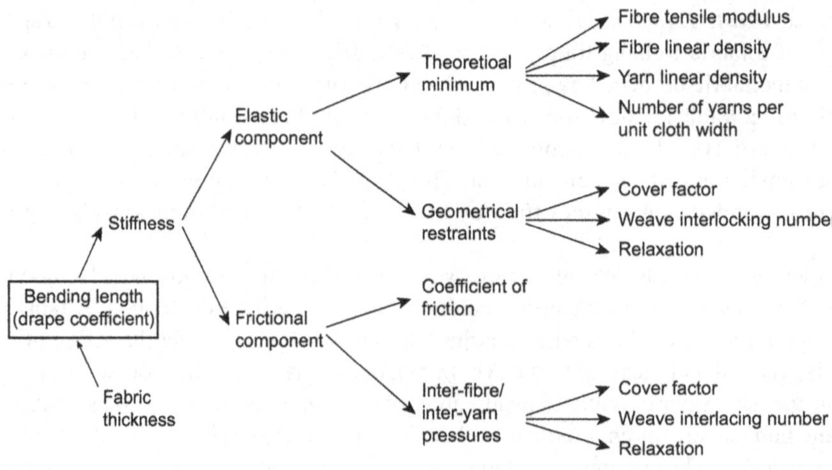

FIGURE 28.5

Some factors contributing to fabric drape behaviour. Direction of arrows indicates whether an increase or decrease in a given parameter will produce an increase or a decrease in the drape coefficient of the fabric.

Source: Anonymous (1981).

most easily and significantly by changing the fibre diametre, and to a certain extent by changing the fibre shape.

28.7.2 YARN PROPERTIES

The effects of changes in yarn properties, within normal practical and commercial limits, will generally be small compared to the effect of changes in fibre diameter and fibre shape and bending modulus. The main yarn parameters that can affect fabric drape are linear density, twist factor (including ply-to-singles twist ratio), bulk/diameter/compactness (volumetric density), hairiness and structure (the latter including factors such as orientation and migration) and wrapper fibres (e.g. Open End (OE) yarns). At constant fibre properties and yarn linear density, the drape coefficient can be increased by a decrease in fibre freedom of movement, which can be brought about by an increase in interfibre friction and adhesion and/or by a decrease in yarn diameter (an increase in yarn compactness), and the presence of surface 'wrapper' fibres.

28.7.3 FABRIC PROPERTIES

Fabric thickness, tightness and weight are the main fabric parameters in determining drape; fabric tightness (cover factor), or compactness, probably has the greatest impact on drape, due to its effect on fibre and yarn freedom of movement, and thus fabric stiffness. As a result, any factor that increases the fibre and yarn freedom of movement within the fabric, such as weave structure or fabric relaxation, will decrease the fabric bending and shear stiffness as well as the fabric drape coefficient. At a constant weight, the fabric drape coefficient will decrease as the float length increases, i.e. number of yarn intersections decreases. Thus, a twill-weave fabric will have a lower drape coefficient than a plain weave fabric of the same mass per unit area.

28.7.4 DYEING AND FINISHING

Fabric dyeing and finishing (including setting and pressing) can reduce fabric bending and shear stiffness, and therefore fabric drape coefficient, significantly. This is mainly due to the associated reduction in the internal stresses and energy in the fabric and the improvement in the fabric state of relaxation, and the ease of fibre and yarn movement within the fabric, together with their effects on interfibre and inter-yarn friction and cohesion. Good examples are the scouring and crabbing of wool fabrics, the scouring and bleaching of cotton fabrics, the heat-setting of fabrics containing thermoplastic fibres, and fabric dyeing, which can promote stress relaxation within the fabric structure. Any treatment that increases interfibre and inter-yarn friction and cohesion (e.g. fibre bonding) will increase fabric bending and shear stiffness, resulting in increased drape coefficient.

28.7.5 GARMENT CONSTRUCTION

The garment construction and type – notably the number, positioning and type of seams, linings and interlinings – as well as cut and style, play significant roles in the draped appearance of the garment, with the presence, nature and bonding of interlinings having a major effect. Fabric colours, depths of shade and patterns also have significant effects, although these are essentially optical in nature and not due to actual changes in the fabric drape per se. Seams mainly affect fabric stiffness (both bending and shear) in their immediate vicinity, the magnitude of the effect depending on their positioning within the garment. For example, bending length tends to increase with the insertion of a vertical seam, while drape coefficient increases with the addition of radial seams, and increasing the seam allowance has little effect.

28.7.6 MEASUREMENT OF DRAPE

Although fabric draping characteristics are often assessed subjectively, various methods for the objective measurement and characterisation of drape have been developed over the years. Because of the predominant effect of fabric stiffness on drape, the instruments were initially designed to measure fabric stiffness, mainly in terms of fabric bending length (the length of fabric that bends to a defined extent under its own weight), and provided a fairly good measure of the two-dimensional (2D) drape. But because fabric shearing properties also play an essential role in determining fabric draping characteristics, the traditional stiffness tests (cantilever method) were unable to accurately reflect fabric drape, and techniques were therefore developed to include fabric shear. At present, the most widely used method is still the traditional one, in which a circular disc of fabric is allowed to drape into folds around the edges of a smaller circular platform or template, and using instruments commonly referred to as drapemeters to measure the relevant parameter – mainly drape coefficient, but also drape profile and nodes. Typical examples of drapemeters include those of Cusick, Fabric Research Laboratory (FRL) and Institut Textile De France (ITF), as well as the MIT Drape-O-Meter.

Other principles of measuring drape include the force to pull a circular fabric sample at a constant speed through a ring, the force being termed the 'drape resistance' of the fabric. Cusick's Drapemeter (Hunter, Fan, & Chau, 2009) is still the standard method of measuring drape and is widely used. The drape of a fabric in this instance is popularly defined as the area of the annular ring covered by the vertical projection of the draped fabric, expressed as a percentage of the area of the flat annular ring of fabric, this being termed the drape coefficient. New techniques such as image analysis have been developed for measuring the size and shape of the draped fabric sample.

28.8 IMPROVING FABRIC AND GARMENT DURABILITY

This Section is from Hunter (2009d). For consumers, durability or serviceability is one of the main requirements for clothing, along with ease-of-care, elegance (aesthetics) and comfort. Durability is a measure of the reasonable wear life of a product and is a complex function of a number of factors that reduce the serviceability and acceptability of the product, including abrasion (flat and flex), tearing, rupturing, bending, stretching, changes in appearance, laundering and cleaning. For example, Stoll (1949) stated that the wear of army uniforms comprised 30% plane (flat) abrasion, 20% edge and projection abrasion, 20% flexing and folding, 20% tear and 10% other mechanical actions. Because of this, wear factors need to be assessed within the context of the requirements of the specific end use for the fabric and garment. With respect to clothing, aspects relating to the overall appearance of the fabric and garment generally dominate. The durability of a garment depends upon the properties of the fabric as well as the conditions it encounters during wear (even the fit of a garment in the case of apparel), and laundering.

For apparel fabrics and garments, resistance to abrasion tends to be more important than strength as an indicator of durability. Abrasion during use not only contributes to the failure of the fabric and garment, but more commonly contributes to changes in fabric appearance such as fuzzing, pilling, frosting (colour change) and 'shine' (making it look old or unattractive). Frequently, the consumer will consider a fabric to have reached the end of its useful life on the basis of appearance rather than mechanical failure such as tearing or rupturing. However, changes in fibre and fabric structure can beneficially affect certain components of durability (e.g. flat abrasion resistance) while adversely affecting others (e.g. tear strength), necessitating compromise and optimisation.

Fabric and garment durability and wear performance are determined by the fibre type and properties, yarn fabric and garment construction, and any chemical and/or mechanical treatment applied to the yarn, fabric or garment. In practice, mechanical damage or breakdown is generally more important than chemical damage (including damage due to laundering and light) in determining wear life. Because of this, wear conditions – including the size, shape, occupation and activities of the wearer, as well as laundering and drying – have a major effect on the wear life of a garment. It is often difficult to distinguish between such factors and fabric shortcomings when attempting to identify the causes of garment failure during use. Frequently, variations amongst wearers and wear conditions are more important than variations in the fabrics or garments themselves in determining wear performance and durability. Such wear generally occurs at localised regions of the garment, such as at the seats of pants, elbows of jackets and jerseys, and collars and cuffs of shirts, depending upon the abrasive forces (magnitude, frequency and duration) imposed on the garment by the wearer, and the wear conditions, including those prevailing during laundering. The moisture and perspiration content of the garment can also affect the degree of wear and abrasion.

28.8.1 FIBRE PROPERTIES

Fibre properties generally play a major role in fabric and garment durability, their relative importance depending to some extent upon the particular type of wear conditions that predominate. Fibre strength, elongation and elastic recovery during and after repeated applications of stress (tensile and bending) are generally important, followed by fibre bending, flexural and shear properties, length and diameter, with fibre shape also playing a role. The energy-absorbing capacity of the fibre assembly also plays a major role in determining durability.

The tensile, elastic, bending and flexing properties of the fibre will be reflected in those of the fabrics and garments and hence in durability. Nylon and polyester are considered highly durable, followed closely by polyolefin fibres, and then acrylic and mod-acrylic fibres, with viscose, rayon and acetate fibres tending to be less durable, probably due to their relatively low strength and elastic recovery, with associated low energy absorption during repeated stress cycles (Galbraith, 1975). In general, combining hard-wearing and strong synthetic fibres, such as polyester and nylon, with natural fibres, such as wool and cotton, should improve the durability of 100% wool or cotton fabrics. However, when fibres that differ greatly in their physical and mechanical properties are blended, differential wear and abrasion can take place, with associated changes in appearance and colour (e.g. frosting). Care must be taken to avoid the problem of 'frosting', where fibres that differ in both colour and abrasion resistance are blended, leading to differential wear and associated changes in colour, as well as problems with pilling.

28.8.2 YARN PROPERTIES

Yarn structure, such as twist, linear density, friction, crimp, number of plies, smoothness and the presence of wrapper or binding fibres, can affect fabric durability, with the magnitude of the effect depending on fabric tightness (cover factor) and structure, it being greater for looser fabric structures and longer float or stitch (loop) lengths. For example, where the yarn structure enables the abrading load to be more evenly spread over a large surface area and more energy to be absorbed, abrasion resistance will be improved.

In general, fabric durability improves as the fibres are better aligned and more securely bound within the yarn structure, for example, by increasing yarn twist factor (up to a point) or using two-ply (as opposed to one-ply) yarn, or by using compact spun yarns. Where laundry-type abrasion (i.e. fibre removal) predominates, increasing yarn twist to increase fibre binding will improve fabric durability. Where a flat abrasion action with low friction predominates, a lower twist may be preferable, since the yarn will deform more easily at the fabric surface and spread the load more evenly over a greater yarn and fabric surface area.

28.8.3 FABRIC PROPERTIES

In general, an increase in fabric tightness, thickness and weight per unit area, and a decrease in float length (i.e. an increase in yarn crossover points) increases fabric durability in terms of strength and flat abrasion, but not necessarily in terms of resistance to flex abrasion and tearing.

Any fabric structural modification that allows the abrasion load to be spread over the largest contact area tends to improve flat abrasion resistance. Fabric structures that improve yarn mobility or sliding will improve resistance to tearing (i.e. tear strength) and possibly flex abrasion. Flat abrasion resistance is more dependent on good fibre binding than on fibre and yarn mobility, whereas the reverse is true for flex abrasion and tear strength. High yarn mobility can be advantageous in edge wear but less important than good fibre binding and balanced distribution of abrasive stress over both warp and weft yarn systems in flex abrasion. Any change in fabric construction that binds fibres more securely should improve fabric durability. The use of yarns of the same linear density, crimp and sett in both warp and weft will generally improve fabric and garment durability.

28.8.4 GARMENT DESIGN AND FIT

The garment type, size and fit, as well as the type, position, strength and abrasion resistance of seams, can influence the performance of garments during wear. The strength characteristics of garments are largely determined by the properties, positioning, type, etc., of seams, together with the fabric properties, while the design and fit of the garment will determine the stress (load and extension) applied to the fabric assembly during wear. Tight-fitting garments containing low-stretch (rigid) fabrics are subjected to large forces, particularly around the knees, elbows and seat regions, and could burst or tear if not sufficiently strong or extensible.

In order to maintain good overall durability, appearance and fit of a garment, the seam strength should be adequate. Different types of garments have different seam strength requirements. Seam failure can be caused by sewing thread wearing out or breaking before the fabric, seam slippage (e.g. when the yarns are smooth or the fabric loose), and the breakage of the yarns by the sewing needle during the sewing operation. Fabric type and weight can affect seam strength, with fabric types that are prone to fraying, such as loosely woven fabrics, always requiring better-constructed and denser seams. Lightweight fabrics require relatively stronger seams and seam strength efficiency than heavy fabrics. The seam direction can also affect the seam quality.

28.8.5 DYEING AND FINISHING

Any dyeing and finishing treatments that damage the fibre (e.g. cause a deterioration in fibre tensile, bending, torsional and flexing properties) will adversely affect fabric and garment wear performance and durability. A good example of this is the resin treatment (cross-linking) of cotton fabrics to improve wrinkle resistance, easy-care and durable-press, which can decrease fabric strength and abrasion resistance, particularly edge, flex and tumbling abrasion, by embrittling and weakening the cotton fibres.

The application of a lubricant or softener can improve both the flat and flex abrasion resistance and tear strength of fabrics, but not tensile strength, as measured in the laboratory. Nevertheless, such improvements are not necessarily reflected in the actual wear performance, particularly after laundering, which can remove much, if not all, of the lubricants and softeners. The effect of dyeing, finishing and laundering on friction, and fibre and yarn mobility and damage, will be reflected in fabric abrasion resistance, flex abrasion in particular, as well as in fabric strength and durability.

28.8.6 MEASUREMENT OF FABRIC DURABILITY

The useful life of a garment is often one of the most important quality factors and being able to predict or engineer durability through laboratory tests would be highly useful. Very many laboratory instruments and test methods, particularly for fabric abrasion resistance, have been developed over many decades in an attempt to simulate and predict, or at the very least estimate, wear performance and durability. Many of these tests have, for various reasons, been discontinued, it being notoriously difficult to simulate the great variety and complexity of the conditions that fabrics and garments experience in the diversity of possible end-use applications.

In apparel, a fabric can be subjected to various mechanical/physical (e.g. tensile, bending, rubbing and flexing) actions and chemical actions, which lead to changes in appearance and functionality of the fabric, and ultimately to the garment no longer being acceptable, either from an appearance or functional point of view. If fabric durability and wear performance are to be reliably estimated in the laboratory, testing would need to measure (or simulate) and appropriately weight all the components of wear and durability, such as abrasion resistance, strength (tensile, tear, bursting), pilling, etc. It is important that the tests subject the fabric to relatively low and random abrasive forces, so including one or more laundering cycles as well usually improves actual wear prediction.

Laboratory tests are probably most useful for comparing different fabrics rather than for prediction of actual durability. It is well known that moisture content, and to a lesser extent also temperature, affects fabric strength and abrasion properties, both in wear and in laboratory tests. As a result, such tests should always be carried out under standard atmospheric conditions (either 20 ± 2 °C and 65 ± 2% RH or 27 ± 2 °C and 65 ± 2% RH), after carrying out the necessary preconditioning and allowing sufficient time for the samples to reach equilibrium in the laboratory.

28.8.6.1 Abrasion resistance

Abrasion tests that involve less severe conditions (i.e. gentle abrasion), but include laundering, usually correlate best with actual wear, but require a prolonged time for testing. Tests developed in recent years have addressed this – for example, by reducing and varying the abrasion forces and including laundering – and testing of abrasion resistance has remained an important measure of fabric durability.

Three forms of abrasion occur and are tested for most frequently, namely flat (plane or surface) abrasion, edge abrasion (i.e. at collars and folds) and flex (flexing and bending) abrasion. Flat abrasion occurs as a result of a rubbing action at the fabric surface; either against itself or against another surface; the latter could be one or more of many widely different materials. Edge abrasion, sometimes also referred to as cuff abrasion, frequently takes place at collars and cuffs, and can be a combination of flat and flex abrasion. Flex abrasion mostly occurs as a result of flexing and bending during use, sometimes also occurring over a sharp edge. It is also referred to as internal abrasion, in which fibres rub against fibres or yarn against yarn within the fabric, although sometimes an external object such as a sharp edge is also involved. Fabric surface properties and lubrication greatly affecting flex abrasion results. A fabric may be subjected to all three forms of abrasion.

Most abrasion test instruments and methods attempt to provide a measure of one or more of the three different types of abrasion. Various types of laboratory testing (and their combinations) are used to measure fabric abrasion resistance:

- Flat (or relatively flat) abrasion (usually the fabric is rubbed against a fabric or another abradant, such as emery paper, under various pressures)
- Flex abrasion
- Combination of flat and flex abrasion
- Edge abrasion
- Tumble abrasion

The tests are not always highly correlated, and changes in fibre, yarn, fabric and finishing parameters often affect them differently, and even oppositely.

28.8.6.2 Flat abrasion

Examples of flat abrasion testers include Martindale (ASTM D4966, BS 5690), Stoll (ASTM D3885 and D3880), Taber (ISO 5470, ASTM D3884) and Schiefer (ASTM D4158). The Martindale abrasion tester is one of the most popular flat abrasion testers, and is also used for testing pilling propensity. A test specimen is rubbed against an abradant (usually an abradant fabric) at a predetermined pressure in a continuously changing direction. Rubbing is continued either for a preset number of cycles, after which the mass loss and change in appearance are determined, or is continued until a predetermined end point is reached – for example, until two fabric threads are ruptured or a hole is formed (in accordance with ASTM 4966 and 4970, BS 3424/5690, BS EN 530, BS EN ISO 12947-1, -2, -3 and -4, and ISO 5470 (rubbed or plastic-coated fabric) and JIS L 1096).

28.8.6.3 Edge abrasion and the tumble test

Mass loss and edge abrasion are often used as a measure of abrasion. Edge abrasion tests include AATCC 119, 120, ASTM D3514, D3885 and D3886. The AATCC 'Accelerotor' test, or AATCC Test Method 93, is a rapid tumble test (for both wet and dry samples), in which the sample is folded and stitched prior to testing to accentuate abrasion of edges. Samples are tumbled in a circular cylinder lined with an appropriate abrasion material, with a rapidly rotating impeller/propeller-shaped rotor creating a tumbling action that beats the sample against the drum wall, causing frictional abrasion, fabric flexing, rubbing, impact, compression, stretching, etc.

28.8.6.4 Flex abrasion

The Stoll-Flex Abrasion Tester applies unidirectional abrasion to a tensioned strip of fabric drawn over an abrasion bar, the fabric being bent or flexed as it is rubbed against (over) the bar (ASTM D3885). Because of the effect of lubricants and softeners on flex abrasion results, which may not be reflected in actual wear performance, it is often advisable to remove such lubricants and softeners from the fabric prior to testing.

28.8.6.5 Strength

Three main types of strength tests are carried out, tensile, bursting and tear. The specific test selected depends on both the type of fabric (e.g. knitted or woven) and the intended end use. Other tests carried out include the peel strength of bonded or laminated fabrics.

Tear strength is generally regarded as a better measure of the serviceability of woven fabrics than tensile or bursting strength. For knitted fabrics, bursting strength is almost solely used to measure fabric strength. Tensile strength and bursting tend to be fairly highly correlated, but fabric elongation has a greater effect on bursting strength than on tensile strength and can therefore affect the correlation. This is not the case for tensile strength and tear strength, which are generally poorly correlated, with fabric cover factor (tightness) affecting the correlation. Although tensile strength is frequently taken as a measure of fabric serviceability, tear strength is preferable, since in many applications fabric tearing is the cause of product failure. In the case of tight fitting nonstretch clothing, the bursting strength of the fabric, whether knitted, woven or nonwoven, is important.

28.8.6.6 Tensile strength

Tensile testing refers to those cases where the force (load) is applied unidirectionally, usually on a strip of fabric – for example, in either the warp or weft direction in the case of a woven fabric. The test could be either a ravelled strip test (ASTM D5035) or a grab test (ASTM D5034), carried out on instruments that enable the fabric extension at break, elastic recovery, etc., to be measured as well. These testers generally operate on the constant rate of extension principle (sometimes constant rate of traverse), with the rate of extension variable according to the test method and the requirements. The tensile strength – also termed 'breaking strength' or 'breaking load' – is the force required to rupture a fabric sample of certain dimensions, with the breaking elongation or breaking extension being the elongation (usually in per cent) at that point. Test methods include, ASTM E-4, D5034, D5035, BS 1610/0.5, BS 2576, BS EN 10002-2, DIN 51221/1 and DIN ISO BS EN 13934-1.

28.8.6.7 Bursting strength

Bursting strength represents a composite and simultaneous measure of the strength of the yarns in all directions (biaxial) when the fabric is subjected to bursting type forces, applied by a ball or an elastic diaphragm (e.g. Mullen Tester). In the diaphragm test, the fabric specimen, mostly circular in shape, is securely clamped over an elastic (rubber) diaphragm in a ring (annular) clamp and subjected to a hydraulic load, the pressure required to burst the fabric being recorded (ASTM D3786 and D3787, DIN 53861, BS 3137, BS 3424-38, and BS 4768, and ISO 13938-1/2960, 2758/2759/3303/3689/13938-2). It is important to relate the values obtained to the specific test conditions, such as testing speed and size of specimen.

28.8.6.8 Tear strength

Tearing strength is defined (ASTM D1682) as the force required to start or to continue to tear a fabric, in either weft or warp direction, under specified conditions. A tear in a fabric or garment generally occurs progressively along a line, and can be initiated by a moving fabric being caught on a sharp object. Several methods are used to measure tear strength, e.g. double tongue rip (tear) test, trapezoid tear test, (ASTM D5587) and single tongue tear test (ASTM D2661, BS 4303).

Another popular method of measuring the tearing strength of a fabric is by using a pendulum-type tester, such as the Elmendorf manual or digital tearing testers (ASTM D1424 and D5734, ISO 1974 and 9290, BS 4253/4468/3424, DIN 53862/53128 and BS EN ISO 13937) to measure the tear energy. Because of the very short tearing time (about 1 s) the Elmendorf test approaches an impact tear test, also referred to as a ballistic (pendulum test method), the test results representing the energy to tear a fixed length of fabric. Specific tear strength is defined as the sum of warp and weft tear strength divided by the fabric mass per unit area.

28.9 RESEARCH AND FUTURE TRENDS

Future trends are likely to include increasing functionalities such as UV protection, self-cleaning, self-healing, anti-odour, aroma-generating, and health and wellness properties in a single garment.

28.10 SUMMARY

Fabric and garment functionality, or multifunctionality, indicates enhanced performance in at least two of the following:

- Handle
- Wrinkle resistance
- Pilling resistance
- Bagging resistance
- Good drape
- Durability, including abrasion resistance and strength

28.11 PROJECT IDEAS

1. Select a range (±20) of commercial fabrics and measure their bending stiffness and flexural rigidity using the cantilever method. Measure the same set of fabrics for drape, on any recognized drape-measuring instrument, using standard test methods. Now relate the cantilever-derived values to the drape coefficient, using graphs to plot the results, and preferably regression analysis. Also, try to explain why the results do not lie on a straight line. Explain which factors play a role in (affect) fabric stiffness and drape.
2. Select a range (±20) of knitted fabrics differing in structure and composition, and measure their pilling propensity using different instruments and the appropriate standard test methods (e.g. random tumble, pill box, Martindale). Compare the results obtained with the different instruments, and determine their agreement and correlation, by plotting the appropriate points and if possible, by correlation and regression analyses. Describe the effect of fibre, yarn and fabric properties on fabric pilling.
3. Select a range (±20) of either commercial knitted or woven fabrics, differing in mass (weight) and composition, and measure their abrasion resistance on a Martindale-type flat abrasion tester, using standard test methods. Determine the abrasion resistance in different ways, for example by measuring the number of cycles (rubs) to end point, the mass loss after a predetermined number of cycles (rubs) and the change in appearance. Relate the different measures to each other, graphically and statistically, and interpret the results. Explain which fibre, yarn and fabric parameters affect fabric abrasion resistance.

28.12 REVISION QUESTIONS

- Describe how fibre, yarn and fabric properties affect fabric handle.
- What are the two types of wrinkles that can occur in fabric?
- List six of the nine factors that influence the extent of wrinkling.
- Describe how pills are formed.
- Describe the three ways bagging can be measured.

REFERENCES

Anonymous. (1972). Methods and finishes for reducing pilling, part 1. *Wool Science Review, 42*, 32–45.

Anonymous. (1981). *Fabric drape (a review)* (IWS E.A.C.T. Technical information letter).

Bhat, A. K. (February 2007). Wrinkle free – processing tips. *Colourage, 54*, 85–86.

Crowther, E. M. (1985). Comfort and fit in 100% cotton-denim jeans. *The Journal of the Textile Institute, 76*, 323–338.

Cusick, G. E. (1965). The dependence of fabric drape on bending stiffness. *The Journal of the Textile Institute, 56*, T596–T606.

De Boos, A. G. (1997). The objective measurement of finished fabric. In P. R. Brady (Ed.), *Finishing and wool fabric properties, a guide to the theory and practice of finishing woven wool fabrics* (44). CSIRO/IWS.

De Boos, A. G., & Tester, D. (1994). *Siro FAST fabric assurance by simple testing: a system for fabric objective measurement and its application in fabric and garment manufacture.* CSIRO Rep No WT 92.02.

Denby, E. F. (1982). Wrinkling. In S. Kawabata, Postle R, & M. Niwa (Eds.), *Proc Japan – Australia joint symposium on objective specification of fabric quality, mechanical properties and processing* (pp. 61–74). Kyoto.

Fan, J., & Hunter, L. (2009). *Engineering apparel fabrics and garments*. Oxford: Woodhead.

Fan, J., Hunter, L., & Liu, F. (2004). Objective evaluation of clothing appearance. In J. Fan, W. Yu, & L. Hunter (Eds.), *Clothing appearance and fit: Science and technology* (pp. 43–71). Cambridge: Woodhead.

Galbraith, R. L. (1975). Abrasion of textile surfaces. In M. J. Schick (Ed.), *Surface characteristics of fibers and textiles* (p. 193). New York: Marcel Dekker.

Grünewald, K. H. (1977). Relationships between laboratory test-results and wearing properties of textiles. *Textil Praxis International*, *32*, 860–863, 947–950.

Hunter, L. (2009a). Wrinkling of fabrics and garments. In J. Fan, & L. Hunter (Eds.), *Engineering apparel fabrics and garments* (pp. 52–70). Oxford: Woodhead.

Hunter, L. (2009b). Pilling of fabrics and garments. In J. Fan, & L. Hunter (Eds.), *Engineering apparel fabrics and garments* (pp. 71–86). Oxford: Woodhead.

Hunter, L. (2009c). Bagging of fabrics and garments. In J. Fan, & Hunter (Eds.), *Engineering apparel fabrics and garments* (pp. 87–101). Oxford: Woodhead.

Hunter, L. (2009d). Durability of fabrics and garments. In J. Fan, & L. Hunter (Eds.), *Engineering apparel fabrics and garments* (pp. 161–200). Oxford: Woodhead.

Hunter, L., Fan, J., & Chau, D. (2009). Fabric and garment drape. In J. Fan, & L. Hunter (Eds.), *Engineering apparel fabrics and garments* (pp. 102–130). Oxford: Woodhead.

Hunter, L., & Hunter, E. L. (2009). Handle and making-up performance of fabrics and garments. In J. Fan, & L. Hunter (Eds.), *Engineering apparel fabrics and garments* (pp. 1–51). Oxford: Woodhead.

Kernaghan, K., Stuart, T., McCall, R. D., & Sharma, S. S. (2007). A review on the development of rapid analytical techniques for assessing physical properties of modified linen fabrics. In R. D. Anandjiwala, L. Hunter, R. Kozlowski, & G. Zaikov (Eds.), *Textiles for sustainable development* (pp. 81–93). New York: Nova Science.

Kirk, J. W., & Ibrahim, S. M. (1966). Fundamental relationship of fabric extensibility to anthropometric requirements and garment performance. *Textile Research Journal*, *36*, 37–47.

Leeder, J. D. (1976). Wrinkling of wool fabrics, part 1: mechanism of wrinkling and methods of assessment. *Wool Science Review*, *52*, 14–29.

Leeder, J. D. (1977). Wrinkling of wool fabrics, part 2: methods of modifying wrinkling performance. *Wool Science Review*, *53*, 18–33.

Morton, W. E., & Hearle, J. W. S. (2008). *Physical properties of textile fibres* (4th ed.). Cambridge: Woodhead. 414–457.

Niwa, M. (December 2001). Clothing science, its importance and prospects. *Textile Asia*, *32*, 35–38.

Potluri, P., Porat, I., & Atkinson, J. (1995). Towards automated testing of fabric. *International Journal of Clothing Science and Technology*, *7*, 11–23.

Smuts, S. (1989). *A review of the wrinkling of wool and wool/polyester fabrics*. Tex Report No 1. South Africa: CSIR.

Stoll, R. G. (1949). An improved multipurpose abrasion tester and its application for the evaluation of wear resistance. *Textile Research Journal*, *19*, 394–415.

Stump, D. M., & Fraser, W. B. (1996). A simplified model of fabric drape based on ring theory. *Textile Research Journal*, *66*, 506–514.

Tomasino, C. (2005). Effect of wet processing and chemical finishing on fabric hand, and Effects of mechanical finishing on fabric hand. In H. M. Behery (Ed.), *Effects of mechanical and physical properties on fabric 'hand'* (p. 289). Cambridge: Woodhead.

Wang, H., Mahar, T. J., & Postle, R. (2013). Instrumental evaluation of orthogonal tactile sensory dimensions of fine lightweight knitted fabrics. *The Journal of the Textile Institute*, *104*, 590–599.

Wilkinson, P. R., & Hoffman, R. M. (1959). The effects of wear and laundry on the wrinkling of fabrics. *Textile Research Journal*, *29*, 652–660.

Zhang, X., Yeung, K. W., Miao, M. H., & Yao, M. (2000). Fabric-bagging: stress distribution in isotropic and anisotropic fabrics. *The Journal of the Textile Institute*, *91*, 563–576.

CHAPTER

IMPROVING THE COMFORT OF GARMENTS 29

L. Hunter[1], J. Fan[2]

[1]*CSIR and Nelson Mandela Metropolitan University, Port Elizabeth, South Africa;*
[2]*Cornell University, New York, NY, USA*

LEARNING OBJECTIVES

At the end of this chapter, you should be able to:

- Describe the principal characteristics of physiological and psychological comfort, as they relate to fabrics and clothing
- Understand the influence of different factors on physiological and psychological comfort
- List and describe the different factors that determine physiological and psychological comfort
- Describe the ways in which the different comfort factors can be measured

29.1 INTRODUCTION

Comfort is a highly subjective phenomenon, with no unanimously agreed quantitative definition. In the Oxford English Dictionary and Webster's Third New International Dictionary, comfort is defined as 'freedom from pain, trouble and anxiety; therefore comfort is a contented enjoyment in physical or mental well-being'. Slater (1986) defined comfort as a pleasant state of physiological, psychological and physical harmony between a human being and his or her environment. Physiological comfort refers to the human body's ability to maintain life. Psychological comfort refers to the mind's ability to keep functioning satisfactorily without external help. Physical comfort refers to the effects of the external environment on the body's physiological and psychological equilibrium (Slater, 1986).

Smith (1986) classified physiological discomfort relating to clothing into three categories: sensorial discomfort, better termed tactile discomfort (i.e. what the fabric/garment feels like when it is worn next to the skin); thermo-physiological discomfort; and garment fit (i.e. restrictions or pressure imposed by the garment). This modified classification is illustrated in Figure 29.1 (Fan, 2009a).

According to Fan (2009a) a physiologically comfortable clothing ensemble should:

- have adequate thermal insulation so as to keep the body and skin temperature of the human body within a narrow limit;
- be highly permeable to moisture transmission and have good liquid water absorption and transport properties so as to keep the skin dry;
- not cause any tactile discomfort;

Textiles and Fashion. http://dx.doi.org/10.1016/B978-1-84569-931-4.00029-5

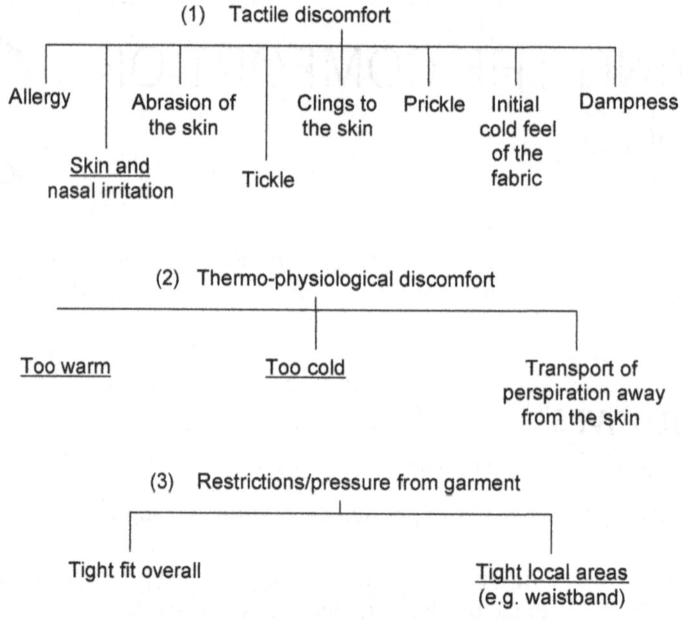

FIGURE 29.1

Classification of clothing physiological discomfort sensations.

Source: Modified from Smith (1986) and Fan (2009a).

- not impose excessive pressure on the body; and
- not restrict movement of the body.

These requirements sometimes compete against each other, and the challenge in engineering physiological comfort of clothing is to find the right balance for different end uses.

29.2 TACTILE COMFORT

Tactile sensations associated with garments arise through the triggering of sensory receptors in or near the skin surface by the contact of the fabric surface with the skin. Tactile comfort is therefore determined by the characteristics of the fabric surface and the protruding fibres (Garnsworthy, Gully, Kenins, Mayfield, & Westerman, 1988).

There are three basic categories of skin-sensory receptors, the touch group, the thermal group and the pain group. The thermal group responds to cold and warmth, and monitors the transfer of heat to or from the skin. The touch group responds to skin tactile sensations ranging from the relatively mild tickle and wet-cling sensations through to the more severe discomfort associated with an allergic reaction. Tactile sensations associated with softness, stiffness and clinginess are examples of information likely to be conveyed by skin-sensory receptors. Tactile sensations and responses include (Fan, 2009a; Mayfield, 1987; Smith, 1985):

- allergies;
- skin and nasal irritation;
- localised irritation;

- skin abrasion;
- tickle;
- clinging;
- prickle;
- warm/cool; and
- dampness/wetness.

Allergies (e.g. rash or other allergic reactions) can be caused by the presence of certain chemical finishes (e.g. formaldehyde) on a fabric surface, while skin and nasal irritation can be caused by loose fibres (fibrous lint) released from the fabric surface and which cause discomfort due to nasal irritation and tickle or prickle sensations on the skin. Localised irritation can be caused by sewn-in garment labels and to a lesser extent by abrasion associated with seams.

Skin abrasion is a common cause of tactile discomfort when there is frequent relative movement between fabric and skin, for example, during physical activity. It is not only related to the fabric–skin contact area, which in turn is related to garment fit, but also to the moisture on the fabric surface, the presence of perspiration, for example, aggravating the situation and promoting abrasion of the skin. Clothing for those who have sensitive skins, particularly babies and the elderly, should therefore be engineered to minimise skin wetness by using water-absorbent and moisture-permeable fabric, and be designed with enhanced ventilation between skin and clothing.

Tickle is caused by protruding fibres on the fabric surface and can be influenced by garment fit (Smith, 1985), which determines the amount of relative movement between the fabric and the body. The more often the fabric moves over the skin, the more frequently a tickle sensation is experienced by the wearer.

Clinging comes in different forms, including wet cling (e.g. perspiration), tacky cling and cling due to static charge, and is often also a function of the area of fabric in contact with the skin, which is related to the fabric structure as well as to the garment fit.

Prickle sensations are caused by stiff, mostly coarse, fibres protruding from the fabric surface, a soft skin (e.g. that of babies) being more sensitive to this kind of discomfort. Prickle is a pain sensation arising through the triggering of the pain group of the skin-sensory receptors. Garnsworthy et al. (1988) found that the subjective magnitude of the prickle sensation depends on the prickle stimulus intensity, which is the number of protruding fibres per 10 cm^2, having buckling loads greater than 75 mN.

$$R_p = 0.54 S_p^{0.66} \quad (29.1)$$

where:

R_p = the subjective magnitude of prickle sensation and
S_p = the prickle stimulus intensity, i.e. the number of protruding fibres per 10 cm^2 having a buckling load greater than 75 mN.

According to buckling theory, the critical buckling load is proportional to Ed^4/l^2 (E: bending modulus, d: diameter of the fibre and l: length of the protruding fibre). It has been shown that, for wool, a fibre diameter of 30 μm and a protruding length of 2 mm will give a critical buckling load of 0.75 mN (Naylor, 1992; Veitch & Naylor, 1992). Therefore, the percentage of coarse fibres (i.e. fibres >30 μm) is important in terms of prickle sensation.

Fibre diameter distribution is a critical factor in terms of the prickliness of a fabric, with moisture also playing a role. Finer fibres, lower fibre-bending modulus, chemical processes that reduce fibre bending modulus, and brushing and raising (which increase the length of protruding fibres), will reduce prickle, while cropping, which shortens the protruding fibres, tends to increase the prickle discomfort.

Initial warm/cool sensations are experienced when a garment is first put on or a fabric is first touched, and are determined by the heat transfer between the skin and the fabric surface. Such sensations are predominantly influenced by the surface contact area with the skin and the fibre characteristics, notably their heat conductivity. In general, fabrics with a hairy surface and bulky fabrics, containing more air, create a warmer feeling; more compact and less hairy fabrics creating a cooler feeling.

Dampness (or wetness) perceptions, due to liquid or moisture on the skin or in clothing, lead to uncomfortable dampness-related sensations during wear, such as clamminess and stickiness, the degree of discomfort being influenced by the amount of moisture at the clothing–skin interface and by the fibre type, notably the fibre hygroscopicity. The sensation of dampness is produced by the loosely held water, i.e. by the water content in excess of the equilibrium regain of the fibre. Highly hygroscopic fibres, such as wool, are therefore perceived to be significantly drier (less damp) than weakly hygroscopic fibres, such as polyester.

29.3 THERMO-PHYSIOLOGICAL (THERMAL) COMFORT

Thermo-physiological comfort, referred to as thermal comfort, is defined by American Society of Heating, Refrigeration and Air-Conditioning Engineers (ASHRAE) as that condition of mind that expresses satisfaction with the thermal environment (Fanger, 1970), and depends on the thermal physiological conditions of the human body. The human being is a homeotherm, which means the central core temperature must be maintained within narrow limits – for a person under extended exposure, maintaining core temperature within 37 ± 0.5 °C is vital for survival. Clothing contributes to thermal comfort by assisting the human body to maintain comfortable thermal physiological conditions over an extended range of environments and periods, and is largely determined by the *thermal resistance* (or insulation), *moisture permeability* and *liquid water transport* properties of the clothing.

29.3.1 FACTORS AFFECTING THE THERMAL INSULATION OF FABRICS AND CLOTHING

29.3.1.1 Fibre and yarn properties

Fan (2009a) drew the following conclusions from work carried out by various researchers over many years:

- The thermal insulation of a textile fabric is primarily a function of the still air contained (trapped) within the fabric and yarn. Factors such as crimping or texturing, which help to trap still air, therefore improve the thermal insulation of fabrics. Another method of trapping still air in fibres is to imitate the medulla in certain animal fibres and create a hollow fibre, which combines good insulation with lightness.
- Due to the relatively low bulk density of textile fabrics, differences in the thermal conductivity of the fibres have little effect on the overall thermal insulation of the fabric.
- Generally, the thermal transmission of textile materials increases with density – the greater the bulk density for a given thickness, the greater the thermal transmission and the lower the warmth,

due to the replacement of air by fibres having a greater heat conductivity. Nevertheless, if the bulk density is very low, or if the fabric construction is sufficiently open, radiant heat from the skin can pass through the garment, thereby reducing its thermal insulation.
- Fabric thickness is the most important factor governing thermal insulation, there being a linear relationship between thermal insulation and fabric thickness. When two fabrics are of equal thickness, the lower density fabric has the greater thermal insulation, although there is a critical density (about 0.6 g/cc) below which convective effects become important and thermal insulation falls.
- The thermal insulation of fabrics increases significantly when two fabrics or garments are layered as compared to a single fabric or garment of similar overall thickness. This is because the trapped air between the two fabrics provides additional thermal insulation.
- Thermal insulation decreases with an increase in the water content of the fabric.
- Thermal insulation can be increased by increasing the air gap between the body and the fabric, but it starts to decrease again beyond a gap of about 7.5–10 mm, because of convection effects.
- Increased wind velocity reduces the thermal insulation of a fabric relative to that in still air. The effect is minimised in closely woven fabrics.
- The thermal insulation of a fabric is improved significantly if it is covered with a fine closely woven outer fabric.
- A slight increase in thermal insulation is observed with an increase in the weight of the fabric.

29.3.1.2 Garment design

Generally, much more heat is lost through openings in a garment than through the fabric layers of the garment. Therefore, garment design, notably in terms of openings and entrapped air, plays a vital role in the overall thermal insulation of clothing. The following design factors affect the thermal insulation of a garment or clothing ensemble (Fan, 2009a):

- The higher the thermal insulation of the fabric used in the garment, the higher that of the garment.
- The greater the amount of the body covered, the better the insulation.
- Looseness or tightness of fit affects the amount of air trapped and therefore thermal insulation.
- An air space between the body and the garment generally has a beneficial effect on thermal insulation.
- A greater ensemble weight provides higher thermal insulation.
- Layering of fabrics in the garment improves thermal insulation.
- Garment openings generally reduce thermal insulation.
- Body movement can reduce thermal insulation, by 10–15%.

29.3.2 FACTORS AFFECTING THE MOISTURE (VAPOUR) TRANSMISSION PROPERTIES OF FABRIC AND CLOTHING

Evaporative heat loss is the most important means of body cooling during high activity or in a hot climate. However, even under indoor sedentary conditions, the human body loses water from the skin by 'insensible perspiration'. If water vapour cannot escape sufficiently fast through clothing, even under normal conditions of atmospheric temperature and humidity and at low levels of body activity, there will be a build-up of moisture at the skin surface and within the clothing, which will result in uncomfortable sensations, such as dampness and clamminess.

The movement of water vapour through a fabric is largely a function of the microporous structure of the material, and this movement can therefore be changed by changing the structure, which is related to fibre properties, fabric construction, yarn texturing, twist, and finishing (chemical and mechanical) treatments. Related information can also be found in Section 29.6.

29.3.2.1 Fibre properties

Moisture transmission through a fabric is affected by the hygroscopicity of the fibres. Hygroscopic fibres, such as wool and cotton, absorb moisture and at the same time release heat of absorption. The absorption of moisture and release of heat change the moisture concentration and temperature profile across the fabric, and consequently change the moisture transfer through the fabric. Fibre swelling and changes in shape can also affect the fabric microstructure and therefore its water vapour transmission.

29.3.2.2 Fabric construction

Generally, the percentage of volume occupied by the fibre in the fabric is a dominant factor in terms of moisture transmission, dense and thick fabrics tending to have a greater resistance to moisture transmission.

29.3.2.3 Types of breathable fabrics

The water vapour transmission properties of waterproof breathable fabrics (or breathability) are affected by the type of coating or membrane, which is discussed in more detail in Section 29.6.

29.3.2.4 Garment design

Garment design has a significant effect on moisture transfer, affecting the amount of body surface area covered, looseness or tightness of fit (air gap), wind penetration and ventilation through openings. The moisture vapour resistance of clothing initially increases with the size of the air gap between the garment and the body, the rate of increase gradually decreasing as the air gap increases, owing to increased convection.

29.3.3 FACTORS AFFECTING LIQUID WATER TRANSPORT PROPERTIES OF FABRICS AND CLOTHING

Liquid water transport properties of fabric and clothing are important, especially for clothing worn under hot and humid conditions or under conditions of high physical activity, when evaporation of perspiration is a major means of body cooling. The ideal fabric should keep the skin dry by allowing perspiration to evaporate and/or flow to the outer layer of the clothing. Wetting and wicking are of major practical importance in the absorption and transportation of liquids in textiles. Wickability can be defined (Harnett & Mehta, 1994) as the ability to sustain capillary flow. Wettability, a prerequisite of wicking, refers to the initial behaviour of the fabric, yarn and fibre, and can be defined as the interaction between the liquid and the substrate before wicking takes place.

29.3.3.1 Fibre properties

Fabrics made from hydrophilic fibres, such as cotton and viscose, tend to have very good wettability, provided any residual oil or wax is removed, since these can greatly affect wettability.

Fibre cross-sectional shape has a significant effect on liquid water transportation properties, the vertical liquid wicking height in a bundle of filament yarns increasing with an increase in the noncircularity of the fibre and with a decrease in void spaces between the filaments, better wicking also taking place when more yarns are bundled together.

29.3.3.2 Fabric construction

Smaller pore sizes in the fabric produce higher capillary pressure and thus enhance liquid spreading. Interfibre pores must have the appropriate dimensions to produce sufficient capillary pressure and interconnective pathways to transport the liquid, and to have sufficient overall porosity to retain the liquid. Recognising the superior water transport system in plants, Fan, Sarkar, Szeto, and Tao (2007) developed textile fabrics which emulate the branching structure of plants and which exhibit excellent initial water absorption and moisture management properties.

29.3.3.3 Surface treatment and finishing

Surface treatment can modify the liquid water transport properties of fabrics. Plasma treatment, for example, improves the wettability, re-wettability and wicking of textile materials. Hydrophilic finishing can improve water absorption, although the durability of the finished product after repeated laundering is a concern.

29.3.4 FACTORS AFFECTING GARMENT FIT AND EASE OF BODY MOVEMENT

Clothing represents the second skin of the human body and must fit the human body and adjust to its movement. The importance of garment fit for clothing comfort has long been recognised – fit determines the amount of relative movement between the fabric and the body and influences tactile sensations.

29.3.4.1 Fabric properties

When people move, the skin elongates and recovers, and this varies for different parts of the body depending on the activity. Consequently, garment fit and ease of body movement are closely related to the extensibility and recovery of fabrics. Most woven fabrics are rigid because the interlacement and crimp of the yarn allows little extension to occur, unless the yarns (or fibres) themselves stretch, as is the case for stretch woven fabrics constructed from elastic fibres and yarns. Knitted fabrics, due to their interlooping yarns, usually possess a minimum of 15% elongation, but can sometimes also be rigid. In general, knitted fabrics are more extensible than woven fabrics, and hence are preferred for tight-fitting garments such as underwear.

29.3.4.2 Garment design

Heavy and bulky garments may provide thermal protection to the wearer in extremely cold or hot conditions. During body movement, the body expands and contracts in the area surrounding the joints, and hence a garment's expansion and contraction should follow the same pattern of body movement. For example, if a garment is too tight-fitting and the fabric is nonstretch, the wearer may have difficulty in bending at the joint. Garments should be designed to facilitate body movement, but be neither too tight nor too loose.

Garments impose pressure on the wearer, which, if excessive, can cause discomfort and detrimental physiological effects. Pressure comfort is therefore an important issue, especially for close-fitting garments such as swimwear, body shapers and some specialised functional garments that can generate excessive pressure, for example, pressure garments for skin healing and baby huggers. To minimise pressure discomfort, it is important to avoid pressure concentration at any position of the body.

Measures of clothing pressure tend to correspond to the sensation of tightness, which is not only related to clothing pressure, but also to factors such as body size, muscular resilience and bone structure. The pressure a garment imposes on the body surface increases with sharpness of the curvature or contour and fabric tension, the latter, for example, increasing with greater negative ease and using fabrics with higher Young's modulus.

29.4 MEASURING PHYSIOLOGICAL COMFORT

Physiological comfort is a complex function of many different factors and no single measurement, except perhaps wearer trials, is available for evaluating the physiological comfort of fabrics and garments. In practice, the different components or aspects of physiological comfort are generally measured separately, as described in the following section.

29.4.1 TACTILE COMFORT

The tactile comfort of clothing can be assessed through subjective wearer trials by using psychological scaling (i.e. rating discomfort on a defined scale) or psychophysical estimation (i.e. measuring a single sensation in relation to its initiating physical stimulus). Psychophysical scaling is considered more reliable than psychological scaling.

Tactile comfort can also be evaluated by measuring objectively the physical stimuli that directly cause a specific tactile discomfort sensation. Many of the tactile discomfort sensations, such as skin irritation, skin abrasion, tickling and prickling, are related to the fabric surface characteristics, which can be evaluated by the Fabric Surface Property Tester of the Kawabata Evaluation System for Fabrics (KESF) system, which measures fabric roughness (or the variation in fabric thickness) and surface frictional forces.

A technique has been devised for measuring the prickle propensity of fabrics objectively. In this method, a fabric under a pressure of 4gf/cm^2 is placed on a thin polytetrafluoroethylene (PTFE) skin spread over a glass slide. The pliability and poor elasticity of the PTFE skin results in fibre ends with buckling loads of about 100 mg or more, leaving imprints on the PTFE skin. The number of imprints are counted under a microscope and used as a measure of prickle propensity (Mayfield, 1987). A more recent development by CSIRO, Australia, has been described (Ramsay, Fox, & Naylor, 2012).

29.4.2 THERMAL CONTACT

The warm–cool contact feeling of the fabric can be physically evaluated by measuring the heat flow from a hot surface to a fabric surface. Examples of two commercially available instruments are KES Thermo Labo-II and Alambeta. The Alambeta uses a heat flux sensor, having a thermal inertia similar to that of the human skin, to measure the heat flow from the measuring head to the fabric specimen, its warm–cool feeling sensitivity approximating that of the human skin.

29.4.3 THERMAL INSULATION

The thermal insulation properties of fabrics can be measured and expressed in terms of either thermal conductivity or thermal resistance. The International System of Units (SI) unit for thermal conductivity is W/mK and for thermal resistance is Km^2/W.

There are two additional popularly used units for thermal resistance or insulation, Tog and clo. A 'Tog' is defined as the approximate insulation of light summer clothing, with 1 Tog = 0.1 km^2/W. The 'clo' is defined as the approximate insulation required to keep a resting person (producing heat at the rate of 58 W/m^2) comfortable in an environment of 21 °C and air movement of 0.1 m/s, or roughly the insulation value of typical indoor clothing. A thermal resistance of one clo is equivalent to a thermal resistance of 1.55 Togs or 0.155 km^2/W.

Several instruments are available for determining the thermal insulation properties of fabrics or fabric assemblies, including:

- Guarded Hot Plate
- KESF Thermo Labo-II
- Alambeta instrument

In general, these methods determine heat flux by measuring the energy required to maintain a set temperature in a heated device when it is covered by a textile, as well as the temperature difference across the textile material.

The thermal insulation properties of **garments or clothing** ensembles can be measured using heated thermal manikins, which can be grouped into three generations: the first generation being standing (not walkable) and nonperspiring, the second being moveable (walkable) and nonperspiring, and the third, most advanced, being moveable and perspiring. A good example of the latter is 'Walter' (see Figure 29.2), developed in Hong Kong.

The thermal insulation of garments or clothing ensembles on the manikin can be determined by:

$$I_t = \frac{A_s (T_s - T_a)}{H} \tag{29.2}$$

where:

I_t = the total thermal insulation of the clothing plus air layer
H = the total the heat loss from the manikin
A_s = the surface area of the manikin
T_s = the mean skin temperature and
T_a = the mean ambient temperature.

29.4.4 WATER VAPOUR PERMEABILITY

The moisture transmission properties of fabrics and clothing may be measured in terms of water vapour transmission rate (WVTR), moisture vapour resistance or moisture permeability index; a summary table of the various test methods is given in Overington and Croskell (2001).

American Society for Testing and Materials (ASTM) E96E describes a Cup Method, which is commonly used for testing the moisture transmission properties of fabrics. It measures the rate of water vapour transmission perpendicularly through a known area of a fabric within a controlled atmosphere. In this method, a cup containing distilled water is covered with a specimen and placed in a controlled environment of 20 °C and 65% relative humidity. The WVTR in grams per hour and per square metre is calculated using the following equation:

$$\text{WVTR} = G/tA \tag{29.3}$$

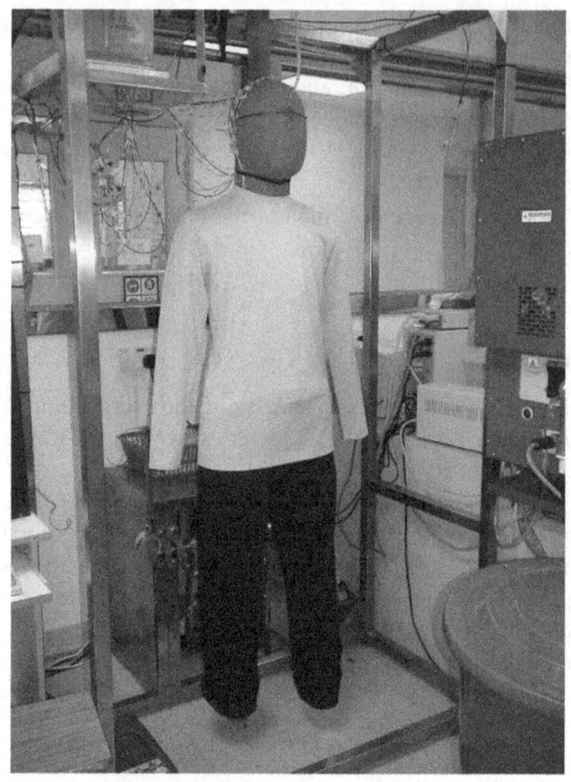

FIGURE 29.2

Sweating fabric manikin Walter.

Source: Fan (2009a).

where:

G = the weight change of the cup with fabric sample in grams
t = the time in hours during which G occurred and
A = the testing area in square metres.

Moisture transmission rate can also be measured by the Moisture Transmission Tester developed by Ludlow Corp., which is a much faster method than the ASTM E96E Cup Method. Using a sweating guarded hot plate (ISO 11092 and ASTM F1868), the total moisture vapour resistance of the fabric sample on the plate, together with the surface air layer, can be determined from the measurement of evaporative heat loss.

The moisture transmission properties of **garments or clothing** ensembles can be measured using a sweating manikin. With the sweating manikin 'Walter', the total moisture vapour resistance is determined using the following formula:

$$R_{et} = \frac{A\,(P_{ss} - P_{sa}H_a)}{H_e} - R_{es} \tag{29.4}$$

where:

A = the surface area of the manikin
P_{ss} = the saturated vapour pressure at the skin temperature
P_{sa} = the saturate vapour pressure at the ambient temperature
H_a = the ambient relative humidity (%)
R_{es} = the moisture vapour resistance of the fabric-skin which is calibrated in advance ($R_{es} = 8.6\,m^2Pa/W$) and
H_e = the evaporative water loss. H_e is calculated from the measurement of evaporative water loss, $H_e = \lambda Q$

where

λ = the heat of evaporation of water at the skin temperature ($\lambda = 0.67\,Wh/g$ at 34 °C) and
Q = the rate of evaporative water loss per hour.

29.4.5 LIQUID WATER TRANSPORT PROPERTIES

Four types of test methods are used to measure the water transport properties of fabrics, namely, **longitudinal wicking 'strip'** tests, **transverse (or transplanar) wicking 'plate'** tests, **areal wicking 'spot'** tests and **syphon** tests, using various instruments.

Two **longitudinal wicking 'strip'** tests are used as industrial standards BS3424 Method 21, Determination of Resistance to Wicking, and DIN 53924, Determination of the Rate of Absorption of Water by Textile Materials (Height of Rise Method). Both these methods use a preconditioned strip of the test fabric, suspended vertically with its lower end immersed in a reservoir of distilled water, to which may be added a dye (of a type known not to affect the wicking behaviour) for tracking the movement of water. After a fixed time has elapsed, the height the water reaches in the fabric above the water level in the reservoir is measured. BS3424 Method 21 specifies a very long time period (24 h) and is intended for coated fabrics with very slow wicking, whereas DIN 53924 specifies a much shorter time for the test (5 min maximum), appropriate for relatively rapid wicking fabrics.

Transverse (or transplanar) wicking 'plate' tests involve measuring the rate water is wicked two-dimensionally along/across a piece of fabric held horizontally. The water can be supplied to the fabric via a sintered or perforated glass plate, or the fabric can come into direct contact with the water. The rate that the water is wicked transplanarly is measured by the mass (weight) loss of the water in the supply reservoir.

There are two published standards in the **areal wicking spot** test category, BS 3554, Determination of Wettability of Textile Fabrics, and American Association of Textile Chemists and Colorists (AATCC) Method 79, Evaluation of Wettability. In these standard tests, a drop of liquid (either distilled water or, for highly wettable fabrics, a 50% sugar solution) is delivered from a height of approximately 6 mm onto a horizontal specimen of the test fabric. The elapsed time between the drop reaching the fabric surface and the disappearance of the reflection from the liquid surface is taken as a measure of how quickly the liquid has spread over, and wetted, the fabric surface. The method is not suitable for highly absorptive fabrics, unless a high-speed camera or electrical sensing device is used for the measurement. A modified procedure uses the reflected light beam to measure the contact angle; alternatively, the drop length–height can be measured. The Moisture Management Tester

(MMT) (Li, Xu, Yeung, & Kwok, 2002) measures the electrical resistance of the upper and lower surface of the fabric and uses this as a measure of spot wicking.

In the **syphon** test, a rectangular strip of the test fabric is used as a syphon by immersing one end in a reservoir of water or saline solution and allowing the liquid to drain from the other end, placed at a lower level, into a collecting beaker. The amount of liquid transferred at successive time intervals can be determined by weighing the collecting beaker. This is a simple test, and does not resemble liquid water transport through clothing during continuous perspiring.

29.4.6 GARMENT FIT AND EASE OF BODY MOVEMENT

Garment fit and ease of body movement are traditionally assessed by means of wearer trials, in which wearers are required to perform a series of activities which normally occur in practice and then asked to rate the ease of body movement on a Likert scale; for example, from one for very stiff to five for very flexible, or from one for very tight to five for very loose. Different assessments are made for different parts of the body under different activities. Sometimes photographs are taken for subsequent visual assessment.

Various objective methods have been developed for evaluating garment fit. These include the fitting index, based on the measurement of the space between the body and clothing, the symmetrised dot pattern technique, based on the measurement of the changes of the dot pattern, and imaging technology through capturing and analysing garment images. Their practical use is very limited due to the difficulty in accurately and efficiently capturing the space between the body and clothing or garment surface, which may be folded or wrinkled.

29.4.7 PRESSURE COMFORT

Pressure comfort may be evaluated by various techniques, including:

- Wearer trials where the wearers are asked to rate the degree of pressure comfort sensation.
- Physiological response testing, such as cardiac output and skin blood flow of human subjects when wearing the pressure garments and undergoing a sequence of activities.
- Measuring the pressure at different body regions by means of pressure sensors when wearing the pressure garments and carrying out a sequence of activities, and then comparing the measured pressure values with the desirable comfortable pressure range.
- Measuring pressure distribution on a soft manikin wearing a pressure garment. The use of a soft manikin that mimics the human body improves the consistency and accuracy of pressure measurement. It also eliminates the need for live models.

A review of various methods for the evaluation of pressure garment is given in Lim (2002).

29.4.8 FORMALDEHYDE CONTENT

Excessive amounts of formaldehyde (CH_2O) present on the fabric or released from the fabric, can cause skin irritation or even skin cancer. National and international standards have been established to determine the formaldehyde content in textiles, including BS EN ISO 14184-1: 2011 and BS EN ISO 14184-2: 2011, for determining the total and free formaldehyde.

29.5 PSYCHOLOGICAL COMFORT

This Section is from Fan (2009B). Psychological comfort may be defined as 'a pleasant state of psychological harmony between a human being and the environment (Slater, 1986) and can perhaps be summarised as a combination of feeling both comfortable and good (a feeling of wellbeing). With respect to clothing, therefore, psychological comfort is the feeling that one is dressed in a style/fashion/manner that is adequate for the purpose of the clothing, and is in accordance with one's view of one's economic, social and functional status vis-à-vis either one's immediate work colleagues or one's wider group of friends, associates and acquaintances (Jeffries, 1988). It can also be when one makes a specific statement or shows allegiance to a specific culture or cause. Whether a garment 'looks right' or not is a careful balance of aesthetics, performance, cost, and whether it conforms to the wearer's perception of what they would like to wear considering the external environment in its totality. Figure 29.3 (Fan, 2009b) summarises the factors related to the psychological comfort of clothing.

Psychological comfort is very much affected by the aesthetic factors of clothing, including colour, texture, garment design elements, garment fit, fashion and prejudice. All these factors should be considered holistically in designing or choosing garments for a targeted end use.

Just like the physical and physiological aspects of comfort, psychological comfort is a neutral sensation and is frequently unnoticed by the wearer. Instead, psychological discomfort is more easily sensed. Since psychological comfort is highly subjective and personal, it is therefore difficult, if not impossible, to indicate how it may be improved. At best, it is possible to indicate how psychological comfort sensations can be reliably measured or assessed and how they are related to fabric and garment design.

29.5.1 FACTORS AFFECTING PSYCHOLOGICAL COMFORT

Various aesthetic and other factors are associated with psychological comfort, including colour, whether the garment style flatters the wearer, shape, fit, sense of fashion, status and/or position, fabric construction and finish, suitability for an occasion, and prejudice. To these could be added factors related to quality and luxury, for example, the 'good or special' feeling when wearing clothing associated with quality and luxury, such as cashmere, silk and leather.

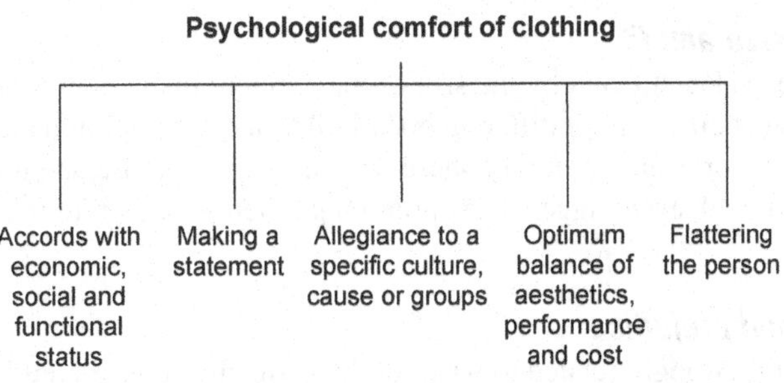

FIGURE 29.3

Factors related to psychological comfort of clothing.

Source: Fan (2009b).

29.5.1.1 Colour

Considerable research has been undertaken on the effect of colour on psychological sensations, these being summarised by Davis (1996). A major influence is culture (Sharpe, 1974). In addition to culture, the association of particular colours with specific feelings and emotions is often different for different age groups. A **warm–cool** perception is influenced by one's experience with the colour of warm or hot objects, and that of cool or cold objects, respectively. Reds, oranges and yellows are associated with a warm feeling; violet, blues and greens are generally associated with a cool feeling; pale values tend to feel cooler, darker and brighter shades warmer, and duller shades cooler.

Colour can also influence human **emotions**, cool hues, dark shades and low intensities making people feel calm, while warm hues, light shades and bright intensities creating a more active feeling. Warm hues, light shades and soft intensities of colour are often associated with femininity, while cool hues, dark shades and bright intensities of colour are often associated with masculinity. Colour can also create **age** illusions, with warm, bright colours tending to project a young and resourceful appearance and dark colours a mature and experienced appearance.

29.5.1.2 Texture

Fabric texture appeals to three senses, namely touch, sight and sound, and greatly affects the mood and psychological perceptions produced by a garment. In general, soft textures, such as flannel and corduroy, suggest casualness and relaxation, whereas firm textures, such as gabardine and other worsteds, are perceived as businesslike, compact and resilient. Texture, colour and garment style should be in harmony to produce a desirable psychological mood.

29.5.1.3 Garment design

Garment design is an integration of all the design elements, including colour, texture, space, lines, pattern silhouette, shape, proportion, balance, emphasis or focal point, rhythm and harmony. Each of these contributes towards the visual perception and psychological comfort of the garment. Principles of illusion can be utilised in garment design to flatter the figure of the wearer (Davis, 1996). For example, the Muller–Lyer illusion (a line with angled extensions at each end appears longer than a line of equal length, but with doubled back angled lines at each end) may be applied to pattern design to lengthen or shorten the perceived figure of the wearer. The lengthening effect may also be created by applying the horizontal–vertical illusion, i.e. a vertical line seems longer than a horizontal one of the same length.

29.5.1.4 Garment size and fit

Perceived body image is also affected by the size of the garment the person wears and how a garment fits the body. The effect is different for different body builds and the optimum garment ease is different for different body sizes. For example, for an obese person, both too loose and too tight-fitting clothing tend to make the person look even bigger. The optimum garment ease to minimise this effect should be around 2–3 cm.

29.5.1.5 Fashion and prejudice

Psychological comfort is strongly related to whether the garment is considered fashionable or not, and whether it fits a certain genre, personal statement, environment and/or age, status, position or other grouping. Fashion is dynamic, changing from season to season and year to year. It is almost impossible to define what makes something become fashionable, except that we know that, for something to be accepted as fashionable, a significant number of people have to acknowledge it as such and give it at least a minimal sign of approval.

Another important factor in psychological comfort is prejudice, defined as an unfair, and often unfavourable, feeling or opinion, not based on reason or adequate knowledge, and sometimes resulting from fear or distrust of ideas different from one's own. A person's sense of fashion and prejudice towards a particular event may affect clothing choice in terms of colour, texture, garment style, etc.

29.5.2 ASSESSING PSYCHOLOGICAL COMFORT

Psychological factors, for which personal idiosyncrasies can be so important, are the hardest of all to measure. There are, however, a number of parameters for measuring the level of psychological discomfort that do not necessarily involve textile products directly (Slater, 1986). These include the determination of interpersonal distance, wish for privacy, level of embarrassment, and effects of stress, and each of these can be estimated by observation, with greater or lesser accuracy. The influence of apparel products on psychological comfort (involving aspects of adornment, status or modesty) is even more difficult to assess directly. Yet, it is possible to observe such indicators as the style of garment bought, the frequency of purchase, or the size of a respondent's wardrobe, in order to check that his or her replies to questioning on the subject are valid.

The perception of one's own body in terms of weight and size is termed 'body image' (Fan, Hunter, & Liu, 2004), with the level of satisfaction with one's own body or body parts being termed 'body cathexis' or body image. Psychological scaling is a common technique used to assess what a wearer thinks or feels. Psychological scales have been developed to assess colour emotions, body image and body cathexis, which are key indicators of psychological comfort.

Bipolar colour–emotion scales are frequently used for quantifying colour emotion. Five-point, seven-point or more commonly nine-point Likert scales have been used to rate body cathexis. The nine-point scale consists of schematic figures or silhouettes of varying sizes, from thin (underweight) to heavy (overweight). The subjects are asked to pick the ideal figure, and the figure that they think most closely matches their own. The difference between the two is a measure of the perception of one's own body (i.e. body image). Body image can be significantly influenced by clothing, with people generally being significantly more satisfied with their clothed bodies than with their nude bodies.

29.6 IMPROVING WATERPROOFING AND BREATHABILITY

This Section is from Hunter & Fan (2009). Consumers, particularly those regularly involved in outdoor activities or subjected to extreme conditions such as snow, rain heat, wind or cold, are increasingly seeking multifunctional clothing to maintain their comfort under such conditions. Ideally, such clothing should keep the wearer dry and comfortable under hot, cold, windy or wet conditions, even under extremes, for example, driving rain and/or extreme temperatures and/or physical activity. In order to do so effectively, the fabric and garment need to have the appropriate heat insulation properties and handle, and preferably be lightweight. Most importantly, such fabrics and garments need to be:

- Breathable (water vapour permeable)
- Waterproof

Although waterproofing and breathability are entirely different concepts, today they are commonly combined – the term 'breathability' is used to imply both 'waterproofing and breathability'. 'Breathability' is taken to mean 'the ability of a textile fabric to allow water vapour (molecules smaller than ≈0.0004 μm in diameter) to pass through it (i.e. from the body surface to the outside) while not allowing liquid water (molecules ≈100 μm in diameter) from the outside to pass through it' (Johnson & Samms, 1997). In recent years, the term 'moisture management' has been used to describe the ability of a garment to transport moisture vapour and liquid (perspiration) away from the skin through the fabric to the outside in a controlled manner, and understanding this property of fabrics has increasingly involved advanced technologies, such as 'body mapping'.

For the body to maintain its core temperature of 37 °C, the heat it produces needs to be removed by conduction, radiation and convection from the skin, and also by evaporation of perspiration released from sweat glands. The first three are adequate for low levels of activity, whereas the fourth predominates at high levels of activity (Holcombe, 1986). To accommodate the latter, the garment must allow moisture to be easily carried away from the point where it is generated, i.e. the skin, so as not to cause the moisture to remain on the skin or to condense on the inside of the garment. This is a critical requirement for maintaining acceptable temperature and comfort, particularly under either very hot or very active conditions, and at various levels and combinations of such conditions.

The challenge is to engineer clothing in such a way that the fabric, each fabric layer, and ultimately the garment, allows air, water vapour and perspiration to pass easily from the inside to the outside, while at the same time providing a barrier to liquid water (as opposed to water vapour) penetrating from the outside. In extreme cases, this is a conflicting requirement, and a compromise must be made. Another important consideration is that the handle and overall comfort properties of the fabric should not be adversely affected by any multifunctional finish or property added.

Some of the relevant terms:

- **Breathability** refers to the ability of a fabric to allow perspiration, evaporated by the body, to escape to the outside (termed moisture vapour transmission), and can be defined as the ability of clothing (and fabric) to allow the transmission or diffusion of moisture vapour, and therefore facilitate evaporative cooling.
- **Water repellency** generally refers to the ability of a fabric to resist wetting. **Water repellent** (also referred to as 'shower resistant') fabrics provide some protection against intermittent rain, but are not suitable in a downpour, as they will become wet through and dampen the wearer. Water will bead and run off the surface (Coffin, 1988) of water repellent fabrics, but under sufficient pressure it will penetrate. The amount of pressure required to do so is a measure of **water resistance**.
- **Water (or rain) resistant** fabrics will resist wetting by water and also prevent it from penetrating or passing through the fabric under most pressures, thereby keeping the wearer dry in moderate to heavy rain (Morris & Morris, 2007).
- **Waterproof** is the extreme case of water resistance, implying complete resistance to water. The fabric will not allow water – even wind-driven rain – through, and such fabrics and garments should keep the wearer dry in prolonged moderate to heavy rain (Morris & Morris, 2007).
- **Windproof** means that air cannot pass through the fabric.

Water resistant fabrics therefore provide better protection against rain, snow and sleet, particularly driving rain than **water repellent** fabrics. Nevertheless, both will eventually become saturated with water and allow water to leak through. Thus, if a fabric or garment needs to keep a person completely dry under virtually all, including prolonged, conditions of rain and weather, it has to be **waterproof**. For a garment to offer such protection, the fabric, seams, zips, etc., all need to be waterproof and the garment must be designed so that there are no openings through which water can penetrate. A waterproof breathable garment needs to satisfy the above waterproof requirements, as well as being able to breathe, i.e. allow water vapour to escape so as to maintain a comfortable microclimate between the skin of the wearer and the garment. The term **weatherproof** stands for both waterproof and/or windproof technologies.

29.6.1 FACTORS AFFECTING FABRIC AND GARMENT BREATHABILITY

Since breathability encompasses two different requirements, namely breathability and waterproofing, both of these characteristics need to be considered and designed into the fabric or garment. This generally involves some compromise between the two extreme cases of total waterproofing and complete breathability. Frequently, the precise balance of properties required for breathable fabrics depends upon the end-use of the fabric and the specific level and balance appropriate to that particular end use.

Moisture diffuses readily through air, whereas fibres present a barrier to such diffusion. Therefore, the resistance of a traditional textile fabric and garment to moisture vapour diffusion will largely depend upon the fabric construction, notably density (compactness or tightness) and thickness, and to a lesser extent on the fibre properties, notably their hydrophilic and hygroscopic properties. Nevertheless, fabrics mostly have a high ratio of air–fibre volume and only in extreme cases do differences in fabric construction per se play a major role in changing moisture vapour diffusion for conventional or traditional fabrics.

In practice, the main ways of achieving waterproof breathable fabrics include:

- using tightly woven or knitted constructions (e.g. involving cotton or microfibres);
- using microporous and/or hydrophilic membranes (a thin polymer membrane laminated to the fabric); and
- using microporous and/or hydrophilic coatings (a thin polymer film, usually polyurethane (PU), applied directly to the fabric).

29.6.1.1 Fibre and yarn properties

Except in the case of closely or tightly woven fabrics, fibre and yarn properties generally play only a secondary role in achieving the required standards for waterproof breathable fabrics. Essentially, the most important fibre properties, within the context of closely woven fabrics, are fibre fineness (the finer the fibre the better) and the ability of the fibre to swell with increasing moisture absorption. Similar considerations apply to the yarn, with yarn twist factor, compactness and smoothness having some effect.

29.6.1.2 Fabric construction

Breathable waterproof fabrics result from the appropriate selection of fabric structure and tightness, together with the appropriate fibre (e.g. cotton or microfibres/filaments), yarn composition and properties, and water repellent fabric finish (e.g. silicone or flourochemicals). The foremost requirement for tightly woven breathable fabrics is that the pore structure is such that water droplets cannot permeate into and through the fabric from the outside. Tightly woven fabrics can be woven from relatively fine synthetic microfibres/filaments such as polyester, polyamide, viscose and acrylic fibres, which produce sufficiently small pores, even when dry.

29.6.1.3 Microporous membranes and coatings

A microporous membrane (or coating) is a thin structure having a morphology of a precisely controlled interconnected network of tiny holes (pores) and an otherwise impermeable polymeric structure, the holes (or pores) being too small to allow water droplets to pass through, but large enough to allow water vapour to pass through. Microporous-based breathable fabrics usually have a surface of hydrophilic PU or water repellent finish (fluorocarbon or silicone), and are generally made from polymers such as PTFE, PU, polyolefins, polyamides, polyester, polyether and polyether-based copolymers, with PTFE and PU being the most popular. In most cases, they can be cast directly onto the fabric (i.e. coated) or formed into a membrane and then laminated to the fabric.

A microporous membrane was first used in the 1970s to produce a breathable fabric called Gore-Tex (W.L. Gore; http://www.gorefabric.com). The membrane (thin film) used was an expanded polytetrafluoroethylene (or Teflon) (ePTFE), having many millions of pores per square centimetre. Today, some of the most durable and high-quality breathable fabrics are produced by laminating (sandwiching) a microporous membrane between two fabrics: an inner, soft and flexible fabric (e.g. warp knitted) and an outer abrasion resistant fabric (e.g. woven nylon fabric), see Figure 29.4 (Ody, 1990). Mid-layer fabrics, i.e. between those in contact with the skin and the outer fabrics, need to offer good wicking and insulation.

29.6.1.4 Hydrophilic membranes and coatings

Hydrophilic (nonporous or nonporometric) membranes and coatings are very thin polymer films, generally of chemically modified polyester or polyurethane, containing essentially no pores, and transmit

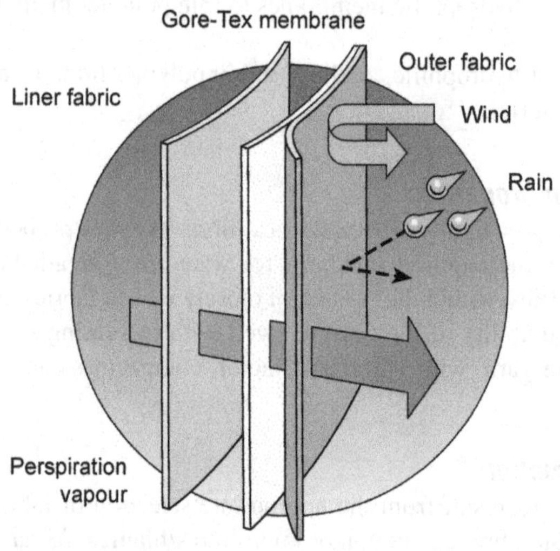

FIGURE 29.4

Gore-Tex®, three-component laminated fabric.

Source: Ody (1990).

water vapour efficiently, but not liquid water. Coatings are generally less expensive and easier to handle than membranes.

29.6.1.5 Combination of microporous and hydrophilic membranes and coatings

Fabrics can also be coated with copolymers (coatings and membranes) combining a nonmicroporous hydrophilic layer with a microporous hydrophobic layer, and this has both advantages and disadvantages compared to a pure microporous film.

29.6.1.6 Smart and biomimetic-based breathable fabrics

Smart (or intelligent) and biomimetic breathable fabrics, as is the case with smart (intelligent) textiles in general, respond to different ambient conditions by changing their properties accordingly, for example, pores that open and close according to water transmission needs. Phase change materials (PCMs) change the breathable properties of the fabric as the environment (microclimate) changes, employing hydrophilic membranes that open up with increasing water vapour and body heat, and close when the converse occurs.

29.6.1.7 Fabric finishes

Generally, waterproof breathable fabrics have a water-repellent (hydrophobic) finish, such as fluoropolymers, fluorochemicals, silicones and waxes applied to the outer layer. It is important that this finish does not adversely affect breathability or any other desirable properties of the fabric, such as handle and comfort. Fluorochemicals (fluorocarbons) are now popular, and provide effective repellency against both aqueous and oil-based substances. Fluorochemicals have good fastness to washing and dry-cleaning, although a heating process (ironing and tumble drying) is often necessary after the cleaning process.

29.6.1.8 Garment construction

The first requirement for producing a waterproof breathable garment is to use fabric(s) with the required performance characteristics; Behara and Singh (2007) have provided a table of commercial breathable fabrics. The design, seaming and seams of waterproof breathable garments need to be carefully engineered to ensure that water cannot enter from the outside, either through seams or through openings in the garment, while still allowing water vapour to escape from the body to the outside. Seams are commonly sealed with special tapes, thermoplastic adhesive films applied (e.g. heat bonded to the laminate) to sewn seams to prevent water from leaking through. Such tapes must be resistant to washing and dry-cleaning. Cuffs, hems, zippers hoods, pockets, etc., all have to be specially designed for water proofing. These are normally specified and compulsory when trademarks are to be applied to garments.

29.6.2 MEASURING WATERPROOFING AND BREATHABILITY

As already mentioned, two distinct concepts are involved, namely waterproofing and breathability, and both need to be measured if a fabric or garment is to be categorised as waterproof and breathable. Two different types of tests are carried out, one to measure the breathability (i.e. water vapour transmission – see also Section 29.4) of the fabric and the other to measure the waterproofing of the fabric. In the case of garments, it is also essential to test the waterproofing of the seams, joins, etc. It should be noted, however, that different test methods and testing conditions can produce widely different values for the water vapour resistance of waterproof fabrics. Furthermore, the different test methods can be differently affected by changes in testing conditions.

Testing for breathability (Water vapour permeability or moisture vapour transport rate – MVTR) is covered in Section 29.4 and will not be dealt with here.

29.6.2.1 Air permeability (wind resistance)

Windproofing (i.e. resistant to wind) is normally assessed by measuring air permeability.

The air permeability of a fabric also has some bearing on its water vapour permeability, but the two are by no means identical. Air permeability can be measured using the following test methods:

- EN ISO 9237: Textile: Determination of the Permeability of Fabrics to Air.
- ASTM D737-96: Standard Test Method for Air Permeability of Textile Fabrics.

29.6.2.2 Water repellence and resistance (shower and rain resistance)

The following test methods can be used to measure water/shower/rain repellence and resistance (Morris & Morris, 2007; Sarkar & Etlers, 2006):

- AATCC TM42: Water Repellency, rain test.
- AATCC TM22: Water Repellency, spray test.
- EN ISO 4920: Textiles – Determination of resistance to surface wetting (spray test) of fabrics
- AATCC TM35: Water Repellency: Impact penetration test.
- EN ISO 9865: Textiles – Determination of Water Repellency of Fabrics by the Bundesmann Rain-Shower Test.
- AATCC Test Method 42-2000: Water Resistance: Impact Penetration Test.
- BS 5066: Method of Test for the Resistance of Fabrics to an Artificial Shower (WIRA Shower).

29.6.2.3 Waterproofing and rainproofing

This Section is from Saini (2004). The following test methods are used to test the water-/rainproofing of fabrics:

- AATCC 127: Water Repellency: Hydrostatic pressure test.
- ISO 811: Textile fabrics – Determination of resistance to water penetration, hydrostatic pressure test.

EMPA concluded that water and rain resistance of materials and garments should be measured dynamically to obtain a more realistic assessment, since the hydrostatic head test is not always a true reflection of what occurs in practice, and has developed new test methods (Weder, 1997) to assess the water and rain resistance of material and garments. The Hohenstein Institute has developed a 'wear comfort quality label' grading system (Anonymous, 2006), from 1 (excellent) to 6 (inadequate), for waterproof breathable garments.

29.7 RESEARCH AND FUTURE TRENDS

Intelligent or smart textiles, which respond to external stimuli such as temperature and moisture, will also become more prevalent, many of these will be based on biomimetic (i.e. imitating nature) principles. Technologies that will increasingly feature in the development of functionalities include biofunctional, nanotechnology, microencapsulation, PCMs, etc. The environmental impact of such technologies and the resulting fabrics and garments will also increasingly come under scrutiny.

29.8 SUMMARY

- Physiological and psychological comfort are important aspects of fabric and garment design.
- There are various ways in which the different factors that determine physiological comfort can be measured, including tests for water repellence, air permeability, insulating properties and garment fit.
- Psychological comfort is harder to measure because it is more subjective, but there are measures for satisfaction with body image, or body cathexis.
- Breathability and waterproofing are increasingly being incorporated into multifunctional clothing.

29.9 CASE STUDY

All traditional garments were breathable in that they allowed, to a lesser or greater extent, perspiration to evaporate and water vapour to escape to the outside, thereby enabling the skin to remain comfortable and dry. Outer garments, such as raincoats or rain jackets, also needed to protect the wearer from rain, even wind-driven rain. Initial attempts to meet this need produced the notorious 'plastic' raincoats, which offered excellent resistance against rain, even torrential rain, but which were extremely uncomfortable, since they did not breathe and therefore perspiration and condensation collected on the inside, causing an extremely uncomfortable wetness and clamminess against the skin. Little, if any, cooling or temperature regulation was possible, since perspiration could not evaporate.

This problem led to the search for, and development of, modern waterproof, breathable fabrics. These fabrics allow the smaller water vapour (i.e. evaporated perspiration) molecules to pass from the inside of the fabric to the outside, while not allowing larger liquid molecules and drops to penetrate from the outside, thereby protecting the wearer from rain while also keeping the microclimate next to the skin comfortable. To achieve this, the fabrics must not only be waterproof and breathable, but the garments must be assembled, by using, for example, water proof tapes and seams and specially designed vents or openings, so that water cannot penetrate through any garment seams or openings.

Waterproof breathable fabrics have been engineered in various ways, such as by using very tightly woven fabrics, incorporating either very fine (micro) fibres or filaments, or fibres, such as cotton, which swell with increasing moisture, or by applying microporous or hydrophilic membranes or coatings to fabrics, sometimes as a multilayered fabric sandwich. In virtually all such waterproof breathable fabrics, a water repellent finish is applied to the outer surface of the fabric or garment. Important issues in creating such products are that they are comfortable in terms of handle, lightweight, can be cleaned a number of times without losing their functionality (or else be re-proofed) and maintain acceptable functionality over an acceptable number of wear and cleaning cycles.

29.10 PROJECT IDEAS

1. Obtain a range (±20) of wool samples, preferably in scoured or top form, but could also be in yarn or fabric form. Accurately measure their mean fibre diameter, CV of fibre diameter and comfort factor (or coarse edge), preferably by means of automatic instruments, such as the Sirolan-Laser-scanTM or the Optical Fibre Diameter Analyser (OFDA), and using the appropriate standard test method (e.g. IWTO, ISO, ASTM). Graphically plot the various results, and interpret the results in terms of their impact on fabric prickle.
2. Source a range (±20) of either knitted or woven fabrics, differing in their mass (weight) and/or thickness. Measure their thickness, mass per unit area and thermal insulation (resistance), and, if possible, also their water vapour resistance, using the appropriate instruments and standard test methods. Relate, graphically and statistically, the different results, for example by plotting thermal insulation and water vapour resistance against fabric mass per unit area and thickness. Interpret and explain the results and trends. Also describe the role of fibre, yarn and fabric properties in determining fabric thermal and water vapour resistance and therefore fabric and garment comfort.
3. Source a range (±20) of woven fabrics that differ in their mass per unit area (weight) and composition. Measure the water transport of the fabrics in terms of their longitudinal wicking, using a standard (e.g. BS 3424 Method 21 and DIN 53,924) longitudinal wicking 'strip' test. Test the fabrics both before and after solvent cleaning (dry-cleaning) and interpret the results in terms of the fabric wicking and wetting behaviour in practice. Explain why the fabrics behave differently before and after solvent cleaning, and describe which factors could influence the wicking and wetting behaviour of fabrics in practice.

29.11 REVISION QUESTIONS

- List nine tactile sensations and responses.
- Give six factors that increase the thermal insulation of fabrics and yarns.
- How does garment design affect moisture transfer?
- What surface treatment improves the wettability, re-wettability and wicking of textile materials?
- Describe the technique used to measure prickle objectively.
- List three instruments used to determine thermal insulation of fabrics.
- List four methods used to test the water transport properties of fabrics.
- What five factors affect psychological comfort?
- List four test methods for water repellence and resistance.

REFERENCES

Anonymous. (2006). Transforming sportswear with functional fabrics. *Africa and Middle East Textiles* (3), 19–20.

Behara, B. K., & Singh, M. K. (September/October 2007). Breathable fabrics: various aspects. *Textile Asia, 38*, 38.

Coffin, D. P. (1988). Stalking the perfect raincoat. *Threads Magazine, 19*, 26.

Davis, M. (1996). *Visual design in dress.* New Jersey: Prentice Hall.

Fan, J., Hunter, L., & Liu, F. (2004). Objective evaluation of clothing appearance. In J. Fan, W. Yu, & L. Hunter (Eds.), *Clothing appearance and fit: Science and technology* (pp. 43–71). Cambridge: Woodhead.

Fan, J., Sarkar, M., Szeto, T., & Tao, X. (2007). Plant structured fabrics. *Materials Letters, 61*, 561–565.

Fan, J. (2009a). Physiological comfort of fabrics and garments. In J. Fan, & L. Hunter (Eds.), *Engineering apparel fabrics and garments* (pp. 201–250). Oxford: Woodhead.

Fan, J. (2009b). Psychological comfort of fabrics and garments. In J. Fan, & L. Hunter (Eds.), *Engineering apparel fabrics and garments* (pp. 251–260). Oxford: Woodhead.

Fanger, P. O. (1970). *Thermal comfort analysis and application in environmental engineering*. New York: McGraw-Hill.

Garnsworthy, R. K., Gully, R. L., Kenins, P., Mayfield, R. J., & Westerman, R. A. (1988). *Journal of Neurophysiology, 59*, 1083–1097.

Harnett, P. R., & Mehta, P. N. (1994). A survey of comparison of laboratory test methods for measuring wicking. *Textile Research Journal, 54*, 471–478.

Holcombe, B. V. (1986). The role of clothing comfort in wool marketing. *Wool Technology and Sheep Breeding, 34*(2), 80–83.

Hunter, L., & Fan, J. (2009). Waterproofing and breathability of fabrics and garments. In J. Fan, & L. Hunter (Eds.), *Engineering apparel fabrics and garments* (pp. 283–308). Oxford: Woodhead.

Jeffries, R. (1988). Work-wear for fire and heat protection. In *Shirley institute conference* (pp. 45–58). Manchester: British Textile Technology Group.

Johnson, L., & Samms, J. (1997). Thermo-plastic polyurethane technologies for the textile industry. *Journal of Coated Fabrics, 27*, 48–62.

Li, Y., Xu, W., Yeung, K. W., & Kwok, Y. L. (2002). Moisture management of textiles, US 6499 338 B2 patent.

Lim, N. Y. (2002). *A study on human perception towards girdle pressure with standard motion* (thesis).Institute of Textiles and Clothing, Hong Kong Polytechnic University.

Mayfield, B. (1987). Preventing prickle. *Textile Horizons, 7*(11), 35–36.

Morris, J., & Morris, P. (2007). *Retail testing standards*. In J. Crimshaw (Ed.). Bradford: World Textile Publications.

Naylor, G. R. S. (1992). The role of coarse fibres in fabric prickle using blended acrylic fibres of fibre diameters. *Wool Technology and Sheep Breeding, 40*(1), 14–18.

Ody, P. (1990). *Learning science through the textile and clothing industry*. Hobsons Publishing PLC.

Overington, Y. H., & Croskell, R. (May 2001). Standards for breathable fabrics. *International Dyer, 186*(5), 23–27.

Ramsay, D. J., Fox, D. B., & Naylor, G. R. S. (2012). An instrument for assessing fabric prickle propensity. *Textile Research Journal, 82*(5), 513–520.

Saini, M. S. (2004). Testing fabrics comfort performance. *Indian Textile Journal, 115*(1), 115–117.

Sarkar, A. K., & Etlers, J. N. (2006). The use of polyacrylamide as an auxiliary in water repellent finishing. *Colourage, 53*(9), 45–46.

Sharpe, D. T. (1974). *The psychology of colour and design*. Chicago: Nelson-Hall.

Slater, K. (1986). The assessment of comfort. *The Journal of the Textile Institute, 77*, 157–171.

Smith, J. E. (1985). The comfort in casuals. *Textile Horizons, 5*, 35–38.

Smith, J. E. (1986). The comfort of clothing. *Textiles, 15*, 23–27.

Veitch, C. J., & Naylor, G. R. S. (1992). The mechanics of fibre buckling in relation to fabric-evoked prickle. *Wool Technology and Sheep Breeding, 40*(1), 31–34.

Weder, M. (1997). How proof is 'proof'? *WSA. Winter, 3*(4), 43–45.

CHAPTER 30

THE MARKETING OF FASHION

K. McKelvey
Northumbria University, Newcastle upon Tyne, UK

LEARNING OBJECTIVES

At the end of this chapter, you should be able to:

- Understand the function of marketing
- Describe the marketing mix
- Explain traditional and new technology media channels
- Understand the value of trends and fashion forecasting
- Understand and use key terminology

30.1 INTRODUCTION

The textile and fashion industry contributes to global and national economies through the purchasing of clothing. How do fashion designers and retailers decide what they will design? How do they know what will sell? The design development process is symbiotic with the marketing process and intelligence data of all kinds are gathered during the process to make sure that there is a market – a consumer, for the product. Here the process is explored in detail and with reference to new technologies as new ways of marketing fashion.

30.2 WHAT IS MARKETING?

Marketing requires a company to think about the customer or potential customer of a particular product. The product needs to fit in with the needs and lifestyle of the customer and generate repeat purchases of the product by that customer. The need to make money is absolutely key.

Marketers achieve this by using a range of 'traditional' approaches, for example, by using market researchers to gather 'intelligence' material and by the advertising of the product. Less obvious approaches are through the 'branding' of the product, the development of the product, the pricing of the product, the promotion of the product, the future possibilities of the product through forecasting and the distribution of the product (Easey, 2009).

Textiles and Fashion. http://dx.doi.org/10.1016/B978-1-84569-931-4.00030-1

30.2.1 THE FOUR P'S: PRODUCT, PRICE, PLACE AND PROMOTION

Marketing was described by James Culliton in 1948 as a 'mixer of ingredients' – which could help to achieve a recipe for success. E. Jerome McCarthy moved this on to invent the 'Four P' concept in 1960 as being the main ingredients in the mix, and this concept is shared broadly in textbooks about marketing.

The four P's are **PRODUCT**, **PRICE**, **PLACE** and **PROMOTION** (http://www.marketingteacher.com/lesson-store/lesson-marketing-mix.html 02/08/11).

There have been further developments of the marketing mix – known as the extended marketing mix, which involves three more P's. These are People, Process and Physical evidence: firstly, People, this would be people who are involved with the consumption of the product; the consumers/customers, what level of the market is the product aimed at and what demographics do the customers share? Secondly, Process, the process of how the product reaches the consumer from concept to realisation and thirdly, Physical, the physical evidence should include the satisfaction of the customer with the product through feedback from the customer.

Markets have developed in recent times, with consumers becoming much more sophisticated, moving from 'homogenised' markets to what became known as 'niche' markets.

Niche markets are markets that supply 'specialist' products to a smaller, specific, target audience. The product focus may be on quality, exclusivity or function, or all of these, but here the product provides a specific need that fits in with the lifestyle of the consumer.

30.2.2 THE FOUR C'S: CONSUMER, COST, CONVENIENCE AND COMMUNICATION

In 1993, Robert F. Lauterborn suggested a Four C's classification to the marketing mix, which is much more consumer oriented and suits 'niche' market development. These four being, firstly, the **CONSUMER**, that is, satisfying the needs of the consumer through regular reference to them through development of the product. Secondly, **COST**, elaborating on the Price by adding consumer ownership and the extension of brand values of the product, such as environmental and sustainability issues. Thirdly, **CONVENIENCE** replaces Place, in terms of locating the product, the ease of finding out information about the product and purchasing the product. Finally, **COMMUNICATION** replaces Promotion by using new technologies to communicate, before and after purchase, with the consumer, through online selling, targeted banner ads, search engine optimisation, pay per click, social media networks, viral marketing and games, videos and email (http://www.customfitfocus.com/marketing-1.htm 03/08/11).

30.3 THE MARKETING OF FASHION

In this 'marketing of fashion' context, the 'consumer product' is the garment or accessory that has been developed by a designer with reference to marketing information. The fashion product allows for a variety of marketing methods to be used and because it is very varied with different customers, different lifestyles, needs and uses, the fashion marketer needs to make a number of informed decisions. These decisions begin with the creation of the product (see Figure 30.1). Design research is undertaken to explore the shape, colour, texture, form and function of the product; also the fashion element is explored – making the product contemporary and desirable so that it fits in with customer

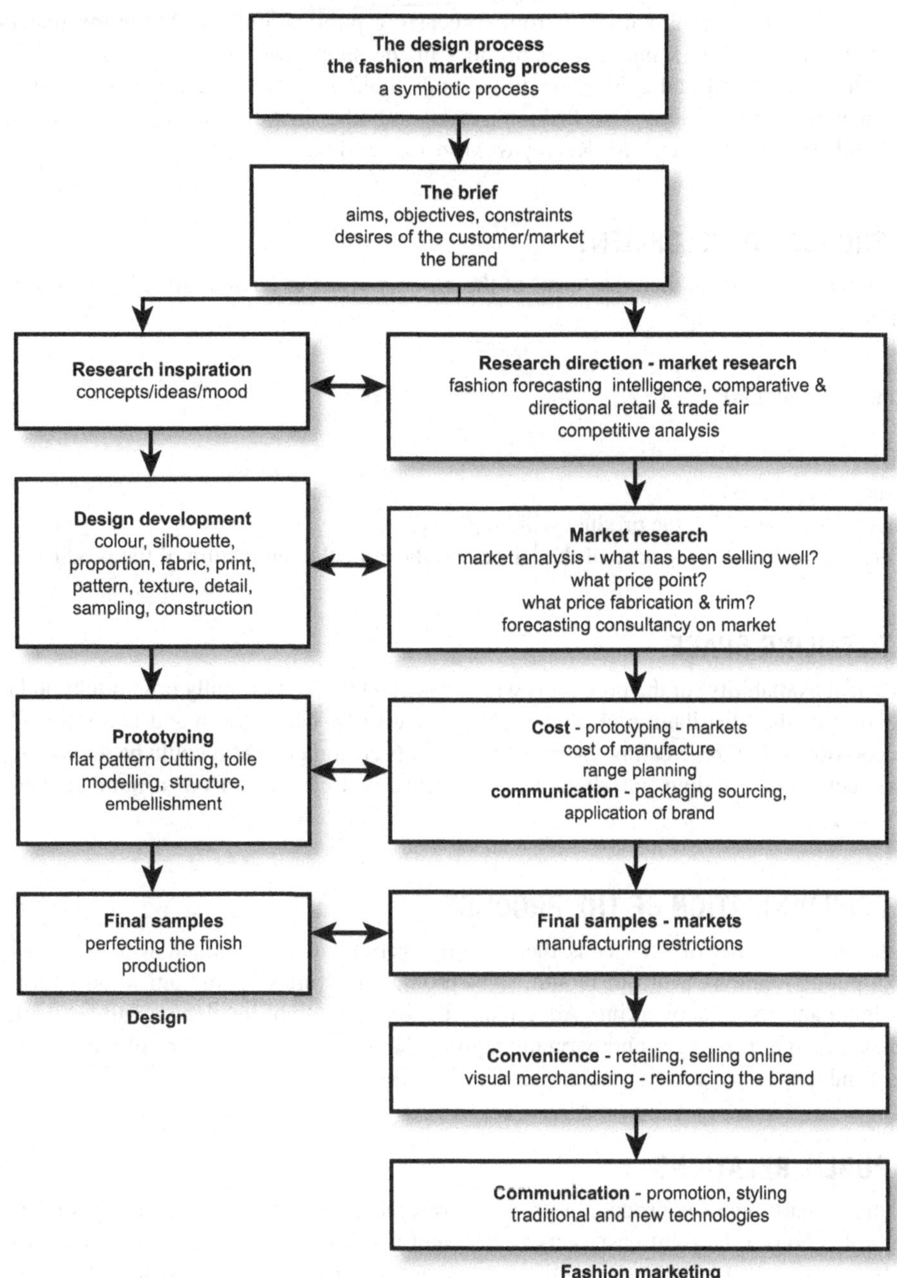

FIGURE 30.1

This chart shows the symbiotic relationship between the fashion design process and fashion marketing.

Source: Created by the author.

needs. Intelligence material is gathered from forecasting publications, retail reports that may be comparative (looking at the same type of product in the shops and comparing fabrication, price points, construction, detail and colour) or directional (looking at new and different products, new directions, new fabric technology, new fashion trends) and other inspirational research material such as cultural and artistic influences (McKelvey & Munslow, 2011).

30.3.1 PRODUCT DEVELOPMENT

Once development has begun then the 'cost' of the product needs to be determined, this is very much dependent upon the following:

- Fabric
- Amount of fabric used
- Trim
- Detail
- Construction complexity
- The market level at which the product will be placed
- Broader issues such as the source of the fabric and the ethical manufacture of the product

30.3.2 RETAILING SPACE

The 'convenient availability' of the product is where it will retail, be it virtually or in bricks and mortar, also referred to as the 'distribution channel'. This may well be where the brand is reinforced in the interior decoration of a store, creating the right mood for its customers. Or, it could be the ease of locating the product online, finding out about its provenance and choosing from a number of delivery options.

30.3.3 COMMUNICATION OF THE PRODUCT

Possibly the broadest area of activity is the 'communication' of the product, which could involve advertising, public relations, word of mouth, sales promotions, further communication with the customer via email and social networking. Advertising the product might involve paying for a television advertising campaign, running a photographic styling shoot to use images in print-based media like newspapers and magazines and on ambient media like billboards.

30.3.4 PUBLIC RELATIONS

Public relations work involves producing press releases and photographic styled images to 'sell' the product or collection to the right customers in the right magazines or newspapers. 'Word of mouth' relies on the satisfied customer spreading the word about the product, and these days this is achieved very successfully through social networking sites. Sales promotion uses strategies like reducing the price of a popular product to encourage the purchase of other products sitting alongside, or customers accumulate rewards points to put towards their next purchase.

30.4 TARGETING A MARKET

Markets can be 'mass' (an overwhelming group of consumers who may use the same household product for example) and homogenised (to make uniform or similar) involving consumers with similar needs and wants. A heterogeneous market (composed of widely dissimilar elements) consists of consumers with diverse needs and wants (Bovee & Thill, 1992).

As 'niche' markets (where there is demand, by the consumer, for a very specialised product) are now frequently targeted there is a necessity to identify 'gaps' or 'opportunities' for a new designer or manufacturer. The market needs to be able to be 'identified', in that it needs to be understood and reached by any marketing message or distribution channel. To achieve a better understanding of this, the tools of segmentation (sub-division of the mass market, for example, for better targeting) and demographics (the statistical data of a population such as the average age or income) are used (Easey, 2009; Jackson & Shaw, 2001).

Geographic Segmentation: This can refer to country regions, city centres or the effect of weather. Fashion is traditionally organised by season, but weather patterns at a given time of year, in holiday regions, for example, will have a strong bearing on the type of clothing required.

Psychographic Segmentation: Here lifestyles are explored, that is, the individual's pattern for living, is expressed through activities, interests and opinions.

Demographic Segmentation: This can refer to gender, occupation, marital status, income, wealth, education, religion, age, size, height, youth subcultures, life-stage of the family, neighbourhood, region, country and climate.

Behavioural Segmentation: The final use of the product, brand loyalty, consumer needs for certain benefits and price sensitivity are considerations here.

Socio-economic Classification: This is considered quite an old system now as it is based on class, but does allow for broad judgements based on occupations and often is used as a reference to the lifestyle of the consumer. See Figure 30.2.

Nowadays, being 'fashionable' has nothing to do with class.

A	Upper class	*Senior professional*
B	Middle class	*Managerial*
C1	Lower middle class	*Supervisors, junior management*
C2	Skilled working class	*Skilled manual workers*
D	Working class	*Semi/unskilled manual workers*
E	Subsistence level - low fixed income	*Pensioners, students, the unemployed*

FIGURE 30.2

Socio-economic classifications.

30.4.1 CUSTOMER PROFILES

These are also known as pen profiles and personas. These can take into account a number of possibilities and tend to tell a story about a potential consumer, analysing lifestyles, attitude, occasion of use for a product, income, age and any other defining characteristics. This tool helps to target advertising and marketing to these potential consumers and so cuts costs. See Customer Profile examples in the Evolution and Stratagem Case studies later in the chapter.

30.4.2 SEASONAL AND OCCASION MARKETS

The traditional spring/summer and autumn/winter fashion seasons have developed into new opportunities such as early spring, spring, early summer, summer sale, high summer, transitional summer, autumn, party wear, Christmas and winter sale. There are a number of reasons for these opportunities, such as climate change and a need for a versatile, perhaps layered wardrobe, many more opportunities to travel abroad and more collections to tempt the consumer whilst waiting for the main season.

Holidays, calendar events, leisure and activities, back to school and weddings all provide further marketing opportunities.

30.5 BRANDING

Branding allows the consumer to differentiate between different products. The brand targets particular consumers.

A brand is a mixture of a number of attributes, many intangible; that is, they are perceived and discerned. Usually these are recognised in a logo or icon forming a trademark – this is the brand identity and allows the consumer to recognise one brand's differentiation from another. Differentiation comes from 'values' that can be built into the brand, these are usually related to human personality traits, such as warmth and reliability, so that consumers can understand exactly what kind of experience they can expect from using a particular product or service (http://www.brandchannel.com/ 18/10/10).

Without branding it is difficult to know how you might market a product or service.

Branding is important in getting your target market to see that you are the only brand that can provide the right solution to their needs, so that you can build loyalty to the brand and, of course, repeat sales.

30.5.1 BRANDING CASE STUDY: APPLE INC.

An example of a well-known brand is Apple who compete in the highly competitive markets of personal computing and the consumer electronics industry and are the largest mobile devices company in the world. The brand is synonymous now with innovation in the form of the iMac and Macbook computers, the iPod, iPhone, iPad, iTunes and applications for these. The Apple brand is about lifestyle; imagination; liberty regained; innovation; passion; hopes, dreams and aspirations and power to the people through technology. These products develop a highly relevant and desirable brand image in the mind of the consumer that they have not previously reached and also promote strong brand loyalty from the consumers that have bought Apple products in the past. Apple prides itself on being a humanistic company with customer experience as its focus. (Marketing Minds, http://www.marketingminds.com.au/branding/apple_branding_strategy.html 03/08/11).

30.6 THE TRADITIONAL MEDIA CHANNELS

With brand values (such human traits as warmth and reliability applied to the product identity) it is easier to work with each element of the promotional mix; the effect will be stronger and will last longer when the 'handwriting' (the visual identity of the brand in the form of a logo and its application) is consistent every time it is used across a variety of media, such as advertisements, on the web, in a store, on the letterhead. It becomes distinctive and is recognisable, it is reinforced in the mind of the consumer and ultimately becomes easier to recall (Easey, 2009).

The traditional channels for promoting products and services are as follows:

Television – This is the most expensive method of advertising, there are many television channels these days with particular markets in mind. Adverts are short though, so there is a limit to how much information can be conveyed, but this is a good medium for consolidating a brand image.

Outdoor advertising – This is known as ambient media and includes billboards, the Underground and other transport systems, street furniture and taxis. This is relatively inexpensive and can target more closely particular users of transport, but it doesn't give much opportunity for lots of information.

Magazine advertising – This is quite an effective, if static, medium as magazines tend to be aimed at a target market in any case. Magazines are often kept for months which means repeat exposure and may lead to sales, there is also the 'pass-along' value to like-minded consumers. However, magazines do tend to require higher production costs and there is limited flexibility in terms of placement of an advertisement.

Celebrity and focussed publications – The 'cult' of celebrity has had an impact on the fashion marketing mix. Celebrity endorsement and sponsorship is a very powerful form of promotion. Having links to the brand with a well-known personality helps to add credibility. An admired celebrity brings their traits of good taste to the brand. Even though they have been paid, if they use the brand it strengthens the brand values. If the celebrity does something damaging in their career, this can have an adverse affect on the brand. Sports brands, in particular, have been very successful in the use of celebrities such as David Beckham.

Radio Advertising – This is not often used by the fashion industry due to the limitations in communicating information as this is obviously not a visual medium, although alerts to sales and events do work.

Sales promotion – This relies on offering discounts on products, or the use of coupons and vouchers, which may then help to convert customers to being brand loyal. Customers may become accustomed to sales if this tool is used too frequently and may not buy at other times. This could also hurt the brand image.

Public relations – These may be in-house or external companies employed to get the brand message across, through launches, parties and events. How they communicate the brand is key, also what response do such events get from the press? Mentioning a product in the press often gives immediate legitimacy. The right public relations company will have strong relationships with press and other media, the wrong one will waste time and money by not getting the brand message across in the right medium.

Personal selling – The knowledgeable sales assistant helps to reinforce the brand. The one-to-one feedback is powerful in personal selling but relies on trust that the assistant will have the customer's best interest at heart.

Direct marketing – This involves mail-outs to targeted customers. This may target the right customer but, as unsolicited mail, may just end up in the waste bin without being read.

30.7 NEW TECHNOLOGIES AS MEDIA CHANNELS

The Internet has become a powerful tool for selling, promoting and feeding back consumer opinion. A full range of products can be reviewed and compared at any time of day and night, products that are normally out of reach in another country can now be purchased and delivered to the customer's door in a very reasonable time frame. Search engines such as Google, one of the best known brands on the Internet, can help to find any product using key words for searching, as well as such sites as the purpose-built product comparison site, Kelkoo, founded in France in 1999. It provides information on products including seller information and price.

Ebay is an American online auction site that is a consumer to consumer business, where a broad range of products and services are bought and sold by businesses and ordinary people and feedback is sought on products and services received, with a view to constantly improving the shopping experience.

The Internet is constantly updating and improving from an original information-based retrieval system to a fully interactive experience, which can also be accessed by a number of devices, such as desktop and laptop computers and tablet computers such as the iPad, mobile devices such as the iPhone and entertainment systems such as the iPod, PSP (Play Station Portable) and Xbox 360. Entertainment can be found by 'surfing' the Internet, downloading music and videos, playing games and downloading entertainment, promotion and information applications. Social networking has also become a popular online pursuit where individuals and organisations connect to each other through interdependency such as friendship, common interest, likes and dislikes, beliefs or knowledge. One of the best known social networking sites is Facebook, where an individual creates a personal profile made available to 'friends' and where friends exchange messages and join common interest groups. Facebook is accessible on any personal computer and on mobile devices such as smartphones, that is mobile phones that combine cameras, media players, touchscreens, Global Positioning System (GPS) navigation, mobile broadband and Wi-Fi.

30.7.1 WEB 2.0 AND OTHER TECHNOLOGICAL DEVELOPMENTS

The expression Web 2.0 is broadly meant to encompass a combination of new technologies (Javascript tools like AJAX) and interaction from the user. The user can be an amateur writer or developer who democratically contributes information and applications freely, and is as credible as traditional written sources and commercial software.

Web 2.0 provides opportunities for users to collaborate with each other online by sharing and interacting with information in what is known as a 'social media dialogue', as creators of user-generated content in a virtual community, in contrast to the passive viewing of content that was created for them (Oreilly http://oreilly.com/web2/archive/what-is-web-20.html). Characteristics such as openness, freedom and collective intelligence are key to Web 2.0.

Examples of such opportunities are social networking sites, like Facebook, wikis like Wikipedia that are editable and extend information that can be 'done' and 'undone', blogs that are personal commentary sites including text, visuals and video, video-sharing sites like YouTube and Vimeo, the ability to 'tag' photographs and text and to use 'extensions' where the web becomes an application platform and document server such as the use of Adobe Flash Player, Adobe Reader, Java, Quicktime and Windows Media Player.

This participatory virtual community also has a 'sharing' phenomenon, which is often free to anyone who wants to contribute, for example, tutorials on using software, software and source code are shared, material on any subject is available for use.

30.7.1.1 Creative Commons

Creative Commons is a non-profit-making organisation that wishes to increase sharing and collaboration legally on the web. Their mantra is 'share, remix, reuse – legally' to fulfil the full potential of the Internet! They release copyright licenses known as Creative Commons licenses to the public free of charge. The licenses allow creators to decide which rights they reserve instead of the default 'all-rights reserved copyright' that was designed pre-Internet. For more information about this initiative visit http://creativecommons.org/about.

30.7.1.2 Facebook

Facebook is what is known as a social networking site, which allows users and companies to create a profile that can be seen online, users can leave 'status' messages and communicate with each other in statements and brief comments. Users can join groups where they have common 'likes'. Companies can leave information about new fashion lines, events and sales. They can pay to advertise on Facebook by closely targeting their customers in banner ads and other smaller adverts. Users can play games and view new content from companies.

30.7.1.3 Twitter

Twitter is a communication platform increasingly used by companies to connect with their customers, as well as individuals and celebrities. There are opportunities to share company information and gather market intelligence that could be invaluable in building relationships and loyalty with customers. Customers can feed back about products that were satisfactory or disappointing, can offer ideas to companies about improvement and can, in return, be informed about special offers, coupon codes can be provided, links to key information or new events can be given, online shopping tips can be shared. Tweets are easy to read at up to 140 characters and offer a 'real time' method of communication.

It is the level of information that makes this useful; for example, it is easy to pass on information about some small defect in a product that would not normally warrant a complaint. It can make the customer feel close to the brand and feel that they are being heard, therefore increasing brand loyalty.

30.7.1.4 Instant messaging

Instant messaging also provides opportunities to inform customers about products, events and 'good deals'. It is a 'non-confrontational' method of promoting, as it is text based and is usually in response to some request from the recipient to receive information. Mobile phones and devices are hugely popular and this is an effective way of informing target markets.

30.7.1.5 Virals

This is the electronic equivalent of 'word of mouth'; an online film or message is created and is designed to be entertaining enough for customers to pass to each other, often through YouTube and email. Therefore making others aware of the brand, at no cost, but at great circulation speed.

30.7.1.6 Blogs

Most blogs (**web logs**) are privately generated and have various uses including showing personal online diaries, sketches, links and videos, but some are developed by companies; these may be developed for branding, marketing and pubic relations.

Blogs are web pages generated by 'users' and published for anyone to see. They are of a simplistic, linear construction and tend to run from oldest to newest postings.

30.7.1.7 YouTube

YouTube is famed for allowing users to post videos online, sharing them so that they can be viewed across the world. A huge range of videos are available to view including virals!

30.7.1.8 Podcast and webcast

A podcast is a series of digital media files that can be downloaded whenever they are posted.

A webcast can be a live or on demand file that uses 'streaming' media technology.

30.7.1.9 Email

Electronic mail or email is ubiquitous nowadays and as well as being a generally informal communication device is also a selling and promotional medium. Free subscriptions to newsletters and Websites allow the recipient to remain up-to-date with developments to their favourite brands. Email may also be the container for viral videos and other marketing material that can be attached for download, and links to Websites can also be contained within the message.

30.7.1.10 Flickr

Flickr hosts an online community that encourages users to share photography and video. The site is used widely by bloggers to host images that are embedded in their blogs and other social media, thereby spreading information. Flickr has an application for the iPhone, Windows 7 phone and the Blackberry.

30.7.1.11 QR codes

QR code is an abbreviation of quick response code and is a two-dimensional barcode that is readable by barcode readers and mobile phone cameras. The code is a square of white with black units (see Figure 30.3). This could hold text, a URL or other information. Users with a camera phone and the correct reader application can scan the codes to get information or go directly to a Website. The technology is ubiquitous in Japan, the Netherlands and South Korea. Applications for reading the code are available for most smartphones.

Personal QR codes can be generated for free and shared. One particular Website is http://delivr.com/qr-code-generator and it generated the QR code shown in Figure 30.3, which is the URL for the author's illustration blog.

30.7.1.12 iPhone and Android smartphones

The iPhone was one of the first mobile phones to be mainly controlled through a multi-touch interface by a touchscreen and can be considered a smartphone.

The App Store can deliver applications directly to the iPhone or iPod Touch over Wi-Fi or cellular networks without requiring a PC. iPhone 4 includes a 960×640 pixel display with a pixel

FIGURE 30.3

Example of a QR code.

density of 326 pixels per inch, a 5 megapixel camera with LED flash capable of recording HD video, a front-facing VGA camera for videoconferencing. This makes it a very versatile and tactile tool. Competition exists from the Android operating system for smartphones, which was released in 2008. Android is an open-source platform supported by Google, along with major developers, such as Intel, Motorola and Samsung; they form the Open Handset Alliance. The software suite included on the Android phone consists of Google Maps, Calendar, Gmail and a full HTML web browser.

30.7.1.13 Apps

Apps or applications can be downloaded from App stores for smartphones. The first App store was for Apple's iPhone. Third-party applications were made available for download, which added considerably to the choice of applications for nominal sums. Applications cover anything from games, entertainment, utilities, social networking, music, productivity, lifestyle, reference, travel, sports, navigation, health and fitness, news, photography, finance, business, education, weather, books and medical and offer great opportunities for promotion. For example, the free All Saints application allows users to shop the full collection with new styles uploaded daily; to search for specific pieces; to create a wish list; to share products with friends via email; to locate the nearest stores to your position using GPS and updates the latest content automatically.

There follows a brief case study of Wickedweb Digital marketing agency, and they use new technologies and social networking as marketing tools. They are responsible for All Saint's marketing strategy, which is discussed in more detail in Case Study 3: All Saints of Spitalfields.

30.7.2 CASE STUDY: WICKEDWEB DIGITAL MARKETING AGENCY

Wickedweb is a web marketing company that sets out to 'build engaging relationships between brands and their audiences using digital communications', they are 'digital marketers' (http://www.wickedweb.co.uk 15/10/10).

They use social media marketing strategies to help companies to improve their brand awareness and loyalty and the sales of product, for example, Facebook, YouTube and Twitter.

The 'conversational' styles of social networking help to break down barriers and build brand perception and loyalty.

Audio and video tools (YouTube), online radio, photographic tools, podcasts and 11webcasts can all help to improve brand loyalty and are utilised by Wickedweb.

Niche markets might require different social networks to communicate; there are many networks on the web to choose from, and Wickedweb could even build a specific network and provide complete management of a company's online presence.

As social networking is a popular tool with millions of users, how does a company make their voice heard? Their solution is to be an expert in the field, so when users can comment, add their own voice, express themselves and share web links, hints and tips on a community site it all helps to improve customer loyalty.

Wickedweb also develops mobile applications for the iPhone that support social networks, provide relevant games, sometimes viral games and information applications that support the brand.

Social networking can attract all kinds of users, so how do you know that you are targeting the right customers? Wickedweb can provide a range of tools that analyse and report on keywords used, target potential customers by similar demographics, the location of customers and positive and negative commentary.

They worked with All Saints of Spitalfields to improve their brand awareness online.

30.8 THE MARKETING PLAN

The marketing mix of consumer, cost, convenience and communication will be used here. A mixture of traditional and new technology marketing methods will be included in a context that relates to two particular fashion concepts.

There follows two case studies exploring the process of the marketing of fashion, one is a womenswear orientated project, the other is menswear; they have different approaches to communication but both show strongly how the 'concept and consumer' is key to creative solutions in the marketing mix.

Both projects follow a process of market research, specifically about the consumer and the target market, competitive analysis, customer profiles, conceptual development, the 'critical path' for achieving the objectives of 'from concept to realisation and promotion', cost, convenience, communication, the traditional marketing approach and the technological marketing approach.

Design development works in tandem with this process.

30.8.1 CASE STUDY 1: 'EVOLUTION' BY KATIE LAY

'A case study to support the evolution of an "Urban-wear" women's-wear collection, which integrates influences from the latest technological advancements with shapes and details inspired by the beauty of nature in the world around us…' (Excerpt from Katie's Marketing Plan; Lay, 2010a).

Market research: Katie's primary objective was to blend the performance and flexibility of sportswear with designs drawn from nature in order to create a practical, comfortable and fashionable range that provided practical solutions to the problems facing the target market. She then organised her thoughts into secondary objectives to undertake her research (Lay, 2010b):

- Use the Internet, library and magazines to thoroughly research the latest urban-wear designs.
- Gather opinion regarding the current understanding of the style by issuing a questionnaire to a significant group of people taken from the proposed target market, using hand filled, email and Facebook.
- Analyse opinions gathered from the returned questionnaires and evaluate the potential for an expanded range of garments.
- Use a number of modes to research current designers in order to understand their thoughts on current and future trends for the introduction of new technologies.
- Understand how new technologies have already been used to influence and direct fashion trends.
- Research new technologies, materials and production methods that could be integrated to create the new look and feel.
- Study aspects of the natural world and extract potential shapes, colours and themes that could be integrated to complement the new look and feel.
- Prepare a number of examples of potential designs and concepts that show how the style can be expanded as proposed for presentation to a focus group.
- Hold a focus group of four people taken from the target market to discuss the proposed integration of themes taken from technology and nature.
- Analyse the feedback from the focus group to determine which ideas received most favourable feedback and should be studied further.

30.8.1.1 Consumer

The Concept

'Evolution' is a new range of 'active-wear' garments, which elaborates on a trend of crossovers between casual, sport and high fashion. This comfortable, wearable range is sophisticated, modern and stylish. The uniqueness of the range is the ability to remove panels and sections from the 'casual' outfit for a transition to 'performance-wear'! The range is aimed at women aged between 18 and 30 who are fashionable and have a strong identity, who desire practical yet stylish clothing.

Second skin-like styling combines with organic, naturalistic patterns and neoprene, high stretch elastane and bonded jersey with contour panelling, with accents in contrast piping and sheer mesh inlays. Quickburst zippers will be used to fit in with the easy-to-remove panel function.

A range of machine-inspired greys with contrasting soft tones emphasises the sophisticated sportswear approach, the fabrics offer high-tech to body-forming solutions. The digital prints are inspired by 'beetles' because of their sleek organic shaping, which fulfils the beauty of nature element (Figures 30.4 and 30.5).

Katie makes her marketing objectives explicit: to design an active lifestyle collection; to study the range of garments offered by major competitors in order to identify niche market areas that are not currently met; to establish realistic price points to attract the target market; to identify potential selling strategies such as where, when and how; to design a unique marketing strategy for the Evolution brand.

The Target Market

A number of 'customer profiles' are created that help to expand the 'target market'; this gives ideas about when and where the collection may be worn and helps to see unforeseen opportunities in developing the range. An analysis of the person's salary, socio-economic class, disposable income after bills,

'EVOLUTION'

The 'Evolution' Collection

*£80

PRINTED JERSEY
HOODED TOP WITH
OVERSIZED COWEL NECK

*£75

*£50

SLOUCH HOODIE
WITH HIGH WAISTED
RIB DETAIL

*£45

OVERSIZED TEE WITH
SLOUCH POCKET AND
BACK DETAIL

*£30 *£20 *£35

CONTRAST PANEL JESREY COVER-UP VESTS

*£50

CONTRAST PRINT SCALE
BATWING JERSEY
JUMPER WITH HIGH
WAISTED HEM DETAIL

FIGURE 30.4

Part of the collection – Evolution.

FIGURE 30.5

The print-based fabric story for Evolution.

wardrobe update allowance, dress size and age is provided. This is followed by a paragraph offering more detail about the lifestyle of the persona.

Competitive Analysis

Usually competitors are attracting the same or similar target markets. Analysis is about understanding what a competitor is developing and how they are reacting to change. This is done by observing and recording the competitor's products and comparing colour, shape, texture, fabrication, price and manufacture in the form of a visual report. It may also help a company to reaffirm what makes them different in terms of their brand message.

The concept requires Katie to look at sportswear and fashion brands, as her collection is a synthesis of the two areas.

Sportswear – Nike

Nike sportswear ranges from the professional to the serious amateur and targets 18–35 year olds. Men, women and children are catered for. The goal is to enhance performance when wearing Nike products. Nike's promotions are diverse and well known, using athletes like Ian Botham, Sebastian Coe, Ronaldo and the England rugby team. Nike innovations include the Triax running watch, which offers a more usable watch with numbers that are readable and buttons that are easily utilised. Nike+ monitors a runner's performance through a radio device in the shoe, which works with the iPod Nano, in collaboration with Apple. They also developed the FIT system, a four-fabric system that copes with heat, cold, snow, wind and sweat.

Sportswear – Adidas

Adidas attempts to develop and create experiences that engage customers in a long-lasting association with their brand. The target market is 14–22 years. Adidas is a mass and niche market provider to maximise consumer reach from shoes, apparel and accessories for sports such as football golf, basketball, to training and fitness. Adidas strive to continuously improve the quality, look and feel of their products. They advertise in the media targeting the youth audience as well as the sporty consumer. Adidas uses new technology to communicate their brand from utilising the world's biggest advertising hoarding in Birmingham during the World Cup, to utilising the Internet and through email campaigns. They try and show at least one improvement in a year and innovate through their cutting edge design, for example, their Dynamic Layering Concept clothing range that provides physiological support for football players.

Katie looks at Top Shop contemporary concessions to see where Evolution fits in, in terms of competition. In this context a concession is a retail business within another retail business which operates under a special license, for example, a department store contains many concessions operated by other retailers.

KTZ – Marjan Pejoski and Sasko Bevzoski design cult streetwear with a feminine angle at Top Shop: this collection was fun filled and had a futuristic feel on American retro looks.

Makin Jan Ma – Makin makes stories and scripts and clothes for his characters then shoots the film, his starting point is always 'love': this was a collection available in Top Shop inspired by 'falling in love'.

Evolution: Inspired by the continuous evolution of technology and nature, this collection is multi-functional, fashionable and jam-packed with performance.

AnneSofie Back – a designer whose signature is based on subverting the human form: this collection was women's daywear with a gothic touch.

Unique – the in-house design team produce strong prints, proportion and embellishment pieces: this range was a sporty, laid-back beach collection.

Ashish – printed tee shirt designer at Top Shop: this was a sportswear range with a studded punk influence.

Boutique – capsule collections at Top Shop with directional and essential pieces: this was a fashion forward range with a neutral colour palette.

Jonathan Saunders – a print designer more recently moving into architectural, clean lines: this collection was about urban luxe with geometric shapes and black.

Danielle Scutt – womenswear designer creating feminine, pretty and often colourful collections: this collection was about bright prints and poolside glamour.

Richard Nicoll – womenswear designer renowned for his strong tailoring: this collection was about city chic through corsetry and menswear details.

Customer Profile

See Section 3.1 Customer Profiles – for an explanation of profiles or personas.

Katie created three customer profiles, the Fashionista, the Sporty Student and the PE Teacher. The Fashionista is described here. Each profile would have a slightly different emphasis, the first is about being fashionable, the second is about the relevance of the high-performance fabrics and the third about the versatility of the garments to quickly change looks and function.

The Fashionista customer profile is loosely based on the movie 'The Devil Wears Prada'; contemporary references, from movies or books, are often helpful in elaborating on customer profiles.

The 'Fashionista' who likes to keep fit.

'Salary: £27,000

Socio-economic class: B

Disposable income: £200 per week

Wardrobe update allowance: £200–400 per month

Dress size: 10

Age: 25

In London, recent graduate Rebecca has just been hired to work as the assistant to the powerful and sophisticated executive fashion editor of a magazine. With help from her colleagues, Rebecca finds herself learning how to dress more appropriately for her demanding new role. Despite being an assistant and living in the centre of London, splitting bills with her friends leaves Rebecca with limited cash for 'affordable' fashion, but with the demand to be constantly 'in fashion' at work, she is finding it difficult affording to keep her wardrobe fresh and up-to-date. With spare time being an issue, Rebecca likes to go to the gym straight after work, but hates having to carry her active wear with her to the office every day.

How did she hear about Evolution?

As an avid fan of Top Shop, Rebecca's eye was caught by the exciting, edgy new concession in the Oxford Street store one Saturday afternoon. Rebecca was captivated by the concept of being able to go to work in her gym kit and still look great.

Why She loves it…

The Evolution garments tick both of Rebecca's main boxes when it comes to fashion – it is affordable and looks great. Not only that, but it allows her to cut out the endless wasted hours trekking between gym, home, work and play and lets her get on with the precious little social life that she is able to enjoy. Evolution's numerous, adjustable accessories allow her to make-over the same outfit several times without breaking the bank by adding or removing attachments as each new season comes round'.

30.8.1.2 Cost

Katie selected a range of her developed garments to demonstrate her costing method, breakdown of materials and price points.

She decided to calculate wholesale prices first as this was relevant for Top Shop. She then did some market research into competitor's pricing to work out a relevant mark-up price for retail purposes, for the brand to sit well amongst its competitors (Figure 30.6).

Materials included the base fabric, mesh, adhesive for the neoprene seams, vinyl, laser cutting, Quickburst zippers, closed end zippers, vinyl and digital prints. Labour was included and a wholesale mark-up of 25% was added. A further recommended retail mark-up of 70% was added to the total wholesale price as was VAT.

Item	Quantity	Unit cost	Total cost
Cotton jersey	2m	£4.95 per m	£ 9.90
Digital printing	2.5m	£ 5.00	£ 12.50
Vinyl	0.05m	£ 9.95	£ 0.50
YKK quickburst zippers	2	£5	£ 10.00
		Total excluding labour	**£ 32.90**
Labour	0.25 day	£20 per day	£ 5.00
		Total cost price	37.90
		Wholesale mark-up @ 25%	9.48
		Wholesale price (excluding VAT)	£ 47.38
		RRP mark-up @ 70%	£ 33.16
		RRP price excluding VAT	£ 80.54
		VAT @ 17.5%	£ 14.09
		RRP including VAT	**£ 94.63**

Item	Quantity	Unit cost	Total cost
Neoprene	1m	£10 per m	£ 10.00
Mesh	0.5m	£10 per m	£ 5.00
Adhesive	1 tube	£3.99	£ 3.99
Vinyl	0.5m	£ 9.95	£ 4.97
Laser cutting	1m	£ 5.00	£ 5.00
YKK quickburst zippers	2	£5	£10
YKK closed end zippers	2	£4.95	£4.95
YKK centre front zipper	1	£9.95	£9.95
		Total excluding labour	**£ 13.96**
Labour	0.5 day	£40 per day	£20
		Total cost price	£ 23.96
		Wholesale mark-up @ 25%	£ 5.99
		Wholesale price (excluding VAT)	£ 29.95
		RRP mark-up @ 70%	£ 20.97
		RRP price excluding VAT	£ 50.92
		VAT @ 17.5%	£ 8.91
		RRP including VAT	**£ 59.83**

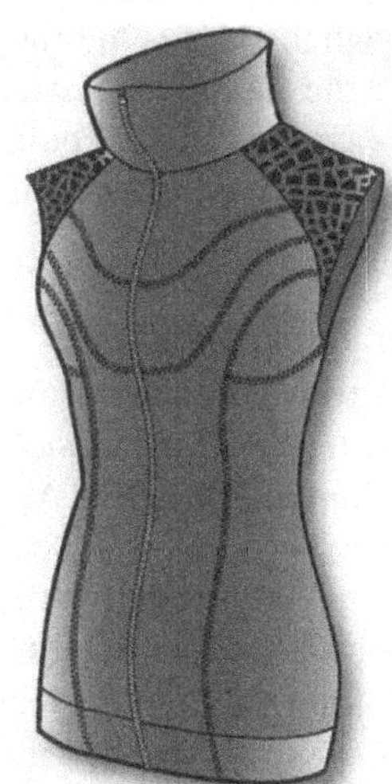

FIGURE 30.6

Two costing sheets for comparison from the Evolution collection.

30.8.1.3 Convenience

The collection would ideally be sold in the Oxford Street Top Shop store in the concessions department. Top Shop is a chain of fashion stores that operates in 20 countries across the world, from Brazil, Canada and the United States across Europe and the Middle East to New Zealand and South East Asia. The flagship store in Oxford Street, London, has a nail bar, a tailoring service, a hair salon and a delivery service delivered by scooter within 1 h of ordering providing that the customer lives in the delivery zone. Top Shop also sells online via its Website www.topshop.co.uk. This would give customers residing outside London the opportunity to buy into the concessions and the Evolution brand.

Top Shop delivers 'basics' to 'edgy trend setting' garments and does not categorise its customer, but relies on its brand values to keep customers. Top Shop has a reputation for supporting new talented young designers. The Top Shop marketing mix is about advertising, word of mouth, constant change in the flagship store and sponsoring of talented designers, which maintains interest with customers in the brand.

The advantage of Evolution being a concession in Top Shop would be low start-up costs, exposure to the right customers and an opportunity to try new and edgy ideas.

As a concession, Evolution would establish its identity by being circled by interactive screens; the screens would allow the customer to choose garments from a revolving hidden rail. The garments would be in 3D and full colour on screen, and garments could be mixed and matched onscreen to create outfits. Once selected, garments would be delivered to the customer and payment could also be made onscreen.

The interactive shopping screen should reinforce the strong technological side of the Evolution brand, create interest in the new shopping experience, attract new customers, deliver the collection concept efficiently and establish the brand.

Data reports could also be created through the screens to feed back information about customers, the brand, sales and consumers' opinions.

Although the idea is to bring the customer into the store, no doubt it could be transposed to the web, or to the smartphone as an application, where merchandise could be viewed and paid for via the phone. The customer could download the application by subscribing and therefore give their details so that garment recommendations could be pushed to them via the phone based on previous sales.

30.8.1.4 Communication

Katie took two approaches to the promotion of Evolution: the traditional and the technological marketing approach using new channels!

The Traditional Marketing Approach

Swing Tags:

Swing tags are attached to the garment; they show the brand message and hold information regarding garment size and price and any special laundering instructions. Every garment would be issued with a swing tag and the customer looks for this when browsing, other associated product's details could be included to increase sales.

The swing tags were developed from the 'beetle' shapes that inspired the print designs (this element would not change with the seasons, unlike the digital print subject matter); the tags were laser-cut out of Perspex to reinforce the technological element of the brand. A layer of the Ventile fabric was attached behind the Perspex with all of the 'point of sale details', such as barcode and size information.

Branding Tags:

Branding tags are more decorative additions to a garment to reinforce the brand; they can be attached permanently or can be hung with the swing tag – they could be made of more permanent material than the swing tag.

This tag was the same as the swing tag in that it was laser cut and a beetle shape, but this was a permanent branding fixture to the garment stitched near the hem (Figure 30.7).

Incentives:

Incentives are given to encourage custom and are usually in the form of a reward such as reductions in price, or giveaways.

To attract customers from the foot traffic outside the store, a series of giveaways would be offered each month, as incentives to investigate the brand, for example, in the summer months a reusable laser-cut lolly stick with the beetle brand would be made into ice lollies. Or a fully branded plastic water bottle would be given away in the summer.

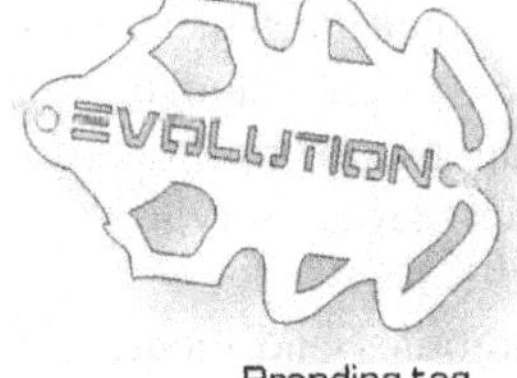

FIGURE 30.7

The laser-cut Perspex swing tag and brand tag for Evolution.

Press Release/Magazine Entries:

The collection needs to be visible in the correct magazine or newspaper with the right readership for the target customer.

New Technologies As Media Channels

The Video wall Experience:

Evolution would have 12 screens linked together to appear like one big screen. The screens would show a promotional video that is made up of shots from the photostyling shoot and print design ideas. It would have a similar kaleidoscopic effect as in the digital prints, ideally pulling the customer in.

The logo would flash up frequently to send a subliminal message to the customer as they are watching.

Interactive Walkway:

As an in-store promotion of Evolution, the interactive walkway would be eye catching, engaging and point any customers at the Evolution collection. When on standby, the walkway would appear like a sticker in the shape of the beetle logo, when stepped upon the beetle would break down into a 1000 digital beetles that will scurry along the walkway to the Evolution stand.

The diagram in Table 30.1 is of the 'critical path', the pulling together of all the necessary design (garments), research, promotion – including the web experience and business tasks – into a time frame.

30.8.2 CASE STUDY 2: 'STRATAGEM' BY LUKE ANTHONY RICHARDSON

The following Marketing Plan is of the menswear case study project, and there are a number of contrasts in this project to Evolution, namely the target market and price points, as this is a more exclusive range of garments (Richardson, 2010). The strong concept allows for another very creative promotional campaign.

30.8.2.1 Consumer

Stratagem is a menswear brand of 'technical' jackets (Figure 30.8). The collection includes a statement that challenges the existence of CCTV and surveillance in this society through concepts such as camouflage and invisibility.

Secondary research suggested that a camouflage print would help the wearer blend into an urban environment. The range is mainly constructed of outerwear garments that will behave as a shield to security cameras and the wearer's identity. The garments, quilted or filled jackets, windcheaters and full body suits, will change the wearer's silhouette. The all-over print is meant to deceive and confuse observers, large hoods and visors hide the face to maintain anonymity. The brand uses disruptive pattern material (DPM) (better known as camouflage print; variations are used by the armed forces worldwide, depending upon the terrain in which they are serving) as an urban option, rather than the typical woodland and desert colours (Figure 30.9).

Inspiration was also taken from traditional and modern workwear, with particular reference to details and function, fabrics and trim from the inside of high-visibility jackets.

Brand Identity

The logo incorporates the word 'rage' in Stratagem, which refers back to the anti-surveillance message. Yellow and black hazard tape is used widely across the promotion and branding and conveys the

Table 30.1 Critical path: case study – Evolution

2010	Research	Garments	Promotion	Business
January	Research into existing brands with similar design philosophies.			
February	Visit action sports retailer trade show, 3rd/4th USA Premiere vision 9th/12th Paris, London fashion week 19th/23rd, London.	Develop trend stories, colour, fabrics and silhouettes.	Begin promotional research.	
March		Begin designing/moulage and developing first prints and textile selections. Initial fabric sourcing, detailing and trim.		Identify key competitors and pricing of garments.
April		Develop collection. Finalise fabric choices and expand prints.		
May		Make up toiles for fittings.		
June		Adjust patterns/toiles for final fits. Work out size grading.	Begin promotional development.	Contact buyers at Top shop to make orders. Set budgets.
July	Visit bread & Butter trade show, 7th/9th Berlin.	Make changes where applicable to details taking from research. Approve final samples for production.		
August		Finalise fabrics and prints for production. Finalise production patterns and size grading. Create detailed technical packs for factories.		Agree on delivery schedule with suppliers.
September		Begin production.	Finalise promotional items and launch details.	
October		Continue production and monitor.		
November		Garment inspection in production and upon delivery.		
December		Final garment checks. Despatch orders to buyers.	Finalise launch and promotion.	Invoice with payment terms as agreed in the contract.
2011				
January	Begin Spring/Summer 2012 research.			Receive payment.
February		Launch collection.		
March				Monitor sales of Spring/Summer 2011 collection.

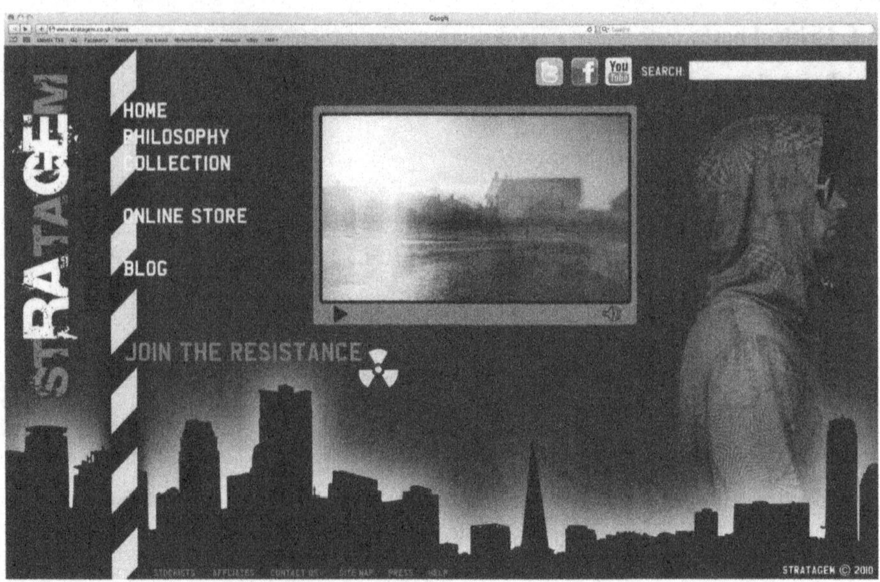

FIGURE 30.8

The branded Stratagem Website.

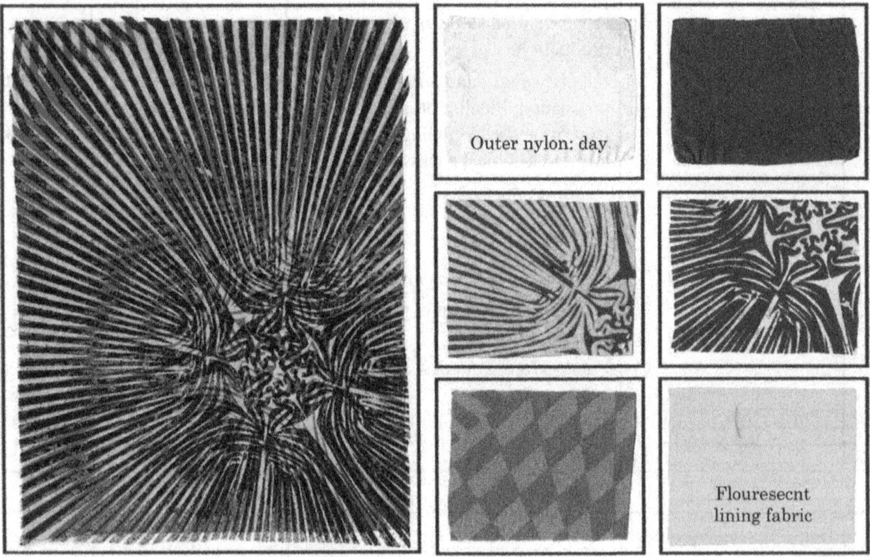

FIGURE 30.9

Moire style prints for the Stratagem collection.

FIGURE 30.10

Swing tags with the Stratagem branding.

message of hazard and exposure in a surveillance society. The garments have no branding on the outside in keeping with the invisibility concept, so swing tags indicate the price, size, colour, print option and garment style number. The collection does not relate to a season but behaves as a fashionable functional item. The range will develop but never to a season.

The range consists of 10 key garments, mostly outerwear, and most have the option of a day or night DPM print. Each item is named after CCTV cameras and systems.

The range consists of moiré prints, which are inspired by the effects that houndstooth style patterns have on television, where they become more obvious, and have an effect known as strobing, which should disrupt and confuse any camera surveillance (Figure 30.10).

Competitive Analysis

Luke looked at competitor brands and analysed their brand values. They were Belstaff, Berghaus, C.P. Company, Griffin, Maharishi, Moncler, The North Face, Stone Island and Superdry.

Customer Profile

Luke approached the customer profile with more specific headings, to get more specific answers and information. For example:

'The Creative':

Age: 25–35

Socio Economic Group: ABC1
Outlook: Appreciates design and conceptual ideas.
Interests: Technology, art & fashion.
Income: Disposable – will spend a lot of money on individual garments if it is something they like.
Lives: In Hoxton, London.
Occupation: A young professional who works in design or a product-based environment.
Travels: By underground or bus to get around town and takes an occasional taxi on nights out. Commutes to the city from work to home to meet friends. Is a regular in the rush hour.
Reads: Magazines – View magazines, Wallpaper, ID magazine.
Shops: In 'exclusive shops' in places such as Kingly Court just off Carnaby Street or Dover Street Market. Makes the occasional trip to Harvey Nicholls and Selfridges. Also buys online for specific pieces.
Brands: Maharishi & Comme des Garcons.
Socialises: Drinks in Soho.

Luke designed two further customer profiles called the ***Activist*** and the ***Functionalist***.

30.8.2.2 Cost

Luke did not include manufacture, labour and shipping costs as he could not estimate these accurately – there is risk in this, but the exercise is hypothetical and the garments are aimed at an expensive level of the market, so if Luke was to go to manufacture he would, of course, be able to complete the costings more accurately. Some trims, such as buckles, are given an average unit price. The price will be wholesale as this is how Stratagem will be sold to stockists. A mark-up will be added to the price to create a recommended retail price. Luke undertook more market research into Stratagem's competitors to create a realistic mark-up for the brand to sit comfortably within its target market.

A typical outerwear piece would require consideration of the outer fabric, lining fabric, silk wool rib trim, stuffing, pocket zips, front zip, invisible zip, Quickburst zipper, magnetic tape, magnetic fastener, visor mesh, reflective tape, webbing, eyelets, shock cord, vinyl, buckles and clips, toggles, grosgrain, branding label and cut vinyl.

Luke added a normal 100% mark-up to the wholesale price and 170% mark-up to the retail price without VAT. Luke's collection ranged from £125 for a jersey shirt to £800 for a jacket (Figure 30.11), all including VAT.

30.8.2.3 Convenience

The main outlet for Stratagem was going to be the Website www.stratagem.co.uk to maintain the philosophy of being inconspicuous. The reach of the brand would be greater on the Internet. Also, a bricks and mortar store was considered too high a risk in the economic climate. The Stratagem customer is a heavy Internet user, who regularly communicates online. The bonus of avoiding the reality of surveillance cameras helps to reinforce the brand.

The site gives the opportunity to customers to subscribe free, to 'Join the Resistance' where events, offers and news is shared.

Luke had a further idea of creating a 'pop up store' in the form of a heavy goods vehicle (HGV). The vehicle would stop off at nine major cities and sell Stratagem clothing: Edinburgh, Newcastle,

Tilt 600 TVL dome jacket				
Item	**Details**	**Quantity**	**Unit cost**	**Total cost**
Fabric	Outer fabric	7	£2.95	£20.65
	Lining fabric	3.5	£3.95	£13.83
Trim	Silk wool rib	0.4	£2.75	£1.10
	Stuffing	1	£6.00	£6.00
	Pocket zip	4	£4.50	£18.00
	Front zip	1	£12.50	£12.50
	Invisible zip	2	£2.50	£5.00
	Quick burst zipper	1	£20.00	£20.00
	Magnetic tape	0.1	£21.95	£2.20
	Magnetic fastener	2	£5.50	£11.00
	Visor mesh	0.25	£13.95	£3.49
	Reflective tape	0.1	£4.95	£0.50
	Webbing	5	£0.50	£2.50
	Eyelets	4	£0.05	£0.20
	Shock cord	0.25	£0.45	£0.11
	Vinyl	0.2	£12.00	£2.40
	Buckles & clips	18	£0.30	£5.40
	Toggles	2	£0.10	£0.20
	Grosgrain	0.05	£0.35	£0.02
Branding	Label	1	£0.50	£0.50
	Cut vinyl	0.05	£12.00	£0.60
			Total cost price	£126.18
			Markup to wholesale price @ 100%	£252.37
			Wholesale to RRP (ex. VAT) @ 170%	£681.39
			VAT @ 17.5%	£119.24
			Total (inc. VAT)	£800.63
			RRP	**£800.00**

FIGURE 30.11

A costing sheet for the Tilt 600 TVL Dome Jacket from the Stratagem collection.

Leeds, Manchester, Liverpool, Birmingham, Bristol, Brighton and London. This idea reinforces the brand, as it would not stay in any city longer than a week, creating interest and distributing promotional material. It also allows for distribution of the brand to its target customer. Some limited Stratagem pieces could be purchased in exclusive stores in the United Kingdom, as long as the brands around the pieces fit in with the Stratagem brand.

30.8.2.4 Communication

The Traditional Marketing Approach

The brand will be featured in Dazed & Confused or Another Man magazine, ideally driving readers to the Website. A promotional incentive can be offered by clicking on links on other sites to the Stratagem site.

New Technologies As Media Channels

Luke planned to use a 'Flash Mob' to launch Stratagem, organised through Facebook and Twitter, where participants can join a group that makes a statement about this surveillance society. This Flash Mob would involve 500 people selected from the group, all wearing Stratagem teeshirts (which they would be allowed to keep); the participants would have to stand still for 10 min in Canonbury Square, Islington, London, the home of George Orwell, author of 1984. CCTV footage of the event would be gathered and made into a film to be played in the Stratagem HGV and also sent out as a viral advertisement. The HGV would be heavily branded (Figure 30.12).

The vehicle would be equipped with CCTV, which would also capture footage from each city location visited, which would then be made into another part of the promotional film that could be viewed

FIGURE 30.12

A branded visualisation of the HGV.

FIGURE 30.13

The videos displayed on 24 screens like a CCTV control room.

online and on the wall of the vehicle. The film will demonstrate how the product works and it will be displayed on 24 screens, like a CCTV control room (Figure 30.13).

The ‘critical path’ for Stratagem is given in Table 30.2. Luke describes each action period as a phase.

Table 30.2 Critical path – Stratagem

Stratagem 2010 collection **Critical path**	**Timing**	**Function**
Trend research into garment styles, shapes and contemporary details.	December–January	Research – Phase 1
Research into competitors' collections.	January–February	Research – Phase 2
Research into promotional ideas.	January	Promotion – Phase 3
Development of initial concept ideas.	January–March	Product – Phase 4
Development of colour palette, silhouettes & fabrications	February–April	Product – Phase 5
Attend London fashion week 15th/23rd Feb. for trend research.	February	Research – Phase 6
Attend Premiere Vision trade show 5th/12th Feb. for sourcing.	February	Research – Phase 7
Begin initial design ideas for garments & prints.	January–June	Product – Phase 8
Start sourcing fabrics and trims.	February–May	Product – Phase 9
Brief fabric mills on prints.	May	Product – Phase 10
Finalise designs and collate range.	May–June	Product – Phase 11
Brief factories on garment designs for sampling.	June	Product – Phase 12
Begin promotional development.	June	Promotion – Phase 13
Set price margins and promotional budget.	June	Business – Phase 14
Receive print samples back and make amendments.	June	Product – Phase 15
Receive garment samples back and conduct fittings.	June–July	Product – Phase 16
Send amendments to factories for secondary sampling.	June–July	Product – Phase 17
Receive secondary samples and produce look-book.	July–August	Product – Phase 18
Agree on costs and delivery schedule with manufacturer.	July	Business – Phase 19
Send potential stockists look-book.	June–July	Product – Phase 20
Attend Bread & Butter trade show with collection 7th/9th July, Berlin.	July	Promotion – Phase 21
Finalise promotion & launch.	August	Business – phase 22
Close books and order fabric quantities.	August	Business – phase 23
Begin production.	July–September	Product – Phase 24
Visit factories to monitor production.	August–September	Product – Phase 25
Distribute garments to stockists.	August - September	Business – Phase 26
Begin next collections trend research (phase 1 and continue phases).	September	Research – Phase 27
Launch Stratagem.	September	Promotion – Phase 28
Flash Mob – stunt.	September	Promotion – Phase 29
Stratagem HGV spends 1 week each in @ Bristol, Brighton, London & Birmingham.	September	Promotion – Phase 30
Stratagem HGV spends 1 week each in @ Liverpool, Manchester, Leeds & Newcastle.	October	Promotion – Phase 31
Stratagem HGV spends 1 week @ Edinburgh.	November	Promotion – Phase 32
Promotional film is launched.	December	Promotion – Phase 33
Monitor sales of 2010 collection.	September–March	Business – Phase 34

30.8.3 CASE STUDY 3: COMMERCIAL CASE STUDY – ALL SAINTS OF SPITALFIELDS

This case study is about a successful commercial enterprise – All Saints of Spitalfields – designed to provide further contrast to the two student case studies Evolution and Stratagem though it does not explore as much detail, it is more about the brand and marketing of the brand.

The distinctive brand All Saints, now a large retail chain, was established in 1998 by Stuart Trevor, and it was taken over around 5 years ago by Kevin Stanford. There are now 90 sites across the United Kingdom, United States and Asia.

Brand Value

There are eight brand values that drive this brand: humility, individuality, honesty, dedication, hunger, decisiveness, distinctiveness and loyalty. Their mission is to 'create a brand that blends culture, fashion and music into a potent formula of desirable clothing that expresses individuality and attitude' (All Saints Brand History).

Customer Type

All Saints sell men's, women's and children's wear, homeware, gifts and vintage garments.

The brand is set in industrial, warehouse style stores, with exposed brickwork, utility lighting and stainless steel fittings. The floors are wooden against distressed wallpaper and bare bulb lighting.

The playing of an eclectic mix of 1970s to contemporary music enhances the atmosphere. The brand is aimed at the 'youthful in spirit', which does not alienate an older customer.

Branding

The branding is that of a line drawing of a ram's skull with 'All Saints of Spitalfields' written in simple, but slightly distressed lettering, reinforcing a feeling of 'provenance' and supporting the 'vintage' feel. This is set on a background of musty looking, aged, but quality, paper carriers with calligraphy print tissue paper wrapping and ram's skull stickers. Online orders arrive in calligraphy print lined boxes with items again wrapped in printed tissue. The price points of garments can be fairly expensive, at around £30 upwards for a tee shirt to over £350 for substantial garments.

They use their store window to sell the strong image, and this remains consistent across the chain. They use 'themes' to tell a story backed up with props, lighting, music and colour in the environment. The aim is to create an entertaining experience for the shopper who can now choose to shop online, by catalogue, as well as instore.

All Saints do not use many traditional marketing methods, other than 'word of mouth'. They took this element of promotion and used new technologies to make the most of what word of mouth they achieve.

The Website offers an 'affiliate' scheme to other companies, to spread the All Saints 'word'. This works by the company adding a text link, banner ad, product feed or content link to the All Saints site. When a sale is made from using this access point to All Saints, the host company can earn a 7% commission (http://www.allsaints.com/ 16/10/10).

All Saints use Webtrends, Analytics 9, to analyse the customer journey from first clicking on the site to the sale. There was a desire to expand into the United States market, so it was important to understand consumer online habits. Analytics 9 works by analysing stock control and presents its data in a visual form, so that it is easy to understand and analyse. It tracks conversion and non-conversion to sales.

Wickedweb, the digital marketing agency, 'a fast paced full service digital marketing agency, combining creative ideas, strategy and technology' (http://www.wickedweb.co.uk/soc

ial-media-marketing/ 15/10/10), provided the social media marketing strategy for All Saints; the brand has a Facebook, Twitter and iPhone application page accessible from the site (many brands use these tools now). It is possible to sign up to the mailing list and get regular emails about sales, new stock and any other events. The 'lookbook' gets emailed out to subscribers. Facebook fans can leave messages about garments and the brand and any other thoughts that provide valuable insight to the company but ultimately drives the consumer back to the ecommerce Website, All Saints' content is shared also, so there are reciprocal relationships when being a Facebook 'friend'. The iPhone application works on the phone, iPod touch and the iPad and updates automatically. The application can be used to locate local stores, browse the range and be aware of new styles, search for products, share 'wish lists' with friends and track any orders!

The Facebook fan base for All Saints grew from 4000 to 10,000 in 3 months, boosting its brand loyalty and sales conversion (Wickedweb)!

These marketing tools work as they provide a 'service' to the consumer so that they can remain loyal to the brand.

30.9 FUTURE TRENDS

Fashion moves very quickly and relies on inspirational intelligence material to inform products for forthcoming seasons. This works in two ways, firstly, a trend forecasting company analyses marketing information and forecasts long-term and short-term changes in consumer patterns. Secondly, a fashion forecasting company offers specifically fashion intelligence material to inspire a new season.

Trend forecasting consultancies use qualitative (this involves the analysis of data obtained from interviews, video or artefacts) and quantitative data (this involves the analysis of numerical data obtained from questionnaires, for example) and statistical data to inform the development of brand strategies for clients (McKelvey & Munslow, 2008). These companies offer insight into consumers and how they evolve in the future. They work across a range of industries such as retail, creative, technology and finance. They report on trends across the globe and analyse social, political, economic, artistic and cultural influences. Anticipating the needs of customers is invaluable to any industry. Developing new brand strategies for companies is undertaken by the trend company. Key companies working in this area include Future Foundation (http://www.futurefoundation.net/), a commercial think tank that looks to the future by forecasting social and consumer trends, based in London. The Future Laboratory (http://www.thefuturelaboratory.com) is a team of trend analysts and ethnographic researchers, based in London, offering clients insights into targeting and understanding future consumers. Faith Popcorn (http://www.faithpopcorn.com/) is a developer of long-term 'lifestyle' trends, aiding clients in developing new relevant products for the future: the service is known as BrainReserve. Trendwatching (http://trendwatching.com/) distributes free monthly briefings by email after subscribing to the service on the Website, giving an insight into new consumer trends. Companies can purchase more in-depth information. Henley centre is a consultancy led by research and intelligence that aims to offer insight and innovation into the future for a range of clients.

30.9.1 FASHION FORECASTING

Fashion forecasting is an essential service to the fashion design industry. Its purpose is to 'forecast' what will happen 18 months to 2 years ahead of any season. This requires a degree of intelligence gathering by these companies, and this consists of looking at the latest stores, designers, brands, trends and business innovations (McKelvey & Munslow, 2008) in leading fashion capitals, such as London, Paris, New York,

Milan and Tokyo. These services create a synthesis between what is happening and what they think is going to happen; much of this feels like intuition, but an experienced forecaster will understand what will happen next due to the cyclical nature of fashion, influences such as social changes, political changes, economic effects, cultural and artistic influences: all have a bearing on what will happen. Forecasting services employ designers and illustrators to realise the 'vision' for a given season.

Such companies have been around for some time, as an identifiable industry in their own right from the 1960s, and sold the information in limited edition hand-swatched books, but they also tailored the intelligence to specific markets dependant upon the client's product area.

Nowadays, much of the information is accessed on the Internet; some companies like Worth Global Style Network (WGSN) offer online information as a 'closed site', which requires a subscription to access the information. The subscription can be a considerable amount of money, for example a full service of 12 publications a year could cost around £5000 a year. Once logged into a site, the resources available can be enormous: WGSN offer information on industry news, business resources, trade shows and fairs, fabrics and trims, catwalk reports, editorial, conceptual inspiration, trend information, retail information – visual and textual, new information in key capitals, beauty information, men's, women's, junior and youth wear, active sportswear, graphic ideas and information about new graduates from fashion courses (this 'showcasing' of a graduate's new ideas and approach to designing for fashion can be inspirational and can also prove to be a good recruitment tool).

This information is analysed and published so that the industry can utilise it at their particular market level. There are a number of companies producing trend information; it remains a fairly small but very significant contributor to the fashion industry.

Also, some larger fashion retailers may well have in-house forecasting where they look specifically at their product and gather their own intelligence by sending buyers and designers on trips across the world.

Some of the companies offering traditional fashion forecasting services (books and consultancy) are Carlin International, Fashion Forecast Services, Here & There, Jenkins Reports Ltd, Milou Ket Styling & Design, Mudpie Ltd, Nelly Rodi, Peclers Paris, Promostyl, Sacha Pasha, Trend Bible (interiors) and Trend Union. These companies often have an online presence, but much of their work is to do with publications and consultancy.

Online forecasting agencies are Fashion Snoops, Infomat, Stylelens, Stylesight, Trendstop Ltd and Trendzine.

Professionals in the forecasting industry have created a group, online, in the professional social network, Linked In. Called Hall5, this is to share information, issues and concepts in this professional forum between each other.

Cool Hunting is a Website that offers inspiration as a designer's personal reference through analysis of innovation in design, technology, art and culture. Updates are available daily and weekly documentaries are available. It is possible to subscribe to the newsletter by RSS feed (Really Simple Syndication – used to publish frequently updated work such as news headlines and blog entries) and by email to Cool Hunting. It is available on iPad and they can be followed on YouTube, Twitter and Facebook (http://www.coolhunting.com/ 04/08/11).

30.9.2 NEW TECHNOLOGIES AND PROCESSES

There are many opportunities to find out what may be the next important future trends by searching on the Internet. General lifestyle trends, marketing trends and the changing consumer, shifting market trends and the impact of new technology trends and contradicting trends can all be found very easily.

Faith Popcorn's site – Brainreserve – offers 17 long-term, evolving trends that tell us about ourselves and how we are changing, what our needs and desires are, for example, 'the way women think and behave is impacting business, causing a marketing shift away from a hierarchical model toward a relational one'; This trend is about marketing to the different needs of women. Popcorn considers it a 'model' rather than a trend. Faith Popcorn (Evolution trend, http://www.faithpopcorn.com/ 04/08/11).

Trendwatching is a leading 'consumer trends firm' that gathers intelligence from across the world, which it pulls together to form a picture of what is going to happen. One of their current 'briefings' suggests that the expansion into global markets will mean that brands will have greater success, but they suggest also that new brands will be developed and the consumer will covet many more into the future; for example, there will be new opportunities emerging from markets in China, India, South Africa and Brazil (http://www.trendwatching.com/briefing/).

The demand for innovation and new technology such as the iPod, iPad and iPhone and gaming consoles such as the Nintendo DS in products and services across all sectors will not stop. The next generation will be 'digital natives' (Marc Prensky (2011) created the term in his publication 'Digital Natives, Digital Immigrants' and refers to people who grew up with twenty-first century new technology, or were born during or after the introduction of digital technology so that it becomes 'second nature' to them. A digital immigrant is an individual born before digital technology existed but has adopted it in their life) and will drive technological developments; however, they will become concerned about green and environmental computing, with the increased use of mobile phones and wireless access (http://ww2.prospects.ac.uk/cms/ShowPage/Home_page/Explore_job_sectors/Information_technology/future_trends/p!eklfif, 04/08/11).

Three forecasts were offered by one Ron Reed (an American entrepreneur, publisher and professional consultant in the real estate and Internet marketing industries; within these two communities he is often referred to as a 'guru'), regarding future trends in technology. His first was about working in 'real time'. In other words, instead of designing, producing and marketing a product, why not put it into the social networking arena and get feedback before you go to all of this trouble. He uses the example of a musician putting out a song for feedback, then reacting to the feedback by developing an album around this.

The second forecast involved the use of augmented reality (AR – combines live video on your mobile phone with computer-generated data and visualisations, perhaps about the buildings in front of you; Figure 30.14).

This works by using GPS on a smartphone, so that at any given moment the satellite system can pinpoint your location and use data from sites like Google and MapQuest to offer information on a restaurant, for example, and maybe get reviews and even menus on your phone. Think how powerful this could be if it was a fashion store with marketing information.

A third suggested forecast and current development (Reed suggests it is already in use in Asia, it is in some use in the United Kingdom by Barclaycard, Orange and by Oyster Card and by the smart card payment symbol on mobile phones) could be the use of the iPhone, or other smartphones, which have a massive market share; as a means to pay for items in a store, instead of using credit/debit cards, or cash, you could scan the phone and add it to your phone bill (http://www.whoisronreed.com/ron-reed/future-trends-in-technology-for-2010, 04/08/11).

FIGURE 30.14

Augmented reality.

Source: Image courtesy of Jack Webber.

30.10 SUMMARY POINTS

- Marketing aims to target customers with relevant products, they use the marketing mix of product, price, place and promotion or more broadly the updated version for 'niche' markets of consumer, cost, convenience and communication.
- Targeting a market can be done in a number of ways, for example, by geographic segmentation, psychographic segmentation, demographic segmentation, behavioural segmentation, socio-economic classification, customer profiles or seasonal and occasion markets.
- Branding is key to achieving a clear identity to be able promote products.
- There are traditional channels to promote products, such as television advertising, ambient media, magazine advertising, celebrity endorsement, radio advertising, sales promotion, public relations, personal selling and direct marketing.
- New technologies are providing new promotional media channels. The development of Web 2.0 has allowed more interaction and social networking on the Internet; Facebook, Twitter, Instant Messaging, virals, blogs, YouTube, podcasts, Webcasts and email are proving to be valuable channels of promotion. 'Digital' marketing agencies create customised strategies using these channels.

- The marketing plan helps to strategically think through the marketing mix.
- Future fashion trends help to inform the marketing plan.
- Future trends regarding the consumer, lifestyles and new technologies help to inform future strategies for marketing fashion products.

30.11 PROJECT IDEAS AND REVISION QUESTIONS

1. What are the fundamental differences between looking at product, price, promotion and place compared to consumer, cost, convenience and communication?
2. List the different methods of targeting customers in the text; can you find two more methods and discuss the advantages and disadvantages of each?
3. Create a brand and brand values for a casual fashion womenswear label aimed at digital immigrants in the middle-market price range.
4. Plan the concept development based on the research from Question 3. Create the critical path.
5. Take your concept: what tools would you use to promote it creatively using traditional methods?
6. Then suggest how you would push the concept further by using new technology channels and give reasons for these suggestions.
7. Discuss how fashion forecasting would inform your concept.
8. Can you think of new ways to use new developments in technologies to promote your brand? Show your understanding by discussing the use of technology in promoting your concept.

30.12 SOURCES OF FURTHER INFORMATION

30.12.1 BOOKS

Bohdanowicz, J., & Clamp, L. (1994). *Fashion marketing*. London: Routledge.
Clifton, R. (2003). *Brands and branding*. Economist Books.
Gobe, M. (2010). *Emotional branding*. New York: Allworth Press.
Goworek, H. (2007). *Fashion buying* (2nd ed.). London: Blackwell Publishing.
Hines, T., & Bruce, M. (2008). *Fashion marketing – Contemporary issues*. Oxford: Butterworth-Heinemann.
Raymond, M. (2010). *The trendforecasters handbook*. London: Laurence King Publishing.
Ries A., & Ries, L. (2000). *The 22 immutable laws of branding*. Profile Business.
Riewoldt, O. (2000). *Retail design*. London: Laurence King Publishing.
Wolfe, M. (2009). *Fashion marketing & merchandising*. Illinois: The Goodheart-Willcox Company, Inc.

30.12.2 TREND FORECASTING COMPANIES

Faith Popcorn – Brainreserve http://www.faithpopcorn.com.
Future Foundation http://www.futurefoundation.net/.
Future Laboratory http://www.thefuturelaboratory.com.

Henley Centre/Headlight Vision http://www.hchlv.com/.
Trendwatching http://www.trendwatching.com.

30.12.3 FASHION FORECASTING COMPANIES

Carlin International http://www.carlin-groupe.com.
Fashion Forecast Services http://www.fashionforecastsevices.com.au.
Fashion Snoops http://www.fashionsnoops.com.
Here & There http://www.doneger.com.
Infomat Inc. http://www.infomat.com.
Jenkins Reports Ltd http://www.jenkinsreports.com.
Milou Ket Styling & Design http://www.milouket.com.
Mudpie Ltd http://www.mudpie.co.uk.
Nelly Rodi http://www.nellyrodi.com.
Peclers Paris http://www.peclersparis.com.
Promostyl http://www.promostyl.com.
Stylelens http://www.stylelens.com.
Stylesight http://www.stylesight.com.
Trend Bible (interiors) http://www.trendbible.co.uk.
Trend Union http://www.trendunion.com.
Trendstop Ltd http://www.trendstop.com.
Trendzine http://www.fashioninformation.com.

30.12.4 MAGAZINES

View Publications – Viewpoint is a magazine about consumers and new markets. http://www.view-publications.com.
WeAr Global Magazine – a magazine that captures the zeitgeist and offers trend reports from major cities, photographs of designer collections, stores, showrooms, interiors and window displays. http://www.wear-magazine.com.

30.12.5 WEBSITES

Branding http://www.brandchannel.com/papers_review.asp?sp_id=1234.
Celebrity Endorsement http://www.oxbridgewriters.com/essays/marketing/consumer-celebrity-endorsement.php.
Guerrilla Marketing ideas http://blogof.francescomugnai.com/2009/11/the-80-best-guerrilla-marketing-ideas-ive-ever-seen/.
Marketing http://inventors.about.com/od/fundinglicensingmarketing/f/Marketing.htm.
News on Design and Technology http://www.fastcompany.com.
What is Marketing? http://www.knowthis.com/principles-of-marketing-tutorials/what-is-marketing/what-is-marketing.
What is Marketing? http://marketingteacher.com/lesson-store/lesson-what-is-marketing.html.
What is Marketing? http://tutor2u.net/business/marketing/what_is_marketing.asp.

REFERENCES

All Saints Brand http://www.guardian.co.uk/business/2010/jun/11/allsaints-spitalfields-store-new-york 16/10/10.

Analytics 9 http://www.marketinguk.co.uk/Marketing/AllSaints-Spitalfields-select-Webtrends-Analytics-9-for-a-tailored-analytics-solution.asp 16/10/10.

Bovee, C. L., & Thill, J. V. (1992). *Marketing*. New York: McGraw-Hill.

Easey, M. (Ed.). (2009). *Fashion marketing* (3rd ed.). Oxford: Wiley-Blackwell.

Jackson, T., & Shaw, D. (2001). *Mastering fashion buying & merchandising management*. London: Macmillan Master Series.

Lay, K. (2010a). Evolution – Marketing plan, Unpublished Degree of BA (Hons) Fashion Marketing @ Northumbria University, School of Design.

Lay, K. (2010b). Major project research document, Unpublished Degree of BA (Hons) Fashion Marketing @ Northumbria University, School of Design.

McKelvey, K., & Munslow, J. (2008). *Fashion forecasting*. Oxford: Wiley-Blackwell.

McKelvey, K., & Munslow, J. (2011). *Fashion design: Process, innovation & practice*. Oxford: Wiley-Blackwell.

Prensky, M. (2001). *Digital natives, digital immigrants* . (Vol. 9) NCB University Press. No. 5.

Richardson, L. A. (2010). Stratagem: Urban camouflage – marketing plan, Unpublished – relevant material is contained within this chapter in the second case study – Stratagem. Degree of BA (Hons) Fashion Marketing @ Northumbria University, School of Design.

Web 2.0 http://oreilly.com/web2/archive/what-is-web-20.html 02/08/11.

CHAPTER

THE CARE OF APPAREL PRODUCTS

31

R.K. Nayak, R. Padhye
RMIT University, Melbourne, VIC, Australia

LEARNING OBJECTIVES

At the end of this chapter, you should be able to:

- Garment performance and the factors that affect serviceable life
- Categories of wear
- Microbial attack and environmental decay
- Shape retention, wrinkle recovery, dimensional stability
- Laundering equipment: types and function
- Laundering aids
- Types of stains and their removal
- Care labelling: care labelling systems, regulations
- Storage of garments: techniques and conditions of temperature and humidity

ABBREVIATIONS

ASTM	American Society for Testing and Materials
BOD	biochemical oxygen demand
COD	chemical oxygen demand
DP	durable press
FTC	Federal Trade Commission
HLCC	Home Laundering Consultative Council
ISO	International Standardisation Organisation
JIS	Japan Industrial Standard
LAS	linear alkyl-benzene sulphonates
NAFTA	North American Free Trade Agreement
O-fading	ozone fading
P/C	polyester/cotton
RH	relative humidity
TC	Technical Committee
UV	ultraviolet
H_2O_2	hydrogen peroxide
NaOCl	sodium hypochlorite
NO_2	nitrogen dioxide

Textiles and Fashion. http://dx.doi.org/10.1016/B978-1-84569-931-4.00031-3

31.1 INTRODUCTION

The performance of garments depends on their initial properties (when new) and the conditions to which they are subjected during use. All garments deteriorate progressively through different mechanisms such as mechanical damage, microbial/insect damage, shrinkage, colour loss and staining. These are the factors that reduce their serviceable life. Damage can be caused by environmental conditions (i.e. weathering) and the conditions of washing, ironing, dry-cleaning and storage. Among these factors, one of the most pervasive is weathering, i.e. the cumulative effect of daylight, temperature, humidity, rain, abrasive dust, reactive gases (pollution) and cosmic radiation (Barnett & Slater, 1991). Weathering results in the decomposition of fabrics that are exposed outdoors for a long period of time. Textile and garment industries around the globe have created a consumer appetite for product performance that will often exceed the capability of the item purchased. Damage caused to garments lead to rejection and consumer dissatisfaction.

Clothing should be capable of useful service. The end-point of usefulness of a garment is governed by the application to which it is subjected and to the standards set by the user. In some cases the end-point is not necessarily determined by the degree of wear but by the failure in other aspects; for example, a garment may be unserviceable even if unworn if the colour is faded. Wear is the net result of a number of agencies that reduce the serviceability of an article. Different types of wear occurring during the use of a garment have been described by Clegg (1949). The paper reported that the failure of textiles in service is mainly due to weakening of the structure caused by mechanical breakdown of individual fibres, although items will also undergo a varying amount of chemical deterioration through laundering, exposure to sunlight and atmospheric conditions.

After being used for a period of time, a garment will no longer fulfil the intended purpose of its particular end-use and ceases to be serviceable. The service life of a garment is reduced by many factors (both physical and chemical) such as:

- dimensional changes or shrinkage of high magnitude such that the garment no longer fits;
- changes in surface appearance, which include formation of pills, pulling out of threads, abrasion of the fabric due to rubbing with external surfaces or body parts;
- internal abrasion and the cutting action of grit particles, which may be ingrained in dirty fabrics;
- biological attack (especially in natural fibre fabrics) by bacteria, fungi and insects;
- colour fading of the garment;
- failure of the seams;
- tearing of the fabrics by sharp objects;
- wearing of the fabrics into holes or wearing away of the surface finish;
- degradation of the fabric caused by normal household items such as bleaches, detergents, soaps, perfumes and antiperspirants;
- wearing of collars, cuff edges and other folded edges to give a frayed appearance.

Garment failures related to design or fabrication were reported by customers within the first 2 or 3 months, whereas those caused by washing were reported over a period of 10 months (Elder, 1978). Design failures included collar distortion, appearance of bubbles, failure of trims, delamination of interlining, fraying of buttonholes, seam breakage and wearing of cuffs. The failure of trims is considered as shrinkage and colour fastness as a result of laundering.

31.2 WEAR OF GARMENTS

Wear is the process of decomposition accelerated by chemical and physical environment or by use, time and exposure. A fabric may be subjected to bending, compression and shearing forces along with stretching, all of which lead to fatigue and wear, a change in fabric properties and to premature failure (Elder, 1978). The change of fabric properties with wear is explained in Figure 31.1. A range of factors contributing to the wear of garments has been investigated by several researchers (Akgun, Becerir, & Alpay, 2006; Alpay, Becerir, & Akgun, 2005; Anderson, Leeder, & Taylor, 1972; Card, Moore, & Ankeny, 2006; Wilcock & Delden, 1985). Different attributes of wear are discussed in the following sections.

31.2.1 PILLING

Pilling is the appearance of small bunches or balls of tangled fibres on the surface of a fabric that are held in place by one or more fibres and gives the garment an unsightly appearance. Before the invention of synthetic fibres, pilling was mainly observed in knitted woollen items made from soft twisted yarns. Both woven and knitted fabrics are prone to pilling. The propensity may be related to the type of fibre used in the fabric, the type and structure of the yarn and the fabric construction. Generally, pills are formed in areas that are especially abraded or rubbed during wear and can be accentuated by laundering and dry-cleaning. The rubbing action causes loose fibres to develop into small spherical bundles anchored to the fabric by few unbroken fibres.

Fabric made from natural fibres is less prone to pilling as the fibres break away and shed the pills. In synthetic fabric, because of higher strength of the fibres, they remain attached to the garment and accumulate to form pills. Pilling is particularly associated with nylon or polyester as may be seen in the collar of men's woven shirts made from polyester/cotton or nylon/cotton blends. Woollen knitted garments with a loose fabric structure made from soft twisted yarn (e.g. jumpers and cardigans) also suffer frequently from pilling. This can be reduced by diminishing the migratory tendency of fibres from constituent yarns in the fabric and is achieved by the use of higher

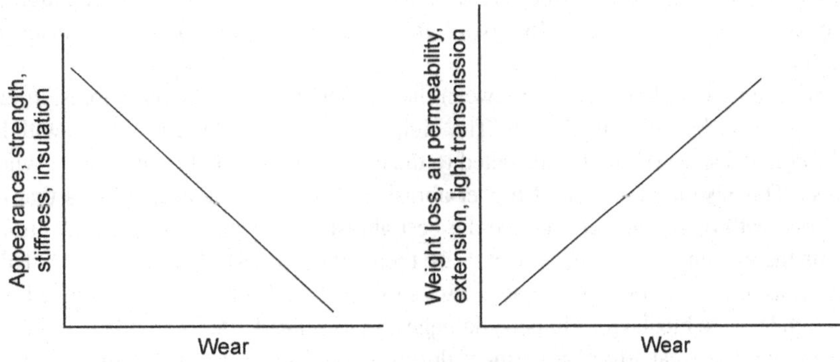

FIGURE 31.1

Change of fabric properties with wear.

twist in the yarn, reduced yarn hairiness, longer fibres and increased inter-fibre friction, a greater number of threads per unit length, brushing and cropping of the fabric surface and specialised chemical finishes (Booth, 1968).

The effect of fabric softeners and cellulase-enzyme containing laundry detergents on pilling was studied (Chiweshe & Crews, 2000). It was found that some softeners were not associated with an increase in pilling and that cellulase-enzyme detergent additives significantly reduced the amount of pilling on all cotton fabrics, except cotton interlock knits.

31.2.2 ABRASION

Abrasion is a progressive loss of fabric caused by rubbing against another surface. It has also been reported to occur through molecular adhesion between surfaces, which may remove material (Tabor, Howell, & Mieszkis, 1959). The hard abradant may also plough into the softer fibre surface (Kalishi, 1958). The breakage of fibres has been reported to be the most important mechanism causing abrasion damage in fabrics (Clegg, 1940). Abrasion can be of three types: flat or plane, edge and flex. In flat abrasion, a flat part of the material is abraded, edge abrasion occurs at collars and folds and flex abrasion rubbing is accompanied by flexing and bending. Abrasion is a series of repeated applications of stress. The selection of suitable yarn and fabric structure can therefore provide high abrasion resistance (Backer & Tanenhaus, 1951).

Abrasion resistance is dependent on several factors such as the fibre type and properties, yarn structure, fabric construction and type and the type and amount of finishing material present. High elongation, elastic recovery and the action of fibre rupture are more important than high strength for good abrasion resistance. Nylon fibre is considered to possess the highest degree of abrasion resistance while viscose and acetates have the lowest (Elder & Ferguson, 1969). Polypropylene and polyester fibres also have good abrasion resistance. The abrasion resistance of wool and cotton can be increased by blending with nylon or polyester. Longer and coarser fibres help to improve the abrasion resistance of a fabric. Increased linear density and balanced twist in a yarn give the best abrasion resistance.

Laundering may cause significant abrasion in fabrics, thus shortening the wear life of a garment. Cotton fabrics laundered in hard water suffered significantly higher edge abrasion than those laundered in soft water and carbonate detergents caused more abrasive damage than phosphate detergents (Morris & Prato, 1976). Neither of these detergents harmed fabrics when used with soft water.

A fabric with evenly distributed crimp between the warp and weft gives good abrasion resistance as damage is spread evenly between the threads. The higher abrasion resistance of fabrics with higher float (such as twill, satin and sateen) may be attributed to the easy relative mobility of threads, which helps in absorbing stress. This also is the cause of higher abrasion resistance in knitted fabrics that have looser structures. Fabrics with optimum sett produce the best abrasion resistance. If the fabric structure is too tight, it prevents the movement of threads, which are then unable to absorb the distortion. This results in lower abrasion resistance. A tight structure also causes the fibres to be stressed and fatigued beyond their yield point, which leads to breakage. Abrasion damage and its related effects are likely to be a significant factor in determining the wear life of a garment during normal use (DeGruy, Carra, Tripp, & Rollins, 1962; Gagliardi & Wehner, 1967; Heap, 1978; Reeves, 1962; Rousselle, Nelson, Hassenboehler, & Legendre, 1976).

31.2.3 COLOUR FADING

One of the major problems garments face is poor colour fastness. A coloured item may encounter a number of agencies during its lifetime that can cause the colour either to fade or to bleed into an adjacent uncoloured or light-coloured item. New garments may experience colour loss due to the removal of excess colour that was not adequately rinsed after dyeing. Colour loss can occur by the migration of weakly bonded dye molecules out of the fibre. Colour loss during washing will stain other materials and this will be influenced by the ratio of coloured to uncoloured items, fibre content of other items, and end-use conditions. A specific hue may be produced by the mixing of two or more dyes. If one component is degraded or lost from the material, the colour will be altered.

The type of dye, the particular shade used, the depth of the shade and the dyeing process all affect the fastness of a colour. Some coloured or printed garments change colour significantly during use. This may be caused by abrasion, rubbing, atmospheric conditions such as ultraviolet (UV) light, oxides of nitrogen or ozone, acid or alkaline substances, laundering or dry-cleaning, ironing, perspiration, rain water, chlorinated water or sea water. Colour loss due to abrasion may be caused by localised wear such as rubbing the elbows against a desk, excessive mechanical agitation during washing or an attempt to remove a stain by rubbing. The exposure of a garment to direct or indirect sunlight may cause colour change resulting in fading because the UV rays in sunlight cause damage to the dye structure.

Atmospheric gas fading or fume fading is the colour change of a fabric caused by acid gases in the atmosphere, which are formed in combustion processes. Garments left hanging for a long period of time will be affected by fume fading. Nitrogen dioxide (NO_2) is primarily responsible for gas fading. Ozone fading (or O-fading) occurs in dyes that are colourfast to fume fading if a high amount of ozone is present in the atmosphere. Disperse and direct dyes are more vulnerable to O-fading, and blue and red dyes are affected to a greater degree than others. Ozone may cause bleaching in acetate, cotton and nylon fabrics.

Colour loss may sometimes be due to acid or alkali present in various products. Certain dyes used in dyeing wool fabrics change colour when exposed to acidic conditions. Alkaline colour change is observed in dark blue and black acetate fabrics. Colour loss through laundering and dry-cleaning has been investigated by several researchers (El-Shishtawy & Nassar, 2002; Mangut, Becerir, & Alpay, 2008; Min, Xiaoli, & Shuilin, 2003). Colourfastness to perspiration is also an important factor for consideration by manufacturers (Schumacher, Heine, & Höcker, 2001; Xie, Hou, & Zhang, 2006). Perspiration is harmful as the bacterial action may lead to a loss or change of colour and finish, loss of fabric strength, odour problems, salt rings and deposits. Alkaline sensitive dyes may be damaged by fresh (acidic) and decomposed (alkaline) perspiration. Many of the direct, basic, acetate and metallic dyes are affected by perspiration. Perspiration may change the hue, cause bleeding of the dye and staining of lighter areas.

Some fabrics change colour in rain water and chlorinated water and some garments may change colour when frequently worn near the ocean. This is normally observed in wool fabrics and is a result of the action of sodium chloride on the sea water.

31.2.4 BREAKING OF YARNS AND FABRICS

Garments may sometimes fail during use because of a loss of strength in the yarns and fabrics, which affects the durability. A garment is subjected to various tensions during wear or washing, and contact with sharp objects which may cause rips or holes in the fabric. This type of damage is affected by the

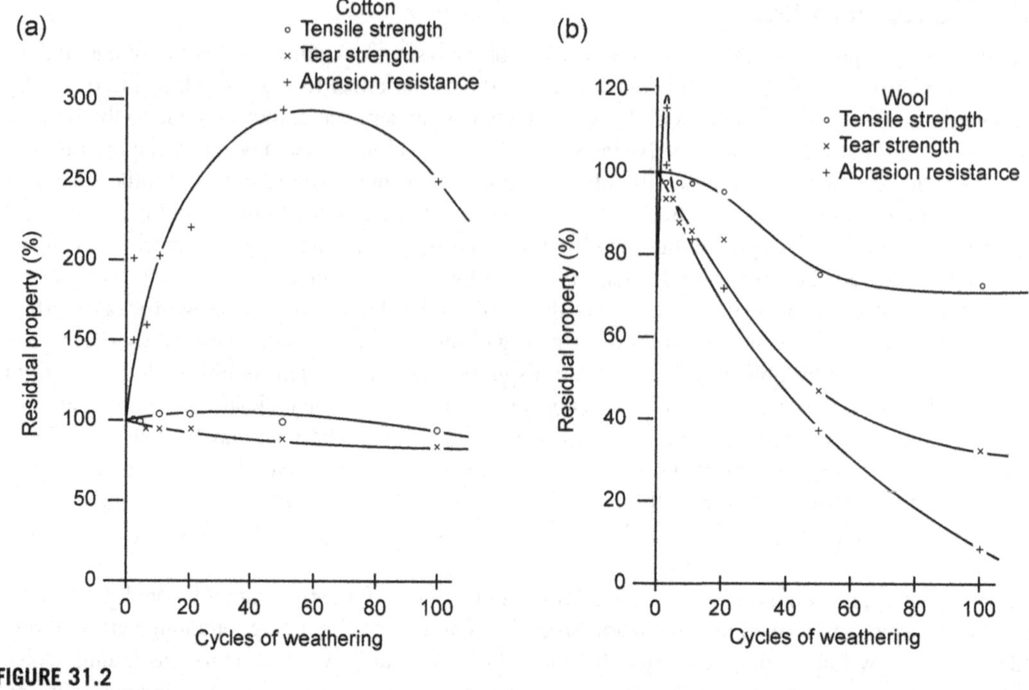

FIGURE 31.2

The effect of weathering time on the tensile strength, tear strength and abrasion resistance of (a) cotton fabric and (b) woollen fabric.

Source: Barnett and Slater (1991).

interrelationships of tensile strength, tear strength and abrasion resistance. Fabrics with a tighter construction have high tensile strength but lower tear strength and vice versa.

The exposure of some clothing to direct or indirect sunlight may cause deterioration in the fibres. The rate of deterioration will vary depending on the fibre content, yarn and fabric construction and the type of dyeing, printing and finishing applied to the fabric. Another significant cause is weathering, i.e. the cumulative effect of daylight, temperature, humidity, rain, abrasive dust, reactive gases (pollution) and cosmetic radiation on fabric. The effect of weathering time on the tensile strength, tear strength and abrasion resistance of cotton and woollen fabrics is explained in Figure 31.2. It can be observed that in the case of cotton, there is an increase in abrasion resistance with the number of weathering cycles, whereas other properties show a decreasing trend. Moisture in the air and grime present in atmospheric acid fumes may reduce the tensile and tear strength of fabrics. Bleaching agents such as hydrogen peroxide (H_2O_2) or sodium hypochlorite (NaOCl) may convert cellulose (in cellulosic fabrics) into oxycellulose, which is much weaker than cellulose. If the bleach is not thoroughly rinsed out, it may cause damage to other fabrics with which it comes in contact.

Laundering effects include the loss of tensile strength, discolouration, overall change in appearance, breakdown of molecular structure and a change in the oxidation state or degree of polymerisation. Hurren, Wilcock, and Slater (1985) reported that the mechanical and chemical degradation of fabric during laundering is mainly due to the abrasion of wet fabric and cleaning agents, respectively. It is reported that durable press (DP) finished fabrics retained a higher proportion of their initial strength in repeated laundering when compared to untreated fabrics (Neelakantan & Mehta, 1981). Lau, Fan, Siu, and Siu (2002) studied the effect of repeated laundering on the performance of wrinkle-free treated garments. They reported that wrinkle-free treatment can reduce the adverse effects of washing on mechanical properties.

The mechanism of breakdown is almost the same for all fibre types, i.e. cotton, wool, silk, linen and rayon. The cause of the breakdown of a large proportion of the fibres is the transverse cracking, which occurs at the position of maximum weakness in the structure as a result of flexing and bending stress suffered during wear. The surface fibres, which are held lightly, undergo gentle abrasion. The breakdown of cotton fibres under abrasive forces in different conditions (dry and wet) during normal wear and laundering has been investigated (Chippindale, 1963). In the dry state, the surface layers are rubbed and eroded with no indication of fibrillar structure. In the wet state, the fibres swell, the fibrillar structure is loosened and the fibrils can be torn out from the fibre surface.

Murdison and Roberts (1949) studied the damage done to cotton fabrics in laundering and storage by measuring the change in tensile strength and fluidity. Samples laundered in 1940 and re-tested in 1948 showed a lower tensile strength and a higher fluidity due to ageing. Uneven cracks also developed in the fibres. FiJan, Šostar Turk, Neral, and Puši (2007) investigated the influence of laundering on the properties of cotton fabrics. It was found that the higher concentrations of H_2O_2 used at higher temperatures in a longer laundering cycle with lower liquor ratios result in a higher degree of chemical and mechanical damage.

31.2.5 SNAGGING

Snagging is the pulling out of warp or weft threads in a woven fabric and coarse or wale threads in a knitted fabric through contact with rough objects, leading to the formation of a loop. Only the appearance of a garment is changed by snagging and its other properties are not affected. Snagging is observed particularly in filament-type fabrics and in extreme cases a single blemish may render an article unserviceable even though unsightly ladders do not necessarily ensue. Soft twisted yarn and loose fabric structure are prone to snagging, which may rupture the yarn and ruin the fabric. Woven fabrics with long floats and fabrics made from bulked continuous filament yarns are susceptible to snagging.

A protrusion is a visible group of fibres or yarn portion that extend above the normal fabric surface. A protrusion is less severe than a snag, which occurs when the fabric surface pulls, plucks, scratches or catches on a component.

31.2.6 SEAM FAILURE

The failure of seams in a sewn garment occurs due to unsuitable selection of sewing thread, stitch type or stitch density; too shallow seam allowance or too tight fit. Although the fabric in a garment may remain in good condition, a failure of the seams reduces serviceability. The failure may be due to slippage or poor seam strength (Saville, 1999).

Slippage is the condition in which a seam sewn in the fabric opens under load and may close on removal of the load, although it may also cause permanent deformation. Seam slippage is a particular problem in fabrics with slippery yarns or an open or loose structure. It is associated with seam allowance, seam type and stitch rate. Tension in the fabric or rubbing of the garment may result in yarn shifting, which causes slippage. Seam strength is the force required to break the sewing thread at the line of stitching.

31.2.7 DIMENSIONAL CHANGE

One of the important properties of a garment is its ability to retain its size and shape. Garments change their size during use, and this is a complex phenomenon with values that may be negative or positive. The negative values are related to the garment shrinkage and the positive values are related to stretch. Changes in fabric dimension may create problems with fit, size, appearance and hence the suitability for its end-use.

31.2.7.1 Shrinkage

A fabric with dimensions that remained constant throughout its useful life would have great technical value. Most fabric production processes involve the application of high tension, which leaves residual strains in the fabric. These residual strains must be removed by the manufacturer before the fabric is converted into a garment or it will lead to shrinkage (i.e. relaxation shrinkage) when the fabric is washed. Several washings are usually required for the complete relaxation of the fabric. The residual strains are relaxed by the hot and wet conditions of washing, which lead to shrinkage. Other forms of shrinkage include hygroscopic or swelling shrinkage; felting shrinkage; thermal or heat shrinkage; and progressive shrinkage.

Hygroscopic or swelling shrinkage is caused by the swelling of fibres due to the absorption of moisture. This shrinkage is highest in the rib weave of wool, cotton, rayon and acetate fibres or their blends and occurs when the garments are washed or dry-cleaned. This group of fabrics may be pre-shrunk and should be cleaned in solvents with low relative humidity (RH). Felting shrinkage is related to wool and other hair fibres and the effect is closely related with their scaly surface features. It is accentuated if woollen items (e.g. socks and underwear) require periodical washing. Felting shrinkage may be caused by excessive mechanical action during washing, high temperature drying and high RH of the solvent during dry-cleaning.

Thermal or heat shrinkage occurs in clothing made of synthetic fibres when subjected to high temperature. The magnitude of the shrinkage depends on fibre morphology and the temperature, tension and washing time. Some knitted garments are made from heat-sensitive fibres that shrink excessively when exposed to heat during drying and finishing. This can be reduced by permanent heat-setting of the fabric during finishing. Progressive shrinkage is the relaxation and swelling shrinkage caused by successive cleaning processes.

Although shrinkage is a common phenomenon in both the length and width of a garment, some may shrink in one dimension and stretch in another. The effects of many aspects of home laundering on the dimensional stability and distortion of cotton knits have been analysed by several researchers (Heap et al., 1983; Heap et al., 1985). Studies on cotton knits showed that tumble drying causes greater levels of shrinkage than line drying in the first few laundering cycles (Collins, 1939; Hearle, 1971; Pierce, 1937). It was also shown that the level of shrinkage increases with successive laundering cycles, reaching a maximum after 5 to 10 cycles.

31.2.7.2 Stretch

There are several accounts of knitted garments increasing in size during wet-cleaning, dry-cleaning and finishing. In some cases, a fabric that shrinks in length will stretch in width. The use of a garment may cause some fabrics to stretch out of shape. Some knitted garments may stretch out of shape if they are hung to dry while still dripping with water or solvent and some may stretch due to manipulation during steam finishing, as the fabric is warm and moist from steam. Stretch can be controlled by suitable yarn, fabric structure and appropriate care during cleaning.

Bagging is a stretch phenomenon commonly observed in knitted garments and occurs at cuffs, ankles and collars (Mehta, 1992). This phenomenon is related to insufficient elastic recovery of the garment. Where a garment is made from fabric with poor elastic recovery properties, it will soon show a baggy appearance, for example, at the knees of trousers and the elbows of jackets. Elastic fabrics containing elastane are capable of stretching far more than conventional fabrics, so increasing the elastic recovery. However, the use of higher amounts of elastane increases fabric stiffness and lowers the tensile and tearing strength (Özdil, 2008).

31.2.8 OTHER PROBLEMS

In addition to the problems listed above, wrinkling, tearing, microbial attack and environmental decay are also associated with garment damage. Garments most commonly become wrinkled during use and are unable to recover from folding deformation. Wrinkling also arises from laundering, especially in cotton fabrics, and causes particular concern to consumers. Wrinkle resistance in a fabric enables it to resist the formation of wrinkles when subjected to folding deformation. The wrinkle behaviour of a fabric depends on the type of fibre and yarn, the fabric construction and the finish applied to the fabric. The resiliency of wool and polyester fibres gives good wrinkle recovery. High twist yarns can improve the wrinkle resistance of a fabric. A tightly woven fabric (having more ends and picks) is more prone to wrinkling than one which is loosely woven. Thinner fabrics are also more prone to wrinkling. A plain woven fabric will wrinkle more than a twill or a 4×4 basket weave. Different finishes can be applied to a fabric or garment to improve the resiliency and therefore the wrinkle recovery.

Tearing is a phenomenon commonly faced by loose-fitting garments. The resistance of a garment to tearing is measured by the force required either to start or to propagate a tear in the fabric. Tearing strength depends on the fibre type, the yarn strength and the fabric construction. Under the action of tearing, the threads in a fabric group close by sliding instead of permitting the successive breakage of individual threads. The grouping of threads becomes easier if the yarns are smooth and able to slip over each other. For this reason, twill and matt weaves exhibit better resistance to tearing than do plain weaves. Fabrics with higher end and pick density prevent the threads from grouping, which reduces the tearing strength. Different finishes, such as anti-crease treatments, may reduce the tearing strength.

Garments may also be damaged by microbial attacks such as insects, mildew and rot. Many insects, including moths, beetles, crickets, roaches and termites, will eat the fibres or any food matter that is allowed to dry on a fabric. Natural fibres are more prone to insect damage though synthetic fibres may also be damaged in this way if they are soiled.

The most common form of insect damage is caused by moth larvae in wool and hair fibres. Such damage may not be noticed prior to cleaning and flexing of the fabric. During cleaning the yarns may be weakened and may break at the point of attack, resulting in a hole in the garment. The risk of insect damage is lowered if garments are cleaned properly and are free from stains before storing. Moth-proofing finishes can also be applied to garments.

Vaeck (1966) has investigated the chemical and mechanical wear of cotton fabric during laundering. Cotton fabric was laundered up to 50 times and the decrease in tensile strength was taken as a measure of wear. The wear was resulted from chemical degradation (caused by oxidising agents or bleaches) and mechanical abrasion. He found the tensile strength loss to be much lower with cold bleach than with hot. It was also reported that in Western Europe, the United Kingdom and the United States, sodium hypochlorite is the most commonly used bleach in commercial laundries, while in Central Europe peroxides are preferred. Hypochlorite bleaching may be done either in a cold rinse or in wash liquor. Most European countries prefer cold rinse, while the United Kingdom and the United States use the second method.

The loss of durability in a garment may sometimes be due to acids or alkalis. Hydrochloric and sulphuric acids are extensively used in industrial plants, dental, medical, photographic, automotive batteries and in all chemical laboratories. Accidental contact with these products may cause fabric damage, i.e. strength loss, disintegration of affected areas and the appearance of holes. Alkaline damage (caused by caustic soda, caustic potash and strong alkaline washing compounds) affects silk, synthetic protein fibres, wool and other hair fibres. Caustic alkalis are common in many household cleaning aids. Sometimes acid or alkali damage will become evident after a garment is cleaned.

It was observed that the repeated laundering of cotton fabrics in alkaline solutions near boiling point and without bleaching left fabric strength unchanged, although linen fabrics suffered 20–30% strength loss (Smit, 1935). After the Second World War, Parisot and Fresco (1954) investigated the mechanical wear of garments by measuring the bursting strength and chemical degradation, which was expressed as degree of polymerisation. Many researchers also reported chemical degradation in laundering, but did not address the loss of tensile strength caused.

Although DP finishes impart shape retention, dimensional stability and wrinkle recovery to cotton and polyester/cotton (P/C) fabrics, other properties such as strength, extensibility and abrasion resistance are adversely affected (Gagliardi, Wehner, & Cicione, 1968; Handu, Sreenivas, & Ranganathan, 1967; Murphy, Margavio, & Welch, 1964; Rollins, Degruy, Hensarling, & Carra, 1970). The extensibility of a wrinkle-free treated garment is significantly reduced after repeated laundering (Lau et al., 2002).

31.3 STAINS

Stains are local deposits of soiling or discolouration that exhibit some degree of resistance to removal by laundering or dry-cleaning, thus creating critical issues in garment care (Kadolph, 2007). The presence of stains on a garment make it dull, stiff and vulnerable to attack by insects. Any attempt to remove stains may cause colour loss or abrasion. Stain removal is affected by the age, extent and type of stain and the type of fabric. The fibre content, fabric construction and the dye and finish characteristics should be considered before stain removal, as the same stain may respond differently in different fabrics. Failure of the cleaning method to remove soiling and stains may lead to product failure.

Stains can be classified according to their characteristics as water or solvent soluble and insoluble. They can also be classified according to the method of removal as protein stains (milk, blood, albumen, pudding, baby food, mud, cream, egg, gelatine, vomit and ice cream), tannin stains (beer, alcoholic beverages, coffee, cologne, fruit juice, soft drinks, tea, tomato juice, berries), oil-based stains (hair oil, automotive oil, grease, salad dressing, butter, lard, suntan lotion, face creams), dye stains (cherry, mustard, colour bleeding in wash) and combination stains (candle wax, ballpoint ink, lipstick, shoe polish, tar, eye make-up, barbecue sauce, gravy, hair spray, tomato sauce).

Identification of the type of stain is important to prevent its removal damaging the fabric. As ageing and heat can set stains permanently, they should be removed as soon as possible. The selection of an unsuitable method may also set the stain permanently. The removal of some stains requires special techniques and solvents and should be removed by laundry professionals.

Polyester fibre and DP finishes retain oily soiling and create cleaning problems. The staining characteristics of resin-treated fabrics have been evaluated by Reeves et al. (Reeves, Summers, & Reinhardt, 1980). It was reported that the removal of soiling from a garment was affected by the fibre content, the type of soil and the process of producing cross links of resin and catalyst. P/C fabrics with and without resin treatment soiled more readily with an oily soil and retained more of the soiling after repeated laundering than did similar cotton fabric. However, with non-oily soil, the P/C fabrics soiled less than cotton fabrics and retained less soil after laundering.

Special care should be taken with regard to temperature during the removal of albuminous stains as higher temperatures may accelerate the coagulation of albumen and fix the stains. Some stains such as glues and paints that contain epoxy resin as a base will damage the fabric during removal. Oily stains should be sponged with a dry solvent and non-greasy stains should be removed with water. Garments packed in polyethylene bags may stain when subjected to excessive heat, e.g. packing the heated garment just after ironing or carrying garments in a car under high-temperature conditions. Ink stains require skill and specialised techniques for complete removal.

31.4 LAUNDERING

Garments may frequently undergo damage during laundering. The laundering of clothes depends on the kind, amount and temperature of water; soaps, laundry aids and detergents. The hardness of water, turbidity, colour, dissolved salts and metals may also affect laundering.

31.4.1 LAUNDERING CHEMICALS

Laundering chemicals (soaps and detergents) are added to water to lower the surface tension for ease of cleaning. Soaps are metallic salts (aluminium, sodium, potassium) of fatty acids and are soluble in water. The soap molecule has two distinct parts: a carboxylate group (attracted to water) and a hydrocarbon chain (repelled by water). A detergent is a chemical composition that removes soiling and is produced by chemical synthesis.

Although both soaps and detergents are surfactants (or surface active agents), they are not the same. Soaps are usually manufactured from natural materials while detergents are made from synthetic materials. Although soaps were the first detergents, they are now being replaced by synthetic detergents. Soap is highly deactivated by hard water. At the early stage of development of non-soap surfactants, the term syndet (short for synthetic detergent) was used to indicate the distinction from natural soap.

Synthetic detergents may be classified as anionic, cationic and non-ionic. Anionic detergents are so called because the detergent portion of the molecule is an anion (negative ion) and the water-soluble portion is a cation (positive ion). Most of the synthetic detergents commonly used in laundering are of the anionic type in which linear alkyl-benzene sulphonates (LAS) are the main anionic compounds. In cationic detergents, the detergent portion is cationic and the water-soluble portion is anionic. The cleansing action of cationic detergents is weaker than that of most anionic detergents. They are used as domestic germicides and fabric softeners. Non-ionic detergents are electrically neutral, having a neutral pH, and are not affected by acids, alkalis or hard water. These detergents are very similar to other detergents having one part of their molecule water soluble and another part solvent soluble. Some non-ionic detergents clean well and have very little lathering action in water.

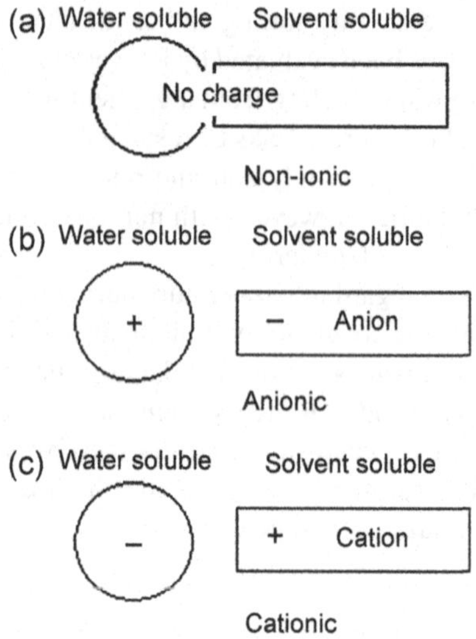

FIGURE 31.3

Three types of detergent molecules: (a) non-ionic, (b) anionic and (c) cationic.

Table 31.1 Components and functions of detergents

Components	Functions
Surfactant	Loosen and disperse soil, provide or control suds; basic 'cleaning' ingredients
Builder	Soften water, aid surfactant in dispersion of soil, buffer detergent solution in alkaline region
Fluorescent whitening agents	Overcome yellowness on fabric and provide whiteness of blue-white hue
Enzymes	Catalyse breakdown of protein- or carbohydrate-based stains to facilitate the removal by surfactant and builder
Anti-redeposition agents	Prevent removed soil from re-depositing on fabric
Sodium silicate, sodium sulphate, water	Carries free-flowing powder

Adapted from 'Home washing products-The technology of home laundry', American Association for Textile Technology, Inc. Monograph No. 108.

All the three types of detergent molecules are explained in Figure 31.3. Synthetic detergents consist of different components that perform different functions. The components and functions of a typical detergent are given in Table 31.1.

Surfactant is a general term for substances such as soluble detergents in a liquid medium, dispersing agents, emulsifying agents, foaming agents, penetrating agents and wetting agents. All these ingredients are necessary in a good detergent. The functions of various ingredients are described in Table 31.2. The laundering process largely depends on the nature of the soiling material. Both alkalinity and temperature are important factors in conserving colour.

Table 31.2 Functions of various components of surfactants

Name	Function
Dispersing agent	Increases the stability of a suspension of particles in a liquid medium
Emulsifying agent	Increases the stability of a dispersion of one liquid in another
Foaming agent	Increases the stability of a suspension of gas bubbles in a liquid medium
Penetrating agent	Increases the penetration of a liquid medium into a porous material
Wetting agent	Increases the spreading of a liquid medium on a surface

Adapted from Lyle (1977).

31.4.2 LAUNDERING AIDS

Laundering aids include bleaches, disinfectants, softeners and starch. Bleaches are used to aid detergents in producing a cleaner and brighter fabric appearance. The laundering bleaches most commonly used are liquid chlorine (strong), powdered chlorine (mild) and oxygen (weak). The hypochlorite ion, which is an oxidising agent, is the chemically active ingredient in liquid chlorine bleach. These bleaches are the most effective stain removers but present a higher risk of damaging clothes. They should not be used for wool, silk, spandex, acetate fibres or their blends. Bleaches should be used in accordance with instructions to prevent damage to fabric dyes and finishes.

The action of powdered chlorine bleach is similar to that of liquid, but is gentler in action. The group of chemicals in these bleaches are *N*-chloro compounds, which release bleaching ingredients more slowly. Powdered chlorine bleaches should never be sprinkled directly on clothes as they can cause damage. Oxygen bleaches are the safest and maintain the brightness of white and coloured items if used regularly.

Disinfectants are used to reduce the amount of bacteria surviving in hot water and laundry detergents (Fijan, Koren, Cencic, & Sostar-Turk, 2007). Bacteria may be spread by laundering if they are not killed by home laundry methods. Quaternary ammonium, phenolic and pine oil disinfectants are commonly used laundry disinfectants. Quaternary ammonium disinfectants should be added to the rinse water and phenolic disinfectants can be added either to the rinse or wash cycles. Pine oil disinfectants and chlorine bleaches should be added to the wash cycle. Liquid chlorine bleaches can also act as disinfectant for fibres on which chlorine can be used. A very small amount of hypochlorite (about 20 ppm) in the washing liquid can destroy the bacteria present in a normal family wash load.

Softeners are used to reduce or eliminate the static charge in synthetic fibres, to make fabric softer, fluffier, and easier to iron, to minimise wrinkling and to help in preventing lint sticking to garments. A lubricating film is created on the fibres by the application of softeners, which allows them to move more readily against each other, thus making the fabric softer and fluffier. In synthetic fibres, the lubricating film absorbs moisture from the air, which helps to reduce the generation of static. Fabric softeners can be added to the wash, rinse or drying cycle. Some softeners are manufactured to be compatible with detergents and other laundering aids. Softeners should be added in the approved concentrations and at the appropriate time of the cycle.

One of the oldest laundering aids is starch, which is still in use in home laundering. Starch is used to (1) obtain a crisp, stiff and shiny fabric appearance; (2) help to keep a garment clean for a longer time; (3) replace the original finish applied to the fabric by the manufacturer and (4) facilitate stain removal as soiling is removed with the starch during washing. Starches may be classified as precooked vegetable starches, starch substitutes and aerosol starches.

Problems may arise from the use of laundering aids. For example, the yellowing of woollen shirt cuffs is due to alkali that is not completely removed after laundering (Petrie, 1939) and which absorbs the borax present in starch.

31.4.3 LAUNDERING EQUIPMENT

This consists of washing and drying equipment. Washing equipment (washer) is designed to wash, rinse and extract water from clothes and to make provision for setting the time, temperature and volume of water. Drying equipment is sometimes combined with washing equipment. The three major functions of washers are (1) removal of soiling from cloth, (2) the rinsing of soap or detergent and soil from the wash process, and (3) the extraction of most of the wash and rinse water prior to drying. Washing equipment can be classified as top-loading or front-loading.

Top-loading machines are fitted with an agitator which transfers mechanical energy from the motor to the clothes. The mechanical agitation, combined with detergent, removes soiling and keeps it in suspension. The central cylinder (solid or perforated) which contains the clothes to be washed is usually steel, coated with porcelain enamel. As the detergent cycle is completed, the water is extracted from the fabric through the perforations of the cylinder. Front-loading machines have a cylinder, but no agitator. The interior of the cylinder consists of baffles that lift the clothes from the water and then drop them back into it.

The water level will depend on the load size and the design of the equipment. Top-loading machines use more water for each cycle, but have fewer cycles than front-loading machines. The temperature of the water generally ranges from cold to 60 °C (140 °F) and is controlled by a temperature unit fitted to the control panel of the washer. In some cases, cold water washing can be used, thus conserving energy. Cold water washing is used for knitted polyester or nylon and permanent press items that wrinkle in hot water. Following the washing cycle, the garments are spun dried or put through a wringer while still hot. This reduces shrinkage, especially in knitted fabrics, chino pants, and some non-sanforised items. It also prevents the setting of certain type of stains (such as milk, egg and blood, which may become permanent if washed in hot water). Hot water washing should be used for the removal of grease and oil stains.

The temperature of cold water can vary from extreme cold to water at body temperature. However, a water temperature of 27 °C (80 °F) or slightly higher gives the best results. Special detergents have been developed for use in cold water which dissolve readily and have good cleaning properties. However, cold water washing will not be bacteria-free. Cold water washing followed by tumble drying at 70 °C (160 °F) will result in little or no bacterial contamination. Investigations by Witt and Warden (1971) showed that the important factors for preventing bacterial growth are water temperature (between 50 and 60 °C (120–140 °F)), detergent and longer washing cycle. Washing followed by steaming or ironing on a hot press have been shown to provide adequate disinfection.

The duration and degree of mechanical agitation in a washing cycle depend on the load, the amount of water used and the extraction of water from the load. A normal washing cycle may be between 30 and 38 min. Some washers may include a pre-soak cycle for heavily soiled items requiring treatment with enzymes, a super wash cycle for heavily soiled items, a longer cycle for permanent press items that require more water and a delicate fabric cycle for delicate items and blankets requiring a higher water level, lower temperature and low mechanical agitation. Some washers may be fitted with an auxiliary device such as a suds saver for areas with a limited water supply. Lightly soiled clothes are washed first, followed by medium to heavily soiled items. During progressive rinse cycles, the rinse water is stored and used again in successive detergent cycles.

Ozone laundering systems have recently been shown to be successful in commercial installations because of reduction in the use of energy, water and chemicals (Rice, Debrum, Cardis, & Tapp, 2009). Ozone enhances the effectiveness of the chemicals by supplying oxygen to the laundry water, thus reducing the need for high temperature washing with lower amount of laundry chemicals. As ozone laundering systems normally require fewer rinse steps, water usage is reduced by an estimated 30–45%. These systems recover most of the water used, so the reductions in water usage may be as high as 70–75%. Ozone oxidises soiling in linen, making it easier to remove from the wash water. It can also reduce the need for harsh, high-pH traditional chemicals. The resultant savings claimed by laundries range from 5 to 30%.

Ozone in water solution performs some of the functions of chlorine bleach. It assists in water softening by helping to remove cations such as calcium and magnesium from the water. Ozone laundering improves the life and quality of textiles because it enables a shorter cycle time and lower temperature. A reduction in the amount of chemicals used also helps to improve fabric life. In addition, the effluent will contain lower levels of biochemical and chemical oxygen demand (BOD and COD) because ozone oxidises bacteria, other microorganisms and some dissolved organic compounds. The reduced washing and rinsing time means the laundry equipment is used more efficiently and the total staff hours per load is reduced.

Various problems such as wrinkling, shrinkage, distortion, colour loss, non-removal of soil, staining, change in texture and other changes in appearance are concerns in laundering (Anand et al., 2002; Higgins, Anand, Holmes, Hall, & Underly, 2003). The appropriate selection of laundering chemicals and washing cycles and attention to care instructions can avoid these problems.

31.5 CARE LABELLING

A care label carries care instructions for cleaning a textile product. Care labels are a series of directions describing procedures for refurbishing a product without adverse effects. Care labelling for garments is essential to identify the product, to assist the consumer in product selection and the retailer in selling the product, and to help the consumer in effective care of the garment (Feltham & Martin, 2006; Lyle, 1977; Seitz, 1988; Shin, 2000). The information on care labels is strongly emphasised as most consumer complaints and claims against apparel products concern colour change, deformation and damage during laundering (Choi & Cha, 1993).

Garment manufacturers or retailers should attach appropriate labels to the apparel and should ensure that all the information provided by the labels is accurate and meets the specified criteria. The information on care labels must be understood by consumers in the post-purchase stage. Care labels carry information on the fibre type, country of origin, registered number (RN number), wash-care instructions, size and the manufacturer or retailer's identification (Chatterjee, Nayak, Bhattacharya, & Kansal, 2006).

The care labelling of apparel sold domestically in the United States is regulated by the Federal Trade Commission (FTC) under its rule (16 CFR Part 423) 'Care Labeling of Textile Wearing Apparel and Certain Piece Goods'. The most recent amendment to the rule (FTC, 2001) states that the manufacturers can use a set of four basic care label symbols developed by the American Society for Testing and Materials (ASTM, 2001) instead of using words. These symbols are a set of graphic images which function like universal symbols on highway signs and do not need to be translated into a variety of languages (Magill, 1998). Following the FTC rules, products sold in the United States can use text only, symbols only, or both text and symbols. Products that are destined for multiple countries should adopt the symbols-only format to avoid the need to label in multiple languages (Meadows, 1999). Consumers with a high need for cognition prefer labels that present care information in text format, while those with a lower need prefer the information in symbol format (Yan, Yurchisin, & Watchravesringkan, 2008).

It was found that a majority of consumers prefer care labels that contain text (text only or text and symbols) (Moore, Doyle, & Thomason, 2001; Yan et al., 2008) because these skills are taught and reinforced from an early age. Care labels that are easily understood by consumers increase their confidence in caring for the apparel and reduce their perceptions of risk concerning the purchase of the item.

31.5.1 CARE LABEL REQUIREMENTS

The care instructions on a permanent label should comply with the following requirements (CS/4, 1957a, 1957b):

1. The label shall be accessible for examination by a prospective consumer, or where that is not possible because of the manner in which the article is packed, displayed or folded, the same instructions shall be provided additionally on a removable label attached to the article.
2. The labels may use symbols in addition to written instructions. Consumers should be able to understand the symbols irrespective of the language. Symbols and letters on the labels must remain legible throughout the useful life of a garment.
3. The label shall carry the entire information of the appropriate care instructions. The wording of the label shall be clearly legible and of medium width lettering of which no individual letter shall be less than 1.5 mm high.
4. Articles with multiple components shall each have separate permanent labels. The care instruction symbols are applicable to whole of the garment, including trimmings, zippers, linings, buttons and sewing thread, unless otherwise mentioned on separate labels.
5. Care labels should be placed accessibly to avoid difficulties in conveying information. The labels for a particular style should be positioned at one place in all garment pieces.
6. All the symbols used in the care labelling system should be placed directly on the article or on a label which shall be affixed in a permanent manner to the article.
7. Care labels should be made of suitable material with resistance to the care treatment indicated in the label at least equal to that of the article on which they are placed.
8. The care symbols selected should give instructions for the most severe process or treatment the garment can withstand while being maintained in a serviceable condition without causing a significant loss of its properties.
9. Care instructions are chosen on the basis of the end-use application of the article and fibre type. In case of fabric with blended fibres, the care instructions should, in most instances, be based on the properties of the most sensitive fibre. The care instructions selected should be verified to ensure that the article complies with the performance requirements prior to sale.

Care instructions may be woven into the labels or printed on labels of mainly rectangular shape. They may be loop labels (sewn at both ends) or fused flat against the fabric without the use of sewing threads. The size of the labels will depend on the amount of information. However, in all cases the wording must be legible. There is no mandatory rule for the positioning of the labels which may be placed at suitable places in different garments. Although the rule states that the organisation or person controlling the manufacture of the finished garment should be responsible for care labelling, the burden of proof falls on the consumer if an item fails to give performance as described in the care label.

The main difficulties with care labels are that (1) some indicate procedures that are far more restrictive than necessary; (2) some instructions make no sense or are difficult to understand; and (3) some abrasive and coarse labels cause skin irritation. A survey found that many people do not fully understand care label information and select more vigorous cleaning methods than those recommended (Hatch & Lane, 1980). Some respondents indicated that they thought bleaching was acceptable, though the instruction warned against it. Similarly, 'line dry' was interpreted incorrectly. Educational programmes are therefore necessary to maximise the number of consumers correctly interpreting the labels. Standardising information on care labels can also minimise misunderstanding (Kadolph, 2007). Considerable evidence also exists to demonstrate that there is no direct relationship between information provision on care labels and information used (Day, 1976).

31.5.2 CARE LABELLING SYSTEMS

At present, there is no universal care labelling system. In the United States, the Wool Products Labelling Act (1938), the Fur Products Labelling Act (1951), the Flammable Fabrics Act (1958) and the Rule on Care Labelling (1972) are in force. The JIS (Japan Industrial Standard) for care labelling came into force in 1962. Similarly, in Korea, the rule on Quality Labelling came into force in 1969, and the use of symbols for care labelling of apparel products was published in 1972.

The ASTM system is accepted in North American Free Trade Agreement (NAFTA) countries. The International Organization for Standardization (ISO or GINETEX) system is accepted in most of Europe and Asia. Japan has its own system. Negotiations are under way to harmonise the two major systems (ASTM and GINETEX) into a universal labelling system for care procedures. An international labelling system can facilitate global trade by avoiding technical or standard barriers. The major systems that are followed worldwide are ASTM, ISO (GINETEX), British (Home Laundering Consultative Council System (HLCC)), Canadian, Dutch and Japanese Care Labelling Systems. Although there are some variations in the symbols, the five basic symbols used in many of the listed systems are discussed in Table 31.3. Table 31.4 indicates some variations to the basic symbols and their explanations.

The Canadian Care Labelling System consists of the five basic symbols (Table 31.3), which are illustrated in three conventional traffic light colours. The red (with a cross superimposed) indicates prohibition, yellow indicates that care is needed and green indicates that no special precautions need to be taken. Similarly in the HLCC system, a number (from 1 to 8) can be indicated in a wash tub indicating the severity. Numbers 1 to 7 relate to articles which may be washed safely either by machine or by hand, and number 8 (with an outline of a hand) always indicates a hand wash process. In some of the systems, dots inside the symbols indicate the severity of the process, i.e. one, two and three dots indicating low, medium and high conditions, respectively. One example of a care label used in a garment is explained in Figure 31.4.

The Technical Committee (TC-38) of ISO handles all types of textile standards through several subcommittees. Subcommittee, SC-11 is concerned with developing standards for care labelling with the primary objective of developing an international symbol system. Manufacturers and retailers follow the ASTM standard (ASTM D 3938 – Determining or confirming care instructions for apparel and other textile consumer products) to ensure correct information is included on care labels. The other standards dealing with care

Table 31.3 Basic symbols used for care instructions in care labels

Symbols	Indicates
	Wash tub: gives instructions about laundering
	Triangle: gives instructions for bleaching
	Square: relates to drying
	Hand iron: provides ironing or pressing instructions
	Circle: gives dry-cleaning instructions

Table 31.4 Explanation of variations of basic symbols used for care instructions in care labels

Washing

	Machine wash, cold		Machine wash, cold, permanent press		Machine wash, cold, gentle cycle
	Machine wash, warm		Machine wash, warm, permanent press		Machine wash, warm, gentle cycle
	Machine wash, hot		Machine wash, hot, permanent press		Machine wash, hot, gentle cycle

Bleaching / **Drying**

	Bleach as needed		Tumble dry, no heat		Tumble dry, permanent press, no heat
	Non-chlorine bleach as needed		Tumble dry, low heat		Tumble dry, permanent press, low heat
			Tumble dry, medium heat		Tumble dry, permanent press, medium heat
	Do not bleach		Tumble dry, high heat		

Table 31.4 Continued

Drying				**Ironing**	
	Tumble dry, gentle cycle, no heat		Do not tumble dry		Iron, steam or dry, with low heat
	Tumble dry, gentle cycle, low heat		Line dry		Iron, steam or dry, with medium heat
	Tumble dry, gentle cycle, medium heat		Drip dry		Iron, steam or dry, with high heat
			Dry flat		Do not iron with steam
					Do not iron
Dry-cleaning					
	Dry-clean				
	Do not dry-clean				

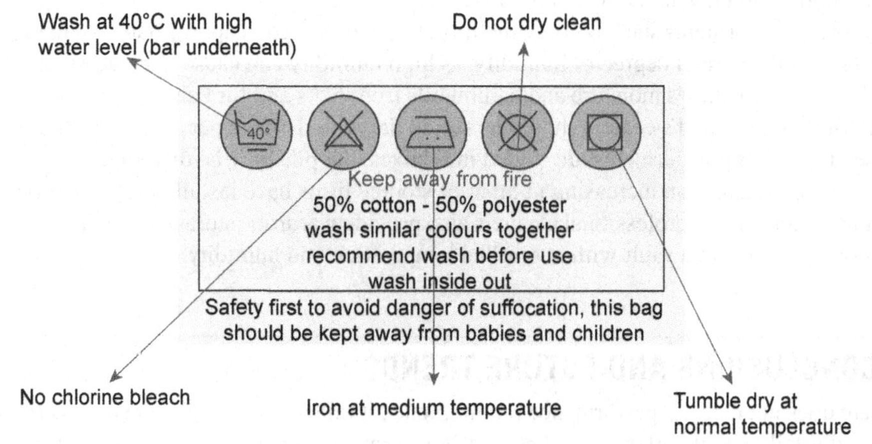

FIGURE 31.4

Example of a care label explaining the meanings of various symbols.

labelling include ASTM D 3136 (Standard terminology relating to care labels for textile and leather products other than textile floor coverings and upholstery); ASTM D 6322 (International test methods associated with textile care procedures) and ASTM D 5489 (Standard guide for care symbols for care instructions).

31.6 CLOTHING STORAGE

The effect of storage on the performance of garments is of special interest to manufacturers, retailers and consumers. The problems associated with the storage of garments are insect, rot and mildew damage or other conditions that may create problems during subsequent use. When textiles are stored in damp or in highly humid conditions, they are vulnerable to insect damage. This will be aggravated if the storage spaces are dark and stagnant with warm and humid areas. Before storing a garment the following points should be noted:

1. The garment should be cleaned and all stains removed.
2. All the fastenings should be closed.
3. Belts should be removed from their loops.
4. The garment should be hung from a coat hanger.
5. A moth preventive should be sprayed if the garment is vulnerable to insect attack.
6. The storage area should be cool and dry.

Many insect species may damage soiled textiles and those made of wool or other animal hair fibres such as mohair, angora and cashmere. Crickets and silverfish may cause irreparable damage to cellulosic garments and a distinct odour may arise, particularly when starch, glue or other attractive materials are present. If synthetic fabrics are stored in a dirty, spotted or stained condition, they may be damaged by insects. A reliable moth preventive such as naphthalene or paradichlorobenzene in the form of balls, flakes or powder should be applied before storing. Storage areas should be cool, dry and away from sunlight to prevent the hatching of insects.

Some expensive garments such as coats and furs should be stored in appropriate storage vaults. Furs must be stored at the correct degree of humidity, as high humidity can cause damage by mould or mildew, and low humidity draws moisture and natural oils from pelts and fur hairs, thus shortening the life of the fur. Knitted items and sweaters should be stored flat with tissue paper stuffing so that fold marks will not be obvious. If pile garments are folded into boxes, the pile may be distorted.

During recent decades, an increasing number of storage firms have installed burglar-proof and fireproof storage vaults. Some professional cleaners also provide wardrobe storage hampers or box-storage and will store garments in a vault with controlled temperature and humidity.

31.7 CONCLUSIONS AND FUTURE TRENDS

If a garment does not meet its performance requirements, it fails to meet its business objectives. Manufacturers and retailers will suffer losses because of returns, complaints and reputational damage with their target market. The durability of a garment depends mainly on its care (i.e. severity in laundering, dry cleaning and ironing). Garment performance can be enhanced by the appropriate selection of fibre, yarn, fabric; combination of production processes and application of finishes. Standard test methods can also be established to check whether the garment meets the performance required for the intended use before it goes to the consumer.

New developments in fibre technology (such as microfibres, nanofibres and specialty fibres) and finishes extend the analytical aspect of garment manufacture. Similar developments in other areas (such as laundering chemicals and techniques) are necessary to cope with these advanced materials. Products with advanced fibres and finishes will also require the development of new care instructions.

31.8 SOURCES OF FURTHER INFORMATION

Further information can be obtained from the following sources:

- Physical Testing and Quality Control, K Slater, Textile Progress, Volume 23, Number 1/2/3.
- Chemical Testing and Analysis, K Slater, Textile Progress, Volume 25, Number 1/2.
- Quality Assurance for Textiles and Apparel, Sara J. Kadolph, Fairchild Publications, NY.
- Physical Testing of Textiles, B P Saville, Textile Institute, UK.
- Performance of Textiles, Dorothy Siegert Lyle, John Wiley and Sons, NY.

31.9 SUMMARY POINTS

All garments are subject to wear. This occurs through a variety of means: atmospheric conditions, usage and laundering.

The appropriate choice of type of textiles, care in laundering, stain removal and storage practises is important in minimising wear.

Care labelling is critical in communicating the above choices.

31.10 PROJECT IDEAS

Describe a collection of garments which cover leisure, everyday and work-wear. Suggest a programme of laundering (including laundering aids) which will optimise the serviceable life of these garments.

What stain/soil removal procedures would you use for garments used in medical situations and procedures?

Design a set of care labels for garments sold outside your own country which will aid in the prevention of dimensional change.

31.11 REVISION QUESTIONS

1. Discuss five factors that reduce the service life of a garment.
2. What are the various causes of colour fading?
3. Highlight the damage caused to apparel by various environmental factors.
4. Discuss laundering chemicals and aids.
5. Discuss the care labelling systems followed worldwide. Give examples of the various symbols and numbers used.
6. How may the proper storage of clothing reduce or avoid damage?

REFERENCES

Akgun, M., Becerir, B., & Alpay, H. R. (2006). Abrasion of polyester fabrics containing staple weft yarns: color strength and color difference values. *AATCC Review, 6*, 40–43.

Alpay, H., Becerir, B., & Akgun, M. (2005). Assessing reflectance and color differences of cotton fabrics after abrasion. *Textile Research Journal, 75*, 357–361.

Anand, S., Brown, K., Higgins, L., Holmes, D., Hall, M., & Conrad, D. (2002). Effect of laundering on the dimensional stability and distortion of knitted fabrics. *Autex Research Journal, 2*, 85.

Anderson, C., Leeder, J., & Taylor, D. (1972). The role of torsional forces in the morphological breakdown of wool fibres during abrasion. *Wear, 21*, 115–127.

ASTM. (2001). *ASTM D5489–01a: Standard guide for care symbols for care instructions on textile products.* West Conshohocken, PA.

Backer, S., & Tanenhaus, S. (1951). The relationship between the structural geometry of a textile fabric and its physical properties. *Textile Research Journal, 21*, 635–654.

Barnett, R., & Slater, K. (1991). The progressive deterioration of textile materials part V: the effects of weathering on fabric durability. *Journal of the Textile Institute, 82*, 417–425.

Booth, J. E. (1968). *Principles of textile testing*. England: J W Arrowsmith Ltd.

Card, A., Moore, M., & Ankeny, M. (2006). Garment washed jeans: impact of launderings on physical properties. *International Journal of Clothing Science and Technology, 18*, 43–52.

Chatterjee, K. N., Nayak, R. K., Bhattacharya, S., & Kansal, N. (2006). Care labeling of apparels. *Indian Textile Journal, 117*, 79–83.

Chippindale, P. (1963). 40—Wear, abrasion, and laundering of cotton fabrics. *Journal of the Textile Institute Transactions, 54*, 445–463.

Chiweshe, A., & Crews, P. (2000). Influence of household fabric softeners and laundry enzymes on pilling and breaking strength. *Textile Chemist and Colorist and American Dyestuff Reporter, 32*, 41–47.

Choi, H. W., & Cha, O. S. (1993). A study on the consumers' dissatisfaction for the clothing product. *Korean Society Clothing & Textile, 17*, 550–564.

Clegg, G. G. (1940). 5—The examination of damaged cotton by the Congo red test: further developments and applications. *Journal of the Textile Institute, 31*, T49.

Clegg, G. G. (1949). A microscopic examination of worn textiles. *Journal of the Textile Institute, 40*, T449–T480.

Collins, G. (1939). Fundamental principles that govern the shrinkage of cotton goods by washing. *Journal of the Textile Institute Proceedings, 30*, 46–61.

CS/4, J. T. C. (1957a). *Textiles-care labelling*. Australia and New Zealand: Standards Australia and New Zealand.

CS/4, J. T. C. (1957b). *Textiles-guide to the selection of correct care labelling instructions from AS/NZS 1957.* Australia/New Zealand: Standards Australia and New Zealand.

Day, G. (1976). Assessing the effects of information disclosure requirements. *The Journal of Marketing, 40*, 42–52.

Degruy, I., Carra, J., Tripp, V., & Rollins, M. (1962). Microscopical observations of abrasion phenomena in cotton. *Textile Research Journal, 32*, 873–882.

El-Shishtawy, R., & Nassar, S. (2002). Cationic pretreatment of cotton fabric for anionic dye and pigment printing with better fastness properties. *Coloration Technology, 118*, 115–120.

Elder, H. (1978). Wear of textiles. *Journal of Consumer Studies & Home Economics, 2*, 1–13.

Elder, H. M., & Ferguson, A. S. (1969). 18—The abrasion-resistance of some woven fabrics as determined by the accelerator abrasion tester. *Journal of Textile Institute, 60*, 251–267.

Feltham, T., & Martin, L. (June 2006). Apparel care labels: understanding consumers' use of information. *Marketing, 27*(3), 231–244. Available at SSRN: http://ssrn.com/abstract=1809865.

Fijan, S., Koren, S., Cencic, A., & Sostar-Turk, S. (2007a). Antimicrobial disinfection effect of a laundering procedure for hospital textiles against various indicator bacteria and fungi using different substrates for simulating human excrements. *Diagnostic Microbiology and Infectious Disease*, *57*, 251–257.

Fijan, S., Šostar Turk, S., Neral, B., & Puši, T. (2007b). The influence of industrial laundering of hospital textiles on the properties of cotton fabrics. *Textile Research Journal*, *77*, 247–255.

FTC. (2001). Care labeling of textile wearing apparel and certain piece goods. In FTC (Ed.), *16 CFR part 423*. Washington, DC.

Gagliardi, D., & Wehner, A. (1967). Influence of swelling and mowosubstitution on the strength of cross-linked cotton. *Textile Research Journal*, *37*, 118–128.

Gagliardi, D., Wehner, A., & Cicione, R. (1968). Improved durable-press cottons produced by conventional pad-dry-cure procedures using pairs of monofunctional and difunctional swelling reactants 1. *Textile Research Journal*, *38*, 426.

Handu, J., Sreenivas, K., & Ranganathan, S. (1967). Chemical and mechanical damage in service wear of cotton apparel fabrics. *Textile Research Journal*, *37*, 997.

Hatch, K., & Lane, S. (1980). Care labels: will more information help consumers? *Family and Consumer Sciences Research Journal*, *8*, 361–368.

Heap, S. (1978). Liquid ammonia treatment of cotton fabrics, especially as a pretreatment for easy-care finishing. *Textile Institute India*, *16*, 387–390.

Heap, S., Greenwood, P., Leah, R., Eaton, J., Stevens, J., & Keher, P. (1983). Prediction of finished weight and shrinkage of cotton knits–the Starfish Project, part I: introduction and general overview. *Textile Research Journal*, *53*, 109–119.

Heap, S., Greenwood, P., Leah, R., Eaton, J., Stevens, J., & Keher, P. (1985). Prediction of finished relaxed dimensions of cotton knits—the Starfish project. *Textile Research Journal*, *55*, 109–119.

Hearle, J. W. S. (1971). The nature of setting. In J. H. E. Al (Ed.), *Setting of fibres and fabrics*. Watford, UK: Merrow Publishing Co.

Higgins, L., Anand, S., Holmes, D., Hall, M., & Underly, K. (2003). Effects of various home laundering practices on the dimensional stability, wrinkling, and other properties of plain woven cotton fabrics. *Textile Research Journal*, *73*, 407.

Hurren, A., Wilcock, A., & Slater, K. (1985). The effects of laundering and abrasion on the tensile strength of chemically treated cotton print cloth. *Journal of the Textile Institute*, *76*, 285–288.

Kadolph, S. J. (2007). *Quality assurance for textiles and apparel*. New York: Fairchild Publications, Inc.

Kalishi, E. J. (1958). *Textile Research Journal*, *25*, 325.

Lau, L., Fan, J., Siu, T., & Siu, L. (2002). Effects of repeated laundering on the performance of garments with wrinkle-free treatment. *Textile Research Journal*, *72*, 931–937.

Lyle, D. (1977). *Performance of textiles*. New York: Wiley.

Magill, R. (1998). Keeping pace with permanent care labeling processes. *Bobbin*, *39*, 36–41.

Mangut, M., Becerir, B., & Alpay, H. (2008). Effects of repeated home launderings and non-durable press on the colour properties of plain woven polyester fabric. *Indian Journal of Fibre & Textile Research*, *33*, 80–87.

Meadows, S. (1999). Taking care labeling to the next level. *Bobbin*, *41*, 38–42.

Mehta, V. P. (1992). *An introduction to quality control for the apparel industry*. New York: Quality Press.

Min, L., Xiaoli, Z., & Shuilin, C. (2003). Enhancing the wash fastness of dyeings by a sol–gel process. Part 1; direct dyes on cotton. *Coloration Technology*, *119*, 297–300.

Moore, C. M., Doyle, S. A., & Thomason, E. (2001). Till shopping us do part – the service requirements of divorced male fashion shoppers. *International Journal of Retail & Distribution Management*, *29*, 399–406.

Morris, M., & Prato, H. (1976). Fabric damage during laundering. *California Agriculture*, *30*, 9.

Murdison, M., & Roberts, J. (1949). 31—a study of the effects of laundering and storage on cotton cloth. *Journal of the Textile Institute Transactions*, *40*, 505–518.

Murphy, A. L., Margavio, M. F., & Welch, C. M. (1964). A method of preventing strength losses during wash-wear finishing of cotton fabric. *American Dyestuff Reporter*, *53*, 25–27.

Neelakantan, P., & Mehta, H. (1981). Wear life of easy-care cotton fabrics. *Textile Research Journal*, *51*, 665–670.

Özdil, N. (2008). Stretch and bagging properties of denim fabrics containing different rates of elastane. *Fibres & Textiles in Eastern Europe*, *16*, 63–67.

Parisot, A., & Fresco, A. (1954). *Bulletin of Institut Textile de France*, *48*, 7.

Petrie, T. (1939). Laundering effects on textile fabrics, with special reference to knitted materials. *Journal of the Textile Institute Proceedings*, *30*, 66–68.

Pierce, F. T. (1937). The geometry of cloth structure. *Journal of the Textile Institute*, *28*, 45–96.

Reeves, W. (1962). Some effects of the nature of cross links on the properties of cotton fabrics. *Journal of the Textile Institute Proceedings*, *53*, 22–34.

Reeves, W., Summers, T., & Reinhardt, R. (1980). Soiling, staining, and yellowing characteristics of fabrics treated with resin or formaldehyde. *Textile Research Journal*, *50*, 711–717.

Rice, R., Debrum, M., Cardis, D., & Tapp, C. (2009). The ozone laundry handbook: a comprehensive guide for the proper application of ozone in the commercial laundry industry. *Ozone: Science & Engineering*, *31*, 339–347.

Rollins, M., Degruy, I., Hensarling, T., & Carra, J. (1970). Abrasion phenomena in durable-press cotton fabrics: a microscopical view. *Textile Research Journal*, *40*, 903–916.

Rousselle, M., Nelson, M., Hassenboehler, C., Jr., & Legendre, D. (1976). Liquid-ammonia and caustic mercerization of cotton fibers: changes in fine structure and mechanical properties. *Textile Research Journal*, *46*, 304–310.

Saville, B. (1999). *Physical testing of textiles*. Cambridge, UK: CRC Press.

Schumacher, K., Heine, E., & Höcker, H. (2001). Extremozymes for improving wool properties. *Journal of Biotechnology*, *89*, 281–288.

Seitz, V. A. (1988). Information needs of catalog consumers. *Journal of Home Economics*, *80*, 39–42.

Shin, S. (2000). Consumers' use of care-label information in the laundering of apparel products. *Journal of the Textile Institute*, *91*, 20–28.

Smit, R. (1935). *Melliand Textilberichte*, *16*, 879.

Tabor, D., Howell, H., & Mieszkis, K. (1959). *Friction in textiles*. London: Butterworth.

Vaeck, S. (1966). Chemical and mechanical wear of cotton fabric in laundering. *Journal of the Society of Dyers and Colourists*, *82*, 374–379.

Wilcock, A., & Delden, E. (1985). A study of the effects of repeated commercial launderings on the performance of 50/50 polyester/cotton momie cloth. *Journal of Consumer Studies & Home Economics*, *9*, 275–281.

Witt, C. S., & Warden, J. (1971). Can home laundering stop the spread of bacteria in clothing? *Textile Chemist and Colorists*, *3*, 55–58.

Xie, K., Hou, A., & Zhang, Y. (2006). New polymer materials based on silicone acrylic copolymer to improve fastness properties of reactive dyes on cotton fabrics. *Journal of Applied Polymer Science*, *100*, 720–725.

Yan, R., Yurchisin, J., & Watchravesringkan, K. (2008). Use of care labels: linking need for cognition with consumer confidence and perceived risk. *Journal of Fashion Marketing and Management*, *12*, 532–544.

GLOSSARY

3D body scanning A body scanner captures information about the surface of an object, creating something called a point cloud. This data can then be used to reconstruct the object as a virtual model, sometimes called an avatar.

Abrasion The progressive loss of fabric caused by friction with another surface.

Absorption The process of one material (the absorbent) taking in another (the absorbate). An example of absorption is a sponge (the absorbent) soaking up water (the absorbate).

Acrylic Polymer that contains the acryloyl group derived from acrylic acid.

Adsorption Adhesion of a thin layer of gas, liquid or dissolved solid (the absorbate) to a surface (the absorbent). Adsorption is an important process in a number of applications, for example for the development of non-stick coatings.

Air-jet spinning A method in which high-pressure air is used to impart twist to drafted fibre strands.

Allocated/allocation The process of distributing garments to individual retail outlets to ensure that stock levels are maintained.

Amorphous The physical state of a material where there is a relatively unorganised molecular structure.

Anionic Describes a substance that exhibits negatively charged ions. Acid, direct and reactive dyes are examples of anionic dyes.

Apron The cloth attached to the front and back beam of the loom, to which the warp is tied.

Aramid Synthetic fibre that is extremely strong and heat resistant. The fibre-forming substance is a long-chain synthetic polyamide. Aramids are commonly used to make heat-resistant clothing.

Art Deco Style of decorative art popularised in the 1920s and 1930s, exemplified by the use of geometric motifs, sharply defined outlines, bold colours and the use of synthetic materials such as plastics.

As-spun (partially drawn) yarn Produced by extrusion of molten polymer filaments that are stretched whilst being cooled.

Atmospheric gas fading Also known as fume fading, this is colour change in a fabric caused by acid gases in the atmosphere.

Augmented reality Interaction with the real world that is enhanced by the use of technology.

Avatar A 3D model or 2D icon that represents a real person on a computer or the Internet.

Ballets Russes (Russian Ballet) The 'Ballets Russes' was a travelling ballet company, directed by Sergei Diaghilev, that performed between 1009 and 1929 in a number of countries – including England, France and the United States.

Barré A fault in a weft-knitted fabric appearing as a light or dark stripe, and arising from differences in lustre, knitting tension or dye affinity of the yarn.

Beat up The pushing of the weft yarns into place with the reed.

Biaxial fabric Fabric manufactured using two sets of yarn (warp and weft).

Bi-component filament yarn A yarn produced from two different polymer components brought together at the fibre-extrusion stage.

Binder An extra warp thread used to tie down weft floats.

Biotextile Textile used in specific biological situations to perform a function within the body. Examples of biotextiles include surgical sutures, artificial arteries, artificial skin and parts of artificial hearts.

Bleaching The process of improving the whiteness of a textile material by oxidation of the colouring matter.

Blended/compound yarn Yarns manufactured through the mixing of different yarns together.

Blog A web log or online journal.

Boat shuttle Used to pass the weft thread through the shed. Made with or without rollers underneath. The hollow cavity in the shuttle holds a purn of weft yarns that are released whilst the shuttle is passed through the shed. The shuttle without rollers is normally used for intricate work, whilst the shuttle with rollers is used to quickly pass the shuttle across the whole width of the warp.

Bobbin or purn Hand made from paper or wood or plastic, it holds wound weft yarns, and the bobbin or purn is then placed into a boat shuttle.
Bobbin winder Hand operated or electric, winds weft yarns evenly onto purn.
Braided structure Structure constructed from three or more interlacing strands. May be two dimensional or solid.
Branding tag Decorative addition to a garment to re-inforce the brand. Can be attached permanently or can be hung with the swing tag – could be made of more permanent material than the swing tag.
Brand value Quality that a brand is trying to communicate to its customers.
Breaking tenacity See Tenacity.
Brightness In terms of describing colour, the converse of dullness.
Bulkiness A measure of size.
Cabled/corded yarn Yarn made by twisting ply yarns together, with the final twist usually in the opposite direction of the ply twist. Where the corded yarn is twisted in the opposite direction to the S-twist of the single-ply yarns, the corded yarn is said to have an SZS form. Otherwise, the corded yarn has an ZSZ form. Corded yarns are commonly used for rope or twine.
Calico A plain-woven cotton fabric, usually unbleached, used for making toiles.
Carbonisation The process of removing cellulosic impurities from wool by treatment with mineral acid.
Carding The conversion of cleaned and blended tufts into individual fibres.
Care label Carries care instructions for cleaning and effective care of a textile product.
Cationic Describes a substance that exhibits positively charged ions. Basic dyes are cationic.
Cationic retarding agent A product used to retard the strike rate of basic (cationic) dyes on acrylic fibres by competing with the dye for access to the fibre.
Chiffon Lightweight sheer fabric with a plain weave.
Chinoiserie Chinoiserie is a decorative style, where art and objects are designed in imitation of traditional Chinese motifs. This style began in the late seventeenth century and was particularly popular in eighteenth-century Europe.
Colour fastness The property of a dyed material in terms of colour loss, change of shade and the staining of other fibres when subjected to various tests designed to simulate appropriate use and aftercare conditions.
Colour match prediction The process of predicting a best available dye recipe from an historical database using either visual or instrumental methods.
Combing The process of arranging filaments into a strict parallel form.
Competitive analysis Attempting to understand what a competitor is developing in terms of product and how they are reacting to change.
Concession A space in a store allocated to a run by a specific brand. The store will take a percentage of the turnover of the concession in return for use of the space.
Continuous-filament yarn Consists of one or more filaments of specified and equal length.
Copyright Laws that protect the rights of the creator of an original work.
Core consumer The target customer that a brand is trying to attract and sell to. All brands will have a core consumer even though there will be other people who will buy their product.
Core-spun yarn Yarn which is made by twisting a fibre around a central filament core, which is often polyester. Core-spun yarn is stronger than normal yarn.
Costing sheet A statement that calculates the final cost of a garment by generally taking into account the following: fabric, trimmings, labour, hangers/packaging and mark-up.
Count The size of yarn (a ratio between the length and the weight of the yarn).
Counter-culture (counterculture) A sociological term used to describe a cultural group, or subculture, whose values and behaviour are in opposition (run 'counter') to those of the social mainstream.
Cover The cover factor is the ratio of the fabric surface occupied by the yarns to the total fabric surface.

Covered yarn A combined yarn consisting of a central fibre with a second fibre wrapped around it.

Craftivism/craftism A dynamic and politicised twenty-first century crafts movement that aims to reclaim hand crafts – knitting, sewing and other crafts traditionally viewed as 'women's work' – using them to make political statements (mostly anti-war, anti-capitalist or pro-freedom) and by giving them significant meaning.

Crimp The crimp of a fibre describes its wavy physical structure. The crimp of a yarn describes the wavy configuration of a yarn interlaced in woven fabric.

Critical path The plan of action that helps to achieve the objectives of 'from concept to realisation and promotion'.

Croissure A technique of twisting silk over itself during reeling in order to consolidate the filaments and remove extra moisture.

Cross-sticks Two sticks tied together at each end, which are placed through the cross of the warp to hold it secure.

Crystalline The physical state of a material where there is a relatively organised molecular structure.

Cubism Defined by its use of geometric forms in highly abstracted works, Cubism was an avant-garde art movement pioneered by Pablo Picasso (1881–1973) and Georges Braque (1882–1963) in Paris between 1907 and 1914. Cubism was one of the most influential visual art styles of the early twentieth century.

Culture A term describing the particular characteristics of a specific geographical region (e.g. British culture), civilisation (e.g. Mayan culture) or group (e.g. youth culture, drug culture), in a specific period of time. It can include the habits, attitudes, belief systems, values, artefacts and politics of a particular group of people at a particular point in time.

Customer profile An analysis of a hypothetical customer where age, salary, socio-economic class, disposable income after bills and lifestyle become the focus with a view to developing relevant products.

Cuticle Skin of the cotton fibre.

Degumming Removal of the gums from bast fibres and silk.

Demographics The statistical data of a population such as the average age or income.

Denier The weight in grams of 9000 m of fibre or yarn.

Dent The space between two wires in the reed.

Desizing The removal of the sizing material from warp yarns after a fabric is constructed.

Detergent A chemical composition that removes soiling and is produced by chemical synthesis.

Digital immigrant An individual born before digital technology existed but who has adopted it in his or her life.

Digital native Refers to someone who grew up with twenty-first century new technology, or was born during or after the introduction of digital technology, so that it became second nature to them.

Directoire style In fashion, a style that mixed ancient classical features with modern elements. For men, trousers and high boots, vests, long, open coats, and top hats; while women dressed in light, long-sleeved chemises with V-shaped necklines and ruffled caps.

Dispersion In dyeing terms, this refers to a state where a relatively insoluble dye is held in an aqueous suspension to facilitate its transfer to the fibre.

Disruptive pattern material Camouflage print, variations used by the armed forces worldwide depending upon the terrain in which they are serving.

Distribution channel The retail environment: online or bricks and mortar.

Doup A long loop above and below the eye of a heddle.

Drape This describes the arrangement of a loosely wrapped cloth.

Draw-down ratio The relation between the rate at which filaments are pulled from the spinneret and their molten flow speed into the spinneret.

Drawing The attenuation of extruded polymer fibres during the spinning process.

Dry spinning Differs from wet spinning in that solidification takes place through evaporation of the solvent.

Dystopia A depressing imagined society where everything is as bad as it can be, characterised by human misery, oppression, disease, squalor and overcrowding.

Edo period (Tokugawa period) Also known as the *sakoku jidai* – the period of isolation. A period in Japanese history established by the shogun Tokugawa Ieyasu. The Tokugawa shogunate ruled over Japan for over 250 years from 1608 to 1868. During this peaceful time in Japanese history, Japan became almost completely isolated from the West, and its culture developed without Western influence.

Elasticity The ability of a material to recover its original shape after deformation.

Elastomer Amorphous polymer existing above its glass transition temperature so that considerable flexibility of motion exists. Elastomers usually display a higher failure strain and lower Young's modulus than other materials.

Elongation The ratio of the extension of a material to its length before stretching.

End Each warp thread is called an end. Multiples are ends.

Epi or epcm The number of warp threads, or ends per inch or cm in a cloth is written as ends per inch/cm (epi or epcm).

Extension A material added to an already existing material in order to enlarge it.

Fabric A cloth produced by knitting, weaving or felting fibres.

Fancy yarn A yarn that has deliberately introduced irregularities in its construction.

Fashion (design) Commonly used to refer to clothing that follows contemporary trends.

Fashion forecasting Forecasts 18 months to 2 years ahead of a season. This is done by gathering intelligence material from the latest stores, designers, brands, trends and business innovations in leading fashion capitals of the world.

Fibre A natural or synthetic filament that may be spun into yarn.

Fibrillation The process of splitting filament, tape or textile film longitudinally into a network of interconnected fibres.

Filament A continuous fibre of long length.

Fineness A relative measure of size, diameter, linear density or mass per unit length expressed in a number of different units.

Flat or untextured yarn Yarn that is straight along its length with little bulkiness.

Flat stick Used when winding on the warp to the back beam. Prevents warp yarns from sticking together and provides even tension across the warp.

Float When a warp or weft yarn crosses more than one other yarn at a time, then floats are formed.

Fluorescent brightening agent A product that has the ability to convert ultraviolet radiation into visible light, thereby increasing the apparent whiteness of a material to which it is applied.

Gel spinning A process in which partially liquid polymer filaments are cooled in a liquid bath. It produces highly oriented and therefore very strong fibres.

Geotextile A permeable textile material used in ground engineering for applications such as stabilisation, reinforcement or drainage.

Gilling The straightening and blending of fibres to improve evenness.

Ginning A process for separating cotton fibres from the seed in order that the fibres may be spun into yarns. Machines used for ginning may be either saw gins or roller gins.

Globalisation Refers to the breaking down of inter-cultural barriers (e.g. language, geographical distance, differences in behaviour) and has brought about a closer integration of the countries and people of the world.

Greige Fabrics that have not received any finishing treatments after being produced by any textile process.

Guerilla Art Unauthorised art that appears surreptitiously (and often suddenly) in a public place, particularly when its intention is to make a political statement. It is a form of street art with a political message, often of dissent, and often created by anonymous artists.

Hand/handle The quality of a textile as perceived by touching or feeling.

Handwriting The visual identity of the brand in the form of a logo and its application.

Harajuku An area of Japan, renowned for its distinctive street fashions.

Haute couture A French term, meaning 'high sewing', or 'high dressmaking'. Refers to fashion that is created specifically for a single individual, at great expense – traditionally fashion for a very wealthy elite.
Heat setting The treatment of thermoplastic fibres or fabrics with heat to set them into a specific shape or form.
Heddle A thin wire loop that is attached to the shaft.
Heterogeneous Dissimilar, composed of widely varying elements.
Homogenised To make uniform or similar, with no diversification.
Hue The attribute of colour whereby it is recognised as being red, green, blue, yellow, violet, brown, etc.
Hydrophilic Materials that have an affinity for water.
Hydrophobic Materials that do not have an affinity for water.
Hygroscopic Tendency of a material to absorb water.
Illuminant A source of light. In dyeing terms, this may include natural daylight and sources such as artificial daylight, tungsten-filament and retailers' specified store lights.
Incentive Reward given to encourage custom, such as reductions in price or giveaways.
Instrumental match prediction system A system for predicting the best available dye recipe using a spectrophotometer to measure the reflectance value of a standard, and comparing this with a database of the reflectance values of proposed dyes.
Isoptropy Having identical properties in all directions.
Japonism, Japonisme, Japonesque Describes the influence of Japanese artistic principles on those of the West. Works that arise from the transfer of Japanese artistic principles onto Western artefacts, particularly those by French artists, are called *japonesque*.
Kimono A loose, wide-sleeved robe, fastened by a wide sash (*obi*) at the waist, characteristic of traditional Japanese costume. The word 'kimono' literally means 'thing to wear'.
Levelness The state of a dyed material in terms of evenness of dye application.
Linear density Mass per unit length expressed as grams per centimetre.
Lumen The central core.
Lustre The amount of light that is reflected off a textile material.
Marketing mix The four P's – product, price, place and promotion; the extended mix adds people, process and physical evidence.
Market research Research that can encompass the latest fashion designs, relevant current designers, new digital technologies, new technologies in terms of materials, methods and production techniques, influences, inspiration, questionnaires with significant consumers, focus groups and feedback on findings.
Mass market An overwhelming group of consumers using, for example, the same household product.
Meiji period The Meiji (meaning 'enlightened rule') era (1868–1912) was a period in Japanese history that followed the Edo period of isolation. During the Meiji era, Japanese society underwent fundamental modernisation to establish Japan as a modern nation-state.
Melt spinning A process in which the polymer is melted, extruded and cooled before the filaments are collected. It enables the production of filaments in a variety of cross-sectional shapes.
Mercerisation The treatment of cellulosic fibres with a strong basic solution to swell the fibres and produce permanent changes.
Metallic yarn Yarn made of monofilament fibres or ply yarns, laminated with a layer of aluminium.
Microfibre Manufactured fibres with a denier of less than 0.5, or weight less than 50 μg per metre.
Microfibril A very fine fibril (fine element bundles of which constitute a fibre).
Micronaire An indicator of the maturity and fineness of cotton fibre.
Migration The movement of a dye or pigment from one part of a material to another.
Modulus The resistance to stretching of a textile.
Moisture regain The amount of moisture measured in material under prescribed conditions and expressed as a percentage of the weight of the moisture-free specimen.

Multicultural A multicultural society is one that represents, or is made up of, several different cultures or cultural elements.

Nanotechnology Nanotechnology refers to the science and technology of manipulating the structure of matter at the molecular level and to the manufacture of objects with dimensions of less than 100 nm. (A nanometre is 1000 millionth part of a metre.)

Nap/nep A small knot of entangled fibres that will not straighten to a parallel position during carding or drafting.

New technology media channel The Internet, Facebook, Twitter, instant messaging, virals, blogs, YouTube, podcasts, webcasts, and e-mail.

Niche market Where there is market demand for a very specialised product. The four C's were devised with 'niche' in mind – consumer, cost, convenience and communication.

Nonionic Describes a substance with neutral ionicity. Some detergents are nonionic.

Non-sanforised Fabric that has not been subject to a process of stretching, shrinking and fixing in length and width before cutting and garment production.

Nonwoven fabric A fabric produced by bonding or interlocking of fibres (as distinct from yarns) into webs. The web structure is stabilised by mechanical, chemical or thermal means, or by treatment with a solvent.

Novelty yarn Used for decorative purposes and seldom used to make an entire fabric. It is usually of the fancy or metallic type.

Number average molecular weight The average molecular weight of a polymer chain in a sample.

Olefin Synthetic fibre manufactured from polyolefins. It has low weight, good strength, colourfastness and comfort. Olefins also have low moisture absorption and good resistance to abrasion, mildew and sunlight.

Op Art An art movement of the 1960s typified by the use of optical illusion, often in black and white.

Opening Breaking down a cotton bale into clumps of fibres.

Orientalism (Orientalist Movement) Imitation or depiction of aspects of the East (or Orient) by Western writers, designers and artists.

Paisley (shawl) A design pattern that originated in India, Pakistan and Persia (now Iran). The Scottish town of Paisley produced shawls for the longest period of time and with the greatest level of economic efficiency. The name of the design therefore became synonymous with the place of manufacture.

Pick Each weft thread is called a pick. Multiples are picks.

Pile fabric Fabric with tufts or loops of fibres standing up from the base fabric.

Pilling The tendency of a synthetic fabric to form small balls of entangled fibre during abrasion or use.

Ply yarn A yarn formed by twisting together two or more single yarns.

Polyamide A polymer in which the short chain (monomer) units are linked together by the amide group –CONH–. An example of a naturally occurring polyamide is silk, and a synthetic polyamide is nylon.

Polycondensation A reaction during which polymers are formed from smaller (monomer) units and that involves removal of simple compounds such as water.

Polyester A polymer in which the monomer units are linked together by the group –COO–.

Polynosic A modified viscose with high tensile strength even when wet.

Pop Art An art movement that emerged in Britain in the 1950s as a response to the mass-consumerism of the post-war period and as a rejection of traditional views of art (particularly abstract art) that Pop Aartists considered pretentious and elitist. Pop Art used commonplace images from popular culture for inspiration, for example comic books, supermarket products, billboards and magazine advertisements.

Popular culture (pop culture) Can be described as culture for mass-consumption. The sum of ideas (e.g. political, or religious), attitudes, values, lifestyle (e.g. music, film and fashion) and items (including art) that surround the everyday lives of people in a particular society, are well known and generally accepted and considered mainstream by informal consensus.

Ppi (ppcm) The number of weft threads, or picks per inch or cm, in a cloth is written as picks per inch/cm (ppi or ppcm).

Qualitative research Involves analysis of data, from interviews, video, or artefacts.

Quantitative research Involves the analysis of numerical data often acquired by filling in questionnaires.

Raddle Used for spacing the warp yarns onto the warp beam before winding the warp on to the loom. It consists of wooden pegs set into a heavy base, with a removable top. The raddle usually has two spaces (dents) to 1 inch (2.5 cm).

Rayon A manufactured fibre composed of regenerated cellulose derived from wood pulp, cotton linters or other vegetable matter.

Reduction clearing An after-treatment to improve the colour fastness of polyester dyed with disperse dyes, using sodium hydroxide and a reducing agent, sodium hydrosulphite.

Reed Spaces the warp yarns after the warp has been wound onto the warp beam and threaded through heddles. The reed is also used to beat up the weft yarn after it has been passed through the shed. Its length varies depending upon the width of the loom, and its size is defined by the number of spaces (dents) between each fine price of wire to one inch, i.e., 30 dents to one inch (2.5 cm). Two, three or four warp yarns are normally placed in each dent.

Reflectance curve A graphic representation of the percentage of light reflected from a substrate at 16 points at wavelengths between 400 and 700 nm.

Resiliency The ability of a textile to recover its original shape after a distortion, such as stretching.

Retting Immersion in water in order to aid separation of the fibres from a woody tissue.

Ring spinning The most widely used method of staple fibre yarn production, in which fibres are twisted around each other to give strength to the yarn.

Roving A process of lengthening and twisting fibres.

RSS feed Really Simple Syndication feed – used to publish frequently updated work such as news headlines and blog entries on websites.

Scouring A wet process for removing dirt or impurities from textiles by application of chemicals or surfactants.

Segmentation The sub-division of the mass market for more accurate market targeting.

Selvedge Runs lengthwise in the warp direction of all fabrics. Formed when the weft yarn turns to go back across the warp.

Sett A series of woven threads crossing at right angles.

Shed The opening formed by lifting the warp threads, or ends, on the loom, through which the weaver passes the next weft thread, or pick. The depth of the shed should be a little more than the depth of the shuttle used. The deeper the shed, the easier it is to weave.

Showa period A period (1926–1989) in Japanese history comprising the reign of the emperor Hirohito. The name Showa means 'bright peace' in Japanese.

Single yarn Single strand composed of fibres held together by at least a small amount of twist.

Sliver A rope of disentangled fibres held together by inter-fibre friction rather than twist.

Slubbing Similar to roving, but without twist.

Snagging The pulling out of warp or weft threads in a woven fabric or of the course or wale threads in a knitted fabric that leads to the formation of a loop.

Soap Metallic salt of fatty acids that is soluble in water.

Social networking Individuals and organisations connecting to each other through interdependency such as friendship, common interest, likes and dislikes, beliefs or knowledge; e.g. Facebook.

Socio-economic classifications A/upper class, B/middle class, C1/lower middle class, C2/skilled working class, D/working class, E/subsistence level-low fixed income.

Softener Creates a lubricating film that allows fibres to move more readily against each other, thus making the fabric softer and fluffier.

Spectral data The values of the percentage of light reflected from a coloured substrate, with the values normally recorded at 20 nm (nm) intervals between 400 and 700 nm.

Spectrophotometer An instrument used to measure the intensity of wavelengths in a spectrum of light.
Staple fibres Short fibres of differing length. Short staple fibres are less than 60 mm long. Long staple fibres are more than 60 mm long.
Staple length The average length of a staple-fibre sample.
Stick shuttle A thin rectangular piece of soft wood with a deep notch at either end, used mainly for rigid heddle and table looms, or for carrying thick weft yarns through the shed.
Strain A measure of material extensibility, expressed as a ratio of elongation to original length.
Strike rate The rate at which a dye is transferred from the application medium to the substrate, the control of which is necessary for the even application of the dye throughout the substrate.
Stress Force per unit area.
Striation Lengthwise marking on the surface of manufactured fibres visible through a microscope.
Sub-culture (subculture) A group with distinctive characteristics that exists within a larger culture. For example, the 'mods' and 'rockers' were sub-cultures existing within the wider context of 1960s youth culture.
Substantivity The attraction between a substrate and a dye whereby the latter is selectively attracted from the application medium by the substrate.
Suint A natural grease formed from dried perspiration found in the fleece of sheep.
Surfactant (surface active agent) An organic compound that can reduce the surface tension of water when used in low concentrations.
Surrealism An art movement depicting imagery from the unconscious mind without making any attempt to rationalise the outcome. Founded in 1924 by André Breton (1896–1966), Surrealism was influenced by the psychoanalytical work of Freud and Jung. The most famous exponent of Surrealism was Salvador Dali (1904–1989).
Sustainability Protection of the environment to support long-term ecological balance and to ensure that the needs of current and future generations are provided for.
Suture A thread used to close a wound by sewing.
Swing tag Attached to garments, showing the brand message, and holding information such as size, price and any special laundering instructions.
Syndet Synthetic detergent.
Taisho period The Taisho period (period of 'great righteousness') in Japan was one of the shortest periods in Japanese history, and corresponded to the reign of the Taishō emperor, Yoshihito (1879–1926). It followed the Meiji period and represented a continuation of Japan's modernisation and rise in international status. During this period, traditional Japanese and Western dress became blended, giving rise to surprising combinations of East and West.
Take up Winding woven fabric onto front beam.
Techno-textiles A term used to define textiles that make use of technological advances.
Tenacity The tensile force a fibre can sustain before rupture. Expressed as the force relative to the fibre density.
Tensile strength The resistance of fibres, yarns or fabrics to a pulling force.
Tension Whilst weaving, both warp and weft threads must be held taught at tension; this tension is held by the use of a loom.
Tenter To dry wet fabrics on a frame on which the fabric is stretched taut and flat.
Tex The weight in grams of 1000 m of continuous-filament yarn. 1000 tex = 1 kilotex (ktex), 0.1 tex = 1 decitex (dtex) and 0.001 tex = 1 millitex (mtex).
Textured yarn Yarn that is processed to impart varied degrees of bulkiness.
Threading hook Device made of metal with a small curved hook on the end, used for threading the warp yarns through the heddles.
Three-dimensional structure Refers to fabrics that are manufactured in 3D form rather than being adapted from two-dimensional sheets.

Tow In bast fibres, the short fibres removed by hackling, or in manufactured fibres, a twistless multifilament strand suitable for conversion in staple fibres or sliver, or direct spinning into yarn.

Traditional media channels Television, advertising, ambient media, magazine advertising, celebrity endorsement, radio advertising, sales promotion, public relations, personal selling and direct marketing.

Trend forecaster Company that analyses marketing information and forecasts long- and short-term changes in consumer patterns.

Triaxial fabric Consists of three sets of yarns, typically intersecting at 60°.

Twist Turns per metre of yarn. Used to hold filaments together.

Vascular graft Materials used to repair infected or damaged sections of arteries, or to replace the entire length of the artery.

Vortex spinning A development of air-jet spinning in which carded cotton yarns can be spun at high speeds.

Wear The result of a number of agencies that reduce the serviceability of an article.

Weathering Damage caused by environmental conditions.

Wet spinning A process in which polymer is dissolved in a solvent before spinning.

Yarn A linear collection of fibres or filaments twisted together or braided by other means.

Yarn count The mass of a standardised length of continuous-filament yarn, expressed in grams per kilometre.

Zeitgeist Spirit of the times.

Index

Note: Page numbers followed by "f" and "t" indicate figures and tables respectively.

A

Abaca, 51
Abrasion resistance, 18, 733
Acrylic jumper, 123, 125f
Acrylics, 7
Actuator, 355, 356f
Adenosine triphosphate (ATP), 368
Adjustable fastenings
  examples of, 454, 455f
  machinery and attachments used to, 455
  materials used to, 454–455
  selecting/applying, safety standards and legislation for, 455–456
  types of, 454
Airflow method, 37
Air jet, 223
Air-jet spinning, 184–185, 185f
Air-jet texturing, 229–231, 229f–230f
Allergies, 741
Amide linkage (CO–NH), 106
Amino acid composition, 59–61, 60t, 67–68, 68t
Amorphous regions, 121, 121f
Android smartphones, 772–773
Animal (protein)-based fibres, 5
Antimicrobials, 314
Antistatic agents, 314
A-POC, 615
App, 685–686
Apparel market, 39
Apps, 773
Aramid fibres, 20
Aramids, 7
Augmented reality, 793, 794f

B

Balanced plain weave, 271–272, 272f
Basket weave, 272, 274f
Batch processing, 460
Behavioural segmentation, 767
Bi-component continuous filament yarns, 167–168
Biotech cotton, 52
Bipolar colour-emotion scales, 753
Bleaching, 552
Bleaching cotton, 463–464
Block draft, 270–271
Blogs, 772
Bombyx mori, 62–63
Bound seams, 385, 386f
Breathability, 754
Buckles fastenings
  examples of, 454, 455f
  machinery and attachments used to, 455
  materials used to, 454–455
  selecting/applying, safety standards and legislation for, 455–456
  types of, 454
Buckling theory, 741
Bulk continuous-filament (BCF) technology, 224
Bursting strength, 735
Button ligne gauge, 429f
Buttons, 422–432
  covered buttons, 424, 424f
  machinery and attachments to apply, 425–430, 430f
  materials used to, 425, 426t–428t
  measurement of, 425, 429f
  multi-material buttons, 425, 425f
  selecting/applying, safety standards and legislation for, 431–432
  sew-through buttons, 423, 423f
  shank buttons, 423–424, 423f
Button-sew machine clamp, 430f

C

CAD. *See* Computer-aided design (CAD)
Calendered thermal bond, 322
CAM. *See* Computer-aided manufacturing (CAM)
Camel hair fibres, 72–73, 72f
Carbon fibre-reinforced composites, 147
Carbon fibre-reinforced plastics (CFRP), 139, 141–142, 153
Cashmere fibres, 71–72, 72f
Celebrity publications, 769
Cellobiose, 30
Cellulose, 214
  fibres, 30
  molecular composition of, 29–30, 30f
  polymer molecule unit, 81, 81f
CFRP. *See* Carbon fibre-reinforced plastics (CFRP)
Chameleon-like camouflage fabrics, 365
Chemical bonding, 311–312
Chemical finishing, 314
Chemical vapour deposition (CVD), 140
Circular braiding machines, 344

Clinging, 741
Coir, 42–43
Colour theory
  colour description and measurement, 476–477
  instrumental colour match prediction and shade assessment, 477
  light and human eye, 476
Computer-aided design (CAD), 298, 301
  case studies, 691–700
  design fashion and textiles products, 674–678
    flats/working drawings, 675
    specification sheets (SPECS), 675–677
    style sheets, 678
  fashion and textile software programs, 673–674
  in fashion design
    App, place of, 685–686
    design presentations, 682
    digital design libraries, 678
    digital design portfolio, 682–685
Computer-aided manufacturing (CAM), 298, 301
  case studies, 691–700
  in fashion and textiles
    digital and virtual fabrication in, 686
    new 3D printing and fabrication in, 688–690
  fashion and textile software programs, 673–674
Computerised loom, 259, 259f
Conductive textile materials, 366
Continuous-filament (CF) yarn
  classification of, 214–215, 214f
  definition, 213–214
  properties of, 234–242
    morphology, 234–237, 234f
    tensile properties, 237–242, 238f
  twisting/plying of, 231–233, 232f
  yarn count system, 215–216
Continuous metallic fibre production, 149, 149f
Continuous processing, 460–461
Cool Hunting, 792
Costing sheets, 677
Counterbalanced loom, 258, 258f
Covered buttons, 424, 424f
Craftivism, 623
Creative Commons, 771
Crimped fibres, 10–11
Crochet fabrics, 342, 343f
Crystalline regions, 121, 121f
Cultivation, 551
Cupro rayon fiber, 80–81
Cut-middles method, 37
CVD. *See* Chemical vapour deposition (CVD)

D

Demographic segmentation, 767
Desizing, 551
Diaplex, 364
Diethylene glycol terephthalate (DGT), 103
Digital portfolio, 682
Dimethyl terephthalate (DMT), 103, 104f
Direct marketing, 769
Direct printing, 508–511
  acid dye printing, 511
  digital inkjet printing, 511
  disperse dye printing, 510
  pigment printing, 509–510
  reactive dye printing, 510
  vat dye printing, 510–511
Dobby loom, 258, 259f
Doffing, 204
Double density triaxial weave, 338, 339f
Double-lap seams, 385, 386f
Double weave, 276–277, 277f
3D printing, 688–690
Draw-down ratio, 216–217
Dry finishing, 313
Drylaid webs, 309
Dry-spinning process, 166, 167f, 221, 222f
3-D solid woven structure, 346, 347f
Dyeing process
  achieving required shade, 478–479
  classes of, 485–493
    acrylic fibres, 492
    cellulosic fibres, 485–487
    fibre blends, 493
    fluorescent brightening agents, 493
    polyamide fibres, 489–491
    polyester fibres, 491–492
    protein fibres, 487–489
  colour fastness, 479–480
  compatibility of, 479
  dyeing conditions, 481
  ensuring quality and effectiveness of, 495–496
    shade assessment, 495–496
  environmental considerations, 479–480
  environmental impact of, 496–498
    air emissions, 497
    effluent emissions, 497–498
    energy consumption, 497
    occupational safety, 498
    safety of dyed products, 498
    water consumption, 497
  further textile colouration, 481–485
  knitted cotton garments, reactive dyeing of, 500–503
  machinery for, 481
  metamerism, 479
  natural/synthetic, strengths and weaknesses of, 494–495
  selection of, 477–480

E

Eco brands, 623
Eco-Chic: The Fashion Paradox, Sandy Black, 567
EDANA. *See* European Disposables and Nonwovens Association (EDANA)
Edge abrasion, 734
Elastomers, 7
Electrochromic materials, 366
Electronic devices, 367
Elongation, 14f, 15–16
Email, 772
English cotton system, 14
Enzymatic retting, 44
Ethical Fashion Forum (EFF), 627
European Disposables and Nonwovens Association (EDANA), 308
Eyelets, 442

F

Fabric feeding, 408
Fabric finishing
  seam, 382–385
    bound seams, 385, 386f
    double-lap seams, 385, 386f
    flat seams, 385, 388f
    quality problems, 402–405
    superimposed seams, 384, 385f
  sewing machines, 385–402
    machine feeding systems, 394–399, 394f
    needle, 390–394, 390f–391f
  stitch formation, 380–382
    class 100 chain stitches, 380–381, 380f
    class 300 lockstitches, 381, 381f
    class 400 multi-thread chain stitches, 381–382, 382f
    class 500 overedge stitches, 382, 383f
    stitch quality, 382, 383f
Fabric preparation processes
  bleaching, 463–464
  carbonization, 465
  desizing, 461–462
  drying, 465
  environmental impact and sustainability of, 467
  heat setting, 465
  mercerization, 464–465
  quality control in, 466–467
  scouring, 462–463
Fabrics, 14
Facebook, 771
FAIRTRADE, 628
False twist, 223
False-twist texturing, 224–228, 225f–226f
Fashion industry
  Asos.com, 642–644
    project ideas, 644–645
  definition, 635–636
  emergence/development and change in, 636–638
    cycle, 637–638
    shapes, 636
    style, 637
  standard fashion-trend cycle, 638–639
  Topman case study
    menswear, 639–642
    project ideas, 642
Fibre alignments, 326
Fibre bending modulus, 709–710
Fibre bundle, 209
Fibre diameter distribution, 741–742
Fibre-extrusion spinning, 216–221
  dry spinning, 221, 222f
  melt-spinning method, 216–219, 217f
  wet spinning, 219–220, 220f–221f
Fibre-grade polyester polymers, 103
Fibre orientation distribution (FOD), 325
Fibre-reinforced composites, 71
Fibres
  blending of, 109–110
  classification of, 98
Fibretronics, 369–370, 370f
Filaments, 7
Filament yarn spinning
  adding functionality to
    dyeability and printability, 243–246
    functional additives, 246–247
    moisture absorption, 242–243
  applications, 247–248
  continuous-filament (CF) yarn
    classification of, 214–215, 214f
    definition, 213–214
    properties of, 234–242
    twisting/plying of, 231–233, 232f
    yarn count system, 215–216
  fibre-extrusion spinning, 216–221
    dry spinning, 221, 222f
    melt-spinning method, 216–219, 217f
    wet spinning, 219–220, 220f–221f
  yarn texturing, 222–231, 223f
    air-jet texturing, 229–231, 229f–230f
    continuous-filament yarns, twisting/plying of, 231–233, 232f
    false-twist texturing, 224–228, 225f–226f
    metallised yarns, 233
Flammability, 20
Flat abrasion, 733–734
Flatbed machine, 294
Flat continuous filament yarns, 166

Flat seam, 385, 388f
Flax, 43–46
Flax count, 14
Flex abrasion, 734
Flickr, 772
Floor/treddle loom, 257, 257f
Focussed publications, 769
Formaldehyde content, 741
Fractured supply chain, 548
Friction spinning, 183, 183f, 206–207, 207f

G

Garment comfort
improving psychological comfort, 751–753
assessing, 753
factors affecting, 751–753, 751f
measuring physiological comfort, 746–750
formaldehyde content, 750
garment fit and body movement, 750
liquid water transport properties, 749–750
pressure comfort, 750
tactile comfort, 746
thermal contact, 746
thermal insulation, 746–747
water vapour permeability, 747–749
tactile comfort, 740–742
thermo-physiological (thermal) comfort, 742–746
garment fit and body movement, 745–746
liquid water transport properties, factors affecting, 744–745
moisture (vapour) transmission, factors affecting, 743–744
thermal insulation, factors affecting, 742–743
waterproofing and breathability
factors affecting, 755–757
improving, 753–758
measuring, 757–758
Garment functionality
fabric and garment drape, 726–729
dyeing and finishing, 729
fabric properties, 728
fibre properties, 727–728
garment construction, 729
measurement of, 729
yarn properties, 728
fabric and garment durability, 730–735
dyeing and finishing, 732
fabric properties, 731
fibre properties, 730–731
garment design and fit, 732
measurement of, 732–735
yarn properties, 731
fabric handle and tailorability, 706–712
dyeing and finishing, 710–711
fabric properties, 710
fibre properties, 707–710
measurement of, 711–712
yarn properties, 710
factors affecting, 706
reducing bagging, 724–726
fabric properties, 725
fibre properties, 724–725
finishing, 725
garment construction, 725
measurement of, 725–726
yarn properties, 725
reducing pilling, 719–724
dyeing and finishing, 722
fabric properties, 722
fibre properties, 720–721
formation of, 720
measurement of, 722–724
yarn properties, 722
reducing wrinkling, 712–719
factors affecting, 713–715
fibre properties, 715–716
measurement of, 717–719
mechanical and chemical finishing to, 717
yarn and fabric parameters, 716–717
Garments wear
abrasion, 802
care labelling, 813–818
requirements, 814–815
systems, 815–818
clothing storage, 818
colour fading, 803
dimensional change, 806–807
laundering, 809–813
pilling, 801–802
seam failure, 805
snagging, 805
stains, 808–809
yarns and fabrics, breaking of, 803–805
Gel-spinning method, 166, 169f
Geographic segmentation, 767
Gill box, 200, 201f
Global culture
cultural exchanges, 607–618
definition, 605–606
European and non-European arenas, 606–607
fashion
ethical fashion, 626–629
globalisation and democratisation of, 620–624
sustainability in, 624–626
Gore-Tex, 756, 756f
Graphite carbon, 143, 143f
Graphite, properties of, 143, 144t
Guerilla art, 578–580

H

Heat of sorption, 19
Hemp, 50–51
Herringbone twill, 274, 275f
High-regain fibres, 19
High-wet-modulus (HWM), 83
Hollow-spindle spinning, 210
Home furnishings, 40
Hosiery and Allied Trades Research Association (HATRA), 299
HWM. *See* High-wet-modulus (HWM)
Hydroentangled fabrics, 317–319, 318f–319f
Hydroentanglement, 311
Hydrophilic fibres, 19
Hydrophobic fibres, 19
Hydroxyl (–OH) groups, 30

I

Impurities, 460
Industrial products, 40
Industrial Revolution, 608
In-rotating-liquid-spinning process, 150, 150f
Instant messaging, 771
Intelligent molecules flow, 367
Intelligent textiles
  biomimetics, 368–369
    examples of, 368
    lotus effect, 368–369, 369f
  future applications of, 369–372, 370f
  future market development, 372–373
  used for, 356–368
    chromic and conductive materials, 365–366
    PCM, 360–363, 362f, 362t
    shape memory materials, 363–365
    smart textile applications, 357–358
    smart textiles, research and development of, 359
    stress-responsive materials, 367
    wearable electronics, 367–368
Interactive Custom Clothes Company (IC3D), 624
International Organization for Standardization (ISO), 307
International Standards Organisation (ISO), 553
International Trade Centre (ITC), 558
Inventory analysis, 554
iPhone, 772–773

J

Jacquard loom, 260, 260f
  designing for, 280
Japonisme, 607–608
Jeans buttons, 442
Joining fabrics
  buckles and adjustable fastenings
    examples of, 454, 455f
    machinery and attachments used to, 455
    materials used to, 454–455
    selecting/applying, safety standards and legislation for, 455–456
    types of, 454
  buttons, 422–432
    covered buttons, 424, 424f
    machinery and attachments to apply, 425–430, 430f
    materials used to, 425, 426t–428t
    measurement of, 425, 429f
    multi-material buttons, 425, 425f
    selecting/applying, safety standards and legislation for, 431–432
    sew-through buttons, 423, 423f
    shank buttons, 423–424, 423f
  cords/ties and belt fastenings, 446–448
    machinery and attachments used to, 447–448
    materials used to, 446–447
    selecting/applying, safety standards and legislation for, 448
  hook-and-bar fasteners
    machinery and attachments used to, 453
    materials used to, 452
    selecting/applying, safety standards and legislation for, 453–454
    types of, 451–452
  hook-and-eye fasteners
    machinery and attachments used to, 450
    materials used to, 449
    selecting/applying, safety standards and legislation for, 450–451
    types of, 449
  hook-and-loop fastenings, 432–435, 432f
    machinery and attachments used to, 434
    materials used to, 434
    selecting/applying, safety standards and legislation for, 434–435
    types of, 433–434, 433f
  press fasteners, 435–446, 436f
    machinery and attachments used to, 439–442
    materials used to, 437
    non-snap components, 442
    selecting/applying, safety standards and legislation for, 442–446
    types of, 436–437
  zips, 413–422
    bonded water-resistant zip fastener, 419, 421f
    components of, 415–416, 416f
    continuous fasteners, 419, 421f
    functions and applications, 416, 417t–418t
    length of opening measurement, 416–419, 417t–418t
    machinery and attachments, 419, 420t
    selecting/applying, safety standards and legislation for, 419–422
Jute, 48

K

Kapok, 40–42
Kenaf, 50
Kinetic energy, 371–372
Knife edge, 223
Knit-de-knit, 223
Knitting methods
  case study, 300–301
  design and technology, impact of computers in, 298–299
  developments, 296–297
  loop formation, 290, 290f
  quality control, 299–300
  terminology of, 291, 291f
  warp knitted structures, 294–295
    warp knitting machines, 294–295, 295f
  weft knitted structures, 292–294, 292f
    weft knitting machinery, 293–294
Knitwear manufacturing unit, 301
Knotting process, 342–343, 343f
Kunit fabric, 320

L

Lace fabrics, 342
Large single engineered image, 523, 524f
Latent heat, 361
Laundering, 809–813
  aids, 811–812
  chemicals, 809–810
  equipment, 812–813
LED devices, 358, 358f
LFW. *See* London Fashion Week (LFW)
Life cycle assessment (LCA), 553
Lifting plan, 267–268, 269f
Limiting Oxygen Index (LOI), 20
Liquid crystal displays (LCD), 103
London Fashion Week (LFW), 640
Long-arm cylinder arm lockstitch machine, 400, 400f
Longitudinal stress-strain curves, 61, 62f
Long-staple cotton, 31
Lyocell rayon fiber, 80

M

Machine feeding systems, 394–399, 394f
  alternating foot, 397
  compound feed, 395–397, 397f
  cup feed, 398, 399f
  differential drop feed, 395, 396f
  different stitching operations, machines for, 400–402
  feeding foot, 397
  four-motion drop feed, 395, 396f
  manual feed, 399
  needle feed, 395
  puller feed, 398, 399f
  unison feed, 398
  variable top and bottom feed, 397, 398f
  wheel feed, 398
Macrame, 341–342, 342f
Magazine advertising, 769
Malivlies fabrics, 319
Maliwatt fabrics, 320, 320f
Man-made fibres, classification of, 98f
Manual fastening machine, 439, 440f
Marketing, 764–766
  branding, 768
    Apple, 768
  fashion forecasting, 791–792
  four C's concept, 764
  four P's concept, 764
  media channels, new technologies as, 770–774
  new technologies and processes, 792–793
  planning, 774–791
  product communication, 766
  product development, 766
  public relations, 766
  retailing space, 766
  targeting, 767
  traditional media channels, 769
  Web 2.0 and technological developments, 770–773
  Wickedweb digital marketing agency, 773–774
Mass customisation, 623
Material culture
  art and society, 564–575
  definition, 563–564
  design/fashion and textiles, impact of culture on, 589–596
    counter-culture, 595
    Cubism and Delaunay, 589
    Op Art, 593–594
    Pop Art movement, 589–592
    pop culture, 594–595
    punk, 595–596
    Surrealism and Schiaparelli, 589
  politics, 575–583
    guerilla art, 578–580
    poster art, 575–576
    Soviet posters, 576–578
    T-shirts, 580–583
  technological advances
    industrial revolution, 564–567
    modern developments, 567–568
    modern period, 575
    orientalism, 569–572
    space race, 572–575
    travel and discovery, 568–569
  war, 584–589
    Bayeux Tapestry, 584, 584f
    fashion and World War II, 585, 586f

Guernica, 584–585, 585f
textile design following, 586–589
Mechanical bonding, 310–311
Medium-staple cotton, 31
Melt-spinning method, 166, 168f, 216–219, 217f, 362–363, 363f
Membrane complex, 59
Mercerising, 552
Meta-aramid, 106, 107f
Metallised yarns, 233
MICREX, 313
Microclimate, 361
Microfuel cells, 359
Micronaire value, 35–36
Mock leno weave, 276, 276f
Modulus, 16f, 17
Mohair fibres, 73–74, 73f
Moisture regain, 18
Moisture transmission rate, 748
Mono ethylene glycol (MEG), 103
Monofilament, 8, 8f
Mulberry silkworm, 62–63
Multiaxial warp knitted structure, 349–350, 349f
Multifilament, 8, 8f
Murata Vortex spinning (MVS), 184
MVS. *See* Murata Vortex spinning (MVS)

N

Nano-Tex, 369
Naturally coloured cotton, 52–54
Natural protein fibres, 57–58
applications of, 74–75
silk fibres, 62–71
amino acid composition, 67–68, 68t
applications, 70–71
fabric manufacture, 65
fine structure of, 66
properties of, 68–70
reeling, 64–65
sericulture and cocoon production, 63–64
specialty hair fibres
camel hair fibres, 72–73, 72f
cashmere fibres, 71–72, 72f
mohair fibres, 73–74, 73f
sustainability and ecological issues, 75
wool fibres, 58–62
amino acid composition, 59–61, 60t
applications, 62
properties of, 61–62
structure of, 58–59, 59f
Natural vegetable fibres
bast fibre
abaca, 51
flax, 43–46
hemp, 50–51
jute, 48
kenaf, 50
pineapple fibre, 51–52
ramie, 46–48
sisal, 51
case studies, 54–55
classifying, 29–30
cellulose, molecular composition of, 29–30
common properties of, 30
cotton
composition of, 32–34
cultivation and ginning, 31–32
definitions and types of, 31
fibre properties, measurement of, 36–37, 39t
physical properties of, 34–36
structure of, 32
textile, application in, 39–40
new technologies/processes, impact of, 55
seed fibres
coir, 42–43
kapok, 40–42
sustainability issues/eco issues, 52–54
bast fibres, 54
biotech cotton, 52
naturally coloured cotton, 52–54
organic cotton, 52
Needle component parts, 390–394, 390f
Needlepunched fabrics, 316–317
Net fabrics, 341, 341f
N-methylmorpholine-N-oxide (NMMO), 80
NMMO. *See* N-methylmorpholine-N-oxide (NMMO)
Non-polymer fibres
carbon fibres
applications, 147
manufacture, 140–142
properties, 143–147
structure, 142–143
ceramic fibres
applications, 152–153
manufacture, 152
structure and properties, 152
glass fibres
applications, 148–149
manufacture, 147–148
properties, 148
structure, 148
metallic fibres
applications, 150
manufacture, 149–150
structure and properties, 150
sporting goods, CFRP in, 153

Nonwoven fabrics
and applications, 329–330
characteristics of, 315–327
chemical bonded fabrics, 322, 323f
hydroentangled fabrics, 317–319, 318f
needlepunched fabrics, 316–317, 317f
stitch-bonded fabrics, 319–321, 320f
thermal bonded fabrics, 321–322, 321f
in fashion, 330–331
formation, technologies for, 308–314
coating and laminating, 314
fibrous web formation, 308t, 309–310
nonwoven fabric finishing and converting techniques, 312–314
web bonding technologies, 309t, 310–312
methods for evaluation, 328–329
standards for, 328–329
standard test methods, 328
properties and performance of, 327
structural parameters, 322–327
fabric porosity and pore size distribution, 326–327
fabric weight, 322–323
fibre orientation distribution, 324–326, 325f
weight uniformity of, 323–324
Novel Temperature Regulating Fibers and Garments (NOTEREFIGA), 363
Nylon 6, 99–100
Nylon 66, 99, 99f
Nylon fabrics, 102
Nylon filament yarn, 100, 100f

O

Olefins, 7
Open-end spinning, 182–184
friction system, 183, 183f
rotor system, 182–183, 182f
vortex spinning, 184, 184f
Organic cotton, 52
Oriental designs, 571–572
Outdoor advertising, 769

P

Pantone, 650–651
Para-aramids, 106, 107f
Paraffin waxes, 362–363
Passive smart textiles, 359
PCM. *See* Phase change material (PCM)
Personal selling, 769
Phase change material (PCM), 360–363, 362f, 362t
Photochromic materials, 365
Photographic manipulation, 523, 524f
Photonics research, 366
Pineapple fibre, 51–52
Plain weave structure, 337, 338f
Podcast, 772
Point-bonded fabrics, 321f
Pointed draft, 268–270, 270f
Point paper, 267, 267f
Point thread, 268
Polyacrylonitrile (PAN), 140, 141f
Polyamides, 7
Polybutylene terephthalate (PBT), 214–215
Polyesters, 7
Polyethylene terephthalate (PET), 214–215
Polytrimethylene terephthalate (PTT), 214–215
Pond retting, 44
Pore connectivity, 327
Porosity, 326
Poster art, 575–576
Press fasteners, 435–446, 436f
machinery and attachments used to, 439–442
materials used to, 437
non-snap components, 442
selecting/applying, safety standards and legislation for, 442–446
types of, 436–437
Prickle sensations, 741
Printing textiles
CAD/CAM in, 523–525
digital inkjet printing, 520–523
design application, 523
technology and characteristics, 522–523
direct printing, 508–511
acid dye printing, 511
digital inkjet printing, 511
disperse dye printing, 510
pigment printing, 509–510
reactive dye printing, 510
vat dye printing, 510–511
screen printing, 515–518
automatic flat-bed screen printing, 516–517, 516f
rotary screen printing, 517, 517f
screen design and production, 518
table screen printing, 515–516
special printing styles
burn-out (devoré) printing, 512
discharge printing, 512
resist printing, 512
traditional printing methods, 513–514, 513f
transfer printing, 519–520
digital paper printing, 520
gravure printing, 519–520
heat transfer press, 520
Processor, 355, 356f

Prong ring press fastener, 435–436, 437f
Prototyping, 372
Psychographic segmentation, 767

Q

QR code, 772, 773f

R

Radio advertising, 769
Radio-frequency identification (RFID), 370
Ramie, 46–48
Raschel machines, 295
Rayon staple fiber, 83
Reed plan, 268, 270f
Regenerated cellulose fiber
  acetate, 81
  cupro rayon fiber, 80–81
  lyocell rayon fiber, 80
  viscose rayon fiber, 80
Regenerated fibres, 4–5
Repco spinning, 208
Resiliency, 18
Resin-transfer moulding, 142
RFID. *See* Radio-frequency identification (RFID)
Rigid heddle loom, 256, 256f
Ring (conventional) spinning, 176–186, 177f
Rivets, 442
Roller drafting system, 184–185
Rotor spinning, 205–206, 206f
Rotor system, 182–183, 182f

S

Sales promotion, 769
Satin weave, 275, 275f
Scattered draft, 271
Scouring, 551
Scouring cotton, 462
Scouring silk, 463
Scouring synthetic fibers, 463
Scouring wool, 463
Screen printing, 515–518
  automatic flat-bed screen printing, 516–517, 516f
  rotary screen printing, 517, 517f
  screen design and production, 518
  table screen printing, 515–516
Seam, quality problems, 402–405
  pucker, 402–405
  thread breakage, 405, 406f
Selfil spinning, 210
Semicrystalline biopolymer, 61
Sensor, 355, 356f
Sew-through buttons, 423, 423f
Shank buttons, 423–424, 423f
Shape memory textile applications, 364
Sheet moulding compound (SMC), 142
Short-staple cotton, 31
Shrinkage, 806
Silica, crystal and glass of, 148, 148f
Silk fibres, 62–71
  amino acid composition, 67–68, 68t
  applications, 70–71
  fabric manufacture, 65
  fine structure of, 66, 66f–67f
  properties of, 68–70
  reeling, 64–65
  sericulture and cocoon production, 63–64
Simple flat braid, 344–345, 345f
Simple triaxial weave, 337–338, 338f
Singeing, 551
Single-stage compact process, 117–118
Sisal, 51
Skin abrasion, 741
Slub yarns, 207
Smart Material Corporation, 359, 360f
SmartShirt, 356–357
SMC. *See* Sheet moulding compound (SMC)
Socio-economic classification, 767
Solar powered jacket, 371–372, 371f
Solid square braid, 345, 346f
Specialist fabric structures
  braided fabrics, 344–345, 345f
  knotted fabrics, 341–343
    crochet, 342, 343f
    knotting process, 342–343, 343f
    lace, 342
    macrame, 341–342, 342f
    net fabrics, 341, 341f
  pile fabrics, 339–341
  three-dimensional fabrics, 346–352
    hollow structures, 347–348
    knitted structures, 349–350
    nonwoven structures, 350–352
    shell structures, 348–349
    solid structures, 346
  triaxial fabrics, 337–339
Spinning line, 216–217
Spinning techniques, 201–208
  open-end spinning
    friction spinning, 206–207, 207f
    rotor spinning, 205–206, 206f
  ring spinning, 202–204, 203f
  self-twist spinning, 208
  twist-spinning methods, 204–205
Spunmelt webs, 309

Stains, 808–809
Staple fibres, 7–8, 8f
Staple-fibre yarns
  manufacturing, 176–186
    air-jet spinning, 184–185
    chenille yarn system, 185–186
    combined systems, 178–180
    doubling system, 180–182
    flocking, 186, 187f
    hollow-spindle spinning, 178, 179f
    mock chenille, 186
    open-end spinning, 182–184
    ring (conventional) spinning, 176–186, 177f
Staple-yarn spinning
  preparation of cotton, 192–196
    blending, 193, 194f
    carding machine, 195, 195f
    combing, 195–196
    drawing, 196
    opening and cleaning, 192–193, 193f
    roving, 196
  spinning techniques for, 201–208
    open-end spinning, friction spinning, 206–207, 207f
    open-end spinning, rotor spinning, 205–206, 206f
    ring spinning, 202–204, 203f
    self-twist spinning, 208
    twist-spinning methods, 204–205
  woolen system, 196–199, 197f, 200f
    blending, 198
    carding, 199
    drying/oiling, 198
    opening, 198
    scouring and carbonising, 198
  worsted system, 197f, 199–201, 200f
    carding, 200
    combing, 201
    gilling, 200–201
  wrap-spinning techniques, 208–210
    air-jet spinning, 208–209, 209f
    filament wrapping techniques, 210
Stitch-bonded fabrics, 315–316, 319–321
Stitchbonding, 311
Storyboards, 682
Straight draft, 268, 270f
Stream retting, 44
Stress-responsive materials, 367
Stretching, 806–807
Stuffer box, 223
Sustainability
  key issues in, 548–549
  socially responsible fashion, 556–558
  textile supply chain, 549–553
    assessing environmental impact of, 553–554
    minimising assessing environmental impact of, 554–555
    natural fibres, fabrics made from, 550–552
    synthetic fibres, 552–553
Synthetic fibres, 4–5
  acrylic fibres, 123–129
    fibre manufacture, 126
    fibre structure, 127, 127f
    production of, 125–126
  aramid fibres, 106–109
    applications, 109
    production of, 106
    structure and properties of, 108–109
  polyamide fibres, 99–103
    applications, 101–103
    production of nylon, 99–100
    structure and properties of, 100–101
  polyester fibres, 103–106
    for apparel applications, 110–112, 111f
    applications, 105–106
    PET fibre formation, 103
    PET polyester, production of, 103
    structure and properties of, 104–105, 105f
  polyolefin fibres, 123
  polypropylene (PP), 116–122
    additives, 119
    applications, 122
    fibre manufacture, 116–117
    fibre properties, 122
    fibre structure, 119–122, 119t, 120f
    production of, 116
    spin finishes, 118–119
    yarn, types of, 117–118
Synthetic filament yarns, 11
Synthetic polymers, 214–215
Syphon test, 750

T

Table loom, 256–257, 257f
Tactile comfort, 746
Tapestry weaving, 281–283, 281f
Tearing strength, 735
Technical textiles
  eco textiles, 538
  industrial textiles and geotextiles, 536
  medical textiles, 537
  smart fabrics and intelligent textiles, 536–537
  wearable textiles and protective clothing, 537–538
Technological advances
  industrial revolution, 564–567
  modern developments, 567–568
  modern period, 575

orientalism, 569–572
space race, 572–575
travel and discovery, 568–569
Techno-textiles, 568
Television, 769
Temperature dependent permeability, 364
Tenacity, 15f, 16, 68, 69f
Tensile testing, 734–735
Textile fibre
chemical reactivity and resistance, 22
colour and lustre, 12–13
electrical properties of, 19–20
fineness, 13–14
length/shape and diameter, 10–12
moisture absorbency, 18–19
properties, 9–10, 10f
strength/flexibility and abrasion resistance, 15–18
abrasion resistance, 18
elasticity, 18
flexibility/stiffness, 16–17
resiliency, 18
tensile strength and extension, 15–16
textile products, properties to, 22–24
thermal properties of, 20
types of, 4–7
yarns and fabrics, 7–9
Textile products
applications of
apparel, 532–534
furnishing/interior textiles, 534–535
technical textiles, 536–538
textile art, 538–539
textile industry, 539–540
Textronics, 369–370, 370f
Textured continuous filament yarns, 167
Textured yarns, 11
Thermal bonding, 310, 312
Thermal conductivity, 17f, 20
Thermal contact, 746
Thermal insulation, 746–747
Thermochromic materials, 365
Threading plan, 267, 267f
Tickle, 741
Tow, 7
Toxic technological waste, 624–625
TRAID, 557
Transfer printing, 519–520
digital paper printing, 520
gravure printing, 519–520
heat transfer press, 520
Trend forecasting consultancies, 791
Triaxial braid, 344–345, 345f
Turbostratic carbon, 143, 143f
Twill weaves, 272–274, 274f
Twitter, 771
Two-spindle wrap system, 178, 180f

U

Unbalanced plain weave, 272
Uniform water absorption, 460

V

Virals, 771
Virtual sampling, 659
Viscose rayon fiber, 80
Visual design techniques
aesthetic qualities in, 655–657
design process, 649–651
collection planning, 650–651
new designs, 651–652
research for, 649–650
samples developing, 651
design tools, 657–658
finding inspiration, 653–655
copyright, 654
creative thinking techniques, 654–655
fashion forecasting, 654
trade shows, 654
garment development, 659–666
market research methods, 652–653
new technologies/processes, 658–659
sample to production, 658
Vortex spinning, 184, 184f

W

Wabi-sabi, 612
War, 584–589
Bayeux Tapestry, 584, 584f
fashion and World War II, 585, 586f
Guernica, 584–585, 585f
textile design following, 586–589
Warp knitted double fabrics, 339, 340f
Waterproof, 754
Water repellency, 754
Water repellents, 314
Water resistant fabrics, 754
Water-retting, 44
Water vapour permeability, 747–749
Weatherproof, 754–755
Weaving
derivative-weave structures
double weave, 276–277, 277f
mock leno, 276, 276f

Weaving (*Continued*)
 documentation
  lifting plan, 267–268, 269f
  point paper, 267, 267f
  reed plan, 268, 270f
  threading plan, 267, 267f
 dressing loom, 262–266, 266f
 finishing, 283–284
 honeycomb woven structures, 283
 Jacquard loom, designing for, 280
 looms, 255–260
  computerised loom, 259, 259f
  counterbalanced loom, 258, 258f
  dobby loom, 258, 259f
  floor/treddle loom, 257, 257f
  Jacquard loom, 260, 260f
  rigid heddle loom, 256, 256f
  table loom, 256–257, 257f
 pattern drafting
  block draft, 270–271
  pointed draft, 268–270, 270f
  scattered draft, 271
  straight draft, 268, 270f
 starting to, 279
 structures
  balanced and unbalanced weave, 271
  balanced plain weave, 271–272, 272f
  basket weave, 272, 274f
  herringbone twill, 274, 275f
  satin weave, 275, 275f
  twill weaves, 272–274, 274f
  unbalanced plain weave, 272
 tapestry weaving, 281–283, 281f
 tips, 284
 warp, 260–266
  calculating, 260–261
  making, 261, 261f
  making a chain, 261–262, 262f
  selecting, 260
 woven textiles, designing for, 279–280
Webcast, 772
Wetlaid webs, 309
Wet spinning, 219–220, 220f–221f
WGSN. *See* Worth Global Style Network (WGSN)
Windproof, 754
Woolen system, 196–199, 197f, 200f
 blending, 198
 carding, 199
 drying/oiling, 198
 opening, 198
 scouring and carbonising, 198
Wool fibres, 58–62
 amino acid composition, 59–61, 60t
 applications, 62
 properties of, 61–62
  size and shape, 61
  tensile properties, 61–62
 structure of, 58–59, 59f
Worsted count, 14
Worsted system, 197f, 199–201, 200f
 carding, 200
 combing, 201
 gilling, 200–201
Worth Global Style Network (WGSN), 792
Woven double fabrics, 339, 340f
Woven I-beam, 346, 347f
Woven shell structure, 348–349, 349f
Woven textiles, designing for, 279–280
Woven tunnel structure, 347–348, 348f

Y

Yarn count, 215–216
Yarns
 classification of
  continuous-filament yarns, 159–160
  high-bulk yarns, 161
  industrial yarns, 161
  novelty yarns, 160
  staple yarns, 159
  stretch yarns, 161
 fancy yarns
  boucle yarn, 171, 171f
  chainette yarn, 174, 174f
  chenille yarn, 174, 174f
  composite yarns, 175
  covered yarns, 175–176, 175f
  diamond yarn, 171, 171f
  fasciated yarn, 173, 173f
  GIMP yarn, 170–171, 170f
  knop yarn, 172–173, 173f
  loop yarn, 171–172, 172f
  marl yarn, 170
  metallic yarns, 176, 176f
  ribbon yarns, 175
  slub yarn, 173, 173f
  snarl yarn, 172, 172f
  spiral/corkscrew yarn, 170
  tape yarn, 174, 174f
 filament yarns
  applications of, 168, 169f
  continuous, structures of, 166–168
  polymer spinning processes, 165–166
  spinning methods, 164–165
 staple-fibre yarns, 161–164
  applications of, 164
  manufacturing, 176–186

operations in, 162
spinning methods, 161–162
yarn structure, 162–164
Yarn texturing, 222–231, 223f
air-jet texturing, 229–231, 229f–230f
continuous-filament yarns, twisting/plying of, 231–233, 232f
false-twist texturing, 224–228, 225f–226f
metallised yarns, 233
Yofuku, 608
Youth tribes, 618
YouTube, 772

Z

Zips, 413–422
bonded water-resistant zip fastener, 419, 421f
components of, 415–416, 416f
continuous fasteners, 419, 421f
functions and applications, 416, 417t–418t
length of opening measurement, 416–419, 417t–418t
machinery and attachments, 419, 420t
selecting/applying, safety standards and legislation for, 419–422

Printed in the United States
By Bookmasters